# Decapod Crustacean Phylogenetics

# Decapod Crustacean Phylogenetics

Edited by
## Joel W. Martin
Natural History Museum of L. A. County
Los Angeles, California, U.S.A.

## Keith A. Crandall
Brigham Young University
Provo, Utah, U.S.A.

## Darryl L. Felder
University of Louisiana
Lafayette, Louisiana, U.S.A.

CRC Press
Taylor & Francis Group
Boca Raton   London   New York

CRC Press is an imprint of the
Taylor & Francis Group, an **informa** business

CRC Press
Taylor & Francis Group
6000 Broken Sound Parkway NW, Suite 300
Boca Raton, FL 33487-2742

© 2009 by Taylor & Francis Group, LLC
CRC Press is an imprint of Taylor & Francis Group, an Informa business

No claim to original U.S. Government works
Printed in the United States of America on acid-free paper
10 9 8 7 6 5 4 3 2

International Standard Book Number-13: 978-1-4200-9258-5 (Hardcover)

This book contains information obtained from authentic and highly regarded sources. Reasonable efforts have been made to publish reliable data and information, but the author and publisher cannot assume responsibility for the validity of all materials or the consequences of their use. The authors and publishers have attempted to trace the copyright holders of all material reproduced in this publication and apologize to copyright holders if permission to publish in this form has not been obtained. If any copyright material has not been acknowledged please write and let us know so we may rectify in any future reprint.

Except as permitted under U.S. Copyright Law, no part of this book may be reprinted, reproduced, transmitted, or utilized in any form by any electronic, mechanical, or other means, now known or hereafter invented, including photocopying, microfilming, and recording, or in any information storage or retrieval system, without written permission from the publishers.

For permission to photocopy or use material electronically from this work, please access www.copyright.com (http://www.copyright.com/) or contact the Copyright Clearance Center, Inc. (CCC), 222 Rosewood Drive, Danvers, MA 01923, 978-750-8400. CCC is a not-for-profit organization that provides licenses and registration for a variety of users. For organizations that have been granted a photocopy license by the CCC, a separate system of payment has been arranged.

**Trademark Notice:** Product or corporate names may be trademarks or registered trademarks, and are used only for identification and explanation without intent to infringe.

---

**Library of Congress Cataloging-in-Publication Data**

---

Decapod crustacean phylogenetics / editors, Joel W. Martin, Keith A. Crandall, Darryl L. Felder.
  p. cm. -- (Crustacean issues)
 Includes bibliographical references and index.
 ISBN 978-1-4200-9258-5 (hardcover : alk. paper)
  1. Decapoda (Crustacea) 2. Phylogeny. I. Martin, Joel W. II. Crandall, Keith A. III. Felder, Darryl L. IV. Title. V. Series.

QL444.M33D44 2009
595.3'8138--dc22                                                              2009001091

---

**Visit the Taylor & Francis Web site at**
http://www.taylorandfrancis.com

**and the CRC Press Web site at**
http://www.crcpress.com

# Contents

Preface ix
JOEL W. MARTIN, KEITH A. CRANDALL & DARRYL L. FELDER

## I  *Overviews of Decapod Phylogeny*

On the Origin of Decapoda 3
FREDERICK R. SCHRAM

Decapod Phylogenetics and Molecular Evolution 15
ALICIA TOON, MAEGAN FINLEY, JEFFREY STAPLES & KEITH A. CRANDALL

Development, Genes, and Decapod Evolution 31
GERHARD SCHOLTZ, ARKHAT ABZHANOV, FREDERIKE ALWES, CATERINA BIFFIS & JULIA PINT

Mitochondrial DNA and Decapod Phylogenies: The Importance of Pseudogenes and Primer Optimization 47
CHRISTOPH D. SCHUBART

Phylogenetic Inference Using Molecular Data 67
FERRAN PALERO & KEITH A. CRANDALL

Decapod Phylogeny: What Can Protein-Coding Genes Tell Us? 89
K.H. CHU, L.M. TSANG, K.Y. MA, T.Y. CHAN & P.K.L. NG

Spermatozoal Morphology and Its Bearing on Decapod Phylogeny 101
CHRISTOPHER TUDGE

The Evolution of Mating Systems in Decapod Crustaceans 121
AKIRA ASAKURA

A Shrimp's Eye View of Evolution: How Useful Are Visual Characters in Decapod Phylogenetics? 183
MEGAN L. PORTER & THOMAS W. CRONIN

Crustacean Parasites as Phylogenetic Indicators in Decapod Evolution 197
CHRISTOPHER B. BOYKO & JASON D. WILLIAMS

The Bearing of Larval Morphology on Brachyuran Phylogeny 221
PAUL F. CLARK

## II  Advances in Our Knowledge of Shrimp-Like Decapods

Evolution and Radiation of Shrimp-Like Decapods: An Overview  245
CHARLES H.J.M. FRANSEN & SAMMY DE GRAVE

A Preliminary Phylogenetic Analysis of the Dendrobranchiata Based on  261
Morphological Characters
CAROLINA TAVARES, CRISTIANA SEREJO & JOEL W. MARTIN

Phylogeny of the Infraorder Caridea Based on Mitochondrial and Nuclear  281
Genes (Crustacea: Decapoda)
HEATHER D. BRACKEN, SAMMY DE GRAVE & DARRYL L. FELDER

## III  Advances in Our Knowledge of the Thalassinidean and Lobster-Like Groups

Molecular Phylogeny of the Thalassinidea Based on Nuclear and  309
Mitochondrial Genes
RAFAEL ROBLES, CHRISTOPHER C. TUDGE, PETER C. DWORSCHAK, GARY C.B. POORE & DARRYL L. FELDER

Molecular Phylogeny of the Family Callianassidae Based on Preliminary  327
Analyses of Two Mitochondrial Genes
DARRYL L. FELDER & RAFAEL ROBLES

The Timing of the Diversification of the Freshwater Crayfishes  343
JESSE BREINHOLT, MARCOS PÉREZ-LOSADA & KEITH A. CRANDALL

Phylogeny of Marine Clawed Lobster Families Nephropidae Dana, 1852,  357
and Thaumastochelidae Bate, 1888, Based on Mitochondrial Genes
DALE TSHUDY, RAFAEL ROBLES, TIN-YAM CHAN, KA CHAI HO, KA HOU CHU, SHANE T. AHYONG & DARRYL L. FELDER

The Polychelidan Lobsters: Phylogeny and Systematics (Polychelida:  369
Polychelidae)
SHANE T. AHYONG

## IV  Advances in Our Knowledge of the Anomura

Anomuran Phylogeny: New Insights from Molecular Data  399
SHANE T. AHYONG, KAREEN E. SCHNABEL & ELIZABETH W. MAAS

## V  Advances in Our Knowledge of the Brachyura

Is the Brachyura Podotremata a Monophyletic Group?  417
GERHARD SCHOLTZ & COLIN L. MCLAY

Assessing the Contribution of Molecular and Larval Morphological  
Characters in a Combined Phylogenetic Analysis of the Superfamily  
Majoidea  
KRISTIN M. HULTGREN, GUILLERMO GUERAO, FERNANDO P.L. MARQUES &  
FERRAN P. PALERO
   437

Molecular Genetic Re-Examination of Subfamilies and Polyphyly in the  
Family Pinnotheridae (Crustacea: Decapoda)  
EMMA PALACIOS-THEIL, JOSÉ A. CUESTA, ERNESTO CAMPOS & DARRYL L.  
FELDER
   457

Evolutionary Origin of the Gall Crabs (Family Cryptochiridae) Based on  
16S rDNA Sequence Data  
REGINA WETZER, JOEL W. MARTIN & SARAH L. BOYCE
   475

Systematics, Evolution, and Biogeography of Freshwater Crabs  
NEIL CUMBERLIDGE & PETER K.L. NG
   491

Phylogeny and Biogeography of Asian Freshwater Crabs of the Family  
Gecarcinucidae (Brachyura: Potamoidea)  
SEBASTIAN KLAUS, DIRK BRANDIS, PETER K.L. NG, DARREN C.J. YEO  
& CHRISTOPH D. SCHUBART
   509

A Proposal for a New Classification of Portunoidea and Cancroidea  
(Brachyura: Heterotremata) Based on Two Independent Molecular  
Phylogenies  
CHRISTOPH D. SCHUBART & SILKE REUSCHEL
   533

Molecular Phylogeny of Western Atlantic Representatives of the Genus  
*Hexapanopeus* (Decapoda: Brachyura: Panopeidae)  
BRENT P. THOMA, CHRISTOPH D. SCHUBART & DARRYL L. FELDER
   551

Molecular Phylogeny of the Genus *Cronius* Stimpson, 1860, with  
Reassignment of *C. tumidulus* and Several American Species of *Portunus*  
to the Genus *Achelous* De Haan, 1833 (Brachyura: Portunidae)  
FERNANDO L. MANTELATTO, RAFAEL ROBLES, CHRISTOPH D. SCHUBART  
& DARRYL L. FELDER
   567

Index
   581

Color Insert

# Preface

JOEL W. MARTIN[1], KEITH A. CRANDALL[2] & DARRYL L. FELDER[3]

[1] *Natural History Museum of Los Angeles County, 900 Exposition Boulevard, Los Angeles, California, U.S.A.*
[2] *Department of Biology, Brigham Young University, Provo, Utah, U.S.A.*
[3] *Department of Biology and Laboratory for Crustacean Research, University of Louisiana, Lafayette, Louisiana, U.S.A.*

Decapods are undoubtedly the most recognizable of all crustaceans. The group includes the well-known "true" crabs (Brachyura), hermit crabs and their relatives (Anomura), shrimps (Dendrobranchiata, Caridea, and Stenopodidea), and lobsters (Astacidea, Thalassinidea), among other lesser known groups. They are the most species-rich and diverse group of the Crustacea, which in turn is the fourth largest assemblage or clade of animals (behind insects, mollusks, and chelicerates) on Earth (e.g., Martin & Davis 2001). Currently, the Decapoda contains an estimated 15,000 species, some of which support seafood and marine industries worth billions of dollars each year to the world's economy. Decapods also are the quintessential group of crustaceans in the public eye. Perhaps more than any other group of marine invertebrates, the crabs, lobsters, and shrimps that make up the Decapoda are familiar to nearly everyone.

In part because of the popularity of the decapods, there has been a long-standing interest in their relationships. Over the years, hypotheses of decapod relationships have relied on sources of information as varied as behavior (such as the early split between swimming or "natant" decapods and crawling or "reptant" forms), adult morphology, larval morphology, and, in more recent years, molecular sequence data. Despite these efforts, we remain largely in the dark as to the evolutionary relationships of the major decapod clades and to the relationships of decapods to other groups of crustaceans. Although there is no shortage of publications reflecting the wide variety of ideas and hypotheses concerning decapod phylogeny, there is also no obvious consensus among carcinologists working today. Additionally, prior to January 2008, the world's leading decapodologists had never assembled with the sole purpose of elucidating relationships among the major decapod lineages and between decapods and other crustaceans.

Toward rectifying this deficit, several key decapod workers (Keith Crandall at Brigham Young University (team leader), Joel Martin at the Natural History Museum of Los Angeles County, Darryl Felder at the University of Louisiana Lafayette, and Rodney Feldmann and Carrie Schweitzer at Kent State University) were funded by the National Science Foundation's "Assembling the Tree of Life" program beginning in the fall of 2005 to work toward elucidating the evolutionary relationships of the decapods. That team has been in contact with other decapod researchers all over the world, many of whom have been supplying fresh and preserved material or fossil material for our combined analysis while also collaborating on a variety of component phylogenetic studies focused on decapods. In short, interest in decapod evolution currently is at an all-time high, with most of the world's carcinologists aware of the ongoing Tree of Life project and eager to contribute in some way.

In January 2008, carcinologists from throughout the world convened at a symposium hosted by the Society of Integrative and Comparative Biology and The Crustacean Society in San Antonio, Texas, in order to (1) present methodological updates for research on the diversity and relationships (phylogeny) of the decapods, (2) present overviews on our understanding of the systematics and relationships within some of the major decapod clades, and (3) work toward assembling and coding molecular and morphological characters toward an overall decapod phylogeny. Invited participants represented a wide variety of backgrounds and included established decapod workers as well as beginning students of decapod phylogeny. Attendees represented fourteen nations (Australia, Belgium, Brazil, China, England, France, Germany, Japan, the Netherlands, New Zealand, Singapore, Spain, Taiwan, and the United States). The chapters that follow are based on contributions to that symposium and on a few additional manuscripts from workers who could not be present at the San Antonio meeting.

The aforementioned meeting on the phylogeny of decapods, as well as this resulting volume, might seem premature at this point, not only because so much remains unknown in general but also because our Tree of Life group is still actively researching the question of decapod evolution from many different angles. Indeed, one of our primary goals is to produce a better-resolved phylogeny of the entire Decapoda than has been published to date. However, the symposium was seen as important for bringing together a majority of the world's preeminent workers, some of whom had not previously met, and for establishing our current state of knowledge with regard to the three major areas outlined above. Thus, the contributions contained herein range rather widely in scope. Some are state-of-the-art reviews of large bodies of literature and/or methodologies for elucidating decapod phylogeny (e.g., Schram on the fossil origin of decapods, Asakura on the evolution of mating and its bearing on phylogeny, Schubart on mitochondrial approaches, Scholtz on decapod "evo-devo" studies, Tudge on decapod spermiocladistics, Palero & Crandall on phylogenetic inference). Others are somewhat preliminary attempts to construct the first known phylogenetic tree for a given group of decapods (e.g., Tavares et al. on the Dendrobranchiata, Tshudy et al. on clawed lobsters, Palacios-Theil et al. on pinnotherid crabs). Several contributions present the most comprehensive analyses to date on major clades of decapods (e.g., Bracken et al. on carideans, Ahyong & Schnabel on anomurans, Robles et al. on thalassinideans, Breinholt et al. on the diversification of the crayfishes, Hultgren et al. on the crab superfamily Majoidea). Still others present data or approaches that, although not widely applied to studies of decapod evolution previously, could be used eventually to help elucidate the phylogeny of the Decapoda (e.g., Porter & Cronin on the evolution of visual elements, Bokyo & Williams on the use of decapod parasites as phylogenetic indicators). All told, we feel that the 29 contributions contained herein constitute both a fascinating overview of where we are currently in our understanding of decapod phylogeny and a tantalizing promise of what's to come.

Many people and several societies participated in supporting the symposium and/or the publication of the resulting volume, and we are indebted to all of them. For financial support of the symposium itself (including the publication of this volume), we thank the U.S. National Science Foundation (NSF grant DEB 072116), the Society of Integrative and Comparative Biology (SICB), the SICB Divisions of Invertebrate Zoology and Evolutionary and Systematic Biology, the American Microscopical Society, the Crustacean Society, and the Society of Systematic Biologists. The decapod crustacean Tree of Life project is also supported by the National Science Foundation via a series of collaborative grants to K. A. Crandall (team leader) and Nikki Hannegan (DEB 0531762), D. L. Felder (DEB 0531603), J. W. Martin (DEB 0531616), and R. Feldmann and C. Schweitzer (DEB 0531670). Our institutions (JWM: Natural History Museum of Los Angeles County; KAC: Brigham Young University; DLF: University of Lousiana, Lafayette) supported us in kind by providing space and facilities for editing the volume and by underwriting some of the research on which it is based. We are extremely grateful to the many conscientious referees who contributed their time to review the chapters on our behalf. Our promise of anonymity prevents us from listing them individually here. We especially thank Dr. Stefan Koenemann, editor of *Crustacean Issues*, for his invitation to publish the proceedings as part of that series and for his help in editing the volume, and John Sulzycki, Senior Editor of CRC Press / Taylor & Francis, for his encouragement and assistance at several stages. We also thank Paul Martin for his invaluable assistance during stages of copy editing and for readying the overall volume for publication, and undergraduate technician Penelope "ChiChi" Boudreaux for support and assistance at ULL.

Finally, we thank Sue Martin, Cindy Crandall, and Jenny Felder for their support and encouragement during the preparation of this volume.

REFERENCES

Martin, J.W. & Davis, G.E. 2001. An updated classification of the Recent Crustacea. *Nat. Hist. Mus. L.A. County, Science Series* 39: 1–124.

# I OVERVIEWS OF DECAPOD PHYLOGENY

# On the Origin of Decapoda

FREDERICK R. SCHRAM

*Burke Museum, University of Washington, Seattle, U.S.A. Contact address: PO Box 1567, Langley WA 98260, U.S.A.*

ABSTRACT

We do not have stem forms in the fossil record for Decapoda, unlike what we have for some groups of crustaceans. Thus, we currently lack a clear understanding concerning the origin of the decapods based on concrete data. Furthermore, several problem areas present themselves: 1) lack of consensus on the sister group to Decapoda, 2) the advanced nature of known Paleozoic decapods, 3) a restricted paleobiogeographic and paleoecologic distribution of these fossils, and 4) possibly incorrect assumptions about what a decapod ancestor should look like. For now the situation seems hopeless, although new data, new lines of evidence, and new perspectives might provide better insight some time in the future.

## 1 INTRODUCTION

Decapoda stands as one of the most diverse orders of crustaceans in terms of expressed variations on its body plan. That plan includes a carapace fused to the underlying thoracic segments, the first three pairs of thoracopods modified as maxillipeds [and thus their name, "deca"-"poda," for their five pairs of pereiopods], a pleon of six segments, and frequently (but not always) a tail fan including a well-developed telson and uropods. It is a very distinctive and easily recognizable body plan. Yet the origin of the order remains obscure. Indeed, comprehending the origin of any crown group is tied to the recognition and interpretation of its stem forms. In order to offer some promise of success, that task requires preservation of such forms in the fossil record.

It is not an unreasonable hope on our part to expect to find such fossils. For some groups of crustaceans, we do in fact possess sufficient knowledge. An example occurs in the unipeltate stomatopods, the mantis shrimp, a group of crustaceans that also exhibit a highly derived, quite distinctive (one might even say extreme) body plan. Calman (1904) recognized mantis shrimp as so idiosyncratic he erected a separate superorder, Hoplocarida, to accommodate them. Unipeltata, the crown stomatopods, have a modest fossil record that indicates the major superfamilies have Mesozoic origins (Hof 1998; Schram & Müller 2004). However, in recent years sufficient fossils in the Paleozoic have come to light that present a transition series that relates to the crown group Unipeltata (Schram 2007). We effectively now have stem forms that allow us to perceive how Unipeltata evolved.

However, no such array of fossil stem taxa exists as yet that would allow us to probe the earliest evolution of Decapoda. Indeed, what we encounter is a series of problems that obscure the ancient derivations of this important order.

## 2 PROBLEM ISSUES

I perceive four major areas of concern. These are: 1) no clear consensus about a sister group to Decapoda [and thus no guidance to orient us toward recognizing or interpreting possible stem forms], 2) the rather derived nature of the currently known Paleozoic decapod fossils, 3) a conundrum concerning the paleobiogeography and paleoecology of Paleozoic malacostracans, and 4) possibly incorrect assumptions concerning an "ancestor" and thus misleading hypotheses about what we might be looking for in a stem form. Let us examine each of these in turn.

## 2.1 *Sister group to Decapoda*

Ever since the first cladistic analysis of eumalacostracan relationships, the issue of the identity of the sister taxon to Decapoda has presented almost too many options. Schram (1981, 1984) found that his shortest trees had the decapods in a clade with Amphionidacea and Euphausiacea, and these in turn had syncarids as a sister group. However, some of the trees had unresolved polychotomies among the major clades. Many researchers consider that Euphausiacea serves as a sister taxon; Calman (1904) assumed such when he placed Euphausiacea and Decapod together within his superorder Eucarida. Some more recent cladistic analyses indeed recovered such an arrangement, e.g., Wills (1998). However, as in Schram (1984), Amphionidacea appeared as the immediate sister group of Decapoda in the analysis of Richter & Scholtz (2001: fig. 7), but in their analysis Euphausiacea emerges as well-embedded within a group they named Xenommacarida, a clade that contains all the other eumalacostracans.

Hence, while Eucarida often finds expression in the cladograms of eumalacostracan relationships, it is not a particularly robust arrangement. In some ways, the amphionidaceans might serve as a stem form, often emerging from phylogenetic analyses between the decapods and the krill. Amphionidaceans do possess a nicely developed maxilliped, and the second and third thoracopods are miniature versions of the more posterior thoracopods but are widely separated from the maxilliped. However, other aspects of their body habitus isolate Amphionidacea as a unique taxon (see Schram 1986).

Schram & Hof (1998) in some of their cladograms obtained a pattern wherein an array of the Late Paleozoic "eocarids," e.g., Belotelsonidea (Fig. 1A) and Waterstonellidea (Fig. 1B), emerge in sister status to decapods (sometimes in combination with Euphausiacea). However, perhaps one should first ask just what is an "eocarid." The group at one time found expression as a formal taxon (Brooks 1962b), but the concept has entailed problems. First, the assemblage is a hodgepodge of often incompletely preserved forms, e.g., lacking complete sets of limbs such as *Eocaris oervigi* Brooks, 1962 (Brooks 1962a: fig. 1C), and *Archangeliphausia spinosa* Dzik, Ivantsov, & Deulin, 2004 (Dzik et al. 2004: fig. 2A). Second, Brooks' definition of the order is ambiguous ["Length of thorax reduced, caridoid facies" (Brooks 1962b: 271)], and the list of implicit characters implied by "caridoid facies" is composed of plesiomorphic features. Third, some of the taxa placed within the order have proven to be highly specialized in their own right, e.g., Belotelsonidea and Waterstonellidea. Finally, some species once placed in the group have proven to be members of other higher taxa. For example, *Palaeopalaemon newberryi* (see below) was once assigned to the eocarids (Brooks 1962b) but has proven to be a true decapod (Schram et al. 1978). Other eocarid taxa yet might be reassigned to more clearly defined groups; for example, the genus *Eocaris* is probably an aeschronectidan hoplocarid, and I suspect that *Archangeliphausia* from the Devonian of northwestern Russia may in fact represent an early eucarid (see below). Hence, the concept of "eocaridacea" is meaningless, a grade rather than a clade, and should not be used.

In regard to the origin of Decapoda, all this is unfortunate. Without a clear consensus on a sister group, we can neither reliably deduce the ground pattern for Decapoda nor derive any well-grounded hypotheses concerning an ancestral form.

## 2.2 *Paleozoic fossils*

A complicating factor in deducing the origins of the decapods resides in the rather derived state of the known Late Paleozoic decapod fossils. Indeed, the earliest definite decapod, the Late Devonian lobster-like *Palaeopalaemon newberryi* Whitfield, 1880 (Fig. 2), is a species that is clearly a reptant

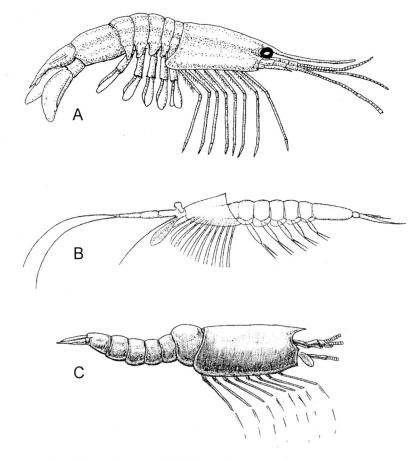

**Figure 1.** Examples of Late Paleozoic "eocarids." (A) *Lobetelson mclaughlinae*, a Middle Pennsylvanian belotelsonid (from Schram 2007). (B) *Waterstonella grantonensis*, the Lower Carboniferous waterstonellid (from Briggs & Clarkson 1983). (C) *Essoidea epiceron*, a Middle Pennsylvanian eumalacostracan of uncertain affinities (from Schram 1974).

(Schram et al. 1978; Hannibal & Feldmann 1984) and that in at least one analysis (Schram & Dixon 2005) emerges high in the decapod tree in a polytomy with Achelata, Anomura, and Brachyura. In any case, it is much too advanced a member of Reptantia to tell us much about decapod origins, let alone be considered an ancestor.

Another intriguing fossil is the Carboniferous genus *Imocaris* Schram & Mapes, 1984 (Fig. 3). Two species are recognized, *I. tuberculata* and *I. colombiensis*. Schram & Mapes (1984) assigned *Imocaris* to Dromiacea, i.e., suggested it belonged among podotreme brachyurans. However, only carapaces are known of this genus, and Racheboef & Villarroel (2003) chose to place *Imocaris* among the pygocephalomorph peracaridans. Resolving the affinities of *Imocaris* is a problem. The pygocephalomorphs bear a single cervical groove on the anterior part of their carapace, and the pattern in *Imocaris* appears more complex, with at least two. In addition, pygocephalomorphs typically bear a long and prominent rostrum, which *Imocaris* lacks. The species of *Imocaris* have a rather ornamented surface, such as one finds in some pygocephalomorphs such as *Tealliocaris* and *Pseudotealliocaris*, but ornamentation is a secondary feature and not particularly useful in phylogenetic comparisons. I still prefer a dromiacean assignment for *Imocaris*, but I am willing to consider other

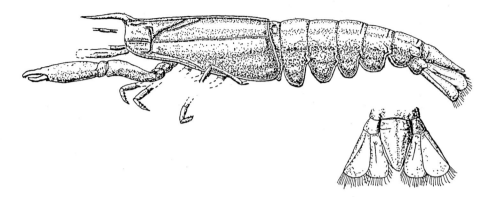

**Figure 2.** Late Devonian *Palaeopalaemon newberryi*, a reptant lobsteroid (modified from Schram et al. 1978; Hannibal & Feldmann 1985).

affinities for it, even with some group other than decapods or pygocephalomorphs. In any case (dromiacean, pygocephalomorph, or some other taxon), *Imocaris* tells us little about decapod origins.

One other set of fossils to consider consists of certain burrows in the Carboniferous of North America; Hasiotis (1999) believes crayfish made these. His interpretation focused on the markings on the walls of these burrows, which led him to conclude that these resemble similar features made by living crayfish in their burrows. There are no actual body fossils recovered from these tunnels. If these burrows do prove to be those of crayfish, they would again only record the presence of yet another rather derived form of reptantian in the Late Paleozoic.

The fossil record for the other major suborders of decapods essentially begins in the Mesozoic. The earliest members of Dendrobranchiata appear during the Triassic (see Garassino & Teruzzi 1995; Garassino et al. 1996), but a good fossil record for the group does not occur until the Jurassic Solnhofen Limestone (see Glaessner 1969). Fossils of Caridea are scarce; the earliest members apparently occur in the Jurassic, although those fossils are poorly preserved and of uncertain affinities (see Glaessner 1969). Reliably identified caridean fossils, however, do appear in the Cretaceous (Bravi & Garassino 1998a, 1998b; Bravi et al. 1999; Garassino 1997) with at least two families (Palaemonidae and Atyidae) represented there. Finally, Stenopodidea until recently had a problematic fossil record; Schram (1986) tentatively suggested that the Lower Jurassic form *Uncina posidoniae* might bear some relationship to the suborder. Subsequently, an apparent spongicolid, *Jilinocaris chinensis*, was identified from the Cretaceous of northern China (Schram et al. 2000), and a stenopodid, *Phoenice pasinii*, occurs in the Cretaceous of Lebanon (Garassino 2001). All of these Mesozoic decapods are more or less easily recognized members of their suborders and have nothing to tell us about decapod origins.

There are some puzzling Devonian fossils that have been recently recognized and bear consideration. Dzik et al. (2004) described *Archangeliphausia spinosa* from the Early Devonian of northeastern-most Europe (Fig. 4A). The fossils lack any preserved thoracic limbs. Nevertheless, the material suggests that the carapace was fused to the underlying thoracic segments. The fossils are flexed ventrally, but the carapaces do not appear to be lifted off the underlying thoracomeres. Furthermore, the segmental boundaries between the thoracic segments are preserved only ventrolaterally and do not extend to include the dorsal tergites—just what one would expect if the carapace were fused to the thoracomeres. The telson is not of the narrow, elongate, subtriangular form we associate with euphausiaceans and dendrobranchiates, but rather resembles the sub-quadrate form we often see in reptantians. I believe *Archangeliphausia spinosa* might in fact be at least a eucarid,

**Figure 3.** Lower Carboniferous *Imocaris tuberculata*, a probable dromiacean (from Schram & Mapes 1984).

and possibly another example of an advanced reptant decapod. We must wait for the collection of fossils with a full set of thoracic limbs.

Finally, another rather well-preserved, middle Paleozoic eumalacostracan is *Angustidontus seriatus* Cooper, 1936. Several species of *Angustidontus* occur in the Late Devonian and early Carboniferous across North America and Europe, and illustrate the difficulties entailed in studying early malacostracans. Originally, only the remarkable terminal segment of the maxilliped was known, and this was interpreted as a jaw of a fish. Rolfe & Dzik (2006) assembled a more extensive collection from Poland and in combination with previously collected material managed to definitively reconstruct this species as eumalacostracan (Fig. 4B). They compared *Angustidontus seriatus* to *Palaeopalaemon newberryi* and even suggested a possible synonymy of these taxa. However, *P. newberryi* is an entirely different animal, clearly a reptant decapod with the first pereiopods bearing chelate claws and the second through fifth pereiopods as walking limbs (Fig. 2). In contrast, *A. seriatus* has seven pairs of rather robust pereiopods and an elongated specialized maxilliped, a distinctly dissimilar body habitus with its singular pair of maxillipeds. What is *Angustidontus*? If we try for a link with decapods, *A. seriatus* evokes Amphionidacea with the first thoracopods as maxillipeds. *Angustidontus*, however, would seem to be a specialized benthic form rather than a mesopelagic creature like *Amphionides*. An alternative assignment of *Angustidontus* might be within Lophogastrida because *A. seriatus* has rather wide thoracic sternites, not unlike those seen in *Gnathophausia* and the pygocephalomorphs. However, no indication of fossilized oöstegites was noted on any of the fossils studied, structures that are known to occur on pygocephalomorph fossils. The wide thoracic sternites on *A. seriatus* might be akin to such sternites seen in decapods such as Achelata. Thus, whether *Angustidontus* is an early eucarid is not certain.

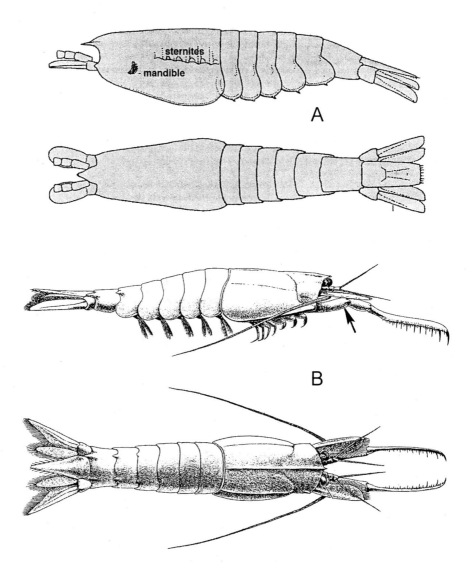

**Figure 4.** Lateral and dorsal reconstructions of Devonian eumalacostracans of uncertain affinities. (A) *Archangeliphausia spinosa*, a possible eucarid (modified from Dzik et al. 2004). (B) *Angustidontus seriatus*; note the large, specialized maxilliped [arrow] (modified from Rolfe & Dzik 2006).

In summary, while the fossil record of the Paleozoic decapods has interesting fossils, at present they tell us little about the origins of the group. The apparently derived nature of *Palaeopalaemon*, and possibly *Imocaris*, does indicate that there possibly was a long history of the order that extended back in time before the earliest fossils in the Late Devonian. *Angustidontus* and *Archangeliphausia* are intriguing in that they appear to indicate occurrences of at least eucarids, if not clear stem decapods, and hold out a promise of even earlier fossils relevant to decapod origins. How far back? Ordovician? Silurian? Cambrian? We cannot now say.

## 2.3 Paleobiogeography and paleoecology

One might feel better about this record if we saw an abundance of fossils from a wide array of localities across the world. However, as is the case for eumalacostracans and hoplocaridans as a whole, the Late Paleozoic record of the decapods has been up to now almost completely restricted to the equatorial island continent of Laurentia (Schram 1977). The Late Devonian *Palaeopalaemon newberryi* occurs in several localities across Ohio and Iowa. The Carboniferous *Imocaris tuberculata* was collected from Arkansas. A singular exception to this Laurentian pattern is *I. colombiensis*, which comes from what is now western Colombia on the Paleozoic continent of Gondwana. However, this site is not far paleogeographically from Arkansas during a time in which the continents were beginning to come together to form Pangaea. In a sense, it is the exception that proves the rule, since Schram (1977) postulated that a dispersal of higher malacostracan crustaceans out from Laurentia began with the formation of Pangaea. Nevertheless, compared to other malacostracans in the late Paleozoic, such as the hoplocaridans and peracaridans, the decapods have a paltry record.

Thus, what we have are three species that are decapods (possibly four, counting the elusive crayfish), from a handful of localities—clearly something is missing.

For instance, where were the decapods before the Devonian, assuming there was not a punctuation event in the Devonian or Late Silurian? The early and middle Paleozoic arthropods of the epicontinental seas of the world are not scarce. The diverse record of the trilobites needs no comment, but there was also an abundant array of xiphosurans, eurypterids, and thylacocephalans in those times. The latter two groups were effective predators. It is tempting to speculate that such an assortment of arthropods simply filled in most of the available niches on the epicontinental seas of those times. Thereafter, the late Devonian through Permian record of malacostracans is marked by an abundance of groups such as Hoplocarida, Syncarida, Peracarida (especially Pygocephalomorpha), Belotelsonidea, and Waterstonellidea. Was there too much competition from these diverse forms to allow the decapods to get established on the epicontinental seas of Laurentia? Such a conclusion would seem peculiar, since we live in a time when decapods have so completely dominated their habitats. Was it an instance of first come, first served?

Of course there are lots of places in the early and middle Paleozoic world where decapods might have lived. The decapods could have been denizens of the deep sea; the Panthallasic and Tethys Oceans were extensive. Or, taking a clue from the amphionidaceans, the decapods of that time may have been in the pelagic realm. Or, it is possible that decapods inhabited extremely cryptic habitats on the continents themselves such as interstitial, groundwater, and cave habitats. In regards to this last possibility, we should not overlook that small, cryptic forms were often important in the origin and early evolution of many groups, even phyla such as the mollusks (Mus et al. 2008). Discovery of the right sort of Lagerstätte in the pre-Devonian might provide us some material of significance in this regard.

## 2.4 Incorrect assumptions concerning "ancestors"

Implicit in all of the above is an assumption that a decapod "ancestor" will essentially be a caridoid with a well-developed pleon of 6 (maybe 7) somites, a carapace fused to the thorax, at least some kind of incipient specialization of the anterior thoracopods towards a maxillipedal condition, and eggs shed freely into the water column. Such an animal, or series of animals, might yet emerge. We do have fossils of caridoids such as *Archangeliphausia, Belotelson, Essoidea, Lobetelson, Waterstonella,* and others, but as mentioned above just what some of these fossils represent is not always clear.

Another deeply embedded assumption about the evolution of Malacostraca is that the 7-segment pleon of the phyllocarids was in some way the precursor of the 6-segment pleon of hoplocaridans and eumalacostracans. However, this supposition seems quite unwarranted. For example, Scholtz (1995) clearly showed in the crayfish *Cherax destructor* that the expression of *engrailed* (a marker for segment boundaries in the arthropod trunk) displays nine, rather than six (or even seven), *engrailed* stripes in the pleon. The ninth stripe is faint and quickly fades to leave eight stripes; the sixth through eighth eventually merge to produce the final 6-segment pleon of the crayfish.

Moreover, this is not a unique pattern. Knopf et al. (2006) recorded in the early development of the amphipod *Orchestia cavimana* eight clearly delineated segmental blocks of cells in the early differentiation of the pleon. In fact, the eighth *Anlage* gives rise to a pair of lateral bulges, and as the seventh and eighth somites are slowly incorporated into the growing sixth pleomere, the bulges continue to grow into distinct lobes that migrate dorsad and mediad to eventually form the so-called bifurcated telson. The adult amphipod pleon clearly begins as a series of eight segmental units.

Finally, in four species of the hermit crab genus *Porcellanopagurus*, a peculiar condition is seen in the urosomal region (cf. McLaughlin 2000). For example, in *P. nihonkaiensis* (Fig. 5), an elongate area of non-sclerotized cuticle separates the tergite of the sixth pleomere and the small telson (Komai & Takeda 2006). This region is clearly not a proximal section of the telson, which retains its characteristic form. From consideration of the larval development of *Porcellanopagurus*, it is obvious that the anus appears initially on the ventral surface of the telson *Anlage* and migrates to a terminal position by the adult stage; hence, this non-sclerotized region has nothing to do with the telson. McLaughlin (personal communication) thinks that this area might somehow be a posterior extension of the sixth pleomere. A similar arrangement is seen in some species of *Solitariopagurus*. Nevertheless, such an extension of a sixth somite posterior to the attachment of the pleopods would be unique. So, what is this? Might this non-sclerotized region be a vestige of additional somites between the sixth pleomere and the telson? The only data that might speak against this as a remnant of such somites are that the area grows in size with growth of an individual. In the examples cited above from *Cherax* and *Orchestia*, the tissues attributed to the putative seventh and eighth somites decrease in size and disappear as the individuals grow. As an alternative hypothesis to consider, I suggest that this tissue does represent remnants of post-sixth somite pleomeres and is worthy of further investigation.

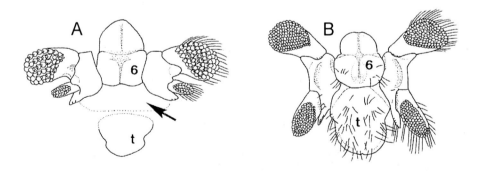

**Figure 5.** Pleon terminus of pagurid hermit crabs of the genus *Porcellanopagurus* (from Komai & Takeda 2006). (A) *P. nihonkaiensis*; note non-sclerotized region [arrow] between uropod-bearing sixth pleomere [6] and telson [t]. (B) *P. japonicus*, with a more typical anatomy of the urosome.

Just how all this impinges on ground patterns within Eumalacostraca is not clear at this time. However, instead of a 7-to-6 pattern long assumed to be the case, there are now alternative hypotheses to be entertained, viz., 8-to-7-to-6, or even separate scenarios of 8-to-7 and 8-to-6. What is clear is that we should not be surprised to find somewhere in the early or middle Paleozoic fossils of eumalacostracan-like creatures with more than the "expected" number of pleomeres.

Another line of evidence that impinges on hypotheses about ancestors arises from a consideration of the central nervous systems of various arthropods. Harzsch (2004) summarizes a series of detailed investigations of brain anatomy. Characteristic patterns of olfactory-globular tracts with chiasmata, olfactory neuropils with glomeruli, and lateral mechano-sensory antenna 1 neuropils suggest a set of synapomorphies shared by Malacostraca and Remipedia. A set of further unique features in regard to the specializations of the protocerebrum and the enervation of the compound eyes draws Hexapoda into this clade. These latter characters would seem to exclude at least the living remipedes, but it is quite possible the fossil enantiopodan remipedes, such as *Tesnusocaris goldichi*, which had very well-developed compound eyes, possessed protocerebral chiasmata as well. Since this complex CNS anatomy could be interpreted as too complicated to be anything other than shared apomorphies, those groups that possess these features might be related. That would mean that the insects, malacostracans, and remipedes form a monophyletic clade, with remipedes and malacostracans as sister groups.

This is a fascinating hypothesis, and it parallels the independent analysis of Schram & Koenemann (2004), which focused on matters of *Bauplan* in crustaceans such as locations of gonopores, *Hox*-gene expression, and numbers and types of trunk segments. They, too, obtained from their cladistic analysis a pattern wherein Remipedia emerged as the sister group to Malacostraca, as well as the core Maxillopoda. In the Schram & Koenemann scenario, we could envision an ancestor with a 16-segment trunk that gave rise to a more derived form bearing an 8-segment thorax and 8-segment pleon, which in turn laid the ground pattern for a line leading to malacostracans.

How all this might bear on the origins of decapods I don't know. On the one hand, the decapods probably emerged after the events suggested above. On the other hand, what comes early has to affect what comes later, and clearly what we had always assumed about caridoid ancestors must be tempered by what we know now. Perhaps we should be willing to consider a non-caridoid ancestor for decapods with weak differentiation between anterior (thorax) and posterior (pleon), a pleon with more than 6 somites, with incipient differentiation of the anterior three thoracopods (putative maxillipeds), and from a cryptic habitat such as groundwater or caves.

## 3 CONCLUSIONS

It would have been nice to suggest a simple little scenario here for the origin of Decapoda with a sequence of fossils at hand that would fill in the details. Unfortunately, this is not now the case. Even when we have such details, such as that seen in the wide array of Paleozoic pre-mantis shrimp relevant to scenarios about the origins of unipeltate Stomatopoda, the pattern derived is not entirely straightforward. In that example, Schram (2007) could arrange the fossils in a row wherein the increasing specialization and enlargement of the ballistic second maxilliped could be explained. However, the actual cladistic analysis of all the scored characters on these fossils indicated that this expected straight-line pattern had to be tempered by information related to the parallel evolution of the stomatopod pleon, and especially the telson.

One has to take the data as they present themselves. I suspect that while we can hope to see fossils someday that display a series of specializations of the maxillipeds toward a decapod condition, we may have to moderate our expectations. As in the stomatopods, we might have to take into account the evolution of the pleon and its urosome, or even some other aspects of the decapod body plan, to arrive at a complete understanding of the origins of this fascinating group.

## ACKNOWLEDGEMENTS

I wish to thank Dr. Pat McLaughlin for showing me the wonders of hermit crab morphology and for reading an early version of the text and making some constructive comments. Prof. Rod Feldmann and Dr. Carrie Schweitzer convinced me that one should express some caution about the possible affinities of *Imocaris*.

## REFERENCES

Bravi, S. & Garassino, A. 1998a. Plattenkalk of the Lower Cretaceous (Albian) of Petina, in the Alburni Mounts (Campania, S Italy) and its decapod crustacean assemblage. *Atti Soc. It. Sci. Nat. Museo Civ. Stor. Nat., Milano* 138: 89–118.

Bravi, S. & Garassino, A. 1998b. New biostratigraphic and palaeoecologic observations on the Plattenkalk of the Lower Cretaceous (Albian) of Pietraroia (Benevento, S Italy), and its decapod crustaceans assemblage. *Atti Soc. It. Sci. Nat. Museo Civ. Stor. Nat., Milano* 138: 119–171.

Bravi, S., Coppa, M.G., Garassino, A. & Patricelli, R. 1999. *Palaemon vesolensis* n. sp. (Crustacea: Decapoda) from the Plattenkalk of Vesole Mount (Salerno, Southern Italy). *Atti Soc. It. Sci. Nat. Museo Civ. Stor. Nat., Milano* 140: 141–169.

Briggs, D.E.G. & Clarkson, E.N.K. 1983. The Lower Carboniferous Granton 'shrimp bed', Edinburgh. *Spec. Pap. Palaeontol.* 30: 161–177.

Brooks, H.K. 1962a. Devonian Eumalacostraca. *Arkiv för Zool.* 2: 307–317.

Brooks, H.K. 1962b. The Paleozoic Eumalacostraca of North America. *Bull. Amer. Paleo.* 44: 163–338.

Calman, W.D. 1904. On the classification of the Crustacea Malacostraca. *Ann. Mag. Nat. Hist.* (7) 13: 144–158.

Cooper, C.L. 1936. Actinopterygian jaws from the Mississippian black shales of the Mississippi Valley. *J. Paleo.* 10: 92–94.

Dzik, J., Ivantsov, A. Yu. & Deulin, Yu. V. 2004. Oldest shrimp and associated phyllocarid from the Lower Devonian of northern Russia. *Zool. J. Linn. Soc. Lond.* 1142: 83–90.

Garassino, A. 1997. The macruran decapod crustaceans of the Lower Cretaceous (Lower Barremian) of Los Hoyas (Cuenca, Spain). *Atti Soc. It. Sci. Nat. Museo Civ. Stor. Nat., Milano* 137: 101–126.

Garassino, A. 2001. New decapod crustaceans from the Cenomanian (Upper Cretaceous) of Lebanon. *Atti Soc. It. Sci. Nat. Museo Civ. Stor. Nat., Milano* 141: 237–250.

Garassino, A. & Teruzzi, G. 1995. Studies on Permo–Trias of Madagascar. 3. The decapod crustaceans of the Ambilobé region (NW Madagascar). *Atti Soc. It. Sci. Nat. Museo Civ. Stor. Nat., Milano* 134: 85–113.

Garassino, A., Teruzzi, G. & dalla Vecchia, F.M. 1996. The macruran decapod crustaceans of the Dolomia di Forni (Norian, Upper Triassic) of Carnia (Undine, NE Italy). *Atti Soc. It. Sci. Nat. Museo Civ. Stor. Nat., Milano* 136: 15–60.

Glaessner, M.F. 1969. Decapoda. In: Moore, R.C. (ed.), *Treatise on Invertebrate Paleontology, Part R, Arthropod 4, Volume 2.* Univ. of Kansas and Gel. Soc. America Lawrence.

Hannibal, J. & Feldmann, R.M. 1985. Newberryi's lobster—the earliest decapod. *Explorer* 27: 10–12.

Harzsch, S. 2004. The tritocerebrum of Euarthropoda: a "non-drosophilocentric" perspective. *Evol. Devo.* 6: 303–309.

Hasiotis, S.T., 1999. The origin and evolution of freshwater crayfish based on crayfish body and trace fossils. *Freshwater Crayfish* 12: 49–70.

Hof, C.H.J. 1998. Fossil stomatopods (Crustacea: Malacostraca) and their phylogenetic impact. *J. Nat. Hist.* 32: 1567–1576.

Knopf, F., Koenemann, S., Schram, F.R. & Wolff, C. 2006. The urosome of the Pan- and Peracarida. *Contrib. Zool.* 75: 1–21.

Komai, T. & Takeda, M. 2006. A review of the pagurid hermit crab (Decapoda: Anomura: Paguroidea) fauna of the Sagami Sea, central Japan. *Mem. Natn. Sci. Mus., Tokyo* 41: 71–144.

McLaughlin, P.A. 2000. Crustacea Decapoda: *Porcellanopagurus* Filhol and *Solitariopagurus* Türkay (Paguridae), from the New Caledonian area, Vanuatu and the Marquesas: new records, new species. *Mem. Mus. Nat. Hist. Nat.* 184: 389–414.

Mus, M.M., Palacios, T. & Jensen, S. 2008. Size of the earliest mollusks: did small helcionellids groom to become large adults? *Geology* 36: 175–178.

Racheboef, P. R. & Villarroel, C. 2003. *Imocaris colombiensis* n. sp. (Crustacea: Decapoda) from the Pensylvanian of Colombia. *N. Jb. Geol. Paläont. Mh.* 2003: 577–590.

Richter, S. & Scholtz, G. 2001. Phylogenetic analysis of the Malacostraca (Crustacea). *J. Zool. Syst. Evol. Res.* 39: 113–136.

Rolfe, W.D.I. & Dzik, J. 2006. *Angustidontus*, a Late Devonian pelagic predatory crustacean. *Trans. Roy. Soc. Edin, Earth Sci.* 97: 75–96.

Scholtz, G. 1995. Expression of the engrailed gene reveals nine putative segment-anlagen in the embryonic pleon of the freshwater crayfish *Cherax destructor* (Crustacea, Malacostraca, Decapoda). *Biol. Bull.* 188: 157–165.

Schram, F.R. 1974. Mazon Creek caridoid Crustacea. *Fieldiana: Geol.* 30:9–65.

Schram, F.R. 1977. Paleozoogeography of Late Paleozoic and Triassic Malacostraca. *Syst. Zool.* 26: 367–379.

Schram, F.R. 1981. On the classification of the Eumalacostraca. *J. Crust. Biol.* 1: 1–10.

Schram, F.R. 1984. Relationships within eumalacostracan crustaceans. *Trans. S.D. Soc. Nat. Hist.* 20: 301–312.

Schram, F.R. 1986. *Crustacea*. Oxford Univ. Press, London.

Schram, F.R. 2007. Paleozoic proto-mantis shrimp revisited. *J. Paleo.* 81: 895–916.

Schram, F.R. & Dixon, C.J. 2005. Decapod phylogeny: addition of fossil evidence to a robust morphological cladistic analysis. *Bull. Mizunami Fossil Mus.* 31: 1–19.

Schram, F.R. & Hof, C.H.J. 1998. Fossils and the interrelationships of major crustacean groups. In: Edgecomb, G. D. (ed.), *Arthropod Fossils and Phylogeny*. Columbia Univ. Press, New York.

Schram, F.R. & Koenemann, S. 2004. Developmental genetics and arthropod evolution: On body regions of Crustacea. In: Scholte, G. (ed.), *Crustacean Issues 15, Evolutionary Developmental Biology of Crustacea*: 75–92. Lisse, Balkema.

Schram, F.R. & Mapes, R.H. 1984. *Imocaris tuberculata* n. gen., n. sp. (Crustacea: Decapoda) from the Upper Mississippian Imo formation, Arkansas. *Trans. San Diego Soc. Nat. Hist.* 20: 165–168.

Schram, F.R. & Müller, H-G. 2004. *Catalog and Bibliography of the Fossil and Recent Stomatopoda*. Backhuys Publ., Leiden.

Schram, F.R., Feldmann, R.M. & Copeland, M.J. 1978. The Late Devonian Palaeopalaemonidae Brooks, 1962, and the earliest decapod crustaceans. *J. Paleo.* 52: 1375–1387.

Schram, F.R., Shen, Y-B., Vonk, R. & Taylor, R.S. 2000. The first fossil stenopodidean. *Crustaceana* 73: 235–242.

Whitfield, R.P. 1880. Notice of new forms of fossil crustaceans from the Upper Devonian rocks of Ohio, with descriptions of new genera and species. *Amer. J. Sci.* 3: 33–42.

Wills, M. 1998. A phylogeny of the recent and fossil Crustacea derived from morphological characters. In: Fortey, R. A. & Thomas, R. H. (eds.), Arthropod Relationships. *Syst. Assc. Spec. Vol. Series* 55: 189–209.

# Decapod Phylogenetics and Molecular Evolution

ALICIA TOON, MAEGAN FINLEY, JEFFREY STAPLES & KEITH A. CRANDALL

*Department of Biology, Brigham Young University, Provo, Utah, U.S.A.*

## ABSTRACT

Decapoda is the most species-rich group of crustaceans, with numerous economically important and morphologically diverse species leading to a large amount of research. Our research groups are attempting to estimate a robust phylogeny of the Decapoda based on molecular and morphological data to resolve the relationships among the major decapod lineages and then to test a variety of hypotheses associated with the diversity of decapod morphological evolution. Thus, we have developed a database of molecular markers for use at different scales of the evolutionary spectrum in decapod crustaceans. We present potential mitochondrial and nuclear markers with an estimation of variation at the genus level, family level, and among infraorders for Decapoda. We provide a methodological framework for molecular studies of decapod crustaceans that is useful at different taxonomic levels.

## 1 MOLECULAR TAXONOMY

There are several competing hypotheses concerning the relationships of the major lineages of Decapoda based on morphological estimates of phylogeny. Early taxonomy of the decapods was largely based on the mode of locomotion; taxa were divided into the swimming lineages (Natantia) and the crawling lineages (Reptantia) (Boas 1880). Morphological and molecular studies suggest Natantia is paraphyletic; it is presently classified based on gill structure (Burkenroad 1963, 1981) dividing Decapoda into the suborders Dendrobranchiata (penaeoid and sergestoid shrimps) and Pleocyemata (all other decapod crustaceans). Relationships within Pleocyemata are still controversial and remain unresolved. As morphological data, both recent and fossil, and genetic data continue to accumulate, we are moving towards phylogenetic resolution of these controversial relationships. Here we present a progress report for the Decapoda Tree of Life effort and the tools with which we will continue our analysis of decapod crustacean phylogenetic relationships.

Several recent hypotheses based on combined analysis of morphological and molecular data or molecular data alone suggest that resolving the systematics of this group is a difficult task (see Fig. 1). There is agreement among these studies that Dendrobranchiata represents a basal lineage within the decapod crustaceans and that within Pleocyemata the Caridea and Stenopodidea are basal infraorders (Porter et al. 2005; Tsang et al. 2008). Molecular research also supports the removal of polychelids from Palinura following Scholtz and Richter (1995) and its establishment as a separate infraorder (Polychelida) (Tsang et al. 2008; Ahyong this volume). Relationships among reptant decapods remain unresolved by the addition of molecular data. Several recent phylogenetic analyses incorporating mitochondrial and nuclear data (Robles et al. this volume) or nuclear data alone (Tsang et al. 2008; Chu et al. this volume) suggest Thalassinidea are not monophyletic but rather may represent several infraorders. The timeline of diversification among the reptant decapods or specifically whether Astacidea (Porter et al. 2005) or the Anomura/Brachyura lineages (Ahyong & O'Meally 2004; Tsang et al. 2008) are the most recently derived lineages remains a question of interest.

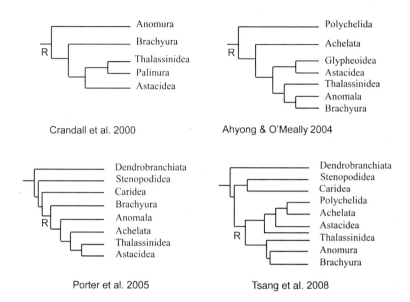

**Figure 1.** Hypotheses of decapod evolutionary relationships based on molecular data. R shows the position of the reptant decapods.

## 2 DEVELOPING GENETIC MARKERS FOR MOLECULAR PHYLOGENY

The order Decapoda includes roughly 175 families (extant and extinct) and more than 15,000 described species. Complicating things further are the estimated 437 million years since the origin of the Decapoda with the major lineages estimated to have been established by 325 million years ago (Porter et al. 2007). Constructing a molecular phylogeny across such breadth of taxa and depth of timescale requires serious consideration of markers that have enough variation to reconstruct relationships at the fine scale (at and within the family level) as well as being conservative enough to be used across infraorders representing these deeper timescales. Our approach is to accumulate molecular sequence data for different gene regions including both mitochondrial and nuclear genes, coding and non-coding. In this way, we will be able to maximize data at deeper nodes where alignment of sequence data is most difficult while retaining information among families and between the most recently diverged taxa.

There are two molecular approaches to amplifying sequence data for use in phylogenetic studies. (1) Isolation of RNA from tissues, coupled with reverse transcription-polymerase chain reaction (RT-PCR) to amplify target genes or gene fragments, reduces problems associated with amplification of pseudogenes (non-coding duplicated gene segments) and sequencing through large introns. The main limitation of RNA work is that fresh tissues, or at least tissues collected in an RNA preserving agent such as RNA*later*, require rapid transfer to $-80°C$ storage. (2) Phylogenetic work using genomic tissue extractions and amplifications is still favored over RNA techniques due to lower costs, ease of field sampling, and the ability to use previously collected specimens in ethanol. To reduce the risk of sequencing multiple copy genes or pseudogenes, gene fragments are first cloned to identify the number of copies that a primer set amplifies. Although this is not the focus of this paper, in the course of looking for useful phylogenetic markers, we have sequenced a number of multigene families such as hemocyanin, actin, and opsins. These markers may be phylogenetically useful if a single gene is isolated and amplified. They also have many uses when looking at genome evolution and the expression of these genes in Decapoda (e.g., Porter et al. 2007; Scholtz this volume). However, one must be certain that the same copy is being amplified across taxa for useful phylogenetic results.

Introns or highly variable regions need to be considered when sequencing as they can be large (greater than 1000 base pairs in length) and include repeat regions in some taxa, making amplification and sequencing difficult. Often there is too much variation in the intron among taxa to be aligned and included in the analysis. Introns can be avoided by first identifying their position and then designing primer sets within the exon to remove the introns. Here we redesigned primers for elongation factor 2 (EF-2) and transmembrane protein (TM9sf4) to exclude regions of high variability of approximately 300 base pairs in EF2 and 500–1000 base pairs in TM9sf4. Although this reduced the total length of sequence amplified, the highly variable regions produce a greater noise-to-signal ratio at the higher phylogenetic relationships, our principal focus. Of course, these more variable introns might become very useful for population genetic and species level phylogenetic work, and we continue to explore their utility at these lower levels of diversity.

## 3 THE GENES AND THEIR DIVERSITY

### 3.1 *Mitochondrial genes: 12S, 16S, and COI*

Mitochondrial ribosomal genes 12S and 16S and coding genes such as COI have been extremely useful in population genetic and systematic studies. Mitochondrial markers have been favored in studies for several reasons (see Schubart, this volume, for details and proposed primer sets for decapod mtDNA amplification). The high copy number of mitochondria in tissues makes them relatively easy to isolate. They are haploid and maternally inherited and consequently are one quarter the effective population size of nuclear genes (Moritz et al. 1987), thus allowing population level studies and systematic studies among recently diverged taxa. Possibly the most important reason to use mitochondrial genes is the availability of universal mtDNA primer sets that have minimized laboratory time in the initial setting up of a project. Finally, there is already an extensive set of nucleotide sequences from these genes in GenBank, as they have been the staple for crustacean molecular phylogenetic work since its inception.

To provide a comparison of gene utility, we have included uncorrected divergence estimates between pairs of taxa: between species, between genera, between families, and between infraorders/suborders for a number of genes. We also included COI on each graph as a reference (see Figs. 2–5). The ribosomal mitochondrial genes show similar levels of divergence to each other across all comparisons. In 12S, divergence estimates range from 3.9% among *Euastacus* species,

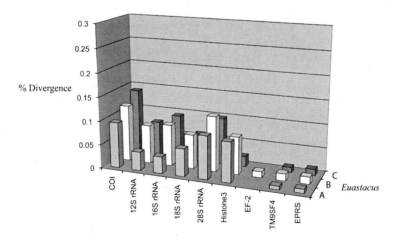

**Figure 2.** Pairwise divergence estimates between species of *Euastacus* (Astacidea) for mitochondrial and nuclear genes. Species are A: *E. eungella* and *E. spinichelatus*, B: *E. robertsi* and *E. eungella*, C: *E. robertsi* and *E. spinichelatus*.

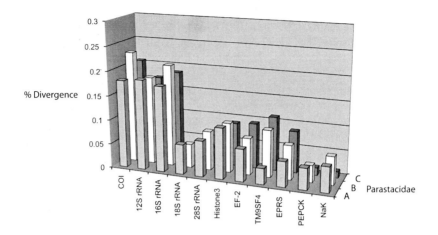

**Figure 3.** Pairwise divergence estimates between species of Parastacidae (Astacidea) for mitochondrial and nuclear genes. For genes COI, 12S, 16S, 18S, 28S, H3, EF-2, TM9SF4, EPRS the species are A: *Euastacus robertsi* and *Astacoides betsileoensis*, B: *E. robertsi* and *Parastacus defossus*, C: *A. betsileoensis* and *P. defossus*. Species for genes PEPCK and NaK are A: *Homarus gammarus* and *Nephropides caribaeus*, B: *H. gammarus* and *Nephropsis stewarti*, C: *N. caribaeus* and *N. stewarti*.

18% among genera within Parastacidae, 18.6% among families of Astacidea, and up to 24.2% among infraorders of Pleocyemata. Divergence of 16S ranges from 3.5% among species, 17.6% among genera, 23.5% among families, and up to 26.2% among infraorders of Pleocyemata. The coding mitochondrial gene COI is highly variable among species, thus making it a good candidate at lower levels. High divergence estimates were found above and including the family level, suggesting that this gene may have problems of nucleotide saturation above this level. This gene may still be useful for phylogenetic inference for resolving deeper nodes; however, it is important to test for

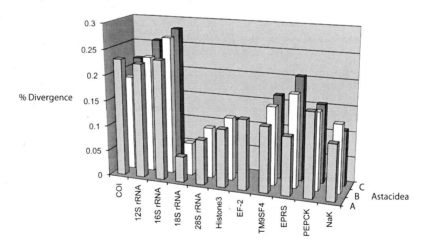

**Figure 4.** Pairwise divergence estimates among family representatives of Astacidea for mitochondrial and nuclear genes. For genes COI, 12S, 16S, 18S, 28S, H3, EF-2, TM9SF4, EPRS the species are A: *E. robertsi* and *Procambarus clarkii* (TM9SF4: *Orconectes virilis*), B: *E. robertsi* and *Nephropsis aculeata* (COI: *Homarus americanus*), C: *P. clarkii* (TM9SF4: *Orconectes virilis*) and *N. aculeate* (COI: *Homarus americanus*). Species for genes PEPCK and NaK are A: *H. gammarus* and *Cherax quadricarinatus*, B: *H. gammarus* and *P. clarkii*, C: *C. quadricarinatus* and *P. clarkii*.

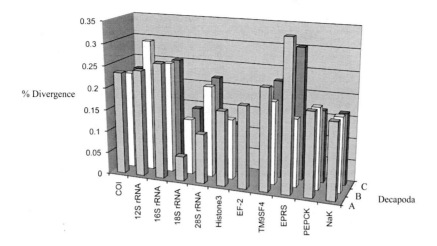

**Figure 5.** Pairwise divergence estimates among representatives of Decapoda for mitochondrial and nuclear genes. For genes COI, 12S, 16S, 18S, 28S, H3, EF-2, TM9SF4, EPRS the species are A: *E. robertsi* and *Calappa gallus* (COI: *Praebebalia longidactyla*), B: *C. gallus* (COI: *P. longidactyla*) and *Penaeus* sp., C: *E. robertsi* and *Penaeus* sp. Species for genes PEPCK and NaK are A: *H. gammarus* and *Calappa philargius*, B: *C. philargius* and *Penaeus monodon*, C: *H. gammarus* and *P. monodon*.

saturation and consider this in the analysis (i.e., use a model of evolution that incorporates multiple mutations at the same site — see Palero & Crandall this volume). A disadvantage of mitochondrial markers is that they are effectively a single locus, and, when used alone, they may not represent the true species tree.

Another problem of some mitochondrial genes such as COI is the presence of pseudogenes (nuclear copies of mitochondrial genes) in some species of decapods (Song et al. 2008).

### 3.2 Nuclear genes

Use of nuclear genes in addition to mitochondrial genes adds to the number of independent markers in a dataset, thus increasing the chances of reconstructing the true species phylogeny. In addition, a larger effective population size, and, on average, a lower substitution rate (Moriyama & Powell 1997), results in nuclear genes evolving slower than mitochondrial genes. Consequently, they may be better at resolving deeper phylogenetic nodes (see Chu et al. this volume). There are several considerations when choosing nuclear markers. There are at least two copies of each gene, although this is not usually a problem for phylogenetic studies as variation within an individual is less than between species. However, as mentioned previously, many genes belong to multigene families where duplications have resulted in genes or domains with a similar nucleotide sequence. In order to establish a single copy or at least the amplification of one dominant copy for new primer sets (EF-2, EPRS, TM9sf4) presented here, we analyzed 16–24 clones in several taxa representing Pleocyemata (Astacidea (*Homarus americanus*), Brachyura (*Cancer* sp.)) and Dendrobranchiata (*Penaeus* sp.). Low variation among some of the clones was observed. This could be attributed to *taq* polymerase error assuming an error rate of $1.6 \times 10^{-6}$ to $2.1 \times 10^{-4}$ per nucleotide per cycle (Hengen 1995) or to very low variation of a diploid gene.

The ribosomal nuclear genes 18S rDNA and 28S rDNA have been extensively used in arthropod systematics including several decapod studies (e.g., Ahyong & O'Meally 2004; Porter et al. 2005; Mitsuhashi et al. 2007; Ahyong et al. 2007). Rates of evolution vary among and within these genes, making them valuable phylogenetic tools at different taxonomic levels (Hillis & Dixon 1991). We found divergence rates for 18S were consistently moderate among species (5.8–7.2%) and

**Table 1.** Gene regions and primer sets selected for reconstructing the phylogeny of decapod crustaceans. For each primer, details of position (3') and a reference sequence are given. NR (nested reaction) refers to the primers used in the first reaction (1) and subsequent hemi-nested reaction (2).

| Gene Region | Primer Name | Primer Sequence (5' – 3') | NR | Position | Reference Sequence | Primer Reference |
|---|---|---|---|---|---|---|
| **Mitochondrial Genes** | | | | | | |
| 12S rRNA | 12sf | GAA ACC AGG ATT AGA TAC CC | | 390 | AY659990 | Mokady et al. 1994 |
| | 12sr | TTT CCC GCG AGC GAC GGG CG | | 778 | AY659990 | Mokady et al. 1994 |
| 16S rRNA | 16s-1472 | AGA TAG AAA CCA ACC TGG | | 99 | AF200829 | Crandall & Fitzpatrick 1996 |
| | 16sf-cray | GAC CGT GCK AAG GTA GCA TAA TC | | 552 | AF200829 | Crandall & Fitzpatrick 1996 |
| COI | LCO1-1490 | GGT CAA CAA ATC ATA AAG ATA TTG | | * | | Folmer et al. 1994 |
| | HCO1-2198 | TAA ACT TCA GGG TGA CCA AAA AAT CA | | * | | Folmer et al. 1994 |
| **Nuclear Genes** | | | | | | |
| 18S rRNA | 18s 1f | TAC CTG GTT GAT CCT GCC AGT AG | | * | | Whiting et al. 1997, Whiting 2002 |
| | 18s b3.0 | GAC GGT CCA ACA ATT TCA CC | | * | | Whiting et al. 1997, Whiting 2002 |
| | 18s a0.79 | TTA GAG TGC TYA AAG C | | * | | Whiting et al. 1997, Whiting 2002 |
| | 18s bi | GAG TCT CGT TCG TTA TCG GA | | * | | Whiting et al. 1997, Whiting 2002 |
| | 18s a2.0 | ATG GTT GCA AAG CTG AAA C | | * | | Whiting et al. 1997, Whiting 2002 |
| | 18s 9R | GAT CCT TCC GCA GGT TCA CCT AC | | * | | Whiting et al. 1997, Whiting 2002 |
| 28S rRNA | 28s-rD1.2a | CCC SSG TAA TTT AAG CAT ATT A | | * | | Whiting et al. 1997, Whiting 2002 |
| | 28s-rD3a | AGT ACG TGA AAC CGT TCA GG | | * | | Whiting et al. 1997, Whiting 2002 |
| | 28s-rd3.3f | GAA GAG AGA GTT CAA GAG TAC G | | * | | Whiting et al. 1997, Whiting 2002 |
| | 28sA | GAC CCG TCT TGA AGC ACG | | * | | Whiting et al. 1997, Whiting 2002 |
| | 28s-rD4.5a | AAG TTT CCC TCA GGA TAG CTG | | * | | Whiting et al. 1997, Whiting 2002 |
| | 28S rD5a | GGY GTT GGT TGC TTA AGA CAG | | * | | Whiting et al. 1997, Whiting 2002 |
| | 28s-rD4b | CCT TGG TCC GTG TTT CAA GAC | | * | | Whiting et al. 1997, Whiting 2002 |
| | 28S B | TCG GAA GGA ACC AGC TAC | | * | | Whiting et al. 1997, Whiting 2002 |
| | 28s-rD5b | CCA CAG CGC CAG TTC TGC TTA C | | * | | Whiting et al. 1997, Whiting 2002 |
| | 28s-rD6b | AAC CRG ATT CCC TTT CGC C | | * | | Whiting et al. 1997, Whiting 2002 |
| | 28S rD7b1 | GAC TTC CCT TAC CTA CAT | | * | | Whiting et al. 1997, Whiting 2002 |
| | 28s3.25a | CAG GTG GTA AAC TCC ATC AAG G | | 602 | AY210833 | this study |
| | 28s4.4b | GCT ATC CTG AGG GAA ACT TCG | | 1594 | AY210833 | this study |

Table 1. continued.

| Gene Region | Primer Name | Primer Sequence (5' – 3') | NR | Position | Reference Sequence | Primer Reference |
|---|---|---|---|---|---|---|
| H3 | H3 AF | ATG GCT CGT ACC AAG CAG ACV GC | | 321 | AB044542 | Colgan et al. 1998 |
| | H3 AR | ATA TCC TTR GGC ATR ATR GTG AC | | 694 | AB044542 | Colgan et al. 1998 |
| EF-2 | EF2a IF2 | TGG GGW GAR AAC TTC TTY AAC | | 824 | EF426560 | Porter ML pers. comm. |
| | EF2a 1R2 | ACC ATY TTK GAG ATG TAC ATC AT | | 1236 | EF426560 | Porter ML pers. comm. |
| | EF2a-F978 | TGG ANA CBC TGA ARA TCA A | 1,2 | 978 | EF426560 | this study |
| | EF2-R1435 | GTT ACC HGC TGG VAC RTC TTC | 2 | 1435 | EF426560 | this study |
| | EF2-R1536 | GAC ACG NWG AAC TTC ATC ACC | 1 | 1536 | EF426560 | this study |
| EPRS | 192fin1f | +GAR AAR GAR AAR TTY GC | | 6874 | U59923 | www.umbi.umd.edu/users/jcrlab/ |
| | 192fin2r | +TCC CAR TGR TTR AAY TTC CA | | 7316 | U59923 | www.umbi.umd.edu/users/jcrlab/ |
| TM9SF4 | 3064fin6f | CAR GAR GAR TTY GGN TGG AA | 1 | 1198 | NM_014742 | www.umbi.umd.edu/users/jcrlab/ |
| | 3064fin7r | AAN CCR AAC ATR TAR TA | | 1841 | NM_014742 | www.umbi.umd.edu/users/jcrlab/ |
| | 3064-F1204 | +GAA TTT GGR TGG AAG CTG GT | 2 | 1204 | NM_014742 | this study |
| | 3064-R1697 | +CTG GGN ATY TGG TTG GTT CG | 1,2 | 1697 | NM_014742 | this study |

" * " see primer reference for primer positions. " + " addition of M13 primers to the 5' end improves PCR amplification (Regier & Shi 2005).

among infraorders (5.6%) within Pleocyemata but were higher among the suborders Pleocyemata and Dendrobranchiata (12.8% and 14.1%). Two hypervariable regions of 28S were identified and removed to avoid inflated estimates of divergence among poorly aligned repeat regions. 28S divergence estimates were higher than 18S among species (9.1–11.6%), within Pleocyemata (11.3%), and among the suborders (20.8–21.8%). Levels of divergence were lower for the intermediate taxon levels, among genera (3.4–8.0%), and among families (7.3–9.9%), and possibly represented a shorter nucleotide alignment due to indels (insertions or deletions) that are absent among species (within a genus).

Two nuclear protein coding genes that are currently used in arthropod systematics are histone 3 (H3) (e.g., Porter et al. 2005) and elongation factor 2 (EF-2) (e.g., Regier & Shultz 2001). Primer sets already developed for H3 (Colgan et al. 1998) amplify the target fragment across a range of decapod crustaceans and show moderate levels of divergence among species (2.2–8.4%), suggesting they are useful nuclear protein coding markers for relationships within a genus. It should be noted that *Euastacus* is relatively older than some decapod genera (see Breinholt et al. this volume) and consequently H3 may not be appropriate for phylogenetic analyses among recently diverged species. Divergence within and among families is also moderate (8.9–12.4%), with a higher level of divergence between *Euastacus robertsi* and *Calappa gallus* within Pleocyemata (17%).

Although we were able to amplify genomic fragments of the EF-2 gene with currently designed primer sets (see Table 1), an intron was located at base pair position 860 relative to mRNA in *Libinia emarginata* (GenBank accession AY305506). The intron may be useful for species/genera level studies, although preliminary analysis suggests it is fewer than 300 base pairs in caridean (Hippolytidae) and brachyuran (Calappidae, Leucosiidae, Goneplacidae, Majidae, Cyclodorippidae) decapods. A new forward primer was designed to exclude the intron, and GenBank sequences were downloaded and aligned to design reverse primers 400–500 base pairs downstream of the forward primer. Using different primer sets, we were able to isolate two copies of EF-2. The two copies were more similar within an individual than between species of *Euastacus* crayfish. Two similar copies of EF-2 are present in *Drosophila melanogaster* (Lasko 2000). The divergence estimates for the longer fragment are presented in figure 2 and were low among species of *Euastacus* (1.3%). Percent divergence within Parastacidae (6.7–9.3%) and between families of Astacidea (13.6%) was moderate. High divergences were noted within Pleocyemata between *E. robertsi* and *C. gallus* (18.7%).

The EPRS locus is a potentially useful nuclear gene for reconstructing phylogenetic relationships among the deeper nodes of decapod crustaceans. The EPRS locus encodes a multifunctional aminoacyl tRNA synthetase, glutamyl–prolyl–tRNA synthetase (Cerini et al. 1991). The two proteins are involved in the aminoacylation of glutamic acid and praline tRNA in *Drosophila* (Cerini et al. 1991; Cerini et al. 1997). Few phylogenetic studies have used EPRS, although a recent study of *Paramysis* (Crustacea: Mysida) demonstrates its usefulness in reconstructing relationships among genera of mysids (Audzijonyte et al. 2008). We found divergence levels were low among species of *Euastacus* (0.8–1.5%) but moderate for within the family Parastacidae (5.2–8.6%) and high between some families of Astacidea (11.3–20.5%). This locus showed high divergences within Pleocyemata between *E. robertsi* and *C. gallus* (33.9%) and between *E. robertsi* and *Penaeus* sp. (15.5–30.1%). The different levels of divergence at different taxonmic levels suggest this marker may be useful among genera up to order level for phylogenetic estimation.

Transmembrane 9 superfamily protein member 4, or TM9sf4, is a small molecule carrier or transporter. Our study is the first to present divergence estimates and phylogenetic results using this gene. Uncorrected pairwise divergence results suggest it has potential as a valuable gene for reconstructing family to order level relationships. Divergence among species within *Euastacus* was low (0.7–1.5%), suggesting this marker may be less informative than other nuclear protein coding markers such as Histone 3 when reconstructing relationships among species. As with EPRS, this marker shows greater divergences (18.8–23%) at the deeper level (among infraorders/suborders) than Histone 3. High levels of divergence are often considered indicative of saturation; however, we found increasing divergence with increasing evolutionary distance, suggesting saturation may not have been reached even among the deeper nodes, indicating the utility of this gene to infer phylogenetic relationships at these higher levels of divergence.

## 4 PHYLOGENY BASED SYSTEMATICS

Reconstructing the evolutionary relationships among decapod crustaceans using molecular data has taken two directions: using only protein coding genes, which are phylogenetically informative at deeper nodes, or incorporating as much molecular information available including both ribosomal RNA and protein coding genes in a family level supertree. We have taken the latter approach and reconstructed Decapoda relationships using a total of eight genes and 46 taxa (see Table 2) including representatives of seven infraorders of Pleocyemata and a representative of Dendrobranchiata (*Penaeus* sp.) as an outgroup. Pleocyemata representatives include Astacidea, Achelata, Polychelida, Thalassinidea, Brachyura, Anomura and Caridea. Non-decapod crustaceans, *Lysiosquillina maculata* (Lysiosquillidae: Stomatopoda), were also included in the analysis as outgroups to all the decapods. Rather than focus on representing all lineages equally, we were interested in reconstructing relationships at many levels from among species within genera, among families, and among infraorders within decapod crustaceans. Therefore, we focused on sampling the Astacidea to demonstrate the usefulness of these genes for reconstructing phylogenies at these various taxonomic levels.

The genes included in our analyses were 12S, 16S, 18S, 28S, H3, EF-2, EPRS, and TM9sf4. A second analysis was run on the four nuclear protein-coding genes. Use of nuclear rRNA 18S and 28S data has been criticized for ambiguities noted in alignments (Tsang et al. 2008). The difficulties in aligning highly variable data may be overcome by using sophisticated methods of alignment employed in recently developed programs such as DIALIGN-T (Subramanian et al. 2005) and MAFFT (Katoh et al. 2002; Katoh et al. 2005). These programs produce more accurate alignments than ClustalW with increasing evolutionary distance (e.g., MAFFT, Nuin et al. 2006) or when gaps are present (indels) in the resulting alignment of sequence data (e.g., DIALIGN-T and MAFFT, Golubchik et al. 2007). To further improve the alignment, GBlocks can be used to identify and exclude ambiguous regions of sequence data (Castresana 2000; Talavera & Castresana 2007). We used MAFFT to align all gene fragments and subsequently ran each dataset through GBlocks (retaining half gap positions) to recover the most useful sequence data. As an example, this reduced the 28S MAFFT alignment from 4489 to 1254 base pairs. Our resulting alignment for the eight-gene dataset was 5104 nucleotides.

Maximum likelihood phylogenies were constructed with RAxML (Stamatakis 2006; Stamatakis et al. 2008) at the CIPRES portal assuming a GTR+G+I model and estimation and optimization of $\alpha$-shape parameters, GTR-rates, and empirical base frequencies for each gene. We allowed the program to choose the number of bootstrap replicates, and for the eight-gene dataset, 150 bootstrap replicates were run before termination. For the smaller nuclear protein coding alignment, 250 bootstrap replicates were run before the program terminated. The estimated parameters are presented in Table 3.

The relationships within Astacidea were well resolved, with bootstrap support in 11 of 14 nodes supported by 95% or greater and all nodes supported greater than 80% (see Fig. 6). As a comparison, the ML phylogeny based on the four-gene dataset (nuclear protein coding) constructed a similar topology within Astacidea although the nodes were not as strongly supported. Only six nodes were supported greater than 95%, with an additional five nodes supported greater than 70%. This result suggests that although the nuclear coding genes have the power to resolve relationships within an infraorder, additional data from ribosomal genes adds to the information available for reconstructing relationships across the whole of decapod diversity. Our group continues to add genes and taxa to achieve our goal of reconstructing a robust phylogenetic estimate for the decapod crustaceans.

**Table 2.** Taxonomy and accession numbers of decapod samples and outgroup included in this study. Accession numbers in bold were obtained from GenBank.

| Taxon | Voucher ID | 12S rRNA | 16S rRNA | 18S rRNA | 28S rRNA | H3 | EF-2 | EPRS | TM9SF4 |
|---|---|---|---|---|---|---|---|---|---|
| Decapoda Latreille, 1802 | | | | | | | | | |
| **Dendrobranchiata Bate, 1888** | | | | | | | | | |
| Penaeoidea Rafinesque, 1815 | | | | | | | | | |
| *Penaeus* sp. Fabricius, 1798 | KCpen | EU920908 | EU920934 | EU920969 | EU921005-EU921006 | EU921075 | — | — | EU921109 |
| Pleocyemata Burkenroad, 1963 | | | | | | | | | |
| **Anomura MacLeay, 1838** | | | | | | | | | |
| Galatheoidea Samouelle, 1819 | | | | | | | | | |
| *Aegla alacalufi* (Jara & López, 1981) | KAC798 | **AY050012** | **AY050058** | EU920958 | **AY595958** | EU921042 | EU921009 | EU910098 | EU921077 |
| *Eumunida funambulus* (Miyake, 1982) | KC3100 | EU920892 | EU920922 | EU920957 | EU920984 | EU921056 | EU921032 | EU910124 | EU921089 |
| *Kiwa hirsute* (Jones & Segonzac, 2005) | KC3116 | — | — | EU920942 | EU920987 | EU921065 | EU921035 | EU910128 | EU921097 |
| *Munidopsis rostrata* (Milne-Edwards, 1880) | KC3102 | EU920898 | EU920928 | EU920961 | EU920985 | EU921066 | EU921034 | EU910126 | EU921100 |
| Lomisoidea Bouvier, 1895 | | | | | | | | | |
| *Lomis hirta* (Lamarck, 1810) | KAClohi | **AY595547** | **AY595928** | **AF436013** | **AY596101** | **DQ079680** | EU921040 | EU910131 | EU921098 |
| Paguroidea Latreille, 1802 | | | | | | | | | |
| *Pomatocheles jeffreysii* (Miers, 1879) | KC3097 | EU920903 | EU920930 | EU920965 | EU920983 | EU921070 | EU921031 | EU910123 | EU921105 |
| **Astacidea Latreille, 1802** | | | | | | | | | |
| Astacoidea Latreille, 1802 | | | | | | | | | |
| *Astacus astacus* (Linnaeus, 1758) | KC702 | EU920881 | **AF235983** | **AF235959** | **DQ079773** | **DQ079660** | EU921008 | — | EU921078 |
| *Barbicambarus cornutus* (Faxon, 1884) | KC1941 | EU920883 | EU920913 | EU920951 | EU920993 | EU921045 | EU921017 | EU910106 | EU921080 |
| *Orconectes virilis* (Hagen, 1870) | KC709 | EU920900 | **AF235989** | **AF235965** | **DQ079804** | **DQ079693** | EU921041 | — | EU921102 |
| *Procambarus clarkii* (Girard, 1852) | KC1497 | EU920901 | **AF235990** | EU920952 | EU920970 | EU921067 | EU921011 | EU910100 | — |
| Parastacoidea Huxley, 1879 | | | | | | | | | |
| *Astacoides betsileoensis* (Petit, 1923) | KC1822 | EU920882 | EU920912 | EU920955 | EU920992 | EU921044 | EU921014 | EU910103 | EU921079 |
| *Cherax cuspidatus* (Riek, 1969) | KC1175 | **DQ006421** | **DQ006550** | EU920960 | EU920996 | EU921048 | EU921010 | EU910099 | EU921083 |
| *Euastacus eungella* (Morgan, 1988) | KC2671 | **DQ006464** | **DQ006593** | EU920964 | EU921000-EU921002 | EU921055 | EU921018 | EU910109 | EU921088 |
| *Euastacus robertsi* (Monroe, 1977) | KC2781 | **DQ006507** | **DQ006633** | EU920962 | EU920988 | EU921058 | EU921019 | EU910110 | EU921091 |
| *Euastacus spinichelatus* (Morgan, 1997) | KC2631 | **DQ006512** | **DQ006638** | EU920963 | EU920989 | EU921059 | — | EU910108 | EU921092 |
| *Gramastacus insolitus* (Riek, 1972) | KC640 | EU920895 | EU920926 | EU920968 | EU920994 | EU921062 | EU921007 | EU910097 | EU921094 |
| *Ombrastacoides huonensis* (Riek, 1967) | KC611 | EU920905 | **AF135997** | EU920956 | EU920995 | EU921072 | — | EU910096 | EU921106 |
| *Parastacus defossus* (Faxon, 1898) | KC1515 | EU920902 | **AF175243** | EU920953 | EU920991 | EU921068 | EU921012 | EU910101 | EU921103 |
| *Parastacus varicosus* (Faxon, 1898) | KC1529 | EU920907 | EU920933 | EU920954 | EU920990 | EU921074 | EU921013 | EU910102 | EU921108 |

Table 2. continued.

| Taxon | Voucher ID | 12S rRNA | 16S rRNA | 18S rRNA | 28S rRNA | H3 | EF-2 | EPRS | TM9SF4 |
|---|---|---|---|---|---|---|---|---|---|
| Nephropoidea Dana, 1852 | | | | | | | | | |
| *Homarus americanus* (Milne-Edwards, 1837) | KAChoam | DQ298427 | HAU11238 | AF235971 | DQ079788 | DQ079675 | — | — | EU921095 |
| *Nephropsis_aculeate* (Smith, 1881) | KC2117 | EU920899 | DQ079727 | DQ079761 | DQ079802 | DQ079691 | — | EU910107 | EU921101 |
| **Brachyura Latreille, 1802** | | | | | | | | | |
| Calappoidea Milne-Edwards, 1837 | | | | | | | | | |
| *Cycloes granulose* (de Haan, 1837) | KC3082 | EU920887 | EU920917 | EU920943 | EU920976 | EU921050 | EU921025 | EU910116 | EU921085 |
| *Calappa gallus* (Herbst, 1803) | KC3083 | EU920886 | EU920916 | EU920947 | EU920977 | EU921049 | EU921026 | EU910117 | EU921084 |
| Dorippoidea MacLeay, 1838 | | | | | | | | | |
| *Ethusa* sp. (Roux, 1830) | KC3088 | — | EU920925 | EU920966 | EU920980 | EU921061 | EU921029 | EU910120 | EU921093 |
| Grapsoidea MacLeay, 1838 | | | | | | | | | |
| *Cyclograpsus cinereus* (Dana, 1851) | KC3417 | EU920884 | EU920914 | EU920945 | EU920997 | EU921046 | EU921038 | EU910130 | EU921081 |
| Leucosioidea Samouelle, 1819 | | | | | | | | | |
| *Ebalia tuberculosa* (Milne-Edwards, 1873) | KC3085 | EU920894 | EU920924 | EU920944 | EU920978 | EU921060 | EU921027 | EU910118 | — |
| *Praebebalia longidactyla* (Yokoya, 1933) | KC3086 | EU920904 | EU920931 | EU920946 | EU920979 | EU921071 | EU921028 | EU910119 | — |
| Majoidea Samouelle, 1819 | | | | | | | | | |
| *Chorilia longipes* (Dana, 1852) | KC3089 | EU920889 | EU920919 | EU920948 | EU920981 | EU921052 | EU921039 | EU910121 | EU921087 |
| Raninoidea de Haan, 1839 | | | | | | | | | |
| *Cosmonotus grayi* (White, 1848) | KC3092 | EU920888 | EU920918 | EU920949 | EU920982 | EU921051 | EU921030 | EU910122 | EU921086 |
| **Caridea Dana, 1852** | | | | | | | | | |
| Palaemonoidea Rafinesque, 1815 | | | | | | | | | |
| *Anchistioides antiguensis* (Schmitt, 1924) | KC3051 | EU920880 | EU920911 | EU920936 | EU920971 | EU921043 | EU921020 | EU910111 | — |
| *Coutierella tonkinensis* (Sollaud, 1914) | KC3068 | EU920890 | EU920920 | EU920937 | EU920975 | EU921053 | EU921024 | EU910115 | — |
| Crangonoidea Haworth, 1825 | | | | | | | | | |
| *Crangon crangon* (Linnaeus, 1758) | KC3052 | EU920885 | EU920915 | EU920938 | EU920972 | EU921047 | EU921021 | EU910112 | EU921082 |
| Bresilioidea Calman, 1896 | | | | | | | | | |
| *Discias* sp. (Rathbun, 1902) | KC3108 | EU920891 | EU920921 | EU920941 | EU920986 | EU921054 | — | EU910127 | — |
| Alpheoidea Rafinesque, 1815 | | | | | | | | | |
| *Hippolyte bifidirostris* (Miers, 1876) | KC3059 | EU920896 | EU920927 | EU920939 | EU920974 | EU921063 | EU921023 | EU910114 | — |
| *Eualus gaimardii* (Milne-Edwards, 1837) | KC3056 | EU920893 | EU920923 | EU920940 | EU920973 | EU921057 | EU921022 | EU910113 | EU921090 |

Table 2. continued.

| Taxon | Voucher ID | 12S rRNA | 16S rRNA | 18S rRNA | 28S rRNA | H3 | EF-2 | EPRS | TM9SF4 |
|---|---|---|---|---|---|---|---|---|---|
| **Achelata Scholtz & Richter, 1995** | | | | | | | | | |
| Palinuroidea Latreille, 1802 | | | | | | | | | |
| *Jasus edwardsii* (Hutton, 1875) | KC3209 | — | DQ079716 | AF235972 | DQ079791 | EU921064 | EU921036 | EU910129 | EU921096 |
| *Palinurus elephas* (Fabricius, 1787) | KC3210 | — | EU920929 | EU920959 | EU920999-EU921000 | EU921069 | EU921037 | — | EU921104 |
| **Polychelida de Haan, 1941** | | | | | | | | | |
| *Polycheles typhlops* (Heller, 1862) | KC3101 | EU920906 | EU920932 | EU920950 | EU921003-EU921004 | EU921073 | EU921033 | EU910125 | EU921107 |
| **Thalassinidea Latreille, 1831** | | | | | | | | | |
| Callianassoidea Dana, 1852 | | | | | | | | | |
| *Lepidophthalmus louisianensis* (Schmitt, 1935) | KAC1852 | EU920897 | DQ079717 | DQ079751 | DQ079792 | DQ079678 | EU921015 | EU910104 | EU921099 |
| *Sergio mericeae* (Manning & Felder, 1995) | KAC1865 | EU920909 | DQ079733 | DQ079768 | DQ079811 | DQ079700 | EU921016 | EU910105 | EU921110 |
| Outgroup | | | | | | | | | |
| Stomatopoda Latreille, 1817 | | | | | | | | | |
| Lysiosquilloidea Giesbrecht, 1910 | | | | | | | | | |
| *Lysiosquillina maculata* (Fabricius, 1793) | KC3832 | EU920910 | EU920935 | EU920967 | EU920998 | EU921076 | — | — | EU921111 |

**Table 3.** Empirical base frequencies for each gene region and associated model parameters estimated from the sequence data in RAxML.

|          | A      | C      | G      | T      | alpha  | pinvar |
|----------|--------|--------|--------|--------|--------|--------|
| 12S rRNA | 0.3670 | 0.0981 | 0.1726 | 0.3622 | 0.6030 | 0.1934 |
| 16S rRNA | 0.3399 | 0.1116 | 0.2027 | 0.3458 | 0.6235 | 0.2879 |
| 18S rRNA | 0.2502 | 0.2342 | 0.2780 | 0.2377 | 0.9231 | 0.4940 |
| 28S rRNA | 0.2501 | 0.2357 | 0.3161 | 0.1981 | 0.7772 | 0.2735 |
| H3       | 0.2152 | 0.3172 | 0.2654 | 0.2022 | 1.0618 | 0.5882 |
| EF-2     | 0.2364 | 0.2469 | 0.2655 | 0.2512 | 1.4067 | 0.4872 |
| EPRS     | 0.2857 | 0.2159 | 0.2523 | 0.2460 | 1.6197 | 0.3690 |
| TM9SF4   | 0.1587 | 0.2784 | 0.2455 | 0.3174 | 0.9592 | 0.4982 |

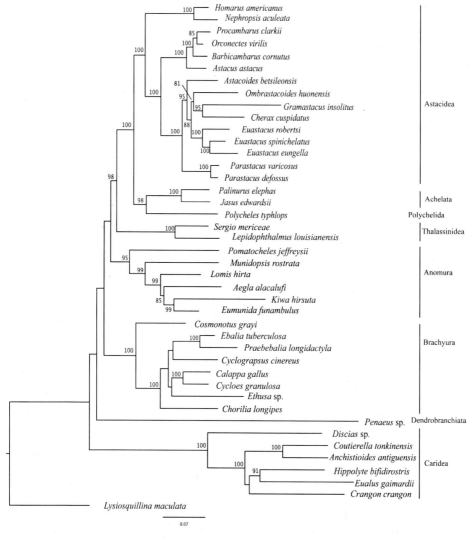

**Figure 6.** Maximum likelihood phylogeny based on two mitochondrial and six nuclear genes constructed in RAxML. Values at nodes represent bootstrap support greater than 70%.

ACKNOWLEDGEMENTS

We thank the wide variety of friends and colleagues who have helped us collect decapod crustaceans from around the world over the past 15 years. Likewise, this study was made possible by the exceptional undergraduates from Brigham Young University who have labored to collect DNA sequence data from decapod crustaceans. Our work was supported by Brigham Young University and a grant from the US NSF EF-0531762 awarded to KAC.

REFERENCES

Ahyong, S.T., Lai, J.C.Y., Sharkey, D., Colgan, D.J. & Ng, P.K.L. 2007. Phylogenetics of the brachyuran crabs (Crustacea: Decapoda): the status of Podotremata based on small subunit nuclear ribosomal RNA. *Mol. Phylogenet. Evol.* 45: 576–86.

Ahyong, S.T. & O'Meally, D. 2004. Phylogeny of the Decapoda Reptantia: resolution using three molecular loci and morphology. *Raffles Bull. Zool.* 52: 673–93.

Audzijonyte, A., Daneliya, M.E. & Vainola, R. 2008. Phylogeny of Paramysis (Crustacea: Mysida) and the origin of ponto-caspian endemic diversity: resolving power from nuclear protein-coding genes. *Mol. Phylogenet. Evol.* 46: 738–59.

Boas, F.E.V. 1880. Studier over decapodernes slaegtskabsforhold. *Dan. Selsk. Skr.* 6: 26–210.

Burkenroad, M.D. 1963. The evolution of the Eucarida (Crustacea, Eumalacostraca) in relation to the fossil record. *Tulane Stud. Geol.* 2: 1–17.

Burkenroad, M.O. 1981. The higher taxonomy and evolution of Decapoda (Crustacea). *Trans. San Diego Soc. Nat. Hist.* 19: 251–68.

Castresana, J. 2000. Selection of conserved blocks from multiple alignments for their use in phylogenetic analysis. *Mol. Biol. Evol.* 17: 540–52.

Cerini, C., Kerjan, P., Astier, M., Gratecos, D., Mirande, M. & Semeriva, M. 1991. A component of the multisynthetase complex is a multifunctional aminoacyl-transfer RNA-synthetase. *Embo J.* 10: 4267–77.

Cerini, C., Semeriva, M. & Gratecos, D. 1997. Evolution of the aminoacyl-tRNA synthetase family and the organization of the Drosophila glutamyl-prolyl-tRNA synthetase gene—intron/exon structure of the gene, control of expression of the two mRNAs, selective advantage of the multienzyme complex. *Eur. J. Biochem.* 244: 176–85.

Colgan, D.J., McLauchlan, A., Wilson, G.D.F., Livingston, S.P., Edgecombe, G.D., Macaranas, J., Cassis G. & Gray, M.R. 1998. Histone H3 and U2 snRNA DNA sequences and arthropod molecular evolution. *Aust. J. Zool.* 46: 419–37.

Crandall, K.A. & Fitzpatrick, J.F. 1996. Crayfish molecular systematics: using a combination of procedures to estimate phylogeny. *Syst. Biol.* 45: 1–26.

Folmer, O., Black, M. Hoeh, W., Lutz, R. & Vrijenhoek, R. 1994. DNA primers for amplification of mitochondrial cytochrome c oxidase subunit I from diverse metazoan invertebrates. *Mol. Mar. Biol. Biotech.* 3: 294–99.

Golubchik, T., Wise, M.J., Easteal, S. & Jermiin, L.S. 2007. Mind the gaps: evidence of bias in estimates of multiple sequence alignments. *Mol. Biol. Evol.* 24: 2433–42.

Hengen, P.N. 1995. Methods and reagents—fidelity of DNA polymerases for PCR. *Trends Biochem. Sci.* 20: 324–25.

Hillis, D.M. & Dixon, M.T. 1991. Ribosomal DNA—molecular evolution and phylogenetic inference. *Q. Rev. Biol.* 66: 411–53.

Katoh, K., Kuma, K., Toh, H. & Miyata, T. 2005. MAFFT version 5: improvement in accuracy of multiple sequence alignment. *Nucleic Acids Res.* 33: 511–18.

Katoh, K., Misawa, K., Kuma, K. & Miyata, T. 2002. MAFFT: a novel method for rapid multiple sequence alignment based on fast Fourier transform. *Nucleic Acids Research* 30: 3059–66.

Lasko, P. 2000. The *Drosophila melanogaster* genome: translation factors and RNA binding proteins. *J. Cell. Biol.* 150: 51–56.

Mitsuhashi, M., Sin, Y.W., Lei, H.C., Chan, T.Y. & Chu, K.H. 2007. Systematic status of the caridean families Gnathophyllidae Dana and Hymenoceridae Ortmann (Crustacea : Decapoda): a preliminary examination based on nuclear rDNA sequences. *Invertebr. Syst.* 21: 613–22.

Mokady, O., Rozenblatt, S., Graur, D. & Loya, Y. 1994. Coral-host specificity of red sea lithophaga bivalves: interspecific and intraspecific variation in 12S mitochondrial ribosomal RNA. *Mol. Mar. Biol. Biotech.* 3: 158–64.

Moritz, C, Dowling, T.E. & Brown, W.M. 1987. Evolution of animal mitochondrial DNA: relevance for population biology and systematics. *Annu. Rev. Ecol. Syst.* 18: 269–92.

Moriyama, E.N. & Powell, J.R. 1997. Synonymous substitution rates in *Drosophila*: mitochondrial versus nuclear genes. *J. Mol. Evol.* 45: 378–91.

Nuin, P.A.S., Wang, Z. & Tillier, E.R.M. 2006. The accuracy of several multiple sequence alignment programs for proteins. *BMC Bioinformatics* 7: 471.

Porter, M.L., Cronin, T.W., McClellan, D.A. & Crandall, K.A. 2007. Molecular characterization of crustacean visual pigments and the evolution of pancrustacean opsins. *Mol. Biol. Evol.* 24: 253–68.

Porter, M.L., Perez-Losada, M. & Crandall, K.A. 2005. Model-based multi-locus estimation of decapod phylogeny and divergence times. *Mol. Phylogenet. Evol.* 37: 355–69.

Regier, J.C. & Shultz, J.W. 2001. Elongation factor-2: a useful gene for arthropod phylogenetics. *Mol. Phylogenet. Evol.* 20: 136–48.

Regier, J.C. & Shi, D. 2005. Increased yield of PCR product from degenerate primers with nondegenerate, nonhomoloogous 5' tails. *BioTechniques*, 38: 34–38.

Scholtz, G. & Richter, S. 1995. Phylogenetic systematics of the reptantian Decapoda (Crustacea, Malacostraca). *Zool. J. Linn. Soc.-Lond.* 113: 289–328.

Song, H, Buhay, J., Whiting, M.F. & Crandall, K.A. 2008. Many species in one: DNA barcoding overestimates the number of species when nuclear mitochondrial pseudogenes are coamplified. *Proc. Nat. Acad. Sci.* In review.

Stamatakis, A. 2006. Raxml-vi-hpc: maximum likelihood-based phylogenetic analyses with thousands of taxa and mixed models. *Bioinformatics* 22: 2688–90.

Stamatakis, A., Hoover, P. & Rougemont, J. 2008. A rapid bootstrap algorithm for the RAxML Web-Servers. *Syst. Biol.* In press.

Subramanian, A.R., Weyer-Menkhoff, J., Kaufmann, M. & Morgenstern, B. 2005. Dialign-T: an improved algorithm for segment-based multiple sequence alignment. *BMC Bioinformatics* 6: 66.

Talavera, G. & Castresana, J. 2007. Improvement of phylogenies after removing divergent and ambiguously aligned blocks from protein sequence alignments. *Syst. Biol.* 56: 564–77.

Tsang, L.M., Ma, K.Y., Ahyong, S.T., Chan, T.-Y. & Chu, K.H. 2008. Phylogeny of Decapoda using two nuclear protein-coding genes: origin and evolution of the Reptantia. *Mol. Phylogenet. Evol.* 48: 359–368.

Whiting, M.F., Carpenter, J.C., Wheeler, Q.D. & Wheeler, W.C. 1997. The Strepsiptera problem: phylogeny of the holometabolous insect orders inferred from 18S and 28S ribosomal DNA sequences and morphology. *Syst. Biol.* 46: 1–68.

Whiting, M.F. 2002. Mecoptera is paraphyletic: multiple genes and phylogeny of Mecoptera and Siphonaptera. *Zool. Scr.* 93–104.

# Development, Genes, and Decapod Evolution

GERHARD SCHOLTZ[1], ARKHAT ABZHANOV[2], FREDERIKE ALWES[1], CATERINA BIFFIS[1] & JULIA PINT[1]

[1] *Humboldt-Universität zu Berlin, Institut für Biologie/Vergleichende Zoologie, Berlin, Germany*
[2] *Department of Organismic and Evolutionary Biology, Harvard University, Cambridge, Massachusetts, U.S.A.*

## ABSTRACT

Apart from larval characters such as zoeal spines and stages, developmental characters are rarely used for inferences on decapod phylogeny and evolution. In this review we present examples of comparative developmental data of decapods and discuss these in a phylogenetic and evolutionary context. Several different levels of developmental characters are evaluated. We consider the influence of ontogenetic characters such as cleavage patterns, cell lineage, and gene expression on our views on the decapod ground pattern, on morphogenesis of certain structures, and on phylogenetic relationships. We feel that developmental data represent a hidden treasure that is worth being more intensely studied and considered in studies on decapod phylogeny and evolution.

## 1 INTRODUCTION

The morphology of decapod crustaceans shows an enormous diversity concerning overall body shape and limb differentiation. On the two extreme ends, we find representatives such as shrimps with an elongated, laterally compressed body, muscular pleon, and limbs mainly adapted to swimming, and groups like the Brachyura exhibiting a dorsoventrally flattened, strongly calcified, broad body with a reduced pleon and uniramous walking limbs. In addition, hermit crabs show a peculiar asymmetric soft and curved pleon, and among all larger decapod taxa there are species with limbs specialized for digging, mollusc shell cracking, and all other sorts and numbers of pincers and scissors. These few examples indicate that the decapod body organization is varied to a high degree. It is obvious that this disparity has been used to establish phylogenetic relationships of decapods and that it is a challenge for considerations of decapod evolution (e.g., Boas 1880; Borradaile 1907; Beurlen & Glaessner 1930; Burkenroad 1981; Scholtz & Richter 1995; Schram 2001; Dixon et al. 2003). One major example for the latter is the controversial discussion about carcinization—the evolution of a crab-like form, which, as the most derived body shape and function, desires an explanation at the evolutionary level (e.g., Borradaile 1916; Martin & Abele 1986; Richter & Scholtz 1994; McLaughlin & Lemaitre 1997; Morrison et al. 2002; McLaughlin et al. 2004).

A closer look at decapod development shows a similarly wide range of different patterns as is found in adult morphology (e.g., Korschelt 1944; Fioroni 1970; Anderson 1973; Schram 1986; Weygoldt 1994; Scholtz 1993, 2000). One can observe decapod eggs with high and low yolk content, with total cleavage and superficial cleavage types, with a distinct cell division and cell lineage pattern, and without these determinations. There are different kinds of gastrulation, ranging from invagination to immigration and delamination, and multiple gastrulation modes and phases within a species. In addition, the growth zone of the embryonic germ band is composed of different numbers of stem cells in the ectoderm, the so called ectoteloblasts (Dohle et al. 2004). Even at the level of gene expression patterns, the few existing publications on decapods reveal some differences between species (e.g., Averof & Patel 1997; Abzhanov & Kaufman 2004). Some groups hatch as a nauplius larva, whereas others hatch at later stages (such as zoea larvae) or exhibit direct development with hatchlings looking like small adults (Scholtz 2000).

With the notable exception of zoeal larval characters (e.g., Gurney 1942; Rice 1980; Clark 2005, this volume), surprisingly little attention has been paid to this developmental diversity and to decapod development in general when the phylogenetic relationships or evolutionary pathways have been discussed.

Here we present some examples of how ontogenetic data, such as cleavage, cell division, and gene expression patterns, can be used to infer phylogenetic relationships and evolutionary pathways among decapod crustaceans. It must be stressed, however, that this is just the beginning. Most relevant data on decapod ontogeny have yet to be described.

## 2 CLEAVAGE PATTERN, GASTRULATION, AND THE DECAPOD STEM SPECIES

It is now almost universally accepted that the sister groups Dendrobranchiata and Pleocyemata form the clade Decapoda (Burkenroad 1963, 1981; Felgenhauer & Abele 1983; Abele & Felgenhauer 1986; Christoffersen 1988; Abele 1991; Scholtz & Richter 1995; Richter & Scholtz 2001; Schram 2001; Dixon et al. 2003; Porter et al. 2005; Tsang et al. 2008). The monophyly of dendrobranchiates is largely based on the putatively apomorphic shape of the gills, which are highly branched, and perhaps on the specialized female thelycum and male petasma (Felgenhauer & Abele 1983). Nevertheless, the monophyly of Dendrobranchiata has been doubted based on characters of eye morphology (Richter 2002). Dendrobranchiata contains sergestoid and penaeoid shrimps, which have a largely similar life style (Pérez Farfante & Kensley 1997). In contrast to this, the pleocyematans include shrimp-like forms, such as carideans and stenopodids, but also the highly diverse reptants, which include lobsters, crayfishes, hermit crabs, and brachyuran crabs among others. When Burkenroad (1963, 1981) established the Pleocyemata, he stressed the characteristic brood-care feature of this group, namely, the attachment of the eggs and embryos to the maternal pleopods. With few exceptions, such as *Lucifer*, which attaches the eggs to the 3rd pleopods (Pérez Farfante & Kensley 1997), dendrobranchiates simply release their eggs into the water column. The monophyly of Pleocyemata is furthermore supported by brain characters (Sandeman et al. 1993).

The early development is quite different between Dendrobranchiata and Pleocyemata. Dendrobranchiates show relatively small, yolk-poor eggs with a total cleavage, a stereotypic cleavage pattern resulting in two interlocking cell bands, a determined blastomere fate, and a gastrulation initiated by two large cells largely following the mode of a modified "invagination" gastrula (e.g., Brooks 1882; Zilch 1978, 1979; Hertzler & Clark 1992; Hertzler 2005; Biffis et al. in prep) (Fig. 1). They hatch as nauplius larvae (Scholtz 2000). Pleocyematans mostly possess relatively large, yolky eggs with a superficial or mixed cleavage, no recognizable cell division pattern, and an immobile embryonized egg-nauplius (see Scholtz 2000; Alwes & Scholtz 2006). There are a few exceptions found in some carideans, hermit crabs, and brachyurans among reptants, which display an initial total cleavage (e.g., Weldon 1887; Gorham 1895; Scheidegger 1976), but these cleavages never show a consistent pattern comparable to that of Dendrobranchiata. The gastrulation is highly variable, and very often it implies immigration and no formation of a proper blastopore (Fioroni 1970; Scholtz 1995). The question is, which of these two types of developmental pathways—the one exhibited by the Dendrobranchiata or the less specified type exhibited by the Pleocyemata—is plesiomorphic within the Decapoda? This can only be answered with an outgroup, since two sister groups with two alternative sets of character states cannnot tell us which states are plesiomorphic. The answer to this question allows inferences on the origin and ground pattern of decapods; in particular, it might inform us as to whether the ancestral decapod was a swimming shrimp-like animal of the dendrobranchiate type or a benthic reptant. A pelagic lifestyle in malacostracan Crustacea is not necessarily

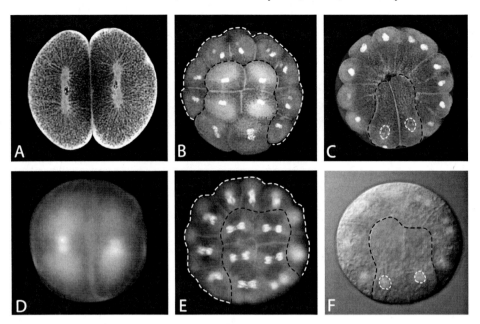

**Figure 1.** Different stages during early development of the dendrobranchiate shrimp *Penaeus monodon* (A-C) and of the euphausiacean *Meganyctiphanes norvegica* (D-F) stained with fluorescent dyes (Sytox A-C; Hoechst D-F). In F the fluorescence is combined with transmission light. The eggs show a low yolk content and total cleavage with a characteristic size and arrangement of the blastomeres. A and D: 2-cell stage. B and E: 32-cell stage. A stereotypic cleavage pattern leads to two interlocking cell bands, a "tennis ball pattern" (surrounded by white and black broken lines each). In B, the mitoses of the previous division are just completed, while in E the cells show the anaphase of the next division. C and F: 62-cell stage. Notice the center of the egg with two differently sized large mesendoderm cells (black broken lines), which arrest their division and initiate gastrulation.

combined with, but facilitates, the absence of brood care, whereas benthic malacostracans always show some degree of investment into the embryos and early larvae.

A comparison with the early development of Euphausiacea helps to polarize the developmental characters of Dendrobranchiata and Pleocyemata. Euphausiacea are either the sister group (Siewing 1956; Christoffersen 1988; Wills 1997; Schram & Hof 1998; Watling 1981, 1999) or are more remotely related to Decapoda (Richter 1999; Scholtz 2000; Jarman et al. 2000; Richter & Scholtz 2001). The Euphausiacea studied show remarkable similarities to dendrobranchiate decapods concerning their early embryonic and larval development (Taube 1909, 1915; Alwes & Scholtz 2004). They also release their eggs into the water column and show no brood care, with some apparently derived exceptions (Zimmer & Gruner 1956). Furthermore, they exhibit a corresponding cleavage pattern, arrangement and fate of blastomeres, and mode of gastrulation (Fig. 1). Like Dendrobranchiata, Euphausiacea hatch as a free nauplius. In particular, the formation of two interlocking germ bands, the origin and fate of the two large mesendoderm cells that initiate the gastrulation, and the formation of distinct cell rings (crown cells) at the margin of the blastopore find a detailed correspondence between dendrobranchiates and euphausiids (Hertzler & Clark 1992; Alwes & Scholtz 2004; Hertzler 2005) (Fig. 1). It must be stressed, however, that the nauplius larvae of dendrobranchiate decapods and Euphausiacea might be the result of convergent evolution (Scholtz 2000). It is furthermore not clear when this type of cleavage and early development evolved within malacostracans. The similarities in early development might indicate that euphausiaceans are the sister group to decapods (see Alwes & Scholtz 2004) (Fig. 2), in agreement with previous suggestions (e.g., Siewing 1956; Christoffersen 1988; Wills, 1997; Schram & Hof 1998; Watling

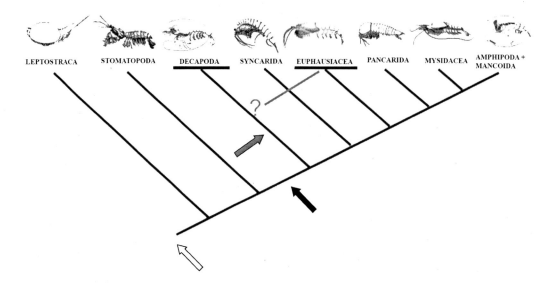

**Figure 2.** Malacostracan phylogeny according to Richter & Scholtz (2001). The arrows indicate the three possibilities for the evolution of the characteristic early development shared by Euphausiacea and Dendrobranchiata (Decapoda). The black arrow shows the possibility that the cleavage pattern evolved in the lineage of Caridoida. The grey arrow indicates a shared evolution of the cleavage pattern for Decapoda and Euphausiacea in combination with the view of a sister group relationship between these two groups (Eucarida), as is indicated with a question mark and light grey line. The white arrow symbolizes an older origin of the developmental pattern, perhaps even in non-malacostracans.

1981, 1999). On the other hand, if we accept the analysis of Richter and Scholtz (2001), the pattern must have evolved in the stem lineage of Caridoida (Fig. 2). However, it might be even older since similar patterns occur in some non-malacostracan crustaceans (Kühn 1913; Fuchs 1914, see Alwes & Scholtz 2004) (Fig. 2).

In either case, this corresponding early development of euphausiids and dendrobranchiate decapods to the exclusion of Pleocyemata strongly suggests that originally decapods did not care for the brood but released their yolk-poor eggs freely into the water. Furthermore, these eggs developed via a stereotypic cleavage pattern with largely determined cell fates and a specific mode of gastrulation. All of this indicates that the early development of Dendrobranchiata is plesiomorphic within Decapoda. In addition, this allows for the conclusion that the ancestral decapod was a more pelagic shrimp-like crustacean.

The oldest known fossil decapod is the late Devonian species *Palaeopalaemon newberryi* (see Schram et al. 1978). According to these authors, this fossil is a representative of the reptant decapods (see also Schram & Dixon 2003). This was disputed by Felgenhauer and Abele (1983), who claimed that the shrimp-like scaphocerite instead indicates an affinity to dendrobranchiates or carideans. Our conclusions, based on ontogenetic data, might lead to reconsidering the affinities of *Palaeopalaemon* as a dendrobranchiate-like decapod. At least there is no morphological structure that contradicts this assumption. This interpretation would furthermore fit with the ideas of Schram (2001) and Richter (2002) who independently concluded, based on eye structure and other arguments, that it is likely that decapods originated in deeper areas of the sea.

## 3 WAS THE ANCESTRAL DECAPOD A DECAPOD?

One of the apomorphies for Malacostraca is the possession of eight thoracic segments and their corresponding eight thoracopods (Richter & Scholtz 2001). In the various malacostracan groups, the thoracopods are diversified to different degrees, with the most conspicuous transformation being

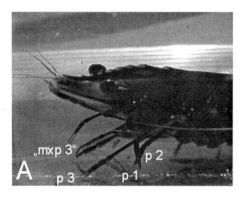 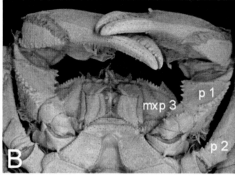

**Figure 3.** Evolution of 3rd maxillipeds in decapods. (A) The dendrobranchiate shrimp *Penaeus monodon* with pediform 3rd maxillipeds (mxp 3), which are not very different from the 1st anterior pereopods (p1 to p3). (B) The 3rd maxilliped (mxp3) of the brachyuran *Eriocheir sinensis* is highly transformed compared to the first two pereopods (p1, p2).

the modification of anterior thoracic limbs to secondary mouthparts, the maxillipeds. Depending on the number of thoracopods transformed to maxillipeds, the number of walking limbs (pereopods) varies. In most malacostracans we find either none (Leptostraca, Euphausiacea), one (e.g., Isopoda, Amphipoda, Anaspidacea) to two (Mysidacea), and sometimes three (Cumacea, most Decapoda) or even five (Stomatopoda) pairs of maxillipeds, which correspondingly means eight, seven, six, five, or three pairs of pereopods (Richter & Scholtz 2001). It is quite safe to assume that the plesiomorphic condition in malacostracans was the absence of any maxillipeds and that the number increased convergently in the course of malacostracan evolution. Only the anteriormost maxilliped might be homologous between those malacostracan taxa that possess it (Richter & Scholtz 2001). Decapods, as the name indicates, are characterized by five pairs of pereopods, which lie posterior to three pairs of maxillipeds. However, the concept of what has to be considered a maxilliped is not very sharp, because it relates to a combination of morphological deviation and different function from a locomotory limb, which is assumed to represent the ancestral throracopod state. Indeed, the locomotory pereopods of malacostracans are often also involved in food gathering and processing of some sort, and the large chelipeds of a lobster, for instance, are seldom used for locomotion. On the other hand, the morphology of some, in particular the posteriormost, maxillipeds is not very different from that of the pereopods. For instance, the 3rd maxillipeds of lobsters are more leg-like than those of most brachyuran crabs in which these form the operculum covering the mouth field (Scholtz & McLay this volume) (Fig. 3).

In particular, in some dendrobranchiates the 3rd maxillipeds are morphologically not really discernible from the pereopods (Fig. 3). They have the same length and segment number as the pereopods and are not kept closely attached to the mouth field. Accordingly, the question arises as to whether the stem species of decapods was equipped with only two pairs of maxillipeds and hence six pairs of pereopods (see Scholtz & Richter 1995; Richter & Scholtz 2001)—in other words, whether it was a dodecapod (dodeka: Greek for twelve) rather than a true decapod.

In their seminal work, Averof and Patel (1997) developed a new molecular criterion for maxillipeds. They found that the Hox gene ultrabithorax (UBX) is expressed in thoracic regions with pereopods, whereas in segments bearing maxillipeds, this gene is not expressed. UBX is needed to differentiate trunk segments, and the absence of UBX expression allows the transformation towards mouthparts (Averof & Patel 1997). This is true for all crustaceans investigated in this respect. Interestingly enough, the two decapod species studied by Averof and Patel (1997) differed slightly in the anterior margin of UBX expression depending on the degree of deviation from a pereopod-like appearance of the 3rd maxillipeds (see Fig. 5). In the lobster, with a more pediform 3rd maxilliped

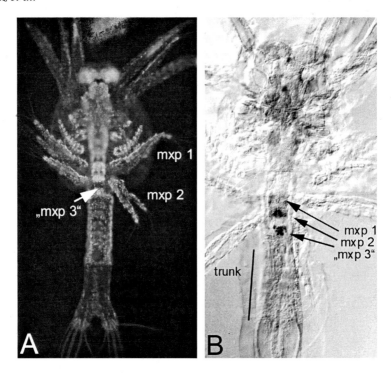

**Figure 4.** Expression of the UBX-AbdA protein in the protozoea of *Penaeus monodon* as seen with the antibody FP6.87. (A) 1st protozoea stained with the nuclear dye Hoechst, showing the overall shape, the limbs, and the central nervous system. The two anterior pairs of maxillipeds (mxp1, 2) are present and the corresponding ganglion anlagen are recognizable. The 3rd maxilliped pair is not yet differentiated but the ganglion is forming (mxp3). (B) 1st protozoea showing UBX expression in the ganglia of the 2nd and 3rd maxillipeds (mxp2, 3) and in the posterior part of the ganglion of the 1st maxilliped segment (mxp1). The anterior expression boundary of UBX is parasegmental. In addition, there is a weak expression in the forming trunk segments. No limbs are stained, which might be due to penetration problems through the well-developed cuticle.

(concerning length, overall shape, and the occurrence of five endopodal articles), the expression, at least in early stages, was also seen in this body segment. However, in the caridean shrimp, with a derived 3rd maxilliped (stout and only three endopodal articles; see, e.g., Bruce 2006), the anterior boundary of UBX expression was always behind the segment bearing the 3rd maxilliped. To test this phenomenon in dendrobranchiate decapods, we used the same antibody against the UBX-AbdA product (FP6.87) as Averof and Patel (1997) to study the expression of UBX in *Penaeus monodon* (Fig. 4). This species is characterized by a pediform 3rd maxilliped that still shows five endopodal segments and that is, compared to most pleocyemate species, still long and slender (Motoh 1981) (Fig. 3). In *Penaeus monodon* protozoea larvae, we find an anterior expression boundary of UBX in the forming nervous system slightly anterior to the 2nd maxilliped segment, which is the anteriormost expression found in a decapod to date (Figs. 4, 5). This result indicates that the specification of the 3rd maxilliped in dendrobranchiates has not reached the degree found in the other decapods and that most likely a 3rd maxilliped in the true sense was absent in the decapod stem species. It furthermore suggests that a true 3rd maxilliped evolved convergently several times within Decapoda. Interestingly enough, a closer look at the situation in the Amphionida, a possible candidate as the sister group to decapods (Richter and Scholtz 2001), supports this conclusion. This group possesses a well-defined maxilliped on the 1st thoracic segment and a reduced 2nd thoracic limb that nevertheless resembles the maxilliped in its overall shape. The 3rd to 8th thoracic appendages are all pereopods with a different morphology (Schram 1986).

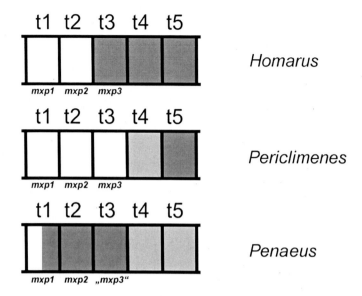

**Figure 5.** Scheme of the anterior expression of the UBX-AbdA protein in three decapod representatives with different degrees of pediform 3rd maxillipeds. *Homarus* and *Penaeus* with more pediform 3rd maxillipeds show a more anterior UBX expression boundary. *Penaeus* with the most pereopod-like 3rd maxilliped reveals the most anterior boundary in the 1st thoracic segment. *Homarus* and *Periclimenes* after Averof & Patel (1997), *Penaeus* this study. Light grey = weak expression, dark grey = strong expression. (mxp1,2,3 =1st to 3rd maxillipeds, t1 to t5 = 1st to 5th thoracic segments).

## 4 THE ORIGIN OF THE SCAPHOGNATHITE

The scaphognathite is a large flattened lobe at the lateral margin of the 2nd maxillae of decapods and amphionids (Fig. 6). The scaphognathite is equipped with numerous plumose setae at its margin and is closely fitted to the walls of the anterior part of the branchial chamber. This allows it to create a water current through the branchial chamber depending on the movement of the 2nd maxilla. This current supplies the gills with fresh oxygen-rich water for breathing. Hence, the scaphognathite is a crucial element of the gill/branchial chamber complex that is apomorphic for Decapoda (including Amphionida). The morphological nature and origin of this important structure, however, have been a matter of debate for more than a century. This relates to the general difficulty in assigning the elements of the highly modified decapod mouthparts to the parts of biramous crustacean limbs, such as the endopod, exopod, or epipods. Accordingly, several authors claim that the scaphognathite is a composite structure formed by the fusion of the exopod and epipod of the 2nd maxilla (Huxley 1880; Berkeley 1928; Gruner 1993). Huxley (1880) even discusses the alternative that it is exclusively formed by the epipod. In contrast to this, carcinologists such as Calman (1909), Giesbrecht (1913), Hansen (1925), Borradaile (1922), and Balss (1940) interpret the scaphognathite as of solely exopod origin. These different traditions are still expressed in recent textbooks (see Gruner 1993; Gruner & Scholtz 2004; Schminke 1996; Ax 1999). But Kaestner (1967: 1073) and Schram (1986: 245), discussing the morphology of decapod 2nd maxillae, state that "Homologie noch unklar!" (homology not clear) and "This appendage is so extensively modified that to suggest homologies with the various components of other limbs is a questionable exercise."

We studied the development of the 2nd maxillae in the embryos of a freshwater crayfish, the parthenogenetic Marmorkrebs (Scholtz et al. 2003; Alwes & Scholtz 2006), applying the means

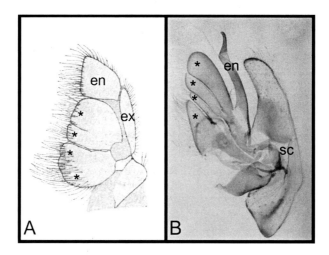

**Figure 6.** The shape and elements of the 2nd maxillae. (A) The 2nd maxilla of the euphausicaean *Meganyctiphanes norvegica* (after Zimmer & Gruner 1956). (B) The 2nd maxilla of the decapod *Axius glyptocereus*. The maxillae of both species show an endopod (en) and four enditic lobes (asterisks). The scaphognathite (sc) characteristic for decapods has such a special shape and function that the homology to the exopod (ex) in euphausiaceans and other malacostracans is controversial.

of histology, scanning electron microscopy, and immunochemistry (Distal-less) to clarify the issue of scaphognathite origins (Fig. 7). The Distal-less gene is involved in the adoption of a distal fate of limb cells in arthropods and is thus a marker for the distal region of arthropod limbs (e.g., Panganiban et al. 1995: Popadic et al. 1998; Scholtz et al. 1998; Williams 1998; Olesen et al. 2001; Angelini & Kaufman 2005). The early limb bud of the 2nd maxilla is undivided. After a short period, the tip of the bud shows a slight cleft that deepens with further development. This process is typical for the early development of crustacean biramous limbs (Hejnol & Scholtz 2004; Wolff & Scholtz 2008). The tips of the undivided limb buds, as well as the later-forming two separate tips, express Distal-less. Again, this is characteristic for biramous crustacean limbs and indicates that the two tips represent the exopod and endopod, since epipods do not express Dll (with the notable exception of the transient expression in epipods of *Artemia* and *Nebalia*, Averof & Cohen 1997; Williams 1998). With further development, the outer branch widens and grows in anterior and posterior directions, eventually adopting the characteristic lobed shape of the adult decapod scaphognathite (Fig. 7). In these later stages endopod and exopod still express Dll (Fig. 7D). A forming epipod is not recognizable at any stage of development, as is also revealed by the comparison to other limb anlagen which are equipped with an epipod.

Our results clearly support the idea that the scaphognathite of decapods is a transformed exopod and that an epipod is not involved in its formation. A comparison with other malacostracans reveals that in no case is the 2nd maxilla equipped with an epipod, but just endopods and exopods with different degrees of deviation from a "normal" limb branch. In addition, the overall shape of the scaphognathite is not so unusual for an exopod if we consider the shape of the exopods of phyllobranchious thoracic limbs in Branchiopoda and Leptostraca (Pabst & Scholtz 2009).

## 5 EMBRYONIC CHARACTERS HELP TO CLARIFY FRESHWATER CRAYFISH MONOPHYLY

Freshwater crayfish, Astacida, show a very disparate geographical distribution. In the Northern Hemisphere, the Cambaridae are found in East Asia and in the eastern part of North America, whereas the Astacidae occur in western Asia, Europe, and in the western parts of North America.

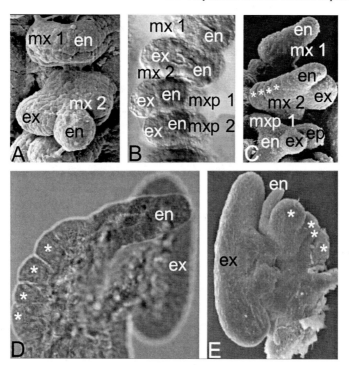

**Figure 7.** Development of the 2nd maxilla and the scaphognathite in the parthenogenetic Marmorkrebs (Astacida). (A) SEM image of the early 1st and 2nd maxillae (mx1, mx2) showing the forming two branches of the endopod (en) and exopod (ex) in the 2nd maxilla. (B) Expression of Distal-less (Dll) in early limb anlagen. Dll is expressed (darker areas) in the tips of the endopods (en) and exopods (ex) of the 2nd maxilla and the maxillipeds (mxp1, 2). The uniramous bud of the 1st maxilla (en) also expresses Dll. (C) SEM image showing the further differentiation of the parts of the 2nd maxilla (mx2). The four enditic lobes are forming (asterisks), and the exopod (ex) begins to form a lobe structure. The 1st maxilliped (mxp1) differentiates an epipod (ep), which finds no correspondence in the two maxillae. (D) Dll expression in an advanced stage. The expression (darker areas) is found in the tip of the endopod and around the margin of the exopod. The asterisks indicate the forming four enditic lobes. (E) SEM image of a 2nd maxilla shortly before hatching. The general shape of the adult maxilla is present (compare with Fig. 6).

Even if both groups, Astacidae and Cambaridae, are not monophyletic as has recently been suggested (Scholtz 1995, 2002; Crandall et al. 2000; Rode & Babcock 2003; Braband et al. 2006; Ahn et al. 2006), this distribution pattern is difficult to explain. The Parastacidae of the Southern Hemisphere live in Australia, New Zealand, some parts of South America, and Madagascar. Crayfish are absent from continental Africa. This is also true for the Indian subcontinent, and in more general terms, there is a crayfish-free circum-tropical zone. To explain this disparate distribution of freshwater crayfish, several hypotheses on the origin and evolution of crayfish have been discussed during the last 130 years. Most authors favored the idea that freshwater crayfish had multiple origins from different marine ancestors, i.e., are polyphyletic, and that they independently invaded freshwater many times (e.g., Huxley 1880; Starobogatov 1995; for review see Scholtz 1995, 2002). This view is based on the fact that freshwater crayfish do not tolerate higher salinities and that an explanation is needed for the occurrence of Astacida on most continents without the possibility of crossing large marine distances. Only Ortmann (1897, 1902) suggested a common origin for freshwater crayfish and a single invasion into freshwater habitats. He hypothesized East Asia as the center of origin from which Astacida spread all over the world, using assumed low sea levels to migrate to other continents (since the concept of continental drift was unknown at that time).

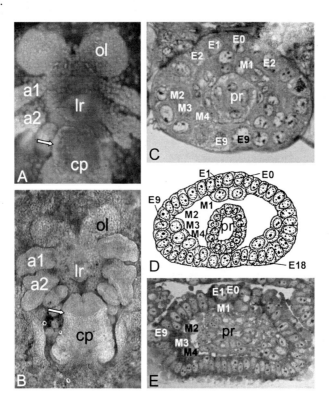

Figure 8. Teloblasts in decapod embryos. (A) Ventral view of the germ band of an embryo of the thalassinid *Callianassa australiensis*. The arrow indicates the area where the teloblasts form a ring (ectoderm and mesoderm) around the ventrally folded caudal papilla (cp). (a1, a2 = 1st and 2nd antennae, lr = labrum, ol = optic lobe). (B) Ventral view of the germ band of an embryo of the crayfish *Cambaroides japonicus* (labels as in A). Note the higher number of cells compared to A. (C) Transverse section through the caudal papilla of the American lobster *Homarus americanus* at the level of the teloblast rings; 19 ectoteloblasts (one unpaired E0 and nine paired E1 to E9 teloblast cells) and 8 mesoteloblast (four pairs in a specific arrangement) surround the forming proctodaeum (pr). (D) Transverse section through the caudal papilla of the Australian crayfish *Cherax destructor* at the level of the teloblast rings. In contrast to *Homarus*, there are about 40 teloblasts in the ectoderm. The mesoteloblasts show the same pattern as in the lobster. (E) Transverse section through the caudal papilla of the Japanese crayfish *Cambaroides japonicus* at the level of the teloblast rings. The pattern in this Northern Hemisphere crayfish is the same as in the Southern Hemisphere representative *Cherax* (after Scholtz 1993; Scholtz & Kawai 2002).

The investigation on cell division patterns in the germ band of embryos of the Australian freshwater crayfish *Cherax destructor* produced the surprising result that the growth zone of this species differs from that of all other malacostracan crustaceans studied so far in this respect (Scholtz 1992). The growth zone of malacostracans is situated in the posterior region of the embryo, immediately anterior to the telson anlage. It is formed by large specialized cells, the teloblasts, which bud off smaller cells only toward the anterior (see Dohle et al. 2004) (Fig. 8). This stem-cell-like cell type occurs in the ectoderm (ectoteloblasts) and the mesoderm (mesoteloblasts), and both sets of teloblasts produce most of the ectodermal and mesodermal material of the post-nauplier germ band. In the ground pattern of Malacostraca, we find 19 ectoteloblasts and 8 mesoteloblasts in circular arrangements (Dohle et al. 2004) (Fig. 8C). These figures are also present in most decapods studied in this respect, such as caridean shrimps, Achelata, Homarida, Thalassinida, Anomala, and Brachyura (Oishi 1959, 1960; Scholtz 1993). In contrast to this, in the freshwater crayfish *Cherax destructor* an individually variable number of more than 40 ectoteloblasts occurs, whereas the 8 mesoteloblasts are conserved (Fig. 8D). Subsequent studies in other crayfish species from the Northern and Southern Hemispheres covering Astacidae, Cambaridae, and Parastacidae revealed that the pattern found in *Cherax* is a general freshwater crayfish character (Scholtz 1993) (Fig. 8E). This different growth zone pattern is hence a clear apomorphy of the Astacida, strongly indicating their monophyly.

This result is corroborated by a number of other developmental, in particular postembryonic, characters (see Scholtz 2002). In addition, phylogenetic analyses based on molecular datasets strongly support the monophyly of Astacida (e.g., Crandall et al. 2000; Ahyong & O'Meally 2004; Tsang et al. 2008). The question of freshwater colonization can now be addressed anew based on the strong support for Astacida monophyly. Monophyly alone is, of course, no proof for a single invasion into freshwater habitats, but parsimony and, in particular, several apomorphic freshwater adaptations strongly argue for a crayfish stem species already living in freshwater (see Scholtz 1995, 2002; Crandall et al. 2000). The modern and almost worldwide distribution of Astacida is thus best explained by the assumption of a freshwater colonization during the Triassic or even earlier before the break-up of Pangaea, which started in the Jurassic (Scholtz 1995, 2002).

## 6 CONCLUSIONS

With these examples, we demonstrate the different levels of impact on our views on decapod evolution resulting from comparative developmental studies (see Scholtz 2004). Including developmental characters in phylogenetic analyses expands our suite of characters for phylogenetic inference. In some cases, ontogenetic characters can be decisive in resolving phylogenetic relationships that cannot be inferred from adult characters alone. An example of this is the resolution of the common origin of astacoidean and parastacoidean crayfish. However, based on ontogenetic data, far-reaching conclusions can be drawn. For instance, the morphological "nature" of adult structures can be clarified with developmental analyses. This touches the core of morphology as a science. Morphological structures are transformed in the course of evolution; they change form and function to various degrees. In addition, new structures (novelties) emerge. These are, however, formed by pre-existing morphological precursors. Developmental analyses offer the possibility to trace these transformations and novelties. The analyses presented here of the 3rd maxillipeds and the scaphognathite of the 2nd maxillae in decapods provide examples for this approach. In the latter case, a century-old controversy was resolved and the evolutionary flexibility of limb structures was shown. In the former case, the correlation between an evolutionary shift of gene expression and altered morphology and function is revealed. Furthermore, evolutionary scenarios can be inferred based on ontogenetic data. This is shown by the timing of the gene expression shift. The transformation of a thoracic limb to a mouthpart takes place at the morphological and functional levels before gene expression has changed to the same degree (see Budd 1999). As is the case in adult structures, several ontogenetic characters are correlated with a certain lifestyle. If these characters are shared between an outgroup and part of the ingroup, it is possible to deduce the ancestral lifestyle of a given taxon. This approach is exemplified by the analysis of the early development of Dendrobranchiata. Yolk-poor eggs with a distinct cleavage pattern are found in shrimp-like crustaceans with a more pelagic lifestyle and a lack of brood care, such as euphausiaceans and, to a certain degree, anaspidaceans. This allows the conclusion that the decapod stem species was a pelagic shrimp-like animal rather than a benthic reptantian and thus strongly corroborates inferences based on the morphology of adults.

## ACKNOWLEDGEMENTS

We thank the organizers of the symposium "Advances in Decapod Crustacean Phylogenetics" at the SICB meeting in San Antonio 2008 for the invitation to GS to present our thoughts and results. Parts of the work presented here were supported by the Deutsche Forschungsgemeinschaft (DFG) and by a grant from the Marie Curie Actions—Early Stage Training Programme (Molmorph) of the European Union.

# REFERENCES

Abele, L.G. 1991. Comparison of morphological and molecular phylogeny of the Decapoda. *Mem. Qld. Mus.* 31: 101–108.

Abele, L.G. & Felgenhauer, B.E. 1986. Phylogenetic and phenetic relationships among the lower Decapoda. *J. Crust. Biol.* 6: 385–400.

Abzhanov, A. & Kaufman, T.C. 2004. Hox genes and tagmatization of the higher Crustacea (Malacostraca). In: Scholtz, G. (ed.), *Crustacean Issues 15 Evolutionary Developmental Biology of Crustacea*: 43–74. Lisse: Balkema.

Ahn, D.-H., Kawai, T., Kim, S.-J. Rho., H.S., Jung, J.W., Kim, W., Lim, B.J., Kim, M. & Min, G.S. 2006. Phylogeny of Northern Hemisphere freshwater crayfishes based on 16S rRNA gene analysis. *Kor. J. Gen.* 28: 185–192.

Ahyong, S. T. & O'Meally, D. 2004. Phylogeny of the Decapoda Reptantia: resolution using three molecular loci and morphology. *Raffl. Bull. Zool.* 52: 673–693.

Alwes, F. & Scholtz, G. 2004. Cleavage and gastrulation of the euphausiacean *Meganyctiphanes norvegica* (Crustacea, Malacostraca). *Zoomorphology* 123: 125–137.

Alwes, F. & Scholtz, G. 2006. Stages and other aspects of the embryology of the parthenogenetic Marmorkrebs (Decapoda, Reptantia, Astacida). *Dev. Genes Evol.* 216: 169–184.

Anderson, D.T. 1973. *Embryology and Phylogeny in Annelids and Arthropods*. Oxford: Pergamon Press.

Angelini, D.R. & Kaufman, T.C. 2005. Insect appendages and comparative ontogenetics. *Dev. Biol.* 286: 57–77.

Averof, M. & Patel, N.H. 1997. Crustacean appendage evolution associated with changes in Hox gene expression. *Nature* 388: 682–686.

Ax, P. 1999. *Das System der Metazoa. II*. Stuttgart: Gustav Fischer Verlag.

Balss, H. 1940. 5. Band, 1. Abteilung, 7. Buch Decapoda, 1. Lieferung. In: Schellenberg, A. (ed.), *Dr. H.G. Bronns Klassen und Ordnungen des Tierreichs 2. Auflage*: 1–165. Leipzig: Akademische Verlagsgesellschaft Becker & Erler.

Berkeley, A.A. 1928. The musculature of *Pandalus danae* Stimpson. *Trans. R. Can. Inst.* 16: 181–231.

Beurlen, K. & Glaessner M.F. 1930. Systematik der Crustacea Decapoda auf stammesgeschichtlicher Grundlage. *Zool. Jb. Syst.* 60: 49–84.

Boas, J.E.V. 1880. Studier over decapodernes Slaegtskabsforhold. *K. Danske.Vidensk. Selsk. Skr*: 6: 163–207.

Borradaile, L.A. 1907. On the classification of the decapod crustaceans. *An. Mag. Nat. Hist.* 19: 457–486.

Borradaile, L.A. 1916. Crustacea. I. Part II. *Porcellanopagurus*; an instance of carcinogenization. British Antarctic ('Terra Nova') Expedition, 1910. *Nat. Hist. Rept. (Zoology), British Museum* 3: 75–126.

Borradaile, L.A. 1922. On the mouth-parts of the shore crab. *Zool. J. Linn. Soc.* 35: 115–142.

Braband, A., Kawai, T. & Scholtz, G. 2006. The phylogenetic position of the East Asian freshwater crayfish *Cambaroides* within the Northern Hemisphere Astacoidea (Crustacea, Decapoda, Astacida) based on molecular data. *J. Zool. Syst. Evol. Res.* 44: 17–24.

Brooks, W.K. 1882. *Leucifer*. A study in morphology. *Phil. Trans. R. Soc. Lond.* 173: 130–137.

Bruce, A.J. 2006. An unusual new Periclimenes (Crustacea, Decapoda, Palaemonidae) from New Caledonia. *Zoosystema* 28: 703–712.

Budd, G.E. 1999. Does evolution in body patterning genes drive morphological change—or vice versa? *BioEssays* 21: 326–332.

Burkenroad, M.D. 1963. The evolution of the Eucarida, (Crustacea, Eumalacostraca), in the relation to the fossil record. *Tulane Stud. Geol.* 2: 1–17.

Burkenroad, M.D. 1981. The higher taxonomy and evolution of Decapoda (Crustacea). *Trans. San Diego Soc. Nat. Hist.* 19: 251–268.

Calman, W.T. 1909. *A Treatise on Zoology, 7: Appendiculata, Crustacea*. London: Adam and Charles Black.

Christoffersen, M.L. 1988. Phylogenetic systematics of the Eucarida (Crustacea, Malacostraca). *Rev. Bras. Zool.* 5: 325–351.

Clark, P.F. 2005. The evolutionary significance of heterochrony in the abbreviated zoeal development of pilumnine crabs (Crustacea: Brachyura: Xanthoidea). *Zool J. Linn. Soc. Lond.* 143: 171–181.

Clark, P.F. (this volume). The bearing of larval morphology on brachyuran phylogeny. In: Martin, J.W., Crandall, K.A. & Felder, D.L. (eds.), *Crustacean Issues: Decapod Crustacean Phylogenetics*. Boca Raton, Florida: Taylor & Francis/CRC Press.

Crandall, K.A., Harris, D.J. & Fetzner, J.W., Jr. 2000. The monophyletic origin of freshwater crayfish estimated from nuclear and mitochondrial DNA sequences. *Proc. R. Soc. Lond. B* 267: 1679–1686.

Dixon, C.J., Ahyong, S.T. & Schram, F.R. 2003. A new hypothesis of decapod phylogeny. *Crustaceana* 76: 935–975.

Dohle, W., Gerberding, M., Hejnol, A. & Scholtz, G. 2004. Cell lineage, segment differentiation, and gene expression in crustaceans. In: Scholtz, G. (ed.), *Crustacean Issues 15 Evolutionary Developmental Biology of Crustacea:* 95–133. Lisse: Balkema.

Felgenhauer, B.E. & Abele, L.G. 1983. Phylogenetic relationships among shrimp-like decapods. In: Schram, F.R. (ed.), *Crustacean Issues 1, Crustacean Phylogeny*: 291–311. Rotterdam, Lisse: Balkema.

Fioroni, P. 1970. Am Dotteraufschluss beteiligte Organe und Zelltypen bei höheren Krebsen; der Versuch zu einer einheitlichen Terminologie. *Zool. Jb. Anat.* 87: 481–522.

Fuchs, F. 1914. Die Keimblätterentwicklung von *Cyclops viridis* Jurine. *Zool. Jb. Anat.* 38: 103–156.

Giesbrecht, W. 1913. II. Klasse: Crustacea. In: Lang, A. (ed.), *Handbuch der Morphologie der wirbellosen Tiere, 4. Bd. Arthropoda, 1. Lieferung*: 9–160. Jena: Gustav Fischer Verlag.

Gorham, F. P. 1895. The cleavage of the egg of *Virbius zostericola* Smith. *J. Morphol.* 11: 741–746.

Gruner, H.-E. 1993. Crustacea. In: Gruner, H.-E. (ed.), Arthropoda (ohne Insecta). *Lehrbuch der Speziellen Zoologie, Bd. I, 4. Teil*: 448–1030. Jena: Gustav Fischer Verlag.

Gruner, H.-E. & Scholtz, G. 2004. Segmentation, tagmata, and appendages. In: Forest, J. & von Vaupel Klein, J.C. (eds.), *Treatise on Zoology: The Crustacea, revised and updated from the Traité de Zoologie, Volume 1*: 13–57, Leiden: Brill.

Gurney, R. 1942. *Larvae of Decapod Crustaceans*. London: Ray Society.

Hansen, H.J. 1925. *Studies on Arthropoda II. On the comparative morphology of the appendages in the Arthropoda. A. Crustacea*. Copenhagen. Gyldendalske Boghandel.

Hejnol, A. & Scholtz, G. 2004. Clonal analysis of *Distal-less* and *engrailed* expression patterns during early morphogenesis of uniramous and biramous crustacean limbs. *Dev. Genes Evol.* 214: 473–485.

Hertzler, P.L. & Clark, W.H.J. 1992. Cleavage and gastrulation in the shrimp *Sicyonia ingentis*: invagination is accombined by oriented cell division. *Development* 116: 127–140.

Hertzler, P.L. 2005. Cleavage and gastrulation in the shrimp *Penaeus* (*Litopenaeus*) *vannamei* (Malacostraca, Decapoda, Dendrobranchiata). *Arthrop. Struct. Dev.* 34: 455–469.

Huxley, T.H. 1880. *The Crayfish: An Introduction in the Study of Zoology*. London: Kegan Paul, Trench & Co.

Jarman, S.N., Nicol, S., Elliot, N.G. & McMinn, A. 2000. 28S rDNA evolution in the Eumalacostraca and the phylogenetic position of Krill. *Mol. Phylogenet. Evol.* 17: 26–36.

Kaestner, A. 1967. Crustacea Krebse. In: Kaestner, A. (ed.), *Lehrbuch der Speziellen Zoologie, Teil I: Wirbellose*: 685–979. Jena: Gustav Fischer Verlag.

Korschelt, E. 1944. 5. Band, 1. Abteilung, 7. Buch Decapoda, 6. Lieferung Ontogenie. In: Schellenberg, A. (ed.), *Dr. H.G. Bronns Klassen und Ordnungen des Tierreichs 2. Auflage*: 671–861. Leipzig: Akademische Verlagsgesellschaft Becker & Erler.

Kühn, A. 1913. Die Sonderung der Keimbezirke in der Entwicklung der Sommereier von *Polyphemus*. *Zool. Jb. Anat.* 35: 243–340.

Martin, J.W. & Abele, L.G. 1986. Phylogenetic relationships of the genus *Aegla* (Decapoda: Anomura: Aeglidae), with comments on anomuran phylogeny. *J. Crust. Biol.* 6: 576–616.

McLaughlin, P.A. & Lemaitre, R. 1997. Carcinization in the Anomura—fact or fiction? *Contr. Zool.* 67: 79–123.

McLaughlin, P.A., Lemaitre, R. & Tudge, C.C. 2004. Carcination in the Anomura—fact or fiction? II. Evidence from larval, megalopal and early juvenile morphology. *Contr. Zool.* 73: 165–205.

Morrison, C.L., Harvey, A.W., Lavery, S., Tieu, K., Huang, Y. & Cunningham, C.W. 2002. Mitochondrial gene rearrangements confirm the parallel evolution of the crab-like form. *Proc. R. Soc. Lond. B* 269: 345–350.

Motoh, H. 1981. *Studies on the fisheries biology of the giant tiger prawn*, Penaeus monodon *in the Philippines*. Tigbauan: Aquaculture Department Southeast Asian Fisheries Development Center.

Oishi, S. 1959. Studies on the teloblasts in the decapod embryo. I. Origin of teloblasts in *Heptacarpus rectirostris* (Stimpson). *Embryologia* 4: 283–309.

Oishi, S. 1960. Studies on the teloblasts in the decapod embryo. II. Origin of teloblasts in *Pagurus samuelis* (Stimpson) and *Hemigrapsus sanguineus* (de Haan). *Embryologia* 5: 270–282.

Olesen, J., Richter, S. & Scholtz G. 2001. The evolutionary transformation of phyllopodous to stenopodous limbs in the Branchiopoda (Crustacea)—Is there a common mechanism for early limb development in arthropods? *Int. J. Dev. Biol.* 45: 869–876.

Ortmann, A.E. 1897. Ueber 'Bipolarität' in der Verbreitung mariner Tiere. *Zool. Jb. Syst.* 9: 571–595.

Ortmann, A.E. 1902. The geographical distribution of fresh-water decapods and its bearing upon ancient geography. *Proc. Amer. Philosp. Soc.* 41 (171): 267–400.

Pabst, T. & Scholtz, G. (2009). The development of phyllopodous limbs in Leptostraca and Branchiopoda. *J. Crust. Biol.* (in press).

Panganiban, G., Sebring, A., Nagy, L.M. & Carroll, S.B. 1995. The development of crustacean limbs and the evolution of arthropods. *Science* 270: 1363–1366.

Pérez Farfante, I. & Kensley, B. 1997. Penaeoid and Sergestoid shrimps and prawns of the world. *Mem. Mus. Nat. Hist. Nat.* 175: 1–233.

Popadic, A., Panganiban, G., Rusch, D., Shear, W.A. & Kaufman, T. C. 1998. Molecular evidence for the gnathobasic derivation of arthropod mandibles and for the appendicular origin of the labrum and other structures. *Dev. Genes Evol.* 208: 142–150.

Porter, M.L., Pérez-Losada, M. & Crandall, K.A. 2005. Model-based multi-locus estimation of decapod phylogeny and divergence times. *Mol. Phylogenet. Evol.* 37: 355–369.

Rice, A.L. 1980. Crab zoeal morphology and its bearing on the classification of the Brachyura. *Trans. Zool. Soc. Lond.* 35: 271–424.

Richter, S. 1999. The structure of the ommatidia of the Malacostraca (Crustacea)—a phylogenetic approach. *Verh. Naturwiss. Ver. Hamburg* (NF) 38: 161–204.

Richter, S. 2002. Evolution of optical design in the Malacostraca (Crustacea). In: Wiese, K. (ed.), *The Crustacean Nervous System*: 512-524. Berlin: Springer Verlag.

Richter, S. & Scholtz, G. 1994. Morphological evidence for a hermit crab ancestry of lithodids (Crustacea, Anomala, Paguroidea). *Zool. Anz.* 233: 187–210.

Richter, S. & Scholtz, G. 2001. Phylogenetic analysis of the Malacostraca (Crustacea). *J. Zool. Syst. Evol. Res.* 39: 113–136.

Rode, A.L. & Babcock, L.E. 2003. Phylogeny of fossil and extant freshwater crayfish and some closely related nephropid lobsters. *J. Crust. Biol.* 23: 418–435.

Sandeman, D.C., Scholtz, G. & Sandeman, R.E. 1993. Brain evolution in decapod Crustacea. *J. Exp. Zool.* 265: 112–133.

Scheidegger, G. 1976. Stadien der Embryonalentwicklung von *Eupagurus prideauxi* Leach (Crustacea, Decapoda, Anomura) unter besonderer Berücksichtigung der Darmentwicklung und der am Dotterabbau beteiligten Zelltypen. *Zool. Jb. Anat.* 95: 297–353.

Schminke, H.K. 1996. Crustacea, Krebse. In: Westheide, W. & Rieger, R. (eds.), *Spezielle Zoologie, Teil 1: Einzeller und wirbellose Tiere*: 501–581. Stuttgart: Gustav Fischer.

Scholtz, G. 1992. Cell lineage studies in the crayfish *Cherax destructor* (Crustacea, Decapoda): germ band formation, segmentation, and early neurogenesis. *Roux's Arch. Dev. Biol.* 202: 36–48.

Scholtz, G. 1993. Teloblasts in decapod embryos: an embryonic character reveals the monophyletic origin of freshwater crayfishes (Crustacea, Decapoda). *Zool. Anz.* 230: 45–54.

Scholtz, G. 1995. Ursprung und Evolution der Flusskrebse (Crustacea, Astacida). *Sitz.-Ber. Ges. Naturf. Freunde Berlin (NF)* 34: 93–115.

Scholtz, G. 2000. Evolution of the nauplius stage in malacostracan crustaceans. *J. Zool. Syst. Evol. Research.* 38: 175–187.

Scholtz, G. 2002. Phylogeny and evolution. In: Holdich, D. (ed.), *Biology of Freshwater Crayfish*: 30–52. Oxford: Blackwell Science.

Scholtz, G. 2004. Baupläne *versus* ground patterns, phyla *versus* monophyla: aspects of patterns and processes in evolutionary developmental biology. In: Scholtz, G. (ed.), *Crustacean Issues 15, Evolutionary Developmental Biology of Crustacea*: 3–16. Lisse: Balkema.

Scholtz, G. & Kawai, T. 2002. Aspects of embryonic and post-embryonic development of the Japanese crayfish *Cambaroides japonicus* (Crustacea, Decapoda) including a hypothesis on the evolution of maternal care in the Astacida. *Acta Zool.* 83: 203–212.

Scholtz, G. & McLay, C.L. (this volume). Is the Brachyura Podotremata a monophyletic group? In: Martin, J.W., Crandall, K.A. & Felder, D.L. (eds.), *Crustacean Issues: Decapod Crustacean Phylogenetics*. Boca Raton, Florida: Taylor & Francis/CRC Press.

Scholtz, G. & Richter, S. 1995. Phylogenetic systematics of the reptantian Decapoda (Crustacea, Malacostraca). *Zool. J. Linn. Soc.* 113: 289–328.

Scholtz, G., Braband, A., Tolley L., Reimann A., Mittmann B., Lukhaup C., Steuerwald F. & Vogt G. 2003. Parthenogenesis in an outsider crayfish. *Nature* 421: 806.

Scholtz, G., Mittmann, B. & Gerberding, M. 1998. The pattern of *Distal-less* expression in the mouthparts of crustaceans, myriapods and insects: new evidence for a gnathobasic mandible and the common origin of Mandibulata. *Int. J. Dev. Biol.* 42: 801–810.

Schram, F.R. 1986. *Crustacea*. Oxford: Oxford University Press.

Schram, F.R. 2001. Phylogeny of decapods: moving towards a consensus. *Hydrobiologia* 449: 1–20.

Schram, F.R. & Dixon, C. 2003. Fossils and decapod phylogeny. *Contr. Zool.* 72: 169–172.

Schram, F.R. & Hof, C.H.J. 1998. Fossils and the interrelationships of major crustaceans groups. In: Edgecombe, G.D. (ed.) *Arthropod Fossils and Phylogeny*: 233–302. New York: Columbia University Press.

Schram, F.R., Feldmann, R.M. & Copeland, M.J. 1978. The late Devonian Palaeopalaemonidae and the earliest decapod crustaceans. *Journ. Paleo.* 52: 1375–1387.

Siewing, R. 1956. Untersuchungen zur Morphologie der Malacostraca (Crustacea). *Zool. Jb. Anat.* 75: 39–176.

Starobogatov, Ya.I. 1995. Taxonomy and geographical distribution of crayfishes of Asia and East Europe (Crustacea: Decapoda: Astacidae). *Arthrop. Sel.* 4: 3–25.

Taube, E. 1909. Beiträge zur Entwicklungsgeschichte der Euphausiden. I. Die Furchung der Eier bis zur Gastrulation. *Z. wiss. Zool.* 92: 427–464.

Taube, E. 1915. Beiträge zur Entwicklungsgeschichte der Euphausiden. II. Von der Gastrula bis zum Furciliastadium. *Z. wiss. Zool.* 114: 577–656.

Tsang, L.M., Ma, K.Y., Ahyong, S.T., Chan, T.-Y. & Chu, K.H. 2008. Phylogeny of Decapoda using two nuclear protein-coding genes: Origin and evolution of the Reptantia. *Mol. Phylogenet. Evol.* 48: 359–368.

Watling, L. 1981. An alternative phylogeny of peracarid crustaceans. *J. Crust. Biol.* 1: 201–210.

Watling, L. 1999. Towards understanding the relationship of the peracaridan orders: the necessity of determining exact homologies. In: Schram, F.R. & von Vaupel-Klein, J.C. (eds.), *Crustaceans and the Biodiversity Crisis. Proc. 4th Int. Crust. Congress, Amsterdam 1998, Vol. I*: 73–89, Leiden: Brill.

Weldon, W.F.R. 1887. The formation of the germ-layers in *Crangon vulgaris*. *Micr. J.* 33: 343–363.

Weygoldt, P. 1994. Le développement embryonaire. In: Forest, J. (ed.) *Traité de Zoologie, tome VII Crustacés Fascicule 1 Morphologie, Physiologie, Reproduction, Systématique*: 807–889. Paris: Masson.

Williams, T.A. 1998. *Distalless* expression in crustaceans and the patterning of branched limbs. *Dev. Genes Evol.* 207: 427–434.

Wills, M.A. 1997. A phylogeny of recent and fossil Crustacea derived from morphological characters. In: Fortey, R.A. & Thomas, R.H. (eds.), *Arthropod Relationships*: 189–209. London: Chapman & Hall.

Wolff, C. & Scholtz, G. 2008. The clonal composition of biramous and uniramous arthropod limbs. *Proc. R. Soc. B* 275: 1023–1028.

Zilch, R. 1978. Embryologische Untersuchungen an der holoblastischen Ontogenese von *Penaeus trisulcatus* Leach (Crustacea, Decapoda). *Zoomorphologie* 90: 67–100.

Zilch, R. 1979. Cell lineage in arthropods? *Fortschr. Zool. Syst. Evolutionsforsch.* 1: 19–41.

Zimmer, C. & Gruner, H.-E. 1956. 5. Band: Arthropoda, 1. Abteilung: Crustacea, 6. Buch, II Teil Euphausiacea 1. Lieferung. In: Gruner, H.-E. (ed.), *Dr. H.G. Bronns Klassen und Ordnungen des Tierreichs*: 1–160. Leipzig: Akademische Verlagsgesellschaft Geest & Portig.

# Mitochondrial DNA and Decapod Phylogenies: The Importance of Pseudogenes and Primer Optimization

CHRISTOPH D. SCHUBART

*Biologie 1, Universität Regensburg, 93040 Regensburg, Germany*

## ABSTRACT

Not much more than fifteen years ago, the first decapod phylogenies based on mitochondrial DNA (mtDNA) sequences revolutionized decapod phylogenetics. Initially, this method was accepted only reluctantly. However, a wider understanding of the methods, and the realization that credibility of specific branching patterns can be measured by statistic confidence values, allowed the recognition of molecular systematics as just another phylogenetic approach, in which homologous characters are compared and interpreted in terms of apomorphic or plesiomorphic status, and best possible trees are calculated based on distances, parsimony, or likelihoods. Similar to morphological characters, some of the shared molecular characters can result from convergence, but the large quantity of potential characters to be compared (15,000–17,000 in mtDNA) promises to reveal phylogenetic signal. For many years, preference was given to mitochondrial genes among the molecular markers, because of the relative ease with which they can be amplified (stable and numerous copies per cell) and interpreted (because they are only maternally inherited and lack introns and recombination), and because of higher mutation rates and thus greater variability than nuclear DNA. More recently, some of these apparent advantages were interpreted as shortcomings of mtDNA, and the discovery of selective sweeps, mitochondrial introgressions, and nuclear copies of mtDNA (numts) have questioned the credibility of phylogenies based exclusively on mtDNA. Here, I revisit the history and importance of mtDNA-based phylogenies of decapods, present two examples of how numts can produce erroneous phylogenies, and emphasize the need for primer optimization for better PCR results and avoidance of numts. Mitochondrial DNA has distinct advantages and disadvantages and, if used in combination with other phylogenetic markers, is still a very effective tool for phylogenetic inference. In most cases, and when used with the necessary care, phylogenies and phylogeographies based on mtDNA will render absolutely reliable results that can be tested and confirmed with other molecular and non-molecular approaches.

## 1 INTRODUCTION

Only a few years after the first publications announced the potential use of mitochondrial DNA for animal phylogenetics and population studies (e.g., Avise et al. 1987; Cann et al. 1987; Moritz et al. 1987) and the mitochondrial genome organization in *Artemia* was described (Batuecas et al. 1988), Cunningham et al. (1992) and Knowlton et al. (1993) published the first mtDNA-based phylogenies for Crustacea. It is noteworthy that these studies were based on sequences of the genes corresponding to the large ribosomal subunit 16S rRNA (16S; Cunningham et al. 1992) and the cytochrome oxidase subunit 1 (Cox1; Knowlton et al. 1993). Up to now, sequences of these genes continue to predominate in molecular phylogenetic studies of Crustacea, even though in many other animal taxa (including humans) other genes, like cytochrome b or the variable mitochondrial control region, have experienced at least a similarly wide use.

The proposal of Cunningham et al. (1992) that king and stone crabs (Anomura: Lithodidae) not only evolved from within the hermit crabs, but from within the genus *Pagurus*, cast a lot of doubt on the methodology and did not help to make the approach very popular among decapod crustacean systematists, causing a lot of skepticism concerning molecular phylogenies in general. For many years, it appeared that evolutionary biologists with molecular methods and taxonomists with morphological methods would continue their research separately. Consequently, there were only a few decapod molecular phylogenies published in the following years, most of them dealing with specific groups with special life history traits (Levinton et al. 1996; Paternello et al. 1996; Sturmbauer et al. 1996; Tam et al. 1996; Kitaura et al. 1998; Schubart et al. 1998a; Tam & Kornfield 1998), rather than with phylogeny and taxonomy per se. Only in Crandall et al. (1995) and Crandall & Fitzpatrick (1996), and in subsequent papers on crayfish systematics and phylogeny (Ponniah & Hughes 1998; Lawler & Crandall 1998), was there an explicit goal to establish molecular systematics, which only Spears et al. (1992) had undertaken previously for decapods, by proposing phylogenetic relationships among brachyuran crabs using nuclear 18S.

This slowly changed as species descriptions became based on, or were accompanied by, mitochondrial DNA data (Daniels et al. 1998; Schubart et al. 1998b, 1999; Gusmão et al. 2000; Macpherson & Machordom 2001, Daniels et al. 2001; Guinot et al. 2002; Guinot & Hurtado 2003; Gillikin & Schubart 2004; Lin et al. 2004, and later papers), when species were synonymized based on mtDNA in the absence of morphological characters (Shih et al. 2004; Robles et al. 2007; Mantelatto et al. 2007), and especially when phylogenetic relationships within genera and families were reconstructed with mtDNA in order to establish new taxonomic classifications (Schubart et al. 2000a, 2002; Kitaura et al. 2002; Tudge & Cunningham 2002; Chu et al. 2003; Lavery et al. 2004; Klaus et al. 2006; Schubart et al. 2006). Only recently, mtDNA has been used as part of multi-locus studies to reconstruct phylogenies at higher levels within decapod Crustacea (Ahyong & O'Meally 2004; Porter et al. 2005; Daniels et al. 2006).

For this kind of higher-level taxonomy, the exclusive use of mitochondrial DNA as a molecular marker is inappropriate (see Schubart et al. 2000b). This is due to the fact that mtDNA is characterized by a relatively high mutation rate, which makes it very useful at low taxonomic levels (intraspecific to intrafamilial levels) but causes increasing saturation when older splits are analyzed. When that occurs, the ratio between "phylogenetic noise," mostly caused by molecular convergence (homoplasy), and phylogenetic signal becomes more and more unfavorable and restricts the use of mtDNA at these levels. Therefore, and because of other potential problems of mtDNA (see Discussion), today the combination of mtDNA with more conserved nuclear markers is essential when reconstructing higher order phylogenies.

mtDNA still has many advantages over nuclear DNA. First, its ring-shaped structure makes it a more stable molecule than the chromosomes in the nucleus. Furthermore, there are hundreds to thousands of mitochondrial genomes per cell (with up to 10 copies per mitochondrion, see Wiesner et al. 1992), whereas there is only one nuclear genome per cell. This makes mtDNA much easier to amplify than nuclear DNA (nDNA), and DNA quality becomes a less critical issue than it is for nDNA. As a result, it is now possible to sequence mtDNA from museum specimens that were preserved in ethanol 150 years ago (e.g., Schubart et al. 2005) or longer, something that would be much more difficult with nDNA. mtDNA is also characterized by the absence of introns, so that basically all DNA is informative. Nevertheless, mutation rates are much higher in mtDNA than in nDNA, allowing phylogenetic signal to accumulate at shorter time frames. The fact that mtDNA appears to not have recombination, and in most cases is only maternally inherited, makes its interpretation much easier and allows for extrapolation, as for example in the calibration of molecular clocks. More recently, the increasing number of multiple gene sequencing of mitochondrial genomes (many of them complete) and their comparison allows the detection of gene rearrangements that may be used to support phylogenetic conclusions (mitogenomics) (e.g., Hickerson & Cunningham 2000; Kitaura et al. 2002; Morrison et al. 2002).

After having listed these well-known and traditionally accepted advantages of mtDNA, below I will discuss potential disadvantages of mtDNA for the reconstruction of decapod crustacean phylogenies. This will be exemplified by the presentation of new data on pseudogenes and a subsequent discussion of their consequences and ways of avoiding them.

## 2 MATERIALS & METHODS

Samples of three species of the genus *Cardisoma* (Brachyura: Thoracotremata: Gecarcinidae) were collected or obtained between 1996 and 2005 from both tropical American coastlines and from western Africa (Table 1). The goal was to establish genetic differentiation between the western African species *C. armatum* Herklots, 1851, and both American species, *C. guanhumi* Latreille, 1828 (western Atlantic), and *C. crassum* Smith, 1870 (eastern Pacific). In a second study, we used single specimens of *Geryon trispinosus* (Herbst, 1803), *G. longipes* (A. Milne-Edwards, 1882), and *Chaceon granulatus* (Sakai, 1978) as part of a study investigating phylogenetic relationships within the Geryonidae and the superfamily Portunoidea (see Schubart & Reuschel this volume). Molecular studies were carried out at the University of Regensburg. DNA was extracted with the Gentra Systems buffer combination. After discovering multiple copies and strongly deviating products in some of our sequencing products, mtDNA enrichment techniques were applied during extractions, such as differential centrifugation in a saccharose gradient and a Triton X-100 treatment (see Burgener & Hübner 1998 and discussion below). This allowed us to work with two separate fractions from the same individual, one with potentially enriched mtDNA, the other with enriched nDNA. Selective amplification of an approximately 580-basepair region of the mitochondrial large ribosomal subunit 16S rRNA was carried out by PCR. Primers used were 16L29, 16L12, 1472, 16H10, 16H12 (see Tables 2, 3). In order to obtain clean sequences from otherwise mixed PCR products in *Cardisoma*, we designed specific primers for the presumed mtDNA (16L13J: 5'-TGTAGATATAAAGAGTTTAA-3') and the presumed nuclear derivate (16L13P: 5'-TGTAGATATAAAGAGTTTAG-3') for PCR and sequencing reactions. These primers differ only in the last nucleotide (3'-end) and should preferentially anneal to one of the two available products.

PCR amplifications were carried out with four minutes denaturation at 94°C, 40 cycles, with 45 s 94°C, 1 min 48°C, 1 min 72°C, and 10 min final denaturation at 72°C. PCR products were purified with Microcon 100 filters (Microcon) or Quick-Clean (Bioline) and then sequenced with the ABI BigDye terminator mix followed by electrophoresis in an ABI Prism 310 Genetic Analyzer (Applied Biosystems, Foster City, USA). Forward and reverse strands were obtained for most products. New sequence data were submitted to the European molecular database EMBL (see Table 1 for accession numbers). In addition, the following sequences from the molecular database were included in our analyses: *Cardisoma guanhumi* (Z79653, from Levinton et al. 1996), *Cardisoma crassum* (AJ130805, from Schubart et al. 2000b), *Chaceon quinquedens* (Smith, 1879) and *C. fenneri* (Manning & Holthuis, 1984) (AY122641 to AY122646 from Weinberg et al. 2003) and *Chaceon affinis* (A. Milne-Edwards & Bouvier, 1894) (AF100914 to AF100916 from Weinberg et al. 2003 and previously unpublished by J. Bautista and Y. Alvarez).

Sequences were aligned and corrected manually with BioEdit (Hall 1999) or XESEE 3.2 (Cabot and Beckenbach 1989). The model of DNA substitution that best fit our data was determined using the software MODELTEST 3.6 (Posada and Crandall 1998). Reconstruction of phylogenetic trees with the corresponding models (TrN+I for *Cardisoma*; TVM+I+G for Geryonidae) in a Bayesian inference analysis (BI) with MrBayes v. 3.0b4 (Huelsenbeck and Ronquist 2001) and without models in a maximum parsimony analysis (MP) with PAUP* (Swofford 2001) revealed that the majority of genetic differences at the interindividual level were so small that the position of most operational taxonomic units was unresolved in major consensus clades. Therefore, a distance-based reconstruction with minimum evolution (ME) (Rzhetsky & Nei 1992) and Maximum Composite Likelihood as implemented in MEGA4 (Tamura et al. 2007) was carried out with 2000 bootstrap pseudoreplicates

**Table 1.** Crab specimens used for phylogenetic reconstruction of pseudogenes with locality of collection, museum catalogue number for vouchers, and genetic database accession numbers.

| Species | Collection Locality | Coll. Date | voucher | mtDNA | numt |
|---|---|---|---|---|---|
| **Cardisoma** | | | | | |
| Cardisoma guanhumi R40 | Jamaica (St. Ann): Priory | 8 Oct. 2000 | SMF 32773 | n.a. | FM 208132 |
| Cardisoma guanhumi CA1 | Jamaica (Hanover): Negril | 14 Oct. 2005 | leg | FM 208123 | FM 208133-35 |
| Cardisoma guanhumi CA2 | Jamaica (Hanover): Negril | 14 Oct. 2005 | SMF 32745 | FM 208123 | FM 208136-37 |
| Cardisoma guanhumi CA3 | Jamaica (St. James): Montego Bay | Oct. 2005 | leg | FM 208124 | FM 208132 |
| Cardisoma guanhumi CA21 | Jamaica (Trelawny): Glistening W. | 22 March 2003 | SMF 32772 | FM 208124 | n.a. |
| Cardisoma guanhumi CA27 | Jamaica (Hanover): Negril | 14 Oct. 2005 | leg | FM 208123 | n.a. |
| Cardisoma guanhumi | Cuba (Pinar de Río): El Rosario | 21 Sept. 1999 | SMF 25747 | FM 208123 | n.a. |
| Cardisoma guanhumi | Honduras (Islas de la Bahía): Utila | 18 Aug. 2000 | SMF 26006 | FM 208123 | n.a. |
| Cardisoma guanhumi | Panama (Caribbean): La Galeta | 3 March 1996 | ULLZ 3796 | FM 208123 | FM 208129-31 |
| Cardisoma armatum tradeSG | West Africa (from aquarium trade) | 1992 | ZRC 1996.121 | FM 208125 | 208127 |
| Cardisoma armatum tradeD | West Africa (from aquarium trade) | 2000 | leg | FM 208126 | 208128 |
| Cardisoma armatum R13 | Ghana: Elmina | 3 July 2001 | SMF 27534 | FM 208125 | n.a. |
| Cardisoma crassum | Costa Rica: Rincón | 18 March 1996 | SMF 24543 | AJ130805 | n.a. |
| **Geryonidae** | | | | | |
| Geryon longipes | Spain (Ibiza): Sta. Eulalia fish market | 28 March 2001 | SMF 32747 | FM 208120 | FM 208119 |
| Geryon trispinosus | North Sea: Flade Grounds | 2000 | SMF 32746 | FM 208121 | |
| Chaceon bicolor | Singapore fish market | 2000 | ZRC 2000.2830 | FM 208122 | |
| Chaceon granulatus | Japan | | SMF 32762 | FM 208775 | |

SMF: Senckenberg Museum, Frankfurt a.M.; ULLZ: University of Louisiana at Lafayette Zoological Collection, Lafayette. ZRC: Zoological Reference Collection, Raffles Museum, National University of Singapore.

**Table 2.** Decapod-specific primers used for amplification of the 16S rRNA–tRNA$_{Leu}$–NDH1 complex and of the Cox1 gene.

16S towards NDH1:
16L2: 5′–TGCCTGTTTATCAAAAACAT–3′ (Schubart et al. 2002)
16L12: 5′–TGACCGTGCAAAGGTAGCATAA–3′ (Schubart et al. 1998)
16L12b: 5′–TGACYGTGCAAAGGTAGCATAA–3′ (new)
16L15: 5′–GACGATAAGACCCTATAAAGCTT–3′ (Schubart et al. 2000c)
16L29: 5′–YGCCTGTTTATCAAAAACAT–3′ (Schubart et al. 2001 as "16L2")
16L6: 5′–TTGCGACCTCGATGTTGAAT–3′ (new)
16L37: 5′–TTACATGATTTGAGTTCARACCGG–3′ (new)
16L11: 5′–AGCCAGGTYGGTTTCTATCT–3′ (new)
16LLeu: 5′–CTATTTTGKCAGATDATATG–3′ (new)

NDH1 towards 16S:
NDH4: 5′–CAAGCYAAATAYATYARCTT–3′ (new)
NDH2: 5′–GCTAAATATATWAGCTTATCATA–3′ (new)
NDH5: 5′–GCYAAYCTWACTTCATAWGAAAT–3′ (new)
NDH1: 5′–TCCCTTACGAATTTGAATATATCC–3′ (new)
16HLeu: 5′–CATATTATCTGCCAAAATAG–3′ (new)
16H10: 5′–AATCCTTTCGTACTAAA–3′ (new)
16H11: 5′–AGATAGAAACCRACCTGG–3′ (new)
16H37: 5′–CCGGTYTGAACTCAAATCATGT–3′ (Klaus et al. 2006)
16H6: 5′–TTAATTCAACATCGAGGTC–3′ (new)
16H12: 5′–CTGTTATCCCTAAAGTAACTT–3′ (new)

Cox1 forward (L) and reverse (H):
COL6: 5′-TYTCHACAAAYCATAAAGAYATYGG-3′ (new, substitute COL1490)
COL14: 5′-GCTTGAGCTGGCATAGTAGG-3′ (Roman & Palumbi 2004, unnamed)
COL19: 5′-ATAGTAGAAAGAGGRGTWGG-3′ (new)
COL7: 5′-GGTGTKGGMACMGGATGAACTGT-3′ (new)
COL8: 5′-GAYCAAATACCTTTATTTGT-3′ (new)
COL4: 5′-TAGCHGGDGCWATYACTAT-3′ (new)
COL12: 5′-GCHATTACTATACTTCTWACWGAYCG-3′ (new)
COL1b: 5′-CCWGCTGGDGGWGGDGAYCC-3′ (new, substitute for COIf)
COL3: 5′-ATRATTTAYGCTATRHTWGCMATTGG-3′ (Reuschel & Schubart 2006)
COH7: 5′-TGWARAGAAAAAATTCCTA-3′ (new)
COH14: 5′-GAATGAGGTGTTTAGATTTCG-3′ (Roman & Palumbi 2004, unnamed)
H7188: 5′-CATTTAGGCCTAAGAAGTGTTG-3′ (Knowlton et al. 1993)
COH6: 5′-TADACTTCDGGRTGDCCAAARAAYCA-3′ (Schubart & Huber, 2006, substitute HCO2198)
COI(10): 5′-TAAGCGTCTGGGTAGTCTGARTAKCG-3′ (Baldwin et al. 1998)
COH3: 5′-AATCARTGDGCAATWCCRSCRAAAAT-3′ (Reuschel & Schubart 2006)
COH8: 5′-TGAGGRAAAAAGGTTAAATTTAC-3′ (new)
COH4: 5′-GGYATACCRTTDARTCCTARRAA-3′ (Mathews et al. 2002)
COH12: 5′-GGYATACCRTTTARTCCTAARAA-3′ (new, substitute for COH4)
COH1b: 5′-TGTATARGCRTCTGGRTARTC-3′ (new, substitute for COIa)
COH18: 5′-CTA TGG AAG ATA CGA TGT TTC-3′ (Reuschel & Schubart 2007)
COH16: 5′-CATYWTTCTGCCATTTTAGA-3′ (new)

and was used for presentation of the phylogenetic relationships as a dichotomous tree (*Cardisoma*) or radiation tree (Geryonidae).

## 3 RESULTS

The aligned region of the 16S rDNA fragment of the three species of *Cardisoma* consisted of 594 basepairs (bp), of which 56 were variable and 39 parsimony-informative, whereas the length of the 16S sequence alignment from the species of *Geryon* and *Chaceon* consisted of 556 bp, of which 34 were variable and 18 parsimony-informative.

Phylogenetic analyses with three reconstruction methods (BI, MP, ME) revealed the evolutionary history of nuclear copies of the mitochondrial 16S rDNA by comparisons of the two products and with closely related species. The resulting topologies were most informative for the ME analysis, which was therefore selected for representation, even if most of the interior branches were not significantly supported. These topologies are not in conflict with the ones produced by BI and MP. In both examples, the successfully recognized numts do not represent the closest related sequence to the mtDNA of the corresponding species, and thus they would confound phylogenetic relationships if erroneously taken for, and treated as, the mitochondrial product.

The phylogenetic tree of the American and West African representatives of the genus *Cardisoma* shows a clear separation (MP bootstraps and BI posterior probabilities 100%) of the mitochondrial sequences, corresponding to three species from different nuclear products of two of the species, the Atlantic *C. guanhumi* and *C. armatum* (see Fig. 1). Clean sequences of numts were

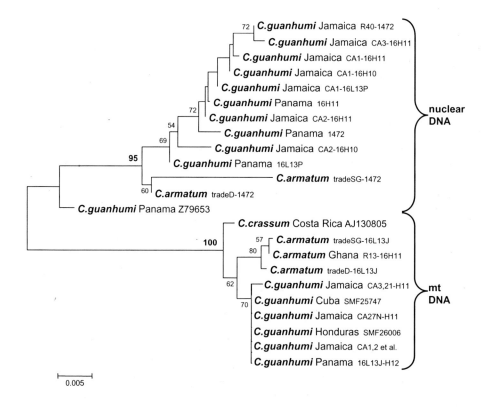

**Figure 1.** Phylogenetic tree of mitochondrial 16S rDNA sequences and nuclear copies obtained from the same individuals of crabs from the genus *Cardisoma* (Brachyura: Thoracotremata: Gecarcinidae). Topology of a Minimum Evolution analysis with confidence values (only $\geq 50$) corresponding to confidence values after 2000 bootstrap pseudoreplicates.

obtained from four freshly preserved specimens of *C. guanhumi* from Jamaica and Panama, especially with the specifically designed primer 16L13P. Older museum specimens like those from Cuba and Honduras never showed signs of the presence of numts, another possible indication of the higher stability of mtDNA compared to nDNA. A pseudogene for the eastern Pacific species *C. crassum* was revealed by double products after PCR, but it has not yet been recovered as a clean sequence. Overall it appears that the evolution of the pseudogenes predates the separation of the mtDNA of the three species involved. Two sequences from GenBank were also included: *C. crassum* AJ130805 fits well within the mitochondrial clade, whereas there are clear indications that *C. guanhumi* Z79653 represents a pseudogene sequence, quite distinct from the other numts from this study, which most likely is the result of the use of different primer combinations (see below).

Phylogenetic reconstruction of all species of the genera *Geryon* and *Chaceon* for which 16S rDNA is available is presented as a radiation tree (unrooted) in Figure 2. This form of representation better demonstrates the phylogenetic position of the nuclear copy of the 16S rDNA from *Geryon longipes*, with respect to not only its mitochondrial counterpart but also to other 16S sequences of the genera *Geryon* and *Chaceon*. Also, the mitochondrial sequence of *G. longipes*

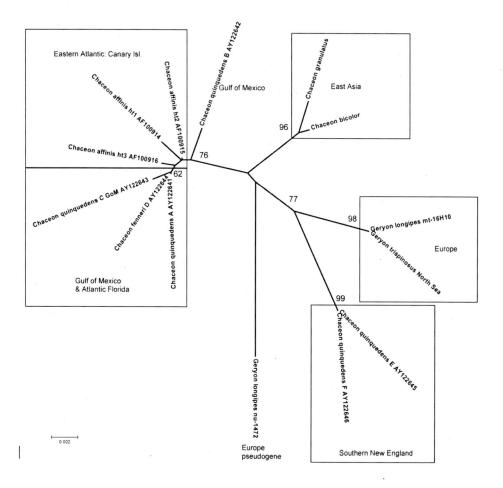

**Figure 2.** Radiation tree (Minimum Evolution, Maximum Composite Likelihood, 2000 bootstrap pseudoreplicates) of representatives from the crab genera *Geryon* and *Chaceon* (Brachyura: Heterotremata: Geryonidae) based on the mitochondrial 16S rDNA sequences and one nuclear copy of the species *G. longipes*.

is more closely related to other species, and even to representatives of another genus, than it is to its corresponding numt. However, available sequences in GenBank for three species of *Chaceon* demonstrate that the taxonomy of this genus is not settled. The American species *Chaceon quinquedens* is especially in need of revision; the North England representatives of this species seem to be more closely related to the genus *Geryon* (two species including the pseudogene) than to their "conspecifics" from the Gulf of Mexico (see also Weinberg et al. 2003). On the other hand, the population of *C. quinquedens* from the Gulf of Mexico is composed of several haplotypes, which do not cluster together but rather cluster with haplotypes of *C. affinis* from the Canary Islands and even share their most common haplotype with *C. fenneri* from Florida (see also Weinberg et al. 2003). If the morphological taxonomy and classification of these species are correct, this represents a case of incomplete lineage sorting, a typical phenomenon following recent speciation events, but a phenomenon that is not unique to mtDNA.

## 4 DISCUSSION

Phylogenies based entirely on mitochondrial DNA have recently and increasingly been criticized, especially because 1) only the maternal evolutionary lineage is considered, 2) there is possible introgression of mtDNA among species (e.g., Llopart et al. 2005), 3) early saturation due to homoplasy in the variable positions is possible (favored by an A&T-bias) (e.g., Chu et al. this volume), and 4) there is the potential for misinterpretation caused by the inclusion of pseudogenes (e.g., Williams & Knowlton 2001). Furthermore, all mitochondrial genes are located on the same molecule and thus cannot be used separately as independent sources of evidence (Moore 1995). I will not list again all the arguments in favor of using mtDNA for phylogenies (already highlighted in the Introduction),

**Table 3.** Large ribosomal subunit 16S rRNA universal primers 16Sbr (Palumbi et al. 1991) and 1472 (Crandall & Fitzpatrick 1996) in 5'-3' direction and the corresponding sequence in selected decapod Crustacea.

| | |
|---|---|
| 16Sbr: | |
| CCGGTCTGAACTCAGATCACGT | 16Sbr (Palumbi et al. 1991) |
| CCGGTCTGAACTCAGATCATGT | *Litopenaeus vannamei* NC 009626 |
| CCGGTCTGAACTCAAATCATGT | *Penaeus monodon* NC 002184 |
| CCGGTCTGAACTCAAATCATGT | *Halocaridina rubra* NC 008413 |
| ATGGTTTGAACTCAAATCATGT | *Macrobrachium rosenbergii* NC 006880 |
| CCGGTCTGAACTCAAATCATGT | *Panulirus japonicus* NC 004251 |
| CCGGTCTGAACTCAAATCATGT | *Cherax destructor* NC 001243 |
| CCGGTCTGAACTCAAATCATGT | *Pagurus longicarpus* NC 003058 |
| CCGGTCTGAACTCAAATCATGT | *Pseudocarcinus gigas* NC 006891 |
| CCGGTCTGAACTCAAATCATGT | *Callinectes sapidus* NC 006281 |
| CCGGTCTGAACTCAAATCATGT | *Portunus trituberculatus* NC 005037 |
| CCGGTTTGAACTCAAATCATGT | *Geothelphusa dehaani* NC 007379 |
| CCGGTTTGAACTCAAATCATGT | *Eriocheir sinensis* NC 006992 |
| CCGGTCTGAACTCAAATCATGT | 16H7 (new) |
| CCGGTTTGAACTCAAATCATGT | 16H3 (Reuschel & Schubart, 2006) |
| | |
| 1472: | |
| AGATAGAAACCAACCTGG | 1472 (Crandall & Fitzpatrick 1996) |
| AGATAGAAACCGACCTGG | *Litopenaeus vannamei* NC 009626 |
| AGATAGAAACCGACCTGG | *Penaeus monodon* NC 002184 |
| AGATAGAAACTAACCTGG | *Halocaridina rubra* NC 008413 |
| AGATAGAAACCAACCTGG | *Macrobrachium rosenbergii* NC 006880 |
| AGATAGAAACCGACCTGG | *Panulirus japonicus* NC 004251 |
| AGATAGAAACCAACCTGG | *Cherax destructor* NC 001243 |
| AGATAGAAACCAACCTGG | *Pagurus longicarpus* NC 003058 |
| AGATAGAAACCAACCTGG | *Pseudocarcinus gigas* NC 006891 |

**Table 3.** (Continued)

| | |
|---|---|
| AGATAGAAACCAACCTGG | *Callinectes sapidus* NC 006281 |
| AGATAGAAACCGACCTGG | *Portunus trituberculatus* NC 005037 |
| AGATAGAAACCGACCTGG | *Carcinus maenas* FM 208763 |
| AGATAGAAACCGACCTGG | *Geryon trispinosus* FM 208776 |
| AGATAGAAACCAACCTGG | *Geothelphusa dehaani* NC 007379 |
| AGATAGAAACCAACCTGG | *Eriocheir sinensis* NC 006992 |
| AGATAGAAACCGACCTGG | *Grapsus grapsus* (unpublished) |
| AGATAGAAACCRACCTGG | 16H11 (new) |

**Table 4.** Cytochrome oxidase subunit I primers LCO1490 and HCO2198 (Folmer et al. 1994) in 5'-3' direction, recommended to be used for barcoding studies and the corresponding sequence in selected decapod Crustacea.

Forward:

| | |
|---|---|
| GGTCAACAAATCATAAAGATATTGG | LCO1490 |
| TTTCTACAAACCACAAAGACATTGG | *Litopenaeus vannamei* NC 009626 |
| TTTCTACAAATCATAAAGACATCGG | *Penaeus monodon* NC 002184 |
| TCTCAACAAACCATAAAGACATTGG | *Halocaridina rubra* NC 008413 |
| TCTCCACCAACCATAAAGATATTGG | *Macrobrachium rosenbergii* NC 006880 |
| TCTCTACTAATCATAAAGACATTGG | *Panulirus japonicus* NC 004251 |
| TTTCAACAAATCATAAAGATATTGG | *Cherax destructor* NC 001243 |
| TCTCTACTAACCACAAAGACATTGG | *Pagurus longicarpus* NC 003058 |
| TTTCTACAAATCATAAAGACATTGG | *Pseudocarcinus gigas* NC 006891 |
| TTTCTACAAATCATAAAGACATTGG | *Callinectes sapidus* NC 006281 |
| TTTCTACAAATCATAAAGATATTGG | *Portunus trituberculatus* NC 005037 |
| TTTCCACAAACCATAAAGATATCGG | *Geothelphusa dehaani* NC 007379 |
| TTTCTACAAATCATAAAGATATTGG | *Eriocheir sinensis* NC 006992 |
| TCWACAAATCATAAAGAYATTGG | COL6a (new) |
| ACAAATCATAAAGATATYGG | COL6b (Schubart & Huber 2006) |
| TYTCHACAAAYCATAAAGAYATYGG | COL6 (new) |

Reverse:

| | |
|---|---|
| TAAACTTCAGGGTGACCAAAAAATCA | HCO2198 |
| TATACTTCTGGGTGACCGAAGAATCA | *Litopenaeus vannamei* NC 009626 |
| TATACTTCAGGATGACCGAAAAATCA | *Penaeus monodon* NC 002184 |
| TAGACTTCTGGGTGGCCGAAAAATCA | *Halocaridina rubra* NC 008413 |
| TATACTTCTGGGTGCCCAAAGAATCA | *Macrobrachium rosenbergii* NC 006880 |
| TAAACTTCGGGATGACCGAAAAACCA | *Panulirus japonicus* NC 004251 |
| TAGACCTCCGGGTGCCCAAAGAATCA | *Cherax destructor* NC 001243 |
| TAAACCTCGGGGTGACCAAAAAACCA | *Austropotamobius torrentium* (unpublished) |
| TAAACTTCTGGGTGGCCGAAAAATCA | *Pagurus longicarpus* NC 003058 |
| TACACTTCAGGGTGTCCAAAAAATCA | *Pseudocarcinus gigas* NC 006891 |
| TAAACTTCAGGATGTCCGAAAAATCA | *Callinectes sapidus* NC 006281 |
| TAGACTTCAGGATGACCAAAAAATCA | *Portunus trituberculatus* NC 005037 |
| TATACTTCGGGATGACCAAAGAACCA | *Pachygrapsus transversus* (unpublished) |
| TAAACTTCTGGGTGACCAAAAAACCA | *Geothelphusa dehaani* NC 007379 |
| TAAACTTCAGGGTGACCGAAAAATCA | *Eriocheir sinensis* NC 006992 |
| TADACTTCDGGRTGDCCAAARAAYCA | COH6 (Schubart & Huber 2006) |

because I think that there are and will be sufficient studies giving evidence of the suitability and credibility of mtDNA-based phylogenies at certain taxonomic levels (see, for example, Schubart & Reuschel this volume). I will also not discuss whether mtDNA or nDNA is the "better" option for reconstructing molecular phylogenies, because this will always depend on the evolutionary time scale to which the respective question refers, and because it is the combination of both that will give us most information (see also Klaus et al. this volume). It is similar to discussions of the potential uses of morphology and genetics when trying to understand evolution of natural lineages; the comparison of both will always increase information content, and it is to no one's advantage to ignore the other source of evidence.

Instead, I will use this discussion to respond to some of the criticisms that mtDNA phylogenies are receiving (e.g., Moore 1995; Zhang & Hewitt 2003; Mahon & Neigel 2008; Tsang et al. 2008; Chu et al. this volume). The topics of introgression and exclusive reconstruction of maternal lineages (criticisms 1 and 2) are important and must be considered in our understanding of the evolution of mtDNA. However, they are biological phenomena and not artifacts. There is nothing that can be done to avoid them, but we need to try to reconstruct and incorporate them in our models of evolution, aided by the independent insights we obtain from other sources of information (e.g., nDNA as, for example, in Shaw 2002). The early saturation of variable positions in mtDNA (criticism 3) may indeed be a problem, when, for example, DNA sequence data of Cox1 are used to reconstruct a phylogeny of the animal kingdom. In those instances, the obvious advantage at low taxonomic levels (i.e., availability of phylogenetic information even for younger differentiation events) becomes a potential problem at higher levels. However, there are ways to avoid this "phylogenetic noise" as a consequence of saturation. In coding genes, third positions can be omitted, as they are the ones most affected by silent mutations; transitions can be omitted, or the translated amino acid sequences used for phylogenetic inference. In their original proposal for implementation of DNA barcodes, Hebert et al. (2003a), for example, presented two independent phylogenetic trees of seven animal phyla and eight insect orders using the amino acid sequences corresponding to the Cox1 gene, while they switched to the DNA sequences (raw data) of the same gene when comparing 200 lepidopteran species. Thus, there are different levels of phylogenetic information that can be obtained from the same mitochondrial marker, depending on the question and on the amount of saturation that may blur the phylogenetic information. Similarly in mitochondrial rRNA genes, exclusion of hypervariable regions in higher-order phylogenies in response to alignment difficulties probably has a similar effect of reducing some of the noise caused by saturation (Schubart et al. 2000a). Nowadays, special software is available to perform these exclusions and avoid subjectivity during the process (Castresana 2000; Talavera & Castresana 2007).

The remaining problem of mtDNA, but also of nDNA, is the occurrence of paralogous copies, such that paralog and homolog DNA sequences may be confounded in comparative studies (criticism 4). The occurrence of non-functional pseudogenes as nuclear copies of mitochondrial genes (numts) is known from the literature and has been demonstrated with two examples in the Results section. Therefore, I would like to dedicate most of the Discussion to this phenomenon, the possibilities of avoiding amplification of paralogs, and the chances that arise when recognizing pseudogenes and possibly using them, together with the functional genes, for phylogenetic reconstruction.

4.1 *Pseudogenes*

The present examples of the occurrence of pseudogenes in the crab genera *Cardisoma* and *Geryon*, and their possible role in confusion of phylogenetic signal, highlight one of the possible problems of mtDNA. Schubart et al. (2000b: 826) noted that the discovery of pseudogenes in 16S rDNA and other mitochondrial genes "suggest[s] that the occurrence of pseudogenes is not an unusual phenomenon and is a potential source of artifacts." In *Menippe mercenaria* and *M. adina*, Schneider-Broussard & Neigel (1997) and Schneider-Broussard et al. (1998) were able to sequence and compare the mitochondrial 16S gene and its nuclear derivative. In this species complex, separation of the two "species" was not possible with both of these sequencing products. In contrast, the South American sister species, *Menippe nodifrons*, represents an outgroup to both the mtDNA and the pseudogene of the *M. mercenaria* complex, when phylogenetically compared with other species (Schubart et al. 2000b), suggesting that the pseudogene evolved relatively recently and after separation of the North and South American forms.

This is not true for other occurrences of pseudogenes, including my examples here, where the nuclear copies must have evolved before the more recent separations within the genus *Cardisoma* (Fig. 1) and before the split of the genera *Geryon* and *Chaceon*, if they are confirmed as monophyletic taxa (Fig. 2). In the case of *Cardisoma*, we provide evidence that more than one nuclear copy of the 16S rDNA may be present in the same individual. Three presumed pseudogenes were obtained from one specimen of *C. guanhumi* from the Caribbean coast of Panama, in addition to the mitochondrial product, depending on the primer combination used for PCR (Fig. 1). Additionally, two specimens from Jamaica, CA1 and CA2, seem to have undergone more than one translocation event with three and two nuclear copies, respectively, detected in our analyses. The only 16S sequence of *C. guanhumi* that had been previously deposited in GenBank (Z79653, from Levinton et al. 1996) also seems to be a very derived pseudogene, not closely related to the pseudogenes obtained in this study (differing in a number of important indels), but also clearly not belonging to the mitochondrial complex of sequences. This can be explained by the fact that Levinton et al. (1996) used the Palumbi et al. (1991) primer combination 16Sar-br, which is suboptimal for most decapod Crustacea (see Table 3 and discussion below) and was not used in our analyses. Weinberg et al. (2003) also noticed "variability in PCR and sequencing results" when using the primers by Palumbi et al. (1991) and designed a new primer for *Chaceon*, thereby considerably shortening the resulting alignment. It is quite possible that this reported "variability" was due to the presence of pseudogenes, since we also detected the existence of such a nuclear copy in the closely related species *Geryon longipes* (Fig. 2). The position of the pseudogene of *G. longipes* in the phylogenetic tree demonstrates how inadvertent amplification of it, and alignment with otherwise mitochondrial products, could easily lead to wrong phylogenetic conclusions, based on the fact that non-homologous evolutionary products would be compared.

The existence of multiple nuclear copies of mitochondrial genes had previously been documented by Williams & Knowlton (2001), who cloned PCR products of the Cox1 gene corresponding to ten species of the snapping shrimp genus *Alpheus*, for which they previously had difficulties in obtaining "good sequences" for Cox1. They found up to seven nuclear copies of the mitochondrial genes per species (from fifteen clones), demonstrating that pseudogenes are a common phenomenon in decapod Crustacea and are often present in more than one copy. Differences among the sequences of pseudogenes from the same individual reached levels of up to 20%. Multiple nuclear Cox1 derivatives have also been found in the ghost crab *Ocypode quadrata* (author's unpublished data).

However, the phenomenon of multiple gene derivates is not unique to mtDNA; it is also a problem in nuclear DNA. By being diploid, there are already at least two copies (maternal and paternal) of all genes present in the nucleus of each individual, and these alleles may differ from each other, complicating the reading of sequences (especially when including length differences) and rendering subsequent analyses more difficult. In addition, many genes are known to be present in multiple copies on different loci throughout the genome. These multiple copies can be functional and on the same chromosome (as, for example, the 28S–5.8S–18S complex) to increase the amount of transcribed DNA, but they can also be nonfunctional and appear as pseudogenes on different chromosomes. This shows that the problem of multiple copies is not unique to mtDNA but is also prevalent in nDNA, where it may be even more difficult to recognize due to the underlying diploidy. Therefore, the challenge for all molecular phylogenetic studies is to recognize pseudogenes and make sure that they, as well as the functional product, are treated independently. Sequences representing pseudogenes do not have to be discarded, but recognized, labelled, and submitted as such. Phylogenies can be built based on functional products as well as on pseudogenes (independently or combined), as long as it is known which sequences are homologous.

There are different approaches to avoid amplification of pseudogenes. One of them would be to generate cDNA through reverse transcriptase out of mRNA (e.g., Palmero et al. 1988; Williams & Knowlton 2001). This would ensure that only DNA that is transcribed, i.e., the functional DNA, is amplified, and that nonfunctional DNA is avoided. However, fresh or frozen material is recommended, or special fixatives like DMSO solutions, to properly preserve the RNA and allow use of this method. It is difficult to apply this method to specimens preserved in ethanol.

Another way to reduce the effect of pseudogenes is enrichment of mtDNA during the extraction process. This can be achieved using mt-rich tissue, by miniprep DNA purification (Beckman et al. 1993) and/or differential centrifugation in a caesium chloride or saccharose gradient (Anderson et al. 1981). Burgener & Hübner (1998) provide a protocol in which the tissue is first exposed to a buffer including Triton-X-100. This commonly used non-ionic detergent makes the mitochondrial membrane soluble, allowing the mtDNA to dissolve in the supernatant, while nDNA stays within the nuclei that remain intact and can be spun down (see also Solignac 1991). However, these methods only allow the enrichment of mtDNA in relation to nDNA and not its isolation. In our study with *Cardisoma* (see above), it was not always possible to obtain clean mtDNA product, even after applying these enrichment methods.

## 4.2 Primer optimization

The best way to avoid pseudogenes is most likely the use of optimized primers. It can be assumed that pseudogenes exist for all mitochondrial genes and maybe for most, if not all, species. Nevertheless, since a normal cell has many more copies of the mitochondrial genome compared to the nuclear genome, the mitochondrial product should be favored in PCRs if both products do not differ in their primer affinities. If, however, the primers have a better fit to the nuclear pseudogene than to the mtDNA, they will preferentially anneal to the nDNA, despite the increased number of mtDNA copies. The result would be a mix of products or a clean sequence corresponding to the pseudogene. In my experience, the occurrence of pseudogenes strongly decreases when using taxon-specific primers. Also, the recorded pseudogenes by Williams and Knowlton (2001) were recovered only from those species "for which good sequences for Cox1 were difficult to obtain from gDNA." Tables 3 and 4 demonstrate how commonly used universal primers are suboptimal for a wide range of decapod Crustacea. The use of these universal primers, which initially were the only ones available, will therefore often result in sequences that have double products or do not represent the mitochondrial product. To help crustacean workers avoid some of the more problematic universal primers, I offer here a list of decapod-specific primers for 16S and Cox1 (Table 2) in addition to comparing the universal primers to "real" DNA sequences in the homologous region of decapod mtDNA (Tables 3, 4).

In 16S rRNA, the primer 16Sar by Palumbi et al. (1991) (formerly considered a forward primer, but according to newest GenBank entries actually the reverse) has a perfect fit to all sequences except for the relatively unimportant first position of the 5'-end, which in most cases is a T instead of C (see primers 16L2 and 16L29 in Table 2). However, the corresponding "reverse" primer 16Sbr (now the forward) has 2 or 3 positions in which it deviates from most decapod sequences. Most critical is a consistent difference at the third from last position, which in the primer is always a T instead of a C as recorded for all known decapod sequences. Since it is relatively close to the 3'-end, which is decisive for primer annealing, it could cause serious problems when amplifying decapod 16S rDNA. I use the primers 16H3, 16H7, or the consensus of the two 16H37 (Tables 2, 3) to avoid this problem when amplifying the corresponding fragment. Probably because of problems adherent to 16Sbr, an alternative forward primer is being frequently used: 1472 by Crandall & Fitzpatrick (1996). This primer normally works very well in combination with 1471 (Crandall & Fitzpatrick 1996), 16Sar (Palumbi et al. 1991), 16L2, or 16L29 (Table 2). However, in some cases it fails to amplify or results in pseudogenes (unpublished observations). After obtaining longer sequences and reading through that primer region, it turned out that in 1472 the seventh position from the 3'-end often is a G instead of an A. This is confirmed with the alignment of a number of decapod sequences for which the entire mitochondrial DNA is known. Therefore, I propose the alternative primer 16H11, which allows easy amplification of sequences with G or A at that position (see Table 3).

For the other most popular mitochondrial marker, Cox1, two regions with a limited overlap have been used for phylogenetic studies: the "Palumbi region" with primers COIa and COIf (Palumbi et al. 1991) (e.g., Knowlton et al. 1993; Schubart et al. 1998a) and the "Folmer region" with primers LCO1490 and HCO 2198 (Folmer et al. 1994) (e.g., Harrison & Crespi, 1999; Trontelj et al. 2005). Subsequent to suggesting the "Folmer region" as a potential molecular barcode gene (Hebert et al. 2003a, b), the number of studies using that region has markedly increased, including the study by Costa et al. (2007) testing the suitability of this Cox1 region for barcoding studies in Crustacea. However, as can be seen in Table 4, the original primers by Folmer et al. (1994) are not optimized for decapod Crustacea, and their usefulness may be limited or could also result in the amplification of pseudogenes. LCO1490 starts with two Gs, which are not found in any of the decapod species with a known sequence of the entire gene. Probably more problematic is that the third position and especially the sixth from last position from the 3'-end (both third positions of the amino acid reading frame) show variability. In LCO1490 they are both Ts, but there are several occasions when they are found to be a C (see Table 4). In Schubart & Huber (2006), an alternative forward primer was suggested that does not include the double G at the beginning and accounts for the possible Cs at the third last position. Alternatively, COL6a can be used, in case the sixth from last position has mutated to C, which is often the case (Table 4). To consider both possible mutations, I propose the primer COL6, which has the same length as the original LCO1490 but accounts for almost all differences that have been observed in decapod crustaceans for which the entire mtDNA has been sequenced (Table 4). Likewise, the primer HCO2198 has some inherent potential problems. In this case, even more decapod species show mutations at the third from last position (C instead of T) and at the sixth from last position (G instead of A), these being again the third positions of the amino acid reading frames, which do not necessarily translate into new amino acids if modified. Also in this case, Schubart & Huber (2006) have proposed the new primer COH6 in their population study of the European crayfish *Austropotamobius torrentium*. This primer fits the sequences of most decapod species much better than the original HCO2198 and, due to its degenerate third and sixth from last positions, is less prone to fail when these mutate (Table 4).

I consider the variability of third positions in coding genes a big disadvantage for their use as universal barcoding genes. Unless taxon-specific primers are used, there is a greater risk of running into amplification problems or generating pseudogenes than in the conserved regions of ribosomal DNA (see Vences et al. 2005). Generation and use of taxon-specific primers should alleviate this problem and make the resulting sequences more trustworthy. In any case, mitochondrial genes will remain the target molecular markers for current and future animal barcoding approaches. They do have a number of advantages, but they must be treated properly. Once genetic barcoding proceeds, there will be a multitude of mitochondrial sequences that can and will be used for reconstructing phylogenies, even if this is not the explicit purpose of the Barcode of Life initiative. Therefore, mitochondrial sequences will continue to be used for molecular phylogenies, and it is easy to predict that there will always be more mitochondrial sequences available for comparisons at different phylogenetic levels than nuclear ones. Nevertheless, it will be important and advisable to complement phylogenies with independent evidence from the nuclear genome (and vice versa) to possibly recognize methodological problems and to distinguish the evolution of maternal lineages from the evolution of entire populations.

ACKNOWLEDGEMENTS

Compilation of this manuscript would not have been possible without the continuing help from my colleagues and students. Especially Nicole Rivera was of great help in the lab when sequencing mitochondrial and nuclear copies of *Cardisoma*. I furthermore would like to thank my companions in the field: José A. Cuesta (Panama & Costa Rica), Tobias Santl, Tobias Weil, Silke Reuschel, René Brodie, Liu Hung-Chang (Jamaica), Klaus Duffner (Ghana), and Carsten Müller (Ibiza). Klaus Anger sent legs from *Cardisoma armatum* from the aquarium trade, Cédric d'Udekem d'Acoz provided the specimen of *Geryon trispinosus*, Ferran Palero confirmed the mt-sequence of *Geryon longipes* with additional sequences, and staff of the Senckenberg Museum in Frankfurt (Michael Türkay et al.) and the ZRC in Singapore (Peter K.L. Ng et al.) allowed access to their collections and tissue extraction from specimens. Special thanks are due to Jody Martin, Darryl Felder, and Keith Crandall and the AToL program for bringing us together during the SICB meeting in San Antonio and for inviting us to contribute to this book. Funding of collections in Jamaica and lab work was provided by the German Science Foundation (DFG) through project SCHU 1460/3.

REFERENCES

Ahyong, S.T. & O'Meally, D. 2004. Phylogeny of the Decapoda Reptantia: resolution using three molecular loci and morphology. *Raffl. Bull. Zool.* 52: 673–693.

Anderson, S., Bankier, A.T., Barrell, B.G., de Bruijn, M.H.L., Coulson, A.R., Drouin, J., Eperon, I.C., Nierlich, D.P., Roe, B.A., Sanger, F., Schreier, P.H., Smith, A.J.H., Staden, R. & Young, I.G. 1981. Sequence and organization of the human mitochondrial genome. *Nature* 290: 457–465.

Avise, J.C., Arnold, J., Ball, R.M., Bermingham, E., Lamb, T., Neigel, J.E., Reeb, C.A. & Saunders, N.C. 1987. Intraspecific phylogeography: the mitochondrial DNA bridge between population genetics and systematics. *Ann. Rev. Ecol. Syst.* 18: 489–522.

Baldwin, J.D., Bass, A.L., Bowen, B.W. & Clark Jr., W.H. 1998. Molecular phylogeny and biogeography of the marine shrimp *Penaeus*. *Mol. Phylogenet. Evol.* 10: 399–407.

Batuecas, B., Garesse, R., Calleja, M., Valverde, J.R. & Marco, R. 1988. Genome organization of *Artemia* mitochondrial DNA. *Nucleic Acids Res.* 16: 6515–6529.

Beckman, K.B., Smith, M.F. & Orrego, C. 1993. Purification of mitochondrial DNA with Wizard$^{TM}$ Minipreps DNA purification system. *Promega Notes Magazine* 43: 10.

Burgener, M. & Hübner, P. 1998. Mitochondrial DNA enrichment for species identification and evolutionary analysis. *Z. Lebensm. Unters. Forsch. A* 207: 261–263.

Cabot, E.L. & Beckenbach, A.T. 1989. Simultaneous editing of multiple nucleic acid and protein sequences with ESEE. *Comput. Appl. Biosci.* 5: 233–234.

Cann, R.L., Stoneking, M. & Wilson, A.C. 1987. Mitochondrial DNA and human evolution. *Nature* 325: 31–36.

Castresana, J. 2000. Selection of conserved blocks from multiple alignments for their use in phylogenetic analysis. *Mol. Biol. Evol.* 17: 540–552.

Chu, K.H., Ho, H.Y., Li, C.P. & Chan, T.Y. 2003. Molecular phylogenetics of the mitten crab species in *Eriocheir, sensu lato* (Brachyura: Grapsidae). *J. Crust. Biol.* 23: 738–746.

Chu, K.H., Tsang, L.M., Ma, K.Y., Chan, T.-Y. & Ng, P.K.L. (this volume). Decapod phylogeny: what can protein-coding genes tell us? In: Martin, J.W., Crandall, K.A. & Felder, D.L. (eds.), *Crustacean Issues: Decapod Crustacean Phylogenetics*. Boca Raton, Florida: Taylor & Francis/CRC Press.

Costa, F.O., deWaard, J.R., Boutillier, J., Ratnasingham, S., Dooh, R.T., Hajibabaei, M. & Hebert, P.D. 2007. Biological identifications through DNA barcodes: the case of the Crustacea. *Can. J. Fish. Aquat. Sci.* 64: 272–295.

Crandall, K.A. & Fitzpatrick Jr., J.E. 1996. Crayfish molecular systematics: using a combination of procedures to estimate phylogeny. *Syst. Biol.* 45: 1–26.

Crandall, K.A., Lawler, S.H. & Austin, C.M. 1995. A preliminary examination of the molecular phylogenetic relationships of some crayfish genera from Australia (Decapoda: Parastacidae). *Freshwater Crayfish* 10: 18–30.

Cunningham, C.W., Blackstone, N.W. & Buss, L.W. 1992. Evolution of king crabs from hermit crab ancestors. *Nature* 355: 539–542.

Daniels, S.R., Cumberlidge, N., Pérez-Losada, M., Marijnissen, S.A. & Crandall, K.A. 2006. Evolution of Afrotropical freshwater crab lineages obscured bymorphological convergence. *Mol. Phylogenet. Evol.* 40: 227–235.

Daniels, S.R., Stewart, B.A. & Gibbons, M.J. 1998. *Potamonautes granularis* sp. nov. (Brachyura: Potamonautidae), a new cyptic species of river crab from the Olifants River system, South Africa. *Crustaceana* 71: 885–903.

Daniels, S.R., Stewart, B.A. & Burmeister, L. 2001. Geographic patterns of genetic and morphological divergence amongst populations of a river crab (Decapoda: Potamonautidae) with the description of a new species from mountain streams in the Western Cape. *Zool. Scri.* 30: 181–197.

Folmer, O., Black, M., Hoeh, W., Lutz, R. & Vrijenhoek, R. 1994. DNA primers for amplification of mitochondrial cytochrome c oxidase subunit I from diverse metazoan invertebrates. *Mol. Mar. Biol. Biotechnol.* 3: 294–299.

Gillikin, D.P. & Schubart, C.D. 2004. Ecology and systematics of the genus *Perisesarma* (Crustacea: Brachyura: Sesarmidae) from East Africa. *Zool. J. Linn. Soc.* 141: 435–445.

Guinot, D., Hurtado, L.A. & Vrijenhoek, R. 2002. New genus and species of brachyuran crab from the southern East Pacific Rise (Crustacea Decapoda Brachyura Bythograeidae). *Compt. Rend. Biol.* 325: 1143–1152.

Guinot, D. & Hurtado, L.A. 2003. Two new species of hydrothermal vent crabs of the genus Bythograea from the southern East Pacific Rise and from the Galapagos Rift (Crustacea Decapoda Brachyura Bythograeidae). *Compt. Rend. Biol.* 326: 423–439.

Gusmão, J., Lazowski, C. & Solé-Cava, A.M. 2000. A new species of Penaeus (Crustacea: Penaeidae) revealed by allozyme and cytochrome oxidase analyses. *Mar. Biol.* 137: 435–446.

Hall, T.A. 1999. BioEdit: a user-friendly biological sequence alignment editor and analysis program for Windows 95/98/NT. *Nucl. Acids Symp. Ser.* 41: 95–98.

Harrison, M.K. & Crespi, B.J. 1999. Phylogenetics of *Cancer* crabs (Crustacea: Decapoda: Brachyura). *Mol. Phylogenet. Evol.* 12: 186–199.

Hebert, P.D.N., Cywinska, A., Ball, S.L. & deWaard, J.R. 2003a. Biological identifications through DNA barcodes. *Proc. R. Soc. Lond. B* 270: 313–321.

Hebert, P.D.N., Ratnasingham, S. & deWaard, R. 2003b. Barcoding animal life: cytochrome c oxidase subunit 1 divergences among closely related species. *Proc. R. Soc. Lond. B* (Suppl.) 270: S96–S99.

Hickerson, M.J. & Cunningham, C.W. 2000. Dramatic mitochondrial gene rearrangements in the hermit crab *Pagurus longicarpus* (Crustacea, Anomura). *Mol. Biol. Evol.* 17: 639–644.

Huelsenbeck, J.P. & Ronquist, F. 2001. MrBayes: Bayesian inference of phylogenetic trees. *Bioinformatics* 17: 754–755.

Jesse, R. Pfenninger, M., Fratini, S., Scalici, M., Streit, B. & Schubart, C.D. Disjunct distribution of the Mediterranean freshwater crab *Potamon fluviatile*—natural expansion or human introduction? *Invas. Biol.*, in press.

Kitaura, J., Wada, K. & Nishida, M. 1998. Molecular phylogeny and evolution of unique mud-using territorial behavior in ocypodid crabs (Crustacea: Brachyura: Ocypodidae). *Mol. Biol. Evol.* 15: 626–637.

Kitaura, J., Wada, K. & Nishida, M. 2002. Molecular phylogeny of grapsoid and ocypodoid crabs with reference to the genera *Metaplax* and *Macrophthalmus*. *J. Crust. Biol.* 22: 682–693.

Klaus, S., Brandis, D., Ng, P.K.L., Yeo, D.C.J. & Schubart, C.D. (this volume). Phylogeny and biogeography of Asian freshwater crabs of the family Gecarcinucidae (Brachyura: Potamoidea). In: Martin, J.W., Crandall, K.A. & Felder, D.L. (eds.), *Crustacean Issues: Decapod Crustacean Phylogenetics*. Boca Raton, Florida: Taylor & Francis/CRC Press.

Klaus, S., Schubart, C.D. & Brandis, D. 2006. Phylogeny, biogeography and a new taxonomy for the Gecarcinucoidea Rathbun, 1904 (Decapoda: Brachyura). *Organ. Div. Evol.* 6: 199–217.

Knowlton, N., Weigt, L.A., Solorzano, L.A., Mills, D.K. & Bermingham, E. 1993. Divergence in proteins, mitochondrial DNA, and reproductive compatibility across the Isthmus of Panama. *Science* 260: 1629–1632.

Lavery, S., Chan, T.Y. & Chu, K.H., 2004. Phylogenetic relationships and evolutionary history of the shrimp genus *Penaeus s.l.* derived from mitochondrial DNA. *Mol. Phylogenet. Evol.* 31: 39–49.

Lawler, S.H. & Crandall, K.A. 1998. The relationship of the Australian freshwater crayfish genera *Euastacus* and *Astacopsis*. *Proc. Linn. Soc. N.S.W.* 119: 1–8.

Levinton, J., Sturmbauer, C. & Christy, J. 1996. Molecular data and biogeography: resolution of a controversy over evolutionary history of a pan-tropical group of invertebrates. *J. Exp. Mar. Biol. Ecol.* 203: 117–131.

Lin, C.-W., Chan, T.-Y. & Chu, K.H. 2004. A new squat lobster of the genus *Raymunida* (Decaopoda: Galatheidae) from Taiwan. *J. Crust. Biol.* 24: 149–156.

Llopart, A., Lachaise, D. & Coyne, J.A. 2005. Multilocus analysis of introgression between two sympatric sister species of *Drosophila*: *Drosophila yakuba* and *D. santomea*. *Genetics* 171: 197–210.

Macpherson, E. & Machordom, A. 2001. Phylogenetic relationships of species of *Raymunida* (Decapoda: Galatheidae) based on morphology and mitochondrial cytochrome oxidase sequences, with the recognition of four new species. *J. Crust. Biol.* 21: 696–714.

Mahon, B.C. & Neigel, J.E. 2008. Utility of arginine kinase for resolution of phylogenetic relationships among brachyuran genera and families. *Mol. Phylogenet. Evol.* 48: 718–727.

Mantelatto, F.L., Robles, R. & Felder, D.L. 2007. Molecular phylogeny of the western Atlantic species of the genus *Portunus* (Crustacea, Brachyura, Portunidae). *Zool. J. Linn. Soc. London* 150: 211–220.

Mathews, L.M., Schubart, C.D., Neigel, J.E. & Felder, D.L. 2002. Genetic, ecological, and behavioural divergence between two sibling snapping shrimp species (Crustacea: Decapoda). *Mol. Ecol.* 11: 1427–1437.

Moore, W.S. 1995. Inferring phylogenies from mtDNA variation: Mitochondrial-gene trees versus nuclear-gene trees. *Evolution* 49: 718–726.

Moritz, C., Dowling, T.E. & Brown, W.M. 1987. Evolution of animal mitochondrial DNA: Relevance for population biology and systematics. *Ann. Rev. Ecol. Syst.* 18: 269–292.

Morrison, C.L., Harvey, A.W., Lavery, S., Tieu, K., Huang, Y. & Cunningham C.W. 2002. Mitochondrial gene rearrangements confirm the parallel evolution of the crab-like form. *Proc. Roy. Soc. Lond. B* 269: 345–350.

Palmero, I., Renart, J. & Sastre, L. 1988. Isolation of cDNA clones coding for mitochondrial 16S ribosomal RNA from the crustacean *Artemia*. *Gene* 68: 239–248.

Palumbi, S.R., Martin, A., Romano, S., Mcmillan, W.O., Stice, L. & Grabowski, G. 1991. *The Simple Fool's Guide to PCR. A Collection of PCR Protocols, Version 2*. Honolulu: University of Hawaii.

Patarnello, T., Bargelloni, L., Varotto, V. & Battaglia, B. 1996. Krill evolution and the Antarctic ocean currents: evidence of vicariant speciation as inferred by molecular data. *Mar. Biol.* 126: 603–608.

Ponniah, M. & Hughes, J.M. 1998. Evolution of Queensland spiny mountain crayfish of the genus *Euastacus* Clark (Decapoda: Parastacidae): preliminary 16S mtDNA phylogeny. *Proc. Linn. Soc. N.S.W.* 119: 9–19.

Porter, M.L., Pérez-Losada, M. & Crandall, K.A. 2005. Model-based multi-locus estimation of decapod phylogeny and divergence times. *Mol. Phylogenet. Evol.* 37: 355–369.

Posada, D. & Crandall, K.A. 1998: MODELTEST: testing the model of DNA substitution. *Bioinformatics* 14: 817–818.

Reuschel, S. & Schubart, C.D. 2006. Geographic differentiation of two Atlanto-Mediterranean species of the genus *Xantho* (Crustacea: Brachyura: Xanthidae) based on genetic and morphometric analyses. *Mar. Biol.* 148: 853–866.

Reuschel, S. & Schubart, C.D. 2007. Contrasting genetic diversity with phenotypic diversity in coloration and size in *Xantho poressa* (Brachyura: Xanthidae), with new results on its ecology. *Mar. Ecol.* 28: 1–10.

Robles, R., Schubart, C.D., Conde, J.E., Carmona-Suárez, C., Alvarez, F., Villalobos, J.L. & Felder, D.L. 2007. Molecular phylogeny of the American *Callinectes* Stimpson, 1860 (Brachyura: Portunidae), based on two mitochondrial genes. *Mar. Biol.* 150: 1265–1274.

Roman, J. & Palumbi, S.R. 2004. A global invader at home: population structure of the green crab, *Carcinus maenas*, in Europe. *Mol. Ecol.* 13: 2891–2898.

Rzhetsky, A. & Nei, M. 1992. A simple method for estimating and testing minimum-evolution trees. *Mol. Biol. Evol.* 9: 945–967.

Schneider-Broussard, R. & Neigel, J.E. 1997. A large subunit mitochondrial ribosomal DNA sequence translocated to the nuclear genome of two stone crabs (*Menippe*). *Mol. Biol. Evol.* 14: 156–165.

Schneider-Broussard, R., Felder, D.L., Chlan, C.A. & Neigel, J.E. 1998. Tests of phylogeographic models with nuclear and mitochondrial DNA sequence variation in the stone crabs, *Menippe adina* and *M. mercenaria*. *Evolution* 52: 1671–1678.

Schubart, C.D., Cannicci, S., Vannini, M. & Fratini, S. 2006. Molecular phylogeny of grapsoid crabs and allies based on two mitochondrial genes and a proposal for refraining from current superfamily classification. *J. Zool. Syst. Evol. Res.* 44: 193–199.

Schubart, C.D., Cuesta, J.A., Diesel, R. & Felder, D.L. 2000a. Molecular phylogeny, taxonomy, and evolution of nonmarine lineages within the American grapsoid crabs (Crustacea: Brachyura). *Mol. Phylogenet. Evol.* 15: 179–190.

Schubart, C.D., Cuesta, J.A. & Felder, D.L. 2002. Glyptograpsidae, a new brachyuran family from Central America: larval and adult morphology, and a molecular phylogeny of the Grapsoidea. *J. Crust. Biol.* 22: 28–44.

Schubart, C.D., Cuesta, J.A. & Felder, D.L. 2005. Phylogeography of *Pachygrapsus transversus* (Gibbes, 1850): the effect of the American continent and the Atlantic Ocean as gene flow barriers and recognition of *Pachygrapsus socius* Stimpson, 1871 as a valid species. *Nauplius* 13: 99–113.

Schubart, C.D., Cuesta, J.A. & Rodríguez, A. 2001. Molecular phylogeny of the crab genus *Brachynotus* (Brachyura: Varunidae) based on the 16S rRNA gene. *Hydrobiologia* 449: 41–46.

Schubart, C.D., Diesel R. & Hedges, S.B. 1998a. Rapid evolution to terrestrial life in Jamaican crabs. *Nature* 393: 363–365.

Schubart, C.D. & Huber, M.G.J. 2006. Genetic comparisons of German populations of the stone crayfish, *Austropotamobius torrentium* (Crustacea: Astacidae). *Bull. Franç. Pêche Piscicult.* 380–381: 1019–1028.

Schubart, C.D., Neigel, J.E. & Felder, D.L. 2000b. Use of the mitochondrial 16S rRNA gene for phylogenetic and population studies of Crustacea. In: von Vaupel Klein, J.C. & Schram, F. (eds.), *Crustacean Issues 12, The Biodiversity Crisis and Crustacea*: 817–830. Rotterdam: Balkema.

Schubart, C.D., Neigel, J.E. & Felder, D.L. 2000c. A molecular phylogeny of mud crabs (Brachyura: Panopeidae) from the northwestern Atlantic and the role of morphological stasis and convergence. *Mar. Biol.* 137: 11–18.

Schubart, C.D., Reimer, J. & Diesel, R. 1998b. Morphological and molecular evidence for a new endemic freshwater crab, *Sesarma ayatum* sp. n., (Grapsidae, Sesarminae) from eastern Jamaica. *Zool. Scr.* 27: 371–380.

Schubart, C.D., Reimer, J. & Diesel, R. 1999. Gonopod morphology and 16S mtDNA sequence alignment of the freshwater crabs *Sesarma bidentatum* and *S. ayatum* (Grapsidae: Sesarminae) from Jamaica. *Zool. Scr.* 28: 367–369.

Schubart, C.D. & Reuschel, S. (this volume). A proposal for a new classification of Portunoidea and Cancroidea (Brachyura: Heterotremata) based on two independent molecular phylogenies. In: Martin, J.W., Crandall, K.A. & Felder, D.L. (eds.), *Crustacean Issues: Decapod Crustacean Phylogenetics*. Boca Raton, Florida: Taylor & Francis/CRC Press.

Shaw, K.L. 2002. Conflict between nuclear and mitochondrial DNA phylogenies of a recent species radiation: What mtDNA reveals and conceals about modes of speciation in Hawaiian crickets. *Proc. Natl. Acad. Sci. USA* 99: 16122–16127.

Shih, H.-T., Ng, P.K.L. & Chang, H.-W. 2004. Systematics of the genus *Geothelphusa* (Crustacea, Decapoda, Brachyura, Potamidae) from southern Taiwan: a molecular appraisal. *Zool. Stud.* 43: 561–570.

Solignac, M., 1991. Preparation and visualization of mitochondrial DNA for RFLP analysis. In: Hewitt, G.M., Johnston, A.W.B. & Young, J.P.W. (eds.), *Molecular Techniques In Taxonomy*: 295–319. Berlin: Springer.

Spears, T., Abele, L.G. & Kim, W. 1992. The monophyly of brachyuran crabs: a phylogenetic study based on 18S rRNA. *Syst. Biol.* 41: 446–461.

Sturmbauer, C., Levinton, J.S. & Christy, J. 1996. Molecular phylogeny analysis of fiddler crabs: test of the hypothesis of increasing behavioral complexity in evolution. *Proc. Natl. Acad. Sci. USA* 93: 10855–10857.

Swofford, D.L. 2001. *PAUP*—Phylogenetic Analysis Using Parsimony (*and Other Methods). Version 4*. Sinauer Associates. Sunderland, Massachusetts.

Talavera, G. & Castresana, J. 2007. Improvement of phylogenies after removing divergent and ambiguously aligned blocks from protein sequence alignments. *Syst. Biol.* 56: 564–577.

Tam, Y.K. & Kornfield, I. 1998. Phylogenetic relationships of clawed lobster genera (Decapoda: Nephropidae) based on mitochondrial 16S rRNA gene sequences. *J. Crust. Biol.* 18: 138–146.

Tam, Y.K., Kornfield, I. & Ojeda, F.P. 1996. Divergence and zoogeography of mole crabs, *Emerita* spp. (Decapoda: Hippidae), in the Americas. *Mar. Biol.* 125: 489–497.

Tamura, K., Dudley, J., Nei, M. & Kumar, S. 2007. MEGA4: Molecular Evolutionary Genetics Analysis (MEGA) software version 4.0. *Mol. Biol. Evol.* 24: 1596–1599.

Trontelj P., Machino, Y. & Sket, B. 2005. Phylogenetic and phylogeographic relationships in the crayfish genus *Austropotamobius* inferred from mitochondrial COX1 gene sequences. *Mol. Phylogenet. Evol.* 34: 212–226.

Tsang L.M., Ma, K.Y., Ahyong, S.T., Chan, T.-Y. & Chu, K.H. 2008. Phylogeny of Decapoda using two nuclear protein-coding genes: origin and evolution of the Reptantia. *Mol. Phylogenet. Evol.* 48: 359–368.

Tudge, C.C. & Cunningham, C.W. 2002. Molecular phylogeny of the mud lobsters and mud shrimps (Crustacea: Decapoda: Thalassinidea) using nuclear 18S rDNA and mitochondrial 16S rDNA. *Invert. Syst.* 16: 839–847.

Vences, M., Thomas, M., Bonett, R.M. & Vieites, D.R., 2005. Deciphering amphibian diversity through DNA barcodes: chances and challenges. *Phil. Trans. R. Soc. B* 360: 1859–1868.

Weinberg, J.R., Dahlgren, T.G., Trowbridge, N. & Halanych, K.M. 2003. Genetic differences within and between species of deep-sea crabs (*Chaceon*) from the North Atlantic Ocean. *Biol. Bull.* 204: 318–326.

Wiesner, R.J., Ruegg, C. & Morano, I. 1992. Counting target molecules by exponential polymerase chain reaction, copy number of mitochondrial DNA in rat tissues. *Biochim. Biophys. Acta* 183: 553–559.

Williams, S.T. & Knowlton, N. 2001. Mitochondrial pseudogenes are pervasive and often insidious in the snapping shrimp genus *Alpheus*. *Mol. Biol. Evol.* 18: 1481–1493.

Zhang, D.-X., Hewitt, G.M. 2003. Nuclear DNA analyses in genetic studies of populations: practice, problems and prospects. *Mol. Ecol.* 12: 563–584.

# Phylogenetic Inference Using Molecular Data

FERRAN PALERO[1] & KEITH A. CRANDALL[2]

[1] *Departament de Genètica, Universitat de Barcelona, Av. Diagonal 645, 08028 Barcelona, Spain*
[2] *Department of Biology, Brigham Young University, Provo, Utah 84602, U.S.A.*

## ABSTRACT

We review phylogenetic inference methods with a special emphasis on inference from molecular data. We begin with a general comment on phylogenetic inference using DNA sequences, followed by a clear statement of the relevance of a good alignment of sequences. Then we provide a general description of models of sequence evolution, including evolutionary models that account for rate heterogeneity along the DNA sequences or complex secondary structure (i.e., ribosomal genes). We then present an overall description of the most relevant inference methods, focusing on key concepts of general interest. We point out the most relevant traits of methods such as maximum parsimony (MP), distance methods, maximum likelihood (ML), and Bayesian inference (BI). Finally, we discuss different measures of support for the estimated phylogeny and discuss how this relates to confidence in particular nodes of a phylogeny reconstruction.

## 1 INTRODUCTION

The main objective of molecular phylogenetic analysis is to infer the evolutionary history of a group of species and represent it as an hierarchical branching diagram, a cladogram, or phylogenetic tree (Edwards & Cavalli-Sforza 1964). The contemporary taxa in that tree (as opposed to the reconstructed ancestral taxa) are called leaves or terminal tips. Internal nodes represent ancestral divergences into two or more (polytomy) genetically isolated groups (Fig. 1). Clades are characterized by shared possession of uniquely derived evolutionary novelties (synapomorphies). Therefore, phylogenetic analysis can be partially regarded as an attempt to recognize the identity and taxonomic distribution of synapomorphies. These could be any kind of inherited phenotypic or genotypic characteristics; it could be the evolutionary appearance of a nauplius larva or the fixation of a change from guanine to adenine at a particular site in a DNA sequence. Thus, phylogenies become essential tools for comparative biology (Harvey & Pagel 1991).

The tree topology is the information on the order of relationships, while the lengths of the branches in the tree can represent the evolutionary distances that separate nodes (phylogram) or not (cladogram). It is important to recognize if branches have been drawn to scale in order to know the relative distance between different species. This is particularly important, since if the sequences do not all evolve at the same rate, it is not possible to have a well-defined time axis on the tree with the standard methods. At this point we should also differentiate between rooted and unrooted trees. Even though biologists tend to think about trees as being rooted and pointing from "lower complexity" to "higher complexity," most phylogenetic methods do not result in a rooted tree (see Modeling Evolution section below). We generally need to define an outgroup by using external evidence not included in the molecular dataset (Weston 1994). Only then can rooted trees inform us about the temporal order of events and about which species have high rates of molecular evolution.

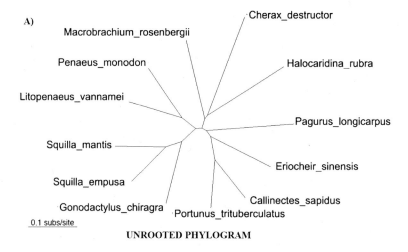
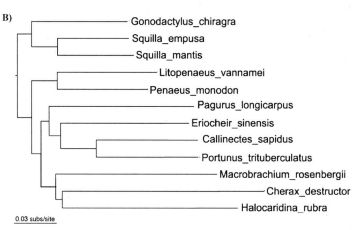
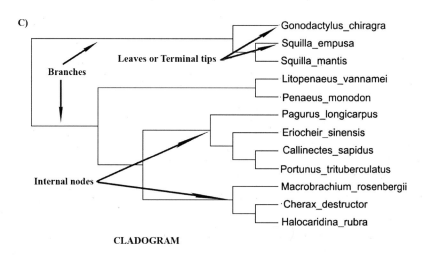

**Figure 1.** Phylogenetic trees obtained using a 966bp segment of the cytochrome B gene of several malacostracan crustaceans. (A) Unrooted phylogram, with distance scale bar indicating substitutions per site. (B) Rooted phylogram; the tree was rooted using Stomatopoda species as the outgroup. (C) Cladogram, showing the tree topology only.

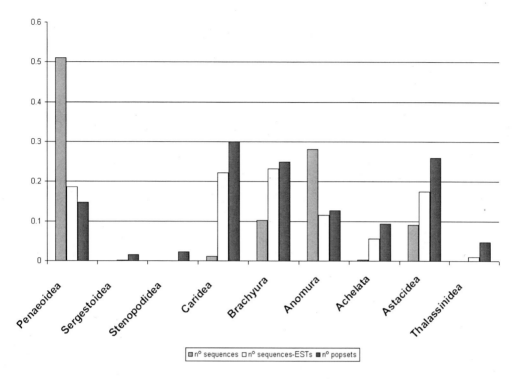

**Figure 2. (See Color Figure 1 in the Color Insert at the end of the book.)** Decapod sequences in GenBank in April 2008, shown as a proportion of the sequences belonging to the different infraorders relative to the total number of sequences available (355,876), the total number of sequences available after excluding ESTs (337,603), and the relative proportion of population study datasets.

## 1.1 Why should we use molecules when we already have morphology-based taxonomies?

Thanks to the popularization of DNA sequencing techniques, the number of decapod crustacean sequences available in GenBank has increased considerably, even though some infra-orders are still underrepresented (Fig. 2). The amplification of long genomic fragments implies that thousands of new, variable characters are made available for the study of phylogenetic relationships among organisms. This is particularly important for groups with very few characters available for developing morphological matrices (e.g., Rhizocephala) or when homology of morphological characters is particularly difficult to establish (Glenner et al. 2003). Moreover, the widespread use of accurate models of evolution and statistical tests allows us to extract a considerable amount of information from molecular sequence data. With the incorporation of closely related species to our group of interest, DNA sequence data allow polarity to be conferred to our phylogenetic reconstruction and allow us to make inferences on the evolution of molecules and/or the morphological characters themselves. An important advantage of molecular data is its objectivity, since results can be independently reproduced from the sequence data that are deposited in public databases.

However, DNA sequences have the same concerns as morphological traits for phylogeny estimation. Homoplasy can be caused by multiple substitutions occurring on a particular site, and character loss can also happen in gene sequences by insertion–deletion events. Phylogeny reconstruction can aid in the homology determination of molecular characters. Homologous genes may be orthologs, if they separated due to a speciation event, or paralogs, if those gene sequences diverged after gene duplication. In fact, gene duplication has been claimed to play a major role in the evolution of the mitochondrial genome of the Japanese freshwater crab *Geothelphusa dehaani* (Segawa & Aotsuka

2005). Furthermore, DNA sequences obtained from PCR products may correspond to pseudogenes, or non-functional copies. Using a mixture of orthologs and paralogs for phylogenetic reconstruction may point to the wrong topology (making distant taxa cluster together), whereas mixing pseudogenes with functional copies (e.g., nuclear copies of mitochondrial genes or numts) also gives the wrong topology but can make even copies from the same individual seem very distant (Song et al. 2008; Schubart this volume). When dealing with molecular sequences, character homology is incorporated with the sequence alignment, so we must be certain about the homology among nucleotide positions in the alignment.

## 2 CHARACTER HOMOLOGY AND THE PROBLEM OF SEQUENCE ALIGNMENT

Phylogenetic analysis attempts to reconstruct evolutionary genealogies of species based on similarities and differences. In an alignment of DNA sequences, each aligned site is a separate character with four character states being four nucleotides (A, C, T, G). Carrying out a multiple alignment means to define positional homology, deciding which nucleotide or amino acid positions are homologous for our sequence data. In order to infer the correct topology, nucleotide or amino acid positions must be aligned correctly. However, alignments of distantly related sequences may not be feasible, and different alignment methods often produce variable results depending on the details of the algorithm (Benavides et al. 2007). The most commonly used algorithms employ dynamic programming procedures seeking to maximize the score of the alignment (Needleman & Wunsch 1970). The score is determined by the choice of a matrix of similarities between nucleotides or amino acids and by the assignment of penalties for opening and extending gaps or insertions (Thompson et al. 1994).

Most dynamic programming methods use a greedy approach for progressively aligning pairs of sequences, but hierarchically aligning pairs of sequences is prone to generate biases and dominance by the most similar sequences. Additionally, the alignment tends to be sensitive to the choice of the similarity matrix and of gap penalties. Alternative approaches for aligning sequences include both dynamic programming and motif-finding algorithms. For example, the alignment program MUSCLE (Edgar 2004) first searches regions of similarity refined through iterations and then optimizes the alignment by applying a dynamic programming procedure locally. Since alignment methods are prone to errors, it is customary to manually adjust the alignment or to eliminate positions that are considered to be uncertain (GBLOCKS: Castresana 2000), a procedure that relies somewhat on the judgment of the investigator. Poorly aligned positions may not be homologous or may have been saturated by multiple substitutions and should be eliminated to increase the reliability of the phylogenetic analysis (Swofford et al. 1996; Castresana 2007). However, misalignments can still go undetected, particularly in large-scale analyses and for distantly related sequences.

### 2.1 Dealing with gaps

DNA sequences of homologous genes from distant species usually have unequal lengths and therefore force us to assume particular insertion and deletion events, defining the location of gaps or indels in the alignment. When dealing with protein coding nucleotide sequences, we could translate to the amino acid sequence, which may be easier to align, and then reverse back to the nucleotide sequence. However, the most commonly used genes for phylogenetic inference are non-protein coding genes (i.e., rDNA), and dealing with gaps remains a problem. Most distance-based analyses and, until recently, most likelihood and Bayesian analyses either treated gaps as unknowns or removed the gap containing column(s) from the analyses for pairs of sequences or for all sequences in an alignment (Lutzoni et al. 2000). The specific treatment of gaps in phylogenetic analysis can affect the results (Ogden & Whiting 2003), and several approaches are available for incorporating indel information into the phylogenetic analysis (Holmes 2005). Indeed, empirical results suggest that incorporating gaps as phylogenetic characters can aid in providing more robust phylogenetic estimates (Egan & Crandall 2008). It has been shown that point estimation of alignment and phylogeny avoids bias that results from conditioning on a single alignment estimate (Lake 1991; Thorne & Kishino 1992).

Within parsimony analysis, gaps may be incorporated as transformations during the cladogram evaluation process (optimization alignment in POY; Varón et al. 2007). It has been shown that in cases where alignment is not totally correct, coding gaps as a fifth state character or as separate presence/absence characters outperforms treating gaps as unknown/missing data nearly 90% of the time (Ogden & Rosenberg 2006). Datasets with higher sequence divergence and polytomies are more affected by gap coding than datasets associated with shallower non-polytomic tree shapes (Ogden & Rosenberg 2007). Redelings & Suchard (2005) describe a statistical method for incorporating indel information into phylogeny estimation under a Bayesian framework. Their method uses a joint reconstruction that simultaneously infers the alignment, tree, and insertion/deletion rates. Estimation proceeds through Markov chain Monte Carlo (MCMC) and naturally accounts for uncertainty in alignments, phylogenies, and other parameters through posterior probabilities. This method is based on a probabilistic model of sequence evolution that contains insertion and deletion events as well as substitution events (Thorne et al. 1991). Gaps are not treated as a fifth character state, since this over-weights the evidence of shared indels by treating an indel of multiple residues as multiple shared indels. Instead, the indel process is separate and independent of the substitution process and allows indels of several residues simultaneously.

## 3 GENETIC DISTANCES AND SATURATION

Theoretically, if the total number of substitutions between any pair of sequences is known, all the distance methods will produce the correct phylogenetic tree. In practice, this number is almost always unknown. In order to estimate a standardized genetic distance between organisms, we could just count the number of nucleotide differences among sequences and divide that number for the total number of nucleotide positions compared (p distance). However, DNA changes usually do not occur randomly along the sequence because of negative selection acting preferentially over some positions (Frank & Lobry 1999). Besides, if two lineages have been evolving separately for a long time, it is likely that multiple nucleotide substitutions have occurred on a particular position (multiple hits). As mutations accumulate, a point is reached at which there is no further divergence between sequences (mutational saturation). From this point on, it becomes impossible to estimate the evolutionary distance from similarity. This point of mutational saturation may occur at any taxonomic level, depending on the pattern of position-specific variability. Variation of mutation rate patterns among sites, functionally constrained sites, rapidly evolving lineages, and ancient evolutionary events will make the estimates of distances uncertain (Philippe & Forterre 1999). Different molecules evolve at different rates, and some of the fast-evolving genes will be saturated with changes even for closely related taxa. Using fast-evolving genes for phylogenetic inference of distantly related species could provide misleading results. A sensible approach for tackling this problem of saturation would be to use molecular markers that present a slower mutation rate and using an appropriate nucleotide substitution model in order to correct the observed distance for the multiple hits. However, if the gene evolves too slowly, there will be very little variation among the sequences, and there will be too little information to construct a phylogeny. Phylogenetic methods are likely to become unreliable if the sequences are too different from one another, and this should be borne in mind when the choice of gene sequences is made initially. Typically, a combination of genes is needed to accurately reconstruct phylogenetic relationships, with faster-evolving genes resolving close relationships and more slowly evolving genes resolving deeper relationships.

## 4 MODELING EVOLUTION AND MODEL SELECTION

More complex models, taking into account a variety of biological phenomena, generally provide more accurate estimates of phylogeny regardless of the method (e.g., parsimony, likelihood, distance, Bayesian) (Huelsenbeck 1995). The most common models of DNA evolution include base frequency, base exchangeability, and rate heterogeneity parameters. The parameter values are usually estimated from the dataset in each particular analysis (model selection). Finally, the evolutionary models are defined by matrices containing the relative rates of all possible replacements (transition probability matrix), which allow us to calculate the probabilities of change from any nucleotide to any other nucleotide (Liò & Goldman 1998). Most models assume reversibility of the transition probability matrix so that no inferences about evolutionary direction can be made unless further information extrinsic to the sequences themselves (e.g., fossil record) is supplied.

The base frequency parameters describe the frequencies of the nucleotide bases averaged over all sequence sites and over the tree. These parameters can be considered to represent constraints on base frequencies due to effects such as overall GC content, and they act as weighting factors in a model by making certain bases more likely to arise when substitutions occur. Base exchangeability parameters describe the relative tendencies of bases to be substituted for one another (Fig. 3). These parameters represent a measure of the biochemical similarity of bases, since transitions (i.e., C↔T or A↔G) usually occur more often than transversions (e.g., C↔G) (Brown et al. 1982; but see also Keller et al. 2007). Furthermore, mutation rates vary considerably among sites of DNA and amino acid sequences or among loci, because of constraints of the genetic code, selection for gene function, etc. In fact, we have to consider that if most of the nucleotide positions in our sequences evolve rather slowly or do not change at all (invariant sites), then base changes will tend to accumulate in a few variable sites, and sequence saturation will be reached much more quickly and at a lower divergence than expected under simpler models that do not

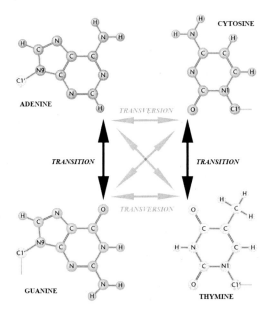

**Figure 3.** Transition versus Transversion mutations. DNA substitution mutations are of two types. Transitions are interchanges of purines (A–G) or pyrimidines (C–T), which involve bases of similar shape. Transversions are interchanges between purine and pyrmidine bases, which involve exchange of one-ring and two-ring structures.

incorporate rate heterogeneity or a proportion of invariant sites. The most widespread approach to modeling rate heterogeneity among sequence sites is to describe each site's rate as a random draw from a gamma distribution (Yang et al. 1994). The shape of the gamma distribution is controlled by a parameter $\alpha$. Large values of $\alpha$ suggest that sites evolve at a similar rate, while small values of the parameter $\alpha$ imply higher levels of rate heterogeneity among sites and the presence of many sites with lower rates of evolution. It is also possible to assign specific rates of substitution to different parts of the sequence in order to account for the heterogeneity on the mutation rate (e.g., to the three codon positions of protein coding sequences or to different domains in rRNA).

We can use the likelihood framework to estimate parameter values and their standard errors from the observed data when selecting the optimal model to perform phylogenetic inference (Yang et al. 1994), since comparisons of two competing models are possible using likelihood ratio tests. Competing models are compared (using their maximized likelihoods) with a statistic that measures how much better an explanation of the data the alternative model gives. When the simpler model is a special case of the more complex model, then the required distribution for the statistic is usually a $\chi^2$ distribution with the number of degrees of freedom equal to the difference in the number of parameters between the two models (Goldman 1993). When the models being compared are not nested, as can often be the case for more complex models of sequence evolution, the required distribution can be estimated by Monte Carlo simulation or by parametric bootstrapping (Huelsenbeck & Rannala 1997). Alternatively, one can use different statistical criteria to evaluate alternative models simultaneously (Posada & Buckley 2004).

Complex models describing selection or structure consistently give significantly improved descriptions of the evolution of protein sequences and are especially valuable in giving new insights into the processes of molecular evolution (Porter et al. 2007). Particularly, codon-based models have been developed that describe the evolution of coding sequences in terms of both DNA substitutions and the selective forces acting on the protein product (Nielsen & Yang 1998; Yang et al. 2000). For example, by studying the relationships between rates of synonymous (amino acid conserving) and nonsynonymous (amino acid altering) DNA substitutions, these models have been used successfully to detect where and when positive selection was important (Zanotto et al. 1999). Other models have attempted to associate the heterogeneity of patterns and rates of evolution among sites with the structural organization of RNA. These complex models accommodating RNA secondary structural elements use 16 states to represent all the possible base pairings in stem regions and four states to model loops (Schöniger & von Haeseler 1994).

Finally, while employing multiple alternative models in phylogenetic analysis might be seen as more rigorous, if this approach is to be meaningful there needs to be some quality control on the models employed (Grant & Kluge 2003). Similarly, all methods of phylogenetic inference assume a model of evolution, either implicitly or explicitly. For example, a strict parsimony analysis assumes all character changes are of equal weight. Thus, it becomes incumbent upon the researcher to justify the choice of model, even if it is an implicit model used to describe character evolution. If there are no restrictions on allowable models, virtually any given phylogeny may be found to be supported by some models and refuted by others. The model averaging approach by Lee & Hugall (2006) addresses both issues: a large number of possible models can be employed, but the results of each model are weighted according to its fit, so that the results of implausible models carry little weight on the final estimate. Likewise, statistically testing alternative models of evolution allows one to determine if the addition of more parameters makes a significant improvement in a likelihood score (Posada & Crandall 2001).

## 5 SEARCHING FOR TREES IN A BROAD TREE SPACE

The reconstruction of a phylogenetic tree using molecular data is an attempt to statistically infer the best estimate of evolutionary relationships given some criterion. While the "true tree" is the goal, what phylogenetic methods actually do is optimize a tree given some model and optimality criterion.

Thus, we are actually searching for not the "true tree" but rather the "optimal tree" and hope that the latter has some relationship to the former. There are two processes involved in this inference: estimation of the topology and estimation of branch lengths for a given tree topology. When a topology is known, statistical estimation of branch lengths is relatively simple, and one can use several statistical methods such as the least squares and the maximum likelihood methods. The problem is the estimation or reconstruction of a topology. The number of possible topologies increases rapidly with the number of sequences (Swofford et al. 1996), and it is generally very difficult to choose the correct topology among them. In phylogenetic inference, a certain optimization principle such as the maximum likelihood (ML) or minimum evolution (ME) principle is often used for evaluating different tree scores and choosing the topology and branch lengths that give an optimal score, so that we need to have tree searching strategies to help us finding the "optimal tree."

**Exhaustive search.** The exhaustive algorithm evaluates all possible trees. Because it examines all possible topologies, exhaustive searches guarantee the most optimal tree(s), but it is very slow (using 12 taxa, more than 600 million trees are evaluated). The advantage of the exhaustive search is the ability to completely explore the tree space and thereby plot the optimality score distribution. This histogram may indicate the "quality" of your matrix, in the sense that there should be a tail to the left such that few short trees are "isolated" from the greater mass of less optimal trees (but see Kitchin et al. 1998).

**Branch and bound.** The branch-and-bound algorithm is guaranteed to find all optimal trees, given some criterion (e.g., maximum parsimony). It discards whole classes of trees that it has determined are suboptimal, without the need to examine all of those one by one. The savings is greater the less homoplasy there is in the data. However, in cases where there are many conflicts between information from different characters and much parallelism and convergence, the branch-and-bound strategy does not perform particularly well. Moreover, branch-and-bound methods still have a complexity that is exponential, and it is not recommended to use the branch-and-bound algorithm for datasets with more than 12 taxa.

**Heuristic searches.** Since most datasets today contain large numbers of sequences, exhaustive and branch-and-bound searches quickly become impractical. We then turn to heuristic searches. Heuristic searches attempt to survey the tree space reasonably well without guaranteeing to find the most optimal tree(s). The key to good heuristic searching is the ability to move around the tree space and spend time exploring reasonable alternative topologies. Thus, a wide variety of branch swapping algorithms has been developed to achieve this goal.

**Nearest-neighbor interchange** (NNI). This heuristic algorithm adds taxa sequentially, in the order they are given in the matrix, to the branch where they will give least increase in tree length (Robinson 1971; Moore et al. 1973). After each taxon is added, all nearest neighbor trees are swapped to try to find an even shorter tree. Like all heuristic searches, this one is much faster than the algorithms above and can be used for large numbers of taxa, but it is not guaranteed to find all or any of the optimal trees. To decrease the likelihood of ending up on a suboptimal local minimum, a number of reorderings can be specified. For each reordering, the order of input taxa could be randomly permutated and another heuristic search attempted.

**Subtree pruning and regrafting** (SPR) is similar to NNI, but with a more elaborate branch swapping scheme. In order to find a shorter tree, a subtree is cut off the tree and regrafted onto all other branches in the tree to find the best alternative (Swofford 2003). This is done after each taxon has been added, and for all possible subtrees. While slower than NNI, SPR will often find shorter trees (Felsenstein 2004).

**Tree bisection and reconnection** (TBR) is similar to SPR, but with an even more complete branch swapping scheme. The tree is divided into two parts, and these are reconnected through every possible pair of branches in order to find a shorter tree. This is done after each taxon is added, and for all possible divisions of the tree (Swofford 2003). TBR will often find shorter trees than SPR and NNI, but it is more time consuming.

**The ratchet**. Different characters in the data may well recommend different trees to us. To prevent the search from becoming focused on a limited set of trees, it may help to use different starting trees as recommended by various subsets of characters. In the ratchet approach, we pick up some characters and increase their representation by increasing their weight (Nixon 1999; Felsenstein 2004). This moves the search to a tree recommended by this reweighted dataset; then we search from that starting point using the full set of characters.

Given the enormously large size of the tree space even for a small dataset, all we can do is hope that if we have searched for a long time without finding any improvement, then we have probably found the best tree. The problem with long-range moves tends to be that they are rather disruptive, moving the search far from the optimal tree. Most real search programs use a combination of NNIs and slightly longer range moves that have been tested and found to be reasonably efficient at finding optimal trees as quickly as possible. The MCMC method (see below) is a way of searching tree space that allows both uphill and downhill moves, allowing for suboptimal tree topologies to be sampled during the search. Regardless of the optimality criterion used, a key aspect of effective heuristic tree searching is to perform the analysis multiple times with different starting positions to be sure the tree space has been reasonably sampled.

## 6 INFERENCE METHODS

Ideally, the inference method used will extract the maximum amount of information available in the sequence data, will combine this with prior knowledge of patterns of sequence evolution (included in the evolutionary model), and will deal with model parameters (e.g., the transition/transversion ratio) whose values are not known a priori. The major inference methods for molecular phylogenetics are maximum likelihood, Bayesian inference, distance methods, and maximum parsimony.

### 6.1 *Maximum likelihood*

Likelihood-based techniques allow a wide variety of phylogenetic inferences from sequence data and a robust statistical assessment of all results. The likelihood of an hypothesis is equal to the probability of observing the data (sequence alignment) if that hypothesis (tree topology) were correct, given the chosen model of sequence evolution (Felsenstein 1981). Thus, a model of nucleotide or amino acid replacement allows the calculation of the likelihood for any possible combinations of tree topology and branch lengths. It permits the inference of phylogenetic trees and also making inferences simultaneously about the patterns and processes of evolution. A great attraction of the likelihood approach in phylogenetics is the existence of a wealth of powerful statistical theory, for example, the ability to perform robust statistical hypothesis tests (see below) and the knowledge that ML phylogenetic estimates are statistically consistent (given enough data and an adequate model, ML will always give the correct tree topology) (Rogers 1997). These strong statistical foundations suggest that likelihood techniques are the most powerful for phylogeny reconstruction and for understanding sequence evolution. Simulation studies show that ML methods generally outperform distance and parsimony methods over a broad range of realistic conditions, and recent developments in distance and parsimony methodology have concentrated on elucidating the relationships of these methods to ML inference and exploiting this understanding to adapt the methods so that they perform more like ML methods (Steel & Penny 2000; Bruno et al. 2000). However, ML suffers from computational intensity, making ML estimation impractical when dealing with several thousands of sequences, but better algorithms are being developed continually that can accommodate an increasingly large number of sequences for ML analyses (Stamatakis et al. 2005).

The ML method is a well-established statistical method of parameter estimation; it gives the smallest variance of a parameter estimate when sample size is large. In the construction of phylogenetic trees, maximization of the likelihood is done for each topology separately by using a different likelihood function, and the topology with the highest (maximum) likelihood is chosen as an estimate of the true topology. Since different topologies represent different probability spaces of parameters, it is not clear whether the maximum likelihood tree is expected to be the true tree unless an infinite number of nucleotides are examined (Felsenstein 2004). Finally, it should be mentioned that the statistical foundation of phylogeny estimation by ML has not been well established, and some authors have pointed out that topologies are parameters, but these parameters are not included in the likelihood function that is being maximized (Yang 1996a).

## 6.2  Bayesian methods

When inferring phylogenies, we should consider methods that deal directly with ensembles of possible trees, rather than chasing after a single best one, and we should be able to consider the information in the data and any prior information about the probabilities of the events. The fundamental importance of evolutionary models is that they contain parameters, and if specific values can be assigned to these parameters based on observations, such as an alignment of DNA sequences, then biologists can learn something about how molecular evolution has occurred. Although both maximum likelihood and Bayesian analyses are based upon the likelihood function, there are fundamental differences in how the two methods treat parameters. ML makes inferences about the parameters of interest while fixing the values for the other parameters (nuisance parameters). However, Bayesians assign a prior probability distribution to the nuisance parameters and the posterior probability is calculated by integrating over all possible values of those nuisance parameters, weighting each by its prior probability. The advantage of this is that inferences about the parameters of interest do not depend upon any particular value for the nuisance parameters. The disadvantage is that it may be difficult to specify a reasonable prior for the parameters. Nevertheless, when there is a large amount of information in the data and the likelihood function changes rapidly as the parameter values are altered, the choice of prior is not so important and it is possible to use uniform or non-informative priors. All branch lengths could be set as equally likely a priori, and a suitable non-informative choice of prior for base frequencies could be to set all sets of frequencies that add up to one as equally probable.

Markov models are routinely used in several domains of science and do not belong specifically to the Bayesian inference methodology; however, they have revolutionized genetic inferences in many aspects (Beaumont & Rannala 2004). A Markov model is a mathematical model for a process with changes of state over time, in which future events occur by chance and depend only on the current state and not on the history of how that state was reached. In molecular phylogenetics, the states of the process are the possible nucleotides or amino acids present at a given time and position in a sequence, and state changes represent mutations in sequences. Therefore, starting from an evolutionary model and a set of nucleotide frequencies, we can get to an equilibrium at which any state has a probability of occurrence that does not depend on the initial state of the process.

Under the MCMC search in a Bayesian framework, the probability of finding a tree will be proportional to its likelihood multiplied by its prior probability. In that case, the new tree is either accepted or rejected, using a rule known as the Metropolis algorithm. If the likelihood of the proposed tree is larger than the likelihood of the current one, the proposed topology is accepted and it becomes the next tree in the sample. If it is rejected, then the next tree in the sample is a repeat of the original tree. It also allows moves that decrease the likelihood, in order to allow for sampling of suboptimal trees. When the MCMC chain reaches the equilibrium, the probability of observing each tree must be constant. This property is known as detailed balance. It is necessary to strike a balance between moves that alter branch lengths and those that alter topology. If changes are very large, then the likelihood ratio of the states will be far from 1, and the likelihood of accepting the downhill move for sampling suboptimal trees will be very small. Finally, failure to diagnose a lack of convergence of the MCMC chain will lead to incorrect tree topology estimates (Huelsenbeck et al. 2002).

## 6.3 *Distance methods*

Distance matrix methods calculate a measure of the distance between each pair of species and then find a tree that predicts the observed set of distances as closely as possible. This leaves out all information from higher-order combinations of character states, reducing the data matrix to a simple table of pairwise distances. Distance methods use the same models of evolution as ML to estimate the evolutionary distance between each pair of sequences from the set under analysis and then try to fit a phylogenetic tree to those distances. The distances will usually be ML estimates for each pair of sequences (considered independently of the other sequences). Disadvantages of distance methods include the inevitable loss of evolutionary information when a sequence alignment is converted to pairwise distances and the inability to deal with models containing parameters for which the values are not known a priori (Steel et al. 1988). We are trying to find the n-species tree that is implied by these distances. The difficulty in doing this is that the individual distances are not exactly the path lengths in the full n-species tree between those two species. Since we are dealing with pairwise distances, we need to be able to find the full tree that does the best job of approximating these individual two-species trees.

In order for distances that are used in these analyses to have the proper expectations, it is essential that they are expected to be proportional to the total branch length between the species. If the distances do not have the linearity property, then wrenching conflicts between fitting the long distances and fitting the short distances arise, and the tree is the worse for them. There are several distance matrix methods available in the literature. Two examples are minimum evolution and neighbor joining.

**Minimum Evolution**. This method seeks to find the tree with the shortest overall branch lengths. First, the least squares trees are determined for different topologies, and the choice is made among them by choosing the one of shortest total length. Rzhetsky & Nei (1993) showed that if the distances were unbiased estimates of the true distance (many distances are not unbiased), then the expected total length of the true tree was shorter than the expected total length of any other. However, that is not the same as showing that the total length is always shorter for the true tree, as the lengths vary along their expectation. Gascuel et al. (2001) have found cases where the minimum evolution is inconsistent when branch lengths are inferred by weighted least squares or by generalized least squares.

**Neighbor Joining**. NJ is a clustering method that produces unrooted trees. It works by successively clustering pairs of sequences together. It is related to the UPGMA method of inferring a branching diagram from a distance matrix. Unlike the UPGMA method, NJ can facilitate contemporary tips of uneven length. This makes it a more appropriate tree reconstruction method than UPGMA in those instances when evolution has not proceeded in a strictly clock-like fashion. NJ is guaranteed to recover the true tree if the distance matrix happens to be an exact reflection of a tree. However, in the real world, distances will not be exactly additive, and therefore NJ is just one approximation. Furthermore, the NJ tree may be misleading. If the input distances are not close to being additive, because pairwise distances were not properly calculated or because sequences were not properly aligned, then NJ will give the wrong tree.

NJ is useful to rapidly search for a good tree that can then be improved by other criteria. Ota & Li (2001) use neighbor joining and bootstrapping to find an initial tree and identify which regions are candidates for rearrangement. They then use ML for further refinement. This results in a substantial improvement in speed over pure likelihood methods. Moreover, modifications of NJ have been developed to allow for differential weighting in the algorithm to take into account differences in statistical noise. Gascuel (1997) has modified the NJ to allow for the variances and covariances of the distances to be proportional to the branch lengths. This is a good approximation provided that the branch lengths are not too long.

## 6.4 *Maximum parsimony*

The theoretical basis of this method is the philosophical idea that the best hypothesis to explain a process is the one that requires the smallest number of assumptions (Occam's Razor). If there are no backward and no parallel substitutions at each nucleotide site (no homoplasy) and the number of informative nucleotides examined is very large, maximum parsimony (MP) methods are expected to provide the correct (realized) tree. MP assumes that maximizing the congruence among characters will be equal to minimizing incongruence (homoplasy) (Farris 1983). Therefore, computing programs will count the number of mutational changes (steps) we need to explain a particular tree and repeat this counting for thousands of trees. The tree or trees that need a minimum number of changes to explain the relationships between species will be accepted as the most parsimonious tree.

There are two main dynamic programming algorithms for counting the number of changes of state. In both cases, the algorithm does not function by actually placing changes or reconstructing states at the nodes of the tree. The **Fitch algorithm** works for characters with any number of states, provided one can change from any one to any other (Kluge & Farris 1969). Fitch characters are reversible and unordered, meaning that all changes have equal cost. This is the criterion with fewest assumptions, and is therefore generally preferable. The Fitch algorithm can be carried out in a number of operations that are directly proportional to the number of species on the tree, and, therefore, the algorithm is less computationally demanding than other methods. The **Sankoff algorithm** starts by assuming that one has a table of the cost of changes between each character state and each other state. In this case, one computes the total cost of the most parsimonious combinations of events by computing it for each character. Given that a node is assigned a particular character state, we will compute the minimal cost of all the events in the subtree that starts from that node and accept it as the most parsimonious result.

Other algorithms allow us to reconstruct character states at the nodes of the tree. The **Camin-Sokal Parsimony** algorithm (C-S) assumes that we know the ancestral state of the character. In its simplest form, only two states are allowed (presence/absence) and reversals are impossible. One application of C-S parsimony is in the evolution of small deletions of DNA, when we have no reason to believe that they could revert spontaneously. In more complex cases, when deletions overlap and we cannot be entirely sure whether any one of them is present or absent, C-S parsimony would not be appropriate. C-S parsimony infers a rooted tree, since it will favor the placement of the root in one particular part of the tree. In its simplest form, **Dollo parsimony** assumes that there are two states (ancestral/derived). The main difference with C-S parsimony is that in this case the derived state is allowed to evolve only once, but it is allowed to revert to the ancestral state multiple times. The number of these reversions is the quantity being minimized, and it is also an inherently rooted method. In "unweighted" (=equal weighting) MP methods, nucleotide or amino acid substitutions are assumed to occur in all directions with equal or nearly equal probability. In reality, however, certain substitutions (e.g., transitional changes) occur more often than other substitutions (e.g., transversional changes). It is therefore reasonable to give different weights to different types of substitutions when the minimum number of substitutions for a given topology is to be computed. MP methods incorporating a weight matrix for the different types of change are weighted MP methods.

Once the most parsimonious phylogenetic tree has been recovered, we can still wonder about the amount of parallelism or reversal that is found on the tree. A particular character state may have evolved independently in two lineages, and multiple hits may cause a particular nucleotide position to return to an ancestral state. Several indices have been developed to measure the relative amount of homoplasy found in a particular tree. For example, the per-character consistency index (ci) is defined as $m/s$, where $m$ is the minimum possible number of character changes (steps) on any tree, and $s$ is the actual number of steps on the current tree. This index hence varies from one (no homoplasy) towards zero (a lot of homoplasy). The ensemble consistency index CI is a similar index, but summed over all characters.

The per-character retention index (ri) is defined as the ratio of (1) the differences between the maximal number of steps for the character on any cladogram and the actual number of steps on the current tree and (2) the differences between the maximal number of steps for the character on any cladogram and the minimum possible number of character changes on any tree (Farris 1989). Therefore, the retention index becomes zero when the site is least informative for MP tree construction, that is, when the difference between the maximal number of steps for the character on any cladogram and the actual number of steps on the current tree is zero.

## 7 NODE SUPPORT AND TREE COMPARISON

Measures of nodal support provide a useful summary of how well data support the relationships defined by a tree. In the MP approach, the Bremer support (decay index) for a clade can be computed as a measure of the confidence on that particular clade. The Bremer support is the number of extra steps you need to construct a tree (consistent with the characters) where that clade is no longer present. When several genes are included in the analysis, the parsimony-based method of partitioned branch support (PBS) estimates the amount that each dataset contributes to a particular clade support, so that we can estimate the extent to which the data partition supports the most parsimonious tree over trees not including a particular clade (Gatesy et al. 1999). An equivalent "partitioned likelihood support" (PLS) can be obtained for each dataset under a likelihood-based approach (Lee & Hugall 2003). Most measures of nodal support attempt to estimate the degree to which an analysis has converged on a stable result. Of course, high support values do not mean that a node is accurate, only that it is well supported by the data. It is well known that model misspecification and taxon sampling can mislead the analysis (Hedtke et al. 2006).

Currently, the nonparametric bootstrap is one of the most widely used methods for assessing nodal support (Felsenstein 1985). The nonparametric bootstrap is a statistical method by which distributions that are difficult to calculate exactly can be estimated by the repeated creation and analysis of artificial datasets. A number of replicates (typically at least 1000) of the original characters (e.g., sites of a DNA sequence alignment) are randomly produced with replacement, obtaining a new dataset in which some characters are represented more than once, some appear once, and some are deleted. The perturbed datasets are each analyzed in the same manner as for the real data, and the number of times that each grouping of species appears in the resulting profile of cladograms is taken as an index of relative support for that grouping.

Perhaps the best interpretation of the bootstrap is that it quantifies the sensitivity of a node to perturbations in the data (Holmes 2005). However, as commonly implemented, the bootstrap gives a biased estimate of accuracy (Hillis & Bull 1993; Holmes 2005), where accuracy is defined as the probability of obtaining a correct phylogenetic reconstruction (Penny et al. 1992). The statistical theory of bootstrap requires that all positions of an alignment are independently and identically distributed, and this assumption does not apply to nucleotide or amino acid sequences. It is worthwhile to point out the difference between nonparametric and parametric bootstraps. In the nonparametric bootstrap, new datasets are generated by resampling from the original data, whereas in the parametric bootstrap, the data are simulated according to the hypothesis being tested. This well-known bias of the bootstrap has led researchers to seek other methods of estimating nodal support, and perhaps the most popular alternative is Bayesian posterior probability (Larget & Simon 1999; Yang & Rannala 1997). A nodal posterior probability is the probability that a given node is found in the true tree, conditional on the observed data, and the model (including both the prior model and the likelihood model). Early observations of Bayesian inference in phylogenetics demonstrated a tendency for posterior probabilities to be more extreme than ML nonparametric bootstrap proportions, although the two tended to be correlated (Buckley et al. 2002). Finally, Lewis et al. (2005) demonstrated that if a polytomy exists but is not accommodated in the prior, resolution of the polytomy will be arbitrary and the nodal support indicated by the posterior probability will appear unusually high compared to ML bootstraps. Because we have little knowledge of the goodness of fit between data and model in typical phylogenetic studies (although goodness of fit tests do exist), we have little idea of the seriousness of the problem of model misspecification in current implementations of Bayesian phylogenetic inference. Goodness of fit tests define how well a statistical model fits a set of observations. Measures of goodness of fit typically summarize the discrepancy between observed values and the values expected under the model in question. The great advantage of the Bayesian posterior probability is that this statistic is drawn from the same distribution that determines the best estimate of tree topology, as opposed to a bootstrap analysis that requires 1000 reruns of the analysis.

## 7.1 Statistical tests of tree topologies

A variety of topology tests has been designed to compare different trees and thereby test alternative hypotheses of phylogenetic relationships. There is a fundamental difference between testing a priori phylogenetic hypotheses versus testing those generated through analyses. The Templeton (1983) test and Kashino-Hasegawa (KH) test (Kishino & Hasegawa 1989) are nonparametric tests designed to compare pairs of topologies selected before a phylogenetic analysis is run, with the Templeton test using a parsimony framework and the KH test using a likelihood framework. However, these approaches may become too liberal when one of the alternative topologies is one estimated from the data (Goldman et al. 2000). In this case, the most widely used parametric test is the Swofford-Olsen-Waddell-Hillis (SOWH) test (Swofford et al. 1996), which uses parametric bootstrapping to simulate replicate datasets that are in turn used to obtain the null distribution. Shimodaira & Hasegawa (1999) have described a non-parametric bootstrap test that directly succeeds the KH test, considering all possible topologies and making the proper allowance for their comparison with the ML topology derived from the same data. Because of the nature of the null hypotheses employed by the nonparametric tests, the Templeton, SH, and KH tests are generally more conservative than the parametric tests (Aris-Brosou 2003; Buckley 2002; Goldman et al. 2000). The more explicit reliance on models of evolution by the parametric tests makes them very powerful tests, yet they are also more susceptible to model misspecification (Buckley 2002; Shimodaira 2002). Bayesian tests of topology are becoming more commonly implemented than the frequentist tests (Aris-Brosou 2003). The Bayesian tests generally rely on Bayes factors to compare marginal likelihoods generated under two hypotheses corresponding to different topologies (Kass & Raftery 1995). The use of Bayes factors in testing topologies will likely receive much greater attention in the future, since it allows for comparison of models that are not hierarchically nested (Nylander et al. 2004).

## 8 USING MULTIPLE GENES

The best phylogenetic estimates come from using robust inference methods coupled with realistic evolutionary models. However, good estimates of phylogeny ultimately depend on good datasets. The two most obvious ways of increasing the accuracy of a phylogenetic inference are to include more sequences in the data and/or to increase the length of the sequences used. Goldman (1998) showed that adding more sequences to an analysis does not increase the amount of information relating to different parts of the tree uniformly over that tree, whereas the use of longer sequences results in a linear increase in information over the whole of the tree. A potentially powerful approach is to analyze the sequences as a concatenated whole or "meta-sequence." The simplest analysis would be to assume that all the genes have the same patterns and rates of evolution (Cao et al. 1994). This naïve method should only be used when there is substantial evidence of a consistent evolutionary pattern across all the genes, which can be assessed by statistical tests of different models (as described above). Otherwise, differences amongst gene replacement patterns or rates can lead to biased results. More advanced analyses of concatenated sequences are possible, which allow for heterogeneity of evolutionary patterns among the genes studied (Yang 1996b). This heterogeneity might be as complex as allowing each gene to evolve with different replacement patterns, and with different rates of replacement in all branches of the gene trees (Yang 1997).

The contradictions in the different phylogenetic reconstructions based on analysis of different protein, gene, or noncoding sequences raise questions concerning the variability of evolutionary processes and the reliability of averaging schemes such as sequence concatenation (Teichmann & Mitchison 1999). Lateral transfer, fusion events, and recombination can make the evolutionary relationships among genes unreliable indicators of the phylogenetic relationships among the species. In that case, the Partition Homogeneity Test or incongruence length difference (ILD) test (Farris et al. 1994) could be used for testing if every gene in the analysis is giving a heterogeneous signal under the maximum parsimony framework. However, this heterogeneity can come solely from branch length differences and is not necessarily indicative of topological differences with different data subsets. Finally, in the so-called "total evidence" approach, genes are concatenated end to end, including also information from morphological characters, and the whole dataset is analyzed using parsimony (Ahyong & O'Meally 2004). This has the great advantage of taking into account the different amounts of sequence in different loci and of combining the evidence in a single tree that does not depend on an arbitrary choice of consensus tree method. Still, if different loci have substantially different rates of change, combining them into one dataset obscures evidence that indicates that one locus should be treated differently from another. In order to include this heterogeneity in the phylogenetic analysis, Kolaczkowski & Thornton (2004) recently presented a new mixture model to account for partitioned sequences. Even though there were some concerns about the computational burdens of implementing more complex evolutionary models, these concerns can be accommodated in a likelihood-based analysis. By using MCMC sampling, mixture models and likelihood-based approaches could be used even when evolution is heterogeneous (Pagel & Meade 2004).

## 9 SUMMARY OF METHODS AND CONCLUSION

> *"The time will come I believe, though I shall not live to see it, when we shall have fairly true genealogical trees of each great kingdom of nature."*
>
> Darwin (1857)

Throughout this review, several methods have been introduced that try to infer phylogenetic relationships between species using molecular data. **(1) Maximum parsimony** seeks to find the tree that is compatible with the minimum number of substitutions among sequences. Finding a maximally parsimonious cladogram is usually a computationally intensive task, but for large problems, fast heuristic algorithms can be employed, even though they cannot guarantee to find the optimal cladogram. Parsimony analysis has been criticized for requiring very stringent assumptions of constancy for substitution rates across sites and similar substitution rates among lineages. It has been found that the performance of MP deteriorates when mutational rates differ between nucleotides or across sites (Yang 1996b) or if evolutionary rates are highly variable among evolutionary lineages (Hendy & Penny 1989; DeBry 1992).

As more divergent sequences are analyzed, the overall degree of homoplasy generally increases, and this implies that the true evolutionary tree becomes less likely to be the one with the least number of changes. Furthermore, when two evolutionary lineages that have undergone a high level of sequence evolution are separated by a short lineage, the long lineages will tend to be spuriously joined in the most parsimonious cladogram produced from the resulting sequence data. Combinations of conditions when this occurs are often called the "Felsenstein zone," and parsimony is particularly affected by this problem because of its inability to deal with homoplasy (Huelsenbeck 1997). Nevertheless, MP methods have some advantages over other tree-building methods. Parsimony analysis is very useful for dealing with morphological characters or some types of molecular data such as insertion sequences and insertion/deletions, and weighted MP methods can be constructed to incorporate information on the evolutionary process.

**(2) Distance methods** such as neighbor joining seek to reconstruct the tree topology that best represents the matrix of distances between pairs of taxonomic units. As with all greedy methods, the NJ algorithm is not guaranteed to find the globally best solution to a general distance matrix with error (Pearson et al. 1999). In an effort to alleviate this problem, some generalizations of the NJ method have been proposed that explore multiple low-error paths in progressively clustering the sequences (Kumar 1996; Pearson et al. 1999). However, the most serious problem with distance methods is that they require a reliable measure of evolutionary distances between sequences. When evolutionary rates vary from site to site in molecular sequences, distances can be corrected for this variation. When variation of rates is large, these corrections become important. In likelihood methods, the correction can use information from changes in one part of the tree to inform the correction in others, but a distance matrix method is inherently incapable of propagating the information in this way. Thus, distance matrix methods must use information about rate variation substantially less efficiently than likelihood methods (Felsenstein 2004).

**(3) Likelihood-based methods** permit the application of mathematical models that incorporate our knowledge on typical patterns of sequence evolution, resulting in more powerful inferences. Furthermore, they use a complete statistical methodology that permits hypothesis tests, enabling validation of the results at all stages: from the values of parameters in evolutionary models, through the comparison of competing models describing the biological factors most important in sequence evolution, to the testing of hypotheses of evolutionary relationship. Computer programs for the robust statistical evolutionary analysis of molecular sequence data are widely available (Table 1).

Nevertheless, ML methods do not directly assign probabilities to the parameters, and if one wants to describe the uncertainty in an estimate, one has to repeat the analysis multiple times (bootstrap), increasing the computational cost. In **Bayesian inference**, information can be drawn directly from the simulated joint distribution of parameters at a reasonable computational cost. On the other hand, a review of the current Bayesian phylogenetic literature indicates that much more emphasis needs to be placed on developing more realistic models, checking the effects of the priors, and monitoring the convergence of posterior distributions.

All in all, it should be pointed out that systematic error will confound any tree reconstruction method. Situations such as long-branch-attraction and base-compositional bias are examples of systematic bias. When inferring phylogenies, we try to define the actual succession of divergence events from the present sampled sequences. This means that the actual genes sampled (gain and loss of genes happens, but we rely only on those genes for which homology can be ascertained), species sampled (extinction of intermediate taxa), selection (causing either among-sites or among-loci rate variation), and the population parameters (mutation rates, recombination rates, effective population sizes, etc.) all may influence the strength of the phylogenetic signal. In conclusion, phylogenetic inference should be approached not as a tool for getting a definitive answer for a taxonomical problem, but rather as a tool for asking new questions on the evolution of molecules and morphology in different species and for trying to uncover the causes of such differences in their evolution.

**Table 1.** A sampling of phylogenetic software to perform evolutionary analyses (see http://evolution.genetics.washington.edu/phylip/software.html for a comprehensive list).

| Name | Methods Implemented | Web | Citation |
|---|---|---|---|
| ClustalW | Progressive multiple sequence alignment | http://www.ebi.ac.uk/clustalw/ | Thompson et al. 1994 |
| MUSCLE | Progressive alignment and refinement using restricted partitioning | http://www.drive5.com/muscle/ | Edgar 2004 |
| POY | Optimization alignment | http://research.amnh.org/scicomp/projects/poy.php | Varón et al. 2007 |
| BAli-Phy | Bayesian inference of alignment and topology | http://www.biomath.ucla.edu/msuchard/bali-phy/index.php | Suchard & Redelings 2006 |
| ModelTest | Model selection | http://darwin.uvigo.es/software/modeltest.html | Posada & Crandall 1998 |
| MrModelTest | Model selection | http://www.abc.se/~nylander/ | Nylander 2004 |
| MEGA | Distance, parsimony and maximum likelihood | http:www.megasoftware.net/index.html | Tamura et al. 2007 |
| PAUP | Maximum parsimony, distance matrix, maximum likelihood | http://paup.csit.fsu.edu/ | Swofford 2003 |
| PHYLIP | Maximum parsimony, distance matrix, maximum likelihood | http://evolution.genetics.washington.edu/phylip.html | Felsenstein 2005 |
| TNT | Maximum parsimony, ratchet | http://www.zmuc.dk/public/phylogeny/TNT/ | Goloboff et al. 2003 |
| Winclada | Maximum parsimony, ratchet | http://www.cladistics.com/aboutWinc.htm | Nixon 2002 |
| PhyML | | http://atgc.lirmm.fr/phyml/ | Guindon & Gascuel 2003 |
| GarLi | Maximum likelihood using genetic algorithms | http://www.bio.utexas.edu/faculty/antisense/garli/Garli.html | Zwickl 2006 |
| PAML | Maximum likelihood | http://abacus.gene.ucl.ac.uk/software/paml.html | Yang 1997 |
| RAxML-HPC | Maximum likelihood, simple maximum parsimony | http://icwww.epfl.ch/~stamatak/ | Stamatakis et al. 2005 |
| MultiDivTime | Dating, molecular clock using Bayes MCMC | http://statgen.ncsu.edu/thorne/multidivtime.html | Thorne & Kishino 2002 |
| BayesPhylo-genies | Bayesian inference | http://www.evolution.rdg.ac.uk/SoftwareMain.html | Pagel & Meade 2004 |
| MrBayes | Bayesian inference | http://mrbayes.csit.fsu.edu/index.php | Ronquist & Huelsenbeck 2003 |

ACKNOWLEDGEMENTS

Thanks are due to P. Abelló, M. Pascual, and E. Macpherson for encouraging the completion of this study. This work was supported by a pre-doctoral fellowship awarded by the Autonomous Government of Catalonia (2006FIC-00082) to FP and by a grant from the US NSF EF-0531762 awarded to KAC. FP is part of the research group 2005SGR-00995 of the Generalitat de Catalunya. Research was funded by project CGL2006-13423 from the Ministerio de Educacion y Ciencia. FP acknowledges EU-Synthesys grant (GB-TAF-1637).

REFERENCES

Ahyong, S.T. & O'Meally, D. 2004. Phylogeny of the Decapoda. Reptantia: resolution using three molecular loci and morphology. *Raffl. Bull. Zool.* 52: 673–693.

Aris-Brosou, S. 2003. Least and most powerful phylogenetic tests to elucidate the origin of the seed plants in presence of conflicting signals under misspecified models. *Syst. Biol.* 52: 781–793.

Beaumont, M. & Rannala, B. 2004. The Bayesian revolution in genetics. *Nat. Rev. Genet.* 5: 251–261.

Benavides, E., Baum, R., McClellan, D. & Sites, J.W. 2007. Molecular phylogenetics of the lizard genus Microlophus (Squamata: Tropiduridae): aligning and retrieving indel signal from nuclear introns. *Syst. Biol.* 56: 776–797.

Brown, W.M., Prager, E.M., Wang, A. & Wilson, A.C. 1982. Mitochondrial DNA sequences of primates: tempo and mode of evolution. *J. Mol. Evol.* 18: 225–239.

Bruno, W.J., Socci, N.D. & Halpern, A.L. 2000. Weighted neighbor-joining: a likelihood-based approach to distance based phylogeny reconstruction. *Mol. Biol. Evol.* 17: 189–197.

Buckley, T.R. 2002. Model misspecification and probabilistic tests of topology: evidence from empirical data sets. *Syst. Biol.* 51: 509–523.

Buckley, T.R., Arensburger, P., Simon, C. & Chambers, G.K. 2002. Combined data, Bayesian phylogenetics, and the origin of the New Zealand cicada genera. *Syst. Biol.* 51: 4–18.

Cao, Y., Adachi, J., Janke, A., Pääbo, S. & Hasegawa, M. 1994. Phylogenetic relationships among eutherian orders estimated from inferred sequences of mitochondrial proteins: instability of a tree based on a single gene. *J. Mol. Evol.* 39: 519–527.

Castresana, J. 2000. Selection of conserved blocks from multiple alignments for their use in phylogenetic analysis. *Mol. Biol. Evol.* 17: 540–552.

Castresana, J. 2007. Topological variation in single-gene phylogenetic trees. *Genome Biol.* 8: 216.

DeBry, R.W. 1992. The consistency of several phylogeny-inference methods under varying evolutionary rates. *Mol. Biol. Evol.* 9: 537–551.

Edgar, R.C. 2004. MUSCLE: multiple sequence alignment with high accuracy and high throughput. *Nucl. Acids Res.* 32: 1792–1797.

Edwards, A.W.F. & Cavalli-Sforza, L.L. 1964. Reconstruction of evolutionary trees. In: McNeill, J. (ed.), *Phenetic and phylogenetic classification*: 67–76. London: Systematics Association Publication.

Egan, A.N. & Crandall, K.A. 2008. Incorporating gaps as phylogenetic characters across eight DNA regions: ramifications for North American Psoraleeae (Leguminosae). *Mol. Phylogenet. Evol.* 46: 532–546.

Farris J.S. 1983. The logical basis of phylogenetic analysis. In: Platnick, N.J. & Funk, V.A. (eds.), *Advances in Cladistics*: 1–36. New York: Columbia Univ. Press.

Farris, J.S. 1989. The retention index and the rescaled consistency index. *Cladistics* 5: 417–419.

Farris, J.S., Källersj, M., Kluge, A.G. & Bult, C. 1994. Testing significance of incongruence. *Cladistics* 10: 315–319.

Felsenstein, J. 1978. The number of evolutionary trees. *Syst. Zool.* 27: 27–33.

Felsenstein, J. 1981. Evolutionary trees from DNA sequences: a maximum likelihood approach. *J. Mol. Evol.* 17: 368–376.

Felsenstein, J. 1985. Confidence limits on phylogenies: an approach using the bootstrap. *Evolution* 39: 783–791.

Felsenstein, J. 2004. Inferring Phylogenies. Sinauer Associates Inc., Massachusetts. 664 pp.

Felsenstein, J. 2005. PHYLIP (Phylogeny Inference Package) version 3.6. Distributed by the author. Department of Genome Sciences, University of Washington, Seattle.

Frank, A.C. & Lobry, J.R. 1999. Asymmetric substitution patterns: a review of possible underlying mutational or selective mechanisms. *Gene* 238: 65–77.

Gascuel, O. 1997. BIONJ: an improved version of the NJ algorithm based on a simple model of sequence data. *Mol. Biol. Evol.* 14: 685–695.

Gascuel, O., Bryant, D. & Denis, F. 2001. Strengths and limitations of the minimum-evolution principle. *Syst. Biol.* 50: 621–627.

Gatesy, J., O'Grady, P. & Baker, R.H. 1999. Corroboration among data sets in simultaneous analysis: hidden support for phylogenetic relationships among higher-level artiodactyl taxa. *Cladistics* 15: 271–313.

Glenner, H., Lützen, J. & Takahashi, T. 2003. Molecular evidence for a monophyletic clade of asexually reproducing parasitic barnacles: *Polyascus*, new genus (Cirripedia: Rhizocephala). *J. Crust. Biol.* 23: 548–557.

Goldman, N. 1993. Statistical tests of models of DNA substitution. *J. Mol. Evol.* 36: 182–198.

Goldman, N. 1998. Phylogenetic information and experimental design in molecular systematics. *Proc. R. Soc. London Ser. B* 265: 1779–1786.

Goldman, N., Anderson, J.P. & Rodrigo, A.G. 2000. Likelihood-based tests of topologies in phylogenetics. *Syst. Biol.* 49: 652–670.

Goloboff, P., Farris, J.S. & Nixon, K. 2003. TNT: Tree analysis using new technology. Program and documentation, available from the authors, and at http://www.zmuc.dk/public/phylogeny.

Grant, T. & Kluge, A.G. 2003. Data exploration in phylogenetic inference: scientific, heuristic, or neither. *Cladistics* 19: 379–418.

Guindon, S. & Gascuel, O. 2003. A simple, fast and accurate algorithm to estimate large phylogenies by maximum likelihood. *Syst. Biol.* 52: 696–704.

Harvey, P.H. & Pagel, M.D. 1991. *The Comparative Method in Evolutionary Biology*. Oxford: Oxford University Press.

Hedtke, S.M., Townsend, T.M. & Hillis, D.M. 2006. Resolution of phylogenetic conflict in large data sets by increased taxon sampling. *Syst. Biol.* 55: 522–529.

Hendy, M.D. & Penny, D. 1989. A framework for the quantitative study of evolutionary trees. *Syst. Zool.* 38: 297–309.

Hillis, D.M. & Bull, J.J. 1993. An empirical test of bootstrapping as a method for assessing confidence on phylogenetic analysis. *Syst. Biol.* 42: 182–192.

Holmes, I. 2005. Using evolutionary expectation maximization to estimate indel rates. *Bioinformatics* 21: 2294–2300.

Huelsenbeck, J.P. 1995. Performance of phylogenetic methods in simulation. *Syst. Biol.* 44: 17–48.

Huelsenbeck, J.P. 1997. Is the Felsenstein zone a fly trap? *Syst. Biol.* 46: 69–74.

Huelsenbeck, J.P. & Rannala, B. 1997. Phylogenetic methods come of age: testing hypotheses in an evolutionary context. *Science* 276: 227–232.

Huelsenbeck, J.P., Larget, B., Miller, R.E. & Ronquist, F. 2002. Potential applications and pitfalls of Bayesian inference of phylogeny. *Syst Biol.* 51: 673–688.

Kass, R.E. & Raftery, A.E. 1995. Bayes factors. *J. Amer. Stat. Assoc.* 90: 773–795.

Keller, I., Bensasson, D. & Nichols, R.A. 2007. Transition-Transversion Bias Is Not Universal: A Counter Example from Grasshopper Pseudogenes. *PLoS Genet* 3(2): e22. doi:10.1371/journal.pgen.0030022

Kishino, H. & Hasegawa, M. 1989. Evaluation of the maximum likelihood estimate of the evolutionary tree topologies from DNA sequence data, and the branching order in Hominoidea. *J. Mol. Evol.* 29: 170–179.

Kitchin, I.J., Forey, P.L., Humphries, C.J. & Williams, D.M. 1998. *Cladistics*. Oxford: Oxford University Press.

Kluge, A.G. & Farris, J.S. 1969. Quantitative phyletics and the evolution of anurans. *Syst. Zool.* 18: 1–32.

Kolaczkowski, B. & Thornton, J.W. 2004. Performance of maximum parsimony and likelihood phylogenetics when evolution is heterogeneous. *Nature* 431: 980–984.

Kumar, S. 1996. A stepwise algorithm for finding minimum evolution trees. *Mol. Biol. Evol.* 13: 584–593.

Lake, J.A. 1991. The order of sequence alignment can bias the selection of tree topology. *Mol. Biol. Evol.* 8: 378-385.

Larget, B. & Simon, D. 1999. Markov chain Monte Carlo algorithms for the Bayesian analysis of phylogenetic trees. *Mol. Biol. Evol.* 16: 750–759.

Lee, M.S.Y. & Hugall, A.F. 2003. Partitioned likelihood support and the evaluation of data set conflict. *Syst. Biol.* 52: 15–22.

Lee, M.S.Y. & Hugall, A.F. 2006. Model type, implicit data weighting, and model averaging in phylogenetics. *Mol. Phylogenet. Evol.* 38: 848–857.

Liò, P. & Goldman, N. 1998. Models of molecular evolution and phylogeny. *Genome Res.* 8: 1233–1244.

Lutzoni, F., Wagner, P., Reeb, V. & Zoller, S. 2000. Integrating ambiguously aligned regions of DNA sequences in phylogenetic analyses without violating positional homology. *Syst. Biol.* 49: 628–651.

Moore, G., Goodman, M. & Barnabas, J. 1973. An iterative approach from the standpoint of the additive hypothesis to the dendrogram problem posed by molecular data sets. *J. Theor. Biol.* 38: 423–457.

Needleman, S.B. & Wunsch, C.D. 1970. A general method applicable to the search for similarities in the amino acid sequence of two proteins. *J. Mol. Biol.* 48: 443–53.

Nielsen, R. & Yang, Z. 1998. Likelihood models for detecting positively selected amino acid sites and applications to the HIV-1 envelope gene. *Genetics* 148: 929–936.

Nixon, K.C. 1999. The Parsimony Ratchet, a new method for rapid parsimony analysis. *Cladistics* 15: 407–414.

Nixon, K.C. 2002. WinClada ver. 1.00.08. Published by the author, Ithaca, NY.

Nylander, J.A.A. 2004. MrModeltest v2. Program distributed by the author. Evolutionary Biology Centre, Uppsala University.

Nylander, J.A., Ronquist, F., Huelsenbeck, J.P. & Nieves-Aldrey, J.L. 2004. Bayesian phylogenetic analysis of combined data. *Syst. Biol.* 53: 47–67.

Ogden, T.H. & Whiting, M. 2003. The problem with "the Paleoptera Problem": sense and sensitivity. *Cladistics* 19: 432–442.

Ogden, T.H. & Rosenberg, M. 2006. How should gaps be treated in parsimony? A comparison of approaches using simulation. *Mol. Phylogenet. Evol.* 42: 817–826.

Ogden, T.H. & Rosenberg, M. 2007. Alignment and topological accuracy of the direct optimization approach via POY and traditional phylogenetics via ClustalW + PAUP*. *Syst. Biol.* 56: 182–193.

Ota, S. & Li, W.H. 2001. NJML+: An extension of the NJML method to handle protein sequence, data and computer software implementation, *Mol. Biol. Evol.* 18: 1983–1992.

Pagel, M. & Meade, A. 2004. A phylogenetic mixture model for detecting pattern-heterogeneity in gene sequence of character-state data. Syst. Biol. 53: 571–581.

Pearson, W.R., Robins, G. & Zhang, T. 1999. Generalized neighbor-joining: more reliable phylogenetic tree reconstruction. *J. Mol. Evol.* 16: 806–816.

Penny, D., Hendy, M.D. & Steel, M.A. 1992. Progress with methods for constructing evolutionary trees. *Trends Ecol. Evol.* 7: 73–79.

Philippe, H. & Forterre, P. 1999. The rooting of the universal tree of life is not reliable. *J. Mol. Evol.* 49: 509–523.

Porter, M.L., Cronin, T., McClellan, D.A. & Crandall, K.A. 2007. Molecular characterization of crustacean visual pigments and the evolution of pancrustacean opsins. *Mol. Biol. Evol.* 24: 253–268.

Posada, D. & Crandall, K.A. 1998. Modeltest: testing the model of DNA substitution. *Bioinformatics* 14: 817–818.

Posada, D. & Crandall, K.A. 2001. Selecting the best-fit model of nucleotide substitution. *Syst. Biol.* 50: 580–601.

Posada, D. & Buckley, T.R. 2004. Model selection and model averaging in phylogenetics: advantages of Akaike Information Criterion and Bayesian approaches over Likelihood Ratio Tests. *Syst. Biol.* 53: 793–808.

Redelings, B. & Suchard, M. 2005. Joint Bayesian estimation of alignment and phylogeny. *Syst. Biol.* 54: 401–418.

Robinson, D.F. 1971. Comparison of labeled trees with Valency Three. *J. Combin. Theor.* 11: 105–119.

Rogers, J.S. 1997. On the consistency of maximum likelihood estimation of phylogenetic trees from nucleotide sequences. *Syst. Biol.* 46: 354–357.

Ronquist, F. & Huelsenbeck, J.P. 2003. MRBAYES 3: Bayesian phylogenetic inference under mixed models. *Bioinformatics* 19: 1572–1574.

Rzhetsky, A. & Nei, M. 1993. Theoretical foundation of the minimum evolution method of phylogenetic inference. *Mol. Biol. Evol.* 10: 1073–1095.

Schöniger, M. & von Haeseler, A. 1994. A stochastic model for the evolution of autocorrelated DNA sequences. *Mol. Phylogenet. Evol.* 3: 240–247.

Segawa, R.D. & Aotsuka, T. 2005. The mitochondrial genome of the Japanese freshwater crab, Geothelphusa dehaani (Crustacea: Brachyura): evidence for its evolution via gene duplication. *Gene* 355: 28–39.

Shimodaira, H. 2002. An approximately unbiased test of phylogenetic tree selection. *Syst. Biol.* 51: 492–508.

Shimodaira, H. & Hasegawa, M. 1999. Multiple comparisons of log-likelihoods with applications to phylogenetic inference. *Mol. Biol. Evol.* 16: 1114–1116.

Song, H., Buhay, J.E. Whiting, M.F. & Crandall, K.A. 2008. DNA barcoding overestimates the number of species when nuclear mitochondrial pseudogenes are coamplified. *Proc. Nat. Acad. Sci. USA*: 105: 13486–13491.

Stamatakis, A., Ludwig, T. & Meier, H. 2005. RAxML-III: a fast program for maximum likelihood-based inference of large phylogenetic trees. *Bioinformatics* 21: 456–463.

Steel, M.A., Hendy, M.D. & Penny, D. 1988. Loss of information in genetic distances. *Nature* 336: 118.

Steel, M. & Penny, D. 2000. Parsimony, likelihood, and the role of models in molecular phylogenetics. *Mol. Biol. Evol.* 17: 839–850.

Suchard, M.A. & Redelings, B.D. 2006. BAli-Phy: simultaneous Bayesian inference of alignment and phylogeny. *Bioinformatics* 22: 2047–2048.

Swofford, D.L. 2003. PAUP*. Phylogenetic Analysis Using Parsimony (*and Other Methods). Version 4. Sinauer Associates, Sunderland, Massachusetts.

Swofford, D.L., Olsen, G.J., Waddell, P.J. & Hillis, D.M. 1996. Phylogenetic inference. In: Hillis, D.M., Moritz, C. & Mable, B.K (eds.), *Molecular Systematics*: 407–514. Sunderland: Sinauer Associates.

Teichmann, S.A. & Mitchison, G. 1999. Making family trees from gene families. *Nat. Genet.* 21: 66–67.

Templeton, A.R. 1983. Convergent evolution and nonparametric inferences from restriction data and DNA sequences. In: Weir, B.S. (ed.), *Statistical Analysis of DNA Sequence Data*: 151–179. New York: Marcel Dekker, Inc.

Thompson, J.D., Higgins, D.G. & Gibson, T.J. 1994. CLUSTAL W: improving the sensitivity of progressive multiple sequence alignment through sequence weighting, positions-specific gap penalties and weight matrix choice. *Nucl. Acids Res.* 22: 4673–4680.

Thorne, J.L., Kishino, H. & Felsenstein, J. 1991. An evolutionary model for maximum likelihood alignment of DNA sequences. *J. Mol. Evol.* 33: 114–124.

Thorne, J.L. & Kishino, H. 1992. Freeing phylogenies from artifacts of alignment. *Mol. Biol. Evol.* 9: 1148–1162.

Thorne, J.L. & Kishino, H. 2002. Divergence time and evolutionary rate estimation with multilocus data. *Syst. Biol.* 51: 689–702.

Varón, A., Vinh, L.S., Bomash, I. & Wheeler, W.C. 2007. POY 4.0 Beta 2635. American Museum of Natural History.

Weston, P.H. 1994. Methods for rooting cladistic trees. In: Scotland, R.W., Siebert, D.J. & Williams, D.M. (eds.), *Models in Phylogeny Reconstruction*: 125–155. Oxford: Oxford Univ. Press.

Yang, Z., Goldman, N. & Friday, A. 1994. Comparison of models for nucleotide substitution used in maximum-likelihood phylogenetic estimation. *Mol. Biol. Evol.* 11: 316–324.

Yang, Z. 1996a. Among-site rate variation and its impact on phylogenetic analysis. *Trends Ecol. Evol.* 11: 367–372.

Yang, Z. 1996b. Maximum-likelihood models for combined analyses of multiple sequence data. *J. Mol. Evol.* 42: 587–596.

Yang, Z. 1997. PAML: a program package for phylogenetic analysis by maximum likelihood. *CABIOS* 13: 555–556.

Yang, Z. & Rannala, B. 1997. Bayesian phylogenetic inference using DNA sequences: Markov chain Monte Carlo methods. *Mol. Biol. Evol.* 14: 717–724.

Yang, Z., Nielsen, R., Goldman, N. & Pedersen, A.-M.K. 2000. Codon-substitution models for heterogeneous selection pressure at amino acid sites. *Genetics* 155: 431–449.

Zanotto, P.M., Kallas, E.Q., de Souza, R.F. & Holmes, E.C. 1999. Genealogical evidence for positive selection in the nef gene of HIV-1. *Genetics* 153: 1077–1089.

Zwickl, D.J. 2006. Genetic algorithm approaches for the phylogenetic analysis of large biological sequence datasets under the maximum likelihood criterion. Ph.D. dissertation, The University of Texas at Austin.

# Decapod Phylogeny: What Can Protein-Coding Genes Tell Us?

K.H. CHU[1], L.M. TSANG[1], K.Y. MA[1], T.Y. CHAN[2] & P.K.L. NG[3]

[1] *Department of Biology, The Chinese University of Hong Kong, Shatin, Hong Kong*
[2] *Institute of Marine Biology, National Taiwan Ocean University, Keelung, Taiwan*
[3] *Department of Biological Sciences, National University of Singapore, Singapore*

ABSTRACT

The high diversity of decapods has attracted the interest of many carcinologists, but there is no consensus on their phylogeny as yet. This is in spite of numerous endeavors using both morphological and molecular approaches. New sources of information are necessary to help elucidate the phylogenetic relationships among decapods. Here we demonstrate the applicability of nuclear protein-coding genes in the phylogenetic analysis of this group. Using only two protein-coding genes, we have successfully resolved most of the infraordinal relationships with good statistical support, indicating the superior efficiency of these markers compared to nuclear ribosomal RNA and mitochondrial genes now commonly used in phylogenetic reconstruction of decapods. Available evidence suggests that these two markers suffer from the problems of alignment ambiguities and rapid saturation, respectively. We have also applied nuclear protein-coding genes in revealing inter- and intrafamilial evolutionary history. Trees with robust support can be obtained using sequences of two to three genes for the infraorders and families tested, including the most species-rich group, the Brachyura. The new genes are also shown to be informative in elucidating interspecific phylogeny. Thus, these nuclear protein-coding genes are applicable at various taxonomic levels and will provide a valuable new source of information for reconstructing the tree of life of Decapoda.

## 1 INTRODUCTION

The Decapoda is one of the most diverse groups of Crustacea. The ecological and morphological diversity of decapods, together with their economic importance, makes them the most studied of all crustaceans (Martin & Davis 2001). A robust phylogeny is therefore crucial to understanding the evolution and diversification in this group of animals. The extraordinary morphological diversity, however, poses substantial challenges to their phylogenetic study. There have been many systematic schemes and phylogenetic hypotheses proposed for Decapoda (reviewed in Martin & Davis 2001; Schram 2001). Morphological cladistic analyses have provided some insights, but they leave many key disputes unsettled, especially concerning the relationship of deeper nodes (e.g., Scholtz & Richter 1995; Dixon et al. 2003; Schram & Dixon 2004). Thus, researchers have recently shifted their attention to new sources of information from the genome to resolve decapod phylogeny.

## 2 MOLECULAR PHYLOGENY OF THE DECAPODA

Mitochondrial genes have been the most commonly used markers in animal phylogenetic studies, including the decapod crustaceans, for many years (Schubart this volume). These markers benefit from the ease of amplification due to relatively higher copy numbers relative to nuclear genes and the availability of many universal primers (Simon et al. 1994). The haploid and non-recombinant nature of mtDNA also presents fewer problems in phylogenetic reconstruction. The rate of nucleotide substitutions among mitochondrial genes is generally more rapid than that among genes in the nuclear genome (Moore 1995). Accordingly, mitochondrial genes could more accurately reflect the relationships among recently diverged taxa. Most of the phylogenetic studies in lower taxonomic levels of decapods rely exclusively on mitochondrial DNA sequences, and these genes do provide us with some insights into the evolutionary history of the Decapoda (reviewed in Schubart et al. 2000; Schubart this volume).

Mitochondrial genes, however, are being criticized for several disadvantages. All of the mitochondrial genes are linked and inherited as a single molecule. Therefore, they share a common evolutionary history and cannot provide an independent phylogenetic inference. The high mutation rate of mitochondrial DNA also limits its utility in the phylogenetics of deep divergences. Furthermore, the highly A/T-biased mitochondrial DNA, especially at the third codon position of the protein-coding genes, suffers from high levels of homoplasy and thus exhibits strong negative effects in phylogenetic analyses. In this regard, decapod molecular systematists have tried to incorporate nuclear rRNA genes, which evolve at a much slower rate, in addition to mitochondrial DNA markers, for decapod phylogeny. Analyses of the 18S rRNA gene have resolved some familial relationships and laid the foundation for further taxonomic revision (e.g., Spears et al. 1992; Pérez-Losada et al. 2002; Ahyong et al. 2007). The nuclear rRNA genes, however, suffer from alignment ambiguities. This poses problems in phylogenetic inference, particularly in nodes with deep divergence (i.e., infraordinal relationships). The two recent studies on the phylogeny of decapod infraorders based primarily on 18S and 28S rRNA gene sequences (combined with morphological characters or the relatively much shorter fragments of mitochondrial 16S rRNA and histone 3) yield contrasting topologies (Ahyong & O'Meally 2004; Porter et al. 2005), suggesting the current markers are insufficient in reconstructing a robust high-level phylogeny of Decapoda.

Consequently, nuclear protein-coding genes could serve as an excellent new source of information. These genes have the clear advantage of being easy to align. Moreover, many potential candidates are present in the genome with diverse evolutionary rates that are suitable to address phylogeny at different taxonomic levels. Despite the apparently high potential utility of protein-coding gene markers, several limitations have restricted the development and application of these markers. First, the protein-coding genes have a much lower number of copies in the genome, compared to highly abundant nuclear rRNA and mitochondrial genes, and therefore are more difficult to amplify through PCR. The degenerate third codon positions further challenge the design of PCR primers, and long stretches of introns might be present, making amplification difficult or even impossible. Furthermore, paralogs might be present, resulting in problems in phylogenetic analyses. Thus, though these genes appear to be informative, their application in decapod phylogenetics has been relatively limited to date (e.g., histone 3: Porter et al. 2005; glyceraldehyde-3-phosphate dehydrogenase: Buhay et al. 2007).

With the recent advances in molecular techniques (e.g., EST) and the accumulation of large amounts of genome sequence data, scientists can search for new molecular markers or apply the existing ones to their target organisms much more easily than before. Accordingly, the protein-coding genes play an increasingly dominant role in phylogenetic studies. This is especially true for the taxonomic groups with more comprehensive genomic information (e.g., vertebrates and insects). New protein-coding gene markers have also been successfully developed for other arthropods (e.g., spider, Ayoub et al. 2007; Mysida, Audzijonyte et al. 2008), and have proved to be informative or even superior to nuclear rRNA and mitochondrial genes in resolving power (Audzijonyte et al. 2008).

Thus, the development and application of these markers in Decapoda molecular systematic studies could be a new strategy in addressing the controversial issues in decapod phylogeny. In this paper, we report recent advances in our laboratory in applying nuclear protein-coding genes to decapod phylogenetics across different taxonomic levels. Their utility was examined by comparing the statistical support in topologies obtained in the present study with those from previous studies using nuclear rRNA and/or mitochondrial genes.

## 3 NEW INSIGHTS INTO THE INFRAORDINAL RELATIONSHIPS AMONG DECAPODA REVEALED BY PROTEIN-CODING GENES

We have employed partial segments of two nuclear protein-coding genes, phosphoenolpyruvate carboxykinase (PEPCK, 570 bp) and sodium-potassium ATPase $\alpha$-subunit (NaK, 534 bp), to reconstruct the phylogeny among 69 decapod species (Tsang et al. 2008a). This analysis has now been extended to 135 species from 60 families (Fig. 6.1). The topology inferred from Bayesian inference reveals that the Reptantia and all but one of its infraorders are monophyletic. The nodal support for most of the infraordinal and inter-familial relationships is high (posterior probability $\geq 0.95$), indicating the high resolving power of the protein-coding genes. Thalassinidea, however, is polyphyletic. This corroborates the results of a previous study based on mitochondrial gene rearrangements and sequences from both mitochondrial and nuclear rRNA genes (Morrison et al. 2002). We recover two distinct lineages in Thalassinidea that correspond to the two strongly supported clades obtained in the previous molecular studies (Tudge and Cunningham 2002; Ahyong and O'Meally 2004; Tsang et al. 2008b). The division of Thalassinidea into the two major groups is also supported by larval morphology, external somatic morphology, and foregut ossicles (Gurney 1938; de Saint Laurent 1973; Sakai 2005; Tsang et al. 2008b).

Within Pleocyemata, Stenopodidea and Caridea form a sister clade to Reptantia, supporting the view of Burkenroad (1981). Anomura and Brachyura show high affinity in concordance with the traditional grouping of Meiura. Enoplometopidae and Thaumastochelidae are found to be closely related to Nephropidae, justifying their placement in Astacidea. Yet Thaumastochelidae is nested within Nephropidae, making the latter paraphyletic, and thus future taxonomic re-evaluation is warranted. An interesting finding is that Polychelidae, long considered to be a basal reptant group, clusters with Achelata and Astacidea, and is therefore more derived than expected. Instead, thalassinidean-like creatures are the stem lineage of Reptantia based on our phylogeny.

All in all, the protein-coding genes apparently provide high resolving power in deeper branches within Decapoda. The phylogenetic positions of several 'problematic' taxa have been clarified and new insights into decapod evolution obtained. We advocate further development and application of these markers for the higher level phylogeny of decapods.

## 4 UTILITY OF PROTEIN-CODING GENES IN SUPERFAMILY/FAMILY LEVEL PHYLOGENETIC STUDIES

### 4.1 *Phylogeny of Penaeoidea*

The penaeoid shrimps constitute a diverse group of marine decapods. This superfamily contains most of the commercially important shrimps, constituting more than one third of the annual crustacean wild catch (FAO fisheries data). A robust phylogenetic tree is, therefore, crucial for creating a stable and natural classification, which would facilitate effective fisheries management and aquaculture. Previous phylogenetic hypotheses concerning Penaeoidea were derived mainly from morphological analyses (e.g., Kubo 1949; Burkenroad 1983; see also Tavares et al. this volume). Recent molecular studies based on mitochondrial markers, however, yielded highly conflicting conclusions. A close association among Aristeidae, Benthesicymidae, and Sicyoniidae was suggested, while Penaeidae was revealed to be paraphyletic due to the incursion of Solenoceridae

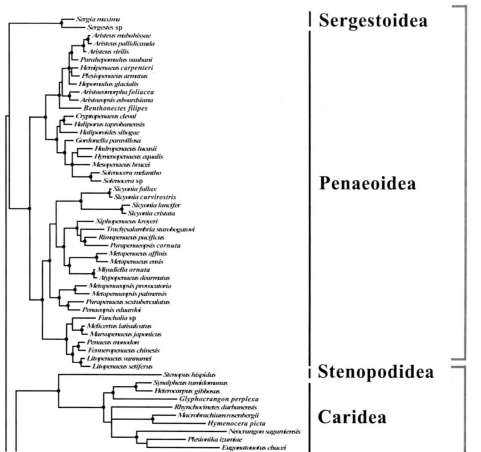

**Figure 1.** continued on next page.

Decapod Phylogeny and Protein-Coding Genes 93

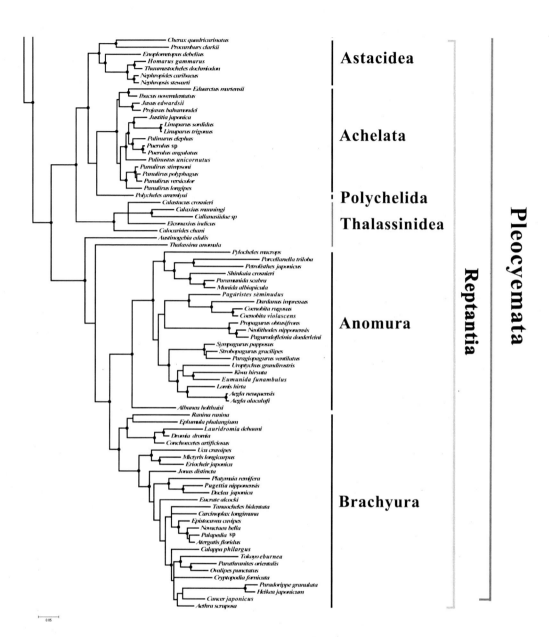

**Figure 1.** Phylogenetic tree of Decapoda (135 species from 60 families) constructed from combined PEPCK and NaK gene sequences (total 1104 bp). The analysis used Bayesian inference under the best-fitting model GTR+I+G. The analysis was run with 5 million generations consisting of four chains, sampled every 500 generations with the first 0.5 million generations discarded as burnin. Three independent runs were performed to confirm the topology. The nodes with posterior probabilities ≥0.95 are denoted by black dots. The infraorder classification of the species is indicated by the bars to the right.

(Vázquez-Bader et al. 2004). Yet these inferred topologies were poorly supported. As a result, it remains unanswered whether the contrasting results represent actual discrepancies between character evolution and speciation or artifacts of gene tree reconstruction.

Using the two nuclear protein-coding genes, PEPCK and NaK, applied in the decapod infraordinal phylogenetic study, we reconstructed a largely resolved, well-supported phylogeny of Penaeoidea (Fig. 2). The monophyly of the superfamily and four out of its five families is evident. Yet the Penaeidae is clearly paraphyletic as Sicyoniidae is nested within it. Two major lineages are recovered in the superfamily, one consisting of Solenoceridae, Aristeidae, and Benthesicymiidae, with the latter two as sister taxa, and the other composed of Penaeidae and Sicyoniidae. This topology is largely congruent with the morphology-inferred phylogeny of the penaeoids. Members from

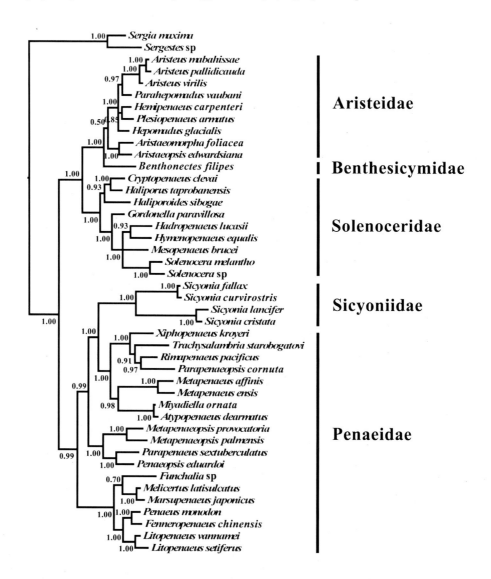

**Figure 2.** Phylogenetic tree of Penaeoidea (42 species + 2 outgroups from Sergestidae) constructed from combined PEPCK and NaK (total 1104 bp) analysis using Bayesian inference under the best-fitting model GTR+I+G. The analysis was run with 5 million generations consisting of four chains, sampled every 500 generations with the first 0.5 million generations discarded as burnin. Three independent runs were performed to confirm the topology. The posterior probability values are indicated on the branches. The bars to the right indicate the five families of Penaeoidea.

the families Penaeidae and Sicyoniidae are predominantly littoral water inhabitants, while those of the Aristeidae, Benthesicymidae, and Solenoceridae are mainly found in bathy- or mesopelagic environments. Our results thus suggest that habitat-associated radiation may play an important role in the diversification of penaeoid shrimps. Moreover, the three tribes of Penaeidae are shown to be monophyletic with strong nodal support, corroborating morphological evidence and the previous molecular study using mitochondrial 16S rDNA sequence data (Chan et al. 2008).

The concordance among sources of information (e.g., between independent genes and morphological characters) and topology with a high statistical support again indicate the superior and high resolving power of protein-coding genes over other markers currently used in decapod molecular systematics.

*4.2 Phylogeny of Brachyura*

With more than 6,500 species, the Brachyura is the most species-rich infraorder of Decapoda (Ng et al. 2008). The large number of species and morphological diversity have led to a large number of phylogenetic hypotheses proposed (reviewed in Martin & Davis 2001). Investigating the phylogeny of Brachyura using nuclear 18S rRNA sequences, Ahyong et al. (2007) found that section Podotremata is paraphyletic, with the Raninidae being more closely related to Eubrachyura than other podotreme crabs. However, the relationships among the families in Eubrachyura are poorly resolved, although the monophyly of the group is strongly supported. These authors attributed the lack of resolution to the insufficient variability in the 18S rRNA sequences in these more recently diverged taxa. More comprehensive taxon sampling and use of more rapidly evolving genetic markers have been advocated (Ahyong et al. 2007).

We tried to reconstruct the phylogeny of Brachyura using three protein-coding genes, NaK, glyceraldehyde-3-phosphate dehydrogenase (GAPDH, 540 bp), and enolase (345 bp), making up a data set of 1419 bp. The topology recovered from Bayesian inference analysis of the combined data set supports the result of Ahyong et al. (2007) that the Podotremata is paraphyletic (Fig. 3), indicating that the gene trees constructed using the two types of markers (nuclear rRNA and protein-coding genes) are congruent. On the other hand, the protein-coding gene tree provides significantly better resolution within the Eubrachyura. The subsections Heterotremata and Thoracotremata are strongly supported to be reciprocally monophyletic, whilst the 18S rRNA gene tree gives little resolution here. Moreover, the close affinities of some of the families are revealed (e.g., Homolidae + Latreilliidae; Xanthidae + Trapeziidae + Goneplacidae; Matutidae + Calapidae + Euryplacidae). The results corroborate the new classification proposed by Ng et al. (2008) to a certain extent (such as most superfamily groupings), suggesting that the protein-coding gene tree is consistent with the morphological patterns observed.

Admittedly, quite a number of internal nodes remain poorly resolved in the present protein-coding gene tree. Yet the number of taxa analyzed here is relatively limited, as many families have not been included and many highly diverse families are only represented by one or two species. This obviously affects the resolution in such a species-rich group. It is worth noting that our data set consists of only 1419 characters, compared to 1830 used by Ahyong et al. (2007). Thus the nuclear protein-coding genes are more efficient in achieving a higher resolving power in comparison with the equivalent length of nuclear rRNA genes. We are confident that a more robust phylogeny of Brachyura could be obtained in future studies with more thorough taxon sampling and additional nuclear protein-coding genes. This study is now ongoing.

## 5 UTILITY OF PROTEIN-CODING GENES IN PHYLOGENETIC RECONSTRUCTION AMONG GENERA/SPECIES: PHYLOGENY OF PALINURIDAE

Spiny lobsters of the family Palinuridae include many economically important species with a high potential in aquaculture. Accordingly, they receive considerable attention in attempts to investigate

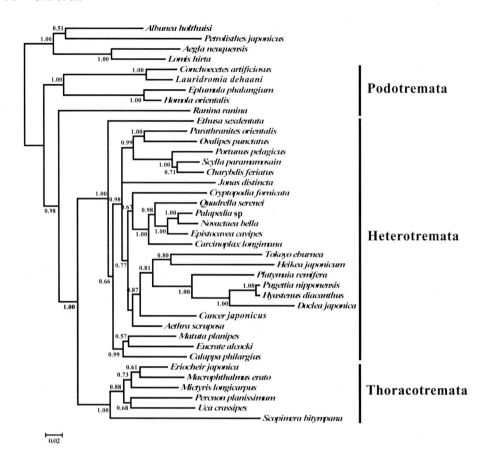

**Figure 3.** Phylogenetic tree of Brachyura (35 species + 4 outgroups from the infraorder Anomura) constructed from combined NaK, GAPDH, and enolase gene sequences (total 1419 bp). The analysis used Bayesian inference under the best-fitting model GTR+I+G. The analysis was run with 2 million generations consisting of four chains, sampled every 100 generations with the first 200,000 generations discarded as burnin. Three independent runs were performed to confirm the topology. The posterior probability values are indicated on the branches. The bars to the right indicate the three sections of Brachyura.

their genetic population structure and phylogeny for fishery management purposes. Morphological analyses recognize two major lineages in the Palinuridae, namely the Silentes and Stridentes, based on whether the lobsters have a stridulating sound-producing organ (George & Main 1967). The evolution of genera within these two groups was proposed to be associated with the invasion of shallow water habitats, formed by past tectonic movement, by ancestral deeper-water inhabitants (Pollock 1995; George 2005, 2006). Modifications in life-history traits are believed to be adaptations for the shallower water habitat (George 2005). Patek and Oakley (2003) investigated the phylogeny of the spiny lobsters using mitochondrial 16S and nuclear 18S and 28S rRNA gene sequences. They found some evidence for the division of Stridentes and Silentes, but most of the internal branches in the rRNA gene tree were poorly resolved, and the reciprocal monophyly of the two groups received very weak support. Moreover, the topologies derived from different gene segments and analytical methods showed conflicts. Thus, the phylogenetic hypotheses proposed could neither be accepted nor rejected confidently.

Using sequences of three nuclear protein-coding genes, PEPCK, NaK, and histone 3, we generated a gene tree of the Palinuridae, with high statistical support for most of the nodes (Fig. 4), which allows us to reconstruct the evolutionary pathway within the family. The reciprocal monophyly of

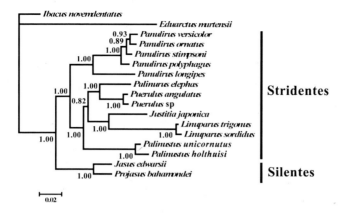

**Figure 4.** Phylogenetic tree of Palinuridae (15 species + 2 outgroups from the family Scyllaridae) constructed from combined PEPCK, NaK, and histone 3 gene sequence (total 1416 bp) analysis using Bayesian inference under the best-fitting model GTR+I+G. The analysis was run with 1 million generations consisting of four chains, sampled every 100 generations with the first 200,000 generations discarded as burnin. Three independent runs were performed to confirm the topology. The posterior probability values are indicated on the branches.

Stridentes and Silentes is strongly supported. Interestingly, the deep-water inhabiting genera of Stridentes (e.g., *Puerulus* and *Linuparus*), which are considered to be primitive (Pollock 1995; George 2006), are revealed to be derived in our tree. *Palinurus* is the basal lineage of the family, supporting the view of Davie (1990). Our present finding based on relatively limited taxa remains preliminary but clearly demonstrates the utility of protein-coding genes in elucidating the phylogeny of Palinuridae, by providing significantly better resolution as compared to previous studies based on similar taxon sampling and sequence data.

Apart from being informative in generic relationships, the protein-coding genes appear to be useful in resolving species level phylogeny as well. The histone 3 gene has already been employed in phylogenetic studies in a number of genera (e.g., Buhay et al. 2007; Page et al. 2008), while the present study represents the first application of the other two genes at this taxonomic level. We found that the five spiny species of *Panulirus* analyzed exhibit up to 6% and 3.5% sequence divergence in PEPCK and NaK, respectively. Moreover, our gene tree indicates the close affinity of *P. ornatus*, *P. versicolor*, *P. stimpsoni*, and *P. polyphagus*, whilst *P. longipes* is more distantly related. This is congruent with the phylogeny inferred from mitochondrial DNA analyses (Ptacek et al. 2001), suggesting the potential of the nuclear protein-coding genes in resolving interspecific relationships.

## 6  CONCLUSIONS

Our analyses using nuclear protein-coding genes indicate that they are highly informative for phylogeny estimation across all taxonomic levels of Decapoda, from infraordinal to interspecific relationships. Some new insights into the higher classifications of decapods are disclosed for the first time (e.g., polyphyly of Thalassinidea), and the phylogenetic positions of selected controversial taxa (e.g., Polychelidae, Enoplometopidae) are also resolved in our gene trees. Thus, these new gene markers are promising for future multi-loci studies on phylogenetic reconstruction of decapods. Our results also demonstrate that a large number of potential candidate genes in the genome remain unexplored for evolutionary studies. It is anticipated that our study will trigger the discovery and application of more protein-coding genes for phylogenetic analysis. The use of these genes as the basic repertoire in the phylogenetic toolkit in analyzing decapod relationships represents a major step towards our goal in assembling the tree of life for Decapoda.

ACKNOWLEDGEMENTS

The work presented in this paper was supported by a grant from the Research Grants Council of the Hong Kong Special Administrative Region, China (project no. 4419/04M), to KHC and grants from the National Science Council, Taiwan, and Center for Marine Bioscience and Biotechnology of the National Taiwan Ocean University to TYC. We are indebted to A. Crosnier and R. Cleva of the Muséum national d'Histoire naturelle, Paris, for the use of materials in the museum for our study. Thanks are also due to Joelle Lai of the National University of Singapore for help with some of the Brachyuran samples.

REFERENCES

Ahyong, S.T. & O'Meally, D. 2004. Phylogeny of the Decapoda Reptantia: resolution using three molecular loci and morphology. *Raff. Bull. Zool.* 52: 673–693.

Ahyong, S.T., Lai, J.C.Y., Sharkey, D., Colgan, D.J. & Ng, P.K.L. 2007. Phylogenetics of the brachyuran crabs (Crustacea: Decapoda): the status of Podotremata based on small subunit nuclear ribosomal RNA. *Mol. Phylogenet. Evol.* 45: 576–586.

Audzijonyte, A., Daneliya, M.E., Mugue, N. & Väinölä, R. 2008. Phylogeny of *Paramysis* (Crustacea: Mysida) and the origin of Ponto-Caspian endemic diversity: resolving power from nuclear protein-coding genes. *Mol. Phylogenet. Evol.* 46: 738–759.

Ayoub, N.A., Garb, J.E., Hedin, M. & Hayashi, C.Y. 2007. Utility of the nuclear protein-coding gene, elongation factor-1 gamma (EF-1γ), for spider systematics, emphasizing family level relationships of tarantulas and their kin (Araneae: Mygalomorphae). *Mol. Phylogenet. Evol.* 42: 394–409.

Buhay, J.E., Moni, G., Mann, N. & Crandall, K.A. 2007. Molecular taxonomy in the dark: evolutionary history, phylogeography, and diversity of cave crayfish in the subgenus *Aviticambarus*, genus *Cambarus*. *Mol. Phylogenet. Evol.* 42: 435–448.

Burkenroad, M.D. 1981. The higher taxonomy and evolution of Decapoda (Crustacea). *Trans. San Diego Soc. Nat. Hist.* 19: 251–268.

Burkenroad, M.D. 1983. Natural classification of Dendrobranchiata, with a key to recent genera. In: Schram, F.R. (ed.), *Crustacean Issues 1, Crustacean Phylogeny:* 279–290. Rotterdam: Balkema.

Chan, T.Y., Tong, J.G., Tam, Y.K. & Chu, K.H. 2008. Phylogenetic relationships among the genera of the Penaeidae (Crustacea: Decapoda) revealed by mitochondrial 16S rRNA gene sequences. *Zootaxa* 1694: 38–50.

Davie, P.J.F. 1990. A new genus and species of marine crayfish, *Palibythus magnificus*, and new records of *Palinurellus* (Decapoda: Palinuridae) from the Pacific Ocean. *Invert. Taxon.* 4: 685–695.

de Saint Laurent, M. 1973. Sur la systématique et la phylogénie des Thalassinidea: définitions des familles des Callianassidae et des Upogebiidae et diagnose de cinq genres nouveaux (Crustacea Decapoda). *CR Acad. Sci. Paris* (D) 277: 513–516.

Dixon, C.J., Ahyong, S.T. & Schram, F.R. 2003. A new hypothesis of decapod phylogeny. *Crustaceana* 76: 935–975.

George, R.W. 2005. Evolution of life cycles, including migration, in spiny lobsters (Palinuridae). *N. Z. J. Mar. Freshw. Res.* 39: 503514.

George, R.W. 2006. Tethys origin and subsequent radiation of the spiny lobster (Palinuridae). *Crustaceana* 79: 397–422.

George, R.W. & Main, A.R. 1967. The evolution of spiny lobster (Palinuridae): a study of evolution in the marine environment. *Evolution* 21: 803–820.

Gurney, R. 1938. Larvae of decapod Crustacea. Part 5. Nephropsidea and Thalassinidea. *Discovery Rep.* 17: 291–344.

Kubo, I. 1949. Studies on the penaeids of Japanese and its adjacent waters. *J. Tokyo Coll. Fish.* 36: 1–467.

Martin, J.W. & Davis, G.E. 2001. An updated classification of the Recent Crustacea. *Nat. Hist. Mus. Los Angeles City Sci. Ser.* 39: 1–124.

Moore, W.S. 1995. Inferring phylogenies from mtDNA variation: mitochondrial-gene trees versus nuclear-gene trees. *Evolution* 49: 718–726.

Morrison, C.L., Harvey, A.W., Lavery, S., Tieu, K., Huang, Y. & Cunningham, C.W. 2002. Mitochondrial gene rearrangements confirm the parallel evolution of the crab-like form. *Proc. R. Soc. Lond.* B 269: 345–350.

Ng, P.K.L., Guinot, D. & Davie, P.J.F. 2008. Systema Brachyuorum: Part I. An annotated checklist of extant brachyuran crabs of the world. *Raffles Bull. Zool.* 17: 1–286.

Page, T.J., Humphreys, W.F. & Hughes, J.M. 2008. Shrimps down under: evolutionary relationships of subterranean crustaceans from Western Australia (Decapoda: Atyidae: *Stygiocaris*). *PloS One* 3: e1618.

Patek, S.N. & Oakley, T.H. 2003. Comparative test of evolutionary trade-offs in a palinurid lobster acoustic system. *Evolution* 57: 2082–2100.

Pérez-Losada, M., Jara, C.G., Bond-Buckup, G., Porter, M.L. & Crandall, K.A. 2002. Phylogenetic position of the freshwater anomuran family Aeglidae. *J Crust. Biol.* 22: 670–676.

Pollock, D.E. 1995. Evolution of life-history patterns in three genera of spiny lobsters. *Bull. Mar. Sci.* 57: 516–526.

Porter, M.L., Pérez-Losada, M. & Crandall, K.A. 2005. Model-based multi-locus estimation of decapod phylogeny and divergence times. *Mol. Phylogenet. Evol.* 37: 355–369.

Ptacek, M.B., Sarver, S.K., Childress, M.J. & Herrnkind, W.F. 2001. Molecular phylogeny of the spiny lobster genus *Panulirus* (Decapoda: Palinuridae). *Mar. Freshw. Res.* 52: 1037–1047.

Sakai, K. 2005. The diphyletic nature of the infraorder Thalassinidea (Decapoda, Pleocyemata) as derived from the morphology of the gastric mill. *Crustaceana* 77: 1117–1129.

Scholtz, G. & Richter, S. 1995. Phylogenetic systematics of the reptantian Decapoda (Crustacea, Malacostraca). *Zool. J. Linn. Soc.* 113: 289–328.

Schubart, C.D., Neigel, J.E. & Felder, D.L. 2000. Use of the mitochondrial 16S rRNA gene for phylogenetic and population studies of Crustacea. In: von Vaupel Klein, J.C. & Schram, F. (eds.), *Crustacean Issues 12, The Biodiversity Crisis and Crustacea*: 817–830. Rotterdam: Balkema.

Schram, F.R. 2001. Phylogeny of decapods: moving towards a consensus. *Hydrobiologia* 449: 1–20.

Schram, F.R. & Dixon, C.J., 2004. Decapod phylogeny: addition of fossil evidence to a robust morphological cladistic data set. *Bull. Mizunami Fossil Mus.* 31: 1–19.

Simon, C., Frati, F., Beckenbach, A., Crespi, B., Liu, H. & Flook, P. 1994. Evolution, weighting, and phylogenetic utility of mitochondrial gene sequences and a compilation of conserved polymerase chain reaction primers. *Ann. Entomol. Soc. Am.* 87: 652–701.

Spears, T., Abele, L.G. & Kim, W. 1992. The monophyly of brachyuran crabs: a phylogenetic study based on 18S rRNA. *Syst. Biol.* 41: 446–461.

Tsang, L.M., Lin, F.-J., Chu, K.H. & Chan, T.-Y. 2008b. Phylogeny of Thalassinidea (Crustacea, Decapoda) inferred from three rDNA sequences: implications for morphological evolution and superfamily classification. *J. Zool. Syst. Evol. Res.* 46: 216–223.

Tsang, L.M., Ma, K.Y., Ahyong, S.T., Chan, T.Y. & Chu, K.H. 2008a. Phylogeny of Decapoda using two nuclear protein-coding genes: origin and evolution of the Reptantia. *Mol. Phylogenet. Evol.* 48: 359–368.

Tudge, C.C. & Cunningham, C.W. 2002. Molecular phylogeny of the mud lobsters and mud shrimps (Crustacea: Decapoda: Thalassinidea) using nuclear 18S rDNA and mitochondrial 16S rDNA. *Invert. Syst.* 16: 839–847.

Vázquez-Bader, A.R., Carrero, J.C., Gárcia-Varela, M., Gracia, A. & Laclette, J.P. 2004. Molecular phylogeny of superfamily Penaeoidea Rafinesque-Schmaltz, 1815, based on mitochondrial 16S partial sequence analysis. *J. Shellfish Res.* 23: 911–917.

# Spermatozoal Morphology and Its Bearing on Decapod Phylogeny

CHRISTOPHER TUDGE

*Biology Department, American University, Washington, D.C. 20016–8007, U.S.A, & Department of Invertebrate Zoology, National Museum of Natural History, Smithsonian Institution, Washington, D.C. 20013–7012, U.S.A.*

## ABSTRACT

The use of spermatozoal characters in elucidating animal phylogeny (spermiocladistics) has been successfully applied in the decapod crustaceans. Most of the studies investigating decapod sperm morphology have been published in the last 18 years and cover 100% of the decapod infraorders, 50% of the families, and approximately 10% of the extant genera, but only 2% of the described, extant species. There is great diversity in sperm morphology within the Crustacea, but overall decapod spermatozoa are quite conservative in comparison. Still, it is difficult to describe a typical decapod sperm cell. Decapod sperm are unusual for several reasons: 1) they are aflagellate (lack a true 9 + 2 flagellum), although microtubular processes are often present; 2) there is no reliable record of motility for any individual sperm cell; 3) the acrosome vesicle is not Golgi-derived as in all other described acrosomes of sperm in the animal kingdom, instead being derived from endoplasmic reticulum vesicles; 4) the decapod sperm nuclear protein is unique, with all other animal sperm nuclear proteins falling into four other categories; 5) the sperm nucleus is composed of diffuse, filamentous, heterogeneous chromatin fibers rather than being uniformly dense; and 6) the mitochondria are degenerate in mature sperm cells. I surveyed spermatozoal characters across the investigated decapod crustaceans, highlighting those of phylogenetic utility, such as acrosome vesicle presence, shape, dimensions and size, and internal complexity; nuclear morphology and shape; and microtubular arm presence, number, and origin. Particular spermatozoal characters, or suites of characters, that define various decapod taxa are provided, and their utility to phylogenetic construction is discussed.

## 1 INTRODUCTION

> *"The sperm seems never to transgress the few rules which govern the production of its fundamental parts, but in the arrangement of these parts every sperm (flagellate or non-flagellate) seems to be a law unto itself."*
>
> Bowen (1925)

Professor Barrie Jamieson coined the term spermiocladistics (Jamieson 1987) and pioneered the use of spermatozoa in decapod phylogenetics (among many other invertebrate and vertebrate groups) using comprehensive datasets based on the ultrastructure of sperm cells from scanning and transmission electron microscopy. Jamieson's contributions to spermiocladistics span two decades, with a significant proportion of this work dedicated to decapod crustaceans. He was not the first to recognize the phylogenetic significance of crustacean spermatozoa, and in fact he was beaten to this claim by 81 years.

The phylogenetic significance of crustacean spermatozoa was first recognized by Koltzoff (1906) and then later by Wielgus (1973). Koltzoff constructed a phylogeny of crustaceans (mostly decapods) based on sperm cell structure observed under the light microscope. He assigned to the

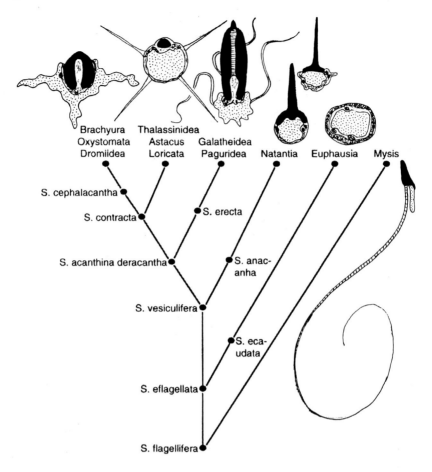

**Figure 1.** Diagram of the sperm phylogeny of Koltzoff (1906) modified to include a representative sperm morphology for the terminal groups. Spermatozoa not to scale.

different sperm types he encountered the "generic" name *Spermia* and a descriptive "species" name. Some of the significant relationships apparent in Koltzoff's phylogenetic tree are shown in Figure 1.

We must also recognize the contributions of others who came before, pioneering the microscopy of spermatozoa in general, including Leeuwenhoek (1678), Swammerdam (1758), Spallanzani (1776), Koltzoff (1906), Retzius (1909), Bowen (1925), Afzelius (1970), and Baccetti (1979), to name a few. We are indebted to their talents, perseverance, foresight, and careful observation.

The considerable decapod sperm literature was ably reviewed by Jamieson (1991), along with the bulk of the crustacean sperm literature to this date. A second review of just the decapod sperm literature from 1991 to 2000 is provided in Jamieson & Tudge (2000). A comprehensive (but not exhaustive) table of subsequent publications (including some missed in the two previous reviews) on spermatozoal descriptions of decapods is provided in Table 1.

Like the animals themselves, spermatozoa of crustaceans are very diverse in their morphology (Pochon-Masson 1983; Jamieson 1989c, 1991). It is therefore difficult to designate sperm features that characterize the entire class. Nevertheless, sperm data are extremely useful in determining relationships among crustacean taxa. Except for the Ascothoracica (Grygier 1982), the Cirripedia (Healy & Anderson 1990), and the Remipedia (Yager 1989), most crustaceans have aflagellate, immotile sperm. The non-caridean, pleocyemate decapods all share a common sperm form consisting of an often large acrosome vesicle (which can be multi-layered), a posterior nucleus of variable

**Table 1.** Decapods investigated for spermatozoal morphology since Jamieson (1991) and Jamieson & Tudge (2000).

| Suborder/Infraorder, SUPERFAMILY & Family | Species | Reference |
|---|---|---|
| **Suborder Dendrobranchiata** | | |
| PENAEOIDEA | | |
| Penaeidae | *Artemesia longinaris* | Scelzo & Medina 2003 |
| | *Fenneropenaeus penicillatus* | Hong et al. 1993, 1999 (both as *Penaeus*) |
| | *Penaeus chinensis* | Lin et al. 1991; Kang et al. 1998; Kang & Wang 2000, Kang et al. 2000 |
| Aristeidae | *Aristaeopsis edwardsiana* | Medina et al. 2006b |
| | *Aristeus varidens* | Medina et al. 2006b |
| Solenoceridae | *Pleoticus muelleri* | Medina et al. 2006a |
| | *Solenocera africana* | Medina et al. 2006a |
| | *Solenocera membranacea* | Medina et al. 2006a |
| SERGESTOIDEA | | |
| Sergestidae | *Peisos petrunkevitchi* | Scelzo & Medina 2004 |
| **Suborder Pleocyemata** | | |
| **Caridea** | | |
| PALAEMONOIDEA | | |
| Palaemonidae | *Macrobrachium nipponense* | Yang et al. 1998 |
| **Palinura** | | |
| PALINUROIDEA | | |
| Scyllaridae | *Thenus orientalis* | Zhu et al. 2002 |
| **Anomura** | | |
| AEGLOIDEA | | |
| Aeglidae | *Aegla longirostri* | Tudge & Scheltinga 2002 |
| HIPPOIDEA | | |
| Albuneidae | *Albunea marquisiana* | Tudge et al. 1999 (as *Albunea* sp.) |
| Hippidae | *Emerita talpoida* | Tudge et al. 1999 |
| | *Hippa pacifica* | Tudge et al. 1999 |
| PAGUROIDEA | | |
| Diogenidae | *Calcinus tubularis* | Tirelli et al. 2006 |
| | *Clibanarius erythropus* | Tirelli et al. 2007 |
| | *Clibanarius vittatus* | Matos et al. 1993 |
| | *Diogenes pugilator* | Manjón-Cabeza & García Raso 2000; Tirelli et al. 2008 |
| | *Loxopagurus loxochelis* | Scelzo et al. 2006 |
| | *Petrochirus Diogenes* | Brown 1966a |
| Paguridae | *Pagurus stimpsoni* | Brown 1966a (as *P. bonairensis*) |
| Pylochelidae | *Pylocheles (Bathycheles)* sp. | Tudge et al. 2001 |
| **Brachyura** | | |
| MAJOIDEA | | |
| Inachidae | *Inachus phalangium* | Rorandelli et al. 2008 |
| PORTUNOIDEA | | |
| Portunidae | *Scylla serrata* | Shang Guan & Li 1994; Wang et al. 1997 |
| Trichodactylidae | *Dilocarcinus septemdentatus* | Matos et al. 1996 |
| POTAMOIDEA | | |
| Gecarcinucidae | *Geithusa pulcher* | Klaus et al. 2008 |
| | *Heterothelphusa fatum* | Klaus et al. 2008 |
| | *Oziothelphusa ceylonensis* | Klaus et al. 2008 |

**Table 1.** continued.

| Suborder/Infraorder, SUPERFAMILY & Family | Species | Reference |
|---|---|---|
| | *Oziothelphusa* sp. | Klaus et al. 2008 |
| | *Parathelphusa convexa* | Klaus et al. 2008 |
| | *Parathelphusa maindroni* | Klaus et al. 2008 |
| | *Phricothelphusa gracilipes* | Klaus et al. 2008 |
| | *Sartoriana spinigera* | Klaus et al. 2008 |
| | *Sayamia bangkokensis* | Klaus et al. 2008 |
| | *Siamthelphusa improvisa* | Klaus et al. 2008 |
| | *Somanniathelphusa* sp. | Klaus et al. 2008 |
| | *Terrathelphusa kuhli* | Klaus et al. 2008 |
| Potamidae | *Geothelphusa albogilva* | Klaus et al. 2008 |
| | *Johora singaporensis* | Klaus et al. 2008 |
| | *Larnaudia beusekomae* | Klaus et al. 2008 |
| | *Malayopotamon brevimarginatum* | Klaus et al. 2008 |
| | *Potamiscus beieri* | Brandis 2000 |
| | *Pudaengon thatphanom* | Klaus et al. 2008 |
| | *Sinopotamon yangtsekiense* | Wang et al. 1999 |
| | *Thaiphusa sirikit* | Klaus et al. 2008 |
| Potamonautidae | *Hydrothelphusa madagascariensis* | Klaus et al. 2008 |
| OCYPODOIDEA | | |
| Ocypodidae | *Uca maracoani* | Benetti et al. 2008 |
| | *Uca thayeri* | Benetti et al. 2008 |
| | *Uca vocator* | Benetti et al. 2008 |
| | *Ucides cordatus* | Matos et al. 2000 |
| GRAPSOIDEA | | |
| Grapsidae | *Metopograpsus messor* | Anilkumar et al. 1999 |
| Varunidae | *Eriocheir sinensis* | Du et al. 1988 |

density, intervening cytoplasm containing some or all of the following organelles — mitochondria, microtubules, lamellar structures and centrioles — and a variable number (from zero to many) of arms or spikes. The arms may be composed of nuclear material, or microtubules, or both. In the Anomura, for example, the arms always contain microtubules, while in the Brachyura they are composed of nuclear material, except for some members of the Majidae that are reported (Hinsch 1969, 1973) to have microtubular elements in the nuclear arms.

Thus, in comparison to the diversity of crustacean spermatozoa, decapods are reasonably conservative, but it is still difficult to describe a typical decapod sperm cell. A taxonomic survey of decapod spermatozoal morphology at this point would be quite extensive, repetitive, and, frankly, dull. Instead, I want to highlight several characteristic and unique spermatozoal characters/features that emphasize the special place that the diverse decapod crustaceans hold within the Crustacea and within the wider animal kingdom.

## 2 THE UNIQUE DECAPOD SPERM

All decapod spermatozoa are unusual for the following six reasons: 1) they are aflagellate (lack a true 9 + 2 flagellum); 2) there is no reliable record of motility of any individual sperm cell; 3) the acrosome vesicle is not Golgi-derived as it is in all other described acrosomes of sperm in the animal kingdom; 4) the decapod sperm nuclear protein is unique; 5) the sperm nucleus is composed of diffuse, filamentous, heterogeneous chromatin fibers rather than being uniformly dense; and 6) the mitochondria are degenerate in mature sperm cells. These unique features will be elaborated below.

## 2.1 Aflagellate sperm cells

Most swimming or flagellate spermatozoa possess a tail(s) with a structured "9 + 2" arrangement of microtubules termed an axoneme. However, in the Crustacea, true flagellate spermatozoa have been recorded only in the Remipedia and in the maxillopodans (Cirripedia, Branchiura, Pentastomida, Mystacocarida, and Ascothoracica). Some apparently flagellate crustacean spermatozoa, such as the long and filamentous ostracod, amphipod, mysid, cumacean, and isopod sperm cells, are considered pseudoflagellate, and their "tail" is most often a long striated extension of the acrosome (see Fig. 1). Jamieson (1987, 1991) referred to this as a pseudoflagellum or striated tail-like appendage and regarded it as a synapomorphy for these peracarids.

Although microtubules are present in many decapod sperm cells, particularly in the long, and often numerous, microtubular arms, no true flagellum has ever been recorded. The entire diverse Decapoda, therefore, possess aflagellate spermatozoa.

## 2.2 Immotile sperm cells

Taking into account the previous character, it is not at all surprising that all recorded sperm cells in the Decapoda are also non-swimming (immotile). Even though the conspicuous arms (often microtubular) seen emanating from sperm cells seem to indicate motility, it has yet to be recorded in decapods. The absence of a true axoneme, with its inherent complexity, in any sperm cells renders them immobile. Some authors have claimed that the extensive and explosive acrosome reaction seen in decapod sperm cells (Brown 1966a, b; Talbot & Chanmanon 1980) constitutes a form of cell motility, but even though it appears to annex new ground for the expanding cell, it does not qualify as independent swimming motion typically associated with sperm cell motility.

## 2.3 Acrosome vesicle

The acrosome vesicle, probably more correctly termed "acrosomal complex" (Baccetti & Afzelius 1976), refers to the often large, concentrically zoned, electron-dense vesicle at the apical end, or constituting the apical portion, of the sperm cell of all decapods (see Figs. 3A, 4A, B). The term acrosome ("akrosoma") was first introduced by Lenhossek (1898) and was later applied to the "capsule" of decapod sperm by Bowen (1925), who also postulated that acrosomal material is formed in close association with the Golgi complex (Figs. 2A, B). Although the typical definition of an acrosome states that its origin is clearly from the Golgi complex, this does not apply to the acrosome of all decapod crustaceans studied to date. It has been shown in a wide range of decapods, including the dendrobranchiate shrimp *Parapenaeus longirostris* (Medina 1994), the caridean shrimp *Palaemonetes paludosus* (Koehler 1979), the crayfish *Procambarus clarkii* (Moses 1961a, b) and *Cambaroides japonicus* (Yasuzumi et al. 1961), the hermit crab *Pagurus bernhardus* (Pochon-Masson 1963, 1968) and the brachyurans *Eriocheir japonicus* (Yasuzumi 1960), *Menippe mercenaria*, *Callinectes sapidus* (Brown 1966a), *Carcinus maenas* (Pochon-Masson 1968), *Uca tangeri* (Medina & Rodriguez 1992), and *Cancer* species (Langreth 1969), that no typical Golgi complex is involved during acrosomal differentiation. Some recent authors (e.g., Yang et al. 1998; Wang et al. 1999) have suggested the presence of Golgi-derived acrosomes in certain decapods, but careful examination of their micrographs indicate that their "Golgi bodies" are complex membrane arrays (admittedly looking remarkably Golgi-like in appearance) and probable extensions of abundant endoplasmic reticulum. The acrosome vesicle of decapods is therefore defined as an acrosome by its position and function and not strictly by its cellular origin.

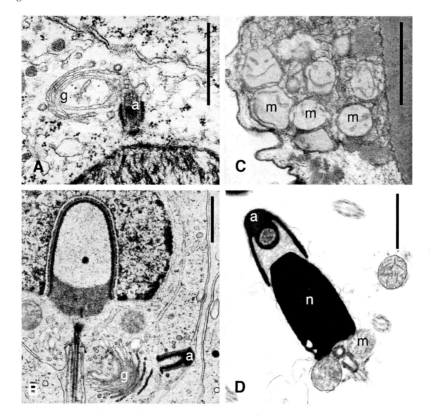

**Figure 2.** Transmission electron micrographs. (A) Golgi body (g) participating in acrosome development during spermiogenesis in the gastropod mollusc *Littorina sitkana*. Modified from Buckland-Nicks & Chia (1976). (B) Golgi body adjacent to developing acrosome (a) in the gastropod mollusc *Nerita picea*. Modified from Buckland-Nicks & Chia (1986). (C) Poorly cristate or acristate mitochondria (m) in the mature spermatozoa of the coconut crab, *Birgus latro*. (D) Typical electron-dense nucleus (n) in the mature spermatozoa of a limpet mollusc, *Cymbula concolor* (note the complex, cristate mitochondria (m) at the base of the nucleus). Photos courtesy of John Buckland-Nicks (A & B) and Alan Hodgson (D). Scale bars = 1 $\mu$m.

2.4  *Sperm nuclear proteins*

In the nucleus of all sperm cells the DNA is closely associated with a collection of proteins referred to as sperm nuclear basic proteins or SNBPs (Bloch 1969). These sperm-specific nuclear proteins appear in late spermiogenesis and are associated with highly compacted and inactive DNA. Unlike the evolutionarily conservative histones in somatic cell nuclei, SNBPs are highly diverse. There are five categories of these SNBPs spread across all the animal kingdom (both protostomes and deuterostomes) (Bloch 1969; Ausio 1995; Kasinsky 1991, 1995). The arthropods, for example, have representatives with all five types of SNBPs: H, P, PL, KP, and O:

- H-type (histones) *Rana* type (named for the animal in which it was first described)
- P-type (protamines) Salmon type (also in plants and the cirripede barnacle, *Balanus*)
- PL-type (protamine-like) *Mytilus* type
- KP-type (keratinous proteins) Mouse type
- O-type (absence of any sperm basic proteins) Crab type

Type "O," as you would expect, is found only in the decapods. Decapods have no SNBPs (but see Kurtz et al. 2008 for new, contrary information) but instead have extra-nuclear basic proteins, first termed "decapodine" by Chevaillier (1967) in *Nephrops*, *Pagurus*, and *Carcinus*. These unique decapodines are found in the large, electron-dense, and often voluminous acrosome vesicle, and migrate there from the nucleus during spermiogenesis (Chevaillier 1968; Vaughn et al. 1969).

## 2.5 Sperm nucleus

Associated with these sperm nuclear basic proteins and their unique absence (once again see Kurtz et al. 2008 for new, contrary information) in the decapod sperm nucleus is the fact that decapod sperm nuclei are also diffuse, electron-translucent, and filamentous in appearance (Fig. 3B) rather than being typically condensed, electron-dense, and granular (Fig. 2D). Condensation of the sperm head (nucleus) is characteristic of most animals regardless of the type of SNBPs they contain, except for decapod crustaceans. In fact, the densest part of the spiked decapod sperm cell is the acrosome vesicle, while the nucleus is electron-lucent and lightly granular or more usually filamentous.

If you look more closely at the structure of the nuclear filaments in decapod sperm under transmission electron microscopy (Fig. 3B), the nucleus has dense fibers ranging from 20 to

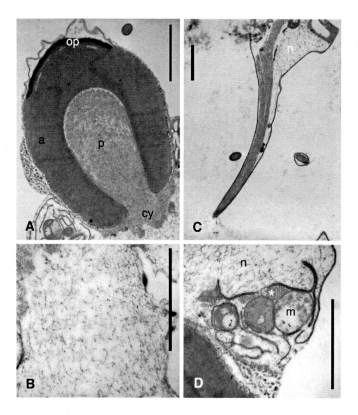

**Figure 3.** Transmission electron micrographs of the mature spermatozoa of the hermit crab *Loxopagurus loxochelis* (Diogenidae). (A) Electron-dense and complexly zoned acrosome vesicle (a). (B) Chromatin fibers in the electron-lucent nucleus (n). (C) Longitudinal section through an external microtubular arm. (D) Internalized microtubular arm (*), in cross-section, adjacent to degenerate mitochondria (m). Other abbreviations: cy, cytoplasm; op, operculum; p, perforatorial chamber. Photos courtesy of Marcelo Scelzo. Scale bars = 1 μm.

200 angstroms (Å) in width. These were argued by Chevaillier (1966b, 1991) to be bare DNA fibers. He also stated that all the SNBPs migrated during spermiogenesis from the sperm nucleus into the acrosome vesicle, where they associated with other proteins to form the characteristic decapodine.

## 2.6 Mitochondria

The last of our six unique decapod sperm characters is the presence in the mature sperm of only degenerate (or nearly so), non-cristate, non-functional mitochondria. In general, decapod sperm have only small amounts of cytoplasm and, therefore, often low numbers of recognizable organelles. Mitochondria can even appear to be totally absent in mature sperm (the Brachyura are a good example of this). What few mitochondria there are usually have few recognizable cristae or are devoid of them (Figs. 2C, 3D).

Studies conducted in the mid-1970's (Pearson & Walker 1975) showed that cytochrome C oxidase activity (an indicator of oxidative phosphorylation and confined to mitochondrial cristae) diminished as mitochondrial morphology changed over spermiogenesis in the crab *Carcinus maenas*. As decapod spermatids mature, most mitochondria are lost or lose their cristae. By the time the sperm cell is mature, it does not show this enzyme activity. This should not be surprising considering that we already established that all decapod sperm are immotile, and so mitochondria are used to power the dynamic process of spermiogenesis only, rather than in cell motility. But aspects of mitochondrial morphology and function in those decapods that store sperm for long periods (e.g., Cheung 1968; Paul 1984) may be worth investigating.

The above six characteristics demonstrate that decapods are unique spermatologically but do not provide much useful information for elucidating phylogenetic relationships within the Decapoda. Of the large suite of spermatozoal characters described in the literature for various decapod sperm, there is only a subset that has any potential phylogenetic utility.

## 3 SPERM AND DECAPOD PHYLOGENY

> *"that one may often safely venture to infer from the specific shape of these elements (spermatozoa) the systematic position and the name of the animals investigated."*
>
> Wagner & Leuckhart (1852)

The use of spermatozoal ultrastructure in taxonomy and phylogeny is well established in various animal groups. Examples include: Oligochaeta (Jamieson 1983); Pentastomida (Storch & Jamieson 1992); Insecta (Jamieson et al. 1999); Anura (Jamieson 2003); Annelida (Rouse & Pleijel 2006); and Aves (Jamieson 2007).

Similarly, in the decapod crustaceans spermatozoal ultrastructure has been successful in elucidating phylogenetic relationships (e.g., Jamieson & Tudge 1990; Jamieson 1994; Jamieson et al. 1995; Medina 1995; Tudge 1997; Medina et al. 1998). Spermatozoal characters have also been used in conjunction with existing morphological character sets in recent phylogenetic analyses (Ahyong & O'Meally 2004) or to support taxonomic or systematic works (Scholtz & Richter 1995; Brandis 2000).

Some spermatozoal characters with relevance to phylogenetic reconstruction of decapod crustaceans include the following, with examples from investigated taxa.

## 3.1 *The acrosome vesicle*

*Presence/absence*: As previously mentioned, the acrosome vesicle is an electron-dense structure, usually used to help define the apical end or pole of the decapod sperm cell, that contains most of the cell's proteins and is therefore often complexly structured. An apical acrosome vesicle (variously sized and shaped) is present in all decapods studied to date, with the notable exception of some of the dendrobranchiate shrimp (families Aristeidae and Sergestidae) and the basal pleocyemate shrimp *Stenopus*, in the family Stenopodidae. Interestingly, several investigated genera in the order Euphausiacea also possess acrosome-less spermatozoa. See Jamieson & Tudge (2000) for a brief review of the supposedly plesiomorphic acrosome-less spermatozoa in the decapods and the novel development and origin of the malacostracan acrosome vesicle (also mentioned above). The loss of the "Golgi-derived" acrosome, common in the rest of the Crustacea, the absence of any acrosome in the above-mentioned basal shrimps, and the independent development of the "ER-derived" malacostracan acrosome vesicle could be important characters for helping to define the early branching patterns in the evolution of the Decapoda.

*Shape*: When present, the decapod acrosome vesicle is either embedded into the sperm cell (Fig. 4) or sits prominently atop the rest of the cell components (cytoplasm and nucleus) (Fig. 3A). The acrosome vesicle also assumes a large variety of shapes including straight spikes, curved spikes, flat discs, hollow domes, ovals (depressed or elongate), hemispheres, spheres (both slightly depressed or slightly elongate), and elongate cones or cylinders. Differences in shape can also occur because of apical perforations (through the operculum, for example) or basal perforations or invaginations, usually termed the perforatorial chamber (Figs. 3A, 4). This term refers to the

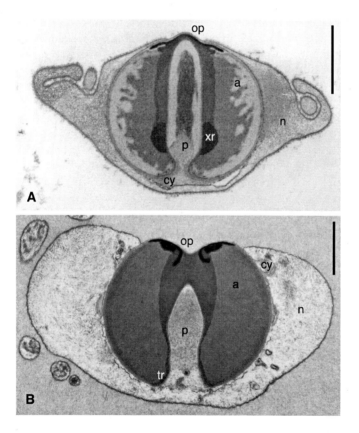

**Figure 4.** Transmission electron micrographs of a longitudinal section through the mature spermatozoa of two brachyuran crabs. (A) *Pilodius areolatus* (Xanthidae). (B) *Camposcia retusa* (Inachidae). Abbreviations: a, acrosome vesicle; cy, cytoplasm; n, nucleus; op, operculum; p, perforatorial chamber; tr, thickened ring; xr, xanthid ring. Photos courtesy of Barrie Jamieson. Scale bars = 1 $\mu$m.

invagination of the posterior end of the acrosome vesicle into a column or tube that penetrates the acrosome vesicle to various depths. The term "perforatorium" was first used by Waldeyer (1870) for a rod of fibrous material between the acrosome and nucleus in an amphibian sperm cell. It was later shown that the vertebrate perforatorium is homologous to equivalent invertebrate acrosomal structures (Dan 1967), and it was convenient to extend the term perforatorium to any subacrosomal material that projects outward at the time of the acrosome reaction (Baccetti 1979). The term perforatorial chamber reflects the fact that it is an invagination of the acrosome vesicle membrane (forming a chamber) that, with its contents, carries out the function of a perforatorium (*sensu* Baccetti 1979) without necessarily being rod-like or fibrous.

*Dimensions and size*: With the diversity of acrosome vesicle shapes comes an equally diverse array of sizes and dimensions for this organelle. A figure plotting acrosome length versus width for a variety of decapod sperm cells was provided by Jamieson (1991: 121), and a similar figure for the Anomura alone can be found in Tudge et al. (2001: 126) showing basic trends of spherical, elongate, or depressed, and any obvious deviations of individual species or groups of taxa. Often, the unusual size and shape differences of some taxa become clearer when plotted in this manner (see the example of the anomuran *Pylocheles* sperm in Tudge et al. 2001).

*Internal complexity*: The decapod acrosome vesicle in its various shapes and sizes also exhibits a range of internal ultrastructural complexity from simple to concentrically arranged in multiple layers or zones, each with its own electron density and morphology (Fig. 3A, 4). The exact biochemical nature and cellular function of most of these acrosomal zones are unknown, beyond their being composed of an array of PAS+ complex polysaccharides (e.g., Pochon-Masson 1965; Brown 1966b; Chevaillier 1966a), migrated sperm nuclear proteins (see above), and cytoskeletal proteins (Jamieson & Tudge 2000). These acrosome vesicle zones are intimately involved in the dynamic acrosome reaction that delivers the posterior nuclear material across the egg membrane at fertilization (see Jamieson & Tudge 2000 for review). Although their exact composition and function are still mysterious, their unique density, granularity, and morphology under TEM have provided a wealth of acrosomal characters for comparison of decapod sperm cells, particularly within the major infraorders. The complexly zoned and morphologically distinct acrosome vesicles have yielded a suite of characteristic and consistent traits identifying and unifying different groups of decapod taxa.

Some notable examples of these acrosome vesicle character traits include: the "dense perforatorial ring" in the hermit crab genus *Clibanarius* (Tudge 1997), the "xanthid ring" (Fig. 4A) common to all investigated members of this heterotreme brachyuran family (Jamieson 1989a, 1991), the distinctive structure of the flattened, centrally depressed, and often perforated majoid operculum (Fig. 4B) (Jamieson 1991; Jamieson et al. 1998; Jamieson & Tudge 2000) seen in this basal eubrachyuran group, and, finally, both the "apical button" perched on top of the operculum and the concentric lamellae present in the outer acrosome zones seen in nearly all thoracotreme crabs (Jamieson & Tudge 2000).

## 3.2 The nucleus

*Membrane-bound*: A defining feature separating the dendrobranchiate shrimp from the remaining pleocyemate decapods is that the nuclear region in the sperm cell of the former is not membrane-bound, while it is always membrane-bound in the latter (Medina 1995; Jamieson & Tudge 2000).

*Morphology and shape*: The basal or posterior sperm nucleus (if the acrosome vesicle is considered apical or anterior) can assume many different shapes throughout the Decapoda. It is spherical or globular in most of the dendrobranchiate shrimp (Medina et al. 1998) and the achelate lobsters (Tudge et al. 1998); triradiate in some of the podotreme brachyuran crabs, such as *Ranina*

(Jamieson 1989b) and *Dromidiopsis* (Jamieson et al. 1993) and the heterotreme brachyurans in the family Leucosiidae (Felgenhauer & Abele 1991; Jamieson & Tudge 2000); amorphous with multiple, pseudopodia-like lateral extensions or arms in many anomurans (Tudge & Jamieson 1991; Tudge 1995) and brachyurans; and secondarily cup-like in overall shape in all the brachyuran crabs where the spherical acrosome vesicle is embedded deeply into the cytoplasm and nuclear material (Jamieson & Tudge 2000).

Sometimes the nucleus is posteriorly extended as a distinct, single, thickened elongation (termed the "posterior median process"), and this has been recorded in the spermatozoa of some homolid and basal heterotreme brachyurans (Hinsch 1973; Jamieson & Tudge 2000). A fundamental difference in spermatozoal nuclear shape has also been used to support a division between the genera within the anomuran family Porcellanidae (Haig 1965; Sankolli 1965; Van Dover et al. 1982). Some genera (e.g., *Petrolisthes*) possess a spherical, more globular nucleus below the large complex acrosome vesicle, while others (e.g., *Aliaporcellana*, *Pisidia* and *Polyonyx*) have the sperm nucleus extended out into a long thick "tail," with a dense microtubular core, splitting terminally to yield multiple microtubular arms (Tudge & Jamieson 1996a, b). This unusual, superficially flagellate, decapod sperm morphology was first illustrated by Retzius (1909) for *Pisidia* (as *Porcellana*).

### 3.3 *Microtubular arms*

*Presence/absence*: As previously stated, all decapod spermatozoa are aflagellate, lacking a true "9+2" flagellum, but many do possess microtubular extensions from the sperm cell, which are often collectively called microtubular arms (Fig. 3C, D). The few decapod groups where no microtubular arms have been recorded include all the dendrobranchiate, caridean, and stenopodidean shrimps and the Brachyura (secondary loss), with the doubtful exception of some lower heterotremes in the majoid group (Jamieson & Tudge 2000). In these latter crabs the lateral arms are nuclear in origin (as they are in all brachyurans) but are said to contain a microtubular core inside them (Hinsch 1973). No independent, "naked," microtubular arms are present in any brachyuran investigated for sperm ultrastructure to date, although microtubules may be evident in sperm cell lateral arms and nuclei under certain conditions (Jamieson & Tudge 2000).

*Number*: In the Decapoda with sperm cells possessing true microtubular arms, the number is highly variable (see Table 2), but it can be simplified into a system whereby four or more arms appear to be plesiomorphic (Astacidea, Thalassinidea, and Palinura). A reduction to three occurs in enoplometopid and nephropid lobsters and most groups in the Anomura (12 of 15 families), and then a further reduction to total loss (as mentioned above) occurs in the Brachyura (Tudge 1997). It is interesting to note that in the podotreme brachyurans, some have sperm cells that exhibit three nuclear arms or extensions (the triradiate condition previously mentioned), and in the few heterotremes with microtubules still present in their nuclei, three lateral nuclear vertices are often apparent (Jamieson & Tudge 2000).

*Origin*: In the Decapoda that have sperm cells possessing true microtubular arms, these are externalized from the cell either from within the cytoplasm or from the nuclear material (Fig. 3C, D). Initially, all microtubules are grown from centrioles in the cytoplasm of the developing sperm cell, but once they become externalized they appear as either originating from the cytoplasm (e.g., all anomurans studied to date) or from the nucleus (e.g., Thalassinidea, Astacidea, and Palinura). This differing "origin" may have some phylogenetic significance (Tudge 1997).

An example of a spermatological character that does not appear to have any phylogenetic significance in the decapods investigated to date is the presence or absence of one or more centrioles in the mature sperm cell. In many decapod sperm cells, the pair (usually) of centrioles is observed

**Table 2.** The number of microtubular arms recorded in spermatozoa across the investigated decapod families, with indications of where the data are not available (NA) or need confirmation (?).

Dendrobranchiata = 0
Pleocyemata
   Stenopodidea = 0

   Caridea = 0

   Astacidea
      Astacidae = 5–8, 15–20
      Cambaridae = 4–7, 20
      Enoplometopidae = 3
      Glypheidae = NA
      Nephropidae = 3
      Parastacidae = 0? (nuclear only?)
      Thaumastochelidae = NA

   Thalassinidea
      Axianassidae = 5
      Axiidae = NA
      Callianassidae = 3?, 4–7
      Callianideidae = NA
      Calocarididae = 4–5
      Ctenochelidae = NA
      Laomediidae = NA
      Micheleidae = NA
      Strahlaxiidae = 4
      Thalassinidae = 3–5?
      Thomassinidae = NA
      Upogebiidae = NA

   Palinura
      Palinuridae = 3–12
      Polychelidae = NA
      Scyllaridae = 6
      Synaxidae = NA

   Anomura
      Aeglidae = 3?
      Albuneidae = >4
      Chirostylidae = 3
      Coenobitidae = 3
      Diogenidae = 3
      Galatheidae = 3
      Hippidae = >4, 3–9
      Kiwaiidae = NA
      Lithodidae = 3
      Lomisidae = 3? (3 nuclear vertices)
      Paguridae = 3
      Parapaguridae = 3
      Porcellanidae = >4
      Pylochelidae = 3
      Pylojacquesidae = NA

   Brachyura = 0 (sometimes 3 nuclear vertices)

in the cytoplasm below the acrosome vesicle in mature spermatozoa, but their occurrence seems erratic and may be more dependent on the state of maturity of the cell, or even on fixation procedures (Jamieson & Tudge 2000). Often, closely related taxa (two species in a genus, for example) will differ in this character state. It should be expected that all sperm cells exhibiting microtubules should have one or more obvious centrioles, but this is not the case, and in fact many brachyuran crab spermatozoa (which mostly do not retain microtubules in the mature sperm cell) show a pair of orthogonally arranged centrioles beneath the acrosome vesicle. Recently, though, the number of centrioles (Benetti et al. 2008) or their unusual arrangement in a parallel pair (Jamieson 1993; Guinot et al. 1997; Klaus et al. 2008) has been suggested to have taxonomic and/or phylogenetic importance in the Ocypodidae and Potamoidea, respectively.

## 4 CONCLUSIONS

Spermatozoal characters have proven to be, and continue to be, useful tools in helping to elucidate phylogenetic relationships in the decapod crustaceans. Their greatest utility, though, does not lie in generating phylogenetic trees using only spermatozoal (and spermatophore) characters (e.g., Jamieson 1994; Tudge 1997), but in providing additional character states for establishing robust nodes and clades in trees generated from more comprehensive datasets. Decapod species investigated for spermatozoal (and spermatophore) morphology will always be a smaller subset (currently 50% of the families, about 10% of genera, and only 2% of species) of those whose somatic morphology or gene sequences are known. Reproductive data, such as spermatozoal structure, can be used to supplement the initial matrices of characters for phylogenetic analysis or can be plotted *a posteriori* onto trees generated by morphological and molecular data to increase support for clades and trace the evolutionary history of the changing reproductive biology of decapod crustaceans. Similar evidence from reproductive biology may also help to confirm the most recent sister group of the Decapoda.

Continued research into the reproductive biology of decapod crustaceans is needed to fill the current gaps in our knowledge of this group, especially representatives from the families and superfamilies whose reproductive biology remains largely or totally unknown (e.g., Glypheidae, Micheleidae, Polychelidae, and Kiwaiidae). Also, further investigation is required on the taxa, and their congeners if available, for which only single species have been investigated for spermatozoal and spermatophore morphology and where they still provide only incomplete or enigmatic results (e.g., *Lomis* and *Aegla* in the Anomura, *Thalassina* in the Thalassinidea, and *Cherax* in the Astacidea).

## ACKNOWLEDGEMENTS

I would like to acknowledge the constant support and guidance of my friend and colleague Professor Barrie Jamieson. His encouragement and academic fervor for this field are directly responsible for my own interest in, and association with, crustacean gamete biology. For this I am eternally grateful, and I dedicate this review paper to him. Without his significant contributions to crustacean and decapod spermatozoal ultrastructure and spermiocladistics, the field would not be in the prominent, dynamic, and advanced state that it currently enjoys. I also wish to acknowledge the support and assistance of Rafael Lemaitre (Chair, Invertebrate Zoology) and his excellent support staff team at the National Museum of Natural History, Smithsonian Institution, Washington, D.C. Barrie Jamieson, Alan Hodgson, John Buckland-Nicks, and Marcelo Scelzo are thanked for providing figures, and Rose Gulledge was invaluable to figure assembly.

## REFERENCES

Afzelius, B.A. 1970. Thoughts on comparative spermatology. In: B. Baccetti (ed.), *Comparative Spermatology*. 565–573. Accademia Nazionale dei Lincei, Rome.

Ahyong, S.T. & O'Meally, D. 2004. Phylogeny of the Decapoda Reptantia: Resolution using three molecular loci and morphology. *Raffles Bull. Zool.* 52: 673–693.

Anilkumar, G., Sudha, K. & Subramoniam, T. 1999. Spermatophore transfer and sperm structure in the brachyuran crab *Metopograpsus messor* (Decapoda: Grapsidae). *J. Crust. Biol.* 19: 361–370.

Ausio, J. 1995. Histone H-1 and the evolution of the nuclear sperm-specific proteins. In: B.G.M. Jamieson, J. Ausio and J.-L. Justine (eds.), *Advances in Spermatozoal Phylogeny and Taxonomy*. *Mém. Mus. natn. Hist. nat.*, Paris 166: 447–462.

Baccetti, B. 1979. The evolution of the acrosomal complex. In: D.W. Fawcett & J.M. Bedford (eds.), *The Spermatozoon. Maturation, Motility, Surface Properties and Comparative Aspects*. Urban and Schwarzenberg, Baltimore-Munich. pp. 305–329.

Baccetti, B. & Afzelius, B.A. 1976. *The Biology of the Sperm Cell*. S. Karger, Basil.

Benetti, A.S., Santos, D.C., Negreiros-Fransozo, M.L. & Scelzo, M.A. 2008. Spermatozoal ultrastructure in three species of the genus *Uca* Leach, 1814 (Crustacea, Brachyura, Ocypodidae). *Micron* 39: 337–343.

Bloch, D.P. 1969. A catalog of sperm histones. *Genetics Suppl.* 61: 93–111.

Bowen, R.H. 1925. Further notes on the acrosome of animal sperm. The homologies of non-flagellate sperms. *Anat. Rec.* 31: 201–231.

Brandis, D. 2000. The taxonomic status of the freshwater crab genus *Potamiscus* Alcock 1909 (Decapoda, Brachyura, Potamidae). *Senckenberg. Biol.* 80(1/2): 57–100.

Brown, G.G. 1966a. *Ultrastructural studies on crustacean spermatozoa and fertilization*. Ph.D. Dissertation, University of Miami, pp. 1–239.

Brown, G.G. 1966b. Ultrastructural studies of sperm morphology and sperm-egg interaction in the Decapod *Callinectes sapidus*. *J. Ultrastr. Res.* 14: 425–440.

Buckland-Nicks, J.A. & Chia, F.S. 1976. Spermatogenesis of a marine snail, *Littorina sitkana*. *Cell Tiss. Res.* 170: 455–475.

Buckland-Nicks, J.A. & Chia, F.S. 1986. Formation of the acrosome and basal body during spermiogenesis in a marine snail, *Nerita picea* (Mollusca: Archaeogastropoda). *Gamete Res.* 15: 13–23.

Cheung, T.S. 1968. Transmoult retention of sperm in the female Stone Crab, *Menippe mercenaria* (Say). *Crustaceana* 15: 117–120.

Chevaillier, P. 1966a. Structure et constitution cytochimique de la capsule du spermatozoïde des Crustacés Décapodes. *C. R. Séanc. Acad. Sci., Paris* 262: 1546–1549.

Chevaillier, P. 1966b. Contribution a l'étude du complexe ADN-histone dans le spermatozoïde du pagure *Eupagurus bernhardus* L. (Crustacé Décapode). *J. Microsc.* 5: 739–758.

Chevaillier, P. 1967. Mise en evidence et etude cytochimique d'une proteine basique extranucleaire dans les spermatozoides des crustaces decapodes. *J. Cell Biol.* 32(3): 547–556.

Chevaillier, P. 1968. Étude cytochimique ultrastructurale des nucléoprotéines dans le spermatozoïde du pagure *Eupagurus bernhardus* L. (Crustacé Décapode). *J. Microsc.* 7: 107–114.

Chevaillier, P. 1991. Nuclear protein transitions during sperm differentiation. In: Baccetti, B. (ed.), *Comparative Spermatology 20 Years After*: New York: Raven Press. pp. 19–25.

Dan, J.C. 1967. Acrosome reaction and lysins. In: Metz, C.B. & Monroy, A. (eds.), *Fertilization: Comparative Morphology, Biochemistry and Immunology*. Vol. 1. Academic Press, New York. pp. 237–293.

Du, N., Xue, L. & Lai, W. 1988. Studies of the sperm of Chinese mitten-handed crab, *Eriocheir sinensis* (Crustacea, Decapoda). Spermatogenesis. *Oceano. Limno. Sinica* 19(1): 71–75.

Felgenhauer, B.E. & Abele, L.G. 1991. Morphological diversity of decapod spermatozoa. In: Bauer, R.T. & Martin, J.W. (eds.), *Crustacean Sexual Biology*. Columbia University Press, New York. pp. 322–341.

Grygier, M.J. 1982. Sperm morphology in Ascothoracida (Crustacea: Maxillipoda): confirmation of generalized nature and phylogenetic importance. *Int. J. Invert. Reprod.* 4: 323–332.

Guinot, D., Jamieson, B.G.M. & Tudge, C.C. 1997. Ultrastructure and relationships of spermatozoa of the freshwater crabs *Potamon fluviatile* and *Potamon ibericum* (Crustacea, Brachyura, Potamidae). *J. Zool., London* 241: 229–244.

Haig, J. 1965. The Porcellanidae (Crustacea, Anomura) of Western Australia with descriptions of four new Australian species. *J. Roy. Soc. West. Austr.* 48: 97–118.

Healy, J.M. & Anderson, D.T. 1990. Sperm ultrastructure in the Cirripedia and its phylogenetic significance. *Rec. Austr. Mus.* 42: 1–26.

Hinsch, G.W. 1969. Microtubules in the sperm of the Spider Crab, *Libinia emarginata*. *J. Ultrastr. Res.* 29: 525–534.

Hinsch, G.W. 1973. Sperm structure of Oxyrhyncha. *Can. J. Zool.* 51: 421–426.

Hong, S., Chen, X., Zhou, S. & Huang, J. 1993. Spermatogenesis in the shrimp *Penaeus penicillatus*: I. The morphology and structure of the sperm. *Acta Zool. Sinica* 39: 243–248.

Hong, S., Xia, C. & Wu, W. 1999. Spermatogenesis in the shrimp *Penaeus penicillatus*. *Oceano. Limno. Sinica* 30: 368–373.

Jamieson, B.G.M. 1983. Spermatozoal ultrastructure: evolution and congruence with a holomorphological phylogeny of the Oligochaeta (Annelida). *Zool. Scripta* 12: 107–114.

Jamieson, B.G.M. 1987. *The Ultrastructure and Phylogeny of Insect Spermatozoa*. Cambridge University Press, Cambridge.

Jamieson, B.G.M. 1989a. The ultrastructure of the spermatozoa of four species of xanthid crabs (Crustacea, Brachyura, Xanthidae). *J. Submicr. Cytol. Pathol.* 21: 579–586.

Jamieson, B.G.M. 1989b. Ultrastructural comparison of the spermatozoa of *Ranina ranina* (Oxystomata) and of *Portunus pelagicus* (Brachygnatha) (Crustacea, Brachyura). *Zoomorphology* 109: 103–111.

Jamieson, B.G.M. 1989c. A comparison of the spermatozoa of *Oratosquilla stephensoni* and *Squilla mantis* (Crustacea, Stomatopoda) with comments on the phylogeny of Malacostraca. *Zool. Scripta* 18: 509–517.

Jamieson, B.G.M. 1991. Ultrastructure and phylogeny of crustacean spermatozoa. *Mem. Queensland Mus.* 31: 109–142.

Jamieson, B.G.M. 1993. Ultrastructure of the spermatozoon of *Potamonautes perlatus sidneyii* (Heterotremata, Brachyura, Crustacea). *S. Afr. J. Zool.* 28: 40–45.

Jamieson, B.G.M. 1994. Phylogeny of the Brachyura with particular reference to the Podotremata: evidence from a review of spermatozoal ultrastructure (Crustacea, Decapoda). *Phil. Trans. Roy. Soc. London*, B. 345: 373–393.

Jamieson, B.G.M. (Ed.) 2003. *Reproductive Biology and Phylogeny of Anura*. Science Publishers, Inc., New Hampshire, USA. pp. 462.

Jamieson, B.G.M. (Ed.) 2007. *Reproductive Biology and Phylogeny of Birds*. Science Publishers, Inc., New Hampshire, USA. pp. 516.

Jamieson, B.G.M. & Tudge, C.C. 2000. Crustacea—Decapoda. In: Jamieson, B.G.M. (ed.), *Reproductive Biology of Invertebrates*. Vol. IX, part C: *Progress in Male Gamete Ultrastructure and Phylogeny*: John Wiley & Sons, Chichester. pp. 1–95.

Jamieson, B.G.M., Guinot, D. & Richer de Forges, B. 1995. Phylogeny of the Brachyura: evidence from spermatozoal ultrastructure (Crustacea, Decapoda). In: Jamieson, B.G.M., Ausio, J. & Justine, J.-L. (eds.), *Advances in Spermatozoal Phylogeny and Taxonomy. Mém. Mus. Natn. Hist. Nat., Paris* 166: 265–283.

Jamieson, B.G.M. & Tudge, C.C. 1990. Dorippids are Heterotremata: evidence from ultrastructure of the spermatozoa of *Neodorippe astuta* (Dorippidae) and *Portunus pelagicus* (Portunidae) (Brachyura: Decapoda). *Mar. Biol.* 106: 347–354.

Jamieson, B.G.M., Tudge, C.C. & Scheltinga, D.M. 1993. The ultrastructure of the spermatozoon of *Dromidiopsis edwardsi* Rathbun, 1919 (Crustacea, Brachyura, Dromiidae): confirmation of a dromiid sperm type. *Austr. J. Zool.* 41: 537–548.

Jamieson, B.G.M., Scheltinga, D.M. & Richer de Forges, B. 1998. An ultrastructural study of spermatozoa of the Majidae with particular reference to the aberrant spermatozoon of *Macropodia longirostris* (Crustacea, Decapoda, Brachyura). *Acta Zool. (Stockholm)* 79: 193–206.

Jamieson, B.G.M., Dallai, R. & Afzelius, B.A. 1999. *Insects Their Spermatozoa and Phylogeny*. Science Publishers, Inc., New Hampshire, USA. pp. 555.

Kang, X. & Wang, S. 2000. Studies on the changes of morphology and structure of the spermatozoon in *Penaeus chinensis*. *Donghai Mar. Sci.* 18: 40–46.

Kang, X., Wang, S. & Du, N. 1998. Studies on cytology spermatogenesis and fertilization of the marine prawn, *Penaeus chinensis* (Osbeck 1765). *J. Hebei Univ.* 18: 399–401.

Kang, X., Wang, S. & Du, N. 2000. Study on the endoplasmic reticulum variation during spermatogenesis in *Penaeus chinensis*. *J. Xiamen Univ.* 39: 849–854.

Kasinsky, H.E. 1991. Sperm basic protein diversity and the biology of fertilization. In: Baccetti, B. (ed.), *Comparative Spermatology 20 Years After*: New York: Raven Press, pp. 41–44.

Kasinsky, H.E. 1995. Evolution and origins of sperm nuclear basic proteins. In: Jamieson, B.G.M., Ausio, J. & Justine, J.-L. (eds.), *Advances in Spermatozoal Phylogeny and Taxonomy. Mém. Mus. Natn. Hist. Nat., Paris* 166: 463–473.

Klaus, S., Schubart, C.D. & Brandis, D. 2008. Ultrastructure of spermatozoa and spermatophores of old world freshwater crabs (Brachyura: Potamoidea: Gecarcinucidae, Potamidae and Potamonautidae). *J. Morph.* 270/2: 175–193.

Koehler, L. 1979. A unique case of cytodifferentiation: spermiogenesis of the prawn, *Palaemonetes paludosus. J. Ultrastr. Res.* 69: 109–120.

Koltzoff, N.K. 1906. Studien über die Gestalt der Zelle. I. Untersuchungen über die Spermien der Decapoden, als Einleitung in das Problem der Zellengestalt. *Arch. Mikrosk. Anat.* 67: 364–572.

Kurtz, K., Martinez-Soler, F., Ausió, J. & Chiva, M. 2008. Histones and nucleosomes in *Cancer* sperm (Decapod: Crustacea) previously described as lacking basic DNA-associated proteins: a new model of sperm chromatin. *Journal of Cellular Biochemistry* 105(2): 574–584.

Langreth, S.G. 1969. Spermiogenesis in *Cancer* crabs. *J. Cell Bio.* 43: 575–603.

Leeuwenhoek, A. van. 1678. Observations D. Anthonii Lewenhoeck, de Natis é semini genitali Animaliculis. *Phil. Trans. Roy. Soc. London* B12: 1040–1043.

Lenhossek, M. von. 1898. Untersuchungen über Spermatogenese. *Arch. Mikrosk. Anat.* 51: 215.

Lin, Q., Liu, R. & Xiang, J. 1991. Study of the morphological structure and physiological and biochemical functions of the spermatozoa in Chinese shrimp *Penaeus chinensis*. I. Ultrastructure of the spermatozoa. *Oceano. Limno. Sinica* 22: 397–401.

Manjón-Cabeza, M.E. & García Raso, J.E. 2000. Morphological reproductive aspects of males of *Diogenes pugilator* (Roux, 1829) (Crustacea, Decapoda, Anomura) from Southern Spain. *Sarsia* 85: 195–202.

Matos, E., Matos, P., Oliveira, E. & Azevedo, C. 1993. Ultrastructural data of the spermatozoa of the hermit crab *Clibanarius vittatus* Bosc, 1802 (Crustacea, Decapoda) of the north littoral of Brazil. *Rev. Brasil. Cienc. Morfol.* 10: 126–131.

Matos, E., Matos, P., Corral, L. & Azevedo, C. 1996. Ultrastructural and morphological aspects of the spermatozoon of *Dilocarcinus septemdentatus* Herbst, 1783 (Crustacea, Decapoda, Trichodactylidae) of the northern sea-coast of Brazil. *Brazil. J. Morphol. Sci.* 13: 31–35.

Matos, E., Matos, P., Corral, L. & Azevedo, C. 2000. Ultrastructure of spermatozoon of the crab *Ucides cordatus* Linnaeus (Crustacea, Decapoda, Ocypodidae) of the northern littoral of Brazil. *Rev. Brasil. Zool.* 17: 753–756.

Medina, A. 1994. Spermiogenesis and sperm structure in the shrimp *Parapenaeus longirostris* (Crustacea, Dendrobranchiata), comparative aspects among decapods. *Mar. Biol.* 119: 449–460.

Medina, A. 1995. Spermatozoal ultrastructure in Dendrobranchiata (Crustacea, Decapoda): taxonomic and phylogenetic considerations. In: Jamieson, B.G.M., Ausio, J. & Justine, J.-L. (eds.), *Advances in Spermatozoal Phylogeny and Taxonomy. Mém. Mus. Natn. Hist. Nat., Paris* 166: 231–242.

Medina, A. & Rodriguez, A. 1992. Spermiogenesis and sperm structure in the crab *Uca tangeri* (Crustacea, Brachyura), with special reference to the acrosome differentiation. *Zoomorphol.* 111: 161–165.

Medina, A., Vila, Y. & Santos, A. 1998. The sperm morphology of the euphausid *Meganyctiphanes norvegica* (Crustacea, Eucarida). *Invert. Reprod. Devel.* 34: 65–68.

Medina, A., Scelzo, M.A. & Tudge, C.C. 2006a. Spermatozoal ultrastructure in three Atlantic solenocerid shrimps (Decapoda, Dendrobranchiata). *J. Morph.* 267: 300–307.

Medina, A., García-Isarch, E., Sobrino, I. & Abascal, F.J. 2006b. Ultrastructure of the spermatozoa of *Aristaeopsis edwardsiana* and *Aristeus varidens* (Crustacea, Dendrobranchiata, Aristeidae). *Zoomorphology* 125: 39–46.

Moses, M.J. 1961a. Spermiogenesis in the crayfish (*Procambarus clarkii*). I. Structural characterization of the mature sperm. *J. Biophys. Biochem. Cytol.* 9: 222–228.

Moses, M.J. 1961b. Spermiogenesis in the crayfish (*Procambarus clarkii*). II. Description of stages. *J. Biophys. Biochem. Cytol.* 10: 301–333.

Paul, A.J. 1984. Mating frequency and validity of stored sperm in the Tanner Crab *Chionoecetes bairdi* (Decapoda, Majidae). *J. Crust. Biol.* 4: 375–381.

Pearson, P.J. & Walker, M.H. 1975. Alteration of cytochrome C oxidase activity during spermatogenesis in *Carcinus maenas*. *Cell Tiss. Res.* 164: 401–410.

Pochon-Masson, J. 1963. Origine et formation de la vésicule du spermatozoïde d'*Eupagurus bernhardus* (Décapode Anomoure). *C. R. Séanc. Acad. Sci., Paris* 256: 2226–2228.

Pochon-Masson, J. 1965. Schéma général du spermatozoide vésiculaire des Décapodes. *C. R. Séanc. Acad. Sci., Paris* 260: 5093–5095.

Pochon-Masson, J. 1968. L'ultrastructure des spermatozoides vésiculaires chez les Crustacés Décapodes avant et au cours de leur dévagination exprimentale. I. Brachyoures et Anomoures. *Ann. Sci. Nat., Zool. Biol. Nat.* 10: 1–100.

Pochon-Masson, J. 1983. Arthropoda—Crustacea. In: Adiyodi, K.G. & Adiyodi, R.G. (eds.), *Reproductive Biology of Invertebrates*. Vol. II. *Spermatogenesis and Sperm Function*: New York: John Wiley & Sons Ltd. pp. 407–409.

Retzius, G. 1909. Die Spermien der Crustaceen. *Biol. Untersuch.* 14: 1–54.

Rorandelli, R., Paoli, F., Cannicci, S., Mercati, D. & Giusti, F. 2008. Characteristics and fate of the spermatozoa of *Inachus phalangium* (Decapoda, Majidae): Description of novel sperm structures and evidence for an additional mechanism of sperm competition in Brachyura. *J. Morph.* 269: 259–271.

Rouse, G. & Pleijel, F. (eds.) 2006. *Reproductive Biology and Phylogeny of Annelida*. Science Publishers, Inc., New Hampshire, USA. pp. 698.

Sankolli, K.N. 1965. Studies on larval development in Anomura (Crustacea, Decapoda)—I. *Proceedings of the Symposium on Crustacea. Held at Ernakulam (January 12–15, 1965). Part II.* Marine Biological Association of India. Bangalore: Bangalore Press. pp. 744–775.

Scelzo, M.A. & Medina, A. 2003. Spermatozoal ultrastructure in *Artemesia longinaris* (Decapoda, Penaeidae). *J. Crust. Biol.* 23: 814–818.

Scelzo, M.A. & Medina, A. 2004. A dendrobranchiate, *Peisos petrunkevitchi* (Decapoda, Sergestidae), with reptant-like sperm: a spermiocladistics assessment. *Acta Zool.* 85: 81–89.

Scelzo, M.A., Medina, A. & Tudge, C.C. 2006. Spermatozoal ultrastructure of the hermit crab *Loxopagurus loxochelis* (Moreira, 1901) (Decapoda: Anomura: Diogenidae) from the southwestern Atlantic. In: Asakura, A. (ed.), *Biology of the Anomura II. Crust. Res.*, Special Number 6: 1–11.

Scholtz, G. & Richter, S. 1995. Phylogenetic systematics of the reptantian Decapoda (Crustacea, Malacostraca). *Zool. J. Linn. Soc.* 113: 289–328.

Shang Guan, B-M. & Li, S-J. 1994. On ultrastructure of the sperm of *Scylla serrata* (Crustacea, Decapoda, Brachyura). *Acta Zool. Sinica* 40: 7–11.

Spallanzani, L. 1776. *Opuscoli di Fisica Animale e Vegetabile*. Modena.

Storch, V. & Jamieson, B.G.M. 1992. Further spermatological evidence for including the Pentastomida (Tongue worms) in the Crustacea. *Int. J. Parasitol.* 22: 95–108.

Swammerdam, J. 1758. *The Book of Nature*. (English translation by Thomas Floyd). C.G. Seyfert, Dean St., Soho, London.

Talbot, P. & Chanmanon, P. 1980. Morphological features of the acrosome reaction of Lobster (*Homarus*) sperm and the role of the reaction in generating forward sperm movement. *J. Ultrastr. Res.* 70: 287–297.

Tirelli, T., Campantico, E., Pessani, D. & Tudge, C.C. 2006. Description of the male reproductive apparatus of the hermit crab *Calcinus tubularis* (Decapoda: Anomura: Diogenidae). In: Asakura, A. (ed.), *Biology of the Anomura II. Crust. Res.*, Special Number 6: 13–21.

Tirelli, T., Campantico, E., Pessani, D. & Tudge, C.C. 2007. Reproductive biology of Mediterranean hermit crabs: male reproductive apparatus of *Clibanarius erythropus* (Decapoda, Anomura). *J. Crust. Biol.* 27: 404–410.

Tirelli, T., Pessani, D., Silvestro, D. & Tudge, C.C. 2008. Reproductive biology of Mediterranean hermit crabs: fine structure of spermatophores and spermatozoa of *Diogenes pugilator* (Decapoda: Anomura) and its bearing on a sperm phylogeny of Diogenidae. *J. Crust. Biol.* 28: 535–543.

Tudge, C.C. 1995. Ultrastructure and phylogeny of the spermatozoa of the infraorders Thalassinidea and Anomura (Decapoda, Crustacea). In: Jamieson, B.G.M., Ausio, J. & Justine, J.-L. (eds.), *Advances in Spermatozoal Phylogeny and Taxonomy*. *Mém. Mus. Natn. Hist. Nat., Paris* 166: 251–263.

Tudge, C.C. 1997. Phylogeny of the Anomura (Decapoda, Crustacea): spermatozoa and spermatophore morphological evidence. *Contrib. Zool.* 67: 125–141.

Tudge, C.C. & Jamieson, B.G.M. 1991. Ultrastructure of the mature spermatozoon of the coconut crab *Birgus latro* (L.) (Coenobitidae, Paguroidea, Decapoda). *Mar. Biol.* 108: 395–402.

Tudge, C.C. & Jamieson, B.G.M. 1996a. Spermatophore and spermatozoal morphology in the Porcellanidae. I. *Aliaporcellana suluensis* and *Pisidia longicornis* (Decapoda, Anomura, Porcellanidae). *Crust. Res.* 25: 73–85.

Tudge, C.C. & Jamieson, B.G.M. 1996b. Spermatophore and spermatozoal morphology in the Porcellanidae. II. The genera *Petrolisthes* and *Polyonyx* (Decapoda, Anomura, Porcellanidae). *J. Crust. Biol.* 16: 535–546.

Tudge, C.C. & Scheltinga, D.M. 2002. Spermatozoal morphology of the freshwater anomuran *Aegla longirostri* Bond-Buckup & Buckup, 1994 (Crustacea: Decapoda: Aeglidae) from South America. *Proc. Biol. Soc. Wash.* 115: 118–128.

Tudge, C.C., Scheltinga, D.M. & Jamieson, B.G.M. 1998. Spermatozoal ultrastructure in the spiny lobster *Jasus novaehollandiae* Holthuis, 1963 (Palinuridae, Palinura, Decapoda). *J. Morph.* 236: 117–126.

Tudge, C.C., Scheltinga, D.M. & Jamieson, B.G.M. 1999. Spermatozoal ultrastructure in the Hippoidea (Anomura, Decapoda). *J. Submicrosc. Cytol. Pathol.* 31: 1–13.

Tudge, C.C., Scheltinga, D.M. & Jamieson, B.G.M. 2001. Spermatozoal morphology in the "symmetrical" hermit crab, *Pylocheles* (*Bathycheles*) sp. (Crustacea, Decapoda, Anomura, Paguroidea, Pylochelidae). *Zoosystema* 23: 117–130.

Van Dover, C.L., Factor, J.R. & Gore, R.H. 1982. Developmental patterns of larval scaphognathites: an aid to the classification of anomuran and brachyuran Crustacea. *J. Crust. Biol.* 2: 48–53.

Vaughn, J.C., Chaitoff, J., Deleon, R., Garland, C. & Thomson, L. 1969. Changing nuclear histone patterns during development. II. Isolation and partial characterization of "Decapodine" from sperm cells of the crab *Emerita analoga*. *Exp. Cell Res.* 54: 362–366.

Wagner, R. & Leuckhart, R. 1852. Semen. In: Todd, R.B. (ed.), *Cyclopaedia of Anatomy* Vol. 4. London: Longman.

Waldeyer, W. 1870. *Eistock und Ei*. Leipzig.

Wang, Y., Zhang, Z. & Li, S. 1997. Ultrastructure of spermatogenesis in the crab *Scylla serrata*. *Acta Zool. Sinica* 43: 249–254.

Wang, L., Du, N.S. & Lai, W. 1999. Studies on spermiogenesis of a freshwater crab *Sinopotamon yangtsekiense* (Crustacea Decapoda). *Acta Hydrobiol. Sinica* 23: 29–33.

Wielgus, E. 1973. The phylogenetic significance of spermatozoa in Decapoda. *Prze. Zool.* 17: 420–426.

Yager, J. 1989. The male reproductive system, sperm, and spermatophores of the primitive, hermaphroditic, remipede crustacean *Speleonectes benjamini*. *Invert. Reprod. Devel.* 15: 75–81.

Yang, W., Du, N. & Lai, W. 1998. Changes of Golgi apparatus during spermatogenesis of *Macrobrachium nipponense*. *Acta Zool. Sinica* 44: 377–383.

Yasuzumi, G. 1960. Spermiogenesis in animals as revealed by electron microscopy. VII. Spermatid differentiation in the crab, *Eriocheir japonicus. J. Biophys. Biochem. Cytol.* 7: 73–78.

Yasuzumi, G., Kaye, G.I., Pappas, G.D., Yamamoto, H. & Tsubo, I. 1961. Nuclear and cytoplasmic differentiation in developing sperm of the crayfish, *Cambaroides japonicus. Z. Zell. Mikrosk. Anat.* 53: 141–158.

Zhu, D., Li, S. & Wang, G. 2002. Ultrastructure of spermatogenesis in the squat lobster (*Thenus orientalis*). *Acta Zool. Sinica* 48: 100–106.

# The Evolution of Mating Systems in Decapod Crustaceans

AKIRA ASAKURA

*Natural History Museum & Institute, Chiba, Japan*

ABSTRACT

The mating systems of decapod crustaceans are reviewed and classified according to general patterns of lifestyles and male–female relations. The scheme employs criteria that focus on ecological, life history, and social determinants of both male and female behavior, and by these criteria nine types of mating systems are distinguished: (1) Short courtship: Both males and females are free-living (= not symbiotic with other organisms), and copulation occurs after brief behavioral interactions between a male and a female. (2) Precopulatory guarding: A male guards a mature female one to several days before copulation; both males and females are generally free-living. (3) Podding: In some large-size decapods, aggregations consisting of an extremely large number of individuals are formed, and mating occurs inside those aggregations. (4) Pair-bonding: In many symbiotic and some free-living species, males and females are found in a heterosexual pair and are regarded as having a monogamous mating system. They may live on or inside other organisms such as sponges, corals, molluscs, polychaetes, sea urchins, ascidians, and algal tubes. (5) Eusocial: In some sponge-dwelling snapping shrimps, a colony of shrimps contains a single reproductive female and many small individuals that apparently never breed. (6) Waving display: In many intertidal and semi-terrestrial crabs inhabiting mudflats or sandy beaches, males conduct visual displays that include species-specific dances to attract females. (7) Visiting: In some hapalocarcinid crabs, females are sealed inside a coral gall, and the male crab normally residing outside the gall is assumed to visit the gall for mating. (8) Reproductive swarm: In some pinnotherid crabs, mating occurs when a female is a free-swimming instar before she enters her definitive host. (9) Dwarf male mating: In some anomuran sand crabs, an extremely small male attaches near the gonopore of a free-living female.

## 1 INTRODUCTION

Decapod crustaceans are a large and diverse assemblage of animals. In most decapods, the sexes live separately and pair briefly as adults. Pairs are formed after a brief display, the sexes remain together for a relatively short period, the sexes separate after copulation, and the females assume all further parental duties such as selecting suitable habitat for egg incubation, aeration, and cleaning (Salmon 1983). However, recent discoveries of often-conspicuous behavior and male–female relations among decapods have shown that their mating system is highly diverse and is sometimes quite similar to mating systems of other animals such as birds, mammals, reptiles, and insects (see Shuster & Wade 2003; Duffy & Thiel 2007 for a review).

As claimed by Emlen & Oring (1977) in their classic work on the relationships among ecological factors, sexual selection, and the evolution of mating system, sexual selection is the driving force that underlies the evolution of male–male competition and female choice. However, ecological factors apparently contribute to the evolution of mating systems as well as to behavioral and morphological differences between the sexes. From this point of view, much study has been conducted recently on the evolution of the mating system of decapods (see section 2 below).

In this paper, I describe the diversity of mating systems of decapods in an attempt to recognize and classify their general patterns from the viewpoints of the ecological, life history, and social determinants of both male and female behavior. Historically, there are two ways of describing mating systems (Shuster & Wade 2003). The first is in behavioral ecology, where mating systems are usually described in terms of the number of mates per male or female, such as monogamy, polygyny, and polyandry. The second is in terms of the genetic relationships between mating males and females, such as random mating, negative assortative mating (outbreeding), and positive assortative mating (inbreeding). My approach to describing mating systems of decapods is a "recognition of general pattern" approach, a kind of a combination of these two approaches that captures variation in the relationship between male and female, from promiscuity to monogamy, as well as the relationship between male guarding and the female tendency to settle down in certain places or to aggregate, and the complex nature of eusociality.

Terminology generally follows Duffy & Thiel (2007). Additionally, some basic terms are redefined here, because these terms are sometimes used in more or less different ways according to taxa, including birds, mammals, and fish:

- **Monogamy** (= pair bonding): One male and one female have an exclusive mating relationship.
- **Polygamy**: One or more males have an exclusive relationship with one or more females. Three types are recognized: **polygyny**, where one male has an exclusive relationship with two or more females; **polyandry**, where one female has an exclusive relationship with two or more males; and **polygynandry**, where two or more males have an exclusive relationship with two or more females (the numbers of males and females need not be equal, and, in vertebrate species studied so far, the number of males is usually fewer).
- **Promiscuity**: Any male within the group mates with any female.
- **Eusociality**: Multigenerational (cohabitation of different generations), cooperative colonies with strong reproductive skew (reproductive division of labor, usually a single breeding female) and cooperative defense of the colony (after Duffy 2003).
- **Symbiosis**: Here defined simply as dissimilar organisms living together.

## 2 HISTORY OF STUDY

The first important review of decapod mating systems was Hartnoll's (1969) publication on brachyuran crabs. He distinguished two types of mating systems. "Soft-female mating" was defined as copulation occurring immediately after molting of the female, usually preceded by a lengthy pre-molt courtship behavior including precopulatory guarding by the male. "Hard-female mating" was defined as mating in which the female copulates during the intermolt stage after a relatively brief courtship behavior.

Through their intensive study of the harlequin shrimp *Hymenocera picta*, Wickler & Seibt (see Reference 16 in Appendix I, Table 10) found that these shrimp form stable heterosexual pairs based on individual recognition by chemical cues at a distance. Wickler & Seibt discussed several similar hypotheses, independently developed in research on crustaceans and humans, for the evolution of monogamy and other mating systems. Individual recognition in the monogamous mating system was intensively studied in the banded shrimp *Stenopus hispidus* by Johnson (1969, 1977).

The report by Emlen & Oring (1977) was influential for studies on crustacean mating systems. They classified the mating system into the following categories:

1. Monogamy
2. Polygyny (subdivided into 2a, resource defense polygyny; 2b, female (or harem) defense polygyny; and 2c, male dominance polygyny (further subdivided into 2c-1, explosive breeding assemblages, and 2c-2, leks))

3. Rapid multiple clutch polygamy
4. Polyandry (subdivided into 4a, resource defense polyandry; and 4b, female access polyandry)

Ridley (1983) intensively reviewed the precopulatory mate guarding behavior in various groups of animals including tardigrades, crustaceans, arachnids, and anurans, and discussed its evolution.

Work on the behavior of the fiddler crabs (genus *Uca*) has contributed greatly to our understanding of the mating systems of brachyuran crabs. These studies include the works of H.O. von Hagen (e.g., von Hagen 1970), J. Crane (e.g., Crane 1975), J. Christy and his coworkers (e.g., Christy et al. 2003a, b), M. Salmon and his coworkers (e.g., Salmon & Hyatt 1979), P. R. Y. Backwell and her coworkers (e.g., Backwell et al. 2000), M. Murai and his coworkers (e.g., Murai et al. 2002), and T. Yamaguchi (e.g., Yamaguchi 2001a, b). Based on the studies of *Uca* and other brachyurans, as well as other decapods, Salmon (1983) reported the diversity of behavioral interactions preceding mating in decapods, and he defined some of the consequences of these interactions in terms of sexual selection, courtship behavior, and mating systems. The book edited by Reback & Dunham (1983), which included Salmon's (1983) work, was a landmark in the study of decapod behavior.

Christy (1987) reviewed the mating systems of brachyuran crabs and classified them, according to modes of competition among males for females, into three major categories and eight subcategories, as follows.

1. Female-centered competition, including: 1a, defense of mobile females following free search; 1b, defense of sedentary females following a restricted search; 1c, capture, carrying, and defense of females at protected mating sites; and 1d, attraction and defense of females at protected mating sites
2. Resource-centered competition, including: 2a, defense of breeding sites; and 2b, defense of refuges
3. Encounter rate competition, including: 3a, neighborhoods of dominance; and 3b, pure search and interception

In their book on crustacean sexual biology, Bauer & Martin (1991) introduced developments in various fields and taxa of crustacean research, including studies on sex attraction, sex recognition, mating behavior, mating system, and structure and function associated with insemination. Bauer and his coworkers have extensively studied the mating behavior, mating system, and hermaphroditism of shrimps (e.g., see Bauer 2004 for a review).

Through their intensive studies on the mating system of the spider crab *Inachus* and of the extended maternal care of semi-terrestrial grapsid crabs of Jamaica, Diesel and his coworker revealed examples of highly specialized mating and social systems in these crabs (see Diesel 1991; Diesel & Schubart 2007 for reviews).

Thiel and his students have conducted intensive research on the mating system of rock shrimps (see Reference 6 in Appendix I, Table 4) and symbiotic anomuran crabs (e.g., Baeza & Thiel 2003). Based on these studies, Thiel & Baeza (2001) and Baeza & Thiel (2007) reviewed factors affecting the social behavior of marine crustaceans living symbiotically with other invertebrates. Similarly, Correa & Thiel (2003) reviewed mating systems in caridean shrimp and their evolutionary consequences for sexual dimorphism and reproductive biology. The book by Duffy & Thiel (2007) on the evolutionary ecology of social and sexual systems of crustaceans is a monumental landmark that synthesizes the state of the field in crustacean behavior and sociobiology and places it in a conceptually based, comparative framework. The relatively recent discovery of eusociality in snapping shrimp by Duffy has opened the door to a new field in social and mating systems of decapods (see Duffy 2007 for a review; see also sections 3.5 *Eusocial type* and 4.5 *Evolution of the eusocial type* below for further explanation).

Asakura (1987, 1990, 1993, 1994, 1995, 1998a, 1998b, 1999, 2001a, b, c), Imazu & Asakura (1994, 2006), and Nomura & Asakura (1998) reported mating systems and various aspects of sexual differences in the ecology and behavior of hermit crabs and other decapods.

## 3 TYPES OF MATING SYSTEMS

### 3.1 *Short courtship type*

This type is generally seen in species whose males and females are free living, that is, not symbiotic with other organisms (Appendix 1, Tables 1, 2). Copulation occurs after a short courtship behavior by the male, or copulation occurs just after brief behavioral interactions between a male and a female. This type of courtship includes very different groups of decapods, from the most primitive group (dendrobranchiate shrimps) to groups specialized for certain habitats such as freshwater crayfishes, intertidal hermit crabs, and semi-terrestrial and terrestrial brachyuran crabs. It is perhaps the most widely seen mating system in decapods.

No intensive aggressive behavior between males (for a female) has been reported in species of dendrobranchiate shrimps of the families Penaeidae and Sicyoniidae, caridean shrimps of the families Palaemonidae, Hyppolytidae, and Pandalidae, or anomuran sand crabs of the family Hippidae. In these species, females are generally similar in size to, or larger than, males. On the other hand, strong aggressive interaction is seen between males in freshwater crayfish species of all three families (Astacidae, Parastasidae and Cambaridae) as well as in brachyuran crabs of the Grapsoidea and Gecarcinidae. In these species, the male body and weaponry (chelipeds) are generally larger than the female.

Among decapods exhibiting this mating system are species whose females molt before copulation (Appendix 1, Table 1) and those whose females do not molt before copulation (Appendix 1, Table 2). In species inhabiting terrestrial and semi-terrestrial habitats, females generally copulate in the hard shell condition; these species include land hermit crabs of the genus *Coenobita* and brachyuran crabs of the Grapsoidea and Gecarcinidae.

In penaeid shrimp, the molting condition of copulating females is determined according to the type of thelycum. The thelycum is the female genital area, i.e., modifications of female thoracic sternites 7 and 8 (sometimes including thoracic sternite 6) that are related to sperm transfer and storage. A female with externally deposited spermatophores is said to have an "open thelycum," which is formed by modifications of the posterior coxae and sternites to which the spermatophores attach. Primitive dendrobranchiate shrimps, including species of the families Aristeidae, Solenoceridae, Benthesicymidae, and the penaeid genus *Litopenaeus,* have open thelyca. In these species, females copulate in the hard shell condition. On the other hand, a "closed thelycum" refers to sternal plates that may (1) enclose a noninvaginated seminal or sperm receptacle, (2) cover a space that leads to spermathecal opening, or (3) form an external shield guarding the spermathecal openings. In the most advanced groups, including the penaeoid genera *Fenneropenaeus, Penaeus, Farfantepenaeus, Melicertus, Marsupenaeus, Trachypenaeus,* and *Xiphopenaeus,* females have closed thelyca. In these species, females molt just before copulation. Since no significant difference is seen in mating behavior between the open thelycum species and the closed thelycum species, Hartnoll's (1969) rule, which predicts a lengthy pre-molt courtship behavior associated with soft-female mating and a relatively brief courtship behavior with hard-female mating, does not hold in the case of the penaeid shrimps.

A sperm plug, which is believed to preclude subsequent insemination by other males, is known in some species of *Farfantepenaeus, Marsupenaeus, Metapenaeus,* and *Rimapenaeus* (Appendix 1, Table 3).

In all the above-mentioned taxa, copulation generally continues only for several minutes. After mating, the male separates from the female and presumably goes on to search for other females.

The habitat of species that exhibit this mating system varies, ranging from terrestrial through intertidal to deep water.

## 3.2 *Precopulatory guarding type*

This mating system also is generally seen in species whose males and females are free living (Appendix 1, Table 4). A male guards a mature female for one to several days before copulation. Generally, males aggressively fight for a female using their cheliped(s) and sometimes also the ambulatory pereopods. In some species, females always molt prior to mating and copulation; in other species, females may or may not molt prior to copulation. There are two types of guarding: (1) contact guarding of hermit crabs and brachyuran crabs, in which a male grasps part of the appendages, the body, or the shell (in the case of hermit crabs) of a mature female, and (2) non-contact guarding, as exhibited in *Macrobrachium* shrimps and *Homarus* lobsters, in which a male keeps a female without grasping her. After mating, postcopulatory guarding by a male for a female is sometimes observed (Appendix 1, Table 5). However, after postcopulatory guarding, or just after copulation, the male and female separate so that both may later mate with other individuals. Generally, in this mating system, the body size of males is larger than that of females, or weaponry (chelipeds) is more developed in males than in females.

Species of the river prawn genus *Macrobrachium* are well known for the extremely long chelipeds in males. A male guards a female for one to several days before copulation and fights with other males using these chelipeds. In some species, such as *M. australiense*, a male has a nest (a saucer-shaped depression on the bottom), beckons a female to the nest, and guards and copulates with her in the nest. In the American lobster *Homarus americanus*, a male guards a female in his shelter, which is dug under rocks, boulders, or eelgrass, and the cohabitation of a male and a female lasts from one to three weeks.

In hermit crabs of the genus *Diogenes* (Diogenidae) and in many species of the family Paguridae, all of which have unequal chelipeds in terms of both size and morphology, a male grasps the rim of the shell inhabited by a mature female by the minor cheliped, guards her for one to several days before copulation, and fights with other males approaching him using the major cheliped. In crab-shaped anomurans, the male *Paralithodes brevipes* conducts both pre-copulatory and post-copulatory guarding. The male claims a female by grasping her chelae or legs with his chelae, or he covers the female with his body. Similarly, the male *Hapalogaster dentata* grasps a female with his left chela and covers the female with his body; these guarding behaviors occur one to three days before copulation.

In the brachyuran crab *Corystes cassivelaunus* (Corystidae), the male carries the female in his chelae, and, while stationary, holds one or both of the female's chelae in his own and holds her carapace close to his sternum. Such behavior continues up to several days before copulation. In species of the Cancridae and Portunidae, males carry the pre-molt female with her carapace or sternum held against the sternum of the male for a period of days; after this period the male releases the female so that she molts, and copulation occurs shortly after the molting. In many species in these two families, the male continues to carry the female after copulation in the pre-molt position until her integument has partially hardened. Sperm plugs, which are regarded as being produced by the males to block the females' genital duct to preclude subsequent insemination by other males (Diesel 1991), also are often reported for species of these families (Appendix 1, Table 6). In *Menippe mercenaria* (Xanthidae), the male guards the entrance to the burrow occupied by the pre-molt female, and they copulate as soon as the female molts. In species of the Majidae and Cheiragonidae, the male guards the female before copulation in a manner similar to what is seen in the Cancridae and Portunidae, where the male grasps the ambulatory pereopods, chelipeds, or body of the female.

Species that exhibit this mating system are from the intertidal through shallow water to deep waters, but they are not found in terrestrial or semi-terrestrial environments.

## 3.3 Podding

In large decapods inhabiting shallow waters, an aggregation consisting of an extremely large number of individuals in certain places is called a "pod." Podding is regarded as a type of behavior that is optional and that is associated with different stages in the species' life history, such as molting, mating, and the incubation period (Appendix 1, Table 7). The pod is also called a "heap" or "mound," according to the locality and/or the species.

The function of the pod may vary depending on the condition of the specimens within it (such as level of maturity, sex, intermolt stage) and possibly on changes in habitat condition, such as water temperature and presence of predators (Sampedro & González-Gurriarán 2004). However, as listed in Appendix 1, Table 7, pods in some species have the function of facilitating mating, so I will treat this as a special kind of mass mating in some species.

Stevens (2003) and Stevens et al. (1994), reporting more than 200 pods with a total of 100,000 crabs of the majid *Chionoecetes bairdi* in an area of only 2 ha off Kodiak Island in Alaska in 1991, observed that the formation of the pods and mating synchronized with the spring tide. Similar observations were made for another majid, *Hyas lyratus*, by Stevens et al. (1992), who reported large aggregations during the mating season from off Kodiak Island. They found 200 mating pairs (males grasping females) among 2000 individuals in one pod. The majid crab *Loxorhynchus grandis*, distributed along the east coast of North America, often forms large aggregations numbering hundreds of animals. The aggregation is composed of crabs of both sexes, and the function is thought to be the attraction of males for mating (Hobday & Rumsey 1999). DeGoursey & Auster (1992) reported large mating aggregations in another majid crab, *Libinia emarginata*, in April and May 1989. Many mating pairs were found in the aggregations, and the percentage of ovigerous females among all females increased from 26% on 1 May to 100% on 14 May. Males paired with females were significantly larger than unpaired males, while the paired and unpaired females were not significantly different in size. Carlisle (1957) monitored a pod consisting of 60–80 individuals of the majid crab *Maja squinado* in shallow waters in the English Channel; 20 were adult males and the rest were juvenile males and females in equal amounts. He observed crabs molting inside the pod and mating between intermolt males and postmolt females, which led him to conclude that the main purpose of podding is to provide protection for newly molted soft crabs against predators and to facilitate mating. However, later behavioral observations by Hartnoll (1969) indicated that copulation occurs between a male and a female in the intermolt stage. Furthermore, Sampedro & González-Gurriarán (2004) found that the gonads of females in the pods were in an early stage of development (= not fully matured) and that the spermathecae were empty, suggesting to them that mating of this species occurs in deeper waters.

In crab-shaped anomurans, large pods of the red king crab *Paralithodes camtschaticus* are well known in the northern Pacific Ocean, with each pod consisting of thousands of crabs in the 2-4 year class (juveniles). Aggregations of adult red king crabs (ovigerous females) also were reported and are thought to be related to mating (Stone et al. 1993), but detailed surveys have not been conducted. Dense aggregations of the southern king crab *Lithodes santolla* have been reported from Chile (South America); however, the crabs forming these aggregations are juveniles, so this behavior is not thought be related to mating (Cardenas et al. 2007).

In summary, podding is known only in large species distributed in temperate or boreal waters in both the Pacific and Atlantic oceans.

## 3.4 Pair-bonding type

Many species of decapods, in particular those that are symbiotic with other animals, have been reported as "found in a heterosexual pair" (Appendix 1, Tables 8–12). Most of these are considered to have a monogamous mating system, which is well known in birds and mammals. In species whose males engage in mate-guarding, temporal heterosexual pairing occurs, where the pair is formed when the female is close to molting or spawning a new batch of unfertilized eggs, and the mate-guarding males abandon the females soon after the eggs are fertilized. However, in pair-bonding species, males cohabit with females, independent of their reproductive status or of the stage of development of the brooded embryos. Nevertheless, the observations for the monogamous nature of these pair-bonding species are often only anecdotal, and how long the pair remains together, and with whom they mate, is rarely recorded. Some well-documented studies include the formation of stable pairing and individual recognition (individuals in a pair can recognize each other as mates), as in the case of the banded shrimp *Stenopus hispidus* (Reference 8 in Appendix 1, Table 10), the scarlet cleaner shrimp *Lysmata debelius* (Reference 12 in Appendix 1, Table 10), and the harlequin shrimp *Hymenocera picta* (Reference 16 in Appendix 1, Table 10).

Detailed observations of the monogamous nature of pairing have been made for several species of snapping shrimps, for example, *Alpheus angulatus* (Reference 97 in Appendix 1, Table 9), *Alpheus heterochaelis* (Reference 99 in Appendix 1, Table 9), *Alpheus armatus* (Reference 28 in Appendix 1, Table 9), and *Alpheus roquensis* (Reference 31 in Appendix 1, Table 9), as well as for the pontoniid shrimp *Pontonia margarita* (Reference 45 in Appendix 1, Table 8), the deep-water sponge-dwelling shrimp *Spongicola japonica* (Reference 1 in Appendix 1, Table 10), a porcelain crab *Polyonyx gibbesi* (Reference 11 in Appendix 1, Table 11), and several species of coral crabs of the genus *Trapezia* (References 2–14 in Appendix 1, Table 12). Many pair-bonding species are known in caridean shrimps of the subfamily Pontoniinae and family Alpheidae, "cleaner" shrimps of the families Stenopodidae and Spongicolidae, crab-shaped anomurans (family Porcellanidae), and brachyuran crabs of the family Trapeziidae.

Most of these species are symbiotic with other animals or live in special habitats. Host animals for these species include sponges, sea anemones, black corals, reef-building corals, gastropods, opistobranch molluscs, bivalves, polychaetes, crinoid feather stars, sea stars, sea urchins, sea cucumbers, and ascidians. The special habitats include gastropod shells used by large hermit crabs; tubes of polychaetes such as *Chaetopterus*; soft, web-like tubes consisting of filamentous algae, sponges, and other debris built by shrimp themselves; burrows excavated in hard dead corals; burrows of gobiid fish; and burrows of the thalassinidean shrimp genus *Upogebia*. However, free-living species are also known, such as stenopodid shrimps inhabiting rocky subtidal zones and many alpheid shrimp species inhabiting rock crevices or found under rubble, around large algae, or in burrows of their own in mudflats and other soft bottoms.

The following generalizations can be made for almost all of these species. They are territorial, and they cooperatively defend their habitats (hosts, special habitats, and burrows) against other conspecific or non-conspecific animals. Thus, the mating system of these species is termed "resource-defense monogamy." The pairs are size-matched (− size-assortative pairing); there is strict preference exerted by either sex for mates of a particular size relative to themselves. Baeza (2008) proposed two possible explanations for this phenomenon in his study on pontoniid shrimps symbiotic with bivalves:

1. The two sexes might choose large individuals of the opposite sex as sexual partners and host companions. In males, a preference for large females should be adaptive, as female size is positively correlated with fecundity in shrimps. In females, sharing a host with a large male might result in indirect benefits (i.e., good genes) or direct benefits (increased protection against predators or competitors).
2. Choice of a certain-size partner could also be a consequence of constraints in the growth rates of shrimps dictated by host individuals. Space limitations for shrimps in hosts are suggested by the tight relationship between shrimp and host size, and by the fact that hosts harboring solitary or no shrimps were among the small hosts.

These species tend to display low sexual dimorphism in weaponry in terms of cheliped size and morphology and often in body size. This is in contrast to the large sexual differences in mate-guarding species in which the weaponry is much more developed and where body size is often much larger in males than in females. Regarding body size, there is a tendency in pair-bonding shrimp for the male to be slightly smaller, in terms of body length, and much more slender than its mate female; in trapeziid crabs the male is often slightly larger than his female mate.

The bathymetric distribution of species with this mating system is generally from intertidal to shallow water, but a few groups of species, such as those of the Spongicolidae, inhabit deep water.

### 3.5 *Eusociality type*

Until the discovery of the eusocial shrimp *Zuzalpheus regalis* (as *Synalpheus regalis*) (Duffy 1996), eusociality was recognized only among social insects, including ants, bees, and wasps (Hymenoptera) and termites (Isoptera); in gall-making aphids (Hemiptera); in thrips (Thysanoptera); and in two mammal species, the naked mole rat (*Heterocephalus glaber*) and the damaraland mole rat (*Cryptomys damarensis*). *Zuzalpheus regalis* lives inside large sponges in colonies of up to >300 individuals, with each colony containing a single reproductive female. Direct-developing juveniles remain in the natal sponge, and allozyme data indicate that most colony members are full siblings. Larger members of the colony, most of whom apparently never breed, defend the colony against heterospecific intruders (Duffy 1996).

Following this initial discovery, Duffy and his coworkers have found several other species of *Zuzalpheus* exhibiting monogynous, eusocial colony organization in the western Atlantic (Appendix 1, Table 13). In the Indo-west Pacific region, Didderen et al. (2006) found a colony of a sponge-dwelling alpheid shrimp, *Synalpheus neptunus neptunus*, with one large ovigerous female or "queen" together with many small individuals, indicating a eusocial colony organization (Appendix 1, Table 13).

Some 20 species of symbiotic decapod species have been reported as found in a group (Appendix 1, Tables 14–15). Among them, examples of *Synalpheus* and *Zuzalpheus* exhibited more than 100 individuals in one aggregation, and, in particular in the case of *Zuzalpheus brooksi*, more than 1000 individuals were recorded from one sponge. These aggregations are regarded either as having a non-social structure (Thiel & Baeza 2001) or with the social structure totally unknown.

### 3.6 *Waving display type*

In many species of the crab families Ocypodidae, Dotillidae, and Macrophthalmidae, and in species of the genus *Metaplax* of the family Varunidae (formerly subfamily Varuninae in the Grapsidae *sensu lato*), males perform waving displays using the chelipeds. As in many other territory advertisement signals in animals, this behavior is commonly thought to have the dual function of simultaneously repelling males and attracting females (e.g., Salmon 1987; Crane 1975). These species typically live in mudflats, tidal creeks, sandbars, and mangrove forests, and each individual has its own burrow with a small territory around it. They often occur in huge numbers, with thousands of individuals living in small, adjacent territories, and with males and females living intermixed. The burrow serves various functions, including a refuge during high tide, an escape from predators, and the site of mating, oviposition, and incubation.

The behavior and mating systems of fiddler crabs (genus *Uca*, Ocypodidae) have been intensively studied (see references in History of Study, above). There are species whose males defend burrows from which they court females and species whose males wander from their burrows and court females on the surface (Christy 1987). For the former group of species, the following generalization is possible (based mainly on P. Backwell and coworkers; see references in History of Study, above). Males wave their enlarged claw, and, when a female is ready to mate (i.e., she matures), she leaves her own burrow and wanders through the population of waving males. The female visits several males before selecting a mate, and a visit consists of a direct approach to the male. Before copulation, both individuals enter the male's burrow, and two behavioral patterns are known: the male enters his burrow first and the female follows him in, or it happens in the reverse order, i.e., the female enters first. The male then gathers up sand or mud to plug the burrow entrance. Mating occurs in the burrow. On the following day, the male emerges, reseals the burrow entrance with the female still underground, and leaves the area. The female remains underground for the following few weeks while she incubates her eggs.

In addition to waving displays, males of some fiddler crab species employ acoustic signals to attract females. In these species, males attract females during the day first by waving and then by producing sounds just within their burrows. At night, the males produce sounds at low rates, but when touched by a female they increase their rate of sound production (Salmon & Atsaides 1968).

Many species of ocypodid crabs build sand structures next to their burrows, some of which function to attract females for mating, such as pillars (*Uca*: Christy 1988a, b), hoods (*Uca*: Zucker 1974, 1981; Christy et al. 2002, 2003a, b), mudballs (*Uca*: Oliveira et al. 1998), and pyramids (*Ocypode*: Linsenmair 1967; Hughes 1973).

## 3.7 Visiting type

An interesting mating system has been suggested for coral gall crabs (family Cryptochiridae), which inhabit cavities in scleractinian corals in (usually) shallow water. However, the information is still anecdotal, based on ecological observations on *Hapalocarcinus marsupialis*, *Troglocarcinus corallicola*, and *Opecarcinus hypostegus* (Potts 1915; Fize 1956; Kropp & Manning 1987; Takeda & Tamura 1981; Hiro 1937; Kotb & Hartnoll 2002; Carricart-Ganivet et al. 2004). In *H. marsupialis* and *T. corallicola*, the male crab normally resides outside the gall, which was constructed by the female, and is thought to visit the gall of the female for mating. The males and females apparently show promiscuity, and male–male aggressive behavior for a female has not been reported. The female is much larger than the male and in some species has a soft body with a very large abdomen. On the other hand, the male is usually hard, with a small abdomen. Geographical distribution includes mostly the tropics (see Wetzer et al. this volume).

In *Opecarcinus hypostegus*, couples were found sharing cavities; ovigerous females and males are recorded inhabiting adjoining cavities on colonies of *Siderastrea stellata* corals (Carricart-Ganivet et al. 2004). This species may have a mating system different from the above.

## 3.8 Reproductive swarm type

This mating system is reported only in pinnotherid crabs that are considered parasitic or co-inhabiting with other animals, including bivalves, gastropods, sea slugs, chitons, polychaetes, echinoderms, burrowing crustaceans, and sea squirts (Cheng 1967; Gotto 1969). In several species of these crabs, mating occurs, or is thought to occur, when the female is in the free-swimming stage before she enters into her definitive host (Appendix 1, Table 16).

The following generalization is possible for these species. Adult females have a soft, membranous carapace, and generally each one lives by itself within its host animal. These females produce broods of planktonic larvae. After development, the larvae metamorphose into the "invasive stage" crab, which is morphologically similar to the later swimming stage in having a flattened shape and ambulatory legs with dense setae adapted for swimming. Following this stage is a stage designated as "prehard"; these crabs invade, and live in, the host invertebrate animals. The crab at this stage is soft, resembling the later posthard stage. These crabs grow and mature into small adults of both sexes and leave their host to join mating swarms in open water. This stage is called the "hard stage," swimming stage, or copulation stage, and it is characterized by a hard body, swimming legs densely fringed with setae, and a thick fringe of setae along the front of the carapace. They copulate at this stage, and, in all reported species (see Appendix 1, Table 16), females copulate in the hard shell condition. After copulation, each female enters the host animal, but the male dies. The female becomes soft and grows much larger in the host, and later the female produces eggs fertilized by sperm from her single mating.

This is a kind of mass mating, with males and females showing promiscuity. In the copulation stage, no intensive aggressive behavior between males for females has been reported. The males in this stage are slightly larger than the females, and the morphology is similar between the sexes. After the female enters the host animal, the female becomes soft and grows much larger and stouter. The species with this mating system are found generally from intertidal to shallow water where their host invertebrates occur. In some pinnotherid species, adult crabs are found in a heterosexual pair in the host animal, although life history and mating systems of these species are mostly unknown.

### 3.9 *Neotenous male type*

Extremely small, neotenous males exist in some species of anomuran sand crabs (genus *Emerita*) inhabiting wave-exposed sandy beaches in tropical and temperate waters (Appendix 1, Table 17). In these species, the males become sexually mature soon after their arrival on the beach as a megalopa. When copulating, a male attaches near one of the female's gonopores, which are located on the coxae of the third pereopods. Surprisingly, the size of the neotenous males is similar to, or smaller than, those coxae.

Protandric hermaphroditism is described in detail in *Emerita asiatica* as it relates to neotenous males (Subramoniam 1981). The neotenous males occur at 3.5 mm carapace length (CL) and above, whereas females acquire sexual maturity at 19 mm CL. The neotenous males, as they continue to grow, gradually lose male functions and reverse sex at about 19 mm CL. In the CL range of 19–22 mm, the male's gonad consists of inactive testicular and active ovarian portions. Androgenic glands, active in the neotenous males, show signs of degeneration in the larger males and disappear in the intersexuals.

The male separates from the female after copulation. Aggressive behavior between males is not reported. As opposed to the female, the neotenous male shows a general simplicity of appendages associated with its small size. Among decapods, this phenomenon is known only in species of *Emerita*.

## 4 EVOLUTION OF MATING SYSTEMS IN DECAPODA

### 4.1 *Introduction*

It is apparent from the above that similar mating systems have evolved independently in different taxa at different times; i.e., convergent evolution is widespread. Species in ecologically similar habitats often display patterns that are strikingly comparable. Here I discuss the possible origin and evolutionary pathway of each mating system and compare them with those of other animals.

### 4.2 *Evolution of the short courtship type and the precopulatory type*

These two mating systems are most dominant among decapods. The mode of life is often quite similar; both males and females are free living (not symbiotic with other organisms), and after mating the male soon separates from the female. However, the habitat is sometimes different; in terrestrial and freshwater species, only the short courtship type has been reported. Therefore, a question arises as to why some groups of species have evolved the prolonged precopulatory mate guarding, whereas others have not.

Precopulatory mate guarding is known in a very broad range of taxa such as tardigrades, crustaceans, arachnids, and anurans (Parker 1974; Ridley 1983; Conlan 1991). It is thought to evolve when male–male competition for females is strong enough and female receptivity is restricted in time (Parker 1974; Jormalainen 1998), or even if receptivity is not time-limited but the guarding costs are low enough (Yamamura 1987). Guarding should be beneficial to the male, if the expected fitness gain achieved by guarding is greater than that expected by continuing to search for other females (Parker 1974). Thus, the optimal guarding duration for the male is determined by the encounter rate of females and the costs of guarding relative to those of searching (Yamamura 1987). The cost of guarding for males includes decreased mobility and feeding (Adams et al. 1985, 1991; Robinson & Doyle 1985), an increase in predation risk while guarding (Verrel 1985; Ward 1986), increased energetic costs associated with carrying females (Sparkes et al. 1996; Plaistow et al. 2003), and an increase in fighting costs through male–male conflict (Benesh et al. 2007; Yamamura & Jormalainen 1996). Additionally, a long guarding time decreases future opportunities to mate with other females (Benesh et al. 2007).

Pelagic dendrobranchiate and caridean shrimps are primarily swimmers, and possibly for that reason they have not evolved prolonged, elaborate behavioral interactions before copulation. However, the above-mentioned energetic cost hypothesis (Sparkes et al. 1996; Plaistow et al. 2003) may be applicable; for males of these species, carrying a swimming female for a long duration requires much more energy than in benthic species. In fact, all species exhibiting a prolonged precoulatory guarding period are benthic species.

In all freshwater crayfish studied, the mating system includes a short courtship without a lengthy precopulatory guarding, even though they have a benthic lifestyle and male–male aggression is often common. They may live in their burrows separately, or underneath boulders or heaps of fallen leaves, and these habitats are quite similar to, or virtually the same as, those of shrimps of the genus *Macrobrachium*. Why males of *Macrobrachium* adopt a precopulatory guarding strategy whereas male crayfish do not is not known.

A similar question arises in intertidal and shallow water decapods. For example, intertidal hermit crab species exhibiting precopulatory guarding have a tendency toward vastly unequal chelipeds, with a well-developed major cheliped particularly in males, who use it for fighting with other males during guarding. Such species include those of the genera *Pagurus* (Paguridae) and *Diogenes* (Diogenidae). On the other hand, species of *Paguristes* have small and similar right and left chelipeds and execute short courtship mating; males do not aggressively fight with other males. Species of *Calcinus*, which conduct short courtship type mating, often have vastly unequal chelipeds, with the well-developed major cheliped similar to those species that display precopulatory guarding. However, males of *Calcinus* species do not aggressively fight with each other during mating. Further study is needed to clarify the relationship between mating behavior and morphology.

In land hermits and land brachyurans, the above-mentioned predation risk hypothesis (Verrel 1985; Ward 1986) may be applicable to those species where mating system is the short-courtship type with hard-female mating. Male–male aggression is common in these taxa, but they have never evolved precopulatory guarding. Prolonged guarding may carry the risk of attack by visual predators such as birds in a terrestrial environment. In these taxa, a strong connection exists between a prolonged precopulatory guarding and soft-female mating as well as between a short courtship and hard-female mating. When marine species adapted to land, the former mating system might have been lost and changed to the latter, i.e., from soft-female to hard-female, to avoid desiccation and to deal with the large and often unpredicted fluctuations in availabilities of females in a terrestrial environment.

The evolution of sperm plugs in species of short-courtship type (penaeid shrimps) and precopulatory type (brachyuran crabs) is interesting. The sperm plug has virtually the same function as the copulation plug (= copulatory plug, mating plug) in mammals (rodents, bats, monkeys, koala), reptiles (snakes and lizards), insects (butterflies, ants, dragonflies, and stinkbugs), spiders, and acanthocephalan worms (Smith 1984). These plugs, secreted by the male after mating, serve to block the female tract for some time to prevent further mating by other males.

### 4.3 *Evolution of the podding type*

Why many animal species (e.g., insects, fish, birds, and herbivorous mammals) group together is one of the most fundamental questions in evolutionary ecology. It is believed that strong selective pressures lead to aggregation rather than to a solitary existence in most of these groups. These pressures include protection against predators, increased foraging efficiency, increased ease of assessing potential mates, and increased information exchange about the location of food (Barta & Giraldeau 2001). Similarly, various ecological reasons for the formation of pods have been proposed, including protection during molting, location of mates, aiding in food capture, and protection from predation (see References in Appendix 1, Table 7). Why some species evolved aggregating behavior and others did not is unknown.

### 4.4 *Evolution of the pair-bonding type*

Heterosexual pairing behavior ("social monogamy," Gowaty 1996; Bull et al. 1998; Gillette et al. 2000; Wickler & Seibt 1981) has evolved many times in a broad range of animal taxa, including mammals, birds, reptiles, amphibians, fish, insects, and crustaceans. For example, a colony of scleractinian coral sometimes yields a pair of goby fish, alpheid shrimps, and trapeziid crabs. Researchers interested in social system evolution must look for ecological and physiological factors (beyond basic sexual differences) that may make social monogamy selectively advantageous to individual males and/or females. Of particular interest are factors that may consistently correlate with such behavior across taxonomic groups. Several hypotheses for the evolution of social monogamy have been developed [see also Mathews (2002b), Baeza (2008), Baeza & Thiel (2007) for a review], as follows.

*Biparental care hypothesis:* Kleiman (1977) argued that the advantages of monogamy in mammals can lead to social monogamy. The hypothesis also implies that both males and females would suffer significantly reduced or zero fitness if they did not cooperate in caring for the offspring. However, this is not the case for marine decapods, where only the females care for the fertilized eggs and where neither parent cares for the larvae.

*Extended mate guarding hypothesis:* If males are under selection to guard females for some time before, during, and/or after courtship and mating, they may be forced into partner-exclusive behavior by some other factor, such as female dispersion (Kleiman 1977; Wickler & Seibt 1981) or female–female aggression (Wittenberger & Tilson 1980). In other words, monogamy can result from males guarding females over one or multiple reproductive cycles, because the female's synchronous receptivity, density, or abundance relative to males renders other male mating strategies (pure searching) less successful (Parker 1970; Grafen & Ridley 1983).

*Territorial cooperation hypothesis:* The fact that most monogamous species are territorial leads to this hypothesis. Territoriality correlates in various ways with social system evolution (Emlen & Oring 1977; Hixon 1987), and cooperation in territorial defense can lead to individual advantages in social groups or pairs (Brown 1982; Davies & Houston 1984; Fricke 1986; Clifton 1989, 1990; Farabaugh et al. 1992). In other words, males and females benefit by sharing a refuge (a territory) as heterosexual pairs because, for example, the risk of being evicted from the territory by intruders decreases (Wickler & Seibt 1981).

Recent intensive behavioral studies in various species shrimps have supported the predictions of the mate-guarding and/or territorial cooperation hypotheses (e.g., in *Hymenocera picta*, Wickler & Seibt 1981; *Alpheus angulatus*, Mathews 2002a, b, 2003; and *Alpheus heterochelis*, Rahman et al. 2002, 2003).

Another hypothesis about social monogamy (Baeza & Thiel 2007) concerns species symbiotic to other organisms (= host). Baeza & Thiel predicted that monogamy evolved when hosts are small enough to support few individuals and are relatively rare, and when predation risk away from the hosts is high. Under these circumstances, movements among hosts are constrained, and monopolization of hosts is favored in males and females due to their scarcity and because of the host's value in offering protection against predators. Because spatial constraints allow only a few adult symbiotic individuals to cohabit in/on the same host, both adult males and females would maximize their reproductive success by sharing "their" dwelling with a member of the opposite sex. This hypothesis was supported by Baeza's (2008) intensive study on a heterosexual pair of *Pontonia margarita*, a species symbiotic to the pearl oyster.

However, as mentioned before, most of observations for this mating system are anecdotal, and further detailed study is needed to clarify actual conditions of monogamous features of those species.

## 4.5 Evolution of the eusocial type

Hypotheses explaining how eusociality has evolved include Trophallaxis Theory (Roubaud 1916), Parental Manipulation Theory (Michener & Brothers 1974), Superorganism Theory (Reeve & Hölldobler 2007), and Inclusive Fitness Theory (Hamilton 1964a, b), of which the last one is most widely accepted. According to the Inclusive Fitness Theory, eusociality may evolve more easily in species exhibiting haplodiploidy, which facilitates the operation of kin selection. Although eusocial mole rats and termites exhibit diploidy, they display high levels of inbreeding by living as a family in a single burrow, such that colony members share more than 50% of their genes, and therefore the same model is considered to apply to these species and also to eusocial *Zuzalpheus* shrimps, in which all members of a colony share a single sponge.

## 4.6 Evolution of the waving display type

As compared to terrestrial species, courtship in aquatic species may be short and may not involve elaborate visual signaling (display) by the males; in aquatic species, chemical or visual cues are more important stimuli. In species of several genera of semi-terrestrial (= upper intertidal) decapods including *Uca* and other ocypodid crabs, visual signalling for prolonged periods is common, and sounds are often emitted by males to "call" females from their burrows to the surface for mating. Salmon & Atsaides (1968) presented ecological arguments to account for these differences in terms of optimal strategy of distance communications in the terrestrial and aquatic environments. Most aquatic decapods are nocturnally active and cryptic and live in an acoustically noisy environment, and this situation virtually eliminates all but the chemical channel for effective distance communication. On the other hand, visual and acoustic signals are effective in terrestrial species and are well developed in most terrestrial animals such as insects, birds, mammals, and also ocypodid and other terrestrial and semi-terrestrial decapods, probably because of the greater visibility in the terrestrial environment.

Waving displays seen in a variety of semi-terrestrial crabs is a case of convergent evolution (Kitaura et al. 2002). Grapsid crabs of the genus *Metaplax* conduct waving displays like species of the ocypodid crab genera *Uca*, *Macrophthalmus*, *Scopimera*, and *Dottila* (Kitaura et al. 2002). Species of *Metaplax*, unlike other grapsid crabs, which generally live along rocky shores, live in mud flats and burrow into the mud like many ocypodids. Salmon & Atsaides (1968) proposed the following factors as advantageous for the evolution of visual signaling in semi-terrestrial crabs: the substrate, which is flat and relatively free from the vegetational obstructions and other discontinuities; diurnal activity of the crabs; and the feeding proximity to their shelters, which leads crabs to live in aggregations so that social contacts are frequent. Therefore, it is assumed that habitat similarity between *Metaplax* and ocypodid crabs resulted in convergent evolution of these displays.

A recent molecular phylogenetic analysis suggested that even the waving display in *Uca* has multiple origins (Sturmbauer et al. 1996). Indo-west Pacific *Uca* species have simpler reproductive social behaviors, are more marine, and were thought to be ancestral to the behaviorally more complex and more terrestrial American species. It was also thought that the evolution of more complex social and reproductive behavior was associated with the colonization of the higher intertidal zones. However, Sturmbauer et al. (1996) demonstrated that species bearing the set of "derived traits" are phylogenetically ancestral, suggesting an alternative evolutionary scenario: the evolution of reproductive behavioral complexity in fiddler crabs may have arisen multiple times during their evolution, possibly by co-opting of a series of other adaptations for high intertidal living and antipredator escape.

This mating system is quite similar to male-territory-visiting polygamy (Kuwamura 1996) in fish, in which many examples are known in intertidal or shallow species; males have a burrow or a territory, and, when a mature female approaches a male, the male changes the color of part of his body and/or conducts species-specific courtship displays, after which the female enters the burrow or territory of the male and spawns (e.g., Miyano et al. 2006). In these fish species, males are brilliantly colored, as are male *Uca* species.

### 4.7 Evolution of the visiting type

A widely recognized tendency among various kinds of animals is that females live in a particular place and have a narrow home range, whereas males have a comparatively wider home range (Clutton-Brock et al. 1982). This "visiting type" mating system (seen in cryptochirid crabs) probably has evolved as one extremity of this tendency, with females living in a very specialized habitat (inside coral galls).

### 4.8 Evolution of the reproductive swarm type

Surprisingly, the function of the reproductive swarm in pinnotherid crabs is very similar to that of the nuptial flight (mating swarm) in ants (Insecta, Formicidae), and indeed their life history is quite similar. In most species of ants, breeding females and males that mature in their mothers' nest have wings and, during the breeding season, fly away from their nests and form swarms. Mating occurs during this period, and the males die shortly afterward. The surviving females land, and each female digs a burrow for the new nest. As eggs are laid in the burrow, stored sperm, obtained during their single nuptial flight, is used to fertilize all future eggs produced.

In the pinnotherids, crabs first grow in their host animals (vs. ants in their initial burrow). Then the crabs with swimming setae leave the hosts and swarm (vs. ants with wings fly away from their nests and conduct the nuptial flight). Mating occurs during this period (in ants, too), after which the female crabs enter the hosts, whereas the males die just after the mating (vs. the female ants make burrows of their own, with males dying just after the mating). As in the case of the ants, the female crabs reproduce by fertilizing their eggs with sperm from a single mating.

### 4.9 Evolution of the neotenous male type

The miniaturization of male mole crabs in the anomuran genus *Emerita* coupled with neoteny is similar to "dwarf males" (parasitic males, complemental males, miniature males), which are tiny males often attached to females. This condition has evolved in various groups of animals, including thoracican barnacles (Yamaguchi et al. 2007), acrothoracican barnacles (Kolbasov 2002), the oyster *Ostrea puelchanas* (Castro & Lucas 1987; Pascual 1997), epicaridean isopods (Mizoguchi et al. 2002), an echiuran *Bonellia* (Berec et al. 2005), anglerfish (Lophiiformes) (Pietsch 2005), blanket octopus (Tremoctopodidae), argonauts (Argonautidae), football octopus (Ocythoidae), and a deeper water octopus *Haliphron atlanticus* (Alloposidae) (Norman et al. 2002). The evolutionary cause for these phenomena has not been fully studied. The neoteny of male *Emerita* is considered to be one rather radical evolutionary solution to the problem of keeping the male and female together in the harsh and turbulent surf zone environment (Salmon 1983; Subramoniam & Gunamalai 2003).

## ACKNOWLEDGEMENTS

I am deeply grateful to Joel W. Martin (Natural History Museum of Los Angeles County, U.S.A.) for giving me the opportunity to present my work at the symposium on decapod phylogenetics at the TCS Winter Meeting in San Antonio, Texas. Thanks are also due to the following persons who provided me with important literature or aided me in my bibliographical survey: Keiichi Nomura (Kushimoto Marine Park, Japan), Tomomi Saito (Port of Nagoya Public Aquarium), Annie Mercier (Memorial University of Newfoundland, Canada), Juan P. Carricart-Ganivet (El Colegio de la Frontera Sur, Unidad Chetumal, Mexico), Jorge Contreras Garduño (Entomología Aplicada, Instituto de Ecología, Mexico), Ana Maria S. Pires-Vanin (Instituto Oceanogrâfico, Universidade de São Paulo, Brazil), J. Antonio Baeza (Smithsonian Tropical Research Institute, Republic of Panama), Martha Nizinski (NOAA/NMFS Systematics Laboratory, Smithsonian Institution, U.S.A.), Pan-wen Hsueh (National Chung Hsing University, Taiwan), Estela Anahí Delgado (Undecimar, Facultad de Ciencias, Uruguay), Michiya Kamio (Georgia State University, U.S.A.), Satoshi Wada (Hokkaido University, Japan), Yoichi Yusa (Nara Women's University, Japan), Tomoki Sunobe (Tokyo University of Marine Science and Technology), Charles H. J. M. Fransen (Nationaal Natuurhistorisch Museum Naturalis, The Netherlands), E. Gaten (University of Leicester, U.K.), Gil G. Rosenthal (Texas A&M University, U.S.A.), and Hiromi Watanabe (JAMSTEC, Japan). Special thanks are due to Raymond Bauer (University of Louisiana Lafayette) and Joel W. Martin for the careful review of an earlier draft of the manuscript.

## REFERENCES

Adams, J., Edwards, A.J. & Emberton, H. 1985. Sexual size dimorphism and assortative mating in the obligate coral commensal *Trapezia ferruginea* Latreille (Decapoda, Xanthidae). *Crustaceana* 48: 188–194.

Adams, J., Greenwood, P., Pollitt, R. & Yonow, T. 1991. Loading constraints and sexual size dimorphism in *Asellus aquaticus*. *Behaviour* 92: 277–287.

Asakura, A. 1987. Population ecology of the sand-dwelling hermit crab, *Diogenes nitidimanus* Terao. 3. Mating system. *Bull. Mar. Sci.* 41: 226–233.

Asakura, A. 1990. Evolution of mating system in decapod crustaceans, with particular emphasis on recent advances in study on precopulatory guarding. *Biol. Sci., Tokyo* 42: 192–200.

Asakura, A. 1993. Recent advances in study on aggressive and agonistic behavior of hermit crabs. I. General introduction and aggressive behavior. *Biol. Sci., Tokyo* 45: 143–160.

Asakura, A. 1994. Recent advances in study on aggressive and agonistic behavior of hermit crabs. II. Shell fighting and evolution of ritualization. *Biol. Sci., Tokyo* 46: 102–112.

Asakura, A. 1995. Sexual differences in life history and patterns of resource utilization by the hermit crab. *Ecology* 76: 2295–2313.

Asakura, A. 1998a. Sociality in decapod crustaceans. I. Relationship between males and females in species found in pair. *Biol. Sci., Tokyo* 49: 228–242.

Asakura, A. 1998b. Sociality in decapod crustaceans. II. Relationship between individuals in species found in group, symbiotic to other organisms. *Biol. Sci., Tokyo* 50: 37–43.

Asakura, A. 1999. Preliminary notes on classification of mating systems in decapod crustaceans. *Aquabiol., Tokyo* 125: 516–521.

Asakura, A. 2001a. Sexual difference and intraspecific competition in hermit crabs (Crustacea: Decapoda: Anomura). I. Morphological aspects. *Aquabiol., Tokyo* 135: 398–403.

Asakura, A. 2001b. Sexual difference and intraspecific competition in hermit crabs (Crustacea: Decapoda: Anomura). II. Difference in growth and survivorship patterns between the sexes. *Aquabiol., Tokyo* 135: 404–410.

Asakura, A. 2001c. Sexual difference and intraspecific competition in hermit crabs (Crustacea: Decapoda: Anomura). II. Behavioral aspects. *Aquabiol., Tokyo* 137: 589–593.

Backwell, P.R.Y., Christy, J.H., Telford S.R., Jennions, M.D. & Passmore, N.I. 2000. Dishonest signalling in a fiddler crab. *Proc. Royal Soc. Ser. B* 267: 719–724.

Baeza, J.A. 2008. Social monogamy in the shrimp *Pontonia margarita*, a symbiont of *Pinctada mazatlanica*, off the Pacific coast of Panama. *Mar. Biol.* 153: 387–395.

Baeza, J.A. & Thiel, M. 2003. Predicting territorial behavior in symbiotic crabs using host characteristics: a comparative study and proposal of a model. *Mar. Biol.* 142: 93–100.

Baeza, J.A. & Thiel, M. 2007. The mating system of symbiotic crustaceans. A conceptual model based on optimality and ecological constraints. In: Duffy, J.E. & Thiel, M. (eds.), *Evolutionary Ecology of Social and Sexual Systems: Crustaceans as Model Organisms*: 249–267. Texas: Oxford Univ. Press.

Barta, Z. & Giraldeau, L.A. 2001. Breeding colonies as information centers: a reappraisal of information-based hypotheses using the producer-scrounger game. *Behav. Ecol.* 12: 121–127.

Bauer, R.T. 2004. *Remarkable shrimps: natural history and adaptations of the carideans*. Norman: Univ. Oklahoma Press.

Bauer, R.T. & Martin, J.W. (eds.) 1991. *Crustacean Sexual Biology*. New York: Columbia Univ. Press.

Benesh, D., Valtonen, T. & Jormalainen, V. 2007. Reduced survival associated with precopulatory mate guarding in male *Asellus aquaticus* (Isopoda). *Ann. Zool. Fennici* 44: 425–434.

Berec, L., Schembri, P.J. & Boukal, D.S. 2005. Sex determination in *Bonellia viridis* (Echiura: Bonelliidae): population dynamics and evolution. *Oikos* 108: 473–484.

Brown, J.L. 1982. Optimal group size in territorial animals. *J. Theor. Biol.* 95: 793–810.

Bull, C.M., Cooper, S.J.B. & Baghurst, B.C. 1998. Social monogamy and extra-pair fertilization in an Australian lizard, *Tiliqua rugosa*. *Behav. Ecol. Sociobiol.* 44: 63–72.

Cardenas, C.A., Canete, J., Oyarzun, S. & Mansilla, A. 2007. Agregaciones de juveniles de centolla *Lithodes santolla* (Molina, 1782) (Crustacea) en asociación con discos de fijación de *Macrocystis pyrifera* (Linnaeus) C. Agardh, 1980. *Investig. Mar., Mayo.* 35: 105–110.

Carlisle, D.B. 1957. On the hormonal inhibition of moulting in decapod Crustacea. II. The terminal anecdysis in crabs. *J. Mar. Biol. Ass. U.K.* 36: 291–307.

Carricart-Ganivet, J.P., Carrera-Parra, L.F., Quan-Young, L.I. & Garcia-Madrigal, M.S. 2004. Ecological note on *Troglocarcinus corallicola* (Brachyura: Cryptochiridae) living in symbiosis with *Manicina areolata* (Cnidaria: Scleractinia) in the Mexican Caribbean. *Coral Reefs* 23: 215–217.

Castro, N.F. & Lucas, A. 1987. Variability of the frequency of male neoteny in *Ostrea puelchana* (Mollusca: Bivalvia). *Mar. Biol.* 96: 359–365.

Cheng, T.C. 1967. Marine molluscs as hosts for symbiosis. *Adv. Mar. Biol.* 5: 1–424.

Christy, J.H. 1987. Competitive mating, mate choice and mating association of brachyuran crabs. *Bull. Mar. Sci.* 41: 177–191.

Christy, J.H. 1988a. Pillar function in the fiddler crab *Uca beebei*. I. Effects on male spacing and aggression. *Ethology* 78: 53–71.

Christy, J.H. 1988b. Pillar function in the fiddler crab *Uca beebei*. II. Competitive courtship signaling. *Ethology* 78: 113–128.

Christy, J.H., Backwell, P.R.Y., Goshima, S. & Kreuter, T.J. 2002. Sexual selection for structure building by courting male fiddler crabs: an experimental study of behavioral mechanisms. *Behav. Ecol.* 13: 366–374.

Christy, J.H., Backwell, P.R.Y. & Schober, U.M. 2003a. Interspecific attractiveness of structures built by courting male fiddler crabs: experimental evidence of a sensory trap. *Behav. Ecol. Sociobiol.* 53 (2): 84–91.

Christy, J.H., Baum, J.K. & Backwell, P.R.Y. 2003b. Attractiveness of sand hoods built by courting male fiddler crabs, *Uca musica*: test of a sensory trap hypothesis. *Anim. Behav.* 66: 89–94.

Clifton, K.E. 1989. Territory sharing by the Caribbean striped parrotfish. *Anim. Behav.* 37: 97–103.

Clifton, K.E. 1990. The costs and benefits of territory sharing for the Caribbean coral reef fish, *Scarus iserti*. *Behav. Ecol. Sociobiol.* 26: 139–147.

Clutton-Brock, T.H., Guinness, F.E. & Albon, S.D. 1982. *Red Deer: Behavior and Ecology of Two Sexes*. Chicago: Univ. Chicago Press.

Conlan, K.E. 1991. Precopulatory mating behavior and sexual dimorphism in the amphipod Crustacea. *Hydrobiol.* 223: 255–282.

Correa, C. & Thiel, M. 2003. Mating systems in caridean shrimp (Decapoda: Caridea) and their evolutionary consequences for sexual dimorphism and reproductive biology. *Rev. Chil. Hist. Nat.* 76: 187–203.

Crane, J. 1975. *Fiddler Crabs of the World: Ocypodidae: Genus Uca*. Princeton: Princeton Univ. Press.

Davies, N.B. & Houston, A.I. 1984. Territory economics. In: Krebs, J.R. & Davies, N.B. (eds.), *Behavioural Ecology, an Evolutionary Approach (2nd ed)*: 148–169. Oxford: Blackwell Publ.

DeGoursey, R.E. & Auster, P.J. 1992. A mating aggregation of the spider crab, *Libinia emarginata*. *J. Northwest Atlantic Fish. Sci.* 13: 77–82.

Didderen, K., Fransen, C.H.J.M. & de Voogd, N.J. 2006. Observations on sponge-dwelling colonies of *Synalpheus* (Decapoda, Alpheidae) of Sulawesi, Indonesia. *Crustaceana* 79: 961–975.

Diesel, R. 1991. Sperm competition and the evolution of mating behavior in Brachyura, with special reference to spider crabs (Decapoda: Majidae). In: Bauer, R.T. & Martin, J.W. (eds.), *Crustacean Sexual Biology*: 145–163, New York: Columbia Univ. Press.

Diesel, R. & Schubart, C.D. 2007. The social breeding system of the Jamaican bromeliads crab, *Metopaulias depressus*. In: Duffy, J.E. & Thiel, M. (eds.), *Evolutionary Ecology of Social and Sexual Systems: Crustaceans as Model Organisms*: 365–386. Oxford: Oxford Univ. Press.

Duffy, J.E. 1996. Eusociality in a coral-reef shrimp. *Nature* 381: 512–514.

Duffy, J.E. 2003. The ecology and evolution of eusociality in sponge-dwelling shrimp. In: Kikuchi, T., Azuma, N. & Higashi, S. (eds.), *Genes, Behavior and Evolution in Social Insects*: 217–252. Sapporo: Hokkaido Univ. Press.

Duffy, J.E. 2007. Ecology and evolution of eusociality in sponge-dwelling shrimp. In: Duffy, J.E. & Thiel, M. (eds.), *Evolutionary Ecology of Social and Sexual Systems: Crustaceans as Model Organisms*: 387–412. Oxford: Oxford University Press.

Duffy, J.E. & Thiel, M. 2007. *Evolutionary Ecology of Social and Sexual Systems: Crustaceans as Model Organisms*. Oxford: Oxford Univ. Press.

Emlen, S.T. & Oring, L.W. 1977. Ecology, sexual selection, and the evolution of mating systems. *Science* 197: 215–223.

Farabaugh, S.M., Brown, E.D. & Hughes, J.M. 1992. Cooperative territorial defense in the Australian Magpie, *Gymnorhina tibicen* (Passeriformes, Cracticidae), a group-living songbird. *Ethology* 92: 283–292.

Fize, A. 1956. Observations biologiques sur les hapalocarcinides. *Ann. Fac. Sci. Univ. Nat. Viet Nam Inst. Oceanogr. Nhat.* 22: 1–30.

Fricke, H.W. 1986. Pair swimming and mutual partner guarding in monogamous butterflyfish (Pisces, Chaetodontidae): a joint advertisement for territory. *Ethology* 73: 307–333.

Gillette, J.R., Jaeger, R.G. & Peterson, M.G. 2000. Social monogamy in a territorial salamander. *Anim. Behav.* 59: 1241–1250.

Gotto, R.V. 1969. *Marine animals: partnerships and other associations*. Amsterdam: Elsevier Publ.

Gowaty, P.A. 1996. Multiple mating by females selects for males that stay: another hypothesis for monogamy in passerine birds. *Anim. Behav.* 51: 482–484.

Grafen, A. & Ridley, M. 1983. A model of mate guarding. *J. Theor. Biol.* 102: 549–567.

Hamilton, W.D. 1964a. The genetical evolution of social behaviour I. *J. Theor. Biol.* 7: 1–16.

Hamilton, W.D. 1964b. The genetical evolution of social behaviour II. *J. Theor. Biol.* 7: 17–52.

Hartnoll, R.G. 1969. Mating in the Brachyura. *Crustaceana* 16: 161–181.

Hiro, F. 1937. Studies on the animals inhabiting reef corals. I. *Hapalocarcinus* and *Cryptochirus*. *Palao Trop. Biol. Stat. Stud.* 1: 137–154.

Hixon, M.A. 1987. Territory area as a determinant of mating systems. *Am. Zool.* 27: 229–247.

Hobday, A. J. & Rumsey, S.M. 1999. Population dynamics of the sheep crab *Loxorhynchus grandis* (Majidae) Stimpson, 1857, at La Jolla California. *Scripps Inst. Oceanogr. Tech. Rep.* 29: 1–32.

Hughes, D.A. 1973. On mating and the "copulation burrows" of crabs of the genus *Ocypode* (Decapoda, Brachyura). *Crustaceana* 24: 72–76.

Imazu, M. & Asakura, A. 1994. Distribution, reproduction and shell utilization patterns in three species of intertidal hermit crabs on a rocky shore on the Pacific coast of Japan. *J. Exp. Mar. Biol. Ecol.* 172: 1–25.

Imazu, M. & Asakura, A. 2006. Descriptions of agonistic, aggressive and sexual behaviors of five species of hermit crabs from Japan (Decapoda: Anomura: Paguridae and Diogenidae). *Crust. Res., Spec. No.* 6: 95–107.

Johnson, V.R., Jr. 1969. Behavior associated with pair formation in the banded shrimp *Stenopus hispidus* (Olivier). *Pac. Sci.* 23: 40–50.

Johnson, V.R., Jr. 1977. Individual recognition in the banded shrimp *Stenopus hispidus*. *Anim. Behav.* 25: 418–428.

Jormalainen, V. 1998. Precopulatory mate guarding in crustaceans: male competitive strategy and intersexual conflict. *Quart. Rev. Biol.* 73: 275–304.

Kitaura, J., Nishida, M. & Wada, K. 2002 Genetic and behavioral diversity in the *Macrophthalmus japonicus* species complex (Crustacea: Brachyura: Ocypodidae). *Mar. Biol.* 140: 1–8.

Kleiman, D.G. 1977. Monogamy in mammals. *Quart. Rev. Biol.* 52: 39–69.

Kolbasov, G.A. 2002. Cuticular structures of some acrothoracican dwarf males (Crustacea: Thecostraca: Cirripedia: Acrothoracica). *Zool. Anz.* 241: 85–94.

Kotb, M.A. & Hartnoll, R.G. 2002. Aspects of the growth and reproduction of the coral gall crab *Hapalocarcinus marsupialis*. *J. Crust. Biol.* 22: 558–566.

Kropp, R.K. & Manning, R.B. 1987. The Atlantic gall crabs, family Cryptochiridae (Crustacea: Decapoda: Brachyura). *Smith. Contrib. Zool.* 462: 1–21.

Kuwamura, T. 1996. An introduction to reproductive strategies of fishes. In: Kuwamura, T. & Nakashima, Y. (eds.), *Reproductive Strategies in Fishes, Vol. 1*: 1–41. Tokyo: Kaiyusha.

Linsenmair, K.E. 1967. Konstruktion und Signalfunktion der Sandpyramide der Reiterkrabbe *Ocypode saratan* Forsk. *Z. Tierpsychol.* 24: 403–456.

Mathews, L.M. 2002a. Territorial cooperation and social monogamy: factors affecting intersexual interactions in pair-living snapping shrimp. *Anim. Behav.* 63: 767–777.

Mathews, L.M. 2002b. Tests of the mate-guarding hypothesis for social monogamy: does population density, sex ratio, or female synchrony affect behavior of male snapping shrimp (*Alpheus angulatus*)? *Behav. Ecol. Sociobiol.* 51: 426–432.

Mathews, L.M. 2003. Tests of the mate-guarding hypothesis for social monogamy: male snapping shrimp prefer to associate with high-value females. *Behav. Ecol.* 14: 63–67.

Michener, C.D. & Brothers, D.J. 1974. Were workers of eusocial Hymenoptera initially altruistic or oppressed? *Proc. Nat. Acad. Sci.* 68: 1242–1245.

Miyano, T., Takegaki, T. & Natsukari, Y. 2006. Spawning and egg-tending behavior of the barred-chin blenny *Rhabdoblennius ellipes*. *Bull. Fac. Fish., Nagasaki Univ.* 87: 1–5.

Mizoguchi, K., Henmi, Y. & Yamaguchi, T. 2002. Parasitic status of epicaridean isopods (Crustacea: Malacostraca) and the effects on their brachyuran crab hosts. *Jpn. J. Benthol.* 57: 79–84

Murai, M., Koga, T. & Yong, H.-S. 2002. The assessment of female reproductive state during courtship and scramble competition in the fidder crab, *Uca paradussumieri*. *Behav. Ecol. Sociobiol.* 52: 137–142.

Nomura, K. & Asakura, A. 1998. The alpheid shrimps (Decapoda: Alpheidae) collected from Kushimoto on the Pacific coast of central Japan, and their spatial distributions, zoogeographical affinities, social structures, and life styles. *Nanki Seibutsu* 40: 25–34.

Norman, M.D., Paul, D., Finn, J. & Tregenza, T. 2002. First encounter with a live male blanket octopus: the world's most sexually size-dimorphic large animal. *New Zealand J. Mar. Freshwt. Res.* 36: 733–736.

Oliveira, R.F., McGregor, P.K., Burford, F.R.L., Custódio, M.R. & Latruffe, C. 1998. Functions of mudballing behaviour in the European fiddler crab *Uca tangeri*. *Anim. Behav.* 55: 1299–1309.

Parker, G.A. 1970. Sperm competition and its evolutionary consequences in the insects. *Biol. Rev.* 45: 525–567.

Parker, G.A. 1974. Courtship persistence and female quarding as male time investment strategies. *Behaviour* 48: 157–184.

Pascual, M.S. 1997. Carriage of dwarf males by adult female puelche oysters: the role of chitons. *J. Exp. Mar. Biol. Ecol.* 212: 173–185.

Pietsch, T.W. 2005. Dimorphism, parasitism, and sex revisited: modes of reproduction among deep-sea ceratioid anglerfishes (Teleostei: Lophiiformes). *Ichthyol. Res.* 52: 207–236.

Plaistow, S.J., Outreman, Y., Moret, Y. & Rigaud, T. 2003. Variation in the risk of being wounded: an overlooked factor in studies of invertebrate immune function? *Ecol. Letters* 6: 489–494.

Potts, F.A. 1915. *Hapalocarcinus*, the gall-forming crab, with some notes on the related genus *Cryptochirus*. *Pap. Dep. Mar. Biol. Carnegie Inst. Wash.* 8: 33–69.

Rahman, N., Dunham, D.W. & Govind, C.K. 2002. Size-assortative pairing in the big-clawed snapping shrimp, *Alpheus heterochelis*. *Behaviour* 139: 1443–1468.

Rahman, N., David, W., Dunham, D.W. & Govind, C.K. 2003. Social monogamy in the big-clawed snapping shrimp, *Alpheus heterochelis*. *Ethology* 109: 457–473.

Reback, S. & Dunham, D.W. (eds.) 1983. *Studies in adaptation: the behavior of higher Crustacea*. New York: J. Wiley and Sons.

Reeve, H.K. & Hölldobler, B. 2007. The emergence of a superorganism through intergroup competition. *Proc. Nat. Acad. Sci.* 104: 9736–9740.

Ridley, M. 1983. *The Explanation of Organic Diversity: The Comparative Method and Adaptations for Mating*. Oxford: Oxford Univ. Press.

Robinson, B.W. & Doyle, R.W. 1985. Trade-off between male reproduction (amplexus) and growth in the amphipod *Gammarus lawrencianus*. *Biol. Bull.* 168: 482–488.

Roubaud, E. 1916. Recherches biologiques sur les guepes solitaires et sociales d'Afrique. La genese de la vie sociale et l'evolution de l'instinct maternel chez les vespides. *Ann. Sci. Nat.* 1: 1–160.

Salmon, M. 1983. Courtship, mating systems, and sexual selection in decapods. In: Rebach, S. & Dunham, D.W. (eds.), *Studies in Adaptation: The Behavior of Higher Crustacea*: 143–169. New York: John Wiley & Sons.

Salmon, M. 1987. On the reproductive behavior of the fiddler crab *Uca thayeri*, with comparisons to *U. pugilator* and *U. vocans*: evidence for behavioral convergence. *J. Crust. Biol.* 7: 2544.

Salmon, M. & Atsaides, S.P. 1968. Visual and acoustical signaling during courtship by fiddler crabs (genus *Uca*). *Am. Zool.* 8: 623–639.

Salmon, M. & Hyatt, G.W. 1979. The development of acoustic display in the fiddler crab *Uca pugilator* and its hybrid with *U. panacea*. *Mar. Behav. Physiol.* 6: 197–209.

Sampedro, M.-P. & González-Gurriarán, E. 2004. Aggregating behaviour of the spider crab *Maja squinado* in shallow waters. *J. Crust. Biol.* 24: 168–177.

Shuster, S.M. & Wade, M.J. 2003. *Mating Systems and Strategies*. New Jersey: Princeton University Press.

Smith, R.L. (ed.) 1984. *Sperm Competition and the Evolution of Animal Mating Systems*. New York: Academic Press.

Sparkes, T.C., Keogh, D.P. & Pary, R.A. 1996. Energetic costs of mate guarding behavior in male stream-dwelling isopods. *Oecologia* 106: 166–171.

Stevens, B.G. 2003. Timing of aggregation and larval release by Tanner crabs, *Chionoecetes bairdi*, in relation to tidal current patterns. *Fish. Res.* 65: 201–216.

Stevens, B.G., Donaldson, W.E. & Haaga, J.A. 1992. First observations of podding behavior for the Pacific lyre crab *Hyas lyratus* (Decapoda: Majidae). *J. Crust. Biol.* 12: 193–195.

Stevens, B.G., Haaga, J.A. & Donaldson, W.E. 1994. Aggregative mating of Tanner crabs, *Chionoectes bairdi*. *Canad. J. Fish. Aquat. Sci.* 51: 1273–1280.

Stone, C.E., O'Clair, C.E. & Shirley, T.C. 1993. Aggregating behavior of ovigerous female red king crab, *Paralithodes camtschaticus*, in Auke Bay, Alaska. *Canada. J. Fish. Aquat. Sci.* 50: 750–758.

Sturmbauer, C., Levinton, J.S. & Christy, J.H. 1996. Molecular phylogeny analysis of fiddler crabs: test of the hypothesis of increasing behavioral complexity in evolution *Proc. Natl. Acad. Sci. U.S.A.* 93: 10855–10857.

Subramoniam, T. 1981. Protandric hermaphroditism in a mole crab, *Emerita asiatica* (Decapoda: Anomura). *Biol. Bull.* 160: 161–174.

Subramoniam, T. & Gunamalai, V. 2003. Breeding biology of the intertidal sand crab, *Emerita* (Decapoda: Anomura). *Adv. Mar. Biol.* 46: 91–182.

Takeda, M. & Tamura, Y. 1981. Coral-inhabiting crabs of the family Hapalocarcinidae from Japan. VIII. Genus *Pseudocryptochirus* and two new genera. *Bull. Biogeogr. Soc. Jap.* 36: 13–27.

Thiel, M. & Baeza, J.A. 2001. Factors affecting the social behaviour of symbiotic Crustacea: a modelling approach. *Symbiosis* 30: 163–190.

Verrel, P.A. 1985. Predation and the evolution of precopula in the isopod *Asellus aquaticus*. *Behaviour* 95: 198–202.

von Hagen, H.O. 1970. Verwandtschaftliche Gruppierung und Verbreitung der Karibischen Winkerkrabben (Ocypodidae, Gattung *Uca*). *Zool. Meded., Leiden* 44: 217–235.

Ward, P.I. 1986. A comparative field study of the breeding behaviour of a stream and a pond population of *Gammarus pulex* (Amphipoda). *Oikos* 46: 29–36.

Wetzer, R., Martin, J.W. & Boyce, S. (this volume). Evolutionary origin of the gall crabs (family Cryptochiridae) based on 16S rDNA sequence data. In: Martin, J.W., Crandall, K.A. & Felder, D.L. (eds.), *Crustacean Issues: Decapod Crustacean Phylogenetics*. Boca Raton, Florida: Taylor & Francis/CRC Press.

Wickler, W. & Seibt, U. 1981. Monogamy in Crustacea and man. *Z. Tierpsychol.* 57: 215–234.

Wittenberger, J.F. & Tilson, R.L. 1980. The evolution of monogamy: hypotheses and evidence. *Ann. Rev. Ecol. Syst.* 11: 197–232.

Yamaguchi, T. 2001a. The breeding period of the fiddler crab *Uca lactea* (Decapoda, Brachyura, Ocypodidae) in Japan. *Crustaceana* 74: 285–293.

Yamaguchi, T. 2001b. Incubation of eggs and embryonic development of the fiddler crab, *Uca lactea* (Decapoda, Brachyura, Ocypodidae). *Crustaceana* 74: 449–458.

Yamaguchi, Y., Ozaki, Y., Yusa, Y. & Takahashi, S. 2007. Do tiny males grow up? Sperm competition and optimal resource allocation schedule of dwarf males of barnacles. *J. Theor. Biol.* 245: 319–328.

Yamamura, N. 1987. A model on correlation between precopulatory guarding and short receptivity to copulation. *J. Theor. Biol.* 127: 171–180.

Yamamura, N. & Jormalainen, V. 1996. Compromised strategy resolves intersexual conflict over pre-copulatory guarding duration. *Evol. Ecol.* 10: 661–680.

Zucker, N. 1974. Shelter building as a means of reducing territory size in the fiddler crab, *Uca terpsichores* (Crustacea: Ocypodidae). *Am. Mid. Nat.* 91: 224–236.

Zucker, N. 1981. The role of hood-building in defending territories and limiting combat in fiddler crabs. *Anim. Behav.* 29: 387–395.

APPENDIX 1

**Table 1.** Species of the short courtship type, in which females molt before copulation (= soft-female mating *sensu* Hartnoll 1969).

---

DENDROBRANCHIATA
Penaeidae: *Marsupenaeus japonicus* (1), *Melicertus kerathurus* (2), *Melicertus brasiliensis* (3), *Melicertus paulensis* (4), *Farfantepenaeus aztecus* (5), *Fenneropenaeus merguiensis* (6), *Penaeus monodon* (7), *Penaeus semisulcatus* (8), *Trachypenaeus similis* (9), *Xiphopenaeus* sp. (10)*,
Sicyoniidae: *Sicyonia dorsalis* (11), *Sicyonia parri* (12), *Sicyonia laevigata* (13)

PLEOCYEMATA
**Caridea**
  Palaemonidae: *Palaemonetes vulgarus* (14), *Palaemonetes varians* (15), *Palaemonetes pugio* (16), *Palaemon serratus* (17), *Palaemon elegans* (18), *Palaemon squilla* (19)
  Alpheidae: *Athanus nitescens* (20), *Alpheus dentipes* (21)
  Hippolytidae: *Heptacarpus picta* (22), *Heptacarpus paludicola* (23)
  Pandalidae: *Pandalus dana* (24), *Pandalus platyceros* (25), *Pandalus borealis* (26)
  Crangonidae: *Crangon crangon* (27), *Crangon vulgaris* (28)
**Astacidea**
  Nephropidae: *Nephrops norvegicus* (29)
**Palinuridea**
  Palinuridae: *Jasus lalandii* (30)*
**Anomura**
  Hippidae: *Emerita asiatica* (31), *Emerita analoga* (32)
  Diogenidae: *Calcinus latens* (33), *Calcinus seurati* (34), *Clibanarius tricolor* (35), *Clibanarius antillensis* (36), *Clibanarius zebra* (37), *Paguristes cadenati* (38), *Paguristes tortugae* (39), *Paguristes anomalus* (40), *Paguristes hummi* (41), *Paguristes oculatus* (42)

---

*Hard-female mating was rarely reported in addition to the soft-female mating. References: (1) Hudinaga (1942 as *Penaeus japonicus*), (2) Heldt (1931 as *Penaeus caramote*), (3) Brisson (1986), (4) de Saint-Brisson (1985), (5)–(6) Aquacop (1977), (7) Primavera (1979), Aquacop (1977), (8) Browdy (1989), (9)–(10) Bauer (1991), (11) Bauer (1992, 1996), (12)–(13) Bauer (1991), (14) Burkenroad (1947), Bauer (1976), (15) Antheunisse et al. (1968), Jefferies (1968), (16) Berg & Sandifer (1984), Bauer & Abdalla (2001), Caskey & Bauer (2005), (17) Nouvel & Nouvel (1937), Forster (1951), Bauer (1976), (18) Hoglund (1943), (19) Hoglund (1943), Bauer (1976), (20) Nouvel & Nouvel (1937), (21) Volz (1938), (22) Bauer (1976), (23) Bauer (1979), (24) Needler (1931), (25) Hoffman (1973), (26) Carlisle (1959), (27) Nouvel (1939), (28) Lloyd & Young (1947), Havinga (1930), Bodekke et al. (1991), (29) Farmer (1974), (30) von Bonde (1936), Silberbauer (1971), McKoy (1979), (31) Menon (1933), Subramoniam (1979), (32) MacGinitie (1938), Efford (1965), (33) Hazlett (1972), (34) Hazlett (1989), (35)–(36) Hazlett (1966), (37) Hazlett (1966, 1989), (38)–(42) Hazlett (1966).

**Table 2.** Species of the short courtship type, in which females do not molt before copulation (= hard-female mating *sensu* Hartnoll 1969).

---

DENDROBRANCHIATA
Penaeoidea: *Litopenaeus vannanmei* (1), *Litopenaeus setiferus* (2), *Litopenaeus stylirostris* (3), *Litopenaeus schmitti* (4)

PLEOCYEMATA
**Astacidea**
   Astacidae: *Pacifastacus trowbridgii* (5), *Pacifastacus leniusculus* (6), *Austropotamobius pallipes* (7), *Austropotamobius italicus* (8), *Austropotamobius torrentium* (9), *Astacus astacus* (10), *Astacus leptodactylus* (11)
   Parastacidae: *Cherax quadricarinatus* (12)
   Cambaridae: *Orconectes nais* (13), *Orconectes limosus* (14), *Faxonella clypeata* (15), *Orconectes rusticus* (16), *Orconectes propinquus* (17), *Orconectes virilis* (18), *Orconectes inermis inermis* (19), *Orconectes pellucidus* (20), *Cambarus blandingi* (21), *Cambaroides japonicus* (22), *Cambarus immunis* (23), *Procambarus alleni* (24), *Procambarus clarkii* (25), *Procambarus hayi* (26)
**Palinuridea**
   Palinuridae: *Panulirus homarus* (27)*, *Panulirus argus* (28)*, *Panulirus longipes cygnus* (29)
**Anomura**
   Diogenidae: *Calcinus verilli* (30), *Calcinus laevimanus* (31), *Calcinus seurati* (32), *Calcinus elegans* (33), *Calcinus hazletti* (34), *Calcinus laurentae* (35)
   Coenobitidae: *Birgus latro* (36), *Coenobita perlatus* (37), *Coenobita clypeatus* (38), *Coenobita compressus* (39)
**Brachyura**
   Leucosiidae: *Philyra scabriuscula* (40), *Ebalia tuberosa* (41)
   Xanthidae: *Lophopanopeus bellus* (42), *Lophopanopeus diegensis* (43), *Paraxanthias taylori* (44), *Pilumnus hirtellus* (45), *Xantho incisus* (46), *Nanopanope sayi* (47), *Eurypanopeus depressus* (48), *Panopeus herbstii* (49)
   Majidae: *Microphrs bicornutus* (50), *Pisa tetraodon* (51), *Pugettia gracilis* (52), *Pugettia producta* (53), *Pleistacantha moseleyi* (54), *Macrocheira kaempferi* (55)
   Grapsoidea: *Aratus pisonii* (56), *Cyclograpsus punctatus* (57), *Cyclograpsus integer* (58), *Cyclograpsus insularum* (59), *Cyclograpsus lavauxi* (60), *Eriocheir sinensis* (61), *Eriocheir japonicus* (62), *Goniopsis cruentata* (63), *Grapsus grapsus* (64), *Leptograpsus variegatus* (65), *Hemigrapsus nudus* (66), *Hemigrapsus crenulatus* (67), *Hemigrapsus oregonensis* (68), *Hemigrapsus sexdentatus* (69), *Pachygrapsus crassipes* (70), *Pachygrapsus gracilis* (71), *Pachygrapsus marmoratus* (72), *Gaetice depressus* (73), *Geograpsus lividus* (74), *Geosesarma percaccae* (75), *Plagusia chabrus* (76), *Planes minutus* (77), *Armases ricordi* (78), *Sesarma reticulatum* (79), *Sesarma bidentatum* (80), *Sesarma verleyi* (81), *Sesarma rectum* (82), *Sesarma eumolpe* (83), *Armases cinereum* (84), *Armases angustipes* (85), *Armases curacaoense* (86), *Helice crassa* (87)
   Gecarcinidae: *Gecarcoidea natalis* (88), *Gecarcoidea lateralis* (89), *Cardisoma guanhumi* (90), *Cardisoma armatum* (91)

**Table 2.** continued.

---

*Soft-female mating was rarely reported in addition to the hard-female mating. References: (1) Yano et al. (1988), Misamore & Browdy (1996), Palacios et al. (2003), (2) Misamore & Browdy (1996), (3) Aquacop (1977), (4) Bueno (1990), (5) Mason (1970a, b), (6) Lowery & Holdich (1988), Stebbing et al. (2003), (7) Ingle & Thomas (1974), Brewis & Bowler (1985), Carral et al. (1994), Villanelli & Gherardi (1998), (8) Galeotti et al. (2007), Rubolini et al. (2006, 2007), (9) Laurent (1988), (10) Cukerzis (1988), (11) Köksal (1988), (12) Barki & Karplus (1999), (13) Pippit (1977), (14) Schone (1968), Holdich & Black (2007), (15) Smith (1953), (16) Berrill & Arsenault (1982), Snedden (1990), Simon & Moore (2007), (17) Tierney & Dunham (1982), (18) Bovbjerg (1953), Rubenstein & Hazlett (1974), Tierney & Dunham (1982), (19)–(20) Bechler (1981), (21) Pearse (1909), (22) Kawai & Saito (2001), (23) Tack (1941), (24) Bovbjerg (1956), Mason (1970a, b), (25) Ameyaw-Akumfi (1981), Corotto et al. (1999), (26) Payne (1972), (27) Berry (1970), Heydon (1969), (28) Sutcliffe (1952, 1953), Kaestner (1970), Lipcius et al. (1983), Lipcius & Herrnkind (1987), (29) Chittleborough (1976), Sheard (1949), (30)–(35) Hazlett (1972), (36) Helfman (1977), (37) Page & Willason (1982), (38) Dunham & Gilchrist (1988), (39) Contreras-Garduño et al. (2007), (40) Naidu (1954), (41) Schembri (1983), (42)–(43) Knudsen (1960, 1964), (44)–(46) Bourdon (1962), (47)–(49) Swartz (1976a, b), (50) Hartnoll (1965a), (51) Vernet-Cornubert (1958a), (52) Knudsen (1964), (57) Boolootian et al. (1959), Grigg personal communication in Hartnoll (1969), Knudsen (1964), (54) Berry & Hartnoll (1970), (55) Arakawa (1964), (56) Warner (1967, 1970), (57) Broekhuysen (1941), (58) Hartnoll (1965b), (59)–(60) Brockerhoff & McLay (2005a, b), (61) Hoestlandt (1948), Peters et al. (1933), (62) Kobayashi & Matsuura (1994), (63) Schone & Schone (1963), Warner (1967, 1970), (64) Kramer (1967), Schone & Schone (1963), (65) Brockerhoff & McLay (2005a, b, c), (66) Knudsen (1964), (67) Yaldwyn (1966b), Brockerhoff (2002), (68) Knudsen (1964), (69) Brockerhoff & McLay (2005a, b, c), (70) Bovbjerg (1960), Hiatt (1948), (71) Brockerhoff & McLay (2005a, b), (72) Vernet-Cornubert (1958b), (73) Fukui (1991, 1994), (74) Hartnoll (1969), (75)–(77) Brockerhoff & McLay (2005a, b), (78) Warner (1967 as *Sesarma ricordi*), (79) Seiple & Salmon (1982), (80)–(81) Hartnoll (1969), (82) von Hagen (1967), (83) Hartnoll (1969), (84) Seiple & Salmon (1982 as *Sesarma cinereum*), (85) Hartnoll (1969 as *Sesarma angustipes*), (86) Hartnoll (1969 as *Sesarma curacaoense*), (87) Nye (1977), Beer (1959), Brockerhoff & McLay (2005a, b), (88) Hicks (1985), (89) Abele et al. (1973), Klassen (1975), Bliss et al. (1978), (90) Gifford (1962), Henning (1975), (91) Ameyaw-Akumfi (1987).

**Table 3.** Penaeid shrimp species in which a sperm plug has been reported.

| | |
|---|---|
| Penaeidae | |
| *Rimapenaeus similis* | (1) |
| *Farfantepenaeus aztecus* | (2) |
| *Rimapenaeus constrictus* | (3) |
| *Marsupenaeus japonicus* | (4) |
| *Metapenaeus joyneri* | (5) |

*References:* (1) Bauer & Min (1993 as *Trachypenaeus similis*), (2) Bauer & Min (1993), (3) Costa & Fransozo (2004), (4) Fuseya (2006), (5) Miyake (1982).

**Table 4.** Species of the precopulatory guarding type, in which males guard females before copulation. S = species in which females molt before copulation. H = species in which females do not molt before copulation. V = species in which both types (S and H) have been reported. ? = molting condition has not been reported.

CARIDEA
Palaemonidae: *Macrobrachium amazonicum* [S](1), *Macrobrachium rosenbergii* [S](2), *Macrobrachium austoraliense* [S](3), *Macrobrachium nipponense* [S](4), *Macrobrachium longipes* [S](5)
Rhynchocinetidae: *Rhynchocinetes typus* [H](6)

ASTACIDEA
Homaridae: *Homarus americanus* [V](7)

ANOMURA
Diogenidae: *Diogenes pugilator* [S](8), *Diogenes nitidimanus* [V](9), *Dardanus punctulatus* [?](10), *Calcinus tibicen* [S?](11)
Paguridae: *Pagurus miamensis* [V](12), *Pagurus pygmaeus* [V](13), *Pagurus bonairensis* [H](14), *Pagurus marshi* [S](15), *Pagurus bernhardus* [S](16), *Pagurus cuanensis* [H](17), *Pagurus anachoretus* [H](18), *Pagurus alatus* [H](19), *Pagurus marshi* [S](20), *Pagurus nigrofascia* [S](21), *Pagurus lanuginosus* [V](22), *Pagurus prideauxi* [H](23), *Pagurus hirsutiuculus* [S](24), *Pagurus maculosus* [?](25), *Pagurus minutus* [V](26), *Pagurus filholi* [V](27), *Pagurus gracilipes* [?](28), *Pagurus middendorffii* [H](29), *Pagurus nigrivittatus* [V](30), *Anapagurus chiroacanthus* [V](31), *Anapagurus breriaculeatus* [V](32), *Pylopagurus* sp. *sensu* Hazlett (1975)[H](33)
Lithodidae: *Paralithodes camtschaticus* [S](34), *Paralithodes brevipes* [S](35), *Lithodes maja* [S](36), *Lithodes santolla* [S](37), *Paralomis granulose* [S](38), *Hapalogaster dentata* [S](39)

BRACHYURA
Leucosiidae: *Philyra laevis* [H](40)
Majidae: *Chionoecetes opilio* [S](41), *Chionoecetes bairdi* [S](42), *Macropodia longirostris* [S](43), *Macropodia rostrata* [S](44)
Hymenosomatidae: *Halicarcinus* sp. [S](45), *Hymenosoma orbiculare* [S](46)
Cancridae: *Cancer gracilis* [S](47), *Cancer irroratus* [S](48), *Cancer magister* [S](49), *Cancer oregonensis* [S](50), *Cancer pagurus* [S](51), *Cancer productus* [S](52), *Cancer borealis* [S](53), *Cancer antennarius* [S](54)
Cheiragonidae: *Telmessus cheiragonus* [S](55), *Erimacrus isenbeckii* [S](56)
Corystidae: *Corystes cassivelaunus* [H](57)
Portunidae: *Callinectes sapidus* [S](58), *Carcinus maenas* [S](59), *Macropipes holsatus* [S](60), *Ovalipes ocellsatus* [S](61), *Portunus pelagicus* [S](62), *Portunus sanguinolentus* [S](63), *Portunus puber* [S](64), *Portunus trituberculatus* [S](65), *Scylla serrata* [S](66)
Xanthidae: *Menippe mercenaria* [S](67)

**Table 4.** continued.

*References:* (1) Guest (1979), (2) Bhimachar (1965), Rao (1967), Ra'anan & Sagi (1985), Kuris et al. (1987), (3) Ruello et al. (1973), Lee & Felder (1983), (4) Ogawa et al. (1981), Mashiko (1981), (5) Shokita (1966), (6) Correa et al. (2000, 2003), Hinojosa & Thiel (2003), Correa & Thiel (2003a, b), Díaz & Thiel (2003), Thiel & Hinojosa (2003), Díaz & Thiel (2004), Thiel & Correa (2004), van Son & Thiel (2006), Dennenmoser & Thiel (2007), (7) Herrick (1909), Templeman (1934, 1936), McLeese (1970, 1973), Hughes & Matthiessen (1962), Aiken & Waddy (1980), Waddy & Aiken (1981), Aiken et al. (2004), (8) Bloch (1935), Hazlett (1968), (9) Asakura (1987), (10) Matthews (1956), (11)–(13) Hazlett (1966), (14)–(17) Hazlett (1968), (18) Hazlett (1968), Hazlett (1975), (19) Hazlett (1968), (20) Hazlett (1975), (21)–(22) Wada et al. (2007), (23) Hazlett (1968), (24) MacGinitie (1935), (25) Imazu & Asakura (2006), (26) Imazu & Asakura (2006), Wada et al. (2007), (27) Imafuku (1986), Goshima et al. (1998), Minouchi & Goshima (1998, 2000), Wada et al. (2007), (28) Imazu & Asakura (2006), (29) Wada et al. (1996, 1999), (30) Wada et al. (2007), (31)–(32) Hazlett (1968), (33) Hazlett (1975), (34) Marukawa (1933), Powell & Nickerson (1965a, b), Gray & Powell (1966), Wallace et al. (1949), McMullen (1969), Matsuura & Takeshita (1976), Takeshita & Matsuura (1989), (35) Wada et al. (1997, 2000), Sato et al. (2005a, b), (36) Pike & Williamson (1959), (37)–(38) Lovrich & Vinuesa (1999), (39) Goshima et al. (1995), (40) Schembri (1983), (41) Watson (1972), (42) Paul (1984), Donaldson & Adams (1989), (43)–(44) Hartnoll (1969), (45) Lucas personal communication in Hartnoll (1969), (46) Broekhuysen (1955), (47) Knudsen (1964), (48) Chidchester (1911), Elner & Elner (1980), Elner & Stasko (1978), Haefner Jr. (1976), (49) Bulter (1960), Cleaver (1949), Snow & Nielsen (1966), (50) Knudsen (1964), (51) Edwards (1966), (52) Knudsen (1964), (53) Elner et al. (1985), (54) Knudsen (1960), (55) Kamio et al. (2000, 2002, 2003), (56) Sasaki & Ueda (1992), (57) Hartnoll (1968), (58) Childchester (1911), Churchill (1919), Hay (1905), Gleeson (1980), Ryan (1966), Gleeson et al. (1984), Christofferson (1970), Teytaud (1971), Jivoff & Hines (1998), (59) Broekhuysen (1936, 1937), Cheung (1966), Childchester (1911), Spalding (1942), Veillet (1945), Williamson (1903), Berrill (1982), Berrill & Arsenault (1982), Jensen (1972), (60) Broekhuysen (1936), (61) Childchester (1911), (62) Delsman & de Man (1925), Broekhuysen (1936), Fielder & Eales (1972), (63) George (1963), Ryan (1966, 1967a, b), Christofferson (1970, 1978), (64) Duteutre (1930), (65) Oshima (1938), (66) Hill (1975), (67) Binford (1913), Cheung (1968), Savage (1971), Porter (1960), Wilber (1989).

Table 5. Duration of guarding time in selected species of decapod crustaceans.

| Species | Precopulatory guarding time | Female condition when copulating | Postcopulatory guarding time | Refe |
|---|---|---|---|---|
| **ANOMURA** | | | | |
| Lithodidae | | | | |
| *Paralithodes brevipes* | 9–84 hrs (mean 38.9±24.9 hrs) | Soft | ? | (1) |
| *Paralithodes brevipes* | | | | |
| 3 males & 3 females | 32.1±44.1 hrs | Soft | ? | (2) |
| 1 male & 5 females | 15.1±20.1 hrs | Soft | ? | (3) |
| *Hapalogaster dentata* | 2–3 days | Soft | ? | (4) |
| **BRACHYURA** | | | | |
| Cancridae | | | | |
| *Cancer pagurus* | 3–21 days | Soft | 1–12 days | (5) |
| *Canner irroratus* | 4.5 days | Soft | 5 days | (6) |
| *Carcinus maenas* | | | | |
| 1 male & 1 female | 2–16 days | Soft | 0–1.5 days | (7) |
| 2 or 3 males – 1 female | 3–10 days | Soft | 1–3.5 days | (8) |
| Majidae | | | | |
| *Chionoecets bairdi* | 1–12 days | Various | ? | (9) |
| *Chionoecets opilio* | 7–9 days | Soft | 8 hrs | (10) |
| Cheiragonidae | | | | |
| *Telmessus cheiragonus* | 11.8 ± 5 SD days | Soft | 4.0 ± 6.6 hrs | (11) |
| Corystidae | | | | |
| *Corystes cassivelaunus* | Up to several days | Hard | 0 | (12) |

*References:* (1) Wada et al. (1997), (2)–(3) Wada et al. (2000), (4) Goshima et al. (1995), (5) Ed▼ (1966), (6) Elner & Elner (1980), (7)–(8) Berrill & Arsenault (1982), (9) Donaldson & Adams (1 (10) Watson (1972), (11) Kamio et al. (2003), (12) Hartnoll (1968).

**Table 6.** Brachyuran crab species, in which a sperm plug has been reported.

| | |
|---|---|
| **Cancridae** | |
|     *Cancer magister* | (1) |
|     *Cancer irroratus* | (2) |
|     *Cancer pagurus* | (3) |
| **Geryonidae** | |
|     *Geryon fenneri* | (4) |
| **Portunidae** | |
|     *Callinectes sapidus* | (5) |
|     *Carcinoplax vestita* | (6) |
|     *Carcinus maenas* | (7) |
|     *Macropipus holsatus* | (8) |
|     *Ovalipes ocellsatus* | (9) |
|     *Portunus sanguinolentus* | (10) |
|     *Necora puber* | (11) |
|     *Liocarcinus depurator* | (12) |
| **Cheiragonidae** | |
|     *Telmessus cheiragonus* | (13) |
| **Eriphiidae** | |
|     *Eriphia smithii* | (14) |

*References:* (1) Oh & Hankin (2004), (2) Childchester (1911), (3) Edwards (1966), (4) Hinsch (1988), (5) Childchester (1911), Wenner (1989), Johnson & Oito (1981), Jivoff (1997), (6) Doi & Watanabe (2006), (7) Broekhuysen (1936, 1937), Spalding (1942), (8) Broekhuysen (1936), (9) Childchester (1911), (10) George (1963), (11) González-Gurriarán & Freire (1994), Norman & Jones (1993), (12) Abelló (1989), (13) Kamio et al. (2003), (14) Tomikawa &Watanabe (1990).

**Table 7.** Species found in large aggregations called a "pod," "heap," or "mound."

| Species | Number of crabs in each aggregation | Reference |
|---|---|---|
| **ANOMURA** | | |
|   **Lithodidae** | | |
|     *Paralithodes camtschaticus* | 1000 or more | (1) |
|     *Lithodes santolla* | 70 ind·m-2 or more | (2) |
| **BRACHYURA** | | |
|   **Majidae** | | |
|     *Maja squinado* | 22-50,000 or more | (3) |
|     *Chionoecetes bairdi* | 100,000s | (4) |
|     *Hyas lyratus* | 2,000 | (5) |
|     *Loxorhynchus grandis* | 100s | (6) |
|     *Libinia emarginata* | 5,000? | (7) |

*References:* (1) Dew (1990), Dew et al. (1992), Powell & Nickelson (1965a, b), Powell et al. (1973), Zhou & Shirley (1997), Stone et al. (1993), (2) Cardenas et al. (2007), (3) Baal (1953), Le Sueur (1954), Carlisle (1957), Sampedro & González-Gurriarán (2004), (4) Stevens (2003), Stevens et al. (1994), (5) Stevens et al. (1992), (6) Debelius (1999), Hobday & Rumsey (1999), (7) DeGoursey & Auster (1992), Hinsch (1968).

**Table 8.** Species of the Pontoniinae reported as "found in pair." Species of shrimps with [host animals in brackets] are listed according to the phyla of the host animals (large capitals).

### PORIFERA
*Apopontonia dubia* [*Spongia* sp.](1), *Onycocaris amakusensis* [*Callyspongia elegans*](2), *Onycocaris oligodentata* [purplish sponge](3), *Onycocaris spinosa* [small sponge](4), *Onycocaridella prima* (5)[*Mycale sulcata*], *Onycocaridella monodoa* (= *Onycocaris monodoa*) [*Pavaesperella hidentata*](6), *Onycocaridites anornodactylus* [sponge] (7), *Orthopontonia ornatus* [*Jaspis stellifera*](8), *Periclimenaeus stylirostris* [sponge](9), *Typton dentatus* [*Reniera* sp.](10)

### CNIDARIA
Antipatharia
*Dasycaris zanzibarica* [black coral, sea whips](11)
Actiniaria
*Periclimenes brevicarpalis* [*Cryptodendron adhaesivum*](12), *Periclimenes colemani* [*Asthenosoma intermedium*](13), *Periclimenes ornatus* [*Entacmaea quadricolor, Heteroactis malu, Parasicyonis actinostroides*](14)
Scleractinia
*Anapontonia denticauda* [*Galaxea fascicularis*](15), *Coralliocaris superba* [*Acropora tubicinaria* and other 15 spp. of *Acropora*](16), *Jocaste lucina* [*Acropora tubicinaria*](17), *Jocaste japonica* [*Acropora* sp., *Acropora humilis, Acropora variabilis, Acropora tubicinaria, Acropora nasuta*](18), *Ischnopontonia lophos* [*Galaxea fascicularis*](19), *Periclimenes lutescens* (20), *Periclimenes koroensis* [*Fungia actiniformis*](21), *Philarius imperialis* [*Acropora* sp., *Acropora millepora*](22), *Vir euphyllius* [*Euphyllia* spp.](23), *Vir philippinensis* [*Plerogyra sinuosa*](24)
Scleractinia [in network of fissures on surface of faviid coral]
*Ctenopontonia cyphastreophila* [*Cyphastrea microphthalma*](25)
Scleractinia [forming galls or bilocular cyst in corals]
*Paratypton siebenrocki* [*Acropora hyacinthus* and other 6 spp. of *Acropora*](26)

### MOLLUSCA
Opistobranchia
*Periclimenes imperator* [*Hexabranchus marginatus*](27)
Bivalvia
*Anchistus demani* [*Tridacna maxima*](28), *Anchistus miersi* [*Tridacna squamosa, Tridacna maxima*](29), *Anchistus pectinis* [*Pecten* sp., *Pecten albicans*], *Anchistus custos* [*Pinna saccata, Pinna* sp.](31), *Chernocaris plaunae* [*Placuna placenta*](32), *Conchodytes biunguiculatus* [*Pinna bicolor*](33), *Conchodytes meleagrinea* [*Meleagrina margaritifera*](34), *Conchodytes monodactylus* [*Pecten* sp., *Atrina* sp.](35), *Conchodytes nipponensis* [*Pinna* sp., *Pecten laquetus, Atrina japonica*](36), *Conchodytes tridacnae* [*Tridacna maxima*](37), *Bruceonia ardeae* (= *Pontonia ardeae*)[*Chama pacifica*](38), *Pontonia domestica* [*Atrina seminuda, Atrina rigida, Pinna muricata*](39), *Pontonia mexicana* [*Pinna cornea, Pinna rigida, Atrina seminuda*](40), *Ascidonia miserabilis* (= *Pontonia miserabilis*)[*Spondylus americanus*](41), ?*Ascidonia miserabilis* (as ?*Pontonia miserabilis*)[*Spondylus americanus*](42), *Pontonia pinnae* [*Pinna rugosa, Atrina tuberculosa*](43), *Pontonia pinnophylax* [*Pinna rudis, Pinna nobilis*](44), *Pontonia margarita* [*Pinctada mazatlanica*](45), *Platypontonia hyotis* [*Pycnodonta hyotis*](46)

**Table 8.** continued.

**ECHINODERMATA**
    Crinoidea: Comatulida
*Palaemonella pottsi* [*Comanthina schlegelii, Comanthus briareus, Stephanometra briareus*](47), *Parapontonia nudirostris* [*Tropiometra afra, Himerometra robustipinna*] (48), *Periclimenes alegrias* [*Lamprometra palmata, Lamprometra klunzingeri, Stephanometra spicata*](49), *Periclimenes attenuatus* [*Comaster multifidus*](50), *Periclimenes novaecaledoninae* [*Lamprometra klunzingeri*](51)
    Echinoidea
*Tuleariocaris holthuisi* [*Astropyge radiata*](52), *Tuleariocaris zanzibarica* [*Astropyge radiata, Diadema setosum*](53)

**CHORDATA**
    Ascidiacea: compound ascidian
*Periclimenaeus diplosomatis* [*Diplosoma ?rayneri*](54), *Periclimenaeus serrula* [*Leptoclinoides incertus*](55), *Periclimenaeus tridentatus* [unidentified ascidian](56), *Ascidonia flavomaculata* (= *Pontonia flavomaculata*)[*Ascidia mentula, Ascidia mammillata, Ascidia involuta, Ascidia interrupta*](57), *Odontonia sibogae* (= *Pontonia sibogae*)[*Styela whiteleggei, Pyura momus, Rhopalaea crassa*](58)
    Ascidiacea: solitary ascidian
    *Dasella ansoni* [*Phallusia depressiuscula*](59)

*References:* (1) Bruce (1983a), (2)–(4) Fujino & Miyake (1969), (5)–(6) Bruce (1981a), (7) Bruce (1987), (8) Bruce (1982), (9) Bruce & Coombes (1995), (10) Bruce & Coombes (1995), Bruce (1980a), (11) Gosliner et al. (1996), (12) Bruce & Svoboda (1983), (13) Bruce (1975), (14) Bruce & Svoboda (1983), Omori et al. (1994), (15) Bruce (1967), (16)–(17) Bruce (1980b), (18) Bruce (1974, 1980b, 1981c), (19) Bruce (1980b, 1981c), Bruce & Coombes (1995), (20) Bruce (1981c), Bruce & Coombes (1995), (21) Bruce & Svobboda (1984), (22) Bruce & Coombes (1995), (23) Martin (2007), (24) Bruce & Svoboda (1984), (25) Bruce (1979), (26) Bruce (1980a, b), (27) Bruce (1972a, 1976a), Bruce & Svoboda (1983), Strack (1993), (28) Bruce (1972a), (29) Bruce (1972a), Debelius (1999), (30) Bruce (1972a), Fujino & Miyake (1967), (31) Bruce (1972a, 1989), Hipeau-Jacquotte (1973), (32) Bruce (1972a), (33) Bruce (1972a), Hipeau-Jacquotte (1973), (34) Bruce (1973), (35)–(36) Bruce (1972a), (37) Bruce (1974), (38) Bruce (1981b), Fransen (2002), (39) Bruce (1972a), Courtney & Couch (1981), Fransen (2002), (40) Bruce (1972a), Criales (1984), Fransen (2002), (41) Fransen (2002), (42) Criales (1984), (43) Bruce (1972a), (44) Debelius (1999), Richardson et al. (1997), (45) Baeza (2008), (46) Hipeau-Jacquotte (1971), (47) Bruce & Coombes (1995), Bruce (1989), (48) Bruce (1992), (49) Bruce (1986), Bruce & Coombes (1995), (50) Bruce (1992), (51) Bruce & Coombes (1995), (52)–(53) Bruce (1967), (54) Bruce (1980b), (55) Bruce & Coombes (1995), (56) Bruce & Coombes (1995), (57) Monniot (1965), Millar (1971), Fransen (2002), (58) Bruce (1972b), Fransen (2002), (59) Bruce & Coombes (1995).

**Table 9.** Species of the Alpheidae reported as "found in pair." Species of shrimps with [host animals in brackets] are listed according to the phyla of host animals (large captals) with higher taxa or habitat when known.

## PORIFERA
*Synalpheus bituberculatus* [sponge](1), *Synalpheus hastilicrassus* [sponge](2), *Synalpheus jedanensis* [sponge](3), *Synalpheus streptodactylus* [sponge](4), *Synalpheus theano* [sponge](5), *Synalpheus fossor* [sponge](6), *Synalpheus harpagatrus* [sponge](7), *Synalpheus nilandensis* [sponge](8), *Synalpheus tumidomanus* [sponge](9), *Zuzalpheus androsi* [*Hyattella intestinalis*](10), *Synalpheus couitere* [sponge](11), *Zuzalpheus bousfield* [*Hymeniacidon* spp.](12), *Zuzalpheus carpenteri* [*Aeglas* spp.](13), *Zuzalpheus goodei* [*Xestospongia wiedenmayeri, Pachypellina podatypa*](14), *Zuzalpheus paraneptunus* [*Hyattella intestinalis, Oceanapia* sp.](15), *Zuzalpheus ruetzleri* [*Hymeniacidon* cf. *caerulea*](16), *Zuzalpheus sanctithomae* [*Hymeniacidon caerulea* etc.](17), *Alpheus parvirostris* [sponge](18), *Alpheus alcyone* [sponge](19), *Alpheus* aff. *eulimene**[sponge](20), *Alpheus paralcyone* [sponge](21), *Alpheus spongiarum* [sponge] (22)

## CNIDARIA
Scyphozoa: Coronatae
*Synalpheus modestus* (23), *Synalpheus* aff. *modestus sensu* Nomura & Asakura (1998) [*Stephanoscyphus racemosus*](24)
Anthozoa: Gorgonacea
*Synalpheus iphinoe* [*Solenocaulon* sp.](25), *Synalpheus trispinosus* [gorgonacean](26)
Anthozoa: Alcyonacea
*Synalpheus neomeris* [*Dendronephthya*](27)
Anthozoa: Actiniaria
*Alpheus armatus* [*Bartholomea annulata*](28), *Alpheus immaculatus* [*Bartholomea annulata*](29), *Alpheus polystuctus* [*Bartholomea annulata*](30), *Alpheus roquensis* [*Heteractis lucida*](31)
Anthozoa: Scleractinia
*Alpheus lottini* [reef coral, *Pocillopora*](32), *Alpheus ventrosus* (33), *Synalpheus charon* [*Pocillopora*, reef coral](34), *Synalpheus scaphoceris* [*Madracis decactis*](35), *Racilius compressus* [*Galaxea fascicularis*](36)
Anthozoa: Scleractinia (in fissures on massive coral)
*Alpheus deuteropus* [*Asteropora, Porites, Acropora, Montipora, Pavona*](37)
Anthozoa: Scleractinia (coral borer, in dead coral head)
*Alpheus saxidomus* (38), *Alpheus simus* (39), *Alpheus schmitti* (40), *Alpheus idiocheles* (41), *Alpheus colluminaus* (42)

## ANNELIDA
Polychaeta
*Alpheus sulcatus* [*Eurythoe complanata*](43)

## CRUSTACEA
Shell used by hermit crab
*Aretopsis amabilis* [*Dardanus sanguinolentus, Dardanus megistos, Dardanus guttatus, Dardanus lagopodes, Clibanarius eurysternus, Calcinus latens*](44), *Aretopsis manazuruensis* [*Aniculus miyakei*](45)
In burrow of thalassinidean shrimps
*Betaeus longidactylus* [*Upogebia pugettensis*](46), *Betaeus harrimani* [*Upogebia pugettensis*](47), *Betaeus ensenadensis* [*Upogebia pugettensis*] (48)
In burrow of mantis shrimp
*Athanas squillophilus* [*Oratosquilla oratoria*](49)

**Table 9.** continued.

## ECHINODERMATA
Crinoidea: Comatulida

*Synalpheus carinatus* [crinoids](50), *Synalpheus comatularum* [*Comanthus timorensis*](51), *Synalpheus demani* [criniod](52), *Synalpheus stimpsoni* [*Comaster multibrachiatus, Comaster multifidus, Comaster gracilis, Comaster alternans*](53), *Synalpheus odontophorus* [crinoid](54)
Echinoidea
*Athanas indicus* [*Echinometra mathaei*](55)

## ECHIURA
*Athanopsis rubricinctuta* [*Ochetostoma erythrogrammon*](56), *Betaeus longidactylus* [*Urechis* sp.](57)

## "PISCES" [in burrow of goby fish]
*Alpheus bellulus* [*Tomiyamichthys* spp, *Amblyeleotris* spp.](58), *Alpheus purpurilenticularis* [*Amblyeleotris steinitzi*], (59) *Alpheus rapacida* [*Myersina* spp., *Vanderhorstia* spp., *Mahidoria* spp.], (60) *Alpheus rapax* [*Cryptocentrus* spp.](61)

## ALGAE TUBE
*Alpheus frontalis* [tube of filamentous blue-green algae such as *Microcoelus* spp.](62), *Alpheus bucephalus* [tube of pure algae or algae with sponges and other material](63), *Alpheus brevipes* [tube of red filamentous alga](64), *Alpheus clypeatus* [tube of red filamentous alga *Acrochaetium*](65), *Alpheus pachychirus* [tube of algae](66)

## FREE LIVING [crack of rock, under rubble, around large algae, burrow in mudflat]
*Alpheopsis chilensis* (67), *Alpheus normanni* (68), *Alpheus euphrosyne richardsoni* (69), *Alpheus strenuus cremnus* (70), *Alpheus diadema* (71), *Alpheus architectus* (72), *Alpheus amirantei* (73), *Alpheus bisincisus* (74), *Alpheus brevicristatus* (75) (might be commensal with goby?), *Alpheus edwardsii* (76), *Alpheus* aff. *gracilipes*\* (77), *Alpheus heeia* (78), *Alpheus* aff. *heeia*\*(79), *Alpheus* aff. *leviusculus* sp. 1\*(80), *Alpheus* aff. *leviusculus* sp. 2\*(81), *Alpheus lobidens* (82), *Alpheus* aff. *lobidens* sp. 1\*(83), *Alpheus* aff. *lobidens* sp. 2\*(84), *Alpheus* aff. *lobidens* sp. 3\*(85), *Alpheus malleodigitus* (86), *Alpheus miersi* (87), *Alpheus obesomanus* (88), *Alpheus pacificus* (89), *Alpheus* aff. *pacificus* (90), *Alpheus paradentipes* (91), *Alpheus parvirostris* (92), *Alpheus polyxo* (93), *Alpheus serenei* (94), *Alpheus suluensis* (95), *Alpheus tenuipes* (96), *Alpheus angulatus* (97), *Alpheus armillatus* (98), *Alpheus heterochaelis* (99), *Alpheus floridanus* (100), *Alpheus inca* (101), *Metalpheus paragracilis* (102)

**Table 9.** continued.

*sensu* Nomura & Asakura (1998). References: (1) Banner & Banner (1975), Nomura & Asakura (1998), (2)–(5) Nomura & Asakura (1998), (6) Didderen et al. (2006), (7) Banner & Banner (1975), (8)–(9) Nomura & Asakura (1998), (10) Rios & Duffy (2007), (11) Nomura & Asakura (1998), (12) Rios & Duffy (2007), (13) Macdonald III et al. (2006), Rios & Duffy (2007), (14)–(17) Rios & Duffy (2007), (18) Banner & Banner (1982), (19)–(27) Nomura & Asakura (1998), (28) Knowlton (1980), Knowlton & Keller (1982, 1983, 1985), Criales (1984), (29)–(31) Knowlton (1980), Knowlton & Keller (1982, 1983, 1985), (32) Vannini (1985), Nomura & Asakura (1998), Abele & Patton (1976), Tsuchiya & Yonaha (1992), (33) Patton (1966), (34) Patton (1966), Nomura & Asakura (1998), (35) Dardeau (1984, 1986), (36) Bruce (1972c), (37) Banner & Banner (1983), (38) Fischer & Meyer (1985), Fischer (1980), (39)–(40) Werding (1990), (41) Kropp (1987), Nomura & Asakura (1998), (42) Banner & Banner (1982), Nomura & Asakura (1998), (43) Banner & Banner (1982), (44) Bruce (1969), Banner & Banner (1973), Kamezaki & Kamezaki (1986), (45) Suzuki (1971), (46)–(48) MacGinitie (1937), (49) Hayashi (2002), (50) Bruce (1989), (51) Banner & Banner (1975), (52) Bruce (1989), Nomura & Asakura (1998), (53) Nomura & Asakura (1998), Van den Spiegel et al. (1998), (54) Nomura & Asakura (1998), (55) Gherardi (1991), (56) Anker et al. (2005), Berggren (1991), (57) MacGinitie (1935), (58) Miya & Miyake (1969), Nomura & Asakura (1998), Nomura (2003), (59) Macnae & Kalk (1962), Karplus (1979), Nomura (2003), (61) Macnae & Kalk (1962), Nomura (2003), (62) Fishelson (1966), Banner & Banner (1982), (63) Banner & Banner (1982), Nomura & Asakura (1998), (64)–(65) Banner & Banner (1982), (66) Cowles (1913), Banner & Banner (1982), (67) Boltana & Thiel (2001), (68) Nolan & Salmon (1970), (69)–(70) Banner & Banner (1982), (71)–(75) Nomura & Asakura (1998), (76) Nomura & Asakura (1998), Jeng (1994), (77)–(96) Nomura & Asakura (1998), (97) Mathews (2002a, b, 2003, 2006, 2007), Mathews et al. (2002), (98) Mathews et al. (2002), (99) Nolan & Salmon (1970), Schein (1975), Obermeier & Schmitz (2003a, b), Rahman et al. (2001, 2002, 2003, 2005), Schmitz & Herberholz (1998), Dworschak & Ott (1993), (100) Dworschak & Ott (1993), (101) Boltana & Thiel (2001), (102) Nomura & Asakura (1998).

Table 10. Species of shrimps other than Pontoniinae and Alpheidae reported as "found in pair." Species of shrimps with [host animals in brackets] are listed according to the phyla of host animals (large capitals) with higher taxa or habitat when known.

**SPONGICOLIDAE**
 PORIFERA
*Spongicola japonica* [*Euplectella oweni*](1), *Spongicola venusta* [*Euplectella aspergillum*](2), *Spongicola levigata* [*Euplectella oweni*?](3), *Spongiocaris semiteres* [hexactinellid sponge], (4) *Spongicoloides iheyaensis* [Euplectellidae & Hyalonematidae](5), *Globospongicola spinulatus* [hexactinellid sponge *Semperella* sp.](6)
 FREE LIVING
*Microprosthema validum* (7)

**STENOPODIDAE**
 FREE LIVING
*Stenopus hispidus* (8), *Stenopus scutellatus* (9), *Stenopus tenuirostris* (10), *Stenopus zanzibaricus* (11)

**HIPPOLIYTIDAE**
 FREE LIVING
*Lysmata debelius* (12), *Lysmata grabhami* (13)
 CNIDARIA
  Actiniaria, Scleractinia
*Thor amboinensis* (14)

**GNATHOPHYLLIDAE**
 ECHINODERMATA
  Holothuroidea
*Pycnocaris chagoae* [*Holothuria cinerascens*](15)
  Asteroidea
*Hymenocera picta* [prey on sea star](16)

*References:* (1) Saito et al. (2001), (2) Miyake (1982), Hayashi & Ogawa (1987), (3) Hayashi & Ogawa (1987), (4) Bruce & Baba (1973), (5) Saito et al. (2006), (6) Komai & Saito (2006), (7) Davie (2002), (8) Johnson (1969, 1977), Castro & Jory (1983), Zhang et al. (1998), Yaldwyn (1964, 1966a), (9) Debelius (1999), (10) Bruce (1976b), (11) Gosliner et al. (1996), (12) Rufino & Jones (2001), Gosliner et al. (1996), (13) Wirtz (1997), Debelius (1999), (14) Stanton (1977), (15) Bruce (1983b), (16) Seibt & Wickler (1972, 1979, 1981), Wickler & Seibt (1970, 1972, 1981), Seibt (1973a, b, 1974, 1980), Wasserthal & Seibt (1976), Wickler (1973), Kraul & Nelson (1986), Fiedler (2002).

**Table 11.** Species of Thalassinidea and Anomura reported as "found in pair." Species with [host animals or habitat in brackets] are listed according to the phyla of host animals (in capitals) with higher taxa or habitat where known.

**THALASSINIDEA**
**Axiidae**
FREE LIVING
*Axiopsis serratifrons* [in burrow in sediments with a higher content of coral rubble](1)
**Laomediidae**
FREE LIVING
*Axianassa australis* [in burrow in mud flat](2)
**Callianassidae**
"PISCES"
*Neotrypaea affinis* [burrow of blind goby *Typhlogobius californiensis*](3)
   FREE LIVING
*Neotrypaea gigas* [burrow in mud](4)
**Upogebiidae**
PORIFERA
*Upogebia synagelas* [*Agelas sceptrum*](5)
   CNIDARIA: Scleractinia
*Pomatogebia rugosa* [inside live colony of *Porites lobata*](6), *Pomatogebia operculata* [inside live coral colony](7), *Upogebia corallifora* [inside dead coral colony](8)
   FREE LIVING
*Upogebia pugettensis* [U- or Y-shaped burrow in mudflat](9), *Upogebia affinis* [burrow in mud](10)

**ANOMURA**
**Porcellanidae**
   CNIDARIA
      Gorgonacea
*Aliaporcellana telestophila* [*Solenocaulon*](11)
      Pennatulacea
*Porcellanella haigae* [*Cavernularia* sp.](12)
   Actiniaria
*Neopetrolisthes oshimai* [*Soichactis* spp.](13), *Neopetrolisthes maculatus* [*Stychodactyla*](14), *Neopetrolisthes alobatus, Neopetrolisthes spinatus* [*Heteroactis malu*](15)
   ANNELIDA
      Polychaeta [in tube of large polychaete species]
*Polyonyx macroheles* [*Chaetopterus variopedatus*](16), *Polyonyx quadriungulatus* [*Chaetopterus variopedatus*](17), *Polyonyx transversus* [*Chaetopterus* sp.](18), *Polyonyx vermicola* [*Sasekumaria selangora*](19), *Polyonyx bella* [*Chaetopterus variopedatus*](20), *Polyonyx gibbesi* [*Chaetopterus variopedatus*](21), *Polyonyx utinomii* [*Chaetopterus* sp.](22), *Heteropolyonyx biforma* [*Chaetopterus* sp.](23), *Polyonyx biunguiculatus* [*Chaetopterus* sp.](24)
   CRUSTACEA [in shell being used by hermit crab]
*Porcellana cancrisocialis* [*Petrochirus californiensis, Dardanus sinistripes, Aniculus elegans, Paguristes digueti*](25), *Porcellana paguriconviva* [*Petrochirus californiensis, Dardanus sinistripes, Aniculus elegans, Paguristes digueti*](26)
   ECHINODERMATA
      Echinoidea
*Clastotoechus vanderhorsti* [*Echinometra lucunter*](27), *Clastotoechus vanderhorsti* [*Echinometra lucunter*](28)
      Asteroida
*Minyocerus angustus* [*Luidia, Astropecten, Tethyaster*](29)

**Table 11.** continued.

---

FREE LIVING
*Pachycheles rudis* [underside of stone, basal portion of large algae](30)
**Galatheidae**
ECHINODERMATA
  Crinoidea
*Galathea inflata* [*Comanthus parvicirrus, Comaster schlehelii*](31)

---

*References:* (1) Dworschak & Ott (1993), (2) Coelho & Rodrigues (1999), Coelho (2001), (3)–(4) Meinkoth (1981), (5) Williams (1987), (6) Fonseca & Cortés (1998), (7) Kleeman (1984), Williams & Ngoc-Ho (1990), Coelho & Rodrigues (1999), Coelho (2001), (8) Williams & Scott (1989), (9) Jensen (1995), (10) Meinkoth (1981), (11) Ng & Goh (1996), (12) Nakasone & Miyake (1972), (13) Seibt & Wickler (1971), (14) Debelius (1984), (15) Osawa & Fujita (2001), (16) Gray (1961), (17) Kudenov & Haig (1974), (18) McNeill & Ward (1930), (19) Ng & Sasekumar (1993), (20) Hsueh & Huang (1998), (21) Rickner (1975), Williams (1984), Grove & Woodin (1996), (22)–(23) Osawa (2001), (24) Macnae & Kalk (1962), (25) Glassell (1936), Parente & Hendrickx (2000), Williams & McDermott (2004), (26) Parente & Hendrickx (2000), Williams & McDermott (2004), (27) Werding (1983), (28) Werding (1983), Schoppe (1991), (29) Werding (1983), Gore & Shoup (1968), (30) Meinkoth (1981), (31) Fujita & Baba (1999).


**Table 12.** Species of brachyuran crabs reported as "found in pair." Species of crabs with [host animals in brackets] are listed within family or superfamily according to the phyla of host animals (in capitals) with higher taxa or habitat where known.

---

**XANTHIDAE**
  CNIDARIA: Scleractinia
*Cymo andreossyi* [*Pocillopora*](1)

**TRAPEZIIDAE**
  CNIDARIA
    Scleractinia: *Pocillopora*
*Trapezia areolata* (2), *Trapezia corallina* (3), *Trapezia cymodoce* (4), *Trapezia dentata* (5),
  *Trapezia digitalis* (6), *Trapezia ferruginea* (7), *Trapezia flavomaculata* (8), *Trapezia guttata* (10),
  *Trapezia intermedia* (11), *Trapezia rufopunctata* (12), *Trapezia tigrina* (13), *Trapezia wardi* (14)
    Antipatharia
*Quadrella maculosa* [*Antipathes*] (15), *Quadrella* spp. [*Cirrhipathes abies, Antipathes* spp.](16),
  *Quadrella reticulata* [*Antipathes* sp.](17)

**TETRALIIDAE**
  CNIDARIA
    Scleractinia: *Acropora*
*Tetralia fulva* (18), *Tetralia nigrolineata* (19), *Tetralia rubridactyla* (20)

**CARPILIIDAE**
FREE LIVING
*Carpilius corallinus* (21)

**Table 12.** continued.

**PINNOTHERIDAE**
  ANNELIDA
    Polychaeta [in tube of large polychaetes]
*Pinnixa tubicola* [terebellids and chaetopterids, *Eupolymnia heterobranchia*, *Amphitrite* sp., *Eupolymnia heterobranchia*, *Neoamphitrite rohusta*, *Thelepus crispus*, *Chaetopterus variopedatus*](22), *Pinnixa chaetopterana* [*Chaetopterida* spp. *Chaetopterus variopedatus*, *Amphitrite ornata*](23), *Pinnixa transversalis* [*Chaetopterus variopedatus*](24)
  MOLLUSCA
    Bivalvia
*Pinnixa faba* [*Tresus capax*, *Tresus nuttalli*](25), *Pinnixa littoralis* [*Tresus capax*](26)
    Gastropoda [inside mantle cavity]
*Orthotheres turboe* [*Turbo* sp.](27), *Orthotheres haliotidis* [*Haliotis asinina*, *Haliotis squamata*](28)
  SIPUNCULA & ECHIURA
*Mortensenella forceps* [*Ochetostoma erythrogrammon*](29)
  ECHINODERMATA
    Echinoidea
*Dissodactylus mellitae* [*Mellita quinguiesperforata*, *Echinarachnius parma*, *Encope michelini*](30), *Dissodactylus crinitichelis* [*Mellita sexiesperforata*](31)
    Holothuroidea
*Holotheres halingi* (= *Pinnotheres halingi*) [*Holothuria scarba*](32), *Holotheres semperi* (= *Pinnotheres semperi*)[*Holothuria fursocinerea*, *Holothuria scabra*](33)
  BURROWS OF OTHER ANIMALS
*Scleroplax granulata* [burrow of echiuroid *Urechis caupo*, mud shrimps *Neotrypaea californiensis*, *Neotrypaea gigas*, *Upogebia pugettensis*, *Upogebia macginiteorum*](34)

**GRAPSOIDEA**
  "REPTILIA": Testudines
*Planes minutus* [loggerhead sea turtle *Caretta caretta*, inanimate flotsam](35)
  ECHINODERMATA
    Echinoidea
*Percnon gibbesi* [*Diadema antillarum*](36)

*References:* (1) Castro (1976), Guinot (1978), Miyake (1983), (2) Miyake (1983), Tsuchiya & Yonaha (1992), Tsuchiya & Taira (1999), (3) Patton (1966), Miyake (1983), Huber (1985), Gotelli et al. (1985), Castro (1996), (4) Patton (1966), Tsuchiya & Yonaha (1992), Tsuchiya & Taira (1999), (5) Patton (1966), Huber (1985), (6) Patton (1966), Preston (1973), Huber (1985, 1987), Huber & Coles (1986), Tsuchiya & Taira (1999), (7) Patton (1966), Preston (1973), Abele & Patton (1976), Finney & Abele (1981), Miyake (1983), Adams et al. (1985), Huber & Coles (1986), Castro (1978, 1996), Tsuchiya & Taira (1999), (8) Patton (1966), Preston (1973), Miyake (1983), (9) Gotelli et al. (1985), Castro (1996), (10) Miyake (1983), Tsuchiya & Yonaha (1992), Tsuchiya & Taira (1999), (11) Preston (1973), Huber & Coles (1986), Huber (1987), (12)–(13) Huber (1985), (14) Preston (1973), Miyake (1983), Huber & Coles (1986), (15) Shih & Mok (1996), (16) Tazioli et al. (2007), (17) Castro (1999), (18) Vytopil & Willis (2001), (19)–(20) Sin (1999), (21) Laughlin (1982), (22) Hart (1982), Wells (1928), Garth & Abbott (1980), Zmarzly (1992), (23) Gray (1961), Grove & Woodin (1996), Grove et al. (2000), McDermott (2005), (24) Baeza (1999), (25) Pearce (1965, 1966a), Hart (1982), Zmarzly (1992), (26) Pearce (1966a), Zmarzly (1992), (27) Sakai (1969), (28) Geiger & Martin (1999), (29) Anker et al. (2005), (30) Bell & Stancyk (1983), Bell (1984), George & Boone (2003), (31) Telford (1978), (32) Hamel et al. (1999), (33) Ng & Manning (2003), (34) Anker et al. (2005), Campos (2006), (35) Dellinger et al. (1997), Frick et al. (2000, 2004, 2006), Carranza et al. (2003), (36) Hayes et al. (1998).

**Table 13.** Eusocial species. All species found inhabiting cavity of sponge.

**Alpheidae**
| | |
|---|---|
| *Zuzalpheus rathbunae* [sponge] | (1) |
| *Zuzalpheus elizabethae* (= *Synalpheus* "*rathbunae* A")[*Lissodendoryx*] | (2) |
| *Zuzalpheus* "*paraneptunus* small" [sponge] | (3) |
| *Zuzalpheus regalis* [*Xestospongia* etc.] | (4) |
| *Zuzalpheus filidigitus* [*Xestospongia* etc.] | (5) |
| *Zuzalpheus chacei* [*Aeglas, Hyattella* etc.] | (6) |
| *Zuzalpheus elizabethae* [*Lissodendoryx* etc.] | (7) |
| *Synalpheus neptunus neptunus* [sponge] | (8) |

*References:* (1) Duffy (2003), (2) Duffy (1996c, 2003), Morrison et al. (2004), (3) Duffy et al. (2000), Duffy (2003), (4) Duffy (1996a, b), Duffy et al. (2002), Rios & Duffy (2007), (5) Duffy (1996c), Duffy & Macdonald (1999), Rios & Duffy (2007), (6) Chace (1972), Duffy (1998), Rios & Duffy (2007),(7) Duffy (1996c), Morrison et al. (2004), Rios & Duffy (2007),(8) Didderen et al. (2006).

**Table 14.** Species found in small groups. Species with [host animals] are listed, according to the phyla of host animals (large capitals) with higher taxa or habitat. One group consists of fewer than 20 individuals on a single host (species, host, number of individuals found, and reference).

### CARIDEA
CNIDARIA
Scyphozoa

| | |
|---|---|
| *Periclimenes holthuisi* [*Cassiopei*] | Max. 8 (various sizes and sexes)(1) |
| Actiniaria | |
| *Periclimenes holthuisi* [sea anemone] | Several individuals (2) |
| *Periclimenes tenuipes* [*Megalactis, Cryptodendron*] | Max. 6 (various sizes and sexes)(3) |
| *Periclimenes longicarpus* [*Entacmaea*] | Max. 7 (various sizes and sexes)(4) |
| *Periclimenes anthophilus* [*Condylactis gigantea*] | Up to 9 (5) |
| Scleractinia | |
| *Thor marguitae* [*Porites andrewsi*] | 10 (2 ♂, 5 ov. ♀, 2 non-ov. ♀, 1 juv.)(6) |
| *Jocaste japonica* [*Acropora divaricata*] | 15 (5 ♂, 6 ov. ♀, 3 non-ov. ♀, 1 juv.)(7) |
| *Periclimenes holthuisi* [corals] | Several individuals (8) |
| *Periclimenes pederosoni* [*Antipathe*] | 7 (2 ♂, 3 ov. ♀, 2 non-ov. ♀)(9) |
| *Anapontonia denticauda* [*Galaxea*] | 5 (1 ♂, 1 ♀, 3 juv.)(10) |

ECHINODERMATA
Echinoidea

| | |
|---|---|
| *Gnathophylloides mineri* [*Tripneustes ventricosus*] | Up to 13, with females greatly outnumbering males (11) |

### GALATHEOIDEA
CNIDARIA
Scleractinia

| | |
|---|---|
| *Lissoporcellana spinuligera* [*Solenocaulon*] | 7 (1 ♂, 3 ov. ♀, 3 juv.)(12) |
| CRUSTACEA: shell used by hermit crab | |
| *Porcellana sayana* [*Dardanus, Petrochirus, Paguristes*] | Max. 11 (several ♂, several ov. ♀)(13) |

**Table 14.** continued.

**BRACHYURA**
   MOLLUSCA
      Bivalvia
*Pinnixa faba* [*Tresus*] More than 3 (1 ♂, 1 ♀, few juv.)(14)

*References:* (1) Bruce & Svoboda (1983), (2) Coleman (1991), (3)–(4) Bruce & Svoboda (1983), (5) Nizinski (1989), (6) Bruce (1978), (7) Bruce (1981b), (8) Coleman (1991), (9) Spotte (1996), (10) Bruce (1967), (11) Patton et al. (1985), (12) Ng & Goh (1996), (13) Gore (1970), (14) Haig & Abbott (1980).

**Table 15.** Species found in large groups. Species with [host animals] are listed, according to the phyla of host animals (large capitals) with higher taxa or habitat. One group consists of more than 20 individuals on a single host.

**CARIDEA**
   PORIFERA

| | |
|---|---|
| *Synalpheus dorae* [*Reiniere*] | 136 (all ♂)(1) |
| *Synalpheus streptodactylus* [sponge] | 105 (68 ♂, 37 ov. ♀, several non- ov. ♀)(2) |
| *Synalpheus crosnieri* [sponge] | 147 (144 ♂, 3 ♀)(3) |
| *Synalpheus paradoxus* [sponge] | 112 (110 ♂, 2 ♀), 132 (130 ♂, 2 ♀)(4) |
| *Zuzalpheus brooksi* [sponge] | 10s to 1000s (5) |
| *Zuzalpheus idios* [*Hymeniacidon* etc.] | Several 10s (including many ov. ♀ & juv.)(6) |
| *Zuzalpheus pectiniger* [*Spheciospongia*] | Few 100s (7) |

   CNIDARIA
      Scyphozoa

| | |
|---|---|
| *Latreutes anoplonyx* [*Nemopilema nomurai*] | More than 100 (8) |

      Scleractinia

| | |
|---|---|
| *Coralliocaris macrophthalma* [*Acropora hyacinthus*] | 24 (including 16 ♀)(9) |
| *Fennera chacei* [*Pocillopora*] | Max. 49 (all adults) (10) |
| *Periclimenes toloensis* [*Lytocarpus philippinensis*] | 110 (including 43 ov. ♀)(11) |

   ECHINODERMATA

| | |
|---|---|
| *Periclimenes affinis* [*Heterometra magnipinna*] | 64 (including 16 ov. ♀)(12) |
| *Periclimenes meyeri* [*Nemaster grandis*] | Max. 25 (various sizes and sexes)(13) |

*References:* (1) Bruce (1988), (2) Banner & Banner (1975, 1982), (3) Banner & Banner (1983), (4) Banner & Banner (1982), (5)–(7) Rios & Duffy (2007), (8) Hayashi et al. (2003), (9) Bruce (1977), (10) Gotelli et al. (1985), (11)–(12) Bruce & Coombes (1995), (13) Criales (1984).

**Table 16.** Selected species of pinnotherid crabs (and their hosts) in which life history has been studied.

| | |
|---|---:|
| MOLLUSCA | |
|   Bivalvia | |
| *Fabia subquadrata* [*Modiohis niodiolus*] | (1) |
| *Tumidotheres maculatus* (= *Pinnotheres maculatus*) [*Mytilus edulis, Argopecten irradians* etc.] | (2) |
| *Pinnotheres ostreum* [*Crassostrea virginica, Mytilus edulis*] | (3) |
| *Pinnotheres pisum* [*Mytilus edulis* etc.] | (4) |
| *Pinnotheres taichungae* [*Laternula marilina*] | (5) |
| *Pinnotheres bidentatus* [*Laternula marilina*] | (6) |
| ANNELIDA: Polychaeta | |
| *Tritodynamia horvathi* [in tube of *Loimia verrucosa*] | (7) |

*References:* (1) Pearce (1962, 1966b), (2) Pearce (1964), Williams (1984), (3) Christensen & McDermott (1958), (4) Atkins (1926), Christensen (1958), Hartnoll (1972), Williams (1984), (5) Hsueh (2003), (6) Hsueh (2001a, b), (7) Matsuo (1998, 1999), Takahashi et al. (1999).


**Table 17.** Species in which neotenous males have been reported.

| | |
|---|---:|
| ANOMURA | |
|   Hippidae | |
|     *Emerita brasiliensis* | (1) |
|     *Emerita asiatica* | (2) |
|     *Emerita emeritus* | (3) |
|     *Emerita holthuisi* | (4) |
|     *Emerita talpoida* | (5) |
|     *Emerita rathbunae* | (6) |

*References:* (1) Delgado & Defeo (2006, 2008), (2) Subramoniam (1981), (3)–(4) Subramoniam & Gunamalai (2003), (5)–(6) Efford (1967).

## APPENDIX 2:

## REFERENCES FOR TABLES OF APPENDIX 1

Abele, L.G. & Patton, W.K. 1976. The size of coral heads and the community biology of associated decapod crustaceans. *J. Biogeogr.* 3: 35–47.

Abele, L.G., Robinson, M.H. & Robinson, B. 1973. Observations on sound production by two species of crabs from Panama (Decapoda, Gecarcinidae and Pseudothelphusidae). *Crustaceana* 25: 147–152.

Abelló, P. 1989. Reproduction and moulting in *Liocarcinus depurator* (Linnaeus, 1758) (Brachyura: Portunidae) in the northwestern Mediterranean Sea. *Sci. Mar.* 53: 127–134.

Adams, J., Edwards, A.J. & Emberton, H. 1985. Sexual size dimorphism and assortative mating in the obligate coral commensal *Trapezia ferruginea* Latreille (Decapoda, Xanthidae). *Crustaceana* 48: 188–194.

Aiken, D.E. & Waddy, S.L. 1980. Reproductive biology. In: Cobb, J.C. & Phillips, B.F. (eds.), *The Biology and Management of Lobsters. Volume 1*: 215–276. New York: Academic Press.

Aiken, D.E., Waddy, S.L. & Mercer, S.M. 2004. Confirmation of external fertilization in the American lobster, *Homarus americanus*. *J. Crust. Biol.* 24: 474–480.

Ameyaw-Akumfi, C. 1981. Courtship in the crayfish *Procambarus clarkii* (Girad) (Decapoda, Astacidea). *Crustaceana* 40: 57–64.

Ameyaw-Akumfi, C. 1987. Mating in the lagoon crab *Cardisoma armatum* Herklots. *J. Crust. Biol.* 7: 433–436.

Anker, A., Murina, G.V., Lira, C., Caripe, J.A.V., Palmer, A.R. & Jeng, M.S. 2005. Macrofauna associated with echiuran burrows: a review with new observations on the innkeeper worm *Ochetostoma erythrogramm* on Leuckartana Riippelin, Venezuela. *Zool. Stud.* 44: 157–190.

Antheunisse, L.J., van den Hoven, N.P. & Jeffries, D.J. 1968. The breeding characters of *Palaemonetes varians* (Leach) (Decapoda, Palaemonidae). *Crustaceana* 14: 259–270.

Aquacop. 1977. Observations sur la maturation et la reproduction en captivité des crevettes pénéides en milieu tropical. *Third Meet. ICES Work. G. Maricult., Brest, France, Actes Colloq. CNEXO* 4: 157–178.

Arakawa, K.Y. 1964. On mating behavior of giant Japanese crab, *Macrocheira kaempferi* De Haan. *Res. Crust.* 1: 40–46.

Asakura, A. 1987. Population ecology of the sand-dwelling hermit crab, *Diogenes nitidimanus* Terao. 3. Mating system. *Bull. Mar. Sci.* 41: 226–233.

Atkins, D. 1926. The moulting stages of the pea crab (*Pinnotheres pisum*). *J. Mar. Biol. Ass. U.K.* 14: 475–493.

Baal, H.J. 1953. Behaviour of spider crabs in the presence of octopuses. *Nature* 171: 887.

Baeza, J.A. 1999. Indicadores de monogamia en el cangrejo comensal *Pinnixa transversalis* (Milne Edwards and Lucas) (Decapoda: Brachyura: Pinnotheridae): distribucion poblacional, asociacion. macho-hembra y dimorfismo sexual. *Anal. Mus. Hist. Nat. Valparaeo (Chile)* 34: 303–313.

Baeza, J.A. 2008. Social monogamy in the shrimp *Pontonia margarita*, a symbiont of *Pinctada mazatlanica*, off the Pacific coast of Panama. *Mar. Biol.* 153: 387–395.

Banner, A.H. & Banner, D.M. 1983. An annotated checklist of the alpheid shrimp from the Western Indian Ocean. *Trav. Doc. ORSTOM* 158: 1–164.

Banner, D.M. & Banner, A.H. 1973. The alpheid shrimp of Australia. Part I: the lower genera. *Rec. Aust. Mus.* 28: 291–382.

Banner, D.M. & Banner, A.H. 1975. The alpheid shrimp of Australia. Part II: the genus *Synalpheus*. *Rec. Aust. Mus.* 29: 267–389.

Banner, D.M. & Banner, A.H. 1982. The alpheid shrimp of Australia. Part III: the remaining alpheids, principally the genus *Alpheus*, and the family Ogyrididae. *Rec. Aust. Mus.* 341: 1–357.

Barki, A. & Karplus, I. 1999. Mating behavior and a behavioral assay for female receptivity in the red-claw crayfish *Cherax quadricarinatus*. *J. Crust. Biol.* 19: 493–497.

Bauer, R.T. 1976. Mating behaviour and spermatophore transfer in the shrimp *Heptacarpus pictus* (Stimpson) (Decapoda: Caridea: Hippolytidae). *J. Nat. Hist.* 10: 315–440.

Bauer, R.T. 1979. Sex attraction and recognition in the caridean shrimp *Heptacarpus paludicola* Holmes (Decapoda: Hippolytidae). *Mar. Behav. Physiol.* 6: 157–174.

Bauer, R.T. 1991. Sperm transfer and storage structures in penaeoid shrimps: a functional and phylogenetic perspective. In: Bauer, R.T. & Martin, J.W. (eds.), *Crustacean Sexual Biology*: 183–207. New York: Columbia Univ. Press.

Bauer, R.T. 1992. Repetitive copulation and variable success of insemination in the marine shrimp *Sicyonia dorsalis* (Decapoda: Penaeoidea). *J. Crust. Biol.* 12: 153–160.

Bauer, R.T. 1996. A test of hypotheses on male mating systems and female molting in decapod shrimp, using *Sicyonia dorsalis* (Decapoda: Penaeoidea). *J. Crust. Biol.* 16: 429–436.

Bauer, R.T. & Abdalla, J.A. 2001. Male mating tactics in the shrimp *Palaemonetes pugio* (Decapoda, Caridea): precopulatory mate guarding vs. pure searching. *Ethology* 107: 185–199.

Bauer, R.T. & Min, L.J. 1993. Spermatophores and plug substances of the marine shrimp *Trachypenaeus similis* (Crustacea: Decapoda: Penaeidae): formation in the male reproductive tract and disposition in the inseminated female. *Biol. Bull.* 185: 174–185.

Bechler, D.L. 1981. Copulatory and maternal-offspring behavior in the hypogean crayfish, *Orconectes inermis inermis* Cope and *Orconectes pellucidus* (Tellkampf) (Decapoda, Astacidea). *Crustaceana* 40: 136–143.

Beer, C.G. 1959. Notes on the behaviour of two estuarine crab species. *Trans. Royal Soc. New Zealand* 86: 197–203.

Bell, J.L. 1984. Changing residence: dynamics of the symbiotic relationship between *Dissodactylus mellitae* (Rathbun) (Pinnotheridae) and *Mellita quinquiesperforata* (Leske) (Echinodermata). *J. Exp. Mar. Biol. Ecol.* 82: 101–115.

Bell, J.L. & Stancyk, S.E. 1983. Population dynamics and reproduction of *Dissodactylus mellitae* (Brachyura: Pinnotheridae) on its sand dollar host *Mellita quinquiesperforata* (Echinodermata). *Mar. Ecol. Prog. Ser.* 13: 141–149.

Berg, A.B. & Sandifer, I.A. 1984. Mating behavior of the grass shrimp *Palaernonetes pugio*. *J. Crust. Biol.* 4: 417–424.

Berggren, M. 1991. *Athanopsis rubricinctuta*, new species (Decapoda: Natantia: Alpheidae), a shrimp associated with an echiuroid at Inhaca Island, Mozambique. *J. Crust. Biol.* 11: 166–178.

Berrill, M. 1982. The life cycle of the green crab *Carcinus maenas* at the northern end of its range. *J. Crust. Biol.* 2: 31–39.

Berrill, M. & Arsenault, M. 1982. Mating behaviour of the green shore crab *Carcinus maenas*. *Bull. Mar. Sci.* 32: 632–638.

Berry, P.F. 1970. Mating behavior, oviposition, and fertilization in the spiny lobster *Panulirus homarus* (Linnaeus). *S. Afr. Oceanogr. Res. Inst. Invest. Rep.* 24: 1–16.

Berry, P.F. & Hartnoll, R.G. 1970. Mating in captivity of the spider crab *Pleistacantha moseleyi* (Miers) (Decapoda, Majidae). *Crustaceana* 19: 214–215.

Bhimachar, B.S. 1965. Life history and behaviour of Indian prawns. *Fish. Technol., Ernakulam* 2: 1–11.

Binford, R. 1913. The germ cells and the process of fertilization in the crab, *Menippe mercenaria*. *J. Morphol.* 24: 147–202.

Bliss, D.E., van Montfrans, J., van Montfrans, M. & Boyer, J.R. 1978. Behavior and growth of the land crab *Gecarcinus lateralis* (Fréminville) in southern Florida. *Bull. Am. Mus. Nat. Hist.* 160: 111–152.

Bloch, D.P. 1935. Contribution à létude des gamètes et de la fécondation chez les Crustacés Décapodes. *Trav. Stn. Zool. Wimereux* 12: 185–270.

Bodekke, R., Bosschieter, J.R. & Goudswaard, P.C. 1991. Sex change, mating, and sperm transfer in *Crangon crangon* (L.). In: Bauer, R.T. & Martin, J.W. (eds.), *Crustacean Sexual Biology*: 164–182. New York: Columbia University Press.

Boltana, S. & Thiel, M. 2001. Associations between two species of snapping shrimp, *Alpheus inca* and *Alpheus chilensis* (Decapoda: Caridea: Alpheidae). *J. Mar. Biol. Ass. U.K.* 81: 633–638.

Boolootian, R.A., Giese, A.C., Farmanfarmaian, A. & Tucker, J. 1959. Reproductive cycles of five west coast crabs. *Physiol. Zool.* 32: 213–220.

Bourdon, R. 1962. Observations préliminaires sur la ponte des Xanthidae. *Bull. Soc. Lorraine Sci.* 2: 3–28.

Bovbjerg, R.V. 1953. Dominance order in the crayfish *Orconectes virilis* (Hagan). *Physiol. Zool.* 26: 173–178.

Bovbjerg, R.V. 1956. Some factors affecting aggressive behavior in crayfish. *Physiol. Zool.* 29: 127–136.

Bovbjerg, R.V. 1960. Courtship behavior of the lined shore crab, *Pachygrapsus crassipes* Randall. *Pac. Sci.* 14: 421–422.

Brewis, J. M. & Bowler, K. 1985. A study of reproductive females of the freshwater crayfish *Austropotamobius pallipes*. *Hydrobiol.* 121: 145–149.

Brisson, S. 1986. Observations on the courtship of *Penaeus brasiliensis*. *Aquacult.* 53: 75–78.

Brockerhoff, A.M. & McLay, C. 2005a. Mating behaviour, female receptivity and male–male competition in the intertidal crab *Hemigrapsus sexdentatus* (Brachyura: Grapsidae). *Mar. Ecol. Prog. Ser.* 290: 179–191.

Brockerhoff, A.M. & McLay, C. 2005b. Factors influencing the onset and duration of receptivity of female purple rock crabs, *Hemigrapsus sexdentatus* (Brachyura: Grapsidae). *J. Exp. Mar. Biol. Ecol.* 314: 123–135.

Brockerhoff, A. & McLay, C. 2005c. Comparative analysis of the mating strategies in grapsid crabs with special reference to the intertidal crabs *Cyclograpsus lavauxi* and *Helice crassa* (Decapoda: Grapsidae) from New Zealand. *J. Crust. Biol.* 23: 507–520.

Brockerhoff, A.M. 2002. *Comparative studies of the reproductive strategies of New Zealand grapsid crabs (Brachyura: Grapsidae) and the effects of parasites on reproductive success*. New Zealand: Ph.D. Thesis, Univ. Canterbury. (Cited from Brockerhoff, A.M. & McLay, C. 2005a.)

Broekhuysen, G.J. 1936. On development, growth and distribution of *Carcinides maenas*. *Arch. Neerl. Zool.* 2: 255–399.

Broekhuysen, G.J. 1937. Some notes on sex recognition in *Carcinides maenas* (L.). *Arch. Neerl. Zool.* 3: 156–164.

Broekhuysen, G.J. 1941. The life history of *Cyclograpsus punctatus*, M. Edw.: breeding and growth. *Trans. Royal Soc. S. Afr.* 28: 331–366.

Broekhuysen, G.J. 1955. The breeding and growth of *Hymenosoma orbiculare* Desm. (Crustacea, Brachyura). *Ann. S. Afr. Mas.* 41: 313–343.

Browdy, C.L. 1989. *Aspects of the reproductive biology of Penaeus semisulcatus*. Dr. Thesis. Tel Aviv Univ., Israel.

Bruce, A.J. 1967. Notes on some Indo-Pacific Pontoniinae, III-IX. Descriptions of some new genera and species from the western Indian Ocean and South China Sea. *Zool. Verhandl., Leiden* 87: 1–73.

Bruce, A.J. 1969. *Aretopsis amabilis* de Man, an alpheid shrimp commensal of pagurid crabs in the Seychelle Islands. *J. Mar. Biol. Ass. India* 11: 175–181.

Bruce, A.J. 1972a. Shrimps that live with molluscs. *Sea Frontiers* 18: 218–227.

Bruce, A.J. 1972b. An association between a pontoniinid shrimp and a rhizostomatous scyphozoan. *Crustaceana* 23: 300–302.

Bruce, A.J. 1972c. On the association of the shrimp *Racilius compressus* Paulson (Decapoda, Alpheidae) with the coral Galaxea clavus (Dana). *Crustaceana* 22: 92–93.

Bruce, A.J. 1973. The pontoniinid shrimps collected by the Yale-Seychelles expedition, 1957–1958 (Decapoda, Palaemoniidae). *Crustaceana* 24: 132–142.

Bruce, A.J. 1974. A report on a small collection of pontoniine shrimps from the Island of Farquhar (Decapoda, Palaemonidae). *Crustaceana* 27: 189–203.

Bruce, A.J. 1975. *Periclimenes colemani* sp. nov., a new shrimp associate of a rare sea urchin from Heron Island, Queensland. *Rec. Aust. Mus.* 29: 485–502.

Bruce, A.J. 1976a. A report on a small collection of shrimps from the Kenya National Marine Parks at Malindi, with notes on selected species. *Zool. Verhandl., Leiden* 145: 1–72.

Bruce, A.J. 1976b. Studies on Indo-West Pacific Stenopodidea, 1. *Stenopus zanzibaricus* sp. nov., a new species from East Africa. *Crustaceana* 31: 90–102.

Bruce, A.J. 1977. A report on a small collection of pontoniine shrimps from Queensland, Australia. *Crustaceana* 33: 167–181.

Bruce, A.J. 1978. *Thor marguitae* sp. nov., a new hippolytid shrimp from Heron Island, Australia. *Crustaceana* 35: 159–169.

Bruce, A.J. 1979. *Ctenopontonia cyphastreophila*, a new genus and species of coral associated pontoniine shrimp from Eniwetok Atoll. *Bull. Mar. Sci.* 29: 423–435.

Bruce, A.J. 1980a. Notes on some Indo-Pacific pontoniinae, XXXIV. Further observations on *Typton dentatus* Fujino & Miyake (Decapoda, Palaemonidae). *Crustaceana* 39: 113–120.

Bruce, A.J. 1980b. Pontoniine shrimps from the Great Astrolabe Reef, Fiji. *Pac. Sci.* 34: 389–400.

Bruce, A.J. 1981a. *Onycocaridella prima*, new genus, new species, a new pontoniine sponge-associate from the Capricorn Islands, Australia (Decapoda, Caridea, Pontoniinae). *J. Crust. Biol.* 1: 241–250.

Bruce, A.J. 1981b. Pontoniine shrimps from Viti Levu, Fijian Islands. *Micronesica* 17: 77–95.

Bruce, A.J. 1981c. Pontoniine shrimps of Heron Island. *Atoll Res. Bull.* 245: 1–33.

Bruce, A.J. 1982. Notes on some Indo-Pacific Pontontinae, XLI. *Orthopontonia*, a new genus proposed for *Periclimenaeus ornatus* Bruce. *Crustaceana* 43:163–176.

Bruce, A.J. 1983a. Further information on *Apopontonia dubia* Bruce (Decapoda Pontoniinae). *Crustaceana* 45: 210–213.

Bruce, A.J. 1983b. A further note on *Pycnocaris chagoae* Bruce (Decapoda, Gnathophyllidae). *Crustaceana* 45: 107–109.

Bruce, A.J. 1986. Three new species of commensal shrimps from Port Essington, Arnhem Land, Northern Australia (Crustacea: Decapoda: Palaemonidae). *Beagle* 3:143–166.

Bruce, A.J. 1987. *Onycocaridites anomodactylus*, new genus, new species (Decapoda: Palaemonidae), a commensal shrimp from the Arafura Sea. *J. Crust. Biol.* 7: 771–779.

Bruce, A.J. 1988. *Synalpheus dorae*, a new commensal alpheid shrimp from the Australian Northwest shelf. *Proc. Biol. Soc. Wash.* 101: 843–852.

Bruce, A.J. 1989. A report on some coral reef shrimps from the Philippine Islands. *Asian Mar. Biol.* 6: 173–192.

Bruce, A.J. 1992. Two new species of *Periclimenes* (Crustaces, Decapoda, Palaemonidae) from Lizard Island, Queensland, with notes on some related taxa. *Rec. Aust. Mus.* 44: 45–84.

Bruce, A.J. & Baba, K. 1973. *Spongiocaris*, a new genus of stenopodidean shrimp from New Zealand and South African waters, with a description of two new species (Decapoda, Natantia, Stenopodidea). *Crustaceana* 25: 153–170.

Bruce, A.J. & Coombes, K.E. 1995. The palaemonoid shrimp fauna (Crustacea: Decapoda: Caridea) of the Cobourg Peninsula, Northern Territory. *Beagle* 12: 101–144.

Bruce, A.J. & Svoboda, A. 1983. Observations upon some pontoniine shrimps from Aqaba, Jordan. *Zool. Verhandl., Leiden* 205: 1–44.

Bruce, A.J. & Svoboda, A. 1984. A report on a small collection of coelenterate-associated pontoniine shrimps from Cebu, Philippines Islands. *Asian Mar. Biol.* 1: 87–99.

Bueno, S.L.S. 1990. Maturation and spawning of the white shrimp *Penaeus schmitti* Burkenroad, 1936, under large scale rearing conditions. *J. World Aquacult. Soc.* 21: 170–179.

Bulter, T.N. 1960. Maturity and breeding of the Pacific edible crab *Cancer magister* Dana. *J. Fish. Res. Bd. Canada* 17: 641–646.

Burkenroad, M.D. 1947. Reproductive activities of decapod Crustacea. *Am. Nat.* 81: 392–398.

Campos, E. 2006. Systematics of the genus *Scleroplax* Rathbun, 1893 (Crustacea: Brachyura: Pinnotheridae). *Zootaxa* 1344: 33–41.

Cardenas, C.A., Canete, J., Oyarzun, S. & Mansilla, A. 2007. Agregaciones de juveniles de centolla *Lithodes santolla* (Molina, 1782) (Crustacea) en asociación con discos de fijación de *Macrocystis pyrifera* (Linnaeus) C. Agardh, 1980. *Investig. Mar., Mayo.* 35: 105–110.

Carlisle, D.B. 1957. On the hormonal inhibition of moulting in decapod Crustacea. II. The terminal anecdysis in crabs. *J. Mar. Biol. Ass. U.K.* 36: 291–307.

Carlisle, D.B. 1959. On the sexual biology of *Pandalus borealis* (Crustacea Decapoda) II. The termination of the male phase. *J. Mar. Biol. Ass. U.K.* 38: 481–491.

Carral, J.M., Celada, J.D., González, J., Sáez-Royuela, M. & Gaudioso, V.R. 1994. Mating and spawning of. freshwater crayfish (*Austropotamobius pallipes* Lereboullet) under laboratory conditions. *Aquacult. Fish. Manag.* 25: 721–727.

Carranza, A., Domingo, A., Verdi, A., Forselledo, R. & Estrades, A. 2003. First report of an association between *Planes cyaneus* (Decapoda: Grapsidae) and loggerhead sea turtles in the southwestern Atlantic Ocean. *Mar. Turtle Newsl.* 102: 5–7.

Caskey, J.L. & Bauer, R.T. 2005. Behavioral tests for a possible contact sex pheromone. *J. Crust. Biol.* 25: 571–576.

Castro, A.D. & Jory, D.E. 1983. Preliminary experiments on the culture of the banded coral shrimp, *Stenopus hispidus* Oliver. *J. Aquacult. Aquat. Sci.* 3: 84–89.

Castro, P. 1976. Brachyuran crabs symbiotic with scleractinian corals: a review of their biology. *Micronesica* 121: 95–110.

Castro, P. 1978. Movements between coral colonies in *Trapezia ferruginea* (Crustacea: Brachyura), an obligate symbiont of scleractinian corals. *Mar. Biol.* 46: 237–245.

Castro, P. 1996. Eastern Pacific species of *Trapezia* (Crustacea, Brachyura: Trapeziidae), sibling species symbiotic with reef corals. *Bull. Mar. Sci.* 58: 531–554.

Castro, P. 1999. Results of the Rumphius Biohistorical Expedition to Ambon (1990). Part 7. The Trapeziidae (Crustacea: Brachyura: Xanthoidea) of Indonesia. *Zool. Meded.* 73: 27–61.

Chace, F.A., Jr. 1972. The shrimps of the Smithsonian-Bredin Caribbean Expeditions with a summary of the West Indian shallow water species (Crustacea: Decapoda: Natantia). *Smith. Contr. Zool.* 98: 1–179.

Cheung, T.S. 1966. An observed act of copulation in the shore crab, *Carcinus maenus* (L.). *Crustaceana* 11: 107–108.

Cheung, T.S. 1968. Trans-molt retention of sperm in the female stone crab, *Menippe mercenaria* (Say). *Crustaceana* 15: 117–120.

Childchester, F.E. 1911. The mating habits of four species of the Brachyura. *Biol. Bull.* 21: 235–248.

Chittleborough, R.G. 1976. Breeding of *Panulirus longipes cygnus* George under natural and controlled conditions. *Aust. J. Mar. Freshwt. Res.* 27: 499–516.

Christensen, A.M. 1958. On the life history and biology of *Pinnotheres pisum*. *Proc. XVth Int. Congr. Zool., London*: 267–270.

Christensen, A.M. & McDermott, J.J. 1958. Life history and biology of the oyster crab, *Pinnotheres ostreum* Say. *Biol. Bull.* 114: 146–179.

Christofferson, J.P. 1970. *An electrophysiological and chemical investigation of the female sex pheromone of the crab* Portunus sanguinolentus. Manoa: Ph.D. Thesis, Univ. Hawaii. (Cited from Dunham 1978.)

Christofferson, J.P. 1978. Evidence for the controlled release of a crustacean sex pheromone. *J. Chem. Ecol.* 4: 633–639.

Churchill, E.P. 1919. Life history of the blue crab. *Bull. U.S. Bur. Fish.* 36: 91–128.

Cleaver, F.C. 1949. Preliminary results of the coastal crab (*Cancer magister*) investigation. *Wash. State Dep. Fish. Biol. Rep.* 49A: 47–82.

Coelho, V.R. 2001. Intraspecific behavior of two pair-bonding thalassinidean shrimp, *Axianassa australis* and *Pomatogebia operculata*. *Fifth Int. Congr. Crust (Abs.), Melbourne, Australia*: 51.

Coelho, V.R. & Rodrigues, S.A. 1999. Comparison between the setal types present on the feeding appendages of two callianassid shrimps. *TCS Summer Meeting (Abs.), Lafayette, Louisiana*: 27.

Coleman, N. 1991. *Encyclopedia of Marine Animals*. New York: Harper Collins Publ.

Contreras-Garduño, J., Osorno, J.L. & Córdoba-Aguilar, A. 2007. Male–male competition and female behavior as determinants of male mating success in the semi-terrestrial hermit crab *Coenobita compressus* (H. Milne Edwards). *J. Crust. Biol.* 27: 411–416.

Corotto, F.S., Bonenberger, D.M., Bounkeo, J.M. & Dukas, C.C. 1999. Antennule ablation, sex discrimination, and mating behavior in the crayfish *Procambarus clarkii*. *J. Crust. Biol.* 19: 708–712.

Correa, C. & Thiel, M. 2003a. Mating systems in caridean shrimp (Decapoda: Caridea) and their evolutionary consequences for sexual dimorphism and reproductive biology. *Rev. Chil. Hist. Nat.* 76: 187–203.

Correa, C. & Thiel, M. 2003b. Population structure and operational sex ratio in the rock shrimp *Rhynchocinetes typus* (Decapoda: Caridea). *J. Crust. Biol.* 23: 849–861.

Correa, C., Baeza, J.A., Dupré, E., Hinojosa, I.A. & Thiel, M. 2000. Mating behavior and fertilization success of three ontogenetic stages of male rock shrimp *Rhynchocinetes typus* (Decapoda: Caridea). *J. Crust. Biol.* 20: 628–640.

Correa, C., Baeza, J.A., Hinojosa, I.A. & Thiel, M. 2003. Male dominance hierarchy and mating tactics in the rock shrimp *Rhynchocinetes typus* (Decapoda: Caridea). *J. Crust. Biol.* 23: 33–45.

Costa, R. C. da. & Fransozo, A. 2004. Reproductive biology of the shrimp *Rimapenaeus constrictus* (Decapoda, Penaeidae) in the Ubatuba Region of Brazil. *J. Crust. Biol.* 24: 274–281.

Courtney, L.A. & Couch, J.A. 1981. Aspects of the host-commensal relationship between a palaemonid shrimp (*Pontonia domestica*) and the Pen Shell (*Atrina rigida*). *Northeast Gulf Sci.* 5: 49–54.

Cowles, R.P. 1913. The habits of some tropical Crustacea. *Philippine J. Sci.* 8 (D): 119–125.

Criales, M.M. 1984. Shrimps associated with coelenterates, echinoderms, and molluscs in the Santa Marta Region, Colombia. *J. Crust. Biol.* 4: 307–317.

Cukerzis, J.M. 1988. *Astacus astacus* in Europe. In: Holdich, D.M. & Lowery, R.S. (eds.), *Freshwater Crayfish: Biology, Management, and Exploitation*: 309–340. London: Champman & Hall.

Dardeau, M.R. 1984. *Synalpheus* shrimps (Crustacea: Decapoda: Alpheidae). I. The *Gambarelloides* group, with a description of a new species. *Mem. Hourglass Cruises 7, Part 2*: 1–125.

Dardeau, M.R. 1986. Redescription of *Synalpheus scaphoceris* Coutiere, 1910 (Decapoda: Alpheidae) with new records from the Gulf of Mexico. *J. Crust. Biol.* 6: 491–496.

Davie, P.J.F. 2002. Crustacea: Malacostraca: Eucarida (Part 2): Decapoda: Anomura, Brachyura. In: Wells, A. & Houston, W.W.K. (eds.), *Zoological Catalogue of Australia 19.3B*: 1–641. Melbourne: CSIRO Publ., Glaessner MF.

de Saint-Brisson, S.C. 1985. The mating behavior of *Penaeus paulensis*. *Crustaceana* 50: 108–110.

Debelius, H. 1984. *Armoured Knights of the Sea*. Essen: Kernen Verlag.

Debelius, H. 1999. *Indian Ocean Reef Guide*. IKAN-Unterwasserarchiv.

DeGoursey, R.E. & Auster, P.J. 1992. A mating aggregation of the spider crab, *Libinia emarginata*. *J. Northwest Atlantic Fish. Sci.* 13: 77–82.

Delgado, D. & Defeo, O. 2006. A complex sexual cycle in sandy beaches: the reproductive strategy of *Emerita brasiliensis* (Decapoda: Anomura). *J. Mar. Biol. Ass. U.K.* 86: 361–368.

Delgado, D. & Defeo, O. 2008. Reproductive plasticity in mole crabs, *Emerita brasiliensis*, in sandy beaches with contrasting morphodynamics. *Mar. Biol.* 153: 1065–1074.

Dellinger, T., Davenport, J. & Wirtz, P. 1997. Comparisons of social structure of Columbus crabs living on loggerhead sea turtles and inanimate flotsam. *J. Mar. Biol. Ass. U.K.* 77: 185–194.

Delsman, C. & de Man, J.G. 1925. On the "Radjungans" of the Bay of Batavia. *Treubia* 6: 308–323.

Dennenmoser, S. & Thiel, M. 2007. Competition for food and mates by dominant and subordinate male rock shrimp, *Rhynchocinetes typus*. *Behaviour* 144: 33–59.

Dew, C.B. 1990. Behavioral ecology of podding red king crab, *Paralithodes camtschatica*. *Canad. J. Fish. Aqua. Sci.* 47: 1944–1958.

Dew, C.B., Cummiskey, P.A. & Munk, J.E. 1992. The behavioral ecology and spatial distribution of red king crab and other target species: Implications for sampling design and data treatment. In: White, L. & Nielson, C. (eds.), *Proceedings of the International Crab Rehabilitation and Enhancement Symposium*: 39–67. Kidiak: Alaska Dept. Fish Game.

Díaz, E. & Thiel, M. 2003. Female rock shrimp prefer dominant males. *J. Mar. Biol. Ass. U.K.* 83: 941–942.

Díaz, E. & Thiel, M. 2004. Chemical and visual communication during mate searching in rock shrimp. *Biol. Bull.* 206: 134–143.

Didderen, K., Fransen, C.H.J.M. & de Voogd, N.J. 2006. Observations on sponge-dwelling colonies of *Synalpheus* (Decapoda, Alpheidae) of Sulawesi, Indonesia. *Crustaceana* 79: 961–975.

Doi, W. & Watanabe, S. 2006. Occurrence of the sperm plugs in *Carcinoplax vestita* (Brachyura: Goneplacidae). *Cancer* 15: 13–15.

Donaldson, W.E. & Adams, A.E. 1989. Ethogram of behavior with emphasis on mating for the Tanner crab *Chionoecetes bairdi* Rathbun. *J. Crust. Biol.* 9: 37–53.

Duffy, J. E. 1996a. *Synalpheus regalis*, new species, a sponge-dwelling shrimp from the Belize Barrier Reef, with comments on host specificity in *Synalpheus*. *J. Crust. Biol.* 16: 564–573.

Duffy, J.E. 1996b. Eusociality in a coral-reef shrimp. *Nature* 381: 512–514.

Duffy, J.E. 1996c. Resource-associated population subdivision in a symbiotic coral-reef shrimp. *Evolution* 50: 360–373.

Duffy, J.E. 1998. On the frequency of eusociality in snapping shrimps with description of a new eusocial species. *Bull. Mar. Sci.* 62: 387–400.

Duffy, J.E. 2003. The ecology and evolution of eusociality in sponge-dwelling shrimp. In: Kikuchi, T., Azuma, N. & Higashi, S. (eds.), *Genes, Behavior and Evolution in Social Insects*: 217–252. Sapporo: Hokkaido Univ. Press.

Duffy, J.E. & Macdonald, K.S. 1999. Colony structure of the social snapping shrimp *Synalpheus filidigitus* in Belize. *J. Crust. Biol.* 19: 283–292.

Duffy, J.E., Morrison, C.L. & Rios, R. 2000. Multiple origins of eusociality among sponge-dwelling shrimps (*Synalpheus*). *Evolution* 54: 503–516.

Duffy, J.E., Morrison, C.L. & Macdonald, K.S. III. 2002. Colony defense and behavioral differentiation in the eusocial shrimp *Synalpheus regalis*. *Behav. Ecol. Sociobiol.* 51: 488–495.

Dunham, D.W. & Gilchrist, S.L. 1988. Behavior. In: Burggren, W.W. & McMahon, B.R. (eds.), *Biology of the Land Crabs*: 97–138. Cambridge: Cambridge Univ. Press.

Duteutre, M. 1930. Mensurations de *Carcinus moenas* en promenade pre-nuptiale. *Ass. Fran. L'avanc. Sci.* 54: 294–250.

Dworschak, P.C. & Ott, J.A. 1993. Decapod burrows in mangrove-channel and back-reef environments at the Atlantic Barrier Reef, Belize. *Ichnos* 2: 277–290.

Edwards, E. 1966. Mating behaviour in the European edible crab (*Cancer pagurus* L.). *Crustaceana* 10: 23–30.

Efford, I.E. 1965. Aggregation in the sand crab, *Emerita analoga* (Stimpson). *J. Anim. Ecol.* 34: 63–75.

Efford, I.E. 1967. Neoteny in sand crabs of the genus *Emerita* (Decapoda, Hippidae). *Crustaceana* 13: 81–93.

Elner, R.W. & Elner, J.K. 1980. Observations on a simultaneous mating embrace between a male and two female rock crabs *Cancer irroratus* (Decapoda, Brachyura). *Crustaceana* 38: 96–98.

Elner, R.W. & Stasko, A.B. 1978. Mating behavior of the rock crab, *Cancer irroratus*. *J. Fish. Res. Bd. Canada* 35: 1385–1388.

Elner, R.W., Gass, C.A. & Campbell, A. 1985. Mating behavior of the Jonah crab, *Cancer borealis* Stimpson (Decapoda, Brachyura). *Crustaceana* 48: 34–39.

Farmer, A.S.D. 1974. Reproduction in *Nephrops novergicus* (Decapoda: Nephroidae). *J. Zool. London* 174: 161–183.

Fiedler, G.C. 2002. The influence of social environment on sex determination in harlequin shrimp (*Hymenocera picta*: Decapoda, Gnathophyllidae). *J. Crust. Biol.* 22: 750–761.

Fielder, P.R. & Eales, A.J. 1972. Observations on courtship, mating and sexual maturity in *Portunus pelagicus* (L., 1766) (Crustacea, Portunidae). *J. Nat. Hist.* 6: 273–277.

Finney, W.C. & Abele, L.G. 1981. Allometric variation and sexual maturity in the obligate coral commensal *Trapezia ferruginea* Latreille (Decapoda, Xanthidae). *Crustaceana* 41: 113–130.

Fischer, R. 1980. Bioerosion of basalt of the Pacific Coast of Costa Rica. *Senckenberg. Marit.* 13: 1–41.

Fischer, R. & Meyer, W. 1985. Observations on rock boring by *Alpheus saxidomus* (Crustacea: Alpheidae). *Mar. Biol.* 89: 213–219.

Fishelson, L. 1966. Observations on the littoral fauna of Israel, V. On the habitat and behaviour of *Alpheus frontalis* H. Milne Edwards (Decapoda, Alpheidae). *Crustaceana* 11: 98–104.

Fonseca, A.C.E. & Cortés, J. 1988. Coral borers of the Eastern Pacific: *Asidosiphon* (*A.*) *elegans* (Sipuncula: Aspidosiphonidae) and *Pomatogebia rugosa* (Crustacea: Upogebiidae). *Pac. Sci.* 52: 170–175.

Forster, G.R. 1951. The biology of the common prawn *Leander serratus* Pennant. *J. Mar. Biol. Ass. U.K.* 30: 333–360.

Fransen, C.H.J.M. 2002. Taxonomy, phylogeny, historical biogeography, and historical ecology of the genus *Pontonia* (Crustacea: Decapoda: Caridea: Palaemonidae). *Zool. Verhandl., Leiden* 336: 1–433.

Frick, M., Williams, K. & Veljacic, D. 2000. Additional evidence supporting a cleaning association between epibiotic crabs and sea turtles: how will the harvest of sargassum seaweed impact this relationship? *Mar. Turtle Newsl.* 90: 11–13.

Frick, M.G., Williams, K.L., Bolten, A.B., Bjorndal, K.A. & Martins, H.R. 2004. Diet and fecundity of columbus crabs, *Planes minutus*, associated with oceanic-stage loggerhead sea turtles, *Caretta caretta*, and inanimate flotsam. *J. Crust. Biol.* 24: 350–355.

Frick, M.G., Williams, K.L., Bresette, M., Singewald, D.A. & Herren, R.M. 2006. On the occurrence of columbus crabs (*Planes minutus*) from loggerhead turtles in Florida, USA. *Mar. Turtle Newsl.* 114:12–14.

Fujino, T. & Miyake, S. 1967. Two species of pontoniid prawns commensal with bivalves (Crustacea, Decapoda, Palaemonidae). *Publ. Seto Mar. Biol. Lab.* 15: 291–296.

Fujino, T. & Miyake, S. 1969. Studies on the genus *Onycocaris* with descriptions of five new species (Crustacea, Decapoda, Palaemonidae). *J. Fac. Agr., Kyushu Univ.* 15: 403–448.

Fujita, Y. & Baba, K. 1999. Two galatheid associates of crinoids from the Ryukyu Islands (Decapoda: Anomura: Galatheidae), with their ecological notes. *Crust. Res.* 28: 112–124.

Fukui, Y. 1991. Mating behavior of brachyuran crabs. *Benthos Res.* 40: 35–46.

Fukui, Y. 1994. Mating behavior of the grapsid crab, *Gaetice depressus* (De Haan) (Brachyura: Grapsidae). *Crust. Res.* 23: 32–39.

Fuseya, R. 2006. Notes on the stopper of the kuruma prawn *Marsupenaeus japonicus*. *Cancer* 15: 7–19.

Galeotti, P., Pupin, F., Rubolini, D., Sacchi, R., Nardi, P.A. & Fasola, M. 2007. Effects of female mating status on copulation behaviour and sperm expenditure in the freshwater crayfish *Austropotamobius italicus*. *Behav. Ecol. Sociobiol.* 61: 711–718.

Garth, J.S. & Abbott, D.P. 1980. Brachyura: the true crabs. In: Morris, R.H., Abbott, D.P. & Haderlie, E.C. (eds.), *Intertidal Invertebrates of California*: 594–630. Stanford: Stanford Univ. Press.

Geiger, D.L. & Martin, J.W. 1999. The pea crab *Orthotheres haliotidis* new species (Decapoda: Brachyura: Pinnotheridae) in the Australian abalones *Haliotis asinina* Linnaeus, 1758 and *H. squamata* Reeve, 1846 (Gastropoda: Vetigastropoda: Haliotidae). *Bull. Mar. Sci.* 64: 269–280.

George, M.J. 1963. The anatomy of the crab *Neptunus sanguinolentus* Herbst. Part IV. Reproductive system and embryological studies. *J. Madras Univ. (Sect. B)* 33: 289–304.

George, S.B. & Boone, S. 2003. The ectosymbiont crab *Dissodactylus mellitae*–sand dollar *Mellita isometra* relationship. *J. Exp. Mar. Biol. Ecol.* 294: 235–255.

Gherardi, F. 1991. Eco-ethological aspects of the symbiosis between the shrimp *Athanas indicus* (Coutiere 1903) and the sea urchin *Echinometra mathaei* (de Blainville, 1825). *Trop. Zool.* 4: 107–128.

Gifford, C.A. 1962. Some observations on the general biology of the land crab, *Cardisoma. guanhumi* (Latreille), in South Florida. *Biol. Bull.* 123: 207–223.

Glassell, S.A. 1936. New porcellanids and pinnotherids from tropical North American Waters. *Trans. San Diego Soc. Nat. Hist.* 8: 227–304.

Gleeson, R.A. 1980. Pheromone communication in the reproductive behavior of the blue crab, *Callinectes sapidus*. *Mar. Behav. Physiol.* 7: 119–134.

Gleeson, R.A., Adams, M.A. & Smith, A.B. III. 1984. Characterization of a sex pheromone in the blue crab, *Callinectes sapidus*: Crustecdysone studies. *J. Chem. Ecol.* 10: 913–921.

González-Gurriarán, E. & Freire, J. 1994. Sexual maturity in the velvet swimming crab *Necora puber* (Brachyura, Portunidae): morphometric and reproductive analyses. *ICES J. Mar. Sci., J. Conseil* 51:133–145.

Gore, R.H. 1970. *Pachycheles cristobalensis*, sp. nov., with notes on the porcellanid crabs of the southwestern Caribbean. *Bull. Mar. Sci.* 20: 957–970.

Gore, R.H. & Shoup, J.B. 1968. A new starfish host and an extension of range for the commensed crab, *Minyocerus angustus* (Dana, 1852) (Crustacea: Porcellanidae). *Bull. Mar. Sci.* 18: 240–248.

Goshima, S., Ito, K., Wada, S., Shimizu, M. & Nakao, S. 1995. Reproductive biology of the stone crab *Hapalogaster dentata* (Anomura: Lithodidae). *Crust. Res.* 24: 8–18.

Goshima S., Kawashima, T. & Wada, S. 1998. Mate choice by males of the hermit crab *Pagurus filholi*: do males assess ripeness and/or fecundity of females? *Ecol. Res.* 13: 151–162.

Gosliner, T.M., Behrens, D.W. & Williams, G.C. 1996. *Coral Reef Animals of the Indo-Pacific: Animal Life from Africa to Hawaii, Exclusive of the Vertebrates*. Monterey: Sea Challengers.

Gotelli, N.J., Gilchrist, S.L. & Abele, L.G. 1985. Population biology of *Trapezia* spp. and other coral-associated decapods. *Mar. Ecol. Prog. Ser.* 21: 89–98.

Gray, G.W. & Powell, G.C. 1966. Sex ratios and distribution of spawning king crabs in Alitak Bay, Kodiak Island, Alaska (Decapoda Anomura, Lithodidae). *Crustaceana* 10: 303–309.

Gray, I.E. 1961. Changes in abundance of the commensal crabs of *Chaetopterus*. *Biol. Bull.* 120: 353–359.

Grove, M.W. & Woodin, S.A. 1996. Conspecific recognition and host choice in a pea crab, *Pinnixa chaetopterana* (Brachyura: Pinnotheridae). *Biol. Bull.* 190: 359–366.

Grove, M.W., Finelli. C.M., Wethey, D.S. & Woodin, S.A. 2000. The effects of symbiotic crabs on the pumping activity and growth rates of *Chaetopterus variopedatus*. *J. Exp. Mar. Biol. Ecol.* 246: 31–52.

Guest, W.C. 1979. Laboratory life history of the palaemonid shrimp *Macrobrachium amazonicum* (Heller) (Decapoda, Palaemonidae). *Crustaceana* 37: 141–152.

Guinot, D. 1978. Principes d'une classification èvolutive des crustacès dècapodes brachyoures. *Bull. Biol. France Belgique* 112: 209–292.

Haefner, P.A., Jr. 1976. Distribution, reproduction and moulting of the rock crab, *Cancer irroratus* Say, 1917, in the mid-Atlantic Bight. *J. Nat. Hist.* 10: 377–397.

Haig, J. & Abbot, D.P. 1980. Macrura and Anomura: the ghost shrimps, hermit crabs, and allies. In: Morris, R.H., Abbott, D.P. & Haderlie, E.C. (eds.), *Intertidal Invertebrates of California*: 577–593. Stanford: Stanford Univ. Press.

Hamel, J.F., Ng, P.K.L. & Mercier, A. 1999. A life cycle of the pea crab *Pinnotheres halingi* sp. nov., an obligate symbiont of the sea cucumber *Holothuria scabra* Jaeger. *Ophelia* 50: 149–175.

Hart, J.F.L. 1982. *Crabs and Their Relatives of British Columbia*. Victoria: British Columbia.

Hartnoll, R.G. 1965a. The biology of spider crabs: a comparison of British and Jamaican species. *Crustaceana* 9: 1–16.

Hartnoll, R.G. 1965b. Notes on the marine grapsid crabs of Jamaica. *Proc. Linn. Soc. London* 176: 113–147.

Hartnoll, R.G. 1968. Reproduction in the burrowing crab, *Corystes cassivelaunus* (Pennant, 1777) (Decapoda, Brachyura). *Crustaceana* 15: 165–170.

Hartnoll, R.G. 1969. Mating in the Brachyura. *Crustaceana* 16: 161–181.

Hartnoll, R.G. 1972. Swimming in the hard stage of the pea crab, *Pinnotheres pisum*. *J. Nat. Hist.* 6: 475–480.

Havinga, B. 1930. Der Granat (*Crangon vulgaris* Fabr.) in den holländischen Gewässern. *J. Cons. Int. Explor. Mer.* 5: 57–87.

Hay, W.P. 1905. The life history of the blue crab (*Callinectes sapidus*). *Rep. U.S. Bur. Fish.* 1904: 395–413.

Hayashi, K.-I. & Ogawa, Y. 1987. *Spongicola levigata* sp. nov., a new shrimp associated with hexactinellid sponge from the East China Sea (Decapoda, Stenopodidae). *Zool. Sci.* 4: 367–373.

Hayashi, K.-I. 2002. A new species of the genus *Athanas* (Decapoda, Caridea, Alpheidae) living in the burrows of a mantis shrimp. *Crustaceana* 75: 395–403.

Hayashi, K.-I., Sakaue, J. & Toyota, K. 2003. *Latreutes anoplonyx* Kemp associated with *Nemopilema nomurai* at Sea of Japan and the Pacific coast of northern Japan. *Cancer* 13: 9–15.

Hayes, F.E., Joseph, V.L., Gurley, H.S. & Wong, B.Y.Y. 1998. Selection by two decapod crabs (*Percnon gibbesi* and *Stenorhynchus seticornis*) associating with an urchin (*Diadema antillarum*) at Tobago, West Indies. *Bull. Mar. Sci.* 63: 241–247.

Hazlett, B.A. 1966. Social behavior of the Paguridae and Doigenidae of Curacao. *Stud. Fauna Curacao* 23: 1–143.

Hazlett, B.A. 1968. The phyletically irregular social behavior of *Diogenes pugilator* (Anomura, Paguridae). *Crustaceana* 15: 31–34.

Hazlett, B.A. 1972. Shell fighting and sexual behaviour in the hermit crab genera *Paguristes* and *Calcinus* with comments on *Pagurus*. *Bull. Mar. Sci.* 22: 806–823.

Hazlett, B.A. 1975. Ethological analysis of reproductive behavior in marine Crustacea. *Pubbl. Staz. Zool. Napoli* 39: 677–695.

Hazlett, B.A. 1989. Mating success of male hermit crabs in shell generalist and shell specialist species. *Behav. Ecol. Sociobiol.* 25: 119–128.

Heldt, J.H. 1931. Observations sur la ponte, la fécondation et les premiers stades du développement de l'œuf chez *Penaeus caramote* Risso. *CR Acad. Sci., Ser. III–vie* 193: 1039–1041.

Helfman, G.S. 1977. Copulatory behavior of the coconut or robber crab *Birgus latio* (L.) (Decapoda, Anomura, Paguridae, Coenobitidae). *Crustaceana* 33: 198–202.

Henning, H.G. 1975. Aggressive, reproductive and molting behavior, growth and maturation of *Cardisoma guanhumi* Latrelle (Crustacea, Brachyura). *Forma Funct.* 8: 463–510.

Herrick, F.H. 1909. Natural history of the American lobster. *Bull. U.S. Bur. Com. Fish.* 29: 149–408.

Heydon, A.E.F. 1969. Notes on the biology of *Panulirus homarus* and on length/weight relationship. *Invest. Rep. Div. Sea. Fish. S. Afr.* 69: 1–19.

Hiatt, R.W. 1948. The biology of the lined shore crab *Pachygrapsus crassipes* Randall. *Pac. Sci.* 2: 135–213.

Hicks, J.W. 1985. The breeding behaviour and migrations of the terrestrial crab *Gecarcoidea natalis* (Decapoda: Brachyura). *Aust. J. Zool.* 33: 127–142.

Hill, B.J. 1975. Abundance, breeding and growth of the crab *Scylla serrata* in two South African estuaries. *Mar. Biol.* 32: 119–126.

Hinojosa, I. & Thiel, M. 2003. Somatic and gametic resources in male rock shrimp, *Rhynchocinetes typus*: effect of mating potential and ontogenetic male stage. *Anim. Behav.* 65: 449–458.

Hinsch, G.W. 1968. Reproductive behavior in the spider crab *Libinia emarginata* (L.). *Biol. Bull.* 135: 273–278.

Hinsch, G.W. 1988. Morphology of the reproductive tract and seasonality of reproduction in the golden crab *Geryon fenneri* from the eastern Gulf of Mexico. *J. Crust. Biol.* 8: 254–261.

Hipeau-Jacquotte, R. 1971. Notes de faunistique et de biologie marines de Madagascar, 5. *Platypontonia hyotis* nov. sp. (Decapoda Natantia, Pontoniinae). *Crustaceana* 20: 125–140.

Hipeau-Jacquotte, R. 1973. Manifestation d'un comportement territorial chez les crevettes Pontoniinae associaées aux mollusques Pinnidae. *J. Mar. Exp. Biol. Ecol.* 13: 63–71.

Hobday, A.J. & Rumsey, S.M. 1999. Population dynamics of the sheep crab *Loxorhynchus grandis* (Majidae) Stimpson, 1857, at La Jolla California. *Scripps Inst. Oceanogr. Tech. Rep.* 29: 1–32.

Hoestlandt, H. 1948. Recherches sur la biologie de l'*Eriocheir sinensis* en France (Crustacé Brachyoure). *Ann. Inst. Océanogr., Monaco* 24: 1–116.

Hoffman, D.L. 1973. Observed acts of copulation in the protandric shrirmp *Pandalus platyceros* Bandt (Decapoda, Pandaliae). *Crustaceana* 24: 242–244.

Hoglund, H. 1943. On the biology and larval development of *Leander squilla* (L.) forma *typica* de Man. *Svenska hydrogr.–biol. Kommn. Skr.* 2: 1–44.

Holdich, D. & Black, J. 2007. The spiny-cheek crayfish, *Orconectes limosus* (Rafinesque, 1817) [Crustacea: Decapoda: Cambaridae], digs into the UK. *Aquat. Invasion* 2: 1–16.

Hsueh, P.-W. 2001a. Intertidal distribution, symbiotic association and reproduction of *Pinnotheres bidentatus* (Brachyura: Pinnotheridae) from Taiwan. *J. Nat. Hist.* 35: 1681–1692.

Hsueh, P.-W. 2001b. Population dynamics of free-swimming stage *Pinnotheres bidentatus* (Brachyura: Pinnotheridae) in tidal waters off the west coast of central Taiwan. *J. Crust. Biol.*: 973–981.

Hsueh, P.-W. 2003. Responses of the pea crab *Pinnotheres taichungae* to the life history patterns of its primary bivalve host *Laternula marilina*. *J. Nat. Hist.* 37: 1453–1462.

Hsueh, P.-W. & Huang, J.F. 1998. *Polyonyx bella*, new species (Decapoda: Anomura: Porcellanidae), from Taiwan, with notes on its reproduction and swimming behaviour. *J. Crust. Biol.* 18: 332–336.

Huber, M.E. 1985. Non-random mating with respect to mate size in the crab *Trapezia* (Brachyura, Xanthidae). *Mar. Behav. Physiol.* 12: 19–32.

Huber, M.E. 1987. Aggressive behavior of *Trapezia intermedia* Miers and *T. digitalis* Latreille (Brachyura: Xanthidae). *J. Crust. Biol.* 7: 238–248.

Huber, M.E. & Coles, S.L. 1986. Resource utilization and competition among the five Hawaiian species of *Trapezia* (Crustacea, Brachyura). *Mar. Ecol. Prog. Ser.* 30: 21–31.

Hudinaga, M. 1942. Reproduction, development and rearing of *Penaeus japonicus* Bate. *Jap. J. Zool.* 10: 305–393.

Hughes, J.T. & Matthiessen, G.C. 1962. Observation on the biology of the American lobster, *Homarus americanus*. *Limnol. Oceanogr.* 7: 414–421.

Imafuku, M. 1986. Sexual discrimination in the hermit crab *Pagurus geminus*. *J. Ethol.* 4: 39–47.

Imazu, M. & Asakura, A. 2006. Descriptions of agonistic, aggressive and sexual behaviors of five species of hermit crabs from Japan (Decapoda: Anomura: Paguridae and Diogenidae). *Crust. Res., Spec. No.* 6: 95–107.

Ingle, R.W. & Thomas, W. 1974. Mating and spawning of the crayfish *Austropotamobius pallipes* (Crustacea: Astacidae). *J. Zool.* 173: 525–538.

Jefferies, D.J. 1968. The breeding characters of *Palaemonetes varians* (Leach) (Decapoda, Palaemonidae). *Crustaceana* 14: 259–270.

Jeng, M.S. 1994. Effect of antennular and antennal ablation on pairing behavior of snapping shrimp *Alpheus edwardsii* (Audouin). *J. Exp. Mar. Biol. Ecol.* 179: 171–178.

Jensen, G.C. 1995. *Pacific Coast Crabs and Shrimps*: 87 pp. Monterey, CA: Sea Challengers.

Jensen, K. 1972. On the agonistic behaviour in *Carcinus maenas* (L.)(Decapoda). *Ophelia* 10: 57–61.

Jivoff, P. 1997. The relative roles of predation and sperm competition on the duration of the postcopulatory association between the sexes in the blue crab, *Callinectes sapidus*. *Behav. Ecol. Sociobiol.* 40: 175–185.

Jivoff, P. & Hines, A.H. 1998. Female behaviour, sexual competition and mate guarding in the blue crab, *Callinectes sapidus*. *Anim. Behav.* 55: 589–603.

Johnson, P.T. & Oito, S.V. 1981. Histology of a bilateral gynandromorph of the blue crab, *Callinectes sapidus* Rathbun (Decapoda: Portunidae). *Biol. Bull.* 161: 236–245.

Johnson, V.R., Jr. 1969. Behavior associated with pair formation in the banded shrimp *Stenopus hispidus* (Olivier). *Pac. Sci.* 23: 40–50.

Johnson, V.R., Jr. 1977. Individual recognition in the banded shrimp *Stenopus hispidus*. *Anim. Behav.* 25: 418–428.

Kaestner, A. 1970. *Invertebrate Zoology* (translated by H. W. Levi & L. R. Levi). New York: Wiley Interscience.

Kamezaki, N. & Kamezaki, Y. 1986. On the ecology of alpheid shrimp *Aretopsis amabilis* de Man. *Nankiseibutu, Nanki Biol. Soc.* 28: 11–15 (in Japanese).

Kamio, M., Matsunaga, S. & Fusetani, N. 2000. Studies on sex pheromones of the helmet crab, *Telmessus cheiragonus*: I. An assay based on precopulatory mate-guarding. *Zool. Sci.* 17: 731–733.

Kamio, M., Matsunaga, S. & Fusetani, N. 2002. Copulation pheromone in the crab, *Telmessus cheiragonus* (Brachyura: Decapoda). *Mar. Ecol. Prog. Ser.* 234: 183–190.

Kamio, M., Matsunaga, S. & Fusetani, N. 2003. Observation on the mating behaviour of the helmet crab *Telmessus cheiragonus* (Brachyura: Cheiragonidae). *J. Mar. Biol. Ass. U.K.* 83: 1007–1013.

Karplus, I. 1979. The tactile communication between *Cryptocentrus steinitzi* (Pisces, Gobiidae) and *Alpheus purpurilenticularis* (Crustacea, Alpheidae). *Z. Tierpsychol.* 49: 173–196.

Kawai, T. & Saito, K. 2001. Observations on the mating behavior and season, with no form alternation, of the Japanese crayfish, *Cambaroides japonicus* (Decapoda, Cambaridae), in Lake Komadome, Japan. *J. Crust. Biol.* 21: 885–890.

Klassen, F. 1975. Ecological and ethological studies on the reproductive biology of *Gecarcinus lateralis* (Decapoda, Brachyura). *Forma Funct.* 8: 101–174.

Kleeman, K. 1984. Lebensspuren von *Upogebia operculata* (Crustacea, Decapoda) in karibischen Steinkorallen (Madreporaria: Anthozoa). *Beitr. Palaeont. Osterreich.* 11: 35–57.

Knowlton, N. 1980. Sexual selection and dimorphism in two demes of a symbiotic, pair-bonding snapping shrimp. *Evolution* 34: 161–173.

Knowlton, N. & Keller, B.D. 1982. Symmetric fights as a measure of escalation potential in a symbiotic, territorial snapping shrimp. *Behav. Ecol. Sociobiol.* 10: 289–292.

Knowlton, N. & Keller, B.D. 1983. A new, sibling species of snapping shrimp associated with the Caribbean Sea anemone *Bartholomea annulata*. *Bull. Mar. Sci.* 33: 353–362.

Knowlton, N. & Keller, B.D. 1985. Two more sibling species of alpheid shrimps localized recruitment in an alpheid shrimp with extended larval development. *Bull. Mar. Sci.* 39: 213–223.

Knudsen, J.W. 1960. Reproduction, life history and larval ecology of the California Xanthidae, the pebble crabs. *Pac. Sci.* 14: 3–17.

Knudsen, J.W. 1964. Observation of the reproductive cycles and ecology of the common Brachyura and crablike Anomura of Puget Sound, Washington. *Pac. Sci.* 18: 3–33.

Kobayashi, S. & Matsuura, S. 1994. Variation in the duration of copulation of the Japanese mitten crab *Eriocheir japonicus*. *J. Ethol.* 12: 73–76.

Köksal, G. 1988. *Astacus leptodactylus* in Europe. In: Holdich, D.M. & Lowery, R.S. (eds.), *Freshwater Crayfish: Biology, Management and Exploitation*: 365–400. London: Chapman & Hall.

Komai, T. & Saito, H. 2006. A new genus and two new species of Spongicolidae (Crustacea, Decapoda, Stenopodidea) from the South-West Pacific. In: De Forges, R.B. & Justine, J.L. (eds.), Tropical Deep-Sea Benthos, 24. *Mém. Mus. Natl. Hist. Nat.* 193. 265–284.

Kramer, P. 1967. The behavior of the rock crab *G. grapsus* on Galapagos. *Noticias Galapagos* 7/8: 18–20.

Kraul, S. & Nelson, A. 1986. The life cycle of the harlequin shrimp. *Freshwt. Mar. Aqua.* 9: 28–31.

Kropp, R.K. 1987. Descriptions of some endolithic habitats for snapping shrimp (Alpheidae) in Micronesia. *Bull. Mar. Sci.* 41: 204–213.

Kudenov, J.D. & Haig, J. 1974. A range extension of *Polyonyx quadriungulatus* Glassell, 1935, new record into the Gulf of California (Decapoda, Anomura, Porcellanidae). *Crustaceana* 26: 105–106.

Kuris, A. M., Ra'anan, Z., Sagi, A. & Cohen, D. 1987. Morphotypic differentiation of male Malaysian giant prawns, *Macrobrachium rosenbergii*. *J. Crust. Biol.* 7: 219–237.

Laughlin, R.A. 1982. Some observations on the occurrence, reproduction and mating of the coral crab *Carpilius corallinus* (Herbst, 1783) (Decapoda, Xanthidae) in the Archipiélago Los Roques, Venezuela. *Crustaceana* 43: 219–221.

Laurent, P.J. 1988. *Austropotamobius pallipes* and *A. torrentium* with observations on their interaction with other species in Europe. In: Holdich, D.M. & Lowery, R.S. (eds.), *Freshwater Crayfish: Biology, Management and Exploitation*: 341–364. London: Croom Helm.

Le Sueur, R.F. 1954. Note on the behaviour of the common spider crab. *Bull. Soc. Jersiaise* 16: 37–38.

Lee, C. L. & Felder, D.R. 1983. Agonistic behaviour and the development of dominance hierarchies in the freshwater prawn, *Macrobrachium australiense* Holthuis, 1950 (Crustacea: Palaemonidae). *Behaviour* 83: 1–16.

Lipcius, R.N. & Herrnkind, W.F. 1987. Control and coordination of reproduction and molting in the spiny lobster *Panulirus argus*. *Mar. Biol.* 96: 207–214.

Lipcius, R.N., Edwards, M.L., Herrnkind, W.F. & Waterman, S.A. 1983. In situ mating behavior of the spiny lobster *Panulirus argus*. *J. Crust. Biol.* 3: 217–222.

Lloyd, A.J. & Young, C.M. 1947. A study of *Crangon vulgaris* in the British Channel and Severn Estuary. *J. Mar. Biol. Assn. U.K.* 26: 626–661.

Lovrich, G.A. & Vinuesa, J.H. 1999. Reproductive potential of the lithodids *Lithodes santolla* and *Paralomis granulosa* (Anomura, Decapoda) in the Beagle Channel, Argentina. *Sci. Mar.* 63 Suppl. 1: 355–360.

Lowery, R.S. & Holdich, D.M. 1988. *Pacifastacus leniusculus* in North America and Europe, with details of the distribution of introduced and native crayfish species in Europe. In: Holdich, D.M. & Lowery, R.S. (eds.), *Freshwater Crayfish: Biology, Management and Exploitation*: 283–308. London: Chapman & Hall.

Macdonald, K.S. III, Rios, R. & Duffy, J.E. 2006. Biodiversity, host specificity, and dominance by eusocial species among sponge-dwelling alpheid shrimp on the Belize Barrier Reef. *Div. Dist.* 12: 165–178.

MacGinitie, G.E. 1935. Ecological aspects of a California marine estuary. *Am. Midl. Nat.* 16: 629–765.

MacGinitie, G.E. 1937. Notes on the natural history of several marine Crustacea. *Am. Midl. Nat.* 18: 1031–1037.

MacGinitie, G.E. 1938. Movements and mating habits of the sand crab (*Emerita analoga*). *Am. Midl. Nat.* 19: 471–481.

Macnae, W. & Kalk, M. 1962. The fauna and flora of sand flats at Inhaca Island, Mozambique. *J. Anim. Ecol.* 31: 93–128.

Martin, I.N. 2007. Notes on taxonomy and biology of the symbiotic shrimp *Vir euphyllius* Martin & Anker, 2005 (Decapoda, Palaemonidae, Pontoniinae), associated with hammer corals *Euphyllia* spp. (Cnidaria, Caryophyllidae). *Inv. Zool.* 4: 15–23.

Marukawa, H. 1933. Taraba-gani chosa [Biological and fishery research on the Japanese king crab *Paralithodes camtschatica* (Tilesius)]. *J. Imp. Fish. Exp. Stn., Tokyo* 4: 1–152.

Mashiko, K. 1981. Sexual dimorphism of the cheliped in the prawn *Macrobrachium nipponense* (de Haan) and its significance in reproductive behavior. *Zool. Mag., Tokyo* 90: 333–337.

Mason, J.C. 1970a. Copulatory behavior of the crayfish, *Pacifastacus trowbridgii* (Stimpson). *Canada. J. Zool.* 48: 969–976.

Mason, J.C. 1970b. Egg-laying in the western north American crayfish, *Pacifastacus trowbridgii* (Stimpson) (Decapoda, Astacidae). *Crustaceana* 19: 37–44.

Mathews, L.M. 2002a. Territorial cooperation and social monogamy: factors affecting intersexual interactions in pair-living snapping shrimp. *Anim. Behav.* 63:767–777.

Mathews, L.M. 2002b. Tests of the mate-guarding hypothesis for social monogamy: does population density, sex ratio, or female synchrony affect behavior of male snapping shrimp (*Alpheus angulatus*)? *Behav. Ecol. Sociobiol.* 51: 426–432.

Mathews, L.M. 2003. Tests of the mate-guarding hypothesis for social monogamy: male snapping shrimp prefer to associate with high-value females. *Behav. Ecol.* 14: 63–67.

Mathews, L.M. 2006. Cryptic biodiversity and phylogeographical patterns in a snapping shrimp species complex. *Mol. Ecol.* 15: 4049–4063.

Mathews, L.M. 2007. Evidence for restricted gene flow over small spatial scales in a marine snapping shrimp *Alpheus angulosus*. *Mar. Biol.* 152: 645–655.

Mathews, L.M., Schubart, C.D., Neigel, J.E. & Felder, D.L. 2002. Genetic, ecological, and behavioural divergence between two sibling snapping shrimp species (Crustacea: Decapoda: *Alpheus*). *Mol. Ecol.* 11: 1427–1437.

Matsuo, M. 1998. Life history of *Tritodynamia horvathi* Nobili (Brachyura, Pinnotheridae). I. *Cancer* 7: 1–8

Matsuo, M. 1999. Life history of *Tritodynamia horvathi* Nobili (Brachyura, Pinnotheridae). II. *Cancer* 8: 3–11.

Matthews, D.C. 1956. The probable method of fertilization in terrestrial hermit crabs based on a comparative study of spermatophores. *Pac. Sci.* 10: 303–309.

Matsuura, S. & Takeshita, K. 1976. Molting and growth of the laboratory-reared king crab, *Paralithodes camtschatica* (Tilesius). *Rep. Fish. Res. Lab., Kyushu Univ.* 3: 1–14.

McDermott, J.J. 2005. Biology of the brachyuran crab *Pinnixa chaetopterana* Stimpson (Decapoda: Pinnotheridae) symbiotic with tubicolous polychaetes along the Atlantic coast of the United States, with additional notes on other polychaete associations. *Proc. Biol. Soc. Wash.* 118: 742–764.

McKoy, J.L. 1979. Mating behaviour and egg laying in captive rock lobster, *Jasus edwardsii* (Crustacea: Decapoda: Palinuridae). *New Zealand J. Mar. Freshwt. Res.* 13: 407–413.

McLeese, D.W. 1970. Detection of dissolved substances by the American lobster (*Homarus americanus*) and olfactory attraction between lobsters. *J. Fish. Res. Bd. Canada* 27: 1371–1378.

McLeese, D.W. 1973. Chemical communication among lobster (*Homarus americanus*). *J. Fish. Res. Bd. Canada* 30: 775–778.

McMullen, J.C. 1969. Effects of delayed mating on the reproduction of king crab, *Paralithodes camtschatica*. *J. Fish. Res Bd. Canada* 26: 2737–2740.

McNeill, F.A. & Ward, M. 1930. Carcinological notes. No. 1. *Rec. Aust. Mus.* 17: 357–383.

Meinkoth, N.A. 1981. *National Audubon Society Field Guide to North American Seashore Creatures*. New York: Alfred A. Knopf.

Menon, M.K. 1933. The life-histories of decapod Crustacea from Madras. *Bull. Madras Govt. Mus., New Ser., Nat. Hist. Sect.* 3: 1–45.

Millar, R.H. 1971. The biology of ascidians. *Adv. Mar. Biol.* 9: 1–100.

Minouchi, S. & Goshima, S. 1998. Effect of male/female size ratio on mating behavior of the hermit crab *Pagurus filholi* (Anomura: Paguridae) under experimental conditions. *J. Crust. Biol.* 18: 710–716.

Minouchi, S. & Goshima, S. 2000. The effect of male size and sex ratio on the duration of precopulatory mate guarding in the hermit crab *Pagurus filholi*. *Benthos Res.* 55: 37–41.

Misamore, M.J. & Browdy, C.L. 1996. Mating behavior in the white shrimps *Penaeus setiferus* and *P. vannamei*: a generalized model for mating in *Penaeus*. *J. Crust. Biol.* 16: 61–70.

Miya, Y. & Miyake, S. 1969. Description of *Alpheus bellulus*, sp. nov., associated with gobies from Japan (Crustacea, Decapoda, Alpheidae). *Publ. Seto Mar. Biol. Lab.* 16: 307–314.

Miyake, S. 1982. *Japanese Crustacean Decapods and Stomatopods in Color; Vol. I. Macrura, Anomura and Stomatopoda*. Osaka: Hoikusha Publ.

Miyake, S. 1983. *Japanese Crustacean Decapods and Stomatopods in Colour; Vol. II. Brachyura (Crabs)*. Osaka: Hoikusha Publ.

Monniot, C. 1965. Étude systematique et evolutive de la famille des Pyuridae (Ascidiacea). *Mem. Mus. Nat. d'Hist. Nat., Paris, Ser. A*, 36: 1–203.

Morrison, C.L., Rios, R. & Duffy, J.E. 2004. Phylogenetic evidence for an ancient rapid radiation of Caribbean sponge-dwelling snapping shrimps (*Synalpheus*). *Mol. Phylogenet. Evol.* 30: 563–58.

Naidu, R.B. 1954. A note on the courtship in the sand crab (*Philyra scabriuscula* (Fabricius)). *J. Bombay Nat. Hist. Soc.* 52: 640–641.

Nakasone, Y. & Miyake, S. 1972. Four unrecorded porcellanid crabs (Anomura: Porcellanidae) from Japan. *Bull. Sci. Eng. Div. Univ. Ryukyu (Nath. Nat. Sci)* 15: 136–147.

Needler, A.B. 1931. Mating and oviposition in *Pandalus danae*. *Canad. Fld. Nat.* 45: 107–108.

Ng, P.K.L. & Goh, N.K.C. 1996. Notes on the taxonomy and ecology of *Aliaporcellana telestophila* (Johnson, 1958) (Decapoda, Anomura, Porcellanidae), a crab commensal on the gorgonian Solenocaulon. *Crustaceana* 69: 652–661.

Ng, P.K.L. & Sasekumar, A. 1993. A new species of *Polyonyx* Stimpson, 1858, of the *P. sinensis* group (Crustacea: Decapoda: Anomura: Porcellanidae) commensal with a chaetopterid worm from Peninsular Malaysia. *Zool. Meded.* 67: 467–472.

Ng, P.K.L. & Manning, R.B. 2003. On two new genera of pea crabs parasitic in holothurians (Crustacea: Decapoda: Brachyura: Pinnotheridae) from the Indo-West Pacific, with notes on allied genera. *Proc. Biol. Soc. Wash.* 116: 901–919.

Nizinski, M.S. 1989. Ecological distribution, demography, and behavioral observations on *Periclimenes anthophilus*, an atypical symbiotic cleaner shrimp. *Bull. Mar. Sci.* 45: 174–188.

Nolan, B.A. & Salmon, M. 1970. The behavior and ecology of snapping shrimp (Crustacea: *Alpheus heterochaelis* and *Alpheus noraonni*). *Forma Funct.* 2: 289–335.

Nomura, K. 2003. A preliminary revision of alpheid shrimps associated with gobiid fishes from the Japanese waters. *Bull. Biogeogr. Soc. Jap.* 58: 49–70.

Nomura, K. & Asakura, A. 1998. The alpheid shrimps (Decapoda: Alpheidae) collected from Kushimoto on the Pacific coast of central Japan, and their spatial distributions, zoogeographical affinities, social structures, and life styles. *Nanki Seibutsu* 40: 25–34.

Norman, C.P. & Jones, M.B. 1993. Reproductive ecology of the velvet swimming crab, *Necora puber* (Brachyura: Portunidae), at Plymouth. *J. Mar. Biol. Ass. U.K.* 73: 379–389.

Nouvel, H. & Nouvel, L. 1937. Recherches sur l'accouplement et la ponte chez les crustacés décapodes Natantia. *Bull. Soc. Zool. France* 62: 208–221.

Nouvel, L. 1939. Observations de l'accouplement chez une espèce de Crevette: *Crangon crangon*. *Comp. Rend. Hebdomadaires Séa. l'Acad. Sci. Paris.* 209: 639–641.

Nye, P.A. 1977. Reproduction, growth and distribution of the grapsid crab *Helice crassa* (Dana, 1851) in the southern part of New Zealand. *Crustaceana* 33: 75–89.

Obermeier, M. & Schmitz, B. 2003a. Recognition of dominance in the big-clawed snapping shrimp (*Alpheus heterochaelis* Say 1818). Part I. Individual or group recognition? *Mar. Freshw. Behav. Physiol.* 36: 1–16.

Obermeier, M. & Schmitz, B. 2003b. Recognition of dominance in the big-clawed snapping shrimp (*Alpheus heterochaelis* Say 1818). Part II. Analysis of signal modality. *Mar. Freshw. Behav. Physiol.* 36: 17–29.

Ogawa, Y., Kakuda, S. & Hayashi, K.-I. 1981. On the mating and spawning behaviour of *Macrobrachium nipponense* (De Haan). *J. Fac. Appl. Biol. Sci. Hiroshima Univ.* 20: 65–69.

Oh, S. J. & Hankin, D.G. 2004. The sperm plug is a reliable indicator of mating success in female dungeness crabs, *Cancer magister*. *J. Crust. Biol.* 24: 314–326.

Omori, K., Yanagisawa, Y. & Hori, N. 1994. Life history of the caridean shrimp *Periclimenes ornatus* Bruce associated with a sea anemone in southwest Japan. *J. Crust. Biol.* 14: 132–145.

Osawa, M. 2001. *Heteropolyonyx biforma*, new genus and new species, from Japan, with a redescription of *Polyonyx utinomii* (Decapoda: Porcellanidae). *J. Crust. Biol.* 21: 506–520.

Osawa, M. & Fujita, Y. 2001. A new species of the genus *Neopetrolisthes* Miyake, 1937 (Crustacea: Decapoda: Porcellanidae) from the Ryukyu Islands, southwestern Japan. *Proc. Biol. Soc. Wash.* 114: 162–171.

Oshima, S. 1938. Biological and fishery research in Japanese blue crab (*Portunus trituberculatus*) (Miers). *Suisan Shikenjyo Hokoku g*: 141–212.

Page, H.M. & Willason, S.W. 1982. Distribution patterns of terrestrial hermit crabs at Enewetak Atoll, Marshall Islands. *Pac. Sci.* 36: 107–117.

Palacios, E., Racotta, I.S. & Villalejo, M. 2003. Assessment of ovarian development and its relation to mating in wild and pond-reared *Litopenaeus vannamei* shrimp in a commercial hatchery. *J. World Aquacult. Soc.* 34: 466–477.

Parente, M.A. & Hendrickx, M.E. 2000. *Pisidia magdalenensis* (Crustacea: Porcellanidae) commensal of the diogenid hermit crab *Petrochirus californiensis* (Decapoda: Diogenidae). *Rev. Biol. Trop.* 48: 265–266.

Patton, W.K. 1966. Decapod Crustacea commensal with Queensland branching corals. *Crustaceana* 10: 271–295.

Patton, W.K., Patton, R.J. & Barnes, A. 1985. On the biology of *Gnathophylloides mineri*, a shrimp inhabiting the sea urchin *Tripneustes ventricosus*. *J. Crust. Biol.* 5: 616–626.

Paul, A.J. 1984. Mating frequency and viability of stored sperm in the Tanner crab Chionoecetes bairdi (Decapoda, Majidae). *J. Crust. Biol.* 4: 375–381.

Payne, J.F. 1972. The life history of *Procambarus hayi*. *Am. Midl. Nat.* 87: 25–35.

Pearce, J.B. 1962. Adaptation in symbiotic crabs of the family Pinnotheridae. *Biologist* 45: 11–15.

Pearce, J.B. 1964. On reproduction in *Pinnotheres maculatus* (Decapoda: Pinnotheridae). *Biol. Bull.* 127: 384.

Pearce, J.B. 1965. On the distribution of *Tresus nuttulli* and *Tresus capax* (Pelecypoda: Mactridae) in the waters of Puget Sound and the San Juan Archipelago. *Veliger* 7: 166–170.

Pearce, J.B. 1966a. On *Pinnixa fubu* and *Pinnixa littoralis* (Decapoda: Pinnotheridae) symbiotic with the clam, *Tresus cupux* (Pelecypoda: Mactridae). In: Barnes, H. (ed.), *Some Contemporary Studies in Marine Science*: 565–589. London: Allen & Unwin.

Pearce, J.B. 1966b. The biology of the mussel crab, *Fabia subquadrata*, from the waters of the San Juan archipelago, Washington. *Pac. Sci.* 20: 3–35.

Pearse, A.S. 1909. Observations on copulation among crayfishes. *Am. Nat.* 43: 746–753.

Peters, N., Panning, A. & Schnakenbeck, W. 1933. Die chinesische Wollhandkrabbe (*Eriocheir sinensis* H.Milne-Edwards) in Deutschland. *Zool. Anz.* 101: 267–271.

Pike, R.B. & Williamson, D.I. 1959. Observations on the distribution and breeding of British hermit crabs and the stone crab (Crustacea: Diogenidae, Paguridae and Lithodidae). *Proc. Zool. Soc. London* 132: 551–567.

Pippett, M.R. 1977. Mating behavior of the crayfish *Orconectes nais* (Faxon 1885). *Crustaceana* 32: 265–271.

Porter, W.J. 1960. Zoeal stages of the stone crab, *Menippe mercenaria* Say. *Chesapeake Sci.* 1: 168–177.

Potts, F.A. 1915. *Hapalocarcinus*, the gall-forming crab with some notes on the related genus *Cryptochirus*. *Pap. Dep. Mar. Biol. Carnegie Inst. Wash.* 8: 33–69.

Powell, G.C. & Nickerson, R.B. 1965a. Reproduction of king crabs (*Paralithodes camtschatica*) (Tilesius). *J. Fish. Res. Bd. Canada* 22: 101–111.

Powell, G.C. & Nickerson, R.B. 1965b. Aggregations among juvenile king crabs (*Paralithodes camtschatica* Tilesius), Kodiac, Alaska. *Anim. Behav.* 13: 374–380.

Powell, G.C., Shafford, B. & Jones, M. 1973. Reproductive biology of young adult king crabs *Paralithodes camtschatica* (Tilesius) at Kodiak, Alaska. *Proc. Natl. Shellfish Assoc.* 63: 78–87.

Preston, E.M. 1973. A computer simulation of competition among five sympatric congeneric species of xanthid crabs. *Ecology* 54: 469–483.

Primavera, J.H. 1979. Notes on the courtship and mating behavior in *Penaeus monodon* Fabricius (Decapoda, Natantia). *Crustaceana* 37: 287–292.

Ra'anan, Z. & Sagi, A. 1985. Alternative mating strategies in male morphotypes of the freshwater prawn *Macrobrachium rosenbergii* (de Man). *Biol. Bull.* 169: 592–601.

Rahman, N., Dunham, D.W. & Govind, C.K. 2001. Mate recognition and pairing in the big-clawed snapping shrimp, *Alpheus herterochelis*. *Mar. Fresh. Behav. Physiol.* 34: 213–226.

Rahman, N., Dunham, D.W. & Govind, C.K. 2002. Size-assortative pairing in the big-clawed snapping shrimp, *Alpheus heterochelis*. *Behaviour* 139: 1443–1468.

Rahman, N., David, W., Dunham, D.W. & Govind, C.K. 2003. Social monogamy in the big-clawed snapping shrimp, *Alpheus heterochelis*. *Ethology* 109: 457–473.

Rahman, N., David, W., Dunham, D.W. & Govind, C.K. 2005. Mate choice in the big-clawed snapping shrimp, *Alpheus heterochaelis* Say, 1818. *Crustaceana* 77: 95–111.

Rao, A.V.P. 1967. Some observations on the biology of *Penaeus indicus* H. Milne-Edwards and *Penaeus monodon* Fabricius from the Chilka Lake. *Indian J. Fish.* 14: 251–270.

Richardson, C.A., Kennedy, H., Duarte, C.M. & Proud, S.V. 1997. The occurrence of *Pontonia pinnophylax* (Decapoda: Natantia: Pontoniinae) in *Pinna nobilis* (Mollusca: Bivalvia: Pinnidae) from the Mediterranean. *J. Mar. Biol. Ass. U.K.* 77: 1227–1230.

Rickner, J.A. 1975. New records of the porcellanid crab, *Polyonyx gibbesi* Haig, from the Texas coast (Decapoda, Anomura). *Crustaceana* 2 : 313–314.

Rios, R. & Duffy, J.E. 2007. A review of the sponge-dwelling snapping shrimp from Carrie Bow Cay, Belize, with description of *Zuzalpheus*, new genus, and six new species (Crustacea: Decapoda: Alpheidae). *Zootaxa* 1602: 1–89.

Rubenstein, D.I. & Hazlett, B.A. 1974. Examination of the agonistic behaviour of the crayfish *Orconectes virilis* by character analysis. *Behaviour* 50: 193–216.

Rubolini, D., Galeotti, P., Ferrari, G., Spairani, M., Bernini, F. & Fasola, M. 2006. Sperm allocation in relation to male traits, female size, and copulation behaviour in freshwater crayfish species. *Behav. Ecol. Sociobiol.* 60: 212–219.

Rubolini, D., Galeotti, P., Pupin, F., Sacchi, R., Nardi, P.A. & Fasola, M. 2007. Repeated matings and sperm depletion in the freshwater crayfish *Austropotamobius italicus*. *Freshw. Biol.* 52: 1898–1906.

Ruello, N.V., Moffitt, P.F. & Phillips, S.G. 1973. Reproductive behaviour in captive freshwater shrimp *Macrobrachium australiense* Holthuis. *Aust. J. Mar. Freshw. Res.* 24: 197–202.

Rufino, M.M. & Jones, D.A. 2001. Binary individual recognition in *Lysmata debelius* (Decapoda: Hippolytidae) under laboratory conditions. *J. Crust. Biol.* 21: 388–392.

Ryan, E.P. 1966. Pheromone: evidence in a decapod crustacean. *Science* 151: 340–341.

Ryan, E.P. 1967a. Structure and function of the reproductive system of the crab *Portunus sanguinolentus* (Herbst) (Brachyura Portunidae). I. The male system. *Symp. Ser. Mar. Biol. Ass. India* 2: 506–521.

Ryan, E.P. 1967b. Structure and function of the reproductive system of the crab *Portunus sanguinolentus* (Herbst) (Brachyura: Portunidae). II. The female system. *Proc. Symp. Crust. Mar. Biol. Ass., India*: 522–544.

Saito, T., Uchida, I. & Takeda, M. 2001. Pair formation in *Spongicola japonica* (Crustacea: Stenopodidea: Spongicolidae), a shrimp associated with deep-sea hexactinellid sponges. *J. Mar. Biol. Ass. U.K.* 81: 789–797.

Saito, T., Tsuchida, S. & Yamamoto, T. 2006. *Spongicoloides iheyaensis*, a new species of deep-sea spongi-associated shrimp from the Iheya Ridge, Ryukyu Islands, southern Japan (Decapoda: Stenopodidea: Spongicolidae). *J. Crust. Biol.* 26: 224–233.

Sakai, T. 1969. Two new genera and twenty-two new species of crabs from Japan. *Proc. Biol. Soc. Wash.* 82: 243–280.

Sampedro, M.-P. & González-Gurriarán, E. 2004. Aggregating behaviour of the spider crab *Maja squinado* in shallow waters. *J. Crust. Biol.* 24: 168–177.

Sasaki, J. & Ueda, Y. 1992. Pairing size of the hair crab, *Erimacrus isenbeckii* (Brandt). *Res. Crust.* 21: 147–152.

Sato, T., Ashidate, M. & Goshima, S. 2005a. Negative effects of delayed mating on the reproductive success of female spiny king crab, *Paralithodes brevipes*. *J. Crust. Biol.* 25: 105–109.

Sato, T., Ashidate, M., Wada, S. & Goshima, S. 2005b. Effects of male mating frequency and male size on ejaculate size and reproductive success of female spiny king crab *Paralithodes brevipes*. *Mar. Ecol. Prog. Ser.* 296: 251–262.

Savage, T. 1971. Mating of the stone crab, *Menippe mercenaria* (Say) (Decapoda, Brachyura). *Crustaceana* 20: 315–316.

Schein, H. 1975. Aspects of the aggressive and sexual behaviour of *Alpheus heterochaelis* Say. *Mar. Behav. Physiol.* 3: 83–96.

Schembri, P.J. 1983. Courtship and mating behaviour in *Ebalia tuberosa* (Pennant) (Decapoda, Brachyura, Leucosiidae. *Crustaceana* 45: 77–81.

Schmitz, B. & Herberholz, J. 1998. Snapping behaviour in intraspecific agonistic encounters in the snapping shrimp (*Alpheus heterochaelis*). *J. Biosci.* 23: 623–632.

Schone, H. 1968. Agonistic and sexual display in aquatic and semiterrestrial brachyuran crabs. *Am. Zool.* 8: 641–654.

Schone, H. & Schone, Y. 1963. Balz und andere Verhaltensweisen der Mangrovenkrabbe *Goniopsis cruentata* Latr. und das Winkverhalten der eulitoralen Brachyuren. *Z. Tierpsychol.* 20: 642–656.

Schoppe, S. 1991. *Echinometra lucunter* (Linnaeus) (Echinoidea, Echinometridae) as host of a complex association in the Caribbean Sea. *Helgoländ. Meeresunter.* 45: 373–379.

Seibt, U. 1973a. Sense of smell and pair-bond in *Hymenocery picta*. *Micronesia* 9: 231–236.

Seibt, U. 1973b. Die beruhigende Wirkung der Partner-Nähe bei der monogamen Garnele *Hymenocera picta*. *Z. Tierpsychol.* 33: 424–427.

Seibt, U. 1974. Mechanismen und Sinnesleistungen fur den Paarzusammenhalt bei der Garnele *Hymenocera picta* Dana. *Z. Tierpsychol.* 35: 337–351.

Seibt, U. 1980. Individuen-Erkennen und Partnerbevorzugung bei der Garnele *Hymenocera picta* Dana. *Z. Tierpsychol.* 52: 321–330.

Seibt, U. & Wickler, W. 1971. New records of the porcellanid crab, *Polyonyx gibbesi* Haig, from the Texas coast (Decapoda, Anomura). *Encycl. Cinematogr. E.* 1723: 1–10.

Seibt, U. & Wickler, W. 1972. Individuen-Erkennen und Partnerbevorzugung bei der Garnele *Hymenocera picta* Dana. *Naturwiss.* 59: 40–41.

Seibt, U. & Wickler, W. 1979. The biological significance of the pair-bond in the shrimp *Hymenocera picta*. *Z. Tierpsychol.* 50: 166–179.

Seibt, U. & Wickler, W. 1981. Paarbildung und Paarbindung bei Krebsen. *Biol. Uns. Zeit* 11: 161–168.

Seiple, W. & Salmon, M. 1982. Comparative social behavior of two grapsid crabs, *Sesarma reticulatum* (Say) and *S. cinereum* (Bosc). *J. Exp. Mar. Biol. Ecol.* 62: 1–24.

Sheard, K. 1949. The marine crayfishes (spiny lobsters), family Palinuridue of Western Australia with particular reference to the fishery on the Western Australian crayfish (*Panulirus longipes*). *CSIRO Aust. Div. Fish. Oceanogr. Bull.* 247: 1–45.

Shokita, S. 1966. Studies on ecology and metamorphosis of the fresh-water shrimp *Macrobrachium longipes* (De Haan). *Biol. Mag., Okinawa* 3: 13–21.

Shih, H.-T. & Mok, H.-K. 1996. *Quadrella maculosa* Alcock, 1898, a genus and species of shallow-water xanthid crab (Brachyura: Xanthoidea: Trapeziidae) new to Taiwan. *Zool. Stud.* 35: 146–148.

Silberbauer, B.I. 1971. The biology of the South African rock lobster, *Jasus lalandli* (H. Milne Edwards) 2. The reproductive organs, mating and fertilization. *S. Afr. Div. Sea Fish. Invest. Rep.* 93: 1–46.

Simon, J.L. & Moore, P.A. 2007. Male–female communication in the crayfish *Orconectes rusticus*: the use of urinary signals in reproductive and non-reproductive pairings. *Ethology* 113: 740–754.

Sin, T. 1999. Distribution and host specialization in *Tetralia* crabs. (Crustacea: Brachyura) symbiotic with corals in the Great Barrier Reef, Australia. *Bull. Mar. Sci.* 65: 839–850.

Smith, E.W. 1953. The life history of the crawfish *Orconectes clypeatus*. *Tulane Stud. Zool.* 1: 79–96.

Snedden, W.A. 1990. Determinants of male mating success in the temperate crayfish *Orconectes rusticus*: chela size and sperm competition. *Behaviour* 115: 100–113.

Snow, C.D. & Nielsen, J.R. 1966. Premating and mating behaviour of the Dungeness crab. (*Cancer magister* Dana). *J. Fish. Res. Bd. Canada* 23: 1319–1323.

Spalding, J.F. 1942. The nature and formation of the spermatophore and sperm plug in *Carcinus maenas*. *Quart. J. Microscop. Sci.* 82/83: 399–422.

Spotte, S. 1996. Supply of regenerated nitrogen to sea anemones by their symbiotic shrimp. *J. Exp. Mar. Biol. Ecol.* 198: 27–36.

Stanton, G. 1977. Habitat partitioning among associated decapods with *Lebrunia danae* at Grand Bahama. In: Taylor D.L. (ed.), *Proceedings of the Third International Coral Reef Symposium, 2 (Biology)*: 169–175. Miami: Univ. Miami Press.

Stebbing, P.D., Bentley, M.G. & Watson, G.J. 2003. Mating behaviour and evidence for a female released courtship pheromone in the signal crayfish *Pacifastacus leniusculus*. *J. Chem. Ecol.* 29: 465–475.

Stevens, B.G. 2003. Timing of aggregation and larval release by Tanner crabs, *Chionoecetes bairdi* in relation to tidal current patterns. *Fish. Res.* 65: 201–216.

Stevens, B.G., Donaldson, W.E. & Haaga, J.A. 1992. First observations of podding behavior for the Pacific lyre crab *Hyas lyratus* (Decapoda: Majidae). *J. Crust. Biol.* 12: 193–195.

Stevens, B.G., Haaga, J.A. & Donaldson, W.E. 1994. Aggregative mating of Tanner crabs, *Chionoectes bairdi*. *Canad. J. Fish. Aquat. Sci.* 51: 1273–1280.

Stone, C.E., O'Clair, C.E. & Shirley, T.C. 1993. Aggregating behavior of ovigerous female red king crab, *Paralithodes camtschaticus*, in Auke Bay, Alaska. *Canada. J. Fish. Aquat. Sci.* 50: 750–758.

Strack, H.L., 1993. Results of the Rumphius Biohistorical Expedition to Ambon (1990). Part. 1. General account and list of stations. *Zool. Verhandl.* 289: 1–72.

Subramoniam, T. 1979. Some aspects of reproductive ecology of a mole crab *Emerita asiatica* Milne Edwards. *J. Exp. Mar. Biol. Ecol.* 36: 259–268.

Subramoniam, T. 1981. Protandric hermaphroditism in a mole crab, *Emerita asiatica* (Decapoda: Anomura). *Biol. Bull.* 160: 161–174.

Subramoniam, T. & Gunamalai, V. 2003. Breeding biology of the intertidal sand crab, *Emerita* (Decapoda: Anomura). *Adv. Mar. Biol.* 46: 91–182.

Sutcliffe, W.H., Jr. 1952. Some observations of the breeding and migration of the Bermuda spiny lobster, *Panulirus argus*. *Proc. Gulf Carib. Fish. Inst., 4th Ann. Sess.*: 64–69.

Sutcliffe, W.H., Jr. 1953. Further observations on the breeding and migration of the Bermuda spiny lobster, *Panulirus argus*. *J. Mar. Res.* 12: 173–183.

Suzuki, H. 1971. Taxonomic review of four alpheid shrimps belonging to the genus *Athanas* with reference to their sexual phenomena. *Sci. Rep. Yokohama Natl. Univ. Sect. II* 17: 1–37.

Swartz, R.C. 1976a. Agonistic and sexual behavior of the xanthid crab, *Neopanope sayi*. *Chesapeake Sci.* 17: 24–34.

Swartz, R.C. 1976b. Sex ratio as a function of size in the xanthid crab, *Neopanope sayi*. *Am. Nat.* 110: 898–900.

Tack, P.I. 1941. The life history and ecology of the crayfish *Cambarus immunis* Hagen. *Am. Midl. Nat.* 25: 420–446.

Takahashi, T., Otani, T. & Matsuura, S. 1999. Swimming behaviour of the pinnotherid crab, *Tritodynamia horvathi* observed during the low temperature season. *J. Mar. Biol. Ass. U.K.* 79: 375–377.

Takeshita, K. & Matsuura, S. 1989. Red king crab in north Pacific. I. Reproduction and growth. *Saibai Giken Tech. Rep. Jpn Sea Branch Program* 18: 35–43.

Tazioli, S., Bo, M., Boyer, M., Rotinsulu, H. & Bavestrello, G. 2007. Ecological observations of some common antipatharian corals in the marine park of Bunaken (North Sulawesi, Indonesia). *Zool. Stud.* 46: 227–241.

Telford, M. 1978. Distribution of two species of *Dissodactylus* (Brachyura: Pinnotheridae) among their echinoid host populations in Barbados. *Bull. Mar. Sci.* 28: 651–658.

Templeman, W. 1934. Mating in the American lobster. *Cont. Canada. Biol. Fish., N. S.* 8: 423–432.

Templeman, W. 1936. Further contributions to mating in the American lobster. *J. Biol. Bd. Canada* 2: 223–226.

Teytaud, A.R. 1971. Laboratory studies of sex recognition in the blue crab *Callinectes sapidus* Rathbun. *Univ. Miami Sea Grant Prog.* 15: 1–63.

Thiel, M. & Correa, C. 2004. Female rock shrimp *Rhynchocinetes typus* mate in rapid succession up a male dominance hierarchy. *Behav. Ecol. Sociobiol.* 57: 62–68.

Thiel, M. & Hinojosa, I. 2003. Mating behavior of female rock shrimp *Rhynchocinetes typus* (Decapoda:caridea)—Indication for convenience polyandry and cryptic female choice. *Behav. Ecol. Sociobiol.* 55: 113–121.

Tierney, A.J. & Dunham, D.W. 1982. Chemical communication in the reproductive isolation of the crayfish *Orconectes propinquus* and *Orconectes virilis* (Decapoda, Cambaridae). *J. Crust. Biol.* 2: 544–548.

Tomikawa, N. & Watanabe, S. 1990. Occurrence of the sperm plugs of *Eriphia smithii* McLeay. *Res. Crust.* 18: 19–21.

Tsuchiya, M. & Yonaha, C. 1992. Community organization of associates of the scleractinian coral *Pocillopora damicornis*: effects of colony size and interactions among the obligate symbionts. *Galaxea* 11: 29–56.

Tsuchiya, M. & Taira, A. 1999. Population structure of six sympatric species of *Trapezia* associated with the hermatypic coral *Pocillopora damicornis* with a hypothesis of mechanisms promoting their coexistence. *Galaxea* 1: 9–18.

Van den Spiegel, Eeckhaut, I. & Jangoux, M. 1998. Host selection by *Synalpheus stimpsoni* (de Man), an ectosymbiotic shrimp of comatulid crinoids, inferred by a field survey and laboratory experiments. *J. Exp. Mar. Biol. Ecol.* 225: 185–196.

Vannini, M. 1985. A shrimp that speaks crab-ese. *J. Crust. Biol.* 5: 160–167.

van Son, T.C. & Thiel, M. 2006. Mating behaviour of male rock shrimp, *Rhynchocinetes typus* (Decapoda: Caridea): effect of recent mating history and predation risk. *Anim. Behav.* 71: 61–70.

Veillet, A. 1945. Recherches sur le parasitisme des crabes et des galathées par les rhizocéphales et les épicarides. *Ann. Inst. Oceanogr.* 22: 194–341.

Vernet-Cornubert, G. 1958a. Biologie générale de *Pisa tetraodon* (Pennant). *Bull. Inst. Océanogr., Monaco* 1113: 1–52.

Vernet-Cornubert, G. 1958b. Recherches sur la sexualité du crabe *Pachygrapsus marmoratus* (Fabricius). *Arch. Zool. Exptl. Gen., Paris* 96: 191–276.

Verrel, P.A. 1985. Predation and the evolution of precopula in the isopod *Asellus aquaticus*. *Behaviour* 95: 198–202.

Villanelli, F. & Gherardi, F. 1998. Breeding in the crayfish, *Austropotamobius pallipes*: mating patterns, mate choice and intermale competition. *Freshw. Biol.* 40: 305–315.

Volz, P. 1938. Studien über das, Knallen" der Alpheiden. Nach Untersuchungen an *Alpheus dentipes* Guérin und *Synalpheus laevimanus* (Heller). *Zoomorph.* 34: 272–316.

von Bonde, C. 1936. The reproduction, embryology and metamorphosis of the Cape crawfish (*Jasus lalandii*) (H. Milne Edwards). *S. Afr. Mar. Biol. Surv. Div. Oceanogr. Invest. Rep.* 6: 1–25.

von Hagen, H.O. 1967. Nachweis einer kinästhetischen Orientierung bei *Uca rapax*. *Z. Morphol. Ökol. Tiere* 58: 301–320.

Vytopil, E. & Willis, B. 2001. Epifaunal community structure in *Acropora* spp. (Scleractinia) on the Great Barrier Reef: implications of coral morphology and habitat complexity. *Coral Reefs* 20: 281–288.

Wada, S., Sonoda, T. & Goshima, S. 1996. Temporal size covariation of mating pairs of the hermit crab *Pagurus middendorffii* (Decapoda: Anomura: Paguridae) during a single breeding season. *Crust. Res.* 25: 158–164.

Wada, S., Tanaka, T. & Goshima, S. 1999. Precopulatory mate guarding in the hermit crab, *Pagurus middendorffii* (Decapoda: Paguridae): effects of population parameters on male guarding duration. *J. Exp. Mar. Biol. Ecol.* 239: 289–298.

Wada, S., Ashidate, M., Yoshino, K., Sato, T. & Goshima, S. 2000. Effects of sex ratio on spawning frequency and mating behavior of the spiny king crab, *Paralithodes brevipes*. *J. Crust. Biol.* 20: 479–482.

Wada, S., Ashidate, M. & Goshima, S. 1997. Observations on the reproductive behavior of the spiny king crab *Paralithodes brevipes* (Anomura: Lithodidae). *Crust. Res.* 26: 56–61.

Wada, S., Ito, A. & Mima, A. 2007. Evolutionary significance of prenuptial molting in female *Pagurus* hermit crabs. *Mar. Biol.* 152: 1263–1270.

Waddy, S.L. & Aiken, D.E. 1981. Natural reproductive cycles of female American lobsters (*Homarus americanus*). In: Clark, W.H., Jr. & Adams, T.S. (eds.), *Developments in Endocrinology. Vol. 11*: 353. New York: Elsevier/North Holland.

Wallace, M., Pertuit, C.J. & Hvatus, A.R. 1949. Contribution to the biology of the king crab *Paralithodes camtschatica* (Tilesius). *U. S. Fish Wildl. Serv. Fish. Leafl.* 340: 1–49.

Warner, G.F. 1967. The life history of the mangrove tree crab, *Aratus pisoni*. *J. Zool., London* 153: 321–335.

Warner, G.F. 1970. Behavior of two species of grapsid crab during intraspecific encounters. *Behaviour* 36: 9–19.

Wasserthal, L.T. & Seibt, U. 1976. Feinstruktur, Funktion und Reinigung der antennalen Sinneshaare der Garnele *Hymenocera picta*. *Z. Tierpsychol.* 42: 186–199.

Watson, J. 1972. Mating behavior in the spider crab, *Chionoecetes opilio*. *J. Fish. Res. Bd. Canada* 29: 447–449.

Wells, W.W. 1928. Pinnotheridae of Puget Sound. *Publ. Puget Sound Biol. Stat., Univ. Wash.* 6: 283–314.

Wenner, E.L. 1989. Incidence of insemination in female blue crabs, *Callinectes sapidus*. *J. Crust. Biol.* 9: 587–594.

Werding, B. 1983. Kommensalische Porzellaniden aus der Karibik (Decapoda, Anomura). *Crustaceana* 45: 1–14.

Werding, B. 1990. *Alpheus schmitti* Chace, 1972, a coral rock boring snapping-shrimp of the tropical western Atlantic (Decapoda, Caridea). *Crustaceana* 58: 88–96.

Wickler, W. 1973. Biology of *Hymenocera picta* Dana. *Micronesia* 9: 225–230.

Wickler, W. & Seibt, U. 1970. Das Verhalten von *Hymenocera picta* Dana, einer Seesterne fressenden Garnele (Decapoda, Natantia, Gnathophyllidae). *Z. Tierpsychol.* 27: 352–368.

Wickler, W. & Seibt, U. 1972. Für den Zusammenhang des Paarsitzens mit anderen Verhaltensweisen bei *Hymenocera picta* Dana. *Z. Tierpsychol.* 31: 163–170.

Wickler, W. & Seibt, U. 1981. Monogamy in Crustacea and man. *Z. Tierpsychol.* 57: 215–234.

Wilber, D.H. 1989. The influence of sexual selection and predation on the mating and postcopulatory guarding behavior of stone crabs (Xanthidae, Menippe). *Behav. Ecol. Sociobiol.* 24: 445–451.

Williams, A.B. 1984. *Shrimps, Lobsters and Crabs of the Atlantic Coast of the Eastern United States, Maine to Florida*. Washington, D.C.: Smith. Inst. Press.

Williams, A.B. 1987. *Upogebia synagelas*, new species, a commensal mud shrimp from sponges in the western central Atlantic (Decapoda: Upogebiidae). *Proc. Biol. Soc. Wash.* 100: 590–595.

Williams, A.B. & Ngoc-Ho, N. 1990. *Pomatogebia*, a new genus of thalassinidean shrimps from western hemisphere tropics (Crustacea: Upogebiidae). *Proc. Biol. Soc. Wash.* 103: 614–616.

Williams, A.B. & Scott, P.J.B. 1989. *Upogebia corallifora*, a new species of coral-boring shrimp from the West Indies (Decapoda: Upogebiidae). *Proc. Biol. Soc. Wash.* 102: 405–410.

Williams, J.D. & McDermott, J.J. 2004. Hermit crab biocoenoses: a worldwide review of the diversity and natural history of hermit crab associates. *J. Exp. Mar. Biol. Ecol.* 305: 1–128.

Williamson, H.C. 1903. On the larval and early young stages and rate of growth of the shore-crab (*Carcinus maenas*). *Rep. Fish. Bd. Scotl., Sci. Invest.* 21: 136–179.

Wittenberger, J.F. & Tilson, R.L. 1980. The evolution of monogamy: hypotheses and evidence. *Ann. Rev. Ecol. Syst.* 11: 197–232.

Wirtz, P. 1997. Crustacean symbionts of the sea anemone *Telmatactis cricoides* at Madeira and the Canary Islands. *J. Zool.* 242: 799–811.

Yaldwyn, J.C. 1964. Pair association in the banded coral shrimp. *Aust. Nat. Hist.* 14: 286.

Yaldwyn, J.C. 1966a. Notes on the behaviour in captivity of a pair of banded coral shrimps, *Stenopus hispidus* (Olivier). *Aust. Zool.* 8: 377–389.

Yaldwyn, J.C. 1966b. Protandrous hermaphroditism in decapod prawns of the families Hippolytidae and Campylonotidae. *Nature* 209: 1366.

Yano, I., Kanna, R.A., Oyarna, R.N. & Wyban, J.A. 1988. Mating behavior in the penaeid shrimp *Penaeus vannamei. Mar. Biol.* 97: 171–175.

Zhang, D., Rhyne, A.L. & Lin, J. 1998. Density-dependent effect on reproductive behaviour of *Lysmata amboinensis* and *L. boggessi* (Decapoda: Caridea: Hippolytidae). *J. Mar. Biol. Ass. U.K.* 87: 517–522.

Zhou, S. & Shirley, T.C. 1997. Distribution of red king crabs and Tanner crabs in the summer by habitat and depth in an Alaskan fjord. *Invest. Mar., Valparaíso* 25: 59–67.

Zmarzly, D.L. 1992. Taxonomic review of pea crabs in the genus *Pinnixa* (Decapoda: Brachyura: Pinnotheridae) occurring on the California Shelf, with descriptions of two new species. *J. Crust. Biol.* 12: 677–713.

# A Shrimp's Eye View of Evolution: How Useful Are Visual Characters in Decapod Phylogenetics?

MEGAN L. PORTER & THOMAS W. CRONIN

*Department of Biological Sciences, University of Maryland Baltimore County, Baltimore, Maryland, U.S.A.*

ABSTRACT

The decapods contain the largest diversity of eye designs and optical types of any group within the Crustacea. This variation has led to debate about the usefulness of visual system characters in the construction of decapod phylogenetic relationships. This debate, however, has not been revisited recently and has never considered the use of molecular aspects of vision. In this paper we review the current understanding of decapod eye anatomy, optics, visual pigments, and evolution. We find that there are many visual system components, including overall optical design and fine structural details, that are potentially useful for reconstructing decapod phylogenetics.

1 INTRODUCTION

Within crustaceans, the decapods are unrivalled in species number, morphological diversity, and ecological distribution. Correspondingly, the decapods also exhibit extraordinary variation in the optical design and morphology of their visual systems. This leads to the simple question: 'Does the observed variation in visual systems contain useful information concerning the evolution of the decapods?' The use of visual system characteristics has been debated throughout the history of decapod taxonomic studies, with just as many decapod researchers arguing for the importance of eye characters as cautioning against their use. In this review we will revisit the debate regarding decapod optical design and phylogenetics. Our goal is to move the debate forward by revising the general question posed above to: 'Does the observed variation in visual systems, both morphological and molecular, have anything useful to tell us about decapod phylogenetics?' In order to investigate this question, we will present the current knowledge regarding the taxonomic and phylogenetic distribution of optical designs and the emerging field of molecular studies on visual system evolution within the decapods.

2 OVERVIEW OF DECAPOD VISUAL SYSTEMS

2.1 *Morphology*

Most Crustacea have compound eyes composed of individual receptive units called ommatidia (Fig. 1). Each ommatidium consists of optical structures (e.g., cornea, lens, crystalline cones) stacked on top of a set of fused retinular cells, which form the photoreceptive rhabdom (Fig. 2). Decapod rhabdoms are formed by eight retinular cells, with seven of these (R1–7) forming the main proximal part of the rhabdom and the eighth (R8), if present, contributing a small distal rhabdomere (Shaw & Stowe 1982). Based on results from a range of methodologies aimed at characterizing visual pigment absorbance and photoreceptor sensitivity (e.g., microspectrophotometry, electrophysiology, intracellular recordings), the spectral characteristics of the R1–7 versus the R8 retinular cells differ. Within the Decapoda, the R1–7 cells of the main rhabdom are sensitive to middle

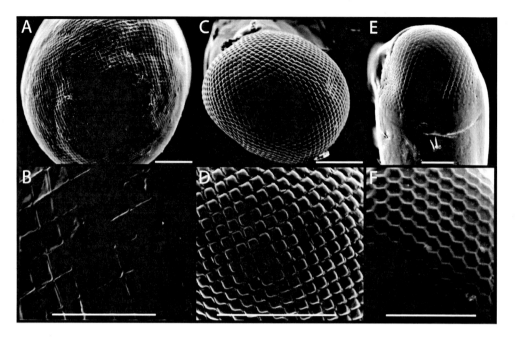

**Figure 1.** Examples of decapod compound eyes demonstrating different facet shapes. Note that some deformation of the shape of the compound eyes has occurred due to the SEM fixation process. (A, B) *Procambarus* sp., illustrating the square facets characteristic of reflecting superposition optics (scale bars: A = 500 μm, B = 200 μm). (C, D) *Stenopus hispidus*, which also contains reflecting superposition optics (scale bars = 500 μm). (E, F) *Clibanarius* sp. (scale bars = 100 μm). Although the underlying optics of this genus have not been investigated specifically, the hexagonal facets imply that this species does not contain reflecting superposition eyes. Other species within the same family (e.g., *Dardanus* sp., Diogenidae) have refracting superposition optics. (Photos by M.L. Porter.)

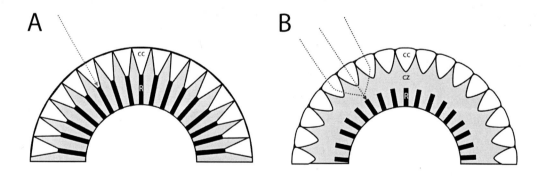

**Figure 2.** Schematics of the two basic compound eye optical designs found in decapod crustaceans: (A) apposition optics, (B) superposition optics. Dashed grey lines represent typical light paths through the crystalline cones to the rhabdoms. Abbreviations: cc = crystalline cone; R = rhabdom; cz = clear zone.

(blue-green) to long (red) wavelengths of light (447–570 nm), while the R8 cells are typically sensitive to violet or UV light (360–440 nm, Fig. 3, Johnson et al. 2002; Porter 2005).

As early as the late 1800s (Exner 1891), it was recognized that the compound eye ground plan can be organized into two optical types: apposition and superposition eyes (Fig. 2). Typically optimized for resolution, apposition eyes contain ommatidia that function as individual units, with screening pigments shielding each individual ommatidium from receiving light from neighboring optical components. In contrast, superposition eyes are commonly optimized for sensitivity, with the optical elements of multiple ommatidia focusing light onto a single rhabdom. Within the Crustacea as a whole, most of the visual systems investigated contain apposition optics, with superposition eyes found only in the Eumalacostraca (Nilsson 1983). In comparison, the decapods contain extraordinary variation in eye design within a single order, exhibiting four fundamentally different optical designs that can be observed among species and different life stages. While all decapod larvae use apposition optics, only a few adult decapods retain apposition eyes, including brachyuran, anomalan, and stenopodidean species (Meyer-Rochow & Reid 1994; Eguchi et al. 1997; Gaten 1998, 2007). Within superposition optics, decapods have evolved three mechanisms for focusing light from multiple ommatidial facets onto a single rhabdom: reflection, refraction, and parabolic optics. Refracting superposition eyes have been found in only two other crustacean groups in addition to the decapods, the Mysida and the Euphausiacea, while reflecting and parabolic superposition eyes are not found outside the Decapoda (Nilsson 1988, 1990).

By far the most widespread design in decapod eyes is reflecting superposition optics, found in the adults of all of the major sub- and infraorders, with the possible exception of the Thalassinidea, where eye design has yet to be rigorously investigated (Table 1). First described in crayfish and deep sea shrimp (Land 1976; Vogt 1977), this optical design uses either mirror boxes lined with a reflective surface or complete internal reflection within the crystalline cone to reflect incoming light to a particular rhabdom. In contrast, the remaining two superposition optical variants are found in only a few decapod families. Refracting superposition optics function using refractive gradients in the crystalline cone and have been described in decapods only from species of deep sea shrimp within the Benthesicymidae and hermit crab species from the genus *Dardanus*, within the Diogenidae (Nilsson 1990, see Table 1). Parabolic superposition optics utilize a combination of structures including lenses, parabolic mirrors, and light guides, and have been characterized only from brachyuran and anomalan crabs (Nilsson 1988).

**Table 1.** Taxonomic distribution of adult decapod compound eye optical designs. Taxonomic designations follow the scheme of Martin & Davis (2001). Question marks indicate uncertainty about eye type. AP = apposition, RFL = reflecting, RFR = refracting, PB = parabolic.

|  | AP | Superposition | | | Reference |
| --- | --- | --- | --- | --- | --- |
|  |  | RFL | RFR | PB |  |
| **Dendrobranchiata** |  |  |  |  |  |
| Benthesicymidae |  |  | X |  | Nilsson 1990 |
| Penaeidae |  | X |  |  | Colin Nicol & Yan 1982; Gaten 1998 |
| Sergestidae |  | X |  |  | Welsh & Chace 1938; Ball et al. 1986 |
| **Caridea** |  |  |  |  |  |
| Crangonidae |  | X |  |  | Gaten 1998 |
| Oplophoridae |  | X |  |  | Welsh & Chace 1937; Land 1976; Gaten et al. 1992 |
| Palaemonidae |  | X |  |  | Doughtie & Rao 1984; Fincham 1984; Meyer-Rochow et al. 1992 |
| Pandalidae |  | X |  |  | Gaten 1998 |
| Pasiphaeidae |  | X |  |  | Gaten 1998 |

**Table 1.** continued.

| | AP | Superposition | | | Reference |
|---|---|---|---|---|---|
| | | RFL | RFR | PB | |
| **Stenopodidea** | | | | | |
| Spongicolidae | X | | | | Gaten 2007 |
| Stenopodidae | | X | | | Richter 2002 |
| **Achelata** | | | | | |
| Palinuridae | | X | | | Eguchi & Waterman 1966; Meyer-Rochow 1975 |
| **Anomala** | | | | | |
| Hippoidea | | | | | |
|   Hippidae | X | | | | Gaten 1998 |
| Galatheoidea | | | | | |
|   Aeglidae | X | | | | Gaten 1998 |
|   Chirostylidae | | X | | | Gaten 1998 |
|   Galatheidae | | X | | | Kampa 1963; Gaten 1994 |
|   Porcellanidae | | X | | | Fincham 1988; Meyer-Rochow et al. 1990 |
| Paguroidea | | | | | |
|   Diogenidae | | | X | | Nilsson 1990 |
|   Paguridae | | | | X | Nilsson 1988 |
| **Astacidea** | | | | | |
| Nephropidae | | X | | | Shelton et al. 1981; Gaten 1988 |
| Astacidae | | X | | | Vogt 1975 |
| Cambaridae | | X | | | Tokarski & Hafner 1984 |
| Parastacidae | | X | | | Bryceson 1981 |
| **Brachyura** | | | | | |
| DROMIACEA | | | | | |
|   Dromiidae | | X | | | Gaten 1998 |
|   Homolidae | | X | | | Gaten 1998 |
|   Latreilliidae | | X | | | Gaten 1998 |
| EUBRACHYURA | | | | | |
| Raninoida | | | | | |
|   Raninidae | | X | | | Gaten 1998 |
| Heterotremata | | | | | |
|   Geryonidae | X? | | | X? | Gaten 1998 |
|   Hymenosomatidae | X | | | | Meyer-Rochow & Reid 1994 |
|   Majidae | | | | X | Nilsson 1988 |
|   Portunidae | X? | | | X | Leggett & Stavenga 1981, Nilsson 1988 |
|   Xanthidae | | | | X | Nilsson 1988 |
| Thoracotremata | | | | | |
|   Grapsidae | X | | | | Arikawa et al. 1987 |
| **Thalassinidea** | | | | | (undescribed) |

On the surface of the eye, either reflecting or parabolic optics can have square ommatidial facets, while apposition, refracting, and parabolic superposition types can all have ommatidial facets ranging from circular to hexagonal. Therefore, the optical design of a visual system cannot be determined without careful investigation of the internal retinal anatomy. As the internal eye structure of only 74 species, representing 32 of ~150 decapod families, has been investigated, the possibility for new discoveries in decapod optical designs still exists.

## 2.2 Evolutionary enigma of eye design

It has been argued that, once evolved, most compound eye designs would not be replaced by another design unless the change rendered a significant optical advantage (Land 1981; Gaten 1998). It is also difficult to conceive how a visual system can move from one eye type to another without going through a near-blind intermediate (Land 1981). This difficulty in moving between states lends support to the stability of eye structure as a phylogenetic character. However, it also makes the evolution of complex eye designs, particularly of superposition optics, an evolutionary enigma.

In comparison with apposition eyes, superposition eyes are optically intricate and a rarity in animal vision (Land 1981). As most crustaceans appear to possess apposition eyes, including all decapod larvae, it is reasonable to postulate that the superposition optics found in adult decapods arose from apposition eyes (Richter 2002). Optically, it is possible to go from apposition to superposition eyes as well, as most decapods make this transition developmentally when changing from larval to adult forms (Meyer-Rochow 1975). In fact, the transparent type of apposition eye found in decapod larvae designed for planktonic life is pre-adapted for superposition optics. Nilsson (1983) showed that the mechanism for superimposing rays is present, but not used, in decapod larval eyes.

Based on taxonomic (Table 1) and phylogenetic distribution (Fig. 4), it is likely that reflecting superposition optics arose early in decapod evolution. Gaten (1998) suggested that reflecting superposition optics are symplesiomorphic for the Decapoda, having evolved only once, probably in the Devonian; however, it has also been hypothesized that Galatheidae (Anomala) independently acquired reflecting superposition eyes based on the presence of a light guide and the formation of the clear zone via elongation of the distal rhabdom (Gaten 1994). The acquisitions of the remaining eye types in decapods, then, represent transitions between superposition types or the paedomorphic retention of apposition eyes (Gaten 2007).

Because reflecting and refracting superposition eyes have approximately similar qualities and brightnesses of the images they produce (Land 1981), it is difficult to imagine the advantage of switching between eye designs. No functional insight is gained from the ecology of the families where refracting optics have been described: the Benthesicymidae, a group of deep-sea shrimp within the Dendrobranchiata, and some species of hermit crabs, e.g., *Dardanus megistos*, found in brightly lit, shallow marine habitats. However, close examination of the structures in these two reflecting eye types indicate different ancestral origins, with the eyes of the Benthesicymidae originating from reflecting optics and the eyes of *Dardanus* being derived from parabolic optics (Nilsson 1990). Furthermore, it is theoretically possible to transform from a parabolic into a refracting superposition eye, and various intermediates between the two types have been found (Nilsson 1990; Gaten 1998). Therefore within the anomalan Paguroidea, it is possible that the ancestral optical state is parabolic superposition, with the *Dardanus* refracting eye representing a derived optical state that was an easier transformation than returning to reflecting optics. Regardless of origin, the taxonomic and phylogenetic distributions of both refracting and parabolic superposition eye types imply that there have been multiple independent acquisitions of these eye designs within the Decapoda (Fig. 4).

## 2.3 Molecular aspects of decapod vision

A considerable amount of research has been devoted previously to decapod visual systems (see reviews by Johnson et al. 2002; Cronin 2005). However, most of this research has investigated the morphological structure (Table 1) and physiological function (Fig. 3) of the eye. Very few molecular studies of the decapod visual system have been undertaken, and none has evaluated the phylogenetic signal of the genes involved in vision.

Sensitivity to light in all animal vision is based on visual pigments, which are composed of a chromophore (vitamin A derivative) bound to an integral membrane protein (opsin) and

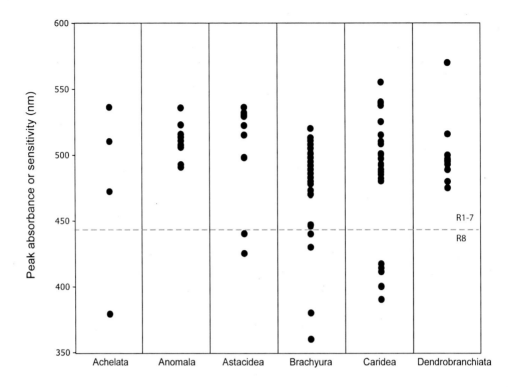

**Figure 3.** Measurements of spectral maxima ($\lambda_{max}$) of visual pigment absorbance and photoreceptor spectral sensitivities recorded from decapod species, separated by major group (suborder Dendrobranchiata, and infraorders Achelata, Anomala, Astacidea, Brachyura, and Caridea). Data were recorded using a variety of methods, including microspectrophotometry, electroretinography, intracellular recordings, and spectroscopy of pigment extracts (for original sources see Johnson et al. 2002; Porter 2005).

characterized by the wavelength of maximal absorption ($\lambda_{max}$). Although there are a number of morphological and physiological methods of controlling the spectral sensitivity of a photoreceptor, the underlying molecular mechanism is the interaction between the particular amino acid sequence of the opsin protein and the type of chromophore. Two different chromophores have been documented from decapod visual pigments, but one of these, the 3-dehydroretinal form, has been found only in crayfish (Suzuki et al. 1984, 1985; Suzuki & Eguchi 1987). All other decapod species studied utilize retinal as the visual pigment chromophore; therefore, the underlying variation in decapod photoreceptor sensitivity is largely determined by the specific amino acid sequence of the opsin protein.

Currently the only available decapod opsin sequences are from two brachyuran crabs (Sakamoto et al. 1996; Kuballa et al. 2007), ten crayfish species (Hariyama et al. 1993; Crandall & Cronin 1997; Crandall & Hillis 1997), one clawed lobster (Porter et al. 2007), and two penaeid shrimp (GenBank accession: DQ825437 and Lehnert et al. 1999). Opsin sequences are notoriously bad for inferring phylogenetic relationships among species due to the high potential for convergence among gene products of a given spectral sensitivity. Because decapods contain only one or two classes of photoreceptors, each tuned to a fairly narrow portion of the visible spectrum, the problem of convergence may be magnified (Fig. 3). However, even given these constraints there are a few important insights regarding the evolution of decapods that can be gleaned from investigating decapod opsin evolution. First, all of the characterized decapod opsin sequences, with the exception of the brachyurans, cluster with insect long- to middle-wavelength sensitive opsins (Fig. 5). However, the decapod sequences do not cluster together and are scattered throughout the crustacean clade. This,

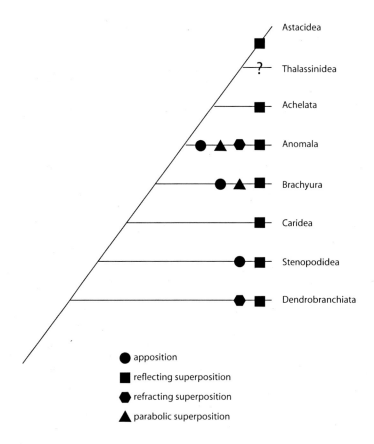

**Figure 4.** Phylogenetic distribution of optical eye designs within the major decapod lineages. Topology of the decapod relationships drawn after Porter et al. (2005).

in conjunction with the identification of three different sequences from a single species (*Penaeus monodon*), implies that opsin gene duplication within the Crustacea has been rampant. Second, the opsin sequences available for brachyuran crabs from *Hemigrapsus sanguinensus* (Sakamoto et al. 1996) and *Portunus pelagicus* (Kuballa et al. 2007) exhibit a distinct phylogenetic placement away from the other decapod sequences. This suggests that in the evolutionary history of opsin gene duplication, diversification, and loss, the brachyuran crabs have co-opted a different copy of the opsin gene from the remaining decapod lineages.

Apart from the admittedly limited information about opsin evolution, little else is known about the network of genes involved in decapod phototransduction. From studies of *Drosophila*, the gene network involved in arthropod phototransduction has been fairly well elucidated (Ranganathan et al. 1991; Zuker 1992, 1996). Few of these interacting genes have been specifically investigated in decapods, and none of the known sequences has been investigated with respect to visual function (Table 2). As opsin is likely to be the most variable gene in the visual signaling cascade due to environmental 'tuning' of the visual pigment spectral absorbance, the remaining genes in the phototransduction network may be more conserved nuclear gene targets for future phylogenetic studies.

## 3 VISUAL SYSTEM COMPONENTS AS PHYLOGENETIC CHARACTERS

Different classification schemes of the decapods have been based on a wide range of characters including behavior (Boas 1880; Borradaile 1907), gill anatomy (Bate 1888; Burkenroad 1963);

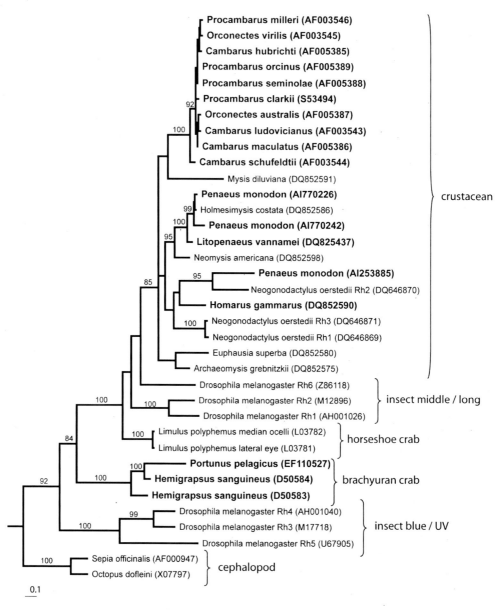

**Figure 5.** Phylogeny of decapod and selected invertebrate opsins based on maximum likelihood analyses of amino acid residues. The phylogeny was reconstructed using PHYML (Guindon & Gascuel 2003) and rooted (not shown) using bovine rhodopsin (NC_007320), chicken pinopsin (U15762), and human melatonin receptor 1A (NM_005958) and GPCR52 (NM_005684). The numbers above each branch indicate the bootstrap proportion from 100 replicates (values less than 70% not shown). The major clusters of opsin sequences are indicated by taxonomic group, and, where possible, the visual pigment spectral sensitivity of each cluster is indicated as middle, long, blue, or ultraviolet (UV) sensitive. Decapod sequences are indicated in bold.

**Table 2.** List of decapod genes known to be involved in phototransduction that are available in GenBank.

| |
|---|
| Dendrobranchiata |
|   Penaeidae |
|     *Penaeus monodon* |
|       Phospholipase C (PLC): AI253804 |
|     *Marsupenaeus japonicus* |
|       Calmodulin: AU175456 |
| Astacidea |
|   Cambaridae |
|     *Procambarus clarkii* |
|       Gq-alpha subunit protein: AAB28122 |
|   Parastacidae |
|     *Cherax quadricarinatus* |
|       Calmodulin: DQ847760, DQ847613 |
|   Nephropidae |
|     *Homarus americanus* |
|       Calmodulin: FD467399, EH116795, CN852450 |
|       Inositol triphosphate: FD467309, EW702750 |
|       Phospholipids phopholipase C beta isoform (PLC): AF128539 |
| Brachyura |
|   Portunidae |
|     *Carcinus maenas* |
|       Gq/11-alpha subunit protein: DV944278, DV642918 |

features of the head, thorax, and carapace (Saint Laurent 1979; Scholtz & Richter 1995); position of the genital openings (Guinot 1978); molecular sequence data (Ahyong & O'Meally 2004; Porter et al. 2005); as well as elements of eye design (Fincham 1980). The utility of visual system components, however, has been debated throughout the history of decapod taxonomic studies. As discussed above (see section 1.1.2), superposition eyes are intricately complex structures, making transitions between different optical types improbable. If this is true, eye structure is a stable character, and therefore the distribution of optical designs in decapods has phylogenetic significance (Fincham 1980; Land 1981; Fincham 1984). Following this line of thinking, elements of the visual system have been used as characters uniting the 'Natantia' or shrimp-like decapods (Fincham 1980) and the 'long bodied' decapods (e.g., shrimp, lobsters, and crayfish) (Land 1981), respectively. In contrast, Nilsson (1983) cautions against the use of visual elements as phylogenetic characters due to repeated, independent gains of similar optical designs.

In fact, visual systems within the decapods exhibit both stable evolutionary characters and independent gains/losses of similar designs. The evolutionary distribution of eye designs within the decapods indicates that the stem lineage most likely contained reflecting superposition optics, at least in adults (Fig. 4, Richter 2002). As the decapods are the only group of crustaceans possessing this unique optical design, reflecting optics serve as a useful character for uniting the decapods. Lineages containing different optical designs, which most assuredly have arisen independently multiple times, may still provide characters for uniting higher-level groups by detailed examination of the optical structures. For example, the refracting optics found in the Benthesicymidae differ from the *Dardanus megistos* refracting eye in fine structural details, including the power of the lens and the origin of the light guide crossing the clear zone (Nilsson 1990). With further detailed investigations of decapod eye structure, these types of details may provide additional visual characters containing strong phylogenetic signal. There are also a number of decapod species that live in light-limited environments (e.g., deep sea, caves, burrows) where eyes are often reduced or lost, and here visual system components may provide little phylogenetic signal (Gaten et al. 1998a, 1998b; Mejia-Ortiz & Hartnoll 2005).

Within decapods, the Anomala and Brachyura contain the greatest diversity of optical designs (Table 1, Fig. 4). This diversity of eye designs has led to multiple interpretations of relationships within the crab infraorders, including removal of the Dromioidea and Homoloidea from the Brachyura based on eye type (Fincham 1980). The validity of the Anomala as a coherent taxon also has been questioned based on the diversity of eye design (Fincham 1980; Gaten 1994). The true evolutionary significance of this variation is still unclear. However, within a phylogenetic context, at least some of the diversity of eye designs found in the crab groups most certainly represents independent acquisitions within specific lineages.

Finally, there are still areas of decapod vision that have not yet been thoroughly investigated, making evaluation of characters for phylogenetic reconstruction difficult. From a molecular perspective, not much is known about the decapod visual system and much work remains. However, the Brachyura appear to use a unique set of opsins not found in other decapods. In some deep sea carideans there is an accessory compound eye on the dorsal margin of the eye (Gaten et al. 1992) that, with further documentation, may provide a useful character within the carideans. Similarly, a number of decapod extraocular photoreceptors have been documented, including intracerebral and caudal photoreceptors (Wilkens & Larimer 1976; Sandeman et al. 1990); investigations of the morphological and molecular components of these extraocular structures also may provide further insight into decapod evolution.

## 4 SUMMARY

The structure and design of decapod compound eyes reveal their function and are influenced by the behavior, ecology, and evolutionary history of the species (Schiff & Hendrickx 1997; Meyer-Rochow 2001). Here we have reviewed the components of the decapod visual system, both structural and molecular, in the hope of providing information that could lead towards a more synthetic phylogenetic reconstruction of decapod relationships. We also highlight some of the critical information still needed to understand visual system evolution within the decapods. Are the optical designs and molecular pathways involved in vision useful for decapod phylogenetic study? Our review of the current data suggests that there are many phylogenetically useful visual system components. However, much work is needed in decapod vision, including investigations of optical design in understudied groups (e.g., Achelata, Thalassinidea, and Stenopodidea) and studies of the phototransduction cascade in general. The overall optical eye designs may be useful characters within, but not among, major lineages, and the fine structural details of each visual system may provide further insights.

## ACKNOWLEDGEMENTS

We would like to thank the decapod ATOL NSF research group for initiating this review. This work was supported by grants from NSF (IOS-0721608) and AFOSR (02NL253).

## REFERENCES

Ahyong, S.T. & O'Meally, D. 2004. Phylogeny of the Decapoda Reptantia: resolution using three molecular loci and morphology. *Raff. Bull. Zool.* 52: 673–693.

Arikawa, K., Kawamata, K., Suzuki, T. & Eguchi, E. 1987. Daily changes of structure, function and rhodopsin content in the compound eye of the crab *Hemigrapsus sanguineus*. *J. Comp. Physiol.* A 161: 161–174.

Ball, E.E., Kao, L.C., Stone, R.C. & Land, M.F. 1986. Eye structure and optics in the pelagic shrimp *Acetes sibogae* (Decapoda, Natantia, Sergestidae) in relation to light-dark adaptation and natural history. *Philos. Trans. R. Soc. Lond. B. Biol. Sci.* 313: 251–270.

Bate, C.S. 1888. Report on the Crustacean Macrura dredged by H.M.S. Challenger during the years 1873–1876. *Rep. Sci. Results Voy. Challenger (Zool)* 24: 1–942.

Boas, J.E.V. 1880. Studier over Decapodernes Slaegtskabsforhold. *Vidensk Selskab Skr.* 6: 25–210.

Borradaile, L.A. 1907. On the classification of the decapod crustaceans. *Ann. Mag. Nat. Hist.* 19: 457–486.

Bryceson, K.P. 1981. Focusing of light by corneal lenses in a reflecting superposition eye. *J. Exp. Biol.* 90: 347–350.

Burkenroad, M.D. 1963. The evolution of the Eucarida (Crustacea, Eumalacostraca), in relation to the fossil record. *Tulane Stud. Geol.* 2: 1–17.

Colin Nicol, J.A. & Yan, H.-Y. 1982. The eye of the grass shrimp *Penaeus monodon*. *Bull. Inst. Zool, Academia Sinica* 21: 27–50.

Crandall, K.A. & Cronin, T.W. 1997. The molecular evolution of visual pigments of freshwater crayfishes (Decapoda: Cambaridae). *J. Mol. Evol.* 45: 524–534.

Crandall, K.A. & Hillis, D.M. 1997. Rhodopsin evolution in the dark. *Nature* 387: 667–668.

Cronin, T.W. 2005. Invertebrate vision in water. In: Warrant, E. & Nilsson, D.-E. (eds.), *Invertebrate Vision*. Cambridge, UK: Cambridge University Press.

Doughtie, D.G. & Rao, K.R. 1984. Ultrastructure of the eyes of the grass shrimp *Palaemonetes pugio*: general morphology and light and dark adaptation at noon. *Cell Tissue Res.* 238: 271–288.

Eguchi, E., Dezawa, M. & Meyer-Rochow, V.B. 1997. Compound eye fine structure in *Paralomis multispina* Benedict, an anomalan half-crab from 1200 m depth (Crustacea; Decapoda; Anomala). *Biol. Bull.* 192: 300–308.

Eguchi, E. & Waterman, T.H. 1966. Fine structure patterns in crustacean rhabdoms. In: Bernhard, C.G. (ed.), *The Functional Organisation of the Compound Eye*: 105–124. Oxford: Pergamon Press.

Exner, S. 1891. *Die Physiologie der facettirten Augen von Krebsen und Insecten*. Leipzig and Vienna, Deuticke.

Fincham, A.A. 1980. Eyes and classification of malacostracan crustaceans. *Nature* 287: 729–731.

Fincham, A.A. 1984. Ontogeny and optics of the eyes of the common prawn *Palaemon (Palaemon) serratus* (Pennant, 1777). *Zool. J. Linn. Soc.* 81: 89–113.

Fincham, A.A. 1988. Ontogeny of anomalan eyes. *Symp. Zool. Soc. Lond.* 59: 123–155.

Gaten, E. 1988. Light-induced damage to the dioptric apparatus in *Nephrops norvegicus* and the quantitative assessment of the damage. *Mar. Behav. Physiol.* 13: 169–183.

Gaten, E. 1994. Geometrical optics of a galatheid compound eye. *J. Comp. Physiol. A* 175: 749–759.

Gaten, E. 1998. Optics and phylogeny: is there an insight? The evolution of superposition eyes in the Decapoda (Crustacea). *Contrib. Zool.* 67: 223–236.

Gaten, E. 2007. Apposition compound eyes of *Spongicoloides koehleri* (Crustacea: Spongicolidae) are derived by neoteny. *J. Mar. Biol. Ass. UK* 87: 483–486.

Gaten, E., Herring, P., Shelton, P.M.J. & Johnson, M.L. 1998a. The development and evolution of the eyes of vent shrimps (Decapoda: Bresiliidae). *Cahiers Biol. Mar.* 39: 287–290.

Gaten, E., Herring, P.J., Shelton, P.M.J. & Johnson, M.L. 1998b. Comparative morphology of the eyes of postlarval bresiliid shrimps from the region of hydrothermal vents. *Biol. Bull.* 194: 267–280.

Gaten, E., Shelton, P.M.J. & Herring, P.J. 1992. Regional morphological variations in the compound eyes of certain mesopelagic shrimps in relation to their habitat. *J. Mar. Biol. Ass. UK* 72: 61–75.

Guindon, S. & Gascuel, O. 2003. A simple, fast, and accurate algorithm to estimate large phylogenies by maximum likelihood. *Sys. Biol.* 52: 696–704.

Guinot, D. 1978. Principes d'une classification évolutive des Crustacés Décapodes Brachyoures. *Bull. Biol. Fr. Belg.* 112: 211–292.

Hariyama, T., Ozaki, K., Tokunaga, F. & Tsukahara, Y. 1993. Primary structure of crayfish visual pigment deduced from cDNA. *FEBS Lett.* 315: 287–292.

Johnson, M.L., Gaten, E. & Shelton, P.M.J. 2002. Spectral sensitivities of five marine decapod crustaceans and a review of spectral sensitivity variation in relation to habitat. *J. Mar. Biol. Ass. UK* 82: 835–842.

Kampa, E.M. 1963. The structure of the eye of a galatheid crustacean, *Pleuroncodes planipes*. *Crustaceana* 6: 69–80.

Kuballa, A.V., Merritt, D.J. & Elizur, A. 2007. Gene expression profiling of cuticular proteins across the moult cycle of the crab *Portunus pelagicus*. *BMC Biol.* 5: 45.

Land, M.F. 1976. Superposition images are formed by reflection in the eyes of some oceanic decapod Crustacea. *Nature* 263: 764–765.

Land, M.F. 1981. Optical mechanisms in the higher Crustacea with a comment on their evolutionary origins. In: Laverack, M.S. & Cosens, D.J. (eds.), *Sense Organs*: 31–48. London: Blackie Press.

Leggett, L.M.W. & Stavenga, D.G. 1981. Diurnal changes in angular sensitivity of crab photoreceptors. *J. Comp. Physiol.* 144: 99–109.

Lehnert, S.A., Wilson, K.J., Byrne, K. & Moore, S.S. 1999. Tissue-specific expressed sequence tags from the black tiger shrimp *Penaeus monodon*. *Mar. Biotechnol. (NY)* 1: 465–476.

Martin, J.W. & Davis, G.E. 2001. An updated classification of the Recent Crustacea. *Nat. Hist. Mus. Los Angeles County, Science Ser.* 39: 1–124.

Mejia-Ortiz, L.M. & Hartnoll, R.G. 2005. Modifications of eye structure and integumental pigment in two cave crayfish. *J. Crust. Biol.* 25: 480–487.

Meyer-Rochow, V.B. 1975. Larval and adult eye of the Western rock lobster (*Panulirus longipes*). *Cell Tissue Res.* 162: 439–457.

Meyer-Rochow, V.B. 2001. The crustacean eye: dark/light adaptation, polarization, sensitivity, flicker fusion frequency, and photoreceptor damage. *Zoolog. Sci.* 18: 1175–1197.

Meyer-Rochow, V.B. & Reid, W.A. 1994. The eye of the New Zealand freshwater crab *Halicarcinus lacustris*, and some eco-physiological predictions based on eye anatomy. *J. R. Soc. of New Zealand* 24: 133–142.

Meyer-Rochow, V.B., Reid, W.A. & Haley, M. 1992. Photomechanical responses in the eye of the jamaican freshwater shrimp *Machrobrachium heterochirus* (Wiegmann). *Res. Crust.* 21: 33–45.

Meyer-Rochow, V.B., Towers, D. & Ziedins, I. 1990. Growth patterns in the eye of *Petrolisthes elongates* (Crustacea; Decapoda; Anomala). *Exp. Biol.* 48: 329–340.

Nilsson, D.-E. 1983. Evolutionary links between apposition and superposition optics in crustacean eyes. *Nature* 302: 818–821.

Nilsson, D.-E. 1988. A new type of imaging optics in compound eyes. *Nature* 332: 76–78.

Nilsson, D.-E. 1990. Three unexpected cases of refracting superposition eyes in crustaceans. *J. Comp. Physiol. A* 167: 71–78.

Porter, M.L. 2005. Crustacean phylogenetic systematics and opsin evolution. Ph.D. Dissertation, Brigham Young University, 189 p.

Porter, M.L., Cronin, T.W., McClellan, D.A. & Crandall, K.A. 2007. Molecular characterization of crustacean visual pigments and the evolution of pancrustacean opsins. *Mol. Biol. Evol.* 24: 253–268.

Porter, M.L., Pérez-Losada, M. & Crandall, K.A. 2005. Model-based multi-locus estimation of decapod phylogeny and divergence times. *Mol. Phylogenet. Evol.* 37: 355–369.

Ranganathan, R., Harris, W.A. & Zuker, C.S. 1991. The molecular genetics of invertebrate phototransduction. *Trends Neurosci.* 14: 486–493.

Richter, S. 2002. Evolution of optical design in the Malacostraca (Crustacea). In: Wiese, K. (ed.), *The crustacean nervous system*: 512–524. Heidelberg: Springer.

Saint Laurent, M.d. 1979. Vers une nouvelle classification des Crustacés Décapodes Reptantia. *Bulletin de l'Office National des Pêches République Tunisienne, Ministere de L'Agriculture* 3: 15–31.

Sakamoto, K., Hisatomi, O., Tokunaga, F. & Eguchi, E. 1996. Two opsins from the compound eye of the crab *Hemigrapsus sanguineus*. *J. Exp. Biol.* 199: 441–450.

Sandeman, D.C., Sandeman, R.E. & DeCouet, H.G. 1990. Extraretinal photoreceptors in the brain of the crayfish *Cherax destructor*. *J. Neurobiol.* 21: 619–629.

Schiff, H. & Hendrickx, M.E. 1997. An introductory survey of ecology and sensory receptors of tropical eastern Pacific crustaceans. *Italian J. Zool.* 64: 13–30.

Scholtz, G. & Richter, S. 1995. Phylogenetic systematics of the reptantian Decapoda (Crustacea, Malacostraca). *Zool. J. Linn. Soc.* 113: 289–328.

Shaw, S.R. & Stowe, S. 1982. Photoreception. In: Atwood, H.L. & Sandeman, D.C. (eds.), *The Biology of Crustacea*: 291–367. New York: Academic Press.

Shelton, P.M.J., Shelton, R.G.J. & Richards, P.R. 1981. Eye development in relation to moult stage in the European lobster *Homarus gammarus* (L.). *J. Cons. Int. Explor.. Mer.* 39: 239–243.

Suzuki, T., Arikawa, K. & Eguchi, E. 1985. The effects of light and temperature on the rhodopsin-porphyropsin visual system of the crafish, *Procambarus clarkii*. *Zoolog. Sci.* 2: 455–461.

Suzuki, T. & Eguchi, E. 1987. Survey of 3-dehydroretinal as a visual pigment chromophore in various species of crayfish and other freshwater crustaceans. *Experientia* 43: 1111–1113.

Suzuki, T., Makino-Tasaka, M. & Eguchi, E. 1984. 3-Dehydroretinal (vitamin A2 aldehyde) in crayfish eye. *Vis. Res.* 24: 783–787.

Tokarski, T.R. & Hafner, G.S. 1984. Regional morphological variations within the crayfish eye. *Cell Tissue Res.* 235: 387–392.

Vogt, K. 1975. Zur Optik des Flußkrebsuages. *Z. Naturforsch* 30c: 691.

Vogt, K. 1977. Ray path and reflection mechanisms in crayfish eyes. *Z. Naturforsch* 32c: 466–468.

Welsh, J.H. & Chace, F.A. 1937. Eyes of deep-sea crustaceans I. Acanthephyridae. *Biol. Bull.* 72: 57–74.

Welsh, J.H. & Chace, F.A. 1938. Eyes of deep-sea crustaceans II. Sergesetidae. *Biol. Bull.* 74: 364–375.

Wilkens, L.A. & Larimer, J.L. 1976. Photosensitivity in the sixth abdominal ganglion of decapod crustaceans: a comparative study. *J. Comp. Physiol.* A 106: 69–75.

Zuker, C.S. 1992. Phototransduction in *Drosophila*: a paradigm for the genetic dissection of sensory transduction cascades. *Curr. Opin. Neurobiol.* 2: 622–627.

Zuker, C.S. 1996. The biology of vision in *Drosophila*. *Proc. Natl. Acad. Sci. USA* 93: 571–576.

# Crustacean Parasites as Phylogenetic Indicators in Decapod Evolution

CHRISTOPHER B. BOYKO[1] & JASON D. WILLIAMS[2]

[1] *Molloy College, Rockville Centre, New York, U.S.A., and American Museum of Natural History, New York, U.S.A.*
[2] *Hofstra University, Hempstead, New York, U.S.A.*

ABSTRACT

The evolutionary history of decapods and their parasites is assessed with particular reference to the use of parasites as proxies for host phylogeny. We focused on two groups of obligate parasites that use decapods as their definitive hosts: parasitic isopods of the family Bopyridae and parasitic barnacles of the superorder Rhizocephala. Bopyrids and rhizocephalans differ in that the rhizocephalans have a direct life cycle whereas bopyrids require an intermediate host. In addition, rhizocephalans cause drastic impacts on hosts (including castration and behavioral modification) whereas bopyrids have less pronounced impacts but often also castrate hosts. The diversity and host specificity of both groups are reviewed and their patterns of association with decapod hosts are analyzed. Aside from the Dendrobranchiata (with 39 bopyrid species) and the Caridea (with 8 rhizocephalan and 203 bopyrid species), the more basal decapods are relatively unparasitized or completely lack representatives of these parasites. In contrast, the most derived decapod taxa (Anomura and Brachyura) host the largest number of parasites (233 rhizocephalan and 282 bopyrid species). Counterintuitively, when the phylogenies of the decapods and parasites are compared, some of the most basal parasite groups are found associated with more derived host groups. Our findings indicate a degree of cospeciation but suggest that host switching has been frequent in these parasites, with colonization of caridean shrimp occurring in both groups. Conclusions based on the coevolutionary analyses are complicated by the fact that comprehensive cladistic analyses of the parasites are presently lacking; our review can act as a catalyst for more directed studies analyzing the coevolution of these groups and testing particular hypotheses on their evolutionary history. Although the value of parasites in the elucidation of the phylogeny of decapods as a whole may be limited due to host switching, parasites may be informative *within* particular decapod taxa. We explore an example of this within the Anomura and indicate how such coevolutionary analyses may show host taxa that we would predict to have parasites but presently appear to be lacking them, likely due to limited sampling or evolution of anti-parasite defenses. In addition, these analyses are important in applied areas of decapod ecology (e.g., fisheries) and a brief discussion is provided on the role of coevolutionary studies in the use of bopyrids and rhizocephalans as biological control agents of invasive and/or pest decapod species.

## 1 INTRODUCTION

Recent attempts to elucidate the phylogenetic relationships among the decapod crustaceans have used a wide variety of characters, both morphological and molecular. However, one character with potentially informative phylogenetic signals has, to date, not been considered in the attempts at reconstructing decapod evolutionary history: parasites. Historically, parasites have been used to infer the phylogeny of diverse host lineages, and within the past two decades methods for coevolutionary analyses have been developed to analyze and reconcile host and parasite lineages (see Brooks 1988; Brooks & McLennan 1993, 2002; Page & Charleston 1998; Legendre et al. 2002; Page 2002; Nieberding & Olivieri 2007; Poulin 2007). More recently, parasites have been used to determine demographic history and movement of their hosts (Whiteman & Parker 2005; Nieberding & Olivieri 2007). In the marine realm, the degree to which the phylogeny of parasites mirrors that of host(s) has been best studied in vertebrates (see review in Hoberg & Klassen 2002); there are few examples of coevolutionary analyses on parasites of invertebrates (e.g., Cribb et al. 2001). To our knowledge there are no coevolutionary studies on marine parasites (protozoan or metazoan) that infest invertebrates as their definitive hosts, although multiple host–parasite lineages have been analyzed separately and are amenable to future studies.

Decapod crustaceans are diverse and numerically dominant components of the marine environment, as well as being well represented in freshwater and terrestrial habitats (Bliss 1990). Many diverse groups of decapods harbor parasitic lineages that may provide phylogenetic signals that support or refute hypotheses of decapod evolution. However, it is essential to study and reveal the phylogenetic patterns within the parasite groups before attempting coevolutionary analyses of the parasites and their hosts. Many different types of organisms parasitize decapods, including bacteria, viruses, fungi (Johnson 1983), protozoans (Couch 1983), and metazoans including platyhelminths, acanthocephalans, nematodes, nematomorphs, and crustaceans (Overstreet 1983; Cressey 1983; Shields et al. 2006; Shields & Overstreet 2007). Within the crustaceans, there are only two parasitic lineages that are known to have evolved with decapod hosts: the rhizocephalan barnacles (Cirripedia) and the "epicaridean" isopods. Note that the classical term "Epicaridea" as a higher-level ranking within the Isopoda is not in current use, and the constituent taxa of Bopyroidea + Cryptoniscoidea are considered to be within the Cymothoida (Brandt & Poore 2003); the term "epicaridean" is used here to refer to both Bopyroidea and Cryptoniscoidea in shorthand, as the monophyly of the Epicaridea has not been demonstrated. However, since the Bopyridae (*sensu stricto*, not including Entoniscidae and Dajidae) is the most speciose and best studied family of epicaridean parasites of decapod hosts, the following analyses will be largely restricted to this group. Rhizocephalans and epicarideans also occur on non-decapod crustaceans, mostly peracarids and cirripedes, but the vast majority of species are known from decapod hosts. Copepods, although containing diverse lineages that parasitize many invertebrates and vertebrates, and being informative in coevolutionary analyses with their teleost hosts (e.g., Paterson & Poulin 1999), have not specialized on decapods. While commensal and mutualistic species also may be informative in coevolutionary analyses (e.g., Griffith 1987; Ho 1988; Cunningham et al. 1991), we focus on the parasitic barnacles and isopods.

Rhizocephalans and bopyrids are obligate parasites of their decapod hosts and are numerically dominant in terms of the parasite fauna on these hosts. As an example, in hermit crabs (Paguroidea), crustacean parasites make up 79% of the described parasite fauna, with bopyrids and rhizocephalans making up 57% and 21%, respectively, of the total number of parasite species (McDermott et al., unpublished data). Additionally, both bopyrids and rhizocephalans are macroparasites and are easily sampled, at least in their adult forms, as they are all either ectoparasitic (most bopyrids) or endoparasitic with an externa (rhizocephalans). This chapter summarizes what is known about the host specificity, diversity, and evolutionary history of rhizocephalans and bopyrids, and uses these data to provide a preliminary investigation of their coevolution with their decapod hosts.

## 1.1  *A brief overview of coevolutionary theory*

Host and parasite phylogenies may be in perfect agreement (i.e., they are congruent and follow Fahrenholz's Rule that the parasites track the phylogeny of hosts), indicating cospeciation of hosts and parasites. However, hosts and parasite lineages often do not exhibit perfect agreement or association by descent (Poulin 2007), and the resulting incongruence can be due to multiple factors, some

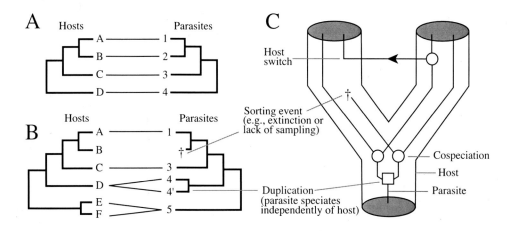

**Figure 1.** Host–parasite coevolution (association by descent) of hypothetical hosts and parasites. (A) Example of perfectly congruent host–parasite phylogenies (cospeciation of hosts and parasites) with all parasites specific to one host. (B) Example of incongruent host–parasite phylogenies, due to: sorting event† (extinction of parasite species or parasite species undetected due to lack of sampling of hosts), duplication (as in parasites 4, 4' that speciated independently of the host lineage and hosts E, F that speciated independently of the parasite lineage). (C) Embedding of a parasite tree inside a host tree. Shown is a duplication event in the parasite lineage and subsequent cospeciation of the resulting two parasite species. One parasite species goes extinct† and another parasite species undergoes a host switch (colonization or horizontal transfer). (A, B: modified from Poulin 2007; C, modified from Page & Charleston 1998.)

of which may represent the true historical associations of these groups (e.g., host switch, intra-host speciation) and others that may reflect our lack of knowledge (e.g., sampling efforts) (Fig. 1). To be able to map the host and parasite phylogenies and determine the degree of congruence present, three data sets must be considered (as indicated in Legendre et al. 2002): 1) association events for hosts and parasites, 2) host phylogenetic tree, and 3) parasite phylogenetic tree.

The first data set is only as good as our knowledge of the associations of the hosts and parasites, and requires accurate identifications of both, as well as reporting of these associations in the literature. To date, there are numerous gaps in our knowledge for this data set pertaining to decapods, as many hosts are reported with undescribed parasites and vice versa. It is important to note that the collection of accurate host/parasite data is essential, as assumptions about parasite occurrences on specific hosts should not be made. Indeed, potential host decapods may have the ability to block infestation by parasites that regularly infest congeners in other parts of the world or by parasite species that are common on sympatric species in the same area (Kuris et al. 2007).

Much progress has been made in the second data set, with many phylogenetic analyses of decapod groups being published in the past several years (e.g., Scholtz & Richter 1995; Pérez-Losada et al. 2002b; Dixon et al. 2003; Ahyong & O'Meally 2004; Porter et al. 2005; Tsang et al. 2008), although most studies have focused on evolutionary patterns above the family level. Although there is still no agreement on the placement of all the decapod constituent groups, a general consensus has developed on the monophyly of some (e.g., Brachyura, Caridea) and the relationships between others (e.g., Anomura+Brachyura, but see Porter et al. 2005).

The third data set is the principal sticking point in terms of generating coevolutionary hypotheses for rhizocephalan and bopyrid parasites and their decapod hosts. In the Bopyridae, no phylogenetic analyses have been performed to identify monophyletic units and there is no cladistic phylogeny for this family, or for the epicarideans as a whole. Cladistic analyses based on molecular and morphological data have shown that bopyrids appear to be derived from the Cymothoida (isopod parasites of fish) (Wägele 1989; Dreyer and Wägele 2001). However, sampling within the Bopyridae was too limited for any conclusions on the relationships of the bopyrid taxa to be made.

Similarly in the Rhizocephala, little work has been done above the species level (e.g., Høeg & Rybakov 1992; Høeg & Lützen 1993), and all of this has been confined to the Akentrogonida. However, two phylogenetic analyses have been published, one purely morphological (Høeg & Lützen 1993) and likewise restricted to the akentrogonids, and one molecular with limited sampling across the Rhizocephala (Glenner & Hebsgaard 2006). The molecular analysis indicated that several traditional groups of rhizocephalans were likely paraphyletic, including the genus *Sacculina*. One other study (Shukalyuk et al. 2007) has used genetic information from rhizocephalans and bopyrids but was conducted so as to produce a phylogeny of select genes, rather than organisms.

## 2 EVOLUTIONARY HISTORY, BIOLOGY, AND DISTRIBUTION OF CRUSTACEAN PARASITES

Several important questions can be asked about the utility of crustacean parasites in understanding decapod host evolution, including: 1) To what degree do the parasites cospeciate with decapod hosts? 2) Do different parasites show similar patterns of coevolution? and 3) Can biogeographic patterns tell us something about the evolutionary history of hosts and parasites?

In order to begin to provide answers to these questions, we summarize below what is known to date regarding relationships between parasites and hosts, both historically and today.

### 2.1 *The history of crustacean parasites of decapods*

Parasitization of decapods by bopyrids is evident from the fossil record and extends at least as far back as the Jurassic (ca. 145–199 mya) (Markham 1986). It is impossible, however, to identify the species of parasites in fossils as only the characteristic swelling of the branchial chambers is evident. Educated speculation about the identity of the parasites is possible (i.e., Ioninae likely in brachyuran fossils) but presently untestable. Some decapod families are known only to have bopyrids in their extant members, possibly due to limitations of fossil preservation, while others with numerous fossil records of parasites, such as the Raninidae (Brachyura), have never been found with bopyrids on members of extant species (Weinberg Rasmussen et al. 2008). The first clear evidence of rhizocephalans in decapods was demonstrated from the Miocene (ca. 5–23 mya) in fossil specimens of *Tumidocarcinus* (Xanthoidea), based on the presence of feminized abdominal segments on otherwise male crabs (Feldmann 1998). However, the origin of rhizocephalans is thought to be much more ancient (Walker 2001). As with bopyrids, there is no way to identify fossil rhizocephalan parasites beyond the higher taxonomic grouping. Although the oldest direct fossil evidences of bopyrids and rhizocephalans are separated by a large span of time, both groups clearly have a long history of association with their hosts.

### 2.2 *Overview of crustacean parasite biology*

Both bopyrids and rhizocephalans use decapods as definitive hosts; however, there are important differences in the two taxa in terms of their life histories. Bopyrids have an indirect life cycle with two hosts being externally parasitized, which is unusual among parasites in that there is no trophic transmission involved. Rhizocephalans, in contrast, have a free-living larval stage before completing their life cycle within a single definitive host. Both bopyrids and rhizocephalans are known to be parasitic castrators of hosts, but rhizocephalans cause more drastic impacts in terms of host modification (physiological and behavioral) through action of hormonal influence (Høeg et al. 2005); the chemical basis for the impacts of bopyrids on hosts remains largely unknown (Lester 2005; Calado et al. 2008). Some bopyrids do not cause "reproductive death" of their hosts, either allowing reduced reproduction by females (smaller clutch sizes) or not interfering with male reproductive ability (Van Wyk 1982; Calado et al. 2005). Although theoretical predictions suggest that parasites with direct life cycles or free-living stages infecting hosts ectoparasitically are expected to exhibit greater congruence with hosts than parasites with indirect life cycles, this is not always the case (Paterson & Poulin 1999). Study of bopyrids and rhizocephalans can provide an additional test for this hypothesis.

### 2.2.1 *Life cycles of the Rhizocephala*

Rhizocephalans either release free-swimming nauplius larvae that develop in the water to the cyprid larval stage (all Akentrogonida and most Kentrogonida) or hatch cyprids directly from the eggs (a few Kentrogonida) that then settle on and initiate parasitism of the crustacean host (Fig. 2). Sexes are separate in rhizocephalans, and although sex determination of some species appears to be environmentally controlled (Walker 2001), the genetic basis for this process is not known for most species (Høeg et al. 2005). Female cyprids settle on new hosts, whereas male cyprids settle on the virgin rhizocephalan externa erupting from hosts. Female cyprids either directly inject

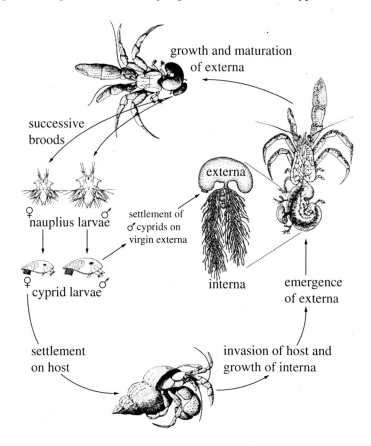

**Figure 2.** Generalized life-cycle diagram for rhizocephalan barnacles (Kentrogonid life cycle shown). Mature externa of parasitized host releases nauplius larvae that develop into cyprids (Akentrogonida lack free-swimming naupliar stages). Female cyprids settle on new hosts, whereas male cyprids settle on juvenile virgin externa. Female cyprids either directly inject inoculum into the host (Akentrogonida) or metamorphose into a kentrogon stage (not shown) that penetrates the host and injects the vermigon (Kentrogonida). Following invasion of the host, the rhizocephalan forms a branched nutrient-absorbing system of rootlets (interna) prior to forming the externa that emerges from the body. Male cyprids that settle on virgin externa will compete to inject generative cells into the female, becoming parasitic males (trichogon stage in the Kentrogonida). (Modified from the life cycle of *Peltogaster paguri* by Høeg (1992); with additional drawings of *P. reticulatus* by Isaeva et al. (2005) and *P. paguri* by Pérez (1937); figures not to scale).

female inoculum into the host (Akentrogonida) or metamorphose into a kentrogon stage that penetrates the host and injects the vermigon (Kentrogonida). The female grows inside the host, forming a branched nutrient-absorbing system of rootlets (interna) prior to forming an externa that emerges from the body. Male cyprids are attracted to settle on these virgin externa, competing to inject male generative cells that invade the female. In the Kentrogonida, a trichogon male stage invades and blocks the female receptacles with its shed cuticle (see fig. 4.21 in Høeg et al. 2005). In the Akentrogonida, the developing ovary or mantle cavity is typically invaded by the male generative cells (a trichogon stage is lacking) (Walker 2001). Eventually the externa matures, producing eggs that are fertilized and develop in the mantle cavity before being released as nonfeeding nauplius or cyprid larvae into the water. Some rhizocephalans produce a single externa while others can undergo asexual reproduction where multiple externae are formed (Isaeva et al. 2005). Because rhizocephalans span both the external and internal environments of their hosts, they are sometimes termed mesoparasites, but their nutrition is taken up by the endoparasitic interna (Høeg 1992). The cues for location and recognition of hosts by rhizocephalans have been investigated (Boone et al. 2004; Pasternak et al. 2004a, b). In addition, cyprid settlement cues and host specificity of some rhizocephalans have been experimentally tested (e.g., Boone et al. 2003; Kuris et al. 2007).

### 2.2.2 *Life cycles of the Bopyridae*

Bopyrid isopods go through three different larval stages in their development. The epicaridium larva hatches from the egg and seeks out an appropriate intermediate host, always a copepod, where it metamorphoses into a microniscus larva and feeds on its hemolymph (Fig. 3). After this period of development on the intermediate host, the microniscus larva transforms into a free-swimming cryptoniscus larva that seeks out an appropriate definitive host, where it typically attaches to the gill filaments inside the branchial chamber or to the abdomen of the host. Species of the subfamily Entophilinae are endoparasites within the thorax or abdomen of hosts; some ectoparasitic species of the subfamily Pseudioninae (*Probopyrus*) are also known to be able to live initially as endoparasites of their hosts (Anderson 1990; Lester 2005). Once attached to their definitive hosts, the isopods transform into a juvenile bopyridium. In some species sex is epigametically or environmentally determined, that is, the first isopod to settle on a host will mature into a female, with any subsequent isopods settling on the same host becoming dwarf males. In some species the females appear to produce a masculinizing substance that reversibly restricts male sex change; when the female dies the males may change sex (Reinhard 1949). In other species determination of sex appears to be genetically controlled (Owens & Glazebrook 1985). Female bopyrids pierce the body of hosts and feed on hemolymph, eventually producing broods of eggs contained within the marsupium and fertilized by the dwarf males.

However, very few species of bopyrids have had their entire life cycles worked out; copepod host choice by epicaridium larva and the patterns of sex determination across the Bopyridae remain unclear. Limited research has investigated the interactions between bopyrids and their intermediate hosts, including the degree of intermediate host specificity (Anderson 1990; Owens & Rothlisberg 1991, 1995). Unfortunately, unlike for the Rhizocephala, little is known about the cues for location or settlement of isopod larvae on definitive hosts.

### 2.3 *Parasite biogeography and host specificity*

Although there have been numerous studies on the taxonomy and biology of bopyrids and rhizocephalans in European waters (e.g., Bourdon 1968; Høeg & Lützen 1985), the geographic ranges and degree of host specificity of many species in both groups are poorly known, especially in areas where sampling has been limited such as the Indo-West Pacific. In such regions, data on the host species may be extensive in taxonomic or ecological publications, but mention of the parasites is often omitted. From the limited worldwide data on the geographic distribution of decapod parasites,

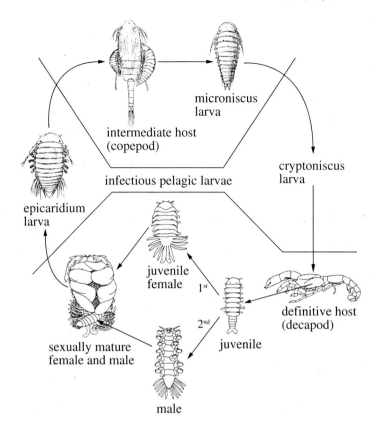

**Figure 3.** Generalized life-cycle diagram for bopyrid isopods. Mature male and female ectoparasitic isopods are typically found in the gill chamber or on the abdomen of decapod definitive hosts (members of the subfamily Entophilinae are endoparasites). Sexually mature females release epicaridium larvae that find a suitable copepod intermediate host, on which they attach and metamorphose into an ectoparasitic microniscus larva. After a period of development the microniscus larva transforms into a cryptoniscus larva that detaches and is free-swimming. The cryptoniscus settles onto suitable definitive hosts (recently settled juveniles are termed bopyridia). The first isopod to settle becomes female; subsequent isopods become dwarf males that live on the female and fertilize the developing eggs in the marsupium. (Modified from the life cycle of *Ione thoracica* by Wägele (1989), with additional drawings by Sars (1899); figures not to scale).

it appears that most species follow the classical pattern of having the parasite occur only within a portion of the range of the host species (Pielou 1974). However, it is clear that some species of bopyrids, at least, can parasitize multiple host species and extend their ranges by this means. As an example, *Athelges takanoshimensis* parasitizes at least 13 species of pagurid and diogenid hermit crabs from Japan, Korea, Hong Kong, and Taiwan (Boyko 2004). Another athelgine bopyrid, *Anathelges hyptius*, may have a range as great as from Massachusetts, USA, to Argentina (Boyko & Williams 2003; Diaz & Roccatagliata 2006) on eight species of pagurid hermit crabs, and perhaps extending all the way around the southern tip of South America to Chile (Diaz & Roccatagliata 2006). In contrast, many other species of bopyrids, as well as most rhizocephalans, appear to be more host-specific and have been found only on a single species of host.

One aspect of the life cycle of bopyrid isopods that may confound our understanding of the factors restricting their distribution is the inclusion of an intermediate copepod host in their life cycle. It is possible that the adaptation to the intermediate copepod host may be the key factor in the distribution of certain species or lineages of bopyrids. Other groups of parasites (e.g., digenean trematodes) have been shown to exhibit a narrower host range in their intermediate hosts than in

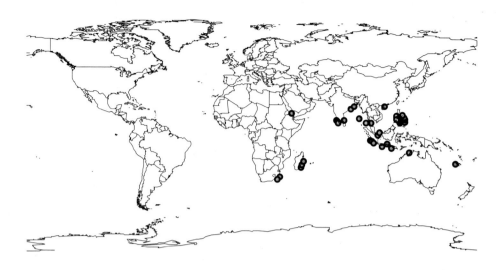

**Figure 4.** Geographic distribution of 36 species of parasitic isopods of the subfamily Orbioninae (each circle represents the type locality of one species; data largely from Bourdon 1979, 1981).

their definitive hosts, but the converse is also true (Cribb et al. 2001). In bopyrids of the subfamily Orbioninae, exclusively parasitic on penaeid shrimp, all of the species are found in the Indo-Pacific region (Fig. 4), despite the fact that penaeid shrimp are widely distributed in all the world's oceans. Bopyrid preference for certain species of copepod hosts may constrain their distribution, rather than the selection of definitive hosts, but this requires investigation. Unfortunately, little is known of the life cycles of Orbioninae or their specificity on copepod hosts. In fact, only a few studies have made direct observations on copepods parasitized by microniscus larvae (see Owens & Rothlisberg 1991, 1995). Coevolutionary analyses involving bopyrids and their intermediate copepod hosts may be informative, but this will require molecular studies to identify the parasites of copepods because bopyrids are typically not identifiable to species based on larval stages. It is notable that the diversity of the Orbioninae is highest in the Philippines, following the general pattern of highest diversity in this region of the Indo-West Pacific for free-living marine species (including invertebrates) (Carpenter & Springer 2005). Other parasitic isopod groups are predicted to exhibit higher diversity in this region (reflecting the diversity of their host groups), but this will require greater efforts in sampling (Markham 1986).

## 3 TAXONOMY AND PHYLOGENY OF DECAPOD CRUSTACEAN PARASITES

Identification of the monophyletic units within the bopyrids and rhizocephalans is essential before any testing of coevolutionary hypotheses can be undertaken. Unfortunately, this has not been done, and the process of identifying them is not simple.

### 3.1 The phylogeny of the Rhizocephala

While there is abundant morphological and developmental evidence supporting the monophyly of the Rhizocephala (Høeg 1992) and its placement within the Cirripedia as sister-taxon to the Thoracica (e.g., Billoud et al. 2000; Pérez-Losada et al. 2002a), there is a less clear picture regarding the relationships of its constituent taxa. An example of this can be seen in the genus *Sacculina*, which contains approximately 115 species, the most of any rhizocephalan genus. Species of *Sacculina*, as well as of the family Sacculinidae (including approximately 50 additional species in six genera), are usually referred to as parasites of brachyurans (e.g., Walker 2001), but two of the species are known only from anomuran hosts (an albuneid and a galatheid) and one from a thalassinidean shrimp. These unusual host records suggest that a closer look at the genus should be undertaken to determine if it is monophyletic in its current configuration (see also Glenner & Hebsgaard 2006) or whether these unusual host associations reflect host switching within this parasitic taxon.

Almost all the species of rhizocephalans have been defined and described based solely on morphological criteria of the mature externae, despite the fact that these animals are among the most morphologically reduced in comparison to their non-parasitic relatives. This has resulted in there being only a limited suite of characters for identification of species, and it is unclear how many described species actually represent distinct taxa. Several recent studies have attempted to unite the limited morphological characters of adults with detailed cyprid morphology and molecular data in order to better define species boundaries and generate larger character selection options for phylogenetic analyses (e.g., Glenner et al. 2003; Chan et al. 2005).

A molecular study using 18S rDNA, 11 species of Sacculinidae, and 11 other rhizocephalans by Glenner & Hebsgaard (2006) resulted in a monophyletic Rhizocephala containing four clades of kentrogonids, with the two most derived being separated by the position of the Akentrogonida, thus rendering the Kentrogonida paraphyletic. The kentrogon stage was shown to be the primitive form of host invasion, with the akentrogonids being derived in their loss of the kentrogon, as well as in reduction in adult externae size. Perhaps most strikingly, *Sacculina carcini*, the type species of the genus, was separated from all other congeners by the position of the Akentrogonida, indicating paraphyly of *Sacculina* even with the limited taxon sampling.

## 3.2 *The phylogeny of the Bopyridae*

The "epicaridean" isopods are currently divided into the two lineages Cryptoniscoidea + Bopyroidea within the Cymothoida (Brandt & Poore 2003). While some of the cryptoniscoids are found parasitizing decapod hosts (e.g., *Danalia ypsilon* on *Galathea* spp.), most (ca. 88%) are known from peracarid, ostracod, or cirripede hosts. Members of the Bopyroidea, in contrast, are primarily known from decapod hosts. With 595 described species, the Bopyridae is the most speciose family in the Bopyroidea, as well as the most speciose family of isopods. Despite this large number of described taxa, the diversity in this group is largely underreported, and evidence for this can be gleaned from the more than 20% increase in the number of known species during the past 20+ years (subsequent to Markham 1986). Many new host records and new taxa await reporting and description, principally from tropical and deep-sea habitats (Bourdon, Markham, pers. communs.; Boyko, Williams, pers. obs.). The other two families of Bopyroidea are the Entoniscidae (ca. 35 spp.), which are endoparasites of decapods, and the Dajidae (ca. 50 spp.), which are ectoparasites of shrimp, mysids, and euphausids. As with the Bopyroidea + Cryptoniscoidea grouping, Bopyridae + Entoniscidae + Dajidae has long been assumed to be monophyletic, based in large part on reproductive biology and the morphology of the males, but no cladistic phylogenetic analyses have ever been conducted for these taxa.

Currently, the Bopyridae is divided into nine subfamilies. A tenth, monotypic subfamily (Bopyrophryxinae) was synonymized with Pseudioninae (Bourdon & Boyko 2005). In the subfamilies Pseudioninae, Bopyrinae, Argeiinae, and Orbioninae, the adult female parasite is located on the decapod host in the right or left branchial chamber. The branchial chamber is also the usual site of attachment for members of the Ioninae, but species of *Rhopalione* are found under the abdomens of their pinnotherid hosts. In the Athelginae, the females are located on the dorsal abdomen of the host hermit or king crab, while in the Phyllodurinae, the female isopod is situated on the ventral surface of the thalassinidean host abdomen. Female isopods of the Hemiarthrinae are found either on the dorsal or ventral surface of the abdomen, laterally on the carapace, or in one species inserted into the mouth region of the host shrimp (Trilles 1999). The two species of Entophilinae are similar in habitat to entoniscid isopods, living as endoparasites in the thoracic or abdominal regions of their anomuran and thalassinidean hosts.

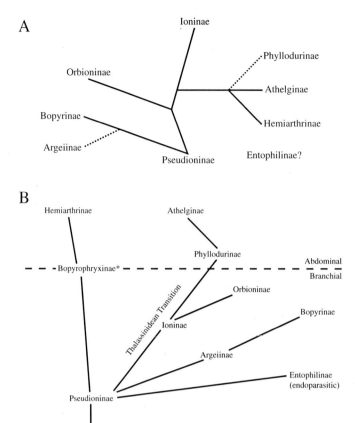

**Figure 5.** Hypotheses of the evolutionary relationships within the Bopyridae. (A) Proposed phylogeny based on Shiino (1965). Dotted lines indicate positioning of subfamilies based on the phylogeny from Shiino (1952); note that Entophilinae was not included in his trees and that subfamily names had not yet been proposed when these trees were originally presented. (B) Proposed phylogeny based on Markham (1986); horizontal dashed line indicates separation of isopods that infest the branchial chamber of hosts (bottom) or their abdomen (top). *Bopyrophryxinae are now members of the Pseudioninae; the Thalassinidean transition refers to those bopyrids that infest callianassid and upogebiid shrimp and are suspected to represent a link between the Pseudioninae and the Ioninae (Markham & Dworschak 2005).

No phylogenetic testing of the monophyly of the Bopyridae or any of its subfamilies has ever been attempted using morphological or molecular data. However, Shiino (1952, 1965) and Markham (1986) proposed evolutionary trees based on their years of research on this group (Fig. 5A & B, respectively). Both Shiino and Markham considered the Pseudioninae to be the basal group, based on morphology and broad range in host use. However, in many other features their trees are quite different. Shiino (1952, 1965) placed Athelginae and Hemiarthrinae (both abdominal parasites) as sister taxa, and showed only two lineages being derived from the Pseudioninae. Markham (1986), in contrast, showed four lineages arising from the basal taxon and placed Athelginae and Hemiarthrinae on two different branches. Additionally, while Shiino's (1952, 1965) trees can be converted into phylogenetic ones, albeit lacking a cladistic analysis, Markham's (1986) trees contain "transitional" taxa that are placed in direct line of descent from one subfamily to another. Specifically, he proposed that those bopyrids infesting callianassid and upogebiid shrimp represent a link between the Pseudioninae and the Ioninae — what he termed the "Thalassinidean transition" (see also Markham &

Dworschak 2005). Whether or not Markham's (1986) transitional forms may represent stem groups is not clear at this time.

Although there is a clear need for phylogenetic analyses of bopyrids, there are many taxonomic problems that need to be sorted out in order to make taxon sampling effective. Given the paucity of specimens for many described species, loss of type specimens, and lack of specimens properly preserved for molecular analysis, a phylogenetic analysis of the Bopyridae based on morphological and/or molecular data is a difficult task. One problem is exemplified by the type species of *Pseudione*, the largest genus in the family. The original description of *P. callianassae* by Kossmann (1881) was based only on an image of the ventral surface of the head of a male bopyrid, with no accompanying descriptive text. There are no useful characters present in the illustration, and this species is, based on this drawing, effectively unidentifiable. Only the choice of host is known (*Callianassa subterranea*), but two species of bopyrids are known from this European host species. On a morphological basis, it has long been suspected that *Pseudione* is paraphyletic, but the lack of an identifiable type species remains a barrier to resolving the taxonomic and phylogenetic issues of this large genus, as well for the Pseudioninae. A second problem is one of limited specimen collection and/or identification, which has resulted in lack of knowledge about the morphological boundaries of many species. In the case of *Metathelges muelleri*, described from a brachyuran host, the species was described from a single female specimen that was later determined to be likely developmentally aberrant (Boyko & Williams 2003). This resulted in the transfer of the genus from the Athelginae, where it was the only species ever reported from a brachyuran host, to the Ioninae, which are predominantly brachyuran parasites. A third potential difficulty, especially important in issues of coevolutionary analysis, is one of identification of the hosts. Usually, the problem is one of consistently recording the host identity and retaining this information with the parasite when it is separated. This has resulted in species' being described with unknown host data, or, occasionally, with incorrect host data, such as *Falsanathelges muelleri* being described by Nierstrasz & Brender à Brandis (1931) as collected from a "*Galathea*" (i.e., Galatheoidea), when it was in fact from a hermit crab collected by the *vessel* "Galathea"!

It is important to choose exemplar taxa for higher-level analyses carefully, as many genera of bopyrids have not been revised and may well be paraphyletic. An example is the genus *Gigantione*, which contains eight species known from brachyuran hosts, including three dromiids, and three species from thalassinoid hosts. This range of hosts may not accurately reflect patterns of host and parasite coevolution; examination of the original descriptions of all *Gigantione* species suggests that the brachyuran parasites and the thalassinoid parasites are not very similar to each other and appear to be currently placed in the same genus principally on the basis of females' having bifurcated uniramous uropods. If this genus is not monophyletic, any discussion of the coevolution of hosts and parasites would be confounded by the paraphyly of the parasite genus.

## 4 DECAPOD HOST AND PARASITE COEVOLUTION: INFERENCES BASED ON CURRENT DATA

Most of our discussion below is based on the decapod phylogeny of Dixon et al. (2003). However, we have also considered the findings of Porter et al. (2005) that present an alternative and dramatically different arrangement for many of the groups. It should be noted that our focus on Dixon et al. (2003) does not imply that we consider their study to be a more accurate representation of decapod phylogeny than other recent works (e.g., Ahyong & O'Meally 2004; Tsang et al. 2008). At this point in time, it is probable that anyone who is "married" to any one particular decapod phylogeny is likely to suffer through a painful divorce at a later date.

In total, there are approximately 244 rhizocephalans and 586 bopyrids that parasitize decapods (representing 2.0 and 4.9% of the total number of decapods being infested by these two groups, respectively). The more derived decapods (Thalassinida + Achelata + Anomura + Brachyura) are host to 575 species of rhizocephalans and bopyrids, the bulk of which (515 species, ~90%) are found on

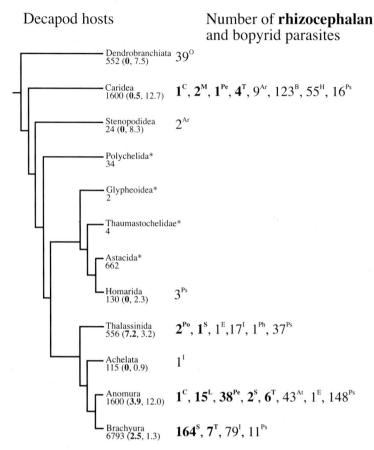

Figure 6. Phylogeny of the Decapoda showing numbers of rhizocephalan and bopyrid species that they host. Numbers under decapod groups indicate current estimates of total number of host species (numbers in parentheses indicate % of host species infested by rhizocephalans and bopyrids, respectively). Decapod phylogeny based on Dixon et al. (2003); *decapod taxa that do not host rhizocephalan or bopyrid parasites; rhizocephalan taxa (in bold): C = Clistosaccidae, L = Lernaeodiscidae, M = Mycetomorphidae, Pe = Peltogastridae, Po = Polysaccidae, S = Sacculinidae, T = Thompsoniidae; Bopyridae taxa: Ar = Argeiinae, At = Athelginae, B = Bopyrinae, E = Entophilinae, H = Hemiarthrinae, I = Ioninae, O = Orbioninae, Ph = Phyllodurinae, Ps = Pseudioninae.

the Anomura and Brachyura (Fig. 6). In contrast, basal decapods (Dendrobranchiata, Caridea, and Stenopodidea) are host to less than half this number (252), the bulk of which (211 species, ~84%) are found on the Caridea. The relative diversity of rhizocephalans compared to bopyrids is low on the more basal decapods (8/252, ~3%) but is slightly less than half that of the bopyrids on the more derived decapods (236/575, ~41%). When the number of parasite species is standardized relative to the diversity of decapod host taxa, the host groups with highest percentages infested are: Anomura (15.9%), Caridea (13.2%), and Thalassinida (10.6%); the rest of the decapods have fewer than 10% infested in each taxon. Lack of parasites in some groups may represent sampling efforts, whereas others can be explained by their evolutionary history. For example, the Astacida harbor no definite parasites (although their commensal ostracods and annelids are thought by some to be parasitic), likely due to their invasion of freshwater habitats that provided a refuge from parasitism. As far as is known, most of the few species of rhizocephalans and bopyrids from hosts collected in freshwater are euryhaline and reproduce at the same time the hosts move towards the ocean to breed (Okada &

Miyashita 1935). There do not appear to be any bopyrids on hosts found in landlocked freshwater habitats, although there are truly freshwater rhizocephalans (Feuerborn 1931, 1933; Andersen et al. 1990). Below we discuss the host relations and coevolution for both these parasite groups.

## 4.1 Rhizocephala

Systematic work on the Rhizocephala subsequent to the contributions of Hildebrand Boschma, who worked on these organisms circa 1925–76, concentrated primarily on the non-Sacculinidae taxa. This has resulted in generation of morphological characters and life cycle data for many species in the Lernaeodiscidae, Peltogastridae, and Akentrogonida (e.g., Ritchie & Høeg 1981; Høeg & Lützen 1985; Lützen & Takahashi 1996). Therefore, there is a greater level of confidence in the monophyly of these groups than in the sacculinid taxa. Members of the Thompsoniidae, one of the most derived taxa in terms of reduced morphology, have the broadest diversity of host selection (four decapod groups plus Stomatopoda) (Fig. 6). This is a case, however, where diversity in host selection is not in conflict with the phylogeny of the group, as Glenner & Hebsgaard (2006) recovered a monophyletic Thompsoniidae. In fact, Glenner & Hebsgaard show a monophyletic Akentrogonida, if the poorly known monotypic *Parthenopea* is included, which generally supports prior morphologically based studies (e.g., Høeg & Rybakov 1992). In the Kentrogonida, Glenner & Hebsgaard (2006) support a monophyletic Peltogastridae + Lernaeodiscidae (which they label as Peltogastridae although there is not enough support or resolution in their tree to combine the two families) and the placement of *Peltogasterella* indicates that it may belong to a separate family. Except for the poorly known *Trachelosaccus* from a caridean, all the other peltogastrids and lernaeodiscids are known from anomuran hosts, a case of basal parasites targeting derived host taxa. The remainder of the Kentrogonida comprising the seven Sacculinidae genera is paraphyletic in Glenner & Hebsgaard's (2006) analysis, which, when combined with their placement of the Peltogastridae + Lernaeodiscidae, makes the Kentrogonida polyphyletic. Based on their results, the evolutionary pattern for the rhizocephalans appears to be: 1) an initial parasitism in anomurans (Peltogastridae + Lernaeodiscidae), 2) parasitism in brachyurans (a basal and a derived lineage of "Kentrogonida"), and 3) a lineage with great reproductive modification (loss of kentrogon) and a corresponding increase in host diversity across much of the Decapoda. One of the main difficulties with the kentrogonids, and the sacculinds in particular, is the high level of species diversity in the group, as compared to all other rhizocephalans. The average number of species per genus in the Sacculinidae is 23.8, but in reality more than 115 species occur in the single genus *Sacculina*. This is in marked contrast to the average number of species per genus in all the other rhizocephalan families that ranges from 1 (Clistosaccidae) to 6.3 (Thompsoniidae). In other words, in all families except the Sacculinidae, the genera are relatively small and better defined.

Several observations can be made from a comparison of the host and parasite phylogenies (Fig. 7), including that the most basal rhizocephalans do not parasitize basal decapods. In fact, none of the dendrobranchiate groups are known to host any rhizocephalans. The carideans are the most basal group to be parasitized, and then only by species of rhizocephalans from the derived akentrogonid genera *Pottsia* and *Sylon*. Species of the derived akentrogonid Mycetomorphidae and one species of the kentrogonid *Trachelosaccus* are also found on carideans, but these taxa were not sampled by Glenner & Hebsgaard (2006). A similar pattern of derived hosts being parasitized by primitive parasites with host-switching leading to invasion of a diverse range of taxa has been found in digenean trematode parasites of molluscs (Cribb et al. 2001).

Although no stenopodideans have been reported with rhizocephalans, one of us (CBB) has examined the stalked "bopyrid parasites" reported from a *Spongicoloides* species by Saito et al. (2006), and they are actually rhizocephalans that appear close to the genus *Trachelosaccus*, a poorly known possible member of the Kentrogonida. None of the polychelid lobsters or the Astacida are known to bear rhizocephalan parasites, making this the largest group of decapods not impacted by parasitic barnacles. Only four species of rhizocephalans are found on thalassinideans, but from three

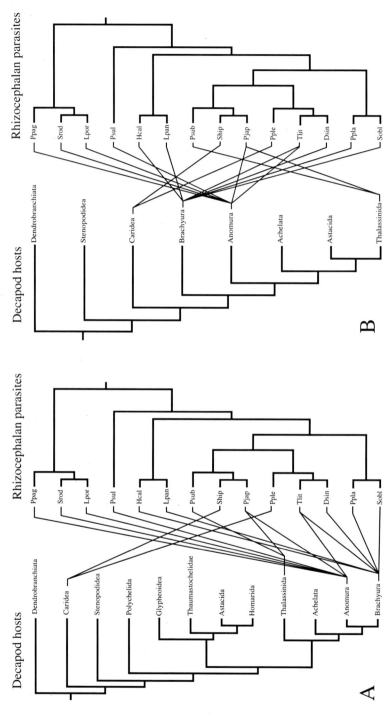

**Figure 7.** Phylogenies of decapod hosts and their rhizocephalan parasites. (A) Comparison of parasite phylogeny based on a subset of taxa from Glenner & Hebsgaard (2006) and host phylogeny based on Dixon et al. (2003). (B) Comparison of parasite phylogeny based on a subset of taxa from Glenner & Hebsgaard (2006) and host phylogeny based on Porter et al. (2005). Ppag = *Peltogaster paguri*, Srod = *Septosaccus rodriguezii*, Lpor = *Lernaeodiscus porcellanae*, Psul = *Peltogasterella sulcata*, Hcal = *Heterosaccus californicus*, Lpan = *Loxothylacus panopaei*, Psub = *Parthenopea subterranea*, Ship = *Sylon hippolytes*, Pjap = *Polysaccus japonicus*, Pple = *Pottsia pleisonikae*, Tlit = *Thompsonia littoralis*, Dsin = *Diplothylacus sinensis*, Ppla = *Polyascus plana*, Sobl = *Sacculina oblonga*.

different families. None of the Achelata have rhizocephalans and, if the spiny and slipper lobsters are indeed rather distant from the clawed lobsters, as born out by some recent analyses (e.g., Dixon et al. 2003), then perhaps the nature of their lobster-type morphology and ecology ("lobsterness") is resistant to rhizocephalan invasion. If, however, achelate lobsters are basal to the Astacura (Glypheidea + Astacidea), this resistance may be based on an evolutionary resistance derived from shared common ancestry. In fact, if considering only the lack of rhizocephalans in Polychelida, Achelata, and Astacura, the tree presented by Ahyong & O'Meally (2004, fig. 3) is more parsimonious in uniting all the taxa above that are known to bear rhizocephalans in a single clade (Lineata) as opposed to that given by Dixon et al. (2003) where Achelata is positioned between Thalassinida and Anomura + Brachyura. This situation indicates the potential utility of parasites in analyzing phylogenetic relationships of host taxa.

## 4.2 *Bopyridae*

Based on host specificity alone, five of the bopyrid subfamilies are likely to be monophyletic: Bopyrinae, Hemiarthrinae, Athelginae, Orbioninae, and the monotypic Phyllodurinae. The diversity of host taxa in the other four subfamilies, especially the Pseudioninae, does not in itself indicate polyphyly but suggests that those subfamilies are in need of rigorous analyses. Indeed, the Argeiinae and Pseudioninae have been suggested as being para- or polyphyletic on the basis of morphological characters (Adkison et al. 1982; Boyko & Williams 2001). However, all of the subfamilies need to have their monophyly tested by both morphological and molecular characters.

As with the Rhizocephala, the bopyrid parasites appear to have invaded relatively derived hosts (anomurans) first and later switched to other decapods (Fig. 8). These findings represent another potential parallel with results obtained by Cribb et al. (2001) in that eco-physiological similarities of hosts may play a role in associations over time. Specifically, the ecological niche of penaeids may have excluded most bopyrids, excepting the ancestral orbionines, from switching to these hosts. Likewise, the distinctive morphology of the relatively exposed abdomens of hermit crabs and caridean may have resulted in either convergent evolution of athelgine and hemiarthrine bopyrids (if they are not sister-taxa as per Markham 1986) or host switching from a putative paguroid host to a caridean one if they are closely related (as per Shiino 1952, 1965). At this juncture, however, the relationship between these abdominal parasite taxa is unclear.

There is a much greater diversity of host range within the Bopyridae than in the Rhizocephala, including several taxa (Dendrobranchiata, Homarida, Achelata) that are known to bear bopyrid parasites but not rhizocephalans. The parasites of homarids and achelates appear to be rather undifferentiated members of large genera (*Pseudione* and *Dactylokepon*), the majority of whose members infest other taxa (anomurans and brachyurans, respectively). In contrast, the parasites of dendrobranchiates are all members of a single lineage (Orbioninae) that has evolved to specialize on these shrimp and whose species are found parasitizing no other types of hosts.

## 4.3 *An example of coevolution within the Bopyridae*

An example of the potential of parasites as a phylogenetic character for decapod evolutionary studies can be seen in the three species of bopyrids found on albuneid crabs (Anomura: Hippoidea) (Fig. 9).

Each of the species of *Albunione* (Pseudioninae) shows the same relationships with respect to each other as their hosts in the genus *Albunea*. *Albunione australiana* is the sister species to the *A. indecora* + *A. yoda* clade, based on morphological characters of both males and females. Likewise, their hosts show the same pattern: *Albunea microps* is the sister species to the clade of *A. groeningi* + *A. paretii* (Boyko & Harvey, unpublished data). Although this analysis suggests some degree of cospeciation between parasites and hosts, reconciling their phylogenies requires the proposal of multiple species of hosts that lack parasites due to sorting events (extinction or

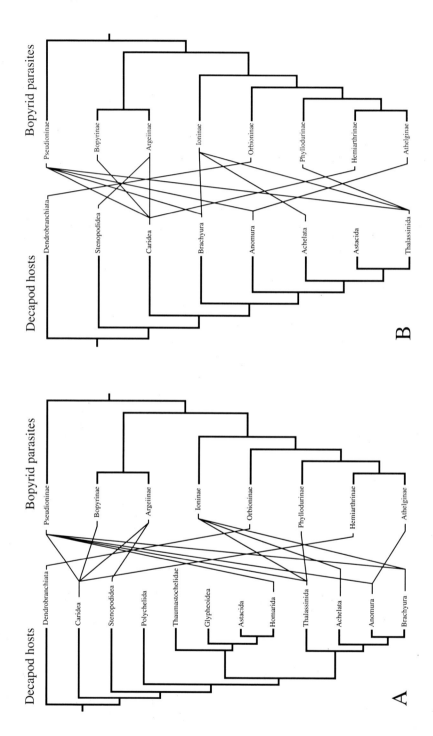

**Figure 8.** Phylogenies of decapod hosts and their bopyrid parasites (minus Entophilinae). (A) Comparison of parasite phylogeny based on Shiino (1952, 1965) and host phylogeny based on Dixon et al. (2003). (B) Comparison of parasite phylogeny based on Shiino (1952, 1965) and host phylogeny based on Porter et al. (2005).

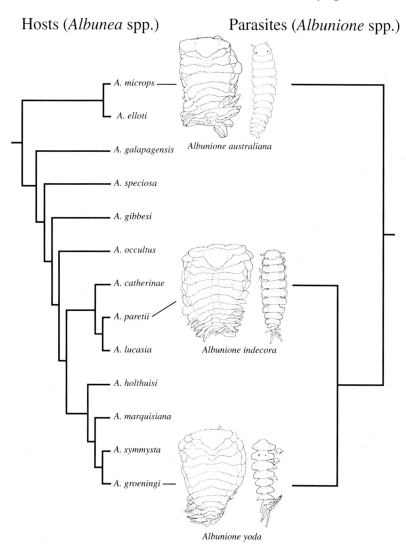

**Figure 9.** Phylogenies of the sand crab genus *Albunea* and the three species of branchial parasitic isopods (genus *Albunione*) that infest them. (Host phylogeny based on Boyko & Harvey, unpublished data; parasite phylogeny based on Markham & Boyko 2003).

lack of sampling). As indicated by Page & Charleston (1998), coevolutionary analyses can lead to hypotheses of hosts that apparently lack parasites but probably do so because they have not been extensively sampled. Given that only these three bopyrids parasitizing species in the Albuneidae are known, it is likely that our knowledge of the diversity of isopod parasites on these anomuran hosts is incomplete. Based on comparison of the analysis of the Shiino/Markham hypotheses with that of the *Albunea/Albunione* relationship, the coevolution of parasites and hosts may be more informative at the genus level than at higher taxonomic levels. However, more data must be gathered and analyzed to draw any general conclusions about this.

## 5 CONCLUSIONS

It is clear from the above discussion that we are only beginning to understand the coevolution between decapods and their crustacean parasites. Cross-phyletic comparisons between rhizocephalans and bopyrids are difficult to interpret due to the fact that the taxonomic levels of the parasites included in the analyses are different (species versus subfamilies). However, one common feature of the rhizocephalan and bopyrid coevolutionary analyses is that anomurans and other more derived host taxa are parasitized by members of basal parasite groups. Because the Anomura is among the more derived groups of decapods, they would be expected to have more derived parasites compared to, for example, penaeids that are more basal. It may be that anomurans, specifically galatheoids, are more susceptible to infestation by parasites than are other decapods. Evidence for this can be found in the robust fossil record for galatheid/bopyrid associations (Markham 1986) and in the large number of extant parasitized anomurans. Although there is clear evidence of some anomurans' having acquired the ability to mechanically resist parasite attack, at least against rhizocephalans (Ritchie & Høeg 1981; Høeg et al. 2005), further study needs to be undertaken to determine if this behavior is found within other decapod groups. Bauer (1981, 1989) hypothesized that selection pressures for natant decapods to remove epifauna that would impede swimming led to efficient mechanisms for removal of parasites, whereas in the more derived, generally non-swimming, decapods (e.g., Anomura and Brachyura) selection pressures to remove these parasites were reduced. One problem with this hypothesis is that, without experimentation, it is not possible to use absence of parasites on hosts as an indicator of their ability to deal with parasites. For example, is the presence of only two species of rhizocephalans (and no bopyrids) on cancrid crabs due to members of the Cancridae having experienced little to no historical parasitic pressure, or have they evolved effective defenses against the parasites? More data need to be collected across the spectrum of decapods in the context of their morphological, physiological, and particularly behavioral adaptations developed in the context of this evolutionary arms race (Ruiz 1991).

Coevolutionary analyses of these parasites of decapods go beyond "ivory tower" research and can inform future studies on the ecology of host–parasite relationships and applied areas of research, including fisheries. As in the *Albunione* example, studies of host/parasite coevolution may allow us to identify host lineages where parasites are unknown but, based on their shared histories, might be expected. Also, a robust understanding of these relationships would allow identification of hosts that are not impacted by parasites (such as the aforementioned cancrid crabs) and suggest the need for further study of the morphological, immunological, and/or behavioral methods they may utilize to resist attack.

These types of coevolutionary studies also can be used in light of the recent attention deservedly given to the problem of invasive species. Rhizocephalan parasites have been suggested as possible biological controls of decapods, in particular the green crab *C. maenas* that has been introduced from Europe to various parts of the world, including the east and west coasts of North America (see Griffen et al. 2007). The rhizocephalan *Sacculina carcini* has been evaluated as a biological control for *C. maenas* (Goddard et al. 2005; Lafferty & Kuris 1996; Thresher et al. 2000; Kuris et al. 2005; Kuris et al. 2007). Along the east coast of the United States, *Carcinus maenas* now competes with the Japanese shore crab *Hemigrapsus sanguineus*, which was first reported from New Jersey in 1988 (McDermott 1991) and has spread from North Carolina to Maine (McDermott 1998, 2000) and has been introduced to Europe and the Mediterranean (Breton et al. 2002; Schubart 2003). In its native habitat of Russia southward to Hong Kong and Japan, this crab is commonly parasitized by the rhizocephalan *Polyascus polygenea*, which sterilizes the crab hosts (Korn et al. 2004), but no rhizocephalans are found impacting the species in its introduced range (McDermott 1998, 2007). The recent rapid spread of *C. maenas* and *H. sanguineus* could reflect their release from parasite pressures (Torchin et al. 2001, 2003). However, introduction of native parasites as biocontrol agents requires detailed studies on host specificity of the parasites (e.g., Goddard et al. 2005; Kuris et al. 2005; Kuris et al. 2007). Given the paucity of our knowledge about the coevolutionary processes that promote or inhibit tightly linked parasite/host relationships, the possibility of decapods being vulnerable to multiple parasite species (Tsuchida et al. 2006), and the ability of some rhizocephalan barnacles to parasitize novel hosts, it would be premature to allow such importation without additional study (Goddard et al. 2005). Even in the event of a controlled release of a rhizocephalan for a target invasive species, there may be little impact on the invader. The rhizocephalan *Heterosaccus dollfusi*, accidentally introduced into the Mediterranean over three decades after its host, has apparently not reduced populations of *Charybdis longicollis* in this region (Innocenti & Galil 2007). The only other known introduced rhizocephalan is *Loxothylacus panopaei*, a parasite of mud crabs that was accidentally introduced to the Chesapeake Bay from the Gulf of Mexico (see Kruse & Hare 2007), which also has limited impacts on host populations (Alvarez et al. 1995). These findings from "natural experiments" suggest that rhizocephalans may have limited utility in combating invasive hosts, and their potential impact on non-target species is far from clear.

The use of parasitic isopods as biological controls has received less attention than rhizocephalans, but some researchers have investigated the potential use of the entoniscid isopod *Portunion maenadis* for the control of *Carcinus maenas* (Høeg et al. 1997; Kuris et al. 2005) and the hyperparasitic cryptoniscid isopod *Cabirops orbionei* for the control of bopyrids on penaeid shrimp (Owens 1993). As indicated by Kuris et al. (2005), the use of epicaridean parasites requires careful evaluation and modeling due to their indirect life cycle and, as in the Rhizocephala, the potential for non-target hosts to be impacted.

In conclusion, parasitic crustaceans may offer insights into the evolutionary histories of their decapod hosts and vice versa. Although host switching among decapod host taxa appears to have occurred in both bopyrids and rhizocephalans, coevolutionary patterns may be more tightly linked at lower taxonomic levels (e.g., species, genera) than at higher ones (infraorders, families). More emphasis should be placed on generating cladistic analyses for parasite lineages, as well as on careful assessment of the status of some, particularly marine, potential host lineages that currently appear parasite-free. Results from these types of studies could be applied across disciplines of interest to crustacean biologists, such as ecological, developmental, and fisheries biology, as well as in the arena of crustacean systematics.

## REFERENCES

Adkison, D.L., Heard, R.W. & Clark, G.T. 1982. Description of the male and notes on the female of *Argeiopsis inhacae* (Crustacea: Isopoda: Bopyridae). *Proc. Biol. Soc. Wash.* 95: 334–337.

Ahyong, S.T. & O'Meally, D. 2004. Phylogeny of the Decapoda Reptantia: resolution using three molecular loci and morphology. *Raffles Bull. Zool.* 52: 673–693.

Alvarez, F., Hines, A.-H. & Reaka-Kudla, M.L. 1995. The effects of parasitism by the barnacle *Loxothylacus panopaei* (Gissler) (Cirripedia: Rhizocephala) on growth and survival of the host crab *Rhithropanopeus harrisii* (Gould) (Brachyura: Xanthidae). *J. Exp. Mar. Biol. Ecol.* 192: 221–232.

Andersen. M.L., Bohn, M., Høeg, J.T. & Jensen, P.G. 1990. Cyprid ultrastructure and adult morphology in *Ptychascus barnwelli*, new species, and *P. glaber* (Cirripedia: Rhizocephala), parasites on semiterrestrial crabs. *J. Crust. Biol.* 10: 20–28.

Anderson, G. 1990. Postinfection mortality of *Palaemonetes* spp. (Decapoda: Palaemonidae) following experimental exposure to the bopyrid isopod *Probopyrus pandalicola* (Packard) (Isopoda: Epicaridea). *J. Crust. Biol.* 10: 284–292.

Bauer, R.T. 1981. Grooming behavior and morphology in the decapod Crustacea. *J. Crust. Biol.* 1: 153–173.

Bauer, R.T. 1989. Functional morphology, adaptive value, and phylogenetic significance of grooming in decapod Crustacea. In: Felgenhauer, B. & Watling, L. (eds.), *Crustacean Issues 6, Functional Morphology of Grooming and Feeding Appendages*: 49–73. Rotterdam: Balkema Press.

Billoud, B., Guerruci, M.-A., Masselot, M. & Deutsch, J.S. 2000. Cirripede phylogeny using a novel approach: molecular morphometrics. *Mol. Biol. Evol.* 17: 1435–1445.

Bliss, D.E. 1990. *Shrimps, Lobsters and Crabs*. New York: Columbia University Press.

Boone, E., Boettcher, A.A., Sherman, T.D. & O'Brien, J.J. 2003. Characterization of settlement cues used by the rhizocephalan barnacle *Loxothylacus texanus*. *Mar. Ecol. Prog. Ser.* 252: 187–197.

Boone, E., Boettcher, A.A., Sherman, T.D. & O'Brien, J.J. 2004. What constrains the geographic and host range of the rhizocephalan *Loxothylacus texanus* in the wild? *J. Exp. Mar. Biol. Ecol.* 309: 129–139.

Bourdon, R. 1968. Les Bopyridae des mers Europennes. *Mém. Mus. Natl. Hist. Nat., n. sér., sér.* A, 50: 77–424.

Bourdon, R. 1979. Sur la taxonomie et l'éthologie de quelques orbioninés (Isopoda Epicaridea). *Internat. Rev. Gesammte Hydrobiol.* 64: 425–435.

Bourdon, R. 1981. Crustacés Isopodes. I. Bopyridae parasites des Pénéides. Résultats des Campagnes MUSORSTOM. I. Philippines (18-28 Mars 1976), volume 1. *Mém.* ORSTOM 91: 237–260.

Bourdon, R. & Boyko, C.B. 2005. Redescription of *Bopyrophryxus branchiabdominalis* Codreanu, 1965 (Crustacea: Isopoda: Bopyridae) with a reappraisal of the subfamily Bopyrophryxinae Codreanu, 1965. *Proc. Biol. Soc. Washington* 118: 108–116.

Boyko, C.B. 2004. The Bopyridae (Crustacea: Isopoda) of Taiwan. *Zool. Stud.* 43: 677–703.

Boyko, C.B. & Williams, J.D. 2001. A review of *Pseudionella* Shiino, 1949 (Crustacea: Isopoda: Bopyridae), with the description of a new species parasitic on *Calcinus* hermit crabs from Easter Island. *Proc. Biol. Soc. Washington* 114: 649–659.

Boyko, C.B. & Williams, J.D. 2003. A revision of *Anathelges* and *Stegophryxus* (Isopoda: Bopyridae: Athelginae) with descriptions of two new genera and one new species. *J. Crust. Biol.* 23: 795–813.

Brandt, A. & Poore, G.C.B. 2003. Higher classification of the flabelliferan and related Isopoda based on a reappraisal of relationships. *Invert. Syst.* 17: 893–923.

Breton, G., Faasse, M., Noël, P. & Vincent, T. 2002. A new alien crab in Europe: *Hemigrapsus sanguineus* (Brachyura: Grapsidae). *J. Crust. Biol.* 22: 184–189.

Brooks, D.R. 1988. Macroevolutionary comparisons of host and parasite phylogenies. *Ann. Rev. Ecol. Syst.* 19: 235–259.

Brooks, D.R. & McLennan, D.A. 1993. *Parascript: Parasites and the Language of Evolution*. Chicago: University of Chicago Press.

Brooks, D.R. & McLennan, D.A. 2002. *The Nature of Diversity*. Chicago: University of Chicago Press.

Calado, R. Bartilotti, C., Goy, J.W. & Dinis, M.T. 2008. Parasitic castration of the stenopodid shrimp *Stenopus hispidus* (Decapoda: Stenopodidae) induced by the bopyrid isopod *Argeiopsis inhacae* (Isopoda: Bopyridae). *J. Mar. Biol. Assoc.* UK 88: 307–309.

Calado, R., Bartilotti, C. & Narciso, L. 2005. Short report on the effect of a parasitic isopod on the reproductive performance of a shrimp. *J. Exp. Mar. Biol. Ecol.* 321:13-18.

Carpenter, K.E. & Springer, V.G. 2005. The center of marine shore fish biodiversity: the Philippine Islands. *Environmental Biol. Fishes* 72: 467–480.

Chan, B.K.K., Poon, D.Y.N & Walker, G. 2005. Distribution, adult morphology, and larval development of *Sacculina sinensis* (Cirripedia: Rhizocephala: Kentrogonida) in Hong Kong coastal waters. *J. Crust. Biol.* 25: 1–10.

Couch, J.A. 1983. Diseases caused by Protozoa. In: Provenzano, A.J., Jr. (ed.), *The Biology of the Crustacea*, Vol. 6, *Pathobiology*: 79–111. New York: Academic Press.

Cressey, R.F. 1983. Crustaceans as parasites of other organisms. In: Provenzano, A.J., Jr. (ed.), *The Biology of the Crustacea, Vol. 6, Pathobiology*: 251–273. New York: Academic Press.

Cribb, T.H., Bray, R.A. & Littlewood, D.T.J. 2001. The nature and evolution of the association among digeneans, molluscs and fishes. *Int. J. Parsitol.* 31: 997–1011.

Cunningham, C.W., Buss, L.W. & Anderson, C. 1991. Molecular and geologic evidence of shared history between hermit crabs and the symbiotic genus *Hydractinia*. *Evolution* 45: 1301–1316.

Diaz, M.V. & Roccatagliata, D. 2006. Remarks on the genus *Anathelges* (Isopoda: Bopyridae), with a new record from the Beagle Channel, Argentina. *J. Crust. Biol.* 26: 331–340.

Dixon, C.J., Ahyong, S.T. & Schram, F.R. 2003. A new hypothesis of decapod phylogeny. *Crustaceana* 76: 935–975.

Dreyer, H. & Wägele, J.-W. 2001. Parasites of crustaceans (Isopoda: Bopyridae) evolved from fish parasites: Molecular and morphological evidence. *Zoology* 103: 157–178.

Feldmann, R.M. 1998. Parasitic castration of the crab, *Tumidocarcinus giganteus* Glaessner, from the Miocene of New Zealand: coevolution within the Crustacea. *J. Paleontol.* 72: 493–498.

Feuerborn, H. 1931. Ein Rhizocephale und zwei Polychaeten aus dem Süsswasser von Java und Sumatra. *Verhandl. Internat. Vereinig. Theoreti. Angewand. Limnologie* 5: 618–660.

Feuerborn, H. 1933. Das Cyprisstadium des Süsswasserrhizocephalen *Sesarmoxenos*. *Verhandl. Deutschen Zoologisch. Gesell.* 35: 127–138.

Glenner, H. & Hebsgaard, M.B. 2006. Phylogeny and evolution of life history strategies of the parasitic barnacles (Crustacea, Cirripedia, Rhizocephala). *Mol. Phylogenet. Evol.* 41: 528–538.

Glenner, H., Lützen, J. & Takahashi, T. 2003. Molecular and morphological evidence for a monophyletic clade of asexually reproducing Rhizocephala: *Polyascus*, new genus (Cirripedia). *J. Crust. Biol.* 23: 548–557.

Goddard, J.H.R., Torchin, M.E., Kuris, A.M. & Lafferty, K.D. 2005. Host specificity of *Sacculina carcini*, a potential biological control agent of the introduced European green crab *Carcinus maenas* in California. *Biol. Invasions* 7: 895–912.

Griffen, B.D., Guy, T. & Buck, J.C. 2007. Inhibition between invasives: a newly introduced predator moderates the impacts of a previously established invasive predator. *J. Animal Ecol.* 77: 32–40.

Griffith, H. 1987. Phylogenetic relationships and evolution in the genus *Dissodactylus* Smith, 1870 (Crustacea: Brachyura: Pinnotheridae). *Canadian J. Zool.* 65: 2292–2310.

Ho, J. 1988. Cladistics of *Sunaristes*, a genus of harpacticoid copepods associated with hermit crabs. *Hydrobiologia* 167/168: 555–560.

Hoberg, E.P. & Klassen, G.J. 2002. Revealing the faunal tapestry: co-evolution and historical biogeography of hosts and parasites in marine systems. *Parasitology* 124: 3–22.

Høeg, J.T. 1992. Rhizocephala. In: Harrison, F.W. & Humes, A.G. (eds.), *Microscopic Anatomy of Invertebrates, vol. 9: Crustacea*: 313–345. New York: Wiley-Liss, Inc.

Høeg, J.T., Glenner, H. & Shields, J.D. 2005. Cirripedia Thoracica and Rhizocephala (barnacles). In: Rohde, K. (ed.), *Marine Parasitology*: 154–165. Wallingford: CSIRO Publishing.

Høeg, J., Glenner, H. & Werner, M. 1997. The epicaridean parasite *Portunion maenadis* as a biological control agent on *Carcinus maenas*. In: *Proceedings of the First International Workshop on the Demography, Impacts and Management of Introduced Populations of the European Crab, Carcinus maenas: 20-21 March 1997*. Hobart, Tazmania: Centre for Research on Introduced Marine Pests. *Technical Report* 11.

Høeg, J.T. & Lützen, J. 1985. Crustacea Rhizocephala. *Mar. Invert. Scandinavia* 6: 1–92.

Høeg, J.T. & Lützen, J. 1993. Comparative morphology and phylogeny of the family Thompsoniidae (Cirripedia, Rhizocephala, Akentrogonida), with descriptions of three new genera and seven new species. *Zool. Scripta* 22: 363–386.

Høeg, J.T. & Rybakov, A.V. 1992. Revision of the Rhizocephala Akentrogonida (Cirripedia), with a list of all the species and a key to the identification of families. *J. Crust. Biol.* 12: 600–609.

Innocenti, G. & Galil, B.S. 2007. *Modus vivendi*: invasive host/parasite relations— *Charybdis longicollis* Leene, 1938 (Brachyura: Portunidae) and *Heterosaccus dollfusi* Boschma, 1960 (Rhizocephala: Sacculinidae). *Hydrobiologia* 590: 95–101.

Isaeva, V.V., Dolganov, S.M. & Shukalyuk, A.I. 2005. Rhizocephalan barnacles—parasites of commercially important crabs and other decapods. *Russian J. Mar. Biol.* 31: 215–220.

Johnson, P.T. 1983. Diseases caused by viruses, Rickettsia, Bacteria and Fungi. In: Provenzano, A.J., Jr. (ed.), *The Biology of the Crustacea, Vol. 6, Pathobiology*: 2–78. New York: Academic Press.

Korn, O.M., Shukalyuk, A.I., Trofimova, A.V. & Isaeva, I.I. 2004. Reproductive stage of the life cycle in the rhizocephalan barnacle *Polyascus polygenea* (Crustacea: Cirripedia). *Russian J. Mar. Biol.* 30: 328–340.

Kossmann, R. 1881. Studien über Bopyriden. *Zeitschr. Wissensch. Zoologie* 35: 652-665, pls. 32–35.

Kruse, I. & Hare, M.P. 2007. Genetic diversity and expanding nonindigenous range of the rhizocephalan *Loxothylacus panopaei* parasitizing mud crabs in the western north Atlantic. *J. Parasitol.* 93: 575–582.

Kuris, A.M., Goddard, J.H.R., Torchin, M.E., Murphy, N., Gurney, R. & Lafferty, K.D. 2007. An experimental evaluation of host specificity: The role of encounter and compatibility filters for a rhizocephalan parasite of crabs. *Int. J. Parasitol.* 37: 539–545.

Kuris, A.M., Lafferty, K.D. & Torchin, M.E. 2005. Biological control of the European green crab, *Carcinus maenas*: natural enemy evaluation and analysis of host specificity. In: Hoddle, M.S. (ed.), *Second International Symposium on Biological Control of Arthropods*: 102–115. Forest Health Technology Enterprise Team, University of California, Riverside.

Lafferty, K.D. & Kuris, A.M. 1996. Biological control of marine pests. *Ecology* 77: 1989–2000.

Legendre, P., Desdevises, Y. & Bazin, E. 2002. A statistical test for host-parasite coevolution. *Syst. Biol.* 51: 217–234.

Lester, R.J.G. 2005. Isopoda (isopods). In: Rohde, K. (ed.), *Ecology of Marine Parasites*: 138–144. CSIRO Publishing, Victoria, Australia.

Lützen, J. & Takahashi, T. 1996. Morphology and biology of *Polysaccus japonicus* (Crustacea, Rhizocephala, Akentrogonida, Polysaccidae, fam. n.), a parasite of the ghost-shrimp *Callianassa japonica*. *Zool. Scripta* 25: 171–181.

Markham, J.C. 1986. Evolution and zoogeography of the Isopoda Bopyridae, parasites of Crustacea. In: Gore, R.H. & Heck, K.L. (eds.), *Crustacean Issues 4, Crustacean Biogeography*: 143–164. Rotterdam: Balkema.

Markham, J.C. & Boyko, C.B. 2003. A new species of *Albunione* Markham & Boyko, 1999 (Crustacea: Isopoda: Bopyridae: Pseudioninae) from Taiwan. *Am. Mus. Novitates* 3410: 1–7.

Markham, J.C. & Dworschak, P.C. 2005. A new species of *Entophilus* Richardson, 1903 (Isopoda: Bopyridae: Entophilidae) from the Gulf of Aqaba. *J. Crust. Biol.* 25: 413–419.

McDermott, J.J. 1991. A breeding population of the western Pacific crab *Hemigrapsus sanguineus* (Crustacea: Decapod: Grapsidae) established on the Atlantic coast of North America. *Biol. Bull.* 181: 195–198.

McDermott, J.J. 1998. The western Pacific brachyuran (*Hemigrapsus sanguineus*: Grapsidae), in its new habitat along the Atlantic coast of the United States: geographic distribution and ecology. *ICES J. Mar. Sci.* 55: 289–298.

McDermott, J.J. 2000. Natural history and biology of the Asian shore crab *Hemigrapsus sanguineus* in the western Atlantic: a review, with new information. In: Peterson, J. (ed.), *Marine Bioinvasions: Proceedings of a Conference January* 24–27, 1999: 193–199. MIT Sea Grant College Program.

McDermott, J.J. 2007. Ectosymbionts of the non-indigenous Asian shore crab, *Hemigrapsus sanguineus* (Decapoda: Varunidae), in the western north Atlantic, and a search for its parasites. *J. Nat. Hist.* 41: 2379–2396.

Nieberding, C.M. & Olivieri, I. 2007. Parasites: proxies for host genealogy and ecology? *Trends Ecol. Evol.* 22: 156–165.

Nierstrasz, H.F. & Brender à Brandis, G.A. 1931. Papers from Dr. Th. Mortensen's Pacific Expedition 1914-16. LVII. Epicaridea II. *Videnskabelige Meddedelser fra den Dansk Naturhistoriske Forening i København* 91: 147–226, pl. 1.

Okada, Y.K. & Miyashita, Y. 1935. Sacculinization in *Eriocheir japonicus* de Haan, with remarks on the occurrence of complete sex-reversal in parasitized male crabs. *Mem. Coll. Sci., Kyoto Imperial Univ., ser. B* 10: 169–208.

Overstreet, R.M. 1983. Metazoan symbionts of crustaceans. In: Provenzano, A.J., Jr. (ed.), *The Biology of the Crustacea, Vol. 6, Pathobiology*: 155–250. New York: Academic Press.

Owens, L. 1993. Prevalence of *Cabirops orbionei* (Epicaridea; Cryptoniscidae) in Northern Australia: a biocontrol agent for bopyrids. *Aust. J. Mar. Freshwater Res.* 44: 381–387.

Owens, L. & Glazebrook, J.S. 1985. Sex determination in the Bopyridae. *J. Parasitol.* 71: 134–135.

Owens, L. & Rothlisberg, P.C. 1991. Vertical migration and advection of bopyrid isopod cryptoniscid larvae in the Gulf of Carpentaria, Australia. *J. Plankton Res.* 13: 779–787.

Owens, L. & Rothlisberg, P.C. 1995. Epidemiology of cryptonisci (Bopyridae: Isopoda) in the Gulf of Carpentaria, Australia. *Mar. Ecol. Progress Ser.* 122: 159–164.

Page, R.D.M. (ed.). 2002. *Tangled Trees: Phylogeny, Cospeciation, and Coevolution*. Chicago: The Chicago University Press.

Page, R.D.M. & Charleston, M.A. 1998. Trees within trees: phylogeny and historical associations. *Trends Ecol. Evol.* 13: 356–359.

Pasternak, Z., Blasius, B. & Abelson, A. 2004a. Host location by larvae of a parasitic barnacle: larval chemotaxis and plume tracking in flow. *J. Plankton Res.* 26: 487–493.

Pasternak, Z., Blasius, B., Achituv, Y. & Abelson, A. 2004b. Host location in flow by larvae of the symbiotic barnacle *Trevathana dentate* using odour-gated rheotaxis. *Proc. R. Soc. London ser. B, Biol. Sci.* 271: 1745–1750.

Paterson, A.M. & Poulin, R. 1999. Have chondracanthid copepods co-speciated with their teleost hosts? *Systematic Parasitol.* 44: 79–85.

Pérez, C. 1937. Sur les racines des Rhizocéphales. *C. R. Int. Congr. Zool. 12 Lisboa* 3: 1555–1563.

Pérez-Losada, M., Høeg, J.T., Kolbasov, G.A. & Crandall, K.A. 2002a. Reanalysis of the relationships among the Cirripedia and the Ascothoracica and the phylogenetic position of the Facetotecta (Maxillopoda: Thecostraca) using 18S rRNA sequences. *J. Crust. Biol.* 22: 661–669.

Pérez-Losada, M., Jara, C.G., Bond-Buckup, G., Porter, M.L. & Crandall, K.A. 2002b. Phylogenetic position of the freshwater anomuran family Aeglidae. *J. Crust. Biol.* 22: 670–676.

Pielou, E.C. 1974. Biogeographic range comparisons and evidence of geographic variation on host-parasite relations. *Ecology* 55: 1359–1367.

Porter, M.L., Pérez-Losada, M. & Crandall, K.A. 2005. Model-based multi-locus estimation of decapod phylogeny and divergence times. *Mol. Phylogenet. Evol.* 37: 355–369.

Poulin, R. 2007. *Evolutionary Ecology of Parasites: Second Edition*. Princeton: Princeton University Press.

Reinhard, E.G. 1949. Experiments on the determination and differentiation of sex in the bopyrid *Stegophryxus hyptius* Thompson. *Biol. Bull.* 96: 17–31.

Ritchie, L.E. & Høeg, J.T. 1981. The life history of *Lernaeodiscus porcellanae* (Cirripedia: Rhizocephala) and co-evolution with its porcellanid host. *J. Crust. Biol.* 1: 334–347.

Ruiz, G.M. 1991. Consequences of parasitism to marine invertebrates: host evolution? *Amer. Zool.* 31: 831–839.

Saito, T., Tsuchida, S. & Yamamoto, T. 2006. *Spongicoloides iheyaensis*, a new species of deep-sea sponge-associated shrimp from the Iheya Ridge, Ryukyu Islands, southern Japan (Decapoda: Stenopodidea: Spongicolidae). *J. Crust. Biol.* 26: 224–233.

Sars, G.O. 1896–1899. *An Account of the Crustacea of Norway with Short Descriptions and Figures of All the Species. Vol. II, Isopoda.* Bergen Museum. (Cryptoniscid text and pls. published in 1899.)

Scholtz, G. & Richter, S. 1995. Phylogenetic systematics of the reptantian Decapoda (Crustacea, Malacostraca). *Zool. J. Linn. Soc.* 113: 289–328.

Schubart, C.D. 2003. The east Asian shore crab *Hemigrapsus sanguineus* (Brachyura: Varunidae) in the Mediterranean Sea: an independent human-mediated introduction. *Scientia Marina* 67: 195–200.

Shields, J.D. & Overstreet, R.M. 2007. Parasites, symbionts, and diseases. In: Kennedy, V. & Cronin, L.E. (eds.), *The Blue Crab Callinectes sapidus*: 299–417. College Park, MD: University of Maryland Sea Grant College.

Shields, J.D., Stephens, F.J. & Jones, J.B. 2006. Chapter 5: Pathogens, parasites and other symbionts. In: Phillips, B.F. (ed.), *Lobsters: Biology, Management, Aquaculture and Fisheries*: 146–204. London: Blackwell Scientific.

Shiino, S.M. 1952. Phylogeny of the family Bopyridae. *Ann. Rep. Prefectural Univ. Mie, sect. 2, Nat. Sci.* 1: 33–56.

Shiino, S.M. 1965. Phylogeny of the genera within the family Bopyridae. *Bull. Mus. Natl. Hist. Nat., sér.* 2, 37: 462–465.

Shukalyuk, A.I., Golovnina, K.A., Baiborodin, S.I., Gunbin, K.V., Blinov, A.G. & Isaeva, V.V. 2007. *vasa*-related genes and their expression in stem cells of colonial parasitic rhizocephalan barnacle [sic] *Polyascus polygenea* (Arthropoda: Crustacea: Cirripedia: Rhizocephala). *Cell Biol. Internatl.* 31: 97–108.

Thresher, R.E., Werner, M., Heg, J.T., Svane, I., Glenner, H., Murphy, N.E. & Wittwer, C. 2000. Developing the options for managing marine pests: specificity trials on the parasitic castrator, *Sacculina carcini*, against the European crab, *Carcinus maenas*, and related species. *J. Exp. Mar. Biol. Ecol.* 254: 37–51.

Torchin, M.E., Lafferty, K.D. & Kuris, A.M. 2001. Release from parasites as natural enemies: increased performance of a globally introduced marine crab. *Biol. Invasions* 3: 333–345.

Torchin, M.E., Lafferty, K.D., Dobson, A.P., McKenzie, V.J. & Kuris, A.M. 2003. Introduced species and their missing parasites. *Nature* 421: 628–630.

Trilles, J.-P. 1999. Ordre des isopodes sous-ordre des épicarides (Epicaridea Latreille, 1825). In: Forest, J. (ed.), *Traité de Zoologie. Anatomie, Systématique, Biologie (Pierre-P. Grass). Tome VII, Fascicule III A, Crustacés Péracarides*: 279–352. *Mem. Inst. Oceanogr., Monaco* 19.

Tsang, L.M., Ma, K.Y., Ahyong, S.T., Chan, T.Y. & Chu, K.H. 2008. Phylogeny of Decapoda using two nuclear protein-coding genes: origin and evolution of the Reptantia. *Mol. Phylogenet. Evol.* 48: 359–368.

Tsuchida, K., Lützen, J. & Nishida, M. 2006. Sympatric three-species infection of *Sacculina* parasites (Cirripedia: Rhizocephala: Sacculinidae) of an intertidal grapsoid crab. *J. Crust. Biol.* 26: 474–479.

Van Wyk, P.M. 1982. Inhibition of the growth and reproduction of the porcellanid crab *Pachycheles rudis* by the bopyrid isopod, *Aporobopyrus muguensis*. *Parasitology* 85: 459–473.

Wägele, J.-W. 1989. Evolution und phylogenetisches System der Isopoda. Stand der Forschung und neue Erkenntnisse. *Zoologica* 140: 1–262.

Walker, G. 2001. Introduction to the Rhizocephala (Crustacea: Cirripedia). *J. Morphol.* 249: 1-8.

Weinberg Rasmussen, H., Jakobsen, H.L. & Collins, J.S.H. 2008. Raninidae infested by parasitic Isopoda (Epicaridea). *Bull. Mizunami Fossil Mus.* 34: 31–49.

Whiteman, N.K. & Parker, P.G. 2005. Using parasites to infer host population history: a new rationale for parasite conservation. *Animal Conservation* 8: 175–181.

# The Bearing of Larval Morphology on Brachyuran Phylogeny

PAUL F. CLARK

*Department of Zoology, The Natural History Museum, Cromwell Road, London, England*

ABSTRACT

Obtaining all developmental stages from an ovigerous decapod female is common in the laboratory. This is a significant advance for larval taxonomic studies, morphological descriptions, systematics, phylogenetics and evolutionary theory. Yet for such studies reliable data must be founded on quality observations and interpretation of setotaxy using a modern high-powered microscope equipped with differential interference contrast. Incorrect setal counts are problematic, especially since first-stage zoeas of congeneric brachyuran species appear to have identical setotaxy. This similarity provides such a high degree of predictability within a taxon that setal differences (incongruence) in a group may suggest incorrect assignment of taxa. However, relationships based on differences and similarities are not necessarily founded on shared derived characters, and instead may be supported by symplesiomorphies. The methodology involved in larval phylogenetics is also problematic. For example, oligomerization is considered to be an evolutionary trend within Crustacea. Decapod larval development suggests that heterochronic processes may provide a dominant evolutionary mechanism influencing loss of characters. Although using an unordered transformation series in a phylogenetic analysis is acknowledged to generate the most parsimonious trees, such an assumption does not necessarily represent a linear evolutionary pathway towards gradual terminal delay of characters as postulated by heterochrony for decapod larvae. A mosaic of heterochronic processes provides a complex evolutionary mechanism influencing oligomerization (reduction and loss) within brachyuran zoeae. This is best captured in a phylogenetic analysis by using "irreversible-up" (terminal delay, not terminal addition) transformation series. Reconstruction of trees using this assumption about character evolution generates longer trees and frequently involves more evolutionary steps to compensate for homoplasy. Yet there is evidence to suggest that homoplasy is common within many brachyuran larval lineages. Nonetheless, larval phylogenetics does appear to have advantages since all decapod zoeal stages are adapted to a planktonic existence, and therefore setal patterns are subject to similar selection pressures. Morphological differences among larvae may provide additional phylogenetic information as compared to possibly convergent adult characters that are more the product of the interaction between genotype and environment.

1 WHY STUDY LARVAE?

Historically, decapod systematics has been established on the basis of adult morphology, but these phenotypic characters are the end product of the interaction between genotype and environment. Consequently, relationships within and between taxa may be postulated on convergence between adults. Another valuable and often-overlooked source of information is the morphology of decapod larvae. Larvae are adapted to the same habitat, a uniform planktonic environment, and as such setal patterns should be subjected to more or less constant selection pressures. Therefore, larval characters may reflect relationships better than the morphology of the adults (see Williamson 1982; Rice 1980; Felder et al. 1985).

The majority of decapod larval studies have addressed relationships within the Brachyura, and these have been based mostly on zoeal characters. As with the adults, larval relationships have normally been established on similarity and difference of morphologically features (e.g., Rice 1980; Martin 1984; Martin et al. 1985; Felder et al. 1985; Ng & Clark 2000; Clark & Ng 2006). But relationships founded on similarities among taxa may be based on ancestral characters and not necessarily those that are shared and derived. With this in mind, several studies have conducted phylogenetic analyses of zoeal characters with a view to confirming or testing relationships based primarily on adult morphology (e.g., Rice 1980; Clark 1983; Clark & Webber 1991; Marques & Pohle 1998; Ng & Clark 2001; Clark & Guerao 2008).

The purpose of this paper is to use a restricted set of data associated with brachyuran (mostly pilumnoid) zoeal stages to review some of the problems identified with constructing phylogenies using setotaxy. The study also aims to show that phylogenetic analysis of Xanthoidea and Pilumnoidea zoeal characters can provide a new insight into a classification traditionally founded on adult convergent morphology.

## 2 COLLECTING LARVAE

Rearing decapod larvae was once considered difficult, but the use of *Artemia* nauplii as a food source has opened up the field. All aspects of larval biology, including biochemistry, ecology, endocrinology, growth, metabolism, moulting, physiology, ultrastructure and other topics (see Anger 2001 for details) can now be more easily studied. Obtaining all developmental stages from an ovigerous female is now common in the laboratory. This is a significant advance for descriptive studies (alpha taxonomy), systematics, phylogenetics and evolutionary theory. However, larval rearing is not without its disappointments and failures. Collecting ovigerous target species still depends on sampling effort and a measure of luck; success is never guaranteed. Once the specimens are safely ensconced in a constant temperature room, rearing is time-consuming, requiring dedication and discipline to see it through to completion. Even then, for no apparent reason, larval cultures occasionally crash. These frustrations aside, there are distinct advantages to rearing larvae in the laboratory as opposed to studying plankton-collected material, such as collecting all life stages with verification from exuvia, providing sufficient specimens for morphological studies, and confirming the identification of the larvae by examining the spent female. The ability to positively identify the species is the distinct advantage that laboratory-reared material has over describing plankton-caught larvae. Confident identification of such larvae to species level is still problematic (e.g., the third and fourth zoeal stages of crab larvae from Atlantic Seamounts described by Rice & Williamson 1977 are still unidentified).

## 3 SETAL OBSERVATIONS

After completing the task of laboratory rearing, many larval morphologists proceed to produce poor descriptions, typically by missing increasing numbers of setal characters during zoeal development. Reliable data are everything, and setotaxy must be founded on high-quality observations and interpretation. Although Rice (1979) and Clark et al. (1998a) made pleas for improved standards in descriptions of crab zoeas, some studies are still inadequate. Zoeal and megalopal characters are still being either overlooked or ignored, for example, the development of the third maxilliped through successive zoeal moults. This situation must be resolved if there is to be progress in brachyuran larval research. A modern-day high-powered microscope equipped with differential interference contrast (DIC) is fundamental to these studies if setal ambiguities are to be resolved. Using lesser microscopes is inadequate for modern larval studies. Additionally, some larval characters, such as the endopod spine on the antennal protopod of xanthoid larvae, may be resolved only by using a scanning electron microscope.

## 4 ZOEAL SIMILARITY

Brachyuran first-stage zoeas of congeneric species appear to have virtually identical setotaxy (Christiansen 1973; Clark 1983, 1984; Ng & Clark 2000). This similarity provides a high degree of predictability within a taxon. Setal differences (incongruence) within a group suggest incorrect assignment of taxa and lack of systematic compatibility. For example, the first stage zoeas of *Chlorodiella nigra* (Forskøal, 1775), *Cyclodius monticulosus* (Dana, 1852), *Pilodius areolatus* (H. Milne Edwards, 1834), *Pilodius paumotensis* Rathbun, 1907 and *P. pugil* Dana, 1852 are similar, if not identical, in terms of setotaxy. Their zoeas cannot be identified to species level. An example shows the usefulness of this similarity: Serène (1984), based on adult features, felt that *Chlorodiella bidentata* (Nobili, 1901) did not belong in *Chlorodiella* and should perhaps be referred to its own genus within the Chlorodiinae Alcock, 1898 (now Chlorodiellinae Ng & Holthuis, 2007). If the hypothesis of Serène (1984) were correct, then the first-stage zoeas of *C. bidentata* would possess a setotaxy identical to those of the other species assigned to the subfamily. According to Ng and Clark (2000), this was not the case. In fact, based on larval characters, especially the antenna, Ng & Clark (2000, table 6) showed that *C. bidentata* was not even a xanthid but a member of the Pilumnidae (now Pilumnoidea Samouelle, 1819; see Ng et al. 2008).

According to Clark & Ng (2004b) there were 72 genera and 408 species of Pilumnoidea known, and of these the zoeas of approximately 30 species (Table 1) are described. The pilumnoid zoeal antenna is a conservative character in that, except for the development of the endopod, its morphology remains unchanged with successive moults and defines all species attributed to this superfamily. It is characteristic of all 30 species listed in Table 1. According to Martin's (1984: 228, Fig. 1H) definition of xanthid group II, pilumnids are characterized by an acutely tipped antennal exopod, about equal in length to or slightly longer than the protopod, armed with small spinules distally, and with a prominent outer seta about halfway along its length; additionally, the antennal protopod is usually longer than the rostrum. However, Martin overlooked a second smaller medial seta on the exopod. Two medial setae on the antennal exopod are diagnostic of this family (Fig. 1A). Furthermore, the exopod is distally bilaterally spinulate, as is the protopod. Interestingly, the antenna exopod of *Aniptumnus quadridentatus* (De Man, 1895) (Fig. 1B) is more elongate than in the other pilumnoids described, but it still retains the two medial setae.

Eumedonic crabs provide another example. Adult eumedonids are associates of echinoderms. Many brachyuran systematists have found their morphology confusing, resulting in their placement in various families, including the Majidae, Parthenopidae, Xanthidae, Pilumnidae, Trapeziidae, Portunidae, Pinnotheridae and Eumedonidae. Ng & Clark (2001) considered the first-stage zoeas of five eumedonid species: *Echinoecus pentagonus* (A. Milne Edwards, 1879), *Harrovia albolineata* Adams & White, 1849, *Permanotus purpureus* (Gordon, 1934), *Rhabdonotus pictus* A. Milne Edwards, 1879 and *Zebrida adamsii* White, 1847. All five possessed the same type of antenna (as in Fig. 1A). On similarity of the zoeal antenna, Ng & Clark (2001) challenged the validity of the Eumedonidae as a distinct (e.g., Martin & Davis 2001) family and suggested that these cryptic crabs were in fact pilumnoids. Their study of eumedonid first-stage zoeas is a classic example of larvae setal patterns resolving the classification of a difficult group of brachyuran species that was previously based on deceptive adult morphology.

Comparisons based on differences and similarities of morphology are of interest because they provide an expectancy (predictability) that the first-stage zoeas of closely related species will share a suite of characters. However, these characters are not necessarily shared derived characters, and therefore relationships founded on similarities among taxa may be based on symplesiomorphic characters.

**Table 1.** References to descriptions of larvae in the brachyuran family Pilumnidae.

| Species | Reference | Stage | Remarks |
|---|---|---|---|
| *Actumnus setifer* (de Haan, 1835) | Aikawa 1937 | ZI | |
| *Actumnus setifer* (de Haan, 1835) | Clark & Ng 2004b | ZI-ZIII, Meg. | |
| *Actumnus squamosus* (de Haan, 1835) | Terada 1988 | ZI-IV, Meg. | |
| *Aniptumnus quadridentatus* (De Man, 1895) | Ng 2002 | ZI | |
| *Aniptumnus quadridentatus* (De Man, 1895) | Ng & Clark 2008 | ZI | |
| *Benthopanope eucratoides* (Stimpson, 1858) | Lim et al. 1986 | ZI-III, Meg. as | *Pilumnopeus eucratoides* |
| *Benthopanope indica* (De Man, 1887) | Takeda & Miyake 1968 | ZI | as *Pilumnopeus indicus* |
| *Benthopanope indica* (De Man, 1887) | Terada 1980 | ZI-IV | as *Pilumnopeus indicus* |
| *Benthopanope indica* (De Man, 1887) | Ko 1995 | ZI-IV, Meg. | |
| *Galene bispinosa* (Herbst, 1794) | Mohan & Kannupandi 1986 | ZI-IV, Meg. | |
| *Halimede fragifer* de Haan, 1835 | Terada 1985 | ZI-II | |
| *Heteropanope glabra* Stimpson, 1858 | Aikawa 1929 | ZI | |
| *Heteropanope glabra* Stimpson, 1858 | Lim et al. 1984 | ZI-IV, Meg. | |
| *Heteropanope glabra* Stimpson, 1858 | Greenwood & Fielder 1984a | ZI-IV, Meg. | |
| *Heteropilumnus ciliatus* (Stimpson, 1858) | Takeda & Miyake 1968 | ZI | |
| *Heteropilumnus ciliatus* (Stimpson, 1858) | Ko & Yang 2003 | ZI-III | |
| *Latopilumnus conicus* Ng & Clark, 2008 | Ng & Clark 2008 | ZI | |
| *Lobopilumnus agassizi* Stimpson, 1871 | Lebour 1950 | ZI | |
| *Pilumnopeus granulata* Balss, 1933 | Ko 1997 | ZI-IV, Meg. | |
| *Pilumnopeus makianus* (Rathbun, 1929) | Lee 1993 | ZI-IV | |
| *Pilumnopeus serratifrons* (Kinahan, 1856) | Wear 1968 | ZI | |
| *Pilumnopeus serratifrons* (Kinahan, 1856) | Greenwood & Fielder 1984b | ZI-III | |
| *Pilumnopeus serratifrons* (Kinahan, 1856) | Wear & Fielder 1985 | ZI | |
| *Pilumnus dasypodus* Kingsley, 1879 | Sandifer 1974 | ZI-IV, Meg. | |
| *Pilumnus dasypodus* Kingsley, 1879 | Bookhout & Costlow 1979 | ZI-IV, Meg. | |
| *Pilumnus hirtellus* (Linnaeus, 1761) | Williamson 1915 | ZI | |
| *Pilumnus hirtellus* (Linnaeus, 1761) | Boraschi 1921 | ZI, | |
| *Pilumnus hirtellus* (Linnaeus, 1761) | Lebour 1928 | ZI-IV, Meg. | |
| *Pilumnus hirtellus* (Linnaeus, 1761) | Bourdillon-Casanova 1960 | ZI | |
| *Pilumnus hirtellus* (Linnaeus, 1761) | Salman 1982 | ZI-IV, Meg. | |

Table 1. continued.

| Species | Reference | Stage | Remarks |
|---|---|---|---|
| *Pilumnus hirtellus* (Linnaeus, 1761) | Ingle 1983 | Meg. | |
| *Pilumnus hirtellus* (Linnaeus, 1761) | Ingle 1991 | ZI-IV, Meg. | |
| *Pilumnus hirtellus* (Linnaeus, 1761) | Ng and Clark 2000 | ZI | |
| *Pilumnus hirtellus* (Linnaeus, 1761) | Clark 2005 | ZI-IV | |
| *Pilumnus kempi* Deb, 1987 | Siddiqui & Tirmizi, 1992 | ZI-II, Meg. | |
| *Pilumnus lumpinus* Bennett, 1964 | Wear 1967 | Meg. | |
| *Pilumnus lumpinus* Bennett, 1964 | Wear & Fielder 1985 | ?ZI Meg. | |
| *Pilumnus longicornis* Hilgendorf, 1879 | Prasad & Tampi 1957 | ZI | |
| *Pilumnus longicornis* Hilgendorf, 1879 | Hashmi 1970 | ?ZI | |
| *Pilumnus longicornis* Hilgendorf, 1879 | Clark & Paula 2003 | ZI | |
| *Pilumnus minutes* de Haan, 1835 | Aikawa 1929 | ZI | |
| *Pilumnus minutes* de Haan, 1835 | Terada 1984 | ZI-IV | |
| *Pilumnus minutes* de Haan, 1835 | Ko 1994b | ZI-IV | |
| *Pilumnus minutes* de Haan, 1835 | Ko 1997 | Meg. | |
| *Pilumnus novaezealandiae* Filhol, 1885 | Wear 1967 | Meg. | |
| *Pilumnus novaezealandiae* Filhol, 1885 | Wear & Fielder 1985 | Meg. | |
| *Pilumnus sayi* Rathbun, 1897 | Bookhout & Costlow 1979 | ZI-IV, Meg. | |
| *Pilumnus scabriusculus* Adams & White, 1849 | Terada 1990 | ZI-IV | |
| *Pilumnus sluiteri* De Man, 1892 | Clark & Ng 2004a | ZI-II, Meg. | |
| *Pilumnus trispinosus* (T. Sakai, 1965) | Terada 1984 | ZI-IV | as *Parapilumnus trispinosus* |
| *Pilumnus trispinosus* (T. Sakai, 1965) | Quintana 1986 | Meg. | as *Parapilumnus trispinosus* |
| *Pilumnus trispinosus* (T. Sakai, 1965) | Ko 1994a | ZI-IV, Meg. | as *Parapilumnus trispinosus* |
| *Pilumnus vespertilio* (Fabricius, 1793) | Aikawa 1929 | ZI | |
| *Pilumnus vespertilio* (Fabricius, 1793) | Lim & Tan 1981 | ZI-III, Meg. | |
| *Pilumnus vespertilio* (Fabricius, 1793) | Terada 1990 | ZI-III | |
| *Pilumnus vespertilio* (Fabricius, 1793) | Clark and Paula 2003 | ZI | |
| *Pilumnus vestitus* Haswell, 1882 | Hale 1931 | Meg. | |
| *Tanaocheles bidentata* (Nobili, 1901) | Ng & Clark 2000 | ZI | |

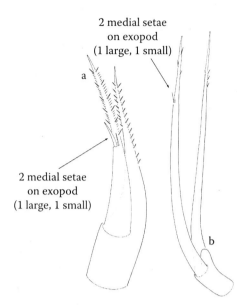

**Figure 1.** Diagnostic characters of the pilumnoid antenna, first-stage zoea. (A) *Pilumnus hirtellus*. (B) *Aniptumnus quadridentatus*.

## 5  HETEROCHRONY

Clark (2001) analyzed patterns in setotaxy and segmentation associated with abbreviated zoeal development in three higher taxa of brachyuran crabs — two portunids, two xanthoids and a number of majids — with different numbers of larval stages. Included were laboratory-reared larvae of species with six zoeal stages [*Charybdis helleri* (A. Milne Edwards, 1867) by Dineen et al. 2001], five stages [*Liocarcinus arcuatus* (Leach, 1814) by Clark 1984], four stages [*Lophozozymus pictor* (Fabricius, 1798) by Clark & Ng 1998], three stages [*Actumnus setifer* (de Haan, 1835) described later by Clark & Ng 2004b], and two stages [*Macrocheira kaempferi* (Temminck, 1838) by Clark & Webber 1991, *Libinia spinosa* H. Milne Edwards, 1834, by Clark et al. 1998b, and *Inachus dorsettensis* (Pennant, 1777) and *Inachus leptochirus* Leach, 1817 both by Clark 1980, 1983]. Comparing these life cycles, Clark (2001) concluded that the development of different characters occurred at different times and/or rates, suggesting that the evolutionary history of brachyuran zoeas provided robust examples of heterochrony. However, Clark (2001) made no attempt to relate his zoeal theory to the heterochronic processes described by McKinney & McNamara (1991).

Heterochrony can be defined as an evolutionary change in the timing of the development of a character between an ancestor and descendant. McKinney & McNamara (1991) illustrated a hierarchical classification of heterochrony, reproduced here in Fig. 2A. They considered that between an ancestor and its descendant, development can be either reduced or increased. Accordingly, a reduction in development resulted in paedomorphosis (child formation), i.e., the retention of juvenile characters of the ancestral forms by adults of their descendants. An increase in development resulted in peramorphosis, i.e., the descendant incorporating all the ontogenetic stages of its ancestor, including the adult stage, in its ontogeny, so that the adult descendant "goes beyond" its ancestor. McKinney & McNamara (1991) recognized three basic types of change for paedomorphosis and peramorphosis: change in rate, change in offset time, and change in onset time. Consequently, six kinds of developmental change were recognized: (1) the rate of change in the descendant can be slower (neoteny) or faster (acceleration) than the ancestor; (2) the onset time in the descendant can be later (postdisplacement) or earlier (predisplacement) than in the ancestor; and (3) the offset time

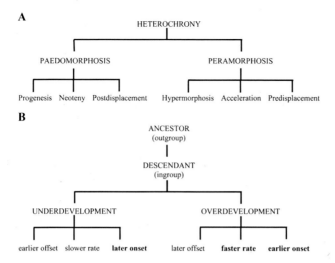

**Figure 2.** Heterochrony. (A) The hierarchical classification of heterochrony (after McKinney & McNamara 1991). (B) Simplified version with the three heterochronic processes associated with brachyuran zoeas highlighted in bold.

in the descendant can be earlier (progenesis) or delayed (hypermorphosis) than in the ancestor. The heterochronic system proposed by McKinney & McNamara (1991) is summarized here in Fig. 2B.

The problem with the hierarchical system of heterochrony as proposed by McKinney & McNamara (1991) in relation to larvae, in particular to zoeal characters, is that three processes are usually associated with sexual maturity, namely progenesis, neoteny and hypermorphosis. Functionally, Decapoda larvae are developmental and dispersal stages and are not influenced by sexual maturity, which develops during the postlarval phase and is continued in the juveniles and adults. Therefore, only three heterochronic mechanisms (see Clark 2005) appear to relate to brachyuran zoeal development (see bold typeface in Fig. 2B): postdisplacement (Table 2), predisplacement (Table 3) and acceleration (Table 4). In addition, the terms onset and offset used by McKinney & McNamara (1991) can be used to describe the presence (expressed) or absence (delayed) of individual setae, segments and even developmental phases/stages.

**Table 2.** Postdisplacement (underdevelopment): four setae are present (expressed, onset) in the ancestor compared to 3 setae (seta 4 absent or delayed) and 2 setae (setae 3 and 4 absent or delayed, offset) in descendants 1 and 2, respectively.

|  | Seta 1 | Seta 2 | Seta 3 | Seta 4 |
|---|---|---|---|---|
| ANCESTOR | present<br>onset<br>expressed | present<br>onset<br>expressed | present<br>onset<br>expressed | present<br>onset<br>expressed |
| DESCENDANT 1 | present<br>onset<br>expressed | present<br>onset<br>expressed | present<br>onset<br>expressed | absent<br>offset<br>delayed |
| DESCENDANT 2 | present<br>onset<br>expressed | present<br>onset<br>expressed | absent<br>offset<br>delayed | absent<br>offset<br>delayed |

*onset of first zoeal stage (hatching)* ↑    *offset of first zoeal stage (molt to second zoeal stage)* ↑

**Table 3.** Predisplacement (overdevelopment): four setae are present (expressed, onset) in the ancestor compared to 5 setae (seta 5 present or expressed) and 6 setae (setae 5 and 6 present or expressed, onset) in descendants 1 and 2, respectively.

|  | Seta 1 | Seta 2 | Seta 3 | Seta 4 | Seta 5 | Seta 6 |
|---|---|---|---|---|---|---|
| ANCESTOR | present onset expressed | present onset expressed | present onset expressed | present onset expressed | absent offset delayed | absent offset delayed |
| DESCENDANT 1 | present onset expressed | present onset expressed | present onset expressed | present onset expressed | present onset expressed | absent offset delayed |
| DESCENDANT 2 | present onset expressed | present onset expressed | present onset expressed | present onset expressed | present onset expressed | present onset expressed |

*onset of first zoeal stage (hatching)* ↑    *offset of first zoeal stage (molt to second zoeal stage)* ↑

**Table 4.** Acceleration (overdevelopment) faster rate: four steps are required in the ancestor to fully develop an appendage from hatching to the offset of the zoeal phase compared to three and two steps in descendants 1 and 2, respectively (see third maxilliped, Clark 2005: 441, fig. 14).

|  | ACCELERATION | | | |
|---|---|---|---|---|
| **ANCESTOR** | UNIRAMOUS | BIRAMOUS | BIRAMOUS with EPIPOD | BIRAMOUS with EPIPOD and ARTHROBRANCH |
| **DESCENDANT 1** | BIRAMOUS | | BIRAMOUS with EPIPOD | BIRAMOUS with EPIPOD and ARTHROBRANCH |
| **DESCENDANT 2** | BIRAMOUS with EPIPOD | | | BIRAMOUS with EPIPOD and ARTHROBRANCH |

*onset of hatching and zoeal phase* ↑    *offset of zoeal phase, onset of megalopal phase* ↑

## 6 POLARITY OF SETAL CHARACTERS

Brachyuran zoeal molts are associated with body growth, division of somites, appearance and development of appendages, and appearance (expression) of setae. On certain body somites and appendage segments, the number of some setae does not increase after successive zoeal moults (stages) and can be considered conservative. For example, the setal patterns on the second maxilliped endopod of xanthoids (Fig. 3A) remain constant (conservative) throughout zoeal development (e.g., *Lophozozymus pictor* as described by Clark & Ng 1998). When analyzing these conservative setal characters for possible phylogenetic significance, a number of brachyuran workers (e.g., Lebour

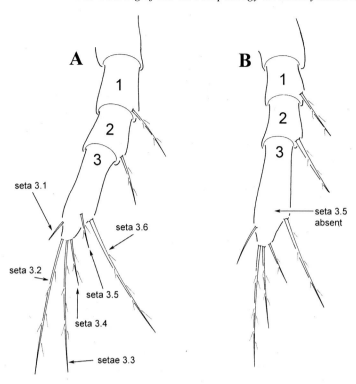

**Figure 3.** First-stage zoea, second maxilliped, setation patterns on the three-segmented endopod. (A) *Pilodius pugil*: seta 3.5 is present (expressed) and is considered to be the ancestral condition. (B) *Banaria subglobosa*: seta 3.5 is lost (absence or delay in appearance) and is regarded as the derived state for this character.

1928, 1931; Bourdillon-Casanova 1960; Kurata 1969; Clark 1980, 1983; Rice 1980, 1983, 1988; Clark & Webber 1991; Ng & Clark 2001) have assumed that zoeal evolution has proceeded by loss or reduction of setae. Under such an assumption, the presence (expression) of a seta would be considered the ancestral state, and its absence (loss or delay in appearance) is considered derived. For example, seta 3.5 is present (expressed) and considered to be the ancestral condition (Fig. 3A), while its loss (absence or delay in appearance) is regarded as the derived state for this character (Fig. 3B).

In contrast to such conservative characters, there are some somites and appendage segments that accumulate setae at successive zoeal moults. Scoring and polarizing these characters is not straightforward. When Clark & Webber (1991) first analyzed majid zoeae using PAUP, they simply counted the setae on each appendage article. As a consequence, five setae on a segment for one species was considered ancestral when compared to the same segment of another species with only four setae (derived). Such an assumption does not take into account which seta had been lost (absent or delayed). Neither did such counting take into account the influence of abbreviated zoeal development on expression of setae (Clark 2005). For example, with reference to the third endopod segment of the first maxilliped in the first stages of *Charybdis helleri* (Portunoidea Rafinesque, 1815; see Ng et al. 2008) and the xanthoid *Chlorodiella nigra*), at first glance a seta is present in ZI of the latter and absent in the former, suggesting that *C. helleri* is the derived condition (compare Fig. 4A with 4E). However, when Dineen et al. (2001) reared *C. helleri* in the laboratory through to stage ZVI, they showed that this seta appeared (was expressed) later (in ZIV) during development (Fig. 4A–D). Reassessing this character now (Fig. 4E), it is clear that the seta on endopod segment 3 has appeared (expressed) early, in ZI, of *Chlorodiella nigra* compared to the outgroup (possible ancestor) of *Charybdis helleri*. From McKinney & McNamara (1991), this early

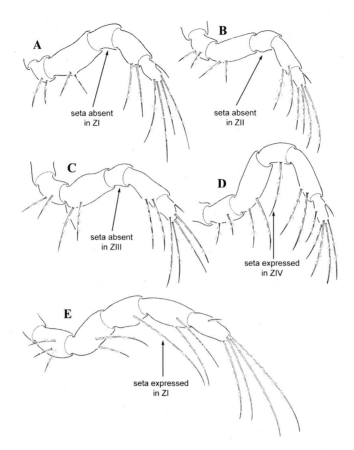

**Figure 4.** First maxilliped, expression (appearance or presence) of the seta on third endopod segment. (A–D) *Charybdis helleri* zoeas I–IV, respectively. (E) *Chlorodiella nigra* zoea I.

expression relates to predisplacement of the seta, overdevelopment (peramorphosis) in *Chlorodiella nigra*, and its early onset is the derived condition. The absence of the seta in ZI of *Charybdis helleri* is therefore the plesiomorphic (ancestral) condition.

Accumulative setae, such as the armature of the maxilla proximal coxal endite in brachyuran zoeas, also are of interest with regard to hetrochrony and polarization. Figure 5A–F illustrates the accumulative setae on the maxilla proximal coxal endite during the development of ZI–VI for *Charybdis helleri* by Dineen et al. (2001); stages ZI to ZVI bear 3,3,3,3,4,5 setae, respectively. Comparison of this accumulation sequence with the zoeal development of *Nanocassiope melanodactyla* (A. Milne Edwards, 1867) by Dornelas et al. (2004), which consists of only four zoeas with setation arranged 4,4,5,6 (Fig. 5G–J), shows that the appearances of 4 (ZI) and 6 (ZIV) setae are both expressed (present) early compared to what is seen in the zoeal stages of *C. helleri* (ZV and ZVI).

Scoring the accumulative setae on the maxilla proximal coxal endite for a phylogenetic analysis with reference to the first-stage zoeas of *C. helleri*, *N. melanodactyla*, *Pilumnus hirtellus* (Linnaeus, 1761) and *Eriphia scabricula* Dana, 1852 is difficult (Fig. 6A–D, respectively). Considering *C. helleri* as the outgroup (ancestor), the character could be scored simply as a multistate character, with the 3 setae of this species being the ancestral condition and accumulation of setae being increasingly more derived.

However, these accumulative setae also could be scored individually with respect to the principles of heterochrony and overdevelopment (peramorphosis). The individual setae can be identified

The Bearing of Larval Morphology on Brachyuran Phylogeny 231

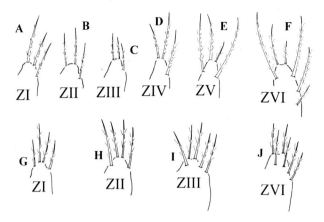

**Figure 5.** Maxilla, setation of proximal coxal endite. (A–F) *Charybdis helleri* (Portunidae). (G–J) *Nanocassiope melanodactyla* (Xanthidae).

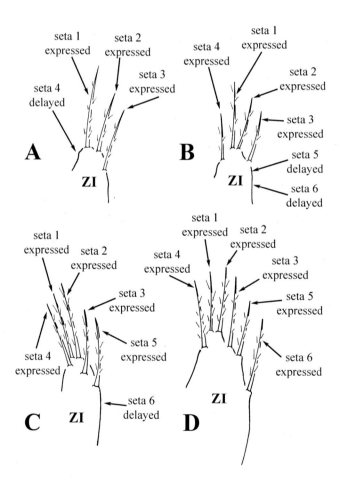

**Figure 6.** Maxilla, setation of proximal coxal endite. (A) *Charybdis helleri*. (B) *Nanocassiope melanodactyla*. (C) *Pilumnus hirtellus*. (D) *Eriphia scabricula*.

and their expression (presence) correlated to an outgroup (possible ancestor) species with a longer zoeal development phase, e.g., *Charybdis helleri* with six zoeal stages. Thus, instead of being a single multistate character, three characters can be scored. In Figure 6A–D, the setae are numbered from 1 to 6. Setae 1–3 are present (expressed) in *C. helleri, N. melanodactyla, P. hirtellus* and *E. scabricula*. Seta 4 is absent (delayed) in *C. helleri* (the outgroup and ancestor), but is expressed (overdeveloped when compared to the ancestor) in *N. melanodactyla, P. hirtellus* and *E. scabricula*. Seta 5 is delayed in *C. helleri* and *N. melanodactyla* but is expressed in *P. hirtellus* and *E. scabricula*, with seta 6 being delayed in *C. helleri, N. melanodactyla* and *P. hirtellus* but expressed in *E. scabricula*. These characters therefore could be scored as delayed (0) vs. expressed (1) for each of the three setae (seta 4, 5 and 6).

## 7 TRANSFORMATION TYPES

The choice of transformation types is important because such decisions affect the number of evolutionary steps in a phylogenetic analysis. Using "irreversible-up" with respect to brachyuran zoeal phylogeny is widely regarded as introducing an element of subjectivity because it does not necessarily produce the shortest (most parsimonious) trees, as postulated by Marques & Pohle (1998).

A problem for the present study is that according to Maddison & Maddison (1992: 79), when using unordered characters, "... a change from any state to any other state is counted as one step" (referred to as "Fitch parsimony"; see Fitch 1971; Hartigan 1973). Thus, a change from 0 to 1, or from 0 to 8 or 7 to 4, is each counted as one step. A five-state unordered character can be represented diagrammatically (Fig. 7A), where change between any two states involves only one step (i.e., only one line has to be traversed in the diagram). An unordered transformation series does not reflect the course of evolution as proposed for decapod larvae and based on heterochrony (Clark 2005). Heterochrony suggests a gradual progressive loss (delayed expression) of characters in a linear transformation series, such as the loss of one seta at a time from the proximal basial endite of the maxilla (Clark 2005: 437, table 19; and fig. 16). Individual setae can be scored (Fig. 6), i.e., the six setae on the proximal basial endite of the maxilla are numbered individually 1 to 6. Empirical observations suggest that seta 6 is lost, then seta 5, then seta 4 and so on in the last zoeal stage of the descendant in relation to the ancestor. Heterochrony within decapod larvae provides no support for the suggestion that any one state can transform to any other state in a single step, e.g., 1 to 4 or 3 to 0. Indeed, heterochrony appears to support a linear transformation series, of which there are two types: ordered and irreversible.

Maddison & Maddison (1992: 79) define an ordered transformation series: "For characters designated as ordered, the number of steps from one state to another state as the (absolute value of the)

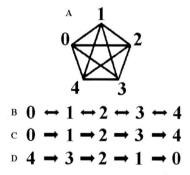

**Figure 7.** Transformation series: (A) unordered. (B) ordered. (C) irreversible-up terminal addition. (D) irreversible-up terminal delay = oligomerization.

difference between their state numbers" ("Wagner parsimony"; Farris [1970]; Swofford and Maddison [1987]). Thus, a change from 0 to 1 is counted as one step, from 0 to 8 as eight steps, from 7 to 4 as three steps. Thus, a five-state ordered character can be represented diagrammatically as shown in Fig. 7B. In this diagram, the number of steps in the change between any two states is equal to the number of lines on the path between the two states; thus, from 1 to 4 is three lines or three steps. The analysis of heterochrony (Clark 2005) provides no support for the existence of ordered transformation of character types in decapod larvae. In the absence of any supporting evidence, it is problematic to accept that zoeal characters once lost in a specific lineage or taxon, e.g., 4 to 3 to 2 to 1 to 0 (Fig. 7B), can then reappear again as 0 to 1 to 2, etc. Within the decapods a number of traits have been lost and not reappeared. For example, the Dendrobranchiata release their eggs directly into the water column, whereas all derived decapods (Pleocyemata) spawn their eggs onto the pleopods, where they remain with parental (female) care until hatching. This strategy, the release of eggs into the sea, has not been reversed in derived decapods. Further, the Dendrobranchiata have a nauplius larval phase, which is lost (present in embryonic development) in the more derived decapods (Pleocyemata) where larvae hatch in a more advanced stage of development as zoeas. Nauplii have not reappeared in the Pleocyemata.

Maddison & Maddison (1992: 79-80) define irreversible as: "For characters designated as irreversible, the number of steps from one state to another state is counted as the difference between their state numbers, with the restriction that decreases in the state number do not occur" ("Camin-Sokal parsimony"; Camin and Sokal [1965]). Thus, a change from 0 to 1 is counted as one step, from 0 to 8 as eight steps, but changes from 1 to 0 or 8 to 0 are impossible. Multiple gains (increases) are allowed, but no losses (decreases) are allowed. A five-state irreversible character can be represented diagrammatically (Fig. 7C). However, this figure represents terminal addition (Clark 2005: 438), whereas the linear transformation series described by Fig. 7D seems to best fit the theories that a mosaic of several heterochronic processes provides a dominant evolutionary mechanism influencing oligomerization within brachyuran zoeae. Terminal delay of characters is represented by Fig. 8 (see also Clark 2005). Once decapod larval characters are lost in any lineage, they are not expressed again.

## 8 HOMOPLASY

Although scoring characters as "irreversible-up" does reflect reduction or abbreviation, ultimately resulting in terminal delay (oligomerization), this option, in general, does not allow reversals in character state changes and forces additional homoplasy. But homoplasy does appear to be extremely widespread in brachyuran zoeal lineages; many derived character states have evolved more than once within different branches (clades). For example, seta 3.5 (Fig. 3B) has been lost (delayed or absent) a number of times in brachyuran zoeal evolution. Examples are found in the Pilumnidae as in *Tanocheles bidentata* (described by Ng & Clark 2000); within the Xanthidae as in *Leptodius exaratus* (H. Milne Edwards, 1834) and *Lybia plumose* Barnard, 1947 (both by Clark & Paula 2003); within the Majidae as in *Inachus* (by Clark 1983) and *Libinia spinosa* H. Milne Edwards, 1834 (by Clark et al. 1998b); and within the Grapsoidea as in *Xenograpsus testudinatus* Ng, Huang & Ho, 2000 (by Min-Shiou et al. 2004). As with the second maxilliped, the expression of the seta on the first endopod segment (Fig. 3) also has been lost (delayed or absent) a number of times in brachyuran zoeal evolution. Examples occur within the Trapezioidea as in *Trapezia richtersi* Galil & Lewinsohn, 1983 (by Clark & Ng 2006); within the Majidae as in *Inachus* (by Clark 1983) and *Libinia spinosa* (by Clark et al. 1998b); and within the Grapsoidea as in *Armases miersii* (Rathbun, 1897) (by Cuesta et al. 1999). Such derived characters have not just evolved once within brachyuran zoeas; they have evolved in many different lineages. Consequently, homoplasy appears to be the norm in the evolution of brachyuran zoeae, not the exception.

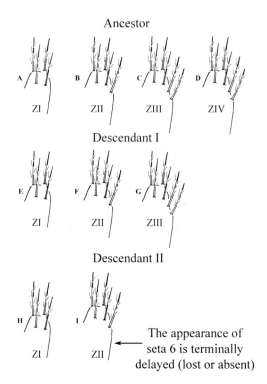

**Figure 8.** Maxilla, proximal basial endite, a representation of terminal delay with respect to seta 6. (A–D) *Pilumnus hirtellus*. (E–F) *Actumnus setifer*. (H–I) *Pilumnus sluiteri* (see Clark 2005).

## 9 PHYLOGENETICS

Our understanding of larval morphology bears not only on classification but also on phylogeny. For example, on the basis of adult morphology, *Tanaocheles bidentata* was originally assigned to the xanthoidean subfamily Chlorodiellinae, and the "Eumedoninae" species have been assigned to various taxa including Eumedonidae, Xanthoidea, Trapezioidea and Portunoidea (for details see Ng & Clark 2000, 2001). However, similarity of the zoeal antenna morphology (Fig. 1) suggests that *T. bidentata* and the "eumedonids" should be assigned to the Pilumnoidea. In order to test this hypothesis, 18 synapomorphic characters of first-stage zoeas from representative taxa were analyzed, including: two xanthids, *Actaea areolatus* (Dana, 1852) and *Chlorodiella nigra*; one tetraiid, *Tetralia cavimana* Heller, 1861; one Portunoidea, *Charybdis helleri* (also the outgroup); four pilumnoids, *Benthopanope indica* (De Man, 1887), *Glabropilumnus edamensis* (De Man, 1888), *Pilumnus hirtellus* and *P. vespertilio* (Fabricius, 1793); and three "eumedonids," *Echinoecus pentagonus*, *Zebrida adamsi* and *Rhabdonotus pictus*. *Rhabdonotus pictus* is used to represent the first-stage zoeas of *Harrovia albolineata* and *Permanotus purpureus* because the setal arrangement of all three larvae is identical.

For this brief example, the data matrix was constructed in MacClade 4.08 OSX (Maddison & Maddison 2000), the trees were generated in PAUP* 4.0b10 (Swofford 2002), and the data set was analyzed using Branch and Bound. One of the 18 characters included in the analysis was treated as unordered because of the difficulty in determining the polarity of exopod antennal spinulation (Clark & Guerao 2008), and the remaining 17 were treated as "irreversible-up." A 50% majority rule consensus was generated from two trees with a consistency index = 0.5714 and tree length of 35.

The resulting tree supported the inclusion of *Tanaocheles bidentata* within the Pilumnoidea (Fig. 9) and in the same clade as *Pilumnus hirtellus*, the type species of the superfamily. There is no

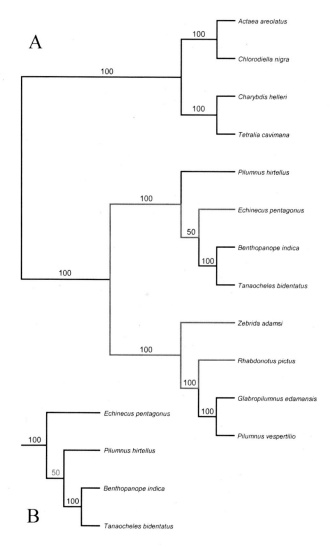

**Figure 9.** Phylogenetic analysis of first-stage zoeas (1) supports the morphological comparisons based on similarity and difference in that *Tanaocheles bidentata* is not a member of a xanthoidean subfamily but should be assigned to the Pilumnoidea Samouelle, 1819; (2) indicates that eumedonid crabs should be assigned to the Pilumnoidea Samouelle, 1819, rather than to a distinct family within the Xanthoidea; and (3) suggests that the Eumedonidae Dana, 1852, may not be a monophyletic taxon because *Echinoecus pentagonus* appears in a separate pilumnoidean clade. Competing topologies for the pilumnoid lineages of tree A are shown in tree B.

phylogenetic support for assigning this species to the Chlorodiellinae, represented in the analysis by the type species *Chlorodiella nigra*. Similarly, there is no support for placing *T. bidentata* in the Trapezioidea Miers, 1886 (represented by *Tetralia cavimana*) as suggested by Kropp (1984) for *Tanaocheles stenochilus* (see Ng and Clark 2000 for details). Although *T. bidentata* possesses some unique larval characters, such as loss of lateral spines and reduced rostral spine, on the basis of this limited analysis there appears to be little support for the assignment of *Tanaocheles* to a new subfamily, Tanaocheleinae (now Tanaocheleidae Ng & Clark 2000, see Ng et al. 2008), as proposed by Ng & Clark (2000). However, more taxa will need to be included to resolve intrafamilial relationships.

In Figure 9, the "eumedonid" taxa represented by *Echinoecus pentagonus*, *R. pictus* and *Z. adamsi* (including *Harrovia albolineata* and *Permanotus purpureus*) were located within the Pilumnoidea clade. There is no support from the first zoeas that the eumedonids were related to the Trapezioidea (represented by *Tetralia cavimana*), the Xanthoidea (represented by *Chlorodiella nigra* and *Actaea areolatus*), or the Portunoidea (represented by *Charybdis helleri*). Furthermore, this analysis suggests that the "eumedonids" may be polyphyletic. These commensal crabs are associated with echinoderms. *Echinoecus pentagonus* is found internally in sea urchins such as *Diadema savignyi*, *Echinothrix calamarix* and *Echinothrix diadema*; *H. albolineata*, *P. purpureus* and *R. pictus* are found on crinoids; and *Zebrida adamsi* is located externally on sea urchins such as *Asthnosoma ijimai* and *Diadema setosum*. From the tree (Fig. 9), *E. pentagonus* and *Z. adamsi* + *R. pictus* (representing *H. albolineata* and *P. purpureus*) are placed in separate clades. Biologically, these two clades correspond to the externally inhabiting eumedonids and the internally associated *E. pentagonus*. Moreover, the externally inhabiting eumedonids appear to be subdivided into those crabs that live on crinoids (*R. pictus* representing *H. albolineata* and *P. purpureus*) and *Z. adamsi*, which is found on sea urchins. More larval descriptions of sea-urchin associates are required to confirm this division. The non-monophyly of the eumedonids also has implications for the subfamily Eumedoninae as proposed by Števčić (2005) and Ng et al. (2008), as two of the genera that they assign to this subfamily, namely *Echinoecus* and *Zebrida*, are in separate clades (Fig. 9). This analysis supports the views expressed by Chia & Ng (1995), who questioned the divisions of the Eumedonidae proposed by Števčić et al. (1988). The larvae of the type species, *Eumedonus niger* H. Milne Edwards, 1835, are not known but are of interest, for if these are similar to those of *Z. adamsi*, *R. pictus*, *H. albolineata* and *P. purpureus*, it would suggest that *E. pentagonus* is not a eumedonine as presently defined. In fact, *E. pentagonus* shares two synapomorphies — absence of dorolateral spines on somites four and five — with the three taxa in the clade (*B. indica*, *T. bidentatus* and *Pilumnus hirtellus*). In summary, this limited phylogenetic analysis of first-stage zoeas supports the inclusion of *T. bidentatus* and the eumedonines within the Pilumnoidea, but suggests the latter taxon may not be monophyletic.

## 10 CONCLUSIONS

Studying only first-stage zoeas or obtaining the complete larvae development from an ovigerous decapod female in the laboratory has one distinct advantage: the species can be subsequently positively identified. A modern high-powered microscope with DIC is essential for basic alpha taxonomy and descriptions of setal patterns.

Brachyuran zoeas of congeneric species appear to have identical setotaxy. This similarity provides a degree of predictability within a taxon. Setal differences (incongruence) within a group are indicative of systematic non-compatibility; they suggest incorrect assignment of taxa. However, similarity does not provide a measure of relationship, which can only be achieved by analyzing shared derived characters.

Oligomerization is considered to be an evolutionary trend within the Crustacea. Study of decapod larval development suggests that heterochronic processes may provide a dominant evolutionary mechanism influencing oligomerization within brachyuran zoeas.

On some body somites and appendage segments, setae do not increase in number after successive zoeal moults, so these are considered conservative characters. When analyzing conservative setal characters for possible phylogenetic significance, their presence (expression) can be considered the ancestral state and their absence (loss or delay) derived. In contrast, there are some somites and segments that accumulate setae; numbers of these setae increase with successive zoeal moults. A method of phylogenetically interpretating these accumulative setae may be to identify individual setae and correlate their expression or delay with respect to an outgroup (possible ancestor) species with a long zoeal development phase.

Unordered characters generate the shortest number of evolutionary steps and produce the most parsimonious trees. However, an unordered transformation series does not represent the linear evolutionary steps toward gradual loss of characters as postulated here by heterochrony. A mosaic of several heterochronic processes provides an evolutionary mechanism influencing oligomerization (reduction and loss) in brachyuran zoeas, and this is best represented by an irreversible transformation series. But reconstruction of trees using "irreversible up" does not necessarily produce the most parsimonious trees and frequently involves more evolutionary steps to compensate for homoplasy. There is evidence that suggests homoplasy is widespread within many brachyuran lineages.

With respect to a classification based on decapod adult morphology, brachyuran larval descriptions can be used to provide an additional perspective on conventional systematics and evolutionary processes.

## ACKNOWLEDGEMENTS

I thank Geoff Boxshall for reading through a draft of this manuscript. I acknowledge the improvements to this paper as suggested by Jody Martin, Joe Goy and an anonymous reviewer. I thoroughly enjoyed the Advances in Decapod Crustacean Phylogenetics symposium held during the SICB Annual Meeting, at San Antonio, Texas, in January 2008. I wish to thank Jody Martin and his team for preparing the program and their organization skills that ensured the meeting ran smoothly. I also thank the NSF consortium "Assembling the Tree of Life" program (Morphological and Molecular Phylogeny of the Decapod Crustaceans) for their invitation and funding.

## REFERENCES

Aikawa, H. 1929. On larval forms of some Brachyura. *Rec. Oceanogr. Wks Japan.* 2: 17–55.

Aikawa, H. 1937. Further notes on brachyuran larvae. *Rec. Oceanogr. Wks Japan.* 9: 87–162.

Anger, K. (ed.) 2001. *Crustacean Issues 14, The Biology of Decapod Crustacean Larvae*: xii + 419 pp. Lisse: Balkema.

Bookhout, C.G. & Costlow, J.D.C. 1979. Larval development of *Pilumnus dasypodus* and *Pilumnus sayi* reared in the laboratory (Decapoda, Brachyura, Xanthidae). *Crustaceana Supplement* 5: 1–16.

Boraschi, L. 1921. Osservazioni sulle larve dei Crostacei Decapodi. *Memorie R. Com. talassogr. ital.* 87: 1–32.

Bourdillon-Casanova, L. 1960. Le méroplancton du Golfe de Marseille: les larves de Crustacés Décapodes. *Rec. Tav. Stn mar. Eudome* 30: 1–286.

Camin, J.H. & Sokal, R.R. 1965. A method for deducing branch sequences in phylogeny. *Evolution* 19: 311–326.

Chia, D.G.B. & Ng, P.K.L. 1995. A revision of the genus *Rhabdonotus* A. Milne Edwards, 1879, with descriptions of two new species and the first zoeal stage of *R. pictus* A. Milne Edwards, 1879 (Brachyura: Eumedonidae). *Crust. Res.* 24: 104–127.

Christiansen, M.E. 1973. The complete larval development of *Hyas araneus* (Linnaeus) and *Hyas coarctatus* Leach (Decapoda, Brachyura, Majidae) reared in the laboratory. *Norw. J. Zool.* 21: 63–89.

Clark, P.F. 1980. British spider crabs of the genus *Inachus*; a morphological study of larval development. M.Sc. Modern Taxonomy (CNAA) thesis. Polytechnic of Central London/City of London Polytechnic. Pp. 140, Figs. 1–29, Tabs 1–15, Pls. 1–14. Unpublished.

Clark, P.F. 1983. The larval and first crab stages of three *Inachus* species (Crustacea: Decapoda: Majidae); a morphological and statistical analysis. *Bull. Brit. Mus. Nat. Hist. (Zool).* 44: 179–190.

Clark, P.F. 1984. A comparative study of zoeal morphology in the genus *Liocarcinus* (Crustacea: Brachyura: Portunidae). *Zool. J. Linn. Soc.* 82: 273–290.

Clark, P.F. 2001. Interpreting patterns in chaetotaxy and segmentation associated with abbreviated brachyuran zoeal development. *Invert. Reprod. Dev.* [2000] 38: 171–181.

Clark, P.F. 2005. The evolutionary significance of heterochrony in the abbreviated zoeal development of pilumnine crabs (Crustacea: Brachyura: Xanthoidea). *Zool. J. Linn. Soc.* 143: 417–446.

Clark, P.F., De Calazans, D.K. & Pohle, G.W. 1998a. Accuracy and standardisation of brachyuran larval descriptions. *Invert. Reprod. Develop.* 33: 127–144.

Clark, P.F., De Calazans, D. & Rodriques, S.S. 1998b. *Libinia spinosa* H. Milne Edwards, 1832 (Decapoda: Brachyura: Majidae): a reappraisal of larval characters from laboratory reared material. *Invert. Reprod. Develop.* 33: 145–157.

Clark, P.F. & Guerao, G. (2008) A description of *Calocarcinus africanus* Calman, 1909 (Brachyura, Xanthoidea) first zoeal stage morphology with implications for Trapeziidae systematics. *Proc. Biol. Soc. Wash.* 121: 475–500.

Clark P.F. & Ng, P.K.L. 1998. The larval development of the poisonous mosaic crab, *Lophozozymus pictor* (Fabricius, 1798) (Crustacea: Decapoda: Brachyura: Xanthidae: Zosiminae), with comments on familial characters for first stage zoeae. *Zoosystema* 20: 201–220.

Clark, P.F. & Ng, P.K.L. 2004a. Two zoeal stages and the megalop of *Pilumnus sluiteri* De Man, 1892 [Crustacea: Brachyura: Xanthoidea: Pilumnidae] described from laboratory reared material. *Invert. Reprod. Dev.* 45: 205–219.

Clark, P.F. & Ng, P.K.L. 2004b. The larval development of *Actumnus setifer* (de Haan, 1835) (Brachyura: Xanthoidea: Pilumnidae) described from laboratory reared material. *Crust. Res.* 33: 27–50.

Clark, P.F. & Ng, P.K.L. 2006. First stage zoeae of *Quadrella* Dana, 1851 [Crustacea: Decapoda: Brachyura: Xanthoidea: Trapeziidae] and their affinities with those of *Tetralia* Dana, 1851, and *Trapezia* Latreille, 1828. *Hydrobiologia*. 560: 267–294.

Clark, P.F. & Paula, J. 2003. Descriptions of ten xanthoidean (Crustacea: Decapoda: Brachyura) first stage zoeae from Inhaca Island, Mozambique. *Raffles Bull. Zool.* 51: 323–378.

Clark, P.F. & Webber, W.R. 1991. A redescription of *Macrocheira kaempferi* (Temminck, 1836) zoeae with a discussion of the classification of the Majoidea Samouelle, 1819 (Crustacea: Brachyura). *J. Nat. Hist.* 25: 1259–1279.

Cuesta, J.A., Schub, M., Diesel, R. & Schubart, D. 1999. Abbreviated development of *Armases miersii* (Grapsidae: Sesarminae), a crab that breeds in supralittoral rock pools. *J. Crust. Biol.* 19: 26–41.

Dineen, J.F., Clark, P.F., Hines, A.H., Reed, S.A. & Walton H.P. 2001. Life history, larval description and natural history of *Charybdis hellerii* (A. Milne Edwards, 1867) (Crustacea, Decapoda, Brachyura, Portunidae), an invasive crab in the western Atlantic. *J. Crust. Biol.* 21: 774–805.

Dornelas, M., Clark, P.F. & Paula, J. 2004. The larval development of *Nanocassiope melanodactyla* (A. Milne-Edwards, 1867) (Crustacea: Decapoda: Brachyura: Xanthidae). *J. Nat. Hist.* 38: 506–535.

Farris, J.S. 1970. Methods of computing Wagner Trees. *Syst. Zool.* 18: 374–385.

Felder, D.L., Martin, J.W. & Goy, J.W. 1985. Patterns in early postlarval development of decapods. In: Wenner, A.M. (ed.), *Crustacean Issues 2, Larval Growth*: 163–225. Rotterdam: A.A. Balkema.

Fitch, W.M. 1971. Toward defining the course of evolution: Minimal change for a specific tree topology. *Syst. Zool.* 20: 406–416.

Greenwood, J.G. & Fielder, D.R. 1984a. The complete larval development, under laboratory conditions, of *Heteropanope glabra* Stimpson, 1858 (Brachyura: Xanthidae), from Australia. *Aust. Zool.* 21: 291–303.

Greenwood, J.G. & Fielder, D.R. 1984b. The zoeal stages of *Pilumnopeus serratifrons* (Kinahan, 1856) (Brachyura: Xanthidae) reared under laboratory conditions. *J. Nat. Hist.* 18: 31–40.

Hale, H.M., 1931. The post-embryonic development of an Australian xanthid crab (*Pilumnus vestitus* Haswell). *Rec. Aust. Mus.* 4: 321–331.

Hartigan, J.A. 1973. Minimum mutation fits to a given tree. *Biometrics* 29: 53–65.

Hashmi, S.S. 1970. Study on larvae of the family Xanthidae (*Pilumnus*) hatched in the laboratory (Deacapoda: Brachyura). *Pakistt. J. Scient. Ind. Res.* 13: 420–426.

Ingle, R.W. 1983. A comparative study of the larval development of *Monodaeus couchi* (Couch), *Xantho incisus* (Leach) and *Pilumnus hirtellus* (Linnaeus) (Crustacea: Brachyura: Xanthidae). *J. Nat. Hist.* 17: 951–978.

Ingle, R.W. 1991. *Larval stages of Northeastern Atlantic Crabs. An Illustrated Key*: xii+1–363, figs 1–2, 40. London: Natural History Museum Publications, Chapman and Hall.

Ko, H.S. 1994a. Larval development of *Parapilumnus trispinosus* Sakai, 1965 (Crustacea, Brachyura, Xanthidae) reared in the laboratory. *Korean J. Zool.* 37: 331–342.

Ko, H.S. 1994b. The zoeal stages of *Pilumnus minutus* de Haan, 1835 (Decapoda: Brachyura: Pilumnidae) in the laboratory. *Korean J. Syst. Zool.* 10: 145–155.

Ko, H.S. 1995. Larval development of *Benthopanope indica* (De Man, 1887) (Decapoda: Brachyura: Pilumnidae) in the laboratory. *J. Crust. Biol.* 15: 280–290.

Ko, H.S. 1997. Larval development of *Pilumnopeus granulata* Balss, 1933 and *Pilumnus minutus* de Haan, 1835 (Crustacea: Brachyura: Pilumnidae), with a key to the known pilumnid zoeae. *Korean J. Biol. Sci.* 1: 31–42.

Ko, H.S. & Yang, H.J. 2003. Zoeal development of *Heteropilumnus ciliatus* (Stimpson, 1858) (Crustacea: Decapoda: Pilumnidae), with a key to the known pilumnid zoeas in Korea and the adjacent waters. *J. Crust. Biol.* 23: 341–351.

Kropp, R.K. 1984. *Tanaocheles stenochilus*, a new genus and species of crab from Guam, Mariana Islands (Brachyura: Xanthidae). *Proc. Biol. Soc. Wash.* 97: 744–747.

Kurata, H. 1969. Larvae of decapod of Brachyura of Arasaki, Sagami Bay—IV. Majidae. *Bull. Tokai reg. Fish. Res. Lab.* 57: 81–127.

Lebour, M.V. 1928. The larval stages of Plymouth Brachyura. *Proc. Zool. Soc. Lond.* 1928: 473–560.

Lebour, M.V. 1931. Further notes on larval Brachyura. *Proc. Zool. Soc. Lond.* 1931: 93–96.

Lebour, M.V. 1950. Notes on some larval decapods (Crustacea) from Bermuda. *Proc. Zool. Soc. Lond.* 120: 369–379.

Lee, D.H. 1993. Larval development of *Macromedaeus distinguendus* (de Haan, 1835) and *Pilumnopeus makiana* (Rathbun, 1929) reared in the laboratory. 53 pp., 9 figs. 4 tabs. Ph. Master Thesis, Busan National University, Korea, Busan, Korea. (unpublished).

Lim, S.L. & Tan, L.W.H. 1981. Larval development of the hairy crab, *Pilumnus vespertilio* (Fabricius) (Bracyura, Xanthidae) in the laboratory and comparisons with larvae of *Pilumnus dasypodus* Kingsley and *Pilumnus sayi* Rathbun. *Crustaceana* 41: 71–88.

Lim, S.L., Ng, P.K.L. & Tan, L.W.H. 1984. The larval development of *Heteropanope glabra* Stimpson, 1858 (Decapoda, Xanthidae) in the laboratory. *Crustaceana* 47: 1–16.

Lim, S.L., Ng, P.K.L. & Tan, L.W.H. 1986. The complete larval development of *Pilumnopeus eucratoides* Stimpson, 1858 (Decapoda, Brachyura, Pilumnidae) in the laboratory. *Crustaceana* 50: 265–277.

Maddison, D.R. & Maddison, W.P. 1992. *MacClade: Analysis of Phylogeny and Character Evolution*. Version 3. xi + 1–398. Sunderland, Massachusetts: Sinauer Associates.

Maddison, D.R. & Maddison, W.P. 2000. *MacClade. Analysis of phylogeny and character evolution. Version 4.0*. Sunderland, Massachusetts: Sinauer Associates.

Marques, F. & Pohle, G. 1998. The use of structural reduction in phylogenetic reconstruction of decapods and a phylogenetic hypothesis for 15 genera of Majidae: testing previous larval hypotheses and assumptions. *Invert. Reprod. Develop.* 33: 241–262.

Martin, J.W. 1984. Notes and bibliography on the larvae of xanthid crabs, with a key to the known xanthid zoeae of the western Atlantic and Gulf of Mexico. *Bull. Mar. Sci.* 34: 220–239.

Martin, J.W. & Davis, E.D. 2001. An updated classification of the Recent Crustacea. *Nat. Hist. Mus. L.A. County, Science Series.* 39: 1–124.

Martin, J.W., Truesdale, F.M. & Felder, D.L.1985. Larval development of *Panopeus bermudensis* Benedict and Rathbun, 1891 (Brachyura, Xanthidae) with notes on zoeal characters in xanthid crabs. *J. Crust. Biol.* 5: 84–105.

McKinney, M.L. & McNamara, K.J. 1991. *Heterochrony: the evolution of ontogeny.* New York, Plenum Press. xix + 1–437.

Ming-Shiou, J., Clark, P.F. & Ng, P.K.L. 2004. The first zoea, megalop and first crab stage of the hydrothermal vent crab, *Xenograpsus testudineus* Ng, Huang & Ho, 2000 (Decapoda: Brachyura: Grapsoidea) and systematic implications for the Varunidae. *J. Crust. Biol.* 24: 188–212.

Mohan, R. & Kannupandi, T. 1986. 20. Complete larval development of the xanthid crab, *Galene bispinosa* (Herbst), reared in the laboratory. In: Thompson, M.-F., Sarojini, R. & Nagabhushanam, R. (eds.), *Biology of Benthic Organisms, Techniques and Methods as Applied to the Indian Ocean*: 193–202. Oxford and IBH Publishing Co., New Delhi.

Ng, P.K.L. 2002. The Indo-Pacific Pilumnidae XVI. On the identity of *Pilumnus cristimanus* A. Milne Edwards, 1873, and the status of *Parapilumnus* Kossmann, 1877 (Crustacea: Decapoda: Brachyura), with description of a new species from rubble beds in Guam. *Micronesica* 34: 209–226.

Ng, P.K.L. & Clark, P.F. 2000. The Indo-Pacific Pilumnidae XI. On the familial placement of *Chlorodiella bidentata* (Nobili, 1901) and *Tanaocheles stenochilus* Kropp, 1984, using adult and larval characters with establishment of a new subfamily, Tanaochelinae (Crustacea: Decapoda: Brachyura). *J. Nat. Hist.* 34: 207–245.

Ng, P.K.L. & Clark, P.F. 2001. The eumedonid file: a case study of systematic compatibility using larval and adult characters (Crustacea: Decapoda: Brachyura). *Invert. Reprod. Develop.* [2000] 38: 225–252.

Ng, P.K.L. & Clark, P.F. 2008. A revision of *Latopilumnus* Türkay & Schuhmacher, 1985, and *Aniptumnus* Ng, 2002 (Crustacea: Decapoda: Brachyura: Pilumnidae). *J. Nat. Hist.* 42: 885–912.

Ng, P.K.L., Guinot, D. & Davie, P.J.F. 2008. Systema Brachyurorum: Part I. An annotated checklist of extant brachyuran crabs of the world. *Raffles Bull. Zool.* Supplement 17: 1–286.

Prasad, R.R. & Tampi, P.R.S. 1957. Notes on some decapod larvae. *J. Zool. Soc. India.* 9: 22–29.

Quintana, R. 1986. On the megalopa and early crab stages of *Parapilumnus trispinosus* Sakai, 1965 (Decapoda, Brachyura, Xanthidae). *Proc. Jap. Soc. Syst. Zool.* 34: 1–18.

Rice, A.L. 1979. A plea for improved standards in crab zoeal descriptions. *Crustaceana* 37: 213–218.

Rice, A.L. 1980. Crab zoeal morphology and its bearing the classification of the Brachyura. *Trans. Zool. Soc. Lond.* 35: 271–424.

Rice, A.L. 1983. Zoeal evidence for brachyuran phylogeny. In: Schram, F.R. (ed.), *Crustacean Issues 1, Crustacean Phylogeny*: 313–329. Rotterdam: Balkema.

Rice, A.L. 1988. The megalopa stage in majid crabs, with a review of spider crab relationships based on larval characters. In: Fincham, A.A. & Rainbow, P.S. (eds.), *Aspects of Decapod Crustacean Biology. Proceedings of a Symposium Held at the Zoological Society of London on 8th and 9th April 1987*: 59: 27–46. Oxford, Clarendon Press. Zoological Society of London Symposia.

Rice, A.L. and Williamson, D.I. 1977, Planktonic stages of Crustacea Malacostraca from Atlantic Seamounts. "Meteor" Forsch.-Ergebnisse, D, 26: 28–64.

Salman, S.D. 1982. Larval development of the crab *Pilumnus hirtellus* (L.) reared in the laboratory (Decapoda, Brachyura, Xanthidae). *Crustaceana* 42: 113–126.

Sandifer, P.A. 1974. Larval stages of the crab, *Pilumnus dasypodus* Kingsley (Crustacea, Brachyura, Xanthidae) obtained in the laboratory. *Bull. Mar. Sci.* 24: 378–391.

Serène, R. 1984. Crustacés Décapodes Brachyoures de l'Océan Indien occidental et de la Mer Rouge, Xanthoidea: Xanthidae et Trapeziidae. Avec un addendum par Crosnier (A): Carpiliidae et Menippidae. *Faune Tropicale* 24: 1–349.

Siddiqui, F.A. & Tirmizi, N.M. 1992. The complete larval development, including the first crab stage of *Pilumnus kempi* Deb, 1987 (Crustacea: Decapoda: Brachyura: Pilumnidae) reared in the laboratory. *Raffles Bull. Zool.* 40: 229–244.

Števčić, Z. 2005. The reclassification of brachyuran crabs (Crustacea: Decapoda: Brachyura). *Natura Croatica, Supplement 1*, 14: 1–159.

Števčić, Z., Castro, P. & Gore, R.H. 1988. Re-establishment of the Family Eumedonidae Dana, 1853 (Crustacea: Brachyura). *J. Nat. Hist.* 22: 1301–1324.

Swofford, D.L. 2002. *PAUP\*. Phylogenetic Analysis Using Parsimony (\* and Other Methods)*. Version 4.0b10. Sinauer Associates, Sunderland, MA, USA.

Swofford, D.L. & Maddison, W.P. 1987. Reconstructing ancestral character states under Wagner parsimony. *Math. Biosci.* 87: 199–229.

Takeda, M. & Miyake, S. 1968. First zoea of two pilumnid crabs of the family Xanthidae. *Sci. Bull. Fac. Agric. Kyushu Univ.* 23: 127–133.

Terada, M. 1980. On the zoeal development of *Pilumnopeus indicus* (De Man) (Brachyura, Xanthidae) in the laboratory. *Res. Crust.* 10: 35–44.

Terada, M. 1984. Zoeal development of two pilumnid crabs (Crustacea, Decapoda). *Proc. Jap. Soc. Syst. Zool.* 28: 29–39.

Terada, M. 1985. Zoeal development of *Halimede fragifer* de Haan (Xanthidae, Xanthinae). *Proc. Jap. Soc. Syst. Zool.* 31: 30–37.

Terada, M. 1988. On the larval stages of *Actumnus squamosus* (de Haan) (Brachyura, Pilumnidae). *Proc. Jap. Soc. Syst. Zool.* 38: 15–25.

Terada, M. 1990. Zoeal development of five species of xanthoid crabs, reared in the laboratory. *Res. Crust.* 18: 23–47.

Wear, R.G. 1967. Life-history studies on New Zealand Brachyura 1. Embryonic and post-embryonic development of *Pilumnus novaezealandiae* Filhol, 1886, and of *P. lumpinus* Bennett, 1964 (Xanthidae, Pilumnidae). *N. Z. J. Mar. Freshw. Res.* 1: 482–535.

Wear, R.G. 1968. Life history studies on New Zealand Brachyura 2. Family Xanthidae. Larvae of *Heterozius rotundifrons* A. Milne Edwards, 1867, *Ozius truncatus* H. Milne Edwards, 1834 and *Heteropanope (Pilumnopeus) serratifrons* (Kinahan, 1856). *N. Z. J. Mar. Freshw. Res.* 2: 293–332.

Wear, R.G. & Fielder, D.R. 1985. The marine fauna of New Zealand: larvae of the Brachyura (Crustacea, Decapoda). *Mem. N. Z. Oceanogr. Inst.* 92: 1–90.

Williamson, D.I. 1982. Larval morphology and diversity. In: Abele, L.G. (ed.), *The Biology of Crustacea*: 43–110. New York: Academic Press.

Williamson, H.C. 1915. VI. Crustacea. Decapoda. Larven. *Nord. Plankt.* 18: 315–588.

# II ADVANCES IN OUR KNOWLEDGE OF SHRIMP-LIKE DECAPODS

# Evolution and Radiation of Shrimp-Like Decapods: An Overview

CHARLES H.J.M. FRANSEN[1] & SAMMY DE GRAVE[2]

[1] *Nationaal Natuurhistorisch Museum Naturalis, Darwinweg 2, 2333 CR Leiden, The Netherlands*
[2] *Oxford University Museum of Natural History, Parks Road, Oxford OX1 3PW, United Kingdom*

ABSTRACT

The shrimp-like Decapoda currently include the suborder Dendrobranchiata and the infraorders Caridea and Stenopodidea within the suborder Pleocyemata. Their phylogenetic relationship with the other Decapoda, as well as previously proposed internal phylogenies, are reviewed. This review shows that only a small percentage of the shrimp-like decapod taxa is incorporated in phylogenetic analyses at higher to lower taxonomic levels and that there remain numerous controversies between and within analyses based on morphological characters and molecular markers. The morphological and molecular characters thus far used in phylogenetic reconstructions are evaluated. It is suggested that when a robust morphological matrix is available, the addition of fossil taxa will be worthwhile, in view of their unique morphology and ecology. A review of potentially phylogenetically informative characters across all caridean families is sorely lacking; such a review needs to be instigated to assess foregut morphology and the mastigobranch–setobranch complex, to name but a few important characters.

## 1 INTRODUCTION

Three groups of shrimp-like decapods are currently recognized (Martin & Davis 2001): the suborder Dendrobranchiata and the infraorders Caridea and Stenopodidea of the suborder Pleocyemata. A count of the number of taxa recognized in these groups shows that the Caridea are by far the largest group with more than 3100 species (Table 1).

The discovery curves in all three groups do not show any sign of reaching a plateau (Fig. 1), suggesting we are a long way off from knowing the true species richness for all groups. Although Stenopodidea are far less species rich than the other two taxa, the median date of description (1978), and the steep incline since then, indicates that many more species remain to be described even in this group—not surprising given the deep-water habitat of many of its constituent species. Focusing on the Caridea, at the end of the 19th century and the beginning of the 20th century, the number of species described increased distinctly to about 25 species per year, mainly due to the publication of the results of major oceanographic expeditions like the "Challenger," "Discovery," and "Siboga." Around 1910, the increment of species slowed down to about 12 species a year until around 1970 when the description rate increased again to a mean of 33 per year. The fossil record of shrimp-like decapods is meager, especially in the Caridea, for which relatively few fossil taxa are known compared to the large number of extant taxa (Crandall et al. in prep).

**Table 1.** Number of extant and extinct (†) taxa within the three shrimp-like decapod groups (current as of August 2008).

| Taxon level | Dendrobranchiata | Caridea | Stenopodidea |
|---|---|---|---|
| Superfamilies | 2 | 16 (1 †) | 0 |
| Families | 9 (2 †) | 36 (1†) | 3 |
| Genera | 56 | 361 | 10 (2†) |
| Species | 505 (74 †) | ca. 3108 (46 †) | 58 (2 †) |

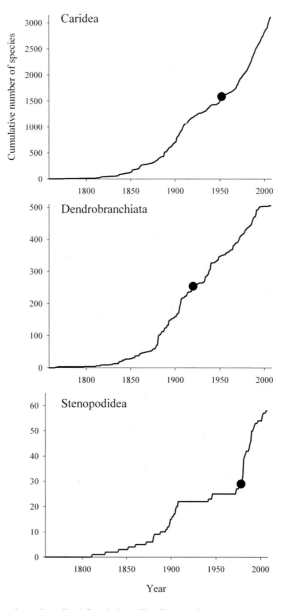

**Figure 1.** Cumulative numbers described for shrimp-like Decapoda per taxon per annum; circle indicates median date of description.

## 2 POSITION OF THE SHRIMP-LIKE DECAPODS WITHIN THE DECAPODA

Ever since Dana (1852) and Huxley (1879) recognized the artificial nature of the Natantia, there has been controversy over the relationships between the shrimp-like decapods as well as their relationship to the remaining groups. Despite this uncertainty, most recent studies demonstrate that the shrimp-like decapods are basal to the other decapod lineages (Richter & Scholtz 2001; Schram 2001; Dixon et al. 2003; Porter et al. 2005). In contrast to these studies, however, the molecular tree presented by Bracken et al. (this volume) indicates that the Stenopodidea might not be as basal as previously assumed.

Earlier classifications, from the 1800s up to 1981, have been succinctly reviewed by Felgenhauer & Abele (1983) and Holthuis (1993), and there appears to be no need to repeat this information here. Burkenroad (1963) firmly established the separate status of the Dendrobranchiata as a suborder, containing the Penaeidae and Sergestidae (now usually treated as the superfamilies Penaeoidea, with 5 families, and the Sergestoidea, with 2 families). Both Burkenroad (1981) and Felgenhauer & Abele (1983) discussed the differences between the Dendrobranchiata and the other shrimp-like decapods, primarily the presence of dendrobranchiate gills, egg broadcasting and the pleonic hinges. Recently Martin et al. (2007) have demonstrated considerable variation in dendrobranch gill morphology. Following on from their study, we recommend that the other distinguishing characters should also be re-studied.

The separate status of the Stenopodidea has long been recognized and is supported by morphological and developmental studies (Felgenhauer & Abele 1983). With the exception of trichobranchiate gills, many of the proposed characters do exhibit some overlap with either Dendrobranchiata or Caridea. Nevertheless, all phylogenetic studies have supported their status as a separate lineage.

The internal classification of the Caridea and their relationship to the other lineages currently appears far from settled, although it is generally accepted that they do constitute a separate lineage (Burkenroad 1963; Felgenhauer & Abele 1983; Abele & Felgenhauer 1986). Of specific interest is the position of the family Procarididae, which remains controversial to date. Prior to the discovery of *Procaris* in 1972, Caridea were characterized by one or both of the two anterior pairs of legs being chelate (Burkenroad 1981), easily differentiating them from the other two lineages, which have the first three pairs nearly always chelate. *Procaris*, and the later discovered *Vetericaris*, not only are achelate but share a number of characters with the Dendrobranchiata (e.g., a well developed gastric mill, L-shaped mastigobranchs, and appendices internae absent) and with Caridea *sensu stricto* (phyllobranchiate gills, wide second abdominal pleuron). Much has been written on whether they should be considered a superfamily within the Caridea (Abele & Felgenhauer 1986; Abele 1991; Chace 1992; Holthuis 1993) or be considered a separate lineage. Felgenhauer & Abele (1983) were the first to address their position, and, although not based on a cladistic analysis, they considered them a separate lineage, branching off earlier than the Caridea. This was opposed by Christoffersen (1988) who, using manual parsimony, considered procaridids as a sister group to the Caridea. Using more objective computer-based methods, Abele & Felgenhauer (1986) reached the same conclusion and considered both taxa closely related, but they did not assign a formal rank to either clade. Bracken et al. (this volume) support the treatment of the Procaridoidea as a sister group to the remaining carideans on the basis of a phylogenetic analysis based on both mitochondrial and nuclear genes.

Both morphological (Dixon et al. 2003; Schram & Dixon 2004) and molecular (Porter et al. 2005) analyses support positioning of the shrimp-like decapods as the most basal clades within the Decapoda. However, the relationships of the three (or four) separate lineages to each other, and indeed to the other Decapoda, are far from settled. All phylogenetic analyses, be they morphological (Abele & Felgenhauer 1986; Dixon et al. 2003; Schram & Dixon 2004) or molecular (Porter et al. 2005), support positioning of the Dendrobranchiata as the most basal clade within the Decapoda. The position of the Stenopodidea and Caridea (including the Procaridoidea or not) remains

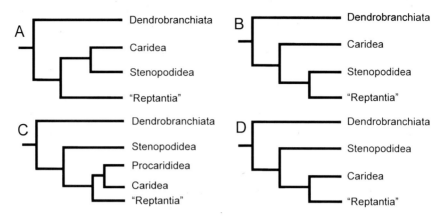

**Figure 2.** Position of the shrimp-like groups within the Decapoda according to (A) Burkenroad (1963), (B) Abele & Felgenhauer (1986), (C) Christoffersen (1988) and (D) Porter et al. (2005).

unsettled. Burkenroad (1963) regarded the Caridea + Stenopodidea as a sister group to the Reptantia (Fig. 2). On the basis of morphological cladistic analyses, two hypotheses have been put forward. Abele & Felgenhauer (1986) considered the Stenopodidea as a sister group to the reptant decapods, preceded by the branching off of the Caridea *sensu lato* (Fig. 2); in contrast, Christoffersen (1988) offered the reverse situation, and considered the Caridea + Procaridoidea as a sister group to the reptant decapods (Fig. 2). The molecular study by Porter et al. (2005), using representatives of all three shrimp-like taxa as well as a score of reptant taxa, resolved a caridean + reptant clade, but it was not statistically different from a stenopodidean + reptant clade (Fig. 2). Interestingly, a caridean + stenopodidean clade, as used by Burkenroad (1963, 1981), was rejected by their analysis (Porter et al. 2005). The analysis by Bracken et al. (this volume) indicates a position of the Stenopodidae within the Repantia, which has been suggested before on the basis of larval development (see Seridji 1990, and references therein). Thus, the exact position of these two shrimp-like taxa in relation to the reptant decapods and indeed to each other remains debated.

From this brief overview, it is evident that more rigorous and more inclusive cladistic analyses are needed to resolve the position of the Caridea and Stenopodidea within the Decapoda.

## 3 PHYLOGENETIC RELATIONSHIPS WITHIN STENOPODIDEA

Saito & Takeda (2003) have published the only phylogeny within the Stenopodidea. Analyzing the family Spongicolidae, they used a morphological matrix composed of 38 characters of 32 species, which resulted in a consensus tree with mainly paraphyletic genera. The phylogeny shows a tendency from primitive "shallow water free living species" towards a more derived group of "deep water sponge-associated" species. All genera and nearly all species in this family are included in this phylogeny. Thus, about half of the genera and species for the infraorder as a whole have been subjected to a cladistic analysis.

## 4 PHYLOGENETIC RELATIONSHIPS WITHIN DENDROBRANCHIATA

In their excellent book on penaeoid and sergestoid shrimps, Pérez Farfante & Kensley (1997) recognized two superfamilies: the Penaeoidea and Sergestoidea, with the Sergestoidea consisting of two families and the Penaeoidea of five distinct families (Table 2). The position of the enigmatic genus *Lucifer* remains problematic (Tavares et al. this volume) due to its aberrant adult morphology. The relation between the two superfamilies has not been treated in any phylogenetic study to date.

Table 2. Number of genera and species in the suborder Dendrobranchiata (as of August 2008).

| Superfamily | Family | Genera | Species |
|---|---|---|---|
| Penaeoidea | Aristeidae | 9 | 26 |
| | Benthesicymidae | 4 | 21 |
| | Penaeidae | 26 | 215 |
| | Sicyoniidae | 1 | 44 |
| | Solenoceridae | 9 | 80 |
| Sergestoidea | Luciferidae | 1 | 9 |
| | Sergestidae | 6 | 90 |
| | | **56** | **505** |

Several phylogenies within the Penaeoidea have appeared in the last four decades (Mulley & Latter 1980; Palumbi & Benzie 1991; Tam & Chu 1993; von Sternberg & Motoh 1995; Baldwin et al. 1998; Tong et al. 2000; Quan et al. 2001; Maggioni et al. 2001; Quan et al. 2004; Lavery et al. 2004; Vazquez-Bader et al. 2004; Voloch et al. 2005; and Chan et al. 2008); however, the relationships within Sergestoidea have not been examined.

Phylogenetic relationships among the five penaeoid families were tackled by Vazquez-Bader et al. (2004), using a partial sequence of about 300 bps of the 16S mitochondrial gene. Their results support monophyly of the superfamily, but they show the Penaeidae to be paraphyletic with regard to the closely related Solenoceridae. This was confirmed by Voloch et al. (2005) using the two mitochondrial markers 16S and COI, although the separate family status of Aristeidae, Benthesicymidae, and Sicyoniidae was questioned, as they form a compact group separated by small genetic distances. These somewhat preliminary results require confirmation based upon more conservative markers, as already acknowledged by Voloch et al. (2005) themselves.

All other phylogenetic studies within the superfamily deal with the family Penaeidae. Crosnier (1987, 1991, 1994a, 1994b) revised the genus *Metapenaeopsis*. He proposed a grouping primarily based on the morphology of the petasma and a subgrouping based on the presence/absence of a stridulating organ. A preliminary phylogeny of selected species within this genus (based on mitochondrial markers) published by Tong et al. (2000) confirms the views of Crosnier. All other studies have focused on the generic division proposed by Pérez Farfante & Kensley (1997), which was, and is, debated by both the fishing industry and the scientific community (Flegel 2007; McLauglin et al. 2008). An overview of molecular research on this topic was published by Dall (2007). He concluded that some of the genera recognised by Pérez Farfante & Kensley (1997) are not monophyletic with regards to the molecular markers used in other analyses (e.g., *Penaeus* and *Melicertus*). More studies using nuclear genes are needed to elucidate the systematic position of these genera and their constituent species groups. In a recent contribution, Chan et al. (2008) studied the phylogenetic relationships of 20 genera of the 26 recognized by Pérez Farfante & Kensley (1997), supporting Burkenroad's (1983) original three-tribe scheme (Peneini, Parapeneini, and Trachypeneini) and synonymizing the genus *Miyadiella* with *Atypopenaeus*. Within the Penaeidae nearly all genera and just over 20% of the species have been the subject of phylogenetic analyses. See also Tavares et al. (this volume) for a preliminary morphological analysis of penaeoid families and genera.

## 5 PHYLOGENETIC RELATIONSHIPS WITHIN CARIDEA

The internal classification of the Caridea by Chace (1992) and Holthuis (1993), which is largely followed by Martin & Davis (2001), is widely used today (Table 3). Minor recent changes are the addition of the family Pseudochelidae (De Grave & Moosa 2004) and the non-recognition of the

Table 3. Number of genera and species in the suborder Dendrobranchiata (as of August 2008).

| Superfamily | Family | Genera | Species |
|---|---|---|---|
| Procaridoidea | Procarididae | 2 | 6 |
| Galatheacaridoidea | Galatheacarididae | 1 | 1 |
| Pasiphaeoidae | Pasiphaeidae | 7 | 97 |
| Oplophoroidea | Oplophoridae | 10 | 73 |
| Atyoidea | Atyidae | 40 | 395 |
| Bresilioidea | Agostocarididae | 1 | 3 |
| | Alvinocarididae | 6 | 18 |
| | Bresiliidae | 3 | 9 |
| | Disciadidae | 3 | 10 |
| | Pseudochelidae | 1 | 3 |
| Nematocarcinoidea | Eugonatonotidae | 1 | 2 |
| | Nematocarcinidae | 4 | 44 |
| | Rhynchocinetidae | 2 | 24 |
| | Xiphocarididae | 1 | 2 |
| Psalidopodoidea | Psalidopodidae | 1 | 2 |
| Stylodactyloidea | Stylodactylidae | 5 | 33 |
| Campylonotoidea | Bathypalaemonellidae | 2 | 11 |
| | Campylonotidae | 1 | 5 |
| Palaemonoidea | Anchistioididae | 1 | 4 |
| | Desmocarididae | 1 | 2 |
| | Euryrhynchidae | 3 | 6 |
| | Gnathophyllidae | 5 | 13 |
| | Hymenoceridae | 2 | 3 |
| | Kakaducarididae | 3 | 3 |
| | Palaemonidae | 116 | 876 |
| | Typhlocarididae | 1 | 3 |
| Alpheoidea | Alpheidae | 43 | 614 |
| | Barbouriidae | 3 | 6 |
| | Hippolytidae | 36 | 302 |
| | Ogyrididae | 1 | 10 |
| Processoidea | Processidae | 5 | 66 |
| Pandaloidea | Pandalidae | 23 | 189 |
| | Thalassocarididae | 2 | 4 |
| Physetocaridoidea | Physetocaridae | 1 | 1 |
| Crangonoidea | Crangonidae | 22 | 190 |
| | Glyphocrangonidae | 1 | 77 |
| | | **360** | **3108** |

Mirocarididae. Studies dealing with phylogenetic relations among the superfamilies and families are scarce. Christoffersen's (1987, 1988, 1989, 1990) contributions, using manually constructed phylogenies, indicate the non-monophyletic nature of the traditional classification. The first comprehensive molecular phylogeny of the group is presented by Bracken et al. (this volume), and suggests polyphyletic and paraphyletic relationships among genera within the families Atyidae, Pasiphaeidae, Oplophoridae, Hippolytidae, Gnathophyllidae, and Palaemonidae. Phylogenetic research has been carried out on 7 of the 36 families within the Caridea, amounting to less than perhaps 3-4% of all species. Christoffersen performed manual and computerized morphological cladistic analyses among the hippolytid (1987), crangonid (1988), and pandaloid (1989) genera.

Within the predominantly freshwater family Atyidae, molecular studies on selected species within genera like *Paratya* (Page et al. 2005; Cook et al. 2006), *Troglocaris* (Zakšek et al. 2007), and *Caridina* (Chenoweth & Hughes 2003; Roy et al. 2006; Page et al. 2007; von Rintelen et al. 2007a, b) in relation to biogeographical issues, as well as the regional study of several genera by Page et al. (2008), have been published.

The phylogenetic relationships among the deep-sea hydrothermal vent shrimp belonging to the Alvinocarididae were analyzed by Shank et al. (1998) using the COI mitochondrial gene. Their molecular phylogeny is consistent with the higher-level taxonomy based on morphology, and demonstrates that the Alvinocarididae form a monophyletic group in relation to the outgroup shrimp taxa used.

A morphological hypothesis about the phylogenetic relationships within the Palaemonoidea (currently containing 910 species) was presented by Pereira (1997), who concluded that both the superfamily Palaemonoidea and the family Palaemonidae (*sensu* Chace 1992) are natural groups, but that a rearrangement of palaemonid subgroups would better reflect their phylogenetic relationships. However, if the classification of Martin & Davis (2001) were to be superimposed upon Pereira's cladogram, the Palaemonidae (*sensu* Martin & Davis 2001) become paraphyletic. Pereira (1997) also indicated that several genera in the subfamily Palaemoninae, such as *Macrobrachium*, *Cryphiops*, *Palaemon*, *Palaemonetes*, and *Pseudopalaemon*, are paraphyletic. The subfamily Pontoniinae remains monophyletic in his view, although several genera, now included in the Palaemoninae (e.g., *Brachycarpus*, *Leander*, *Leandrites*), should be transferred to the Pontoniinae. Page et al. (2008) showed the genera *Kakaducaris* and *Leptopalaemon* (currently in the family Kakaducarididae) as a strongly supported clade within the Palaemoninae that is closely related to the genus *Macrobrachium*. This result is confirmed by Bracken et al. (this volume).

Recent work by Mitsuhashi et al. (2007), using the nuclear 18S rRNA and 28S rRNA genes, showed the families Hymenoceridae and Gnathophyllidae to be closely related and nested within the Pontoniinae, which is also confirmed by the study of Bracken et al. (this volume). This clade is clearly distinct from the clade with representatives of the Palaemoninae, in accordance with the relationships among the families as suggested by larval characters (Bruce 1986; Yang & Ko 2002). A review of the literature on the first zoea shows that the characters suggested by Yang & Ko (2002) to separate palaemonine and pontoniine genera hold true except for five genera: *Leander*, *Leandrites*, *Harpilius*, *Kemponia*, and *Philarius*. Such a shift of several genera from the Palaemoninae to the Pontoniinae is in line with the ideas put forward by Pereira (1997). Future molecular work including representatives of these genera should elucidate the boundaries between the Pontoniinae and Palaemoninae and their relationship to the other palaemonoid clades, including the Anchistioididae, with its peculiar larval development.

Within the Pontoniinae, a phylogeny of 72 genera based on 80 morphological characters was published by Li and Liu (1997). They regard the subfamily, as currently defined, to be a monophyletic group but suggest that the status of some newly erected genera should be reexamined. They further conclude that commensal Pontoniinae are evolved from free-living Palaemoninae, and they propose the genus *Periclimenes* to be the evolutionary link between free-living and commensal taxa. As currently much taxonomic work is focused around the paraphyletic genus *Periclimenes sensu lato*, this conclusion seems premature. Fransen (2002) published a morphological phylogeny of the genus *Pontonia* s.l., splitting the genus into six genera, with species in these genera associating either with bivalves or ascidians. Molecular work on selected genera using 16S and COI mitochondrial genes in relation to certain host groups is in progress, providing building blocks for a molecular phylogeny within this subfamily.

Within the Palaemoninae, several phylogeographical studies on *Macrobrachium rosenbergii* have been published in recent years by de Bruyn and coworkers (2004a, 2004b, 2005, 2007). Additionally, Murphy & Austin (2002, 2004) studied the origin and classification of Australian species of *Macrobrachium* using the 16S gene.

Anker et al. (2006) presented the first phylogenetic hypothesis of relationships among 36 extant genera of alpheid shrimps based on a cladistic analysis of 122 morphological characters from 56 species. In that study there is strong support for the monophyly of the family. Nodes defining genera were relatively well supported, though many basal nodes showed weak support. Six genera appeared paraphyletic, the large genus *Alpheus* (276 species) being amongst these. As suggested by the authors, the remaining uncertainties in the phylogenetic relations among the genera would benefit from tests with independent larval and molecular data.

Molecular phylogenies of alpheids also have been produced as a component of studies on eusociality among species of *Synalpheus* by Duffy et al. (2000) and Morrison et al. (2004). Williams et al. (2001) used one mitochondrial (COI) and two nuclear genes (GPI, EF-1$\alpha$) to analyze the status of the 7 morphological groups within the genus *Alpheus* recognized by Coutière (1905). This analysis showed the existence of three major clades within the genus; these clades showed no particular relationship to the groupings of Coutière (1905). Finally, a morphological phylogeny of the genus *Athanopsis* was presented by Anker & Ahyong (2007).

## 6 MORPHOLOGICAL CHARACTERS

The monophyly of both the Dendrobranchiata and the Stenopodidea is uncontroversial and is supported by several characters, of which the following can be considered to be of phylogenetic significance: the dendrobranchiate gill, male petasma, nauplius egg eclosion, and pleonic hinge structure in the Dendrobranchiata (Felgenhauer & Abele 1983; Abele & Felgenhauer 1986; Abele 1991; Dixon et al. 2003); and the enlarged third pereiopod and spherical spermatozoa in the Stenopodidea (Felgenhauer & Abele 1983; Abele & Felgenhauer 1986; Abele 1991; Dixon et al. 2003). As Martin et al. (2007) recently described considerable variation in dendrobranch gill morphology, a fresh look at some of the generally accepted characters may reveal further incongruities.

The monophyly of the Caridea is harder to address, as it is based on a large number of variable morphological characters (Felgenhauer & Abele 1983). Bracken et al. (this volume) consider the taxon as monophyletic, but perhaps excluding Procarididae. The true position of the family Procarididae, although unquestionably closely related to other carideans, remains unresolved. Procaridids share only one character with the other caridean families, the second abdominal pleuron overlapping the first and third somites, which is however variable in Glyphocrangonidae and Psalidopodidae. Procaridids differ from carideans in the attachment position of the phyllobranch gills, which is precoxal in *Procaris* versus higher on the body wall in Caridea, whereas other characters are similar to Dendrobranchiata (e.g., the foregut; see Felgenhauer & Abele 1983).

Currently the family level classification of Caridea is based primarily on the structure of the propodus and dactylus of the first two pereiopods, non- or multi-articulated carpus of the second pereiopod, features of the mandible, second and third maxilliped, and the number of epipods and branchial formula (Chace 1992; Holthuis 1993) Although these characters are of considerable use in the identification of Caridea, their phylogenetic significance at the family level appears uncertain. It is far beyond the current review to highlight all discrepancies, and we can only discuss a few salient ones. The chelae of carideans come in a bewildering variety of shapes and sizes, ranging from the relatively unspecialised examples in Palaemoninae, Processidae, and Pandalidae (the latter two with a multiarticulated carpus) to the specialized structures in Alpheidae, Atyidae, and Disciadidae, the homologies of these structures remaining unclear. Burkenroad (1981) proposed that the plesiomorphic gill formula in Caridea is one arthrobranch and one pleurobranch on thoracic segments 3 to 7, which is reduced in various ways to a minimum formula of a single pleurobranch each on thoracic segments 4 to 7, considered the most derived condition (Bauer 2004). However, within families there exists much variation in this character, especially in the Atyidae, and its phylogenetic usefulness remains to be proven.

Several authors (Thompson 1967; Felgenhauer & Abele 1983; Christoffersen 1990; Bauer 2004) have offered their opinion on which characters could be phylogenetically useful. Thompson (1967) placed much emphasis on the mandible, considering a fused molar and incisor process, combined with a 3-segmented palp, to be ancestral. Although there exists considerable variation at the generic level in some families, this could indeed be a valuable phylogenetic character. Felgenhauer & Abele (1983) and Abele & Felgenhauer (1986) discussed the protocephalon, pleonic hinges, and the gastric mill. These characters also may prove to be of value, but a survey of their variation across all families is still lacking. Christoffersen (1990) used a combination of previously highlighted characters (e.g., mandible, telson armature), with a score of "new" characters (e.g., corneal ocellus, bifid dorsal carina on the third abdominal somite, and a distolateral tooth on the basicerite) in his new superfamily/family arrangement. Many of Christoffersen's characters do, however, appear to be of low phylogenetic value. Finally, Bauer (2004) reviewed some of the above characters and emphasized the mandible, first to third maxillipeds, first and second pereiopods, pereiopodal exopods, gills, and the mastigobranch-setobranch complex. Currently, there is not enough information on the evolutionary polarity and indeed on even the mere occurrence of many of these characters across (and within) all families to address their phylogenetic usefulness, although work on this is now in progress by one of the authors.

## 7 MOLECULAR MARKERS

Several mitochondrial genes have been used for phylogenetic studies of shrimp-like decapods. Cytochrome C Oxidase Subunit I (COI) is a protein coding gene that has been used in more than 30 studies. COI is especially informative at low taxonomic levels with good resolution among populations of a species and sometimes at the family level. The protein coding gene Cytochrome B has been used in a few studies at the species and infraspecific levels of, for instance, *Typhlatya* (Webb 2003; Hunter et al. 2008). The non-protein coding 16S ribosomal RNA (16S) gene is slightly more conservative than COI with good resolution at species to family levels. The 12S ribosomal RNA (12S) gene has been applied to study infraspecific variation in a penaeid species (Palumbi & Benzie 1991; Bouchon et al. 1994). The complete mitochondrial genome of 6 shrimps has been sequenced: *Penaeus monodon* by Wilson et al. (2000), *Marsupenaeus japonicus* by Yamauchi et al. (2004), *Litopenaeus vannamei* by Xin Shen et al. (2007), *Fenneropenaeus chinensis* by Xin Shen et al. (2007), *Macrobrachium rosenbergii* by Miller et al. (2005), and *Halocaridina rubra* by Ivey & Santos (2007). As only a few complete mitochondrial sequences of species from different higher taxa are yet available, phylogenetic analyses have been performed only on these taxonomic levels.

Nuclear genes have been applied in a few phylogenetic studies of shrimp-like decapods so far. The following protein coding genes have so far been used: Myosin Heavy Chain (MyHC) for cryptic diversity and phylogeography in an *Alpheus* species-complex (Mathews, 2006); Glucose-6-phosphate isomerase (GPI) to analyze the status of the species-groups within the genus *Alpheus* (Williams et al. 2001); Elongation factor-1α (EF-1α) for infraspecific variation in penaeid species (Duda & Palumbi 1999; France et al. 1999); and the analysis of *Alpheus* species-groups (Williams et al. 2001). Histone H3 was used by Porter et al. (2005) in combination with 3 other genes for the elucidation of phylogenetic relations among the higher Decapod taxa. Non-coding nuclear genes used are: Internal Transcribed Spacer (ITS), applied in analysis of infraspecific variation in penaeid species (Chu et al. 2001; Wanna et al. 2006); 18S ribosomal DNA gene, used at higher taxonomic levels among families to orders (Kim & Abele 1990; Porter et al. 2005; Mitsuhashi et al. 2007; Bracken at al. this volume); and the 28S ribosomal DNA gene, also used at higher taxonomic levels (Porter et al. 2005; Mitsuhashi et al. 2007), although Zakšek et al. (2007) used it within the cave-shrimp genus *Troglocaris*.

## 8 FOSSILS

The fossil record of the shrimp-like decapods is particularly scant, due to their poorly calcified exoskeleton and perhaps also to their mode of life. Of the three groups, the Dendrobranchiata has the best fossil record with 74 fossil taxa known. Examples of extant families extend only as far back as the lower Cretaceous (100 mya), but the extinct Aegeridae range from the upper Triassic to the upper Jurassic, and a few species of the extinct Carpopenaeidae are present in the mid-Cretaceous. Two families of Stenopodidea contain a single extinct species each, both of lower Cretaceous age, one of which is a freshwater form. The Caridea have an extraordinarily poor fossil record, with a mere 46 extinct species compared to more than 3100 extant taxa. Taxa positively assigned to extant families occur only from the lower Cretaceous and later. In contrast to these confirmed ages, Porter et al. (2005) estimate the origin of the Dendrobranchiata to be in the early Silurian (437 mya) and the origin of the Caridea to be in the Devonian (417–423 mya), leaving a considerable gap in the historical record between the appearance of fossils and the estimated origin of the major lineages.

Although a good proportion of fossil taxa can be placed confidently within extant families, several remain enigmatic. This is particularly the case in the Caridea, with 9 fossil genera unplaced within any recent family, whilst the Udorellidae cannot be assigned to a superfamily (Crandall et al. in prep.). Interestingly, the achelate first and second pereiopods of the Udorellidae have led to speculation that they are related to the Procarididae (Abele & Felgenhauer 1983).

Several positively assigned fossil taxa exhibit features that are not present in modern-day lineages. For instance, the Carpopenaeidae, currently assigned to the Dendrobranchiata, harbor a multiarticulate carpus on the second and third pereiopods. Equally incongruous, the recently erected caridean superfamily Pleopteryxoidea (erected for *Pleopteryx kuempeli*) differs from all known carideans by the multiarticulate first pereiopod combined with achelate second pereiopods (Schweigert & Garassino 2006).

A robust, combined cladistic analysis of extant and extinct taxa in the shrimp-like decapods currently appears difficult to achieve, as classification of extant forms is largely based on rarely fossilized structures such as mouthparts, epipods, and gill structure/formulae (Holthuis 1993). Such studies are further hindered by the current lack of a robust phylogeny for the extant forms themselves. When a robust phylogeny of recent forms does become available, it would be instructive to pursue experimental analyses akin to Schram & Dixon (2004), by incorporating selected fossil taxa. Certainly, Solnhofen-type taxa (the origin of many fossil shrimp) may be of sufficient preservation status to circumvent the "vraagteken effect" (see Schram & Hof 1998). Equally, the addition of characters lacking in extant taxa may shed light on evolutionary pathways, whilst the addition of non-extant ecological niches (such as the freshwater Dendrobranchiata and Stenopodidea) could contribute interesting information.

## 9 CONCLUSION

This overview shows that relatively few representatives of shrimp-like decapod taxa thus far have been incorporated into phylogenetic analyses at higher to lower taxonomic levels and that controversies remain between the outcomes of various morphological and molecular analyses.

A survey of many morphological characters across (and within) families is sorely needed. These surveys should target characters previously suggested to be of phylogenetic importance, such as the mandible, the mastigobranch-setobranch complex, and pleonic hinges, but they should also include other characters known to vary among genera and families, such as the carpo-propodal brush and the setal brush on the fifth pereiopod in carideans. Additionally, the homology of certain characters needs to be put on a firmer footing, such as the L-shaped mastigobranch in Dendrobranchiata, Procarididae, and basal Caridea. Certain characters have been dismissed as being of phylogenetic value and should be re-appraised, including the structure of the gastric mill. This structure is generally assumed to be lacking in all carideans, but Felgenhauer & Abele (1983) discuss its occurrence in several families. Comparative morphological studies across all taxa, both at the family level within the Caridea and across all shrimp-like taxa, are urgently needed for morphological phylogeny to progress and to keep pace with the predictable flood of molecular phylogenies.

Currently, molecular phylogenetic work lags behind the amount of effort devoted to the Brachyura, but it is rapidly gaining momentum, with a score of new studies appearing in print each year. Nevertheless, the range of taxa included in molecular work, and their systematic breadth and scope, must be further expanded.

In other decapod groups, an interesting body of literature exists on various systematically informative biological attributes, such as larval development, spermatozoan ultrastructure, and even evo-devo processes. Works of this nature in shrimp-like Decapoda are few and far between. These will need to be integrated with molecular and morphological studies, underpinned by continued morphological studies, in order for the decapod Tree of Life to fully embrace available technologies for integrative systematics.

## REFERENCES

Abele, L.G. 1991. Comparison of morphological and molecular phylogeny of the Decapoda. *Mem. Queensland Mus.* 31: 101–108.

Abele, L.G. & Felgenhauer, B.E. 1986. Phylogenetic and phenetic relationships among the lower Decapoda. *J. Crust. Biol.* 63: 385–400.

Anker, A. & Ahyong, S.T. 2007. A rediagnosis of *Athanopsis australis* Banner & Banner, 1982, a rare alpheid shrimp from southern Australia, with a phylogeny of *Athanopsis* Coutière, 1897 and remarks on antitropical distributions in the Alpheidae (Decapoda, Caridea). *Crustaceana* 80: 685–698.

Anker, A., Ahyong, S.T., Noël, P.Y. & Palmer, A.R. 2006. Morphological phylogeny of alpheid shrimps: parallel preadaptation and the origin of a key morphological innovation, the snapping claw. *Evolution* 60: 2507–2528.

Baldwin, J.D., Bass, A.L., Bowen, B.W. & Clark, W.H. 1998. Molecular phylogeny and biogeography of the marine shrimp *Penaeus. Mol. Phylogenet. Evol.* 10: 399–407.

Bauer, R.T. 2004. *Remarkable shrimp: adaptations and natural history of the carideans*. University of Oklahoma Press, Norman.

Bouchon, D., Souty-Grosset, C. & Raimond, R. 1994. Mitochondrial DNA variation and markers of species identity in two Penaeid shrimp species: *Penaeus monodon* Fabricius and *P. japonicus* Bate. *Aquaculture* 127: 131–144.

Bracken, H., De Grave, S. & Felder, D.L. (this volume). Phylogeny of the infraorder Caridea based on mitochondrial and nuclear genes (Crustacea: Decapoda). In: Martin, J.W., Crandall, K.A. & Felder, D.L. (eds.), *Crustacean Issues: Decapod Crustacean Phylogenetics*. Boca Raton, Florida: Taylor & Francis/CRC Press.

Bruce, A.J. 1986. Observations on the family Gnathophyllidae Dana, 1852 (Crustacea: Decapoda). *J. Crust. Biol.* 6: 463–470.

Burkenroad, M.D. 1963. The evolution of the Eucarida (Crustacea, Eumalocostraca) in relation to the fossil record. *Tulane Stud. Geol.* 2: 3–16.

Burkenroad, M.D. 1981. The higher taxonomy and evolution of Decapoda (Crustacea). *Trans. San Diego Soc. Nat. Hist.* 19: 251–268.

Burkenroad, M.D. 1983. Natural classification of Dendrobranchiata, with a key to recent genera. In: Schram, F.R. (ed.), *Crustacean Issues 1, Crustacean Phylogeny*: 279–290. Rotterdam: Balkema.

Chace, F.A. 1992. On the classification of the Caridea (Decapoda). *Crustaceana* 63: 70–80.

Chan, T-Y, Tong, J., Tam, Y.K. & Chou, K.H. 2008. Phylogenetic relationships among the genera of the Penaeidae (Crustacea: Decapoda) revealed by mitochondrial 16S rRNA gene sequences. *Zootaxa* 1694: 38–50.

Chenoweth, S.F. & Hughes, J.M. 2003. Speciation and phylogeography in *Caridina indistincta*, a complex of freshwater shrimps from Australian heathland streams. *Mar. Freshw. Res.* 54: 807–812.

Christoffersen, M.L. 1987. Phylogenetic relationships of hippolytid genera, with an assignment of new families for the Crangonoidea and Alpheoidea (Crustacea, Decapoda, Caridea). *Cladistics* 3: 348–362.

Christoffersen, M.L., 1988. Phylogenetic systematics of the Eucarida (Crustacea Malacostraca). *Rev. Bras. Zool.* 5: 325–351.

Christoffersen, M.L., 1989. Phylogeny and classification of the Pandaloidea (Crustacea, Caridea). *Cladistics* 5: 259–274.

Christoffersen, M.L., 1990. A new superfamily classification of the Caridea (Crustacea: Pleocyemata) based on phylogenetic pattern. *Zeitschr. Zool. Syst. Evolutionsforsch.* 28: 94–106.

Chu, K.H., Li, C.P., & Ho, H.Y. 2001. The first Internal Transcribed Spacer (ITS-1) of ribosomal DNA as a molecular marker for phylogenetic and population analyses in Crustacea. *Mar. Biotechnol.* 3: 355–361.

Cook, B.D., Baker, A.W., Page, T.J., Grant, S.C., Fawcett, J.H., Hurwood, D.A. & Hughes, J.M. 2006. Biogeographic history of an Australian freshwater shrimp, *Paratya australiensis* (Atyidae): the role life history transition in phylogeographic diversification. *Mol. Ecol.* 15: 1083–1093.

Coutière, H. 1905. Les Alpheidae. In: Gardiner, J.S. (ed.), *The Fauna and Geography of the Maldive and Laccadive Archipelagoes, vol. 2*: 852–921. Cambridge, U.K: Cambridge University Press.

Crosnier, A. 1987. Les espèces indo-ouest-pacifiques d'eau profonde du genre *Metapenaeopsis* (Crustacea Decapoda Penaeidae). *Mém. Mus. Nat. Hist. Nat.* 2: 409–453.

Crosnier, A. 1991. Crustacea Decapoda: Les *Metapenaeopsis* indo-ouest-pacifiques sans appareil stridulant (Penaeidae). Deuxième partie. In: Crosnier, A. (ed.), *Résultats des Campagnes MUSORSTOM. Vol. 2. Mém. Mus. Nat. Hist. Nat.* 152: 155–297.

Crosnier, A. 1994a. Crustacea Decapoda: Les *Metapenaeopsis* indo-ouest-pacifiques avec un appareil stridulant (Penaeidae). Deuxième partie. In: Crosnier, A. (ed.), *Résultats des Campagnes MUSORSTOM. Vol. 9. Mém. Mus. Nat. Hist. Nat.* 161: 255–337.

Crosnier, A. 1994b. Crustacea Decapoda: Les *Metapenaeopsis* indo-ouest-pacifiques sans appareil stridulant (Penaeidae). Description de deux espèces nouvelles. In: Crosnier, A. (ed.), *Résultats des Campagnes MUSORSTOM. Vol. 9. Mém. Mus. Nat. Hist. Nat.* 161: 339–349.

Dall, W. 2007. Recent molecular research on *Penaeus* sensu lato. *J. Crust. Biol.* 27: 380–382.

Dana, J.D. 1852. Crustacea, Part 1. In: United States exploring expedition during the years 1838, 1839, 1840, 1841, 1842, under the command of Charles Wilkes, U.S.N., 13: i- viii, 1–685.

de Bruyn, M. & Mather, P.B. 2007. Molecular signatures of Pleistocene sea-level changes that affected connectivity among freshwater shrimp in Indo-Australian waters. *Mol. Ecol.* 16: 4295–4307.

de Bruyn, M., Nugroho, E., Mokarrom Hossain, Md., Wilson, J.C. & Mather, P.B. 2005. Phylogeographic evidence for the existence of an ancient biogeographic barrier: the Isthmus of Kra Seaway. *Heredity* 94: 370–378.

de Bruyn, M., Wilson, J.A. & Mather, P.B. 2004a. Huxley's line demarcates extensive genetic divergence between eastern and western forms of the giant freshwater prawn, *Macrobrachium rosenbergii*. *Mol. Phylogenet. Evol.* 30: 251–257.

de Bruyn, M., Wilson, J.A. & Mather, P.B. 2004b. Reconciling geography and genealogy: phylogeography of giant freshwater prawns from the Lake Carpentaria region. *Mol. Ecol.* 13: 3515–3526.

De Grave, S. & Moosa, M.K. 2004. A new species of the enigmatic shrimp genus *Pseudocheles* (Decapoda: Bresiliidae) from Sulawesi (Indonesia), with the designation of a new family Pseudochelidae. *Crust. Res.* 33: 1–9.

Dixon C.J., Ahyong, S.T. & Schram, F.R. 2003. A new hypothesis of decapod phylogeny. *Crustaceana* 76: 935–975.

Duda, T.F.J. & Palumbi, S.R. 1999. Population structure of the black tiger prawn, *Penaeus monodon*, among western Indian Ocean and western Pacific populations. *Mar. Biol.* 134: 705–710.

Duffy, J.E., Morrison, C.L. & Ríos, R. 2000. Multiple origins of eusociality among sponge-dwelling shrimps (*Synalpheus*). *Evolution* 54: 503–516.

Felgenhauer, B.E. & Abele, L.G. 1983. Phylogenetic relationships among shrimp-like decapods. *Crustacean Issues* 1: 291–311.

Flegel, T.W. 2007. The right to refuse revision in the genus *Penaeus*. *Aquaculture* 264: 2–8.

France, S.C., Tachino, N., Duda, T.F., Jr., Shleser, R.A., and Palumbi, S.R. (1999). Intraspecific genetic diversity in the marine shrimp *Penaeus vannamei*: multiple polymorphic elongation factor-1a loci revealed by intron sequencing. *Mar. Biotechnol.* 1: 261–268.

Fransen, C.H.J.M. 2002. Taxonomy, phylogeny, historical biogeography, and historical ecology of the genus *Pontonia* Latreille (Crustacea: Decapoda: Caridea: Palaemonidae). *Zool. Verh.* 336: 1–433.

Holthuis, L.B. 1993. *The recent genera of the caridean and stenopodidean shrimps (Crustacea, Decapoda) with an appendix on the order Amphionidacea*. Nationaal Natuurhistorisch Museum Leiden.

Hunter, R.L., Webb, M.S., Iliffe, T.M., & Bremer, J.R.A. 2008. Phylogeny and historical biogeography of the cave-adapted shrimp genus *Typhlatya* (Atyidae) in the Caribbean Sea and western Atlantic. *J. Biogeogr.* 35: 65–75.

Huxley, T.H. 1879. On the classification and the distribution of the Crayfishes. *Proc. Zool. Soc. London* 1878: 752–788.

Ivey, J.L. & Santos, S.R. 2007. The complete mitochondrial genome of the Hawaiian anchialine shrimp *Halocaridina rubra* Holthuis, 1963 (Crustacea: Decapoda: Atyidae). *Gene* 394: 35–44.

Kim, W. & Abele, L.G. 1990. Molecular phylogeny of selected decapod crustaceans based on 18s rRNA nucleotide sequences. *J. Crust. Biol.* 10: 1–13.

Lavery, S., Chan, C.H., Tam, Y.K. & Chu, K.H. 2004. Phylogenetic relationship and evolutionary history of the shrimp genus *Penaeus* s.l. derived from mitochondrial DNA. *Mol. Phylogenet. Evol.* 31: 39–49.

Li, X. & Liu, J.Y 1997. A preliminary study on the phylogeny of Pontoniinae (Decapoda: Palaemonidae). *Oceanol. Limnol. Sin.* 28: 383–393.

Maggioni, R., Rogers, A.D., Maclean, N. & D'Incao, F., 2001. Molecular phylogeny of Western Atlantic *Farfantepenaeus* and *Litopenaeus* shrimp based on mitochondrial 16S partial sequences. *Mol. Phylogenet. Evol.* 18: 66–73.

Martin, J.W. & Davis, G.E. 2001. An updated classification of the Recent Crustacea. *Natural History Museum of Los Angeles County, Science Series* 39: 1–124.

Martin, J.W., Liu, E.M. & Striley, D. 2007. Morphological observations on the gills of dendrobranchiate shrimps. *Zool. Anz.* 246: 115–125.

Mathews, L.M. 2006. Cryptic biodiversity and phylogeographical patterns in a snapping shrimp species complex. *Mol. Ecol.* 15: 4049–4063.

McLaughlin, P.A., Lemaitre, R., Ferrari, F.D., Felder, D.L. & Bauer, R.T. 2008. A reply to T.W. Flegel. *Aquaculture* 2175: 370–373.

Miller, A.D., Murphy, N.P, Burridge, C.P. & Austin, C.M. 2005. Complete mitochondrial DNA sequences of the decapod crustaceans *Pseudocarcinus gigas* (Menippidae) and *Macrobrachium rosenbergii* (Palaemonidae). *Mar. Biotechnol.* 7: 339–349.

Mitsuhashi, M., Sin, Y.W., Lei, H.C., Chan, T.-Y. & Chu, K.H. 2007. Systematic status of the caridean families Gnathophyllidae Dana and Hymenoceridae Ortmann (Crustacea: Decapoda): a preliminary examination based on nuclear rDNA sequences. *Inv. Syst.* 21: 613–622.

Morrison, C.L., Ros, R. & Duffy, J.E. 2004. Phylogenetic evidence for an ancient rapid radiation of Caribbean sponge-dwelling snapping shrimps (*Synalpheus*). *Mol. Phylogenet. Evol.* 30: 563–581.

Mulley, J.C. & Latter, B.D.H. 1980. Genetic variation and evolutionary relationships within a group of thirteen species of penaeid prawns. *Evolution* 34: 904–916.

Murphy, N.P. & Austin, C.M. 2002. A preliminary study of 16S rRNA sequence variation in Australian *Macrobrachium* shrimps (Palaemonidae: Decapoda) reveals inconsistencies in their current classification. *Inv. Syst.* 16: 697–701.

Murphy, N.P. & Austin, C.M. 2004. Multiple origins of the endemic Australian *Macrobrachium* (Decapoda: Palaemonidae) based on 16S rRNA mitochondrial sequences. *Aust. J. Zool.* 52: 549–559.

Page, T.J., Baker, A.M., Cook, B.D. & Hughes, J.M. 2005. Historical transoceanic dispersal of a freshwater shrimp: the colonization of the South Pacific by the Genus *Paratya* (Atyidae). *J. Biogeogr.* 32: 581–593.

Page, T.J., von Rintelen, K. & Hughes, J.M. 2007. Phylogenetic and biogeographic relationships of subterranean and surface genera of Australian Atyidae (Crustacea: Decapoda: Caridea) inferred with mitochondrial DNA. *Inv. Syst.* 21: 137–145.

Page, T.J., Short, J.W., Humphrey, C.L., Hillyer, M.J. & Hughes, J.M. 2008a. Molecular systematics of the Kakaducarididae (Crustacea: Decapoda: Caridea). *Mol. Phylogenet. Evol.* 46: 1003–1014.

Page, T.J., Cook, B.D., von Rintelen, T., von Rintelen, K. & Hughes, J.M. 2008b. Evolutionary relationships of atyid shrimps imply both ancient Caribbean radiations and common marine dispersal. *J. N. Am. Benth. Soc.* 27: 68–83.

Palumbi, S.R. & Benzie, J. 1991. Large mitochondrial DNA differences between morphologically similar penaeid shrimp. *Mol. Mar. Biol. Biotechn.* 1: 27–34.

Pereira, G. 1997. A cladistic analysis of the freshwater shrimps of the family Palaemonidae (Crustacea, Decapoda, Caridea). *Acta Biol. Venez.* 17: 1–69.

Pérez Farfante, I. & Kensley, B.F. 1997. Penaeoid and sergestoid shrimps and prawns of the world. Keys and diagnoses for the families and genera. *Mém. Mus. Nat. His. Nat.* 175: 1–233.

Porter, M.L., Pérez-Losada, M. & Crandall, K.A. 2005. Model-based multi-locus estimation of decapod phylogeny and divergence times. *Mol. Phylogenet. Evol.* 37: 355–369.

Quan, J., Lü, X.-M., Zhuang, Z., Dai, J., Deng, J. & Zhang, Y.-P. 2001. Low genetic variation of *Penaeus chinensis* as revealed by mitochondrial COI and 16S rRNA gene sequences. *Biochem. Gen.* 39: 297–284.

Quan, J., Lü, X.-M., Zhuang, Z., Dai, J., Deng, J. & Zhang, Y.-P. 2004. Phylogenetic relationships of 12 Penaeoidea shrimp species deduced from mitochondrial DNA sequences. *Biochem. Gen.* 42: 331–345.

Richter, S. & Scholtz, G. 2001. Phylogenetic analysis of the Malacostraca (Crustacea). *J. Zool. Syst. Evol. Res.* 39: 113–136.

von Rintelen, K., von Rintelen, T. & Glaubrecht, M. 2007a. Molecular phylogeny and diversification of freshwater shrimps (Decapoda, Atyidae, *Caridina*) from ancient Lake Poso (Sulawesi, Indonesia)—the importance of being colourful. *Mol. Phylogenet. Evol.* 45: 1033–1041.

von Rintelen, K., von Rintelen, T., Meixner, M., Lüter, C., Cai, Y. & Glaubrecht, M. 2007b. Freshwater shrimp–sponge association from an ancient lake. *Biol. Letters* 3: 262–264.

Roy, D., Kelly, D.W., Fransen, C.H.J.M., Heath, D.D. & Haffner, G.D. 2006. Evidence of small-scale vicariance in *Caridina lanceolata* (Decapoda: Atyidae) from the Malili Lakes, Sulawesi. *Evol. Ecol. Res.* 8: 1087–1099.

Saito, T. & Takeda, M. 2003. Phylogeny of the Spongicolidae (Crustacea: Stenopodidae): evolutionary trend from shallow-water free-living to deep-water sponge-associated habitat. *J. Mar. Biol. Assoc. U.K.* 83: 119–131.

Schram, F.R. 2001. Phylogeny of decapods: moving towards a consensus. *Hydrobiologia* 449: 1–20.

Schram, F.R. & Dixon, C.J. 2004. Decapod phylogeny: addition of fossil evidence to a robust morphological cladistic data set. *Bull. Mizunami Fossil Mus.* 31: 1–19.

Schram, F.R. & Hof, C.H.J. 1998. Fossils and the interrelationships of major crustacean groups. In: G.D. Edgecombe (ed.), *Arthropod Fossils and Phylogeny*: 233–302. New York: Columbia Univ. Press.

Schweigert, G. & Garassino, A. 2006. News on *Pleopteryx kuempeli* Schweigert & Garassino, an enigmatic shrimp (Crustacea: Decapoda: Caridea: Pleopteryxoidea superfam. nov.) from the Upper Jurassic of S Germany. *N. Jahrb. Geol. Pal.* 6: 449–461.

Seridji, R. 1990. Description of some planktonic larval stages of *Stenopus spinosus* Risso, 1826: notes on the genus and the systematic position of the Stenopodidae as revealed by larval characters. *Sc. Mar.* 54: 293–303.

Shank, T.M., Lutz, R.A. & Vrijenhoek, R.C. 1998. Molecular systematics of shrimp (Decapoda; Bresiliidae) from deep-sea hydrothermal vents: I. Enigmatic "small orange" shrimp from the Mid-Atlantic Ridge are juvenile *Rimicaris exoculata*. *Mol. Mar. Biol. Biotech.* 7: 88–96.

Shen, X., Ren, J., Cui, Z., Sha, Z., Wang, B., Xiang, J. & Liu, B. 2007. The complete mitochondrial genomes of two common shrimps (*Litopenaeus vannamei* and *Fenneropenaeus chinensis*) and their phylogenetic considerations. *Gene* 403: 98–109.

Tam, Y.K. & Chu, K.H. 1993. Electrophoretic study on the phylogenetic relationships of some species of *Penaeus* and *Metapenaeus* (Decapoda: Penaeidae) from the South China Sea. *J. Crust. Biol* 13: 697–705.

Tavares, C.T., Serejo, C. & Martin, J.W. (this volume). A preliminary phylogenetic analysis of the Dendrobranchiata based on morphological characters. In: Martin, J.W., Crandall, K.A. & Felder, D.L. (eds.), *Crustacean Issues: Decapod Crustacean Phylogenetics*. Boca Raton, Florida: Taylor & Francis/CRC Press.

Thompson, J.R. 1967. Comments on phylogeny of section Caridea (Decapoda, Natantia) and the phylogenetic importance of the Oplophoridea. *Mar. Biol. Soc. India Symp. Crustacea* 1: 314–326.

Tong, J.G., Chan, T.-Y. & Chu, K.H. 2000. A preliminary phylogenetic analysis of *Metapenaeopsis* (Decapoda: Penaeidae) based on mitochondrial DNA sequences of selected species from the Indo-West Pacific. *J. Crust. Biol.* 20: 541–549.

Vazquez-Bader, A.R., Carrero, J.C., Garcia-Varela, M., Garcia, A. & Laclette, J.P. 2004. Molecular phylogeny of superfamily Penaeoidea Rafinesque-Schmaltz, 1815, based on mitochondrial 16S partial sequence analysis. *J. Shellf. Res.* 23: 911–916.

Voloch, C.M., Freire, P.R. & Russo, C.A.M. 2005. Molecular phylogeny of penaeid shrimps inferred from two mitochondrial markers. *Gen. Mol. Res.* 4: 668–674.

von Sternberg, R. & Motoh, H. 1995. Notes on the phylogeny of the American *Penaeus* shrimps (Decapoda: Dendrobranchiata: Penaeidae). *Crust. Res.* 24: 146–156.

Wanna, W., Chotigeat, W. & Phongdara, A. 2006. Sequence variations of the first ribosomal internal transcribed spacer of *Penaeus* species in Thailand. *J. Exp. Mar. Biol. Ecol.* 331: 64–73.

Webb, M.S. 2003. *Intraspecific relationships among the stygobitic shrimp*, Typhlatya mitchelli, *by analyzing sequence data from mitochondrial DNA*. Master of Science Thesis, Texas A&M University.

Williams, S.T., Knowlton, N., Weigt, L.A. & Jara, J.A. 2001. Evidence for three major clades within the snapping shrimp genus *Alpheus* inferred from nuclear and mitochondrial gene sequence data. *Mol. Phylogenet. Evol.* 20: 375–389.

Wilson, K., Cahill, V., Ballment, E. & Benzie, J. 2000. The complete sequence of the mitochondrial genome of the crustacean *Penaeus monodon*: are malacostracan crustaceans more closely related to insects than to branchiopods? *Mol. Biol. Evol.* 17: 863–874.

Yamauchi, M.M., Miya, M.U., Machida, R.J. & Nishida, M. 2004. A PCR–based approach for sequencing the mitochondrial genomes of decapod crustaceans, with a practical example from the kuruma prawn *Marsupenaeus japonicus*. *Mar. Biotechnol.* 6: 419–429.

Yang, H.J. & Ko, H.S. 2002. First zoea of *Palaemon ortmanni* (Decapoda, Caridea, Palaemonidae) hatched in the laboratory, with notes on the larval morphology on the Palaemonidae. *Korean J. Syst. Zool.* 18: 181–189.

Zakšek, V., Sket, B. & Trontelj, P. 2007. Phylogeny of the cave shrimp *Troglocaris*: evidence of a young connection between Balkans and Caucasus. *Mol. Phylogenet. Evol.* 42: 223–235.

# A Preliminary Phylogenetic Analysis of the Dendrobranchiata Based on Morphological Characters

CAROLINA TAVARES,[1] CRISTIANA SEREJO[1] & JOEL W. MARTIN[2]

[1] *Museu Nacional/UFRJ, Quinta da Boa Vista, s/n, Rio de Janeiro, Brazil 20940-040*
[2] *Natural History Museum of Los Angeles County, 900 Exposition Boulevard, Los Angeles, California, U.S.A.*

ABSTRACT

Dendrobranchiata currently is composed of two superfamilies, Penaeoidea (families Aristeidae, Benthesicymidae, Penaeidae, Sicyoniidae, and Solenoceridae) and Sergestoidea (families Sergestidae and Luciferidae). Although the monophyly of Dendrobranchiata is rather firmly established, little is known about the relationships among its families. We analyzed 24 taxa of Dendrobranchiata using three different combinations of outgroups, with differing results. In the majority of the most parsimonious trees, Dendrobranchiata, Penaeoidea, and Sergestoidea appear monophyletic, as do the families Aristeidae, Solenoceridae, Sicyoniidae, Sergestidae, and Luciferidae. The families Penaeidae and Benthesicymidae are not monophyletic. Dendrobranchiata is defined by having dendrobranchiate gills, prominent pleonic hinges, larvae hatching as nauplii or protozoeae, and the presence of a petasma in males. Sergestoidea is defined primarily by "lost" characters, including the loss of the exopod on maxilliped 3, the absence of a dactyl on P1, and the related absence of a P1 chela. Penaeoidea is defined by the presence of a tubercle on the terminal article of the eyestalk and the presence of a branchiocardiac carina. There are no clear synapomorphies defining the Aristeidae. Solenoceridae is defined by the presence of a postorbital spine and the presence of a distolateral projection on the male pleopod 2. Sicyoniidae is defined by many characters, including the presence of an ocular stylet. Sergestidae and Luciferidae also are defined by many characters, such as the presence of a clasper organ on the male antenna 1 in the sergestids and the brooding of eggs on the female pereopods in luciferids.

## 1 INTRODUCTION

The decapod suborder Dendrobranchiata contains some 500 species of shrimps, including most of the 10–15 commercially important species worldwide. Dendrobranchiates also play important ecological roles in estuaries and other marine systems. Species range from shallow waters in the tropics to depths of 1000 m or more on the continental slopes (Pérez Farfante & Kensley 1997).

These shrimps have had a somewhat confusing taxonomic history. Boas (1880) divided the Decapoda into the Natantia, a "swimming" group that included all shrimps and shrimp-like forms, and the Reptantia for the remaining (crawling) species of decapods. Bate (1888) first recognized the different types of gills among the Natantia and divided the group into three subgroups: Dendrobranchiata, Phyllobranchiata, and Trichobranchiata. Bate (1888) also divided the "tribe Penaeidea" into the families Penaeidae and Sergestidae. Calman's (1909) treatment of the Dendrobranchiata (as Tribe Penaeidea) included the family Penaeidae (with the subfamilies Aristeinae, Sicyoninae, and Penaeinae) and the family Sergestidae (with subfamilies Sergestinae and Leuciferinae). Much later, Crosnier (1978) treated Penaeidae as consisting of two families: Aristeidae, containing the subfamilies Aristeinae, Benthesicyminae, and Solenocerinae, and Penaeidae containing the subfamilies Penaeinae, and Sicyoninae. Crosnier (1978) also suggested that most or all of the penaeid subfamilies should be raised to familial level, an action finally taken by Pérez Farfante & Kensley (1997).

Currently, the suborder Dendrobranchiata contains two superfamilies: Penaeoidea and Sergestoidea. The Penaeoidea includes the families Aristeidae, Benthesicymidae, and Solenoceridae, species of which are found in the deep sea, and the Penaeidae and Sicyoniidae, found more often on the continental shelf. The Sergestoidea includes only two families, the Sergestidae (mostly in the deep sea but with some freshwater species) and the highly aberrant and exclusively planktonic Luciferidae.

The first phylogenetic hypothesis for any dendrobranchiate taxa was proposed in 1983, when Burkenroad (1983) presented a more or less intuitively based hypothesis, unfortunately without a corresponding character matrix. Since then there have been many papers published on the relationships of these shrimp, and nearly all of these studies have agreed that the Dendrobranchiata is a basal group among the Decapoda and is the sister group to the Pleocyemata (e.g., Burkenroad 1981; Felgenhauer & Abele 1983; Schram 1984; Abele & Felgenhauer 1986; Abele 1991; Wills 1997; Richter & Scholtz 2001; Dixon et al. 2003). Reviewing the details of all of these studies is beyond the scope of this paper, but noteworthy contributions include Felgenhauer & Abele's (1983) recognition of the Dendrobranchiata as a natural group and their addition of other important characters to the diagnosis of the suborder; Abele's (1991) first molecularly derived phylogeny of the Dendrobranchiata and his comparison of that tree to a morphology-based phylogeny, strongly supporting the monophyly of the dendrobranchs; and Wills's (1997) support of dendrobranchiate monophyly in his analysis of all major crustacean taxa (extant and fossil). Most recent studies have assumed or supported monophyly of the Dendrobranchiata, such as Dixon et al. (2003), who considered monophyly of the group probable from their analysis of ordered characters, while at the same time emphasizing that the clade was not recovered in all of the most parsimonious trees in that study.

Defining morphological characters of the Dendrobranchiata (based primarily on the works of Pérez Farfante & Kensley 1997; Burkenroad 1981, 1983; Dixon et al. 2003) are: 1) the presence of gills that are "dendrobranchiate" (defined as "secondarily branching;" see Martin et al. 2007); 2) the presence of chelae on the first three pairs of pereopods (with some exceptions); 3) the pleura of the second abdominal somite not overlapping those of the first (as opposed to the situation in the caridean shrimps); 4) the presence of prominent hinges between the pleonic somites; 5) the direct release of eggs into the water (as opposed to being carried on the female pleopods) and the subsequent hatching of the eggs as nauplii or protozoeae; 6) the presence of a petasma in males; and 7) the absence of an appendix interna on the pleopods (with the exception of a vestigial structure found in some males). Here, we use morphological characters and cladistic methods to establish a preliminary phylogeny of the Dendrobranchiata and to test the monophyly of the two superfamilies and seven families currently treated as dendrobranchiates.

## 2 MATERIALS AND METHODS

The material used in this study was obtained from three institutions: Museu Nacional/UFRJ, Brazil; FURG (Fundação Universitária Rio Grande), Brazil; and NMNH (National Museum of Natural History, Smithsonian Institution), USA (Appendix 1). For the ingroup, 24 species distributed among the seven families of Dendrobranchiata were examined. For the outgroups, 3 species of Caridea, one of Stenopodidea, and one of Nephropidea were examined, in three different combinations: one with Caridea alone, another with Caridea and Stenopodidea, and a third with Caridea, Stenopodidea, and Nephropidea.

For selection of the morphological characters, specimens of Dendrobranchiata were examined using compound and stereoscope microscopes. Drawings of most of the phylogenetically

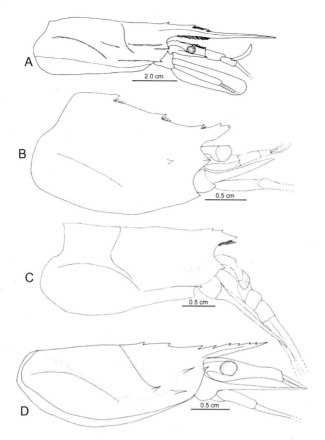

**Figure 1.** Selected morphological characters. Rostrum length. (A) *Plesiopenaeus coruscans*, surpassing antennular peduncle. (B) *Sicyonia typica*, not surpassing antennular peduncle, reaching cornea. (C) *Benthesicymus bartletti*, not surpassing antennular peduncle, not reaching cornea. (D) *Hymenopenaeus debilis*, not surpassing antennular peduncle, surpassing cornea.

informative characters are provided (Figs. 1–4). A total of 102 morphological characters was selected. When appropriate, characters were combined into multistate groupings to avoid overly dependent characters. This combining into multistate characters resulted in a matrix of 68 binary characters and 34 multistate characters. Of the 34 multistate characters, 8 were regarded as continuous characters. These characters were split into multistate characters following an arbitrary method in which we took the range between the lowest and the highest values and divided that range into three equal parts; each of these parts was then treated as one character state. All characters were unordered.

The data matrix was assembled using the program Delta (Dallwitz et al. 1993, 1998). This program allows users to prepare a dataset and export it as a nexus format. The cladistic analysis was performed using PAUP 4.0 Beta version (Swofford 2000), with a heuristic search option, in stepwise addition, with 1000 replicates. Bootstrap analysis and Bremer support (Bremer 1994) also were performed using PAUP 4.0.

For character optimization we used the tool trace character of MacClade 4.03 (Maddison & Maddison 2001). For character polarization we followed Nixon & Carpenter (1993) for outgroup comparisons.

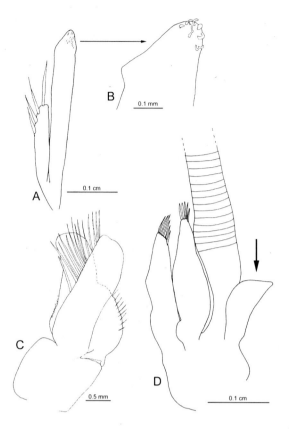

**Figure 2.** Selected morphological characters. Hook setae on male appendix interna. (A) *Pasiphaea princeps*, present. (B) *Pasiphaea princeps*, present, detailed. (C) *Benthesicymus bartletti*, absent. Disto-lateral projection on male pleopod 2. (D) *Hymenopenaeus debilis*, present.

## 3 RESULTS

### 3.1 *Description and optimization of characters*

Characters used and explanations of their distribution and polarity are given in Appendix 2. Because Dendrobranchiata is widely recognized as a basal group within Decapoda, it is difficult to find true synapomorphies for the group. For this reason, character optimization was performed by comparison with the three outgroups, meaning that some characters appearing here as "apomorphic" to (or within) the Dendrobranchiata may in fact be plesiomorphic in the Decapoda as a whole. One example is the second abdominal pleuron overlapping the first, a character that is clearly derived (occurring only in the Caridea) but that appears "plesiomorphic" here when the Caridea is used as the outgroup for the dendrobranchs. The same problem occurs with characters 38 (releasing eggs freely into the water as opposed to carrying them on the pleopods), 40 (hatching as nauplius larvae), and 83 (absence of hook setae on the male appendix interna), in which states treated in this analysis as apomorphic for the Dendrobranchiata are actually plesiomorphic among the Decapoda as a whole.

### 3.2 *Analysis 1 - Caridea as the outgroup*

Sixty-nine equally most parsimonious trees were found (for indices see Table 1), and from these two consensus trees were calculated (strict and majority rule) (Figs. 5, 6). Character states

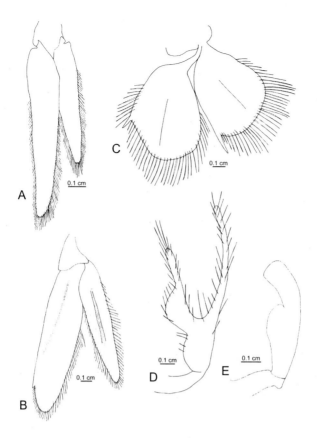

**Figure 3.** Selected morphological characters. Uropods. (A) *Sergestes armatus*. (B) *Artemesia longinaris*. (C) *Nephropsis agassizi*. Epipod shape. (D) *Haliporoides sibogae sibogae*, bifid. (E) *Litopenaeus schmitti*, foliaceous.

considered non-homoplastic are depicted in bold. For the majority rule consensus tree, we obtained the following results:

Dendrobranchiata (clade 3) is a monophyletic group. The suborder is defined by 6(2), 30(1), **31(2)**, **39(1)**, 79(1), **80(2)**, **83(1)**, **84(1)**. Luciferidae (clade 5) is a monophyletic group, defined by **6(1)**, 26 (2), **38 (3)**, **41(1)**, **49(1)**, **51(1)**, 58(1), 59(1), **64(1)**, **65(1)**, **72(1)**, **76(1)**. Sergestoidea is not a natural group. Sergestidae (clade 7) is a natural group defined by 32(4), **42(2)**, 58(1), 59(1), **73(1)**, **77(1)**. Penaeoidea (clade 8) is a natural group defined by 9(2); 23(2); 44(2); 54(2); 99(4). Benthesicymidae is not a natural group. Sicyoniidae (clade 17) is a natural group defined by **8(2)**, **32(3)**, **37(2)**, **81(4)**, **101(2)**. Penaeidae is not resolved, with members of the family in a trichotomy with Sicyoniidae in clade 12. Solenoceridae (clade 19) is a natural group defined by **14(2)**, **90 (2)**. Aristeidae (clade 24) is a natural group. Characters 22 (3), 61 (1), 67 (1), 85(1) characterize the family, but it is not possible to determine plesiomorphic vs. apomorphic states.

Other clades (most of which are currently not defined taxonomically) resulting from the analysis were: Clade 2: All species except *Pasiphaea princeps*. This clade is characterized by 8(1), 16(2), 22(1), 47(2), 48(2), 53(1), 54(2), 61(3), 67(2), 92(2), 97(2). Clade 6: All Dendrobranchiata except the family Luciferidae, defined by 24(2), **55(1)**, **70(2)**. Clade 9: Penaeoidea except for *Benthesicymus* sp., defined by 1(2), 4(2), 28(2). Clade 10: Penaeoidea except for *Benthesicymus bartletti* and *Benthesicymus* sp., defined by 17(2), 32(4). Clade 11: Penaeoidea except for Benthesicymidae and Aristeidae, defined by 10(2), 23(1), **40(2)**, 93(1). Clade 12: Sicyoniidae and

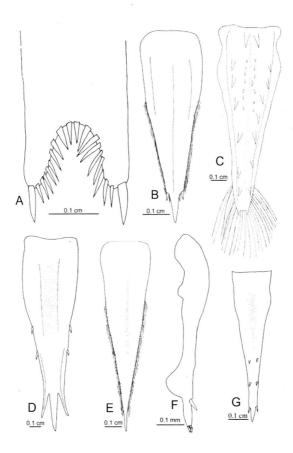

**Figure 4.** Selected morphological characters. Telson posterior margin. (A) *Pasiphaea princeps*, cleft. (B) *Rimapenaeus constrictus*, with robust setae only. (C) *Stenopus hispidus*, truncate. Telson ornamentation. (D) *Penaeopsis serrata*, with spines and robust setae. (E) *Xiphopenaeus kroyeri*, unarmed, with spines only. Telson robust setae position. (F) *Lucifer typus*, lateral and terminal. (G) *Oplophorus spinosus*, lateral and dorsal.

Penaeidae, defined by 99(3), 100(1). Clade 13: *Penaeopsis serrata* and *Artemesia longinaris*, defined by 46(2), 93(3). Clade 14: *Farfantepenaeus paulensis*, *Litopenaeus schmitti*, *Parapenaeus americanus*, *Xiphopenaeus kroyeri* and *Rimapenaeus constrictus*, defined by 19(1). Clade 15: *Farfantepenaeus paulensis* and *Litopenaeus schmitti*, defined by 1(3). Clade 16: *Parapenaeus americanus*, *Xiphopenaeus kroyeri* and *Rimapenaeus constrictus*, defined by 13(2).

### 3.3 Analysis 2 - Caridea and Stenopodidea as outgroups

Ninety-three equally most parsimonious trees were found (for indices see Table 1), and from these two consensus trees were calculated (strict and majority rule) (Fig. 7). Character states considered non-homoplastic are depicted in bold. For the majority rule consensus tree, we obtained the following results:

Dendrobranchiata (clade 3) is monophyletic, defined by 6(2), **30(1)**, 38(1), **39(1)**, 45(1), **80(2)**, 90(2). Luciferidae (clade 5) is monophyletic, defined by **6(1)**, 19(1), 26(2), **38(3)**, **41(1)**, **51(1)**, **64(1)**, **65(1)**, **72(1)**, **76(1)**. Sergestoidea (clade 28) is a natural group, now with the families Luciferidae and Sergestidae in a monophyletic clade, defined by 57(1), **58(1)**, **59(1)**. Sergestidae (clade 7) is a natural group defined by 32(4), **42(2)**. Penaeoidea (clade 8) is a natural group defined by 9(2), 23(2), 33(4), 82(2). Benthesicymidae is not a natural group. Sicyoniidae (clade 17) is a natural

*Morphological Phylogeny of the Dendrobranchiata* 267

**Table 1.** Some values of the three different analyses. NT = total number of trees; Tl = total length; CI = consistency index; RI = retention index; RC = rescaled consistency index.

| Analysis | NT | TI | CI | RI |
|---|---|---|---|---|
| 1 | 69 | 290 | 0.50 | 0.64 |
| 2 | 93 | 304 | 0.49 | 0.63 |
| 3 | 69 | 319 | 0.49 | 0.63 |

group defined by the following apomorphies: **8(2)**, **33(3)**, **37(2)**, 57(1), **81(4)**, **101(2)**. Penaeidae (clade 29) is a natural group characterized by 81(4), although optimization is not possible. Solenoceridae (clade 19) is a natural group defined by **14(2)**, **90(2)**. Aristeidae (clade 24) is a natural group. As in analysis 1, characters 22(3), 61(1), 67(1), 85(1) characterize the family but cannot be optimized.

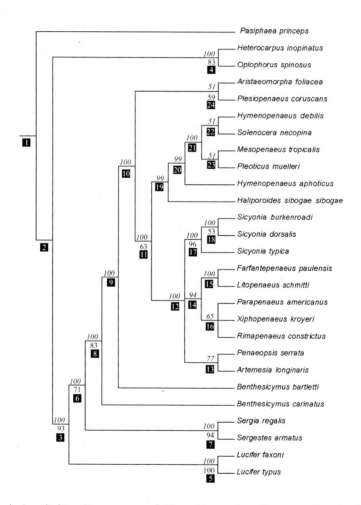

**Figure 5.** Analysis 1 majority rule consensus of 69 equally parsimonious trees (length = 290), with clade numbers (black squares), bootstrap and MR (percentage of appearance of each clade in all original trees, in italics) values.

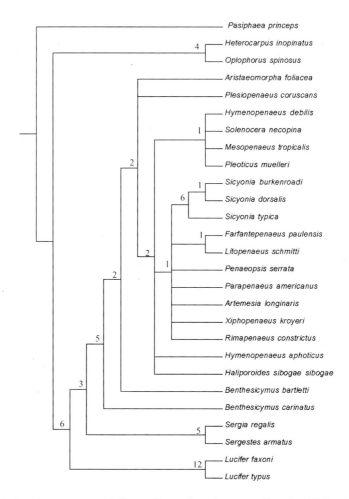

**Figure 6.** Analysis 1, strict consensus of 69 equally parsimonious trees (length = 290), with Bremer support index values.

Other clades (not taxonomically defined or named) in the analysis are: Clade 9: Penaeoidea except for *Benthesicymus* sp., defined by 1(2), 4(2), 28(2). Clade 10: Penaeoidea except for *Benthesicymus bartletti* and *Benthesicymus* sp., defined by 17(2), 32(4). Clade 11: Penaeoidea except for Benthesicymidae and Aristeidae, defined by 10(2), **40(2)**, 93(1). Clade 12: Sicyoniidae and Penaeidae, defined by 82(1), **99(3)**, 100(1). Clade 13: *Penaeopsis serrata* and *Artemesia longinaris*, defined by 46(2), **93(3)**. Clade 14: *Farfantepenaeus paulensis, Litopenaeus schmitti, Parapenaeus americanus, Xiphopenaeus kroyeri* and *Rimapenaeus constrictus*, characterized by 19(1). Clade 15: *Farfantepenaeus paulensis* and *Litopenaeus schmitti*, defined by 1(3). Clade 16: *Parapenaeus americanus, Xiphopenaeus kroyeri* and *Rimapenaeus constrictus*, defined by 13(2). Clade 27: Dendrobranchiata and *Stenopus hispidus*, defined by 31(2), 55(1), 70(2), 79(1), 82 (1), 84(1).

## 3.4 *Analysis 3 - Caridea, Stenopodidea and Nephropidae as outgroups*

Sixty-nine equally most parsimonious trees were found (for indices see Table 1) and, from these, two consensus trees were calculated (strict and majority rule) (Figs. 8, 9). Character states considered non-homoplastic are depicted in bold. For the majority rule consensus tree, we obtained the following results:

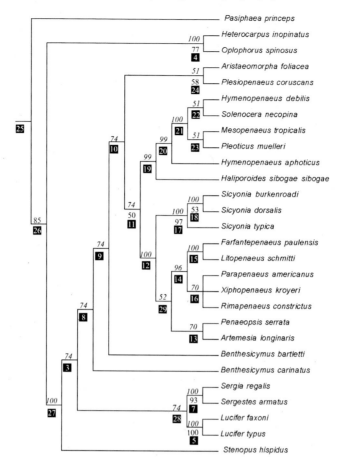

**Figure 7.** Analysis 2, majority rule consensus of 93 equally parsimonious trees (length = 304), with clade numbers (black squares), bootstrap and MR (percentage of appearance of each clade in all original trees, in italic) values.

Dendrobranchiata (clade 3) is monophyletic, defined by 6(2), **30(1)**, 38(1), **39(1)**, 45(1), **80(2)**. Luciferidae (clade 5) is a monophyletic group, defined by **6(1)**, 19(1), 26(2), **38(3)**, **41(1)**, **49(1)**, **51(1)**, **64(1)**, **65(1)**, **72(1)**, **76(1)**. Sergestoidea (clade 28) is a natural group. As in analysis 2, the families Luciferidae and Sergestidae constitute a monophyletic clade defined by 57(1), **58(1)**, **59(1)**. Sergestidae (clade 7) is a natural group defined by 32(4), **42(2)**. Penaeoidea (clade 8) is a natural group defined by 9(2), 15(2), 25(2), 46(2), 82(2). Benthesicymidae is not a natural group. As in analysis 1 and 2, the benthesicymid species do not appear together. Sicyoniidae (clade 17) is a natural group defined by **8(2)**, 32(3), **37(2)**, 57(1), **81(4)**, **101(2)**. Penaeidae could not be evaluated (as in analysis 1). Solenoceridae (clade 19) is a natural group defined by **14(2)**, **90(2)**. Aristeidae (clade 24) is a natural group characterized (as in analyses 1 and 2) by 22(3), 61(1), 67(1), 85(1), but optimization of characters is not possible.

Other clades depicted in this analysis are: Clade 9: Penaeoidea except for *Benthesicymus* sp., defined by 1(2), 4(2), 28(2). Clade 10: Penaeoidea except for *Benthesicymus bartletti* and *Benthesicymus* sp., defined by 17(2), 32(4). Clade 11: Penaeoidea except for Benthesicymidae and Aristeidae, defined by 10(2), **40(2)**, 93(1). Clade 12: Sicyoniidae and Penaeidae, defined by 82(1), **99(3)**, 100(1). Clade 13: *Penaeopsis serrata* and *Artemesia longinaris*, defined by 46(2), **93(3)**. Clade 14: *Farfantepenaeus paulensis, Litopenaeus schmitti, Parapenaeus americanus, Xiphopenaeus kroyeri*

**Figure 8.** Analysis 3, majority rule consensus of 69 equally parsimonious trees (length = 319), with clade numbers (black squares), bootstrap and MR (percentage of appearance of each clade in all original trees, in italic) values.

and *Rimapenaeus constrictus*, characterized by 19(1). Clade 15: *Farfantepenaeus paulensis* and *Litopenaeus schmitti*, defined by 1(3). Clade 16: *Parapenaeus americanus*, *Xiphopenaeus kroyeri* and *Rimapenaeus constrictus*, defined by 13(2). Clade 33: Dendrobranchiata and *Nephropsis agassizi*, defined by 31(2), 55(1), 70(2), 79(1), 82(1), 84(1).

## 4 DISCUSSION

### 4.1 *Choice of outgroup and different analyses*

Selecting the best outgroup for phylogenetic analysis is often a difficult decision, and this was true in our case as well. Although Pleocyemata is often depicted as the sister group to Dendrobranchiata in the literature, that group (Pleocyemata) is highly diverse, and it is unclear which group among the Pleocyemata should be used. Consequently, we prepared three different analyses using different Pleocyemata groups. Interestingly, although some topologies are similar, all three analyses differed. When we compared clades that appeared in two or all three analyses, sometimes character polarity differed. Analysis 3 is perhaps the most realistic in that more pleocyemata taxa are included,

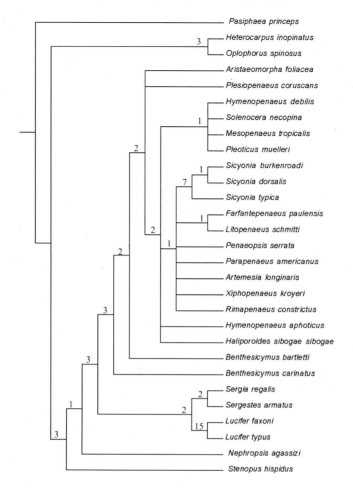

**Figure 9.** Analysis 3, strict consensus of 69 equally parsimonious trees (length = 319), with Bremer support index values.

although all three analyses are valuable in highlighting characters and polarities that might be important in dendrobranchiate phylogeny. Euphausiids, another potential outgroup choice, were not used in this study.

4.2 *Dendrobranchiata as a monophyletic group*

We began with the hypothesis that the suborder is monophyletic, as indicated in the literature (e.g., Burkenroad 1983; Felgenhauer & Abele 1983; Dixon et al. 2003), and with the suborder defined by the presence of 1) dendrobranchiate gills (but see Martin et al. 2007); 2) the first three pairs of pereopods usually chelate; 3) the pleura of the second abdominal somite not overlapping those of the first; 4) prominent hinges between their pleonic somites; 5) eggs released directly into water (rather than carried by females); 6) larvae hatching as nauplii or protozoea; 7) a petasma in males; and 8) pleopods without an appendix interna, except for some vestigial structure found in males. In our analyses, only the following characters proved to be synapomorphies of Dendrobranchiata: dendrobranchiate gills [6(2)], prominent pleonic hinges [30(1)], larvae hatching as nauplii or protozoea [39(1)], and the presence of a petasma in males [80(2)]. All species we examined have the first two pereopods chelate (except for Sergestoidea). A distinctive character of the dendrobranchiates (as noted in previous studies) is the presence of a chelate third pereopod [70(2)]. However, although this character is "typical" of Dendrobranchiata, in analysis 1 it appears as a synapomorphy of clade 6 (Penaeoidea + Sergestidae), and in analyses 2 and 3 as a synapomorphy of clades 27 and 32, respectively. The pleura of the second abdominal somite not overlapping those of first [31(2)] is apomorphic only in analysis 1; in analyses 2 and 3 this character appears as a synapomorphy of clades 27 and 32. Eggs released directly into the water [38(1)] vs. being retained on the female pereopods [38(3)] is a synapomorphy for Dendrobranchiata only in analyses 2 and 3. The ratio between scaphocerite and antennae 1 peduncle [45(1)] is a synapomorphy for Dendrobranchiata also in analyses 2 and 3. The absence of hook setae on the male appendix interna [83(1)] is apomorphic only in analysis 1; in analyses 2 and 3 this character is a synapomorphy of clades 27 and 32.

Despite the fact that there is much evidence to indicate that the suborder is monophyletic, in the strict consensus of analysis 2, the dendrobranchiate species appear as a non-monophyletic clade, grouped with *Stenopus hispidus* as the sister group to the Caridea. Similarly, Dixon et al. (2003) did not recover Dendrobranchiata in the most parsimonious trees in their ordered analysis. Yet we think it unlikely that Dendrobranchiata is non-monophyletic, with most of the above discrepancies explained by outgroup choice or character polarity. Here, we accept the monophyly and current classification of Dendrobranchiata, divided into two superfamilies, Sergestoidea and Penaeoidea, as discussed below.

The position of Luciferidae is a salient question in any consideration of dendrobranchiate phylogeny. The family is extremely different from other Dendrobranchiata, with most of the differences assumed to be modifications for a planktonic life. Although the inclusion of Luciferidae within Dendrobranchiata by Bate (1988) was not based on cladistic methods, it was assumed (then and now) that most of the family's unusual features represented simple character loss. In all of our analyses, the family clustered with the other families of Dendrobranchiata; for this reason we feel that Luciferidae should be maintained for now as a Dendrobranchiata family.

### 4.3 *Sergestoidea as a natural group*

Sergestoidea includes two families, Sergestidae and Luciferidae. Traditionally, the superfamily has been poorly defined, often by such different character states as having pereopods 4 and 5 reduced or absent and/or having the antennular flagellum modified or absent. In analysis 1, Sergestoidea appears as non-monophyletic. However, in analyses 2 (except for strict consensus) and 3 these families appear together in clade 28, defined by the absence of the exopod on maxilliped 3 [57(1)], the absence of a dactyl on P1 [58(1)], and the absence of a chela on P1 [59(1)]. The absence of a P1 dactyl and consequently the chela is scored here as non-homoplastic, but from the literature we know that this is indeed homoplastic, as other sergestid genera not treated here (e.g., *Acetes*, *Peisos*, *Sicyonella*) possess a minute chela on P1. Although these characters have been described in the literature, they were never used to define the superfamily.

### 4.4 *Penaeoidea as a natural group*

The superfamily Penaeoidea contains five families: Aristeidae, Benthesicymidae, Penaeidae, Sicyoniidae, and Solenoceridae. In all analyses, the superfamily was monophyletic (clade 8), with the exception of the strict consensus of analysis 2. In the literature the superfamily is defined by having all five pereopods well developed, at least some somites with three branchiae on each side, and at least 11 well-developed gills on each side. None of these characters was found as a synapomorphy here, where the superfamily is defined instead by the presence of a tubercle on the terminal article of the eyestalk [9(2)] and the presence of the branchiocardiac carina [23(2)].

## 4.5 Benthesicymidae as a non-natural group

In all trees, this family did not appear as a monophyletic clade. Characters used in the literature to define the family (e.g., the presence of an open petasma [82(1)] and the presence of a tubercle on the eyestalk [9(2)]) are not synapomorphies, as they are shared by other species within the Penaeoidea. It is important to notice that the two species used in this study belong to two different groups among the genus *Benthesicymus*. The first group is defined in the literature by the following characters: presence of marginal branchiostegal spine, with branchiostegal carina not sharp; exopods of first maxilliped narrowing abruptly to tip; merus of second maxilliped expanded laterally; dactylus of third maxilliped triangular, with only one spine at tip; exopods of all pereopods small but easily perceptible. The second group is defined by the following characters: presence of non-marginal branchiostegal spine, with very sharp branchiostegal carina; exopods of first maxilliped tapering to tip; merus of second maxilliped not expanded laterally; dactylus of third maxilliped subrectangular, distal margin bearing more than 1 strong spine; exopods of all pereopods minute (Burkenroad 1936; Kikuchi & Nemoto 1991; Dall 2001). Our study suggests that this morphological separation is in accordance with evolutionary patterns within the genus *Benthesicymus*. However, very few species of the family, which includes some 40 species, were used in our analyses, so our results have to be considered preliminary.

## 4.6 Penaeidae as a non-natural group

Most studies on penaeid phylogeny have indicated that the family is not monophyletic (Quan et al. 2004; Vázquez-Bader et al. 2004; Voloch et al. 2005). Characters previously used to diagnose the family are not always synapomorphs; e.g., the presence of an ocular scale [10(2)] is synapomorphic to clade 11, not to Penaeidae only. Similarly, the exopods of maxilliped 2 [54(2)] and maxilliped 3 [57(2)] are characteristic of clade 10, not just the Penaeidae. Other characters are "one time" occurrences with no phylogenetic signal, such as the semi-open petasma [80(3)] found only in *Litopenaeus schmitti* (a semi-closed petasma [81(3)] is characteristic of clade 29). Analyses 1 and 3 resulted in a trichotomy (clade 12) of two groups of Penaeidae (clades 13, 14) and a group of Sicyoniidae (clade 17); no further resolution was possible here. On the other hand, in analysis 2 the majority rule consensus Penaeidae clades are nested in a monophyletic clade 29, characterized by the presence of a semi-closed petasma [81(3)]; however, clade 29 is not supported by either Bremer index or bootstrap analysis. Regardless of whether Penaeidae is monophyletic, two groups emerged consistently: clade 14 (*Farfantepenaeus paulensis, Litopenaeus schmitti, Parapenaeus americanus, Xiphopenaeus kroyeri*, and *Rimapenaeus constrictus*), defined by the absence of a branchiocardiac carina [19(1)], and clade 13 (*Penaeopsis serrata* + *Artemesia longinaris*), defined by the presence of the parapenaeid spine [46(2)] and a telson armed with spines and robust setae [93(3)]. The close relationship between sicyoniids and penaeids shown here was suggested earlier by both Crosnier (1978) and Burkenroad (1983).

## 4.7 Solenoceridae as a natural group

Although this clade is present only in majority rule consensus trees (99%) and additionally was not supported by Bremer index and bootstrap analysis, we continue to consider the family monophyletic based on two non-homoplastic synapomorphies (presence of a postorbital spine [14(2)] and presence of a distolateral projection on male pleopod 2 [90(2)]), as has been noted previously in the literature. The position of the family among Dendrobranchiata in all analyses obtained here showed solenocerids closer to penaeids and sicyoniids (as in clade 11), in contrast with some previous authors (e.g., Crosnier 1978; Burkenroad 1983) who placed solenocerids closer to aristeids.

## 4.8 Status of the Aristeidae

This clade is present in all majority rule consensus trees, although it was not supported by Bremer index and has a low bootstrap value (58%). Additionally, no synapomorphies were found to define or characterize the family. Characters used to describe the family in the past, such as the presence of an ocular tubercle and an open petasma, are present also in clades 8 and 9. Because of the preliminary nature of this analysis, we are leaving the question of aristeid monophyly unanswered for now.

## ACKNOWLEDGEMENTS

We thank Dr. Fernando D'Incao (FURG) and Dr. Rafael Lemaitre (NMNH) for providing access to the material examined and Dr. Charles Fransen (Nationaal Natuurhistorisch Museum) for critically reviewing this manuscript. The first author also thanks CAPES (Coordenação de Aperfeiçoamento de Pessoal de Nvel Superior) for the fellowship. This work was supported in part by grant number DEB 0531616 from the National Science Foundation's "Assembling the Tree of Life" program to J. W. Martin, in conjunction with collaborative awards to Keith Crandall and Nikki Hannegan (Brigham Young University), Darryl Felder (University of Louisiana Lafayette), and Rodney Feldmann and Carrie Schweitzer (Kent State University). The symposium during which these results were first presented was funded by NSF grant DEB 0721146, with additional support from the American Microscopical Society, the Crustacean Society, the Society of Systematic Biologists, the Society of Integrative and Comparative Biology (SICB), and the SICB Divisions of Invertebrate Zoology and Systematics & Evolutionary Biology. The participation of J. Martin was also made possible by funding from the U.S. National Oceanographic and Atmospheric Administration NOAA for systematic work on crabs of the Hawaiian Islands.

# APPENDIX 1

List of species examined in this study with specimen collection numbers. MNRJ = UFRJ collection, Museu Nacional, Brazil; FURG = Fundação Universitária Rio Grande, Brazil; USNM = National Museum of Natural History, Smithsonian Institution, USA.

Suborder Dendrobranchiata Bate, 1888
Superfamily Penaeoidea Rafinesque-Schmaltz, 1815
  Family Aristeidae Wood-Mason, 1891
    *Aristaeomorpha foliacea* (Risso, 1827) MNRJ 13775, MNRJ14561
    *Plesiopenaeus coruscans* (Wood-Mason, 1891) MNRJ 14522, MNRJ 14577
  Family Benthesicymidae Wood-Mason, 1891
    *Benthesicymus bartletti* Smith, 1882 MNRJ 19167, MNRJ19164
    *Benthesicymus carinatus* Smith, 1884 MNRJ 14731
  Family Penaeidae Rafinesque-Schmaltz, 1815
    *Artemesia longinaris* Bate, 1888 MNRJ 1653
    *Farfantepenaeus paulensis* (Pérez Farfante, 1967) MNRJ 28
    *Litopenaeus schmitti* (Burkenroad, 1936) MNRJ 15835
    *Parapenaeus americanus* Rathbun, 1901 MNRJ 14815
    *Penaeopsis serrata* Bate, 1881 MNRJ 14784
    *Rimapenaeus constrictus* (Stimpson, 1874) MNRJ 1680
    *Xiphopenaeus kroyeri* (Heller, 1862) MNRJ 49
  Family Sicyoniidae Ortmann, 1898
    *Sicyonia burkenroadi* Cobb, 1971 MNRJ 14632
    *Sicyonia dorsalis* Kingsley, 1878 MNRJ 68, MNRJ 1656
    *Sicyonia typical* (Boeck, 1864) MNRJ 63, MNRJ 1692
  Family Solenoceridae Wood-Mason, 1891
    *Haliporoides sibogae sibogae* (De Man, 1907) USNM 261459
    *Hymenopenaeus aphoticus* Burkenroad, 1936 FURG 1609, FURG 2681
    *Hymenopenaeus debilis* Smith, 1882 - MNRJ 14794, MNRJ 14796, MNRJ 14798, MNRJ, 14807
    *Mesopenaeus tropicalis* (Bouvier, 1905) FURG 220
    *Pleoticus muelleri* (Bate, 1888) - MNRJ 39
    *Solenocera necopina* Burkenroad, 1939 MNRJ 14631, MNRJ 14630
Superfamily Sergestoidea Dana, 1852
  Family Luciferidae Thompson, 1829
    *Lucifer typus* H. Milne Edwards, 1837 MNRJ 18048, MNRJ 18050
    *Lucifer faxoni* Borradaile, 1915 MNRJ 18046, MNRJ 18054
  Family Sergestidae Dana, 1852
    *Sergestes armatus* Kroyer, 1855 MNRJ 15505
    *Sergia regalis* (Gordon, 1939) MNRJ 15507, MNRJ 15508, MNRJ 15509
Suborder Pleocyemata Burkenroad, 1963
Infraorder Caridea Dana, 1852
Superfamily Pandaloidea Haworth, 1825
  Family Pandalidae Haworth, 1825
    *Heterocarpus inopinatus* Tavares, 1999 MNRJ 14693
Superfamily Oplophoroidea Dana, 1852
  Family Oplophoridae Dana, 1852
    *Oplophorus spinosus* (Brullé, 1839) MNRJ 14874
Superfamily Pasiphaeoidea Dana, 1852
  Family Pasiphaeidae Dana, 1852
    *Pasiphaea princeps* Smith, 1884 MNRJ 19525, MNRJ 19522
Infraorder Stenopodidea Bate, 1888
  Family Stenopodidae Claus, 1827
    *Stenopus hispidus* (Olivier, 1811) MNRJ 2288
Infraorder Astacidea Latreille, 1802
Superfamily Nephropoidea Dana, 1852
  Family Nephropidae Dana, 1852
    *Nephropsis agassizii* A. Milne-Edwards, 1880 MNRJ 19232

## APPENDIX 2

Morphological characters used in the analyses. Although some characters listed below proved to be uninformative, we have listed them here for informational purposes and the possibility of future analyses.

1. Rostral spines: (1) absent; (2) present, dorsal only; (3) present, dorsal and ventral; (4) present, dorsal and lateral; (5) present, lateral only.
2. Number of dorsal rostral spines: (1) up to 5; (2) 6–9; (3) 10 or more.
3. Number of ventral rostral spines: (1) up to 4; (2) 5–7; (3) 8 or more.
4. Post-rostral spines: (1) absent; (2) present.
5. Number of post-rostral spines: (1) up to 2; (2) 3; (3) 4 or more.
6. Gills: (1) absent; (2) dendrobranch; (3) phyllobranch; (4) trichobranch.
7. Number of gills: (1) at least 11 on each side of the body; (2) from 1 to 8 on each side of the body.
8. Ocular stylet: (1) absent; (2) present.
9. Ocular tubercle: (1) absent; (2) present.
10. Ocular scale: (1) absent; (2) present.
11. Ocelo on eye: (1) absent; (2) present.
12. Rostrum length: (1) surpassing antennular peduncle (Fig. 1A); (2) not surpassing antennular peduncle, reaching cornea (Fig. 1B); (3) not surpassing antennular peduncle, not reaching cornea (Fig. 1C); (4) not surpassing antennular peduncle, surpassing cornea (Fig. 1D).
13. Orbital spine: (1) absent; (2) present.
14. Post-orbital spine: (1) absent; (2) present.
15. Adrostral carina: (1) absent; (2) present.
16. Antennal spine: (1) absent; (2) present.
17. Antennal carina: (1) absent; (2) present.
18. Cervical sulci: (1) absent; (2) present; (3) reduced.
19. Branchiocardiac carina: (1) absent; (2) present.
20. Hepatic sulci: (1) absent; (2) present.
21. Hepatic spine: (1) absent; (2) present.
22. Branchiostegal spine: (1) absent; (2) present, marginal; (3) present, not marginal.
23. Branchiostegal carina: (1) absent; (2) present.
24. Post-cervical sulci: (1) absent; (2) present.
25. Gastro-orbital sulcus: (1) absent; (2) present.
26. Pterygostomian spines: (1) absent; (2) present.
27. Longitudinal carina on carapace: (1) absent; (2) present.
28. Thoracic sternites width: (1) sternites 3–8 narrow; (2) sternites 3–5 narrow; (3) sternites 3–6 narrow.
29. Pleon: (1) laterally compressed; (2) dorso-ventral compressed.
30. Pleonic hinges: (1) prominent; (2) hidden; (3) slight.
31. Second abdominal pleura: (1) overlapping first; (2) not overlapping first.
32. Posterior spines on abdominal pleura: (1) absent; (2) present on somites 3–6; (3) present on somites 5–6; (4) present on somite 6.
33. Dorso-abdominal carina: (1) absent; (2) present on somites 2–6; (3) present on somites 3–6; (4) present on somites 4–6; (5) present on somite 6; (6) present on somites 1–6; (7) present on somites 3–5; (8) present on somites 5–6.
34. Dorso-posterior spines on abdominal somites: (1) absent; (2) 3–6; (3) 4–6; (4) 6; (5) 5; (6) 1,5,6.
35. Abdominal somite 6 with posterior dorso-lateral spines: (1) absent; (2) present.
36. Ventral projections on male abdominal somite 6: (1) absent; (2) present two rounded big projections and without a small disto-ventral projection; (3) present two sharp-pointed big projections and with a small disto-ventral projection.
37. Pleopods 3–5: (1) biramous; (2) uniramous.
38. Eggs: (1) released free in water; (2) brooded in female pleopods; (3) brooded in female pereopods.
39. Larvae: (1) hatch as nauplius; (2) hatch as protozoea.
40. Antenna 1 prosartema: (1) absent; (2) present.
41. Antenna 1: (1) uniflagellate; (2) biflagellate.
42. Male antenna 1: (1) without clasper organ; (2) with clasper organ.
43. Scaphocerite: (1) absent; (2) present.

44. Antenna 1 first article: (1) without disto-lateral spine on outer margin; (2) with disto-lateral spine on outer margin.
45. Ratio scaphocerite/antenna 1 peduncle: (1) up to 1.39; (2) 1.4–1.98; (3) 1.99 or more.
46. Ventromesial (parapenaeid) spine: (1) absent; (2) present.
47. Mandible: (1) only with incisor process; (2) with molar and incisor processes together; (3) with molar and incisor processes separated.
48. Mandibular palp: (1) absent; (2) present.
49. Maxilla 1 palp: (1) absent; (2) present.
50. Maxilla 2: (1) with two bilobed setose endites; (2) with one bilobed and one unilobed setose endites; (3) with reduced endites; (4) with one bilobed and one reduced endites.
51. Maxilla 2 palp: (1) absent; (2) present.
52. Number of maxillipeds: (1) 0; (2) 3.
53. Maxilliped 1 endite: (1) oval; (2) reduced, no defined sharp; (3) absent.
54. Maxilliped 2 exopod: (1) absent; (2) present.
55. Articles of maxilliped 3 endopod: (1) separated; (2) fused.
56. Maxilliped 3 dactyl: (1) with only one article; (2) with 5 articles.
57. Maxilliped 3 exopod: (1) absent; (2) present.
58. Pereopod 1 dactyl: (1) absent; (2) present.
59. Pereopod 1: (1) without chela; (2) with chela.
60. Pereopod 1 without chela: (1) with a subchela formed by a row of strongly flexed robust setae present on distal margin of carpus and proximal margin of propodus; (2) without subchela.
61. Pereopod 1 merus: (1) with a sub-distal robust setae; (2) with a sub-distal spine; (3) unarmed; (4) with a row of 5 spines; (5) with a sub-distal robust setae and a row of 3 spines.
62. Pereopod 1 ischium: (1) unarmed; (2) with a mesial spine; (3) with a distal spine.
63. Right and left pereopod 2: (1) of equal size; (2) of unequal size.
64. Pereopod 2 dactyl: (1) absent; (2) present.
65. Pereopod 2: (1) without chela; (2) with chela.
66. Pereopod 2 carpus: (1) divided; (2) entire.
67. Pereopod 2 merus: (1) with a sub-distal robust seta; (2) unarmed; (3) with a disto-lateral row of 5–7 robust setae.
68. Pereopod 2 ischium: (1) unarmed; (2) with one spine.
69. Pereopod 3 dactyl: (1) absent; (2) present.
70. Pereopod 3: (1) without chela; (2) with chela.
71. Pereopod 3 merus: (1) with a robust setae row; (2) without a robust setae row.
72. Pereopod 4: (1) absent; (2) present.
73. Pereopod 4 dactyl: (1) absent; (2) present.
74. Pereopod 4 merus: (1) with a robust setae row; (2) without a robust setae row.
75. Ratio P4/ P3: (1) up to 1.1; (2) 1.11–1.6; (3) 1.61 or more.
76. Pereopod 5: (1) absent; (2) present.
77. Pereopod 5 dactyl: (1) absent; (2) present.
78. Ratio P5/ P3: (1) up to 1.19; (2) 1.2–1.98; (3) 1.99 or more.
79. Exopods on pereopods: (1) absent; (2) present, reduced; (3) present, not reduced.
80. Petasma: (1) absent; (1) present.
81. Petasma present: (1) open; (2) semi-open; (3) semi-closed; (4) closed.
82. Male appendix interna: (1) absent; (2) present only on pleopod 2; (3) present on pleopods 2–5.
83. Hook setae on male appendix interna: (1) absent (Fig. 2C); (2) present on pleopods 2–5 (Fig. 2A,B).
84. Female appendix interna: (1) absent; (2) present on pleopods 2–5.
85. Appendix masculina: (1) smaller than appendix interna; (2) about the same size as appendix interna; (3) bigger than appendix interna.
86. Appendix masculina size: (1) longer than wide; (2) as long as wide.
87. Appendix interna size: (1) as long as wide; (2) longer than wide.
88. Thelycum: (1) absent; (2) present.
89. Thelycum present: (1) open; (2) closed.
90. Disto-lateral projection on male pleopod 2: (1) absent; (2) present, near appendix interna and appendix masculina (Fig. 2D).
91. Uropods: (1) exopod and endopod unarmed (Fig. 3A); (2) exopod with an outer lateral spine, endopod unarmed (Fig. 3B); (3) endopod and exopod with an outer lateral spine both (Fig. 3C).
92. Telson posterior margin: (1) cleft (Fig. 4A); (2) pointed (Fig. 4B,D,E,G); (3) truncate (Fig. 4C,F).
93. Telson ornamentation: (1) only with spines (Fig. 4C ); (2) only with robust setae (Fig. 4A,B,F,G); (3) with spines and robust setae (Fig. 4D); (4) unarmed (Fig. 4E).

94. Robust setae position: (1) lateral (Fig. 4B); (2) terminal (Fig. 4A); (3) lateral and terminal (Fig. 4F); (4) lateral and dorsal (Fig. 4G).
95. Number of robust setae on each side of telson: (1) up to 4; (2) 4.1–7.1; (3) 7.2 or more.
96. Number of spines on each side of telson: (1) up to 3.6; (2) 3.7–6.3; (3) 6.4 or more.
97. Photophores: (1) absent; (2) present.
98. Pesta organ: (1) absent; (2) present. (uninformative)
99. Epipods on pereopods 1–5: (1) absent; (2) present on P1–P5; (3) present on P1–P3; (4) present on P1–P4.
100. Epipods on pereopods 1–5 shape: (1) bifid (Fig. 3D); (2) foliaceous (Fig. 3E).
101. Abdominal somites with antero-dorsal spines: (1) absent; (2) present on somite 1.
102. Abdominal pleurae with lateral carina: (1) absent; (2) present.

REFERENCES

Abele, L. 1991. Comparison of morphological and molecular phylogeny of the Decapoda. *Mem. Qld. Mus.* 31: 101–108.

Abele. L. & Felgenhauer, B.E. 1986. Phylogenetic and phenetic relationships among the lower Decapoda. *J. Crust. Biol.* 6: 385–400.

Bate, C.S. 1888. Report on the Crustacea Macrura collected by the H.M.S. Challenger during the years 1873–1876. *Challenger Rept. Zool.* 24: 1–942.

Boas, J.E.V. 1880. Studier over Decapodernes Slaegtskabsforhold. *Vidensk. Selsk. Skrifter (Nat.)* 6, 1: 25–210.

Bremer, K. 1994. Branch support and tree stability. *Cladistics* 10: 295–304.

Burkenroad, M.D. 1936. The Aristeinae, Solenocerinae and pelagic Penaeinae of the Bingham oceanographic collection. *Bull. Bingham Oceanogr. Collect.* 5: 1–151.

Burkenroad, M.D. 1981. The higher taxonomy and evolution of Decapoda (Crustacea). *Trans. San Diego Soc. Nat. Hist.* 19: 251–268.

Burkenroad, M.D. 1983. Natural classification of Dendrobranchiata, with a key to recent genera. In: Schram, F.R. (ed.), *Crustacean Issues 1, Crustacean Phylogeny*: 279–290. Rotterdam: Balkema.

Calman, W.T. 1909. *A Treatise on Zoology, 7: Appendiculata, Crustacea*. London: Adam & Charles Black.

Crosnier, A. 1978. Crustacés Décapodes Pénéides Aristeidae (Benthesicymidae, Aristeidae, Solenoceridae). *Faune Madag.* 46: 1–197.

Dall, W. 2001. Australian species of Aristeidae and Benthesicymidae (Penaeoidea: Decapoda). *Mem. Qld. Mus.* 46: 409–441.

Dallwitz, M.J., Paine, T.A. & Zurcher, E.J. 1993 and onwards. User's guide to the DELTA System: a general system for processing taxonomic descriptions. 4th edition. Available at: http://biodiversity.uno.edu/delta/

Dallwitz, M.J., Paine, T.A. & Zurcher, E.J. 1998. Interactive keys. In: Bridge, P., Jeffries, P., Morse, D.R. & Scott, P.R. (eds.), *Information Technology, Plant Pathology and Biodiversity*: 201–212. Wallingford: CAB International.

Dixon, C.J., Ahyong, S.T. & Schram, F.R. 2003. A new hypothesis of decapod phylogeny. *Crustaceana* 76 (8): 935–975.

Felgenhauer, B.E. & Abele, L.G. 1983. Phylogenetic relationships among shrimp-like decapods. In: *Crustacean Issues 1, Crustacean Phylogeny*: 291–311. Rotterdam: Balkema.

Kikuchi, T. & Nemoto, T. 1991. Deep-sea shrimps of the genus *Benthesicymus* (Decapoda: Dendrobranchiata) from the western north Pacific. *J. Crust. Biol.* 11: 64–89.

Maddison, D.R. & Maddison, W.P. 2001. *MacClade: Analysis of Phylogeny and Character Evolution*, version 4.03. Sunderland, Massachusetts: Sinauer Assoc.

Martin, J.W., Liu, E.M. & Striley, D. 2007. Morphological observations on the gills of dendrobranchiate shrimps. *Zool. Anz.* 246: 115–125.

Nixon, K. & Carpenter, J. 1993. On outgroups. *Cladistics* 9: 413–426.

Pérez Farfante, I. & Kensley, B.F. 1997. Penaeoid and sergestoid shrimps and prawns of the world. Keys and diagnoses for the families and genera. *Mém. Mus. Nat. Hist. Nat.* 175: 1–233.

Quan, J., Zhang, Z., Deng, J., Dai, J. & Zhang, Y. 2004. Phylogenetic relationships of 12 Penaeoidea shrimp species deduced from mitochondrial DNA sequences. *Biochemical Genetics* 42(9–10) 331–345.

Richter, S. & Scholtz, G. 2001. Phylogenetic analysis of the Malacostraca (Crustacea). *J. Zoolog. Syst. & Evo. Res.* 39: 113–136.

Schram, F.R. 1984. Relationships within Eumalacostracan Crustacea. *Trans. San Diego Soc. Nat. Hist.* 20: 301–312.

Swofford, D.L. 2000. *PAUP: Phylogenetic Analysis Using Parsimony*. Version 4.0 Beta. Sunderland, Massachusetts: Sinauer Assoc.

Vázquez-Bader, A.R., Carrero, J.C., Garca-Varela, M., Gracia, A. & Laclette, J.P. 2004. Molecular phylogeny of superfamily Penaeoidea Rafinesque-Schmaltz, 1815, based on mitochondrial 16S partial sequence analysis. *J. Shellfish Res.* 23: 911–917.

Voloch, C.M., Freire, P.R. & Russo, C.A.M. 2005. Molecular phylogeny of penaeid shrimps inferred from two mitochondrial markers. *Genet. Mol. Res.* 4: 668–674.

Wills, M.A. 1997. A phylogeny of recent and fossil Crustacea derived from morphological characters. In: Fortey, R.A. & Thomas, R.H. (eds.), *Arthropod Relationships*: 189–209. London: Chapman & Hall.

# Phylogeny of the Infraorder Caridea Based on Mitochondrial and Nuclear Genes (Crustacea: Decapoda)

## HEATHER D. BRACKEN[1], SAMMY DE GRAVE[2] & DARRYL L. FELDER[3]

[1] *University of Louisiana at Lafayette, Department of Biology, Lafayette, Louisiana, U.S.A.*
[2] *Oxford University Museum of Natural History, Parks Road, Oxford OX1 3PW, United Kingdom*

## ABSTRACT

Shrimps of the infraorder Caridea occur commonly throughout marine and freshwater habitats. Despite general knowledge of the group, phylogenetic relationships within the infraorder remain poorly known. The few studies that have focused specifically on the classification and evolutionary history within the Caridea have relied entirely on morphological characters and suggest conflicting phylogenetic relationships. Robust molecular analysis is required to test current hypotheses. We present the first comprehensive molecular phylogeny of the group, combining nuclear and mitochondrial gene sequences, to evaluate the relationships among 14 superfamilies and 30 families. Bayesian and likelihood analyses were conducted on a concatenated 18S/16S alignment composed of 1835 basepairs. Results indicated no evidence contrary to hypotheses of monophyly within the families Alpheidae, Processidae, and Alvinocarididae. Ogyrididae is resolved as a sister clade to the Alpheidae, as has been previously suggested. Our findings raise questions as to the systematic placement of the Procarididae within Caridea and suggest polyphyletic and paraphyletic relationships among genera within the families Atyidae, Pasiphaeidae, Oplophoridae, Hippolytidae, Gnathophyllidae, and Palaemonidae, as currently defined. Our results in some cases confirm and in others reject placements of controversial taxa within higher-level phylogeny and provide new insights for classifications within the Caridea.

## 1 INTRODUCTION

The range of adaptation and biological diversity within the infraorder Caridea is remarkable among the decapod crustaceans. While many caridean families inhabit marine shallow tropical and subtropical waters, some can be found associated with hydrothermal vents and hydrocarbon seeps, while others occur in freshwater lakes, mountain streams, anchialine caves, and deep-sea basins (Shank et al. 1999; Anker & Iliffe 2000; Komai & Segonzac 2003; Cai & Anker 2004; Martin & Wicksten 2004; Alvarez et al. 2005; Richardson & Cook 2006; Komai et al. 2007; Page et al. 2007; De Grave et al. 2008). With approximately 36 families, 361 genera, and 3,108 species (Fransen & De Grave this volume), carideans dominate the natantian decapods in terms of morphological and ecological diversity (Martin & Davis 2001; Bauer 2004; De Grave & Moosa 2004).

Members of the infraorder Caridea are abundant in epifaunal and fouling communities and contribute to the structure and function of aquatic ecosystems (Richardson & Cook 2006). They commonly establish temporary or lifelong associations with other organisms including cnidarians, sponges, molluscs, echinoderms, echiurans, stomatopods, fish, and other crustaceans (Knowlton 1980; Knowlton & Keller 1983; Pratchett 2001; Duffy 2002; Hayashi 2002; Khan et al. 2003; Silliman et al. 2003; Bauer 2004; Marin et al. 2005; Macdonald et al. 2006). Many aspects of these unique associations make caridean shrimps ideal organisms for studies of symbiosis, communication, behavioral ecology, and evolutionary biology.

## 1.1 *Evolutionary history of the Caridea*

Over the last five decades, several studies have addressed the systematic placement of the infraorder Caridea within the decapods (Burkenroad 1963, 1981; Abele & Felgenhauer 1982; Christoffersen 1988a; Abele 1991; Chace 1992; Porter et al. 2005), but phylogenetic relationships within the infraorder remain poorly known. Few studies have specifically examined the systematic arrangements and evolutionary relationships among superfamilies and families within the Caridea (Holthuis 1955; Thompson 1967; Christoffersen 1986, 1987, 1988b, 1989, 1990; Chace 1992; Holthuis 1993). Although these studies were crucial in contributing to an evolutionary understanding of the group, they relied entirely on morphological characters and resulted in conflicting patterns of phylogeny.

Difficulties in determining relationships among carideans have been attributed to inconsistent and insufficient coding of morphological characters, lack of comparative larval and molecular studies, a limited fossil record (Thompson 1967; Schram 1986; Christoffersen 1990), and a general dearth of phylogenetic work. One study examined evolutionary relationships using 16S data but lacked sufficient taxon sampling (n = 20) and showed little support for the resulting phylogeny (Xu et al. 2005). Some workers have attempted classifications at the superfamilial and familial levels with relative trepidation, all acknowledging that further work is necessary to validate current hypotheses (Holthuis 1955; Thompson 1967; Christoffersen 1990; Chace 1992; Holthuis 1993). Here we acknowledge a few studies that were essential to constructing the currently applied classification of the Caridea (for a further summary of early studies, see Christoffersen 1987).

Early comparative work by Thompson (1967) divided the Caridea into 10 superfamilies and 23 families on the basis of adult morphology. In this account, he suggested a suite of evolutionarily informative characters, such as chelae adaptations, mandible shape, telson armature, and branchial formula, and proposed an updated classification of Caridea. Thompson assumed the group to be a monophyletic unit, and his hypothesized evolutionary tree suggested an early branching of the families Pasiphaeidae, Stylodactylidae, Glyphocrangonidae, and Crangonidae, while postulating that the remaining families arose from an oplophorid-like ancestor. Thompson's diagram included what are now regarded as some unnatural groupings, such as the polyphyly of Heterocarpodoidea, Bresilioidea, and Oplophoroidea, but did provide hypotheses for subsequent testing and called attention to morphological characters later used in cladistic analyses.

During the 1980s and early 1990s, Christoffersen conducted a series of cladistic analyses examining the phylogenetic relationships within the Caridea (Christoffersen 1986, 1987, 1988a, 1988b, 1989, 1990). During the course of his work, he resurrected, revalidated, rejected, restricted, and reassigned many groups to construct a new superfamily and family level classification of the Caridea. In his final contribution, he divided the Caridea into eight superfamilies and 36 families using 19 adult and larval synapomorphies (Christoffersen 1990). Unfortunately, this classification was based on a limited number of characters. Furthermore, the characters for a number of species were scored using available literature only, which even the author conceded to be inadequate and subject to possible misinterpretation. Christoffersen's work was not accepted at the time but is slowly gaining some recognition. He was the first to attempt a true phylogenetic analysis of the group, using cladistic methods and establishing polarities for morphological characters. As did Thompson (1967), he offered a potential explanation for the evolutionary transition from a pelagic to benthic lifestyle, proposing a suite of morphological characters that were derived from this adaptation.

Two years later, a strikingly different classification of the Caridea was presented, which grouped superfamilies and families on the basis of morphological similarity (Chace 1992). Primarily based on the three anterior pairs of pereopods and six pairs of mouthparts, the infraorder was divided into 15 superfamilies and 28 families. It was acknowledged that this arrangement might not necessarily indicate relationships, since superfamilial and familial arrangements were constructed using relative similarity. However, with minor alterations, the currently used caridean classification stems from this work, and it has yet to be challenged by molecular systematists or morphological cladists.

A recently published consensus on classification divided the Caridea into 36 families (Martin & Davis 2001) after a review of varied morphologically based analyses (Holthuis 1955; Thompson 1967; Christoffersen 1986, 1987, 1988a, 1988b, 1989, 1990; Chace 1992; Holthuis 1993), which we follow as our frame of reference, with two minor revisions. It should be noted that since this publication the family Mirocarididae has been synonymized with Alvinocarididae, and a new family, Pseudochelidae, has been described (De Grave & Moosa 2004).

The current subdivision of the infraorder may not reflect phylogenetic relationships, given aforementioned limitations of cladistic morphological analyses and the lack of previous studies examining higher-level caridean relationships on the basis of molecular data. Here, we present the first comprehensive molecular phylogenetic analysis for the infraorder Caridea, combining nuclear and mitochondrial sequences, to investigate relationships among 30 families, 75 genera, and 104 species. It is intended to identify monophyletic and polyphyletic groups and highlight congruence or incongruence between molecular phylogenies and currently applied classifications.

## 2 MATERIALS AND METHODS

### 2.1 *Ingroup taxa and outgroup selection*

Representatives from 30 families, 75 genera, and 104 species of caridean shrimp were used in this analysis. Families containing a greater number of genera and species were sampled more extensively than others. Sequences of the families Galatheacarididae, Bresiliidae, Pseudochelidae, Campylonotidae, Barbouriidae, and Physetocarididae were not available for inclusion in the analyses because material was unattainable. Specimens were collected during cruise and field expeditions or requested on loan from various museums (National Museum of Natural History—Smithsonian Institution, Oxford University Museum of Natural History, Universidad Nacional Autónoma de México). Sequences from 18 of the 104 caridean species used in this study were obtained from GenBank (Table 1). Fresh specimens were either frozen in glycerol at $-80°C$ and later transferred to 80% ethyl alcohol (EtOH) or placed directly into 80% EtOH. Identifications of all materials were confirmed by two or more authors to limit the chance of misidentifications.

Since the identity of the sister group to the Caridea remains debatable, we included 10 outgroup taxa to represent all of the other presently recognized decapod suborders, infraorders, and superfamilies (Penaeoidea, Sergestoidea, Anomura, Brachyura, Stenopodidea, Astacidea, Palinuroidea, and Thalassinidea). Additionally, we included one representative of the order Euphausiacea, putative sister order to the Decapoda within the superorder Eucarida. Sequences representing the putative sister order Amphionidacea were not available for inclusion in the analysis. Sequences for eight of the ten outgroup taxa were obtained from GenBank (Table 1).

### 2.2 *DNA extraction, PCR, and sequencing*

Total genomic DNA was extracted from the abdomen, gills, pereopods, and pleopods under one of three different extraction protocols. Extraction kits included the Genomic DNA Extraction Kit for Arthropods (Cartagen Cat. No. 20810-050) and Qiagen DNeasy® Blood and Tissue Kit (Cat. No. 69504). For some extractions, we used an isopropanol precipitation as follows: Muscle was ground and then incubated for 12h in 600 $\mu$l of lysis buffer (100 mM EDTA, 10 mM tris pH 7.5, 1% SDS) at 65°C; protein was separated by the addition of 200 $\mu$l of 7.5 M ammonium acetate and subsequent centrifugation. DNA was precipitated by the addition of 600 $\mu$l of cold isopropanol followed by overnight refrigeration (4°C) and later centrifugation (10–30 min at 14,000 rpm); the

**Table 1.** Taxonomy, voucher catalog numbers, and GenBank accession numbers for gene sequences used in study. An "N/A" designates gene sequences we were unable to acquire. ULLZ = University of Louisiana at Lafayette Zoological Collection; USNM = National Museum of Natural History, Smithsonian Institute Invertebrate Collection; OUMNH = Oxford University Museum of Natural History, Zoological Collection; CNCR = Colección Nacional de Crustáceos, Universidad Nacional Autónoma de México. Catalog numbers accompanied by asterisk (*) represent cataloged tissue specimens (isolated appendages, gills, eggs, or abdomens) originating from presently uncataloged specimens at OUMNH.

|  |  | GenBank Nos. | |
|---|---|---|---|
| Taxon | Voucher Cat. No. | 16S | 18S |
| **Outgroups** | | | |
| Euphausiacea Dana, 1852 | | | |
| Euphausiidae Dana, 1852 | | | |
| *Euphausia* sp. | ULLZ 8093 | EU868655 | EU868746 |
| Decapoda Latreille, 1802 | | | |
| Dendrobranchiata Bate, 1888 | | | |
| Penaeoidea Rafinesque, 1815 | | | |
| *Penaeus semisulcatus* de Hann, 1844 | GenBank | DQ079731 | DQ079766 |
| Sergestoidea Dana, 1852 | | | |
| *Sergia* sp. | ULLZ 8089 | EU868710 | EU868807 |
| Pleocyemata Burkenroad, 1963 | | | |
| Brachyura Latreille, 1802 | | | |
| *Dromia dehaani* Rathbun, 1923 | GenBank | AY583899 | AY583972 |
| Stenopodidea Claus, 1872 | | | |
| *Stenopus hispidus* (Olivier, 1811) | GenBank | AY583884 | AY743957 |
| Astacidea Latreille, 1802 | | | |
| *Enoplometopus occidentalis* (Randall, 1840) | GenBank | AY583892 | AY583966 |
| *Procambarus clarkii* (Girard, 1852) | GenBank | DQ666844 | AF436001 |
| Anomura MacLeay, 1838 | | | |
| *Pagurus longicarpus* Say, 1817 | GenBank | NC_003058 | AF436018 |
| Achelata Scholtz & Richter, 1995 | | | |
| *Panulirus argus* (Latreille, 1804) | GenBank | AF337966 | AY743955 |
| Thalassinidea Latreille, 1831 | | | |
| *Upogebia affinis* (Say, 1818) | GenBank | AF436047 | AF436007 |

Table 1. continued.

| Taxon | Voucher Cat. No. | GenBank Nos. 16S | 18S |
|---|---|---|---|
| **Ingroups** | | | |
| Decapoda Latreille, 1802 | | | |
| Pleocyemata Burkenroad, 1963 | | | |
| Caridea Dana, 1852 | | | |
| Alpheoidea Rafinesque, 1815 | | | |
| Alpheidae Rafinesque, 1815 | | | |
| *Alpheopsis trigonus* (Rathbun, 1901) | ULLZ 7283 | EU868633 | EU868723 |
| *Alpheus packardii* Kingsley, 1880 | ULLZ 7248 | EU868630 | EU868720 |
| *Alpheus vanderbilti* Boone, 1930 | ULLZ 7461 | EU868639 | EU868730 |
| *Automate rectifrons* Chace, 1972 | ULLZ 7303 | EU868631 | EU868721 |
| *Automate* sp. | ULLZ 7754 | EU868635 | EU868725 |
| *Betaeus* sp. | CNCR 16850 | N/A | EU868726 |
| *Coronalpheus natator* Wicksten, 1999 | ULLZ 8938 | EU868636 | EU868727 |
| *Coutieralpheus* sp. | ULLZ 8939 | EU868637 | EU868728 |
| *Fenneralpheus chacei* Felder & Manning, 1986 | ULLZ 4559 | EU868638 | EU868729 |
| *Leptalpheus forceps* Williams, 1965 | ULLZ 5594 | EU868670 | EU868763 |
| *Leptalpheus axianassae* Dworschak & Coelho, 1999 | ULLZ 5913 | EU868671 | EU868764 |
| *Synalpheus bousfieldi* (Chace, 1972) | ULLZ 7137 | EU868646 | EU868737 |
| *Synalpheus fritzmuelleri* Coutière, 1909 | ULLZ 7136 | EU868642 | EU868733 |
| *Synalpheus hemphilli* Coutière, 1909 | ULLZ 7147 | EU868643 | EU868734 |
| *Synalpheus pandionis* (Coutière, 1909) | ULLZ 7241 | EU868647 | EU868738 |
| *Yagerocaris cozumel* Kensley, 1988 | ULLZ 8883 | EU868645 | EU868736 |
| Hippolytidae Dana, 1852 | | | |
| *Hippolyte varians* Leach, 1814 | ULLZ 6970 | EU868662 | EU868753 |
| *Hippolyte obliquimanus* Dana, 1852 | ULLZ 9137 | EU868661 | EU868752 |
| *Hippolyte pleuracanthus* (Stimpson, 1871) | GenBank | N/A | AY743956 |
| *Latreutes fucorum* (Fabricius, 1798) | ULLZ 9135 | EU868664 | EU868755 |
| *Lysmata* cf. *wurdemanni* | ULLZ 7433 | EU868666 | EU868757 |

Table 1. continued.

| Taxon | Voucher Cat. No. | GenBank Nos. 16S | GenBank Nos. 18S |
|---|---|---|---|
| *Lysmata* sp. | ULLZ 8931 | EU868665 | EU868756 |
| *Lysmata boggessi* Rhyne & Lin, 2006 | GenBank | DQ079719 | DQ079753 |
| *Lysmata debelius* (Bruce, 1983) | GenBank | DQ079718 | DQ079752 |
| *Thoralus cranchii* (Leach, 1817) | ULLZ 6969 | EU868667 | EU868758 |
| *Tozeuma* cf. *carolinense* | ULLZ 7445 | EU868669 | EU868760 |
| *Tozeuma serratum* A. Milne-Edwards, 1881 | ULLZ 7446 | EU868668 | EU868759 |
| *Trachycaris rugosa* (Bate, 1888) | ULLZ 7425 | N/A | EU868761 |
| *Trachycaris* sp. | ULLZ 7749 | N/A | EU868762 |
| Ogyrididae Holthuis, 1955 | | | |
| *Ogyrides* sp. | ULLZ 7755 | EU868679 | EU868772 |
| *Ogyrides* sp. | ULLZ 7756 | EU868680 | EU868773 |
| Atyoidea de Haan, 1849 | | | |
| Atyidae de Haan, 1849 | | | |
| *Antecaridina* sp. | | EF173754 | EF173850 |
| *Atya scabra* Leach, 1815 | CNCR 17094 | EU868632 | EU868722 |
| *Atyoida bisulcata* (Randall, 1840) | GenBank | DQ079704 | DQ079738 |
| *Atyopsis* sp. | ULLZ 9174 | EU868634 | EU868724 |
| *Halocaridina rubra* Holthuis, 1963 | GenBank | EF173749 | EF173848 |
| *Halocaridinides trigonophthalma* (Fujino & Shokita, 1975) | GenBank | EF173752 | EF173849 |
| *Paratya australiensis* Kemp, 1917 | USNM 1073432 | EU868640 | EU868731 |
| *Potimirim mexicana* (De Saussure, 1857) | CNCR 17140 | EU868641 | EU868732 |
| *Typhlatya mitchelli* Hobbs & Hobbs, 1976 | CNCR 22696 | EU868644 | EU868735 |
| *Typhlatya pearsei* Creaser, 1936 | GenBank | DQ079735 | DQ079770 |
| Bresilioidea Calman, 1896 | | | |
| Agostocarididae Hart & Manning, 1986 | | | |
| *Agostocaris* sp. | USNM 1014071 | EU868626 | EU868716 |
| Alvinocarididae Christoffersen, 1986 | | | |
| *Alvinocaris muricola* Williams, 1988 | CNCR 24875 | EU868627 | EU868717 |

Table 1. continued.

| Taxon | Voucher Cat. No. | GenBank Nos. 16S | GenBank Nos. 18S |
|---|---|---|---|
| *Alvinocaris muricola* Williams, 1988 | CNCR 24873 | EU868628 | EU868718 |
| *Chorocaris chacei* (Williams & Rona, 1986) | GenBank | AM087922 | AM087653 |
| *Rimicaris exoculata* (Williams & Rona, 1986) | GenBank | AM076958 | AM087652 |
| Disciadidae Rathbun, 1902 | | | |
| *Discias atlanticus* Gurney, 1939 | ULLZ 8953 | EU868652 | EU868743 |
| Campylonotoidea Sollaud, 1913 | | | |
| Bathypalaemonellidae de Saint Laurent, 1985 | | | |
| *Bathypalaemonella* sp. | ULLZ 8929* | EU868648 | EU868739 |
| Crangonoidea Haworth, 1825 | | | |
| Crangonidae Haworth, 1825 | | | |
| *Crangon crangon* (Linnaeus, 1758) | ULLZ 6967 | EU868649 | EU868740 |
| *Crangon franciscorum* Stimpson, 1856 | GenBank | N/A | AY859567 |
| *Pontophilus gracilis* Smith, 1882 | ULLZ 8287 | EU868650 | EU868741 |
| Glyphocrangonidae Smith, 1884 | | | |
| *Glyphocrangon alispina* Chace, 1939 | ULLZ 7878 | EU868656 | EU868747 |
| *Glyphocrangon alispina* Chace, 1939 | ULLZ 8084 | EU868657 | EU868748 |
| Nematocarcinoidea Smith, 1884 | | | |
| Eugonatonotidae Chace, 1937 | | | |
| *Eugonatonotus chacei* Chan & Yu, 1991 | ULLZ 8880* | EU868653 | EU868744 |
| Nematocarcinidae Smith, 1884 | | | |
| *Nematocarcinus cursor* A. Milne-Edwards, 1881 | ULLZ 8044 | EU868673 | EU868766 |
| *Nematocarcinus rotundus* Crosnier & Forrest, 1973 | ULLZ 7736 | EU868672 | EU868765 |
| *Nematocarcinus rotundus* Crosnier & Forrest, 1973 | ULLZ 7736 | EU868674 | EU868767 |
| Rhynchocinetidae Ortmann, 1890 | | | |
| *Cinetorhynchus manningi* Okuno, 1996 | ULLZ 7414 | N/A | EU868805 |
| Xiphocarididae Ortmann, 1895 | | | |
| *Xiphocaris elongata* (Guérin-Méneville, 1856) | ULLZ 8882* | EU868714 | EU868809 |
| Oplophoroidea Dana, 1852 | | | |

Table 1. continued.

| Taxon | Voucher Cat. No. | GenBank Nos. 16S | 18S |
|---|---|---|---|
| Oplophoridae Dana, 1852 | | | |
| *Acanthephyra* sp. | ULLZ 8026 | EU868675 | EU868768 |
| *Acanthephyra curtirostris* Wood-Mason, 1891 | ULLZ 6702 | EU868676 | EU868769 |
| *Acanthephyra purpurea* A. Milne-Edwards, 1881 | ULLZ 7579 | EU868677 | EU868770 |
| *Ephyrina figueirai* Crosnier and Forest, 1973 | GenBank | AM076960 | AM087654 |
| *Meningodora* sp. | ULLZ 7738 | EU868678 | EU868771 |
| *Systellaspis debilis* (A. Milne-Edwards, 1881) | ULLZ 7854 | EU868682 | EU868775 |
| *Systellaspis debilis* (A. Milne-Edwards, 1881) | ULLZ 6713 | EU868678 | EU868771 |
| Palaemonoidea Rafinesque, 1815 | | | |
| Anchistioididae Borradaile, 1915 | | | |
| *Anchistiodes antiguensis* (Schmitt, 1924) | ULLZ 7454 | EU868629 | EU868719 |
| Desmocarididae Borradaile, 1915 | | | |
| *Desmocaris* sp. | ULLZ 8358 | EU868651 | EU868742 |
| Euryrhynchidae Holthuis, 1950 | | | |
| *Euryrhynchus wrzesniowskii* Miers, 1878 | ULLZ 9070 | EU868654 | EU868745 |
| Gnathophyllidae Dana, 1852 | | | |
| *Gnathophylloides mineri* Schmitt, 1933 | ULLZ 8596 | EU868658 | EU868749 |
| *Gnathophylloides mineri* Schmitt, 1933 | ULLZ 8932 | EU868659 | EU868750 |
| *Gnathophyllum americanum* Guérin-Méneville, 1855 | ULLZ 8597 | EU868660 | EU868751 |
| Hymenoceridae Ortmann, 1890 | | | |
| *Hymenocera picta* Dana, 1852 | ULLZ 8595 | EU868663 | EU868754 |
| Kakaducarididae Bruce, 1993 | | | |
| *Leptopalaemon gagadjui* Bruce & Short, 1993 | ULLZ 9120 | EU868693 | EU868787 |
| Palaemonidae Rafinesque, 1815 | | | |
| *Brachycarpus biunguiculatus* (Lucas, 1846) | ULLZ 7382 | EU868685 | EU868778 |
| *Brachycarpus biunguiculatus* (Lucas, 1846) | ULLZ 7430 | EU868686 | EU868779 |
| *Brachycarpus biunguiculatus* (Lucas, 1846) | ULLZ 7426 | EU868684 | EU868777 |
| *Coralliocaris graminea* (Dana, 1852) | GenBank | N/A | AM083319 |

Table 1. continued.

| Taxon | Voucher Cat. No. | GenBank Nos. | |
|---|---|---|---|
| | | 16S | 18S |
| *Creaseria morleyi* (Creaser, 1936) | CNCR 22720 | EU868687 | EU868780 |
| *Creaseria morleyi* (Creaser, 1936) | CNCR 22732 | EU868688 | EU868781 |
| *Cryphiops caementarius* (Molina, 1782) | GenBank | DQ079711 | DQ079747 |
| *Kemponia americana* (Kingsley, 1878) | ULLZ 7431 | EU868701 | EU868795 |
| *Leander tenuicornis* (Say, 1818) | ULLZ 7765 | EU868690 | EU868783 |
| *Macrobrachium ohione* (Smith, 1874) | ULLZ 8715 | EU868694 | EU868788 |
| *Macrobrachium potiuna* (Müller, 1880) | GenBank | DQ079721 | DQ079756 |
| *Palaemon elegans* Rathke, 1837 | ULLZ 6968 | EU868696 | EU868790 |
| *Palaemonetes pugio* Holthuis, 1949 | ULLZ 7458 | EU868697 | EU868791 |
| *Palaemonetes vulgaris* (Say, 1818) | GenBank | N/A | AY743941 |
| *Periclimenaeus wilsoni* (Hay, 1917) | ULLZ 7384 | EU868702 | EU868797 |
| *Periclimenes pedersoni* Chace, 1958 | GenBank | N/A | AY743954 |
| *Pontonia* sp. | ULLZ 8886 | EU868706 | EU868801 |
| *Pontonia manningi* Fransen, 2000 | ULLZ 8536 | EU868705 | EU868800 |
| Typhlocarididae Annandale & Kemp, 1913 | | | |
| *Typhlocaris salentina* Caroli, 1924 | ULLZ 9152* | EU868713 | EU868808 |
| Pandaloidea Haworth, 1825 | | | |
| Pandalidae Haworth, 1825 | | | |
| *Heterocarpus ensifer* A. Milne-Edwards, 1881 | ULLZ 8362 | EU868689 | EU868782 |
| *Heterocarpus ensifer* A. Milne-Edwards, 1881 | GenBank | AMO76962 | AMO83320 |
| *Pandalus montagui* Leach, 1814 | ULLZ 6966 | EU868698 | EU868792 |
| *Parapandalus richardi* (Coutière, 1905) | ULLZ 6706 | N/A | EU868793 |
| *Plesionika holthuisi* Crosnier & Forrest, 1968 | ULLZ 7953 | EU868703 | EU868798 |
| *Plesionika longipes* (A. Milne-Edwards, 1881) | ULLZ 8363 | EU868704 | EU868799 |
| Thalassocarididae Bate, 1888 | | | |
| *Thalassocaris crinita* (Dana, 1852) | ULLZ 8359 | EU868712 | EU868810 |
| Pasiphaeoidea Dana, 1852 | | | |
| Pasiphaeidae Dana, 1852 | | | |

Table 1. continued.

| Taxon | Voucher Cat. No. | GenBank Nos. | |
|---|---|---|---|
| | | 16S | 18S |
| *Leptochela carinata* Ortmann, 1893 | ULLZ 7232 | EU868692 | EU868786 |
| *Leptochela bermudensis* (Gurney, 1939) | ULLZ 7888 | EU868691 | EU868785 |
| *Leptochela papulata* Chace, 1976 | ULLZ 8614 | N/A | EU868784 |
| *Pasiphaea merriami* Schmitt, 1931 | ULLZ 6703 | EU868700 | EU868796 |
| *Pasiphaea merriami* Schmitt, 1931 | ULLZ 8088 | EU868699 | EU868794 |
| Procaridoidea Chace & Manning, 1972 | | | |
| Procarididae Chace & Manning, 1972 | | | |
| *Procaris mexicana* Sternberg & Schotte, 2004 | ULLZ 9224 | EU868715 | EU868811 |
| Processoidea Ortmann, 1890 | | | |
| Processidae Ortmann, 1890 | | | |
| *Ambidexter symmetricus* Manning & Chace, 1971 | ULLZ 6432 | EU868683 | EU868776 |
| *Nikoides schmitti* Manning & Chace, 1971 | ULLZ 7441 | EU868695 | EU868789 |
| *Processa guyanae* Holthuis, 1959 | ULLZ 7378 | EU868707 | EU868802 |
| *Processa guyanae* Holthuis, 1959 | ULLZ 7150 | EU868708 | EU868803 |
| Psalidopodoidea Wood Mason & Alcock, 1892 | | | |
| Psalidopodidae Wood Mason & Alcock, 1892 | | | |
| *Psalidopus barbouri* Chace, 1939 | ULLZ 7805 | EU868709 | EU868804 |
| Stylodactyloidea Bate, 1888 | | | |
| Stylodactylidae Bate, 1888 | | | |
| *Stylodactylus multidentatus* Kubo, 1942 | ULLZ 8881* | EU868711 | EU868806 |
| *Stylodactylus libratus* Chace, 1983 | GenBank | AM076943 | AM083323 |

resulting pellet was rinsed in 70% EtOH, dried in a speed vacuum system (DNA110 Speed Vac®), and resuspended in 10–50 µl of nanopure water (Robles et al. 2007).

One mitochondrial gene and one nuclear gene were selected due to their utility in resolving phylogenetic relationships at different taxonomic levels (Spears et al. 1992; Spears et al. 1994; Giribet et al. 1996; Schubart et al. 2000; Stillman & Reeb 2001; Tudge & Cunningham 2002; Porter et al. 2005; Mantelatto et al. 2006; Mantelatto et al. 2007; Robles et al. 2007). The 16S large ribosomal subunit (~550 bps) was selected as our mitochondrial gene, and the complete 18S, large ribosomal subunit (~1850 bps) was selected as the nuclear gene. Targeted sequences were amplified by means of the polymerase chain reaction (PCR). The mitochondrial gene, 16S, was amplified with the primers 16SL2, 16S-ar, and 1472 to create one overlapping region of approximately 550 basepairs in length (Palumbi et al. 1991; Crandall & Fitzpatrick 1996; Schubart et al. 2002). The nuclear gene, 18S, was amplified with the primers A–L, C–Y, and O–B to yield three overlapping regions of approximately 600–700 basepairs in length each (Medlin et al. 1988; Apakupakul et al. 1999). Additionally, slightly shorter internal 18S primers (B–D18s1R, D18s2F–D18s2R, D18s3F–D18s3R, D18s4F–D18s4R, and D18s5F–A) were designed to yield five overlapping regions ranging from approximately 450–600 basepairs in length each (all primers listed in Table 2).

Reactions were performed in 25 µl volumes containing 0.5 µM forward and reverse primer for each gene, 200 µM each dNTP, PCR buffer, magnesium chloride, 5 M betaine, 1 unit AmpliTaq-GOLD® polymerase, and 30–50 ng extracted DNA. The thermal cycling profile conformed to the following parameters: initial denaturation for 10 min at 94°C followed by 40 cycles of 1 min at 94°C, 1.5 min at 46–58°C, 1.5 min at 72°C, and a final extension of 10 min at 72°C. PCR products were purified using filters (Microcon-100® Millipore Corp., Billerica, MA, USA or EPOCH GenCatch PCR Clean-up Kit Cat. No. 13-60250) and sequenced with ABI BigDye® terminator mix (Applied Biosystems, Foster City, CA, USA). A Robocycler 96 cycler was used in all PCR and cycle sequencing reactions and sequencing products were run (forward and reverse) on a 3100 Applied Biosystems automated sequencer.

**Table 2.** 16S and 18S primers used in this study.

| Gene | Primer | Primer Pair | Sequence 5' → 3' | Ref. |
| --- | --- | --- | --- | --- |
| 16S | 16S-ar | 1472 | CGC CTG TTT ATC AAA AAC AT | (1) |
| 16S | 16S-L2 | 1472 | TGC CTG TTT ATC AAA AAC AT | (2) |
| 16S | 1472 | 16S-ar/16S-L2 | AGA TAG AAA CCA ACC TGG | (3) |
| 18S | 18S-A | 18S-L | AAC CTG GTT GAT CCT GCC AGT | (4) |
| 18S | 18S-L | 18S-A | CCA ACT ACG AGC TTT TTA ACT G | (5) |
| 18S | 18S-C | 18S-Y | CGG TAA TTC CAG CTC CAA TAG | (5) |
| 18S | 18S-Y | 18S-C | CAG ACA AAT CGC TCC ACC AAC | (5) |
| 18S | 18S-O | 18S-B | AAG GGC ACC ACC AGG AGT GGA G | (5) |
| 18S | 18S-B | 18S-O | TGA TCC TTC CGC AGG TTC ACC T | (4) |
| 18S | D18s1R | 18S-B | CTT AAT TCC GAT AAC GAA CGA GAC TCT G | New |
| 18S | D18s2F | D18s2R | TCT AAG GGC ATC ACA GAC CTG | New |
| 18S | D18s2R | D18s2F | AGA TAC CGC CCT AGT TCT AAC C | New |
| 18S | D18s3F | D18s3R | GGT TAG AAC TAG GGC GGT ATC | New |
| 18S | D18s3R | D18s3F | TGG AGG GCA AGT CTG GTG | New |
| 18S | D18s4F | D18s4R | GCA ACA AAC TTT AAT ATA CG | New |
| 18S | D18s4R | D18s4F | TGG TAA TTC TAG AGC TAA TAC | New |
| 18S | D18s5F | 18S-A | GTT ATT TTT CGT CAC TAC CTC CC | New |

*References:* (1) Palumbi et al. 1991, (2) Schubart et al. 2002, (3) Crandall & Fitzpatrick 1996, (4) Medlin et al. 1988, (5) Apakupakul et al. 1999.

## 2.3 Phylogenetic analyses

Sequences were assembled using the computer program Sequencher 4.7 (GeneCodes, Ann Arbor, MI, USA). Once assembled, sequences were aligned using MUSCLE (multiple sequence comparison by log-expectation), a computer program found to be more accurate and faster than other alignment algorithms (Edgar 2004). Since many regions within the 16S and 18S datasets were extremely divergent and difficult to align, we used GBlocks v0.91b (Castresana 2000) to omit poorly aligned positions (GBlocks parameters optimized for dataset and modeled after previous studies (Porter et al. 2005): minimum number of sequences for a conserved position = 62/57; minimum number of sequences for a flanking position = 104/95; maximum number of contiguous non-conserved positions = 8/8; minimum length of a block = 6/6; allowed gap positions = half/half). GBlocks pruned approximately 400 and 170 basepairs from the 18S and 16S alignments, resulting in two datasets composed of 1458 and 377 characters, respectively. Recent studies have shown an increase in phylogenetic resolution when multiple genes are combined in phylogenetic analyses. These approaches have gained popularity over single gene studies because of their potential to resolve phylogenies at different taxonomic levels (Ahyong & O'Meally 2004; Porter et al. 2005). For these reasons, we concatenated our 18S and 16S datasets into a single alignment consisting of 1835 basepairs and 122 sequences. We conducted a partition test of heterogeneity (incongruence length difference test (ILD)) (Bull et al. 1993), as implemented in PAUP* (Swofford 2003), and results indicated that the two gene regions could be combined. Before concatenation, we generated single gene trees (16S and 18S). Although we observed similar patterns of phylogeny, the 18S tree showed better resolution at the deeper nodes, while the 16S tree showed higher resolution between species.

The model of evolution that best fit the individual datasets (18S, 16S) was determined by MODELTEST 3.06 (Posada & Crandall 1998) before conducting maximum likelihood (ML) and Bayesian Inference (BAY) analyses. The ML analysis was conducted using RAxML (Randomized Axelerated Maximum Likelihood) (Stamatakis et al. 2005) with computations performed on the computer cluster of the Cyberinfrastructure for Phylogenetic Research Project (CIPRES) at the San Diego Supercomputer Center. The BAY analysis was conducted in MrBayes v3.0b4 (Huelsenbeck & Ronquist 2001). Each analysis was run three times to evaluate the consistency among runs.

Likelihood settings followed the General Time Reversible Model (GTR) with a gamma distribution and invariable sites and RAxML estimated all free parameters following a partitioned dataset. Confidence in the resulting topology was assessed using non-parametric bootstrap estimates (Felsenstein 1985) with 1000 replicates. Values > 50% are presented on the BAY phylogram (Fig. 1). The BAY analysis was performed using parameters selected by MODELTEST. A Markov chain Monte Carlo (MCMC) algorithm ran for 2,000,000 generations, sampling one tree every 100 generations. Preliminary analyses and observation of the log likelihood ($L$) values allowed us to determine burn-ins and stationary distributions for the data. Once the values reached a plateau, a 50% majority rule consensus tree was obtained from the remaining saved trees. Clade support was assessed with posterior probabilities (p$P$), and values > 0.5 are presented on the BAY phylogram (Fig. 1). Trees were initially generated as unrooted phylograms to help designate outgroup taxa. Ten taxa showed a clear separation from the Caridea and were selected as outgroups (Table 1).

---

**Figure 1.** (Opposite Page) Bayesian (BAY) phylogram for the infraorder Caridea (n = 112) and selected outgroups (n = 10) based on 18S (rDNA) and 16S (rDNA) concatenated dataset. ML bootstrap values and BAY posterior probabilities are noted above branches (ML/BAY). Values < 50% are not shown. Vertical black bars indicate 8 major clades within the Caridea. Clades I–IV and VIII represent multiple families and Clades V–VII represent a single family or genus. * = node for each clade.

Phylogeny of the Infraorder Caridea 293

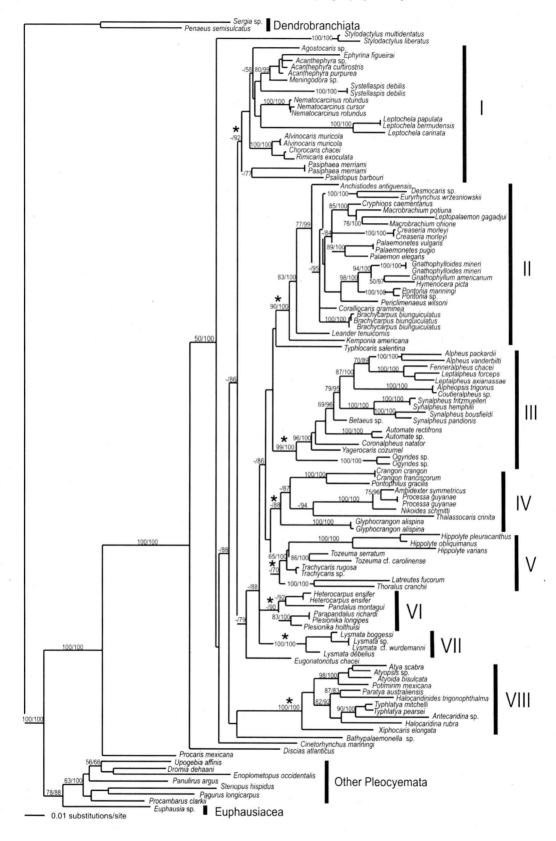

## 3 RESULTS

Our study included representatives from 14 of the 16 superfamilies and 30 of the 36 families presently encompassed in the infraorder Caridea. In total, we generated 87 new complete 18S (~1850 bps), 7 new partial 18S (~700-1450 bps), and 88 new partial 16S sequences (~550 bps) (Table 1). Missing data were designated as a "?" for partial sequences. The ILD test showed no significant incongruence ($P = 0.65$) between datasets, so the 18S and 16S alignments were combined. After the 18S and 16S alignments were run through GBlocks, they were concatenated; of the 1835 basepairs for 122 sequences used in the phylogenetic analyses, 1458 were for 18S and 377 for 16S gene sequences. The optimal model of evolution selected in MODELTEST for the individual datasets was the General Time Reversible (GTR) model (18S) with gamma-distributed among-site rate heterogeneity and invariant sites (base frequencies = 0.2639, 0.2217, 0.2725, 0.2419; Rmat = 1.4462, 2.6478, 1.2472, 1.1228, 4.5836; gamma shape parameter = 0.4927; proportion of invariable sites = 0.3884) and the Transition (TIM) model (16S) with gamma-distributed among-site rate heterogeneity and invariant sites (base frequencies = 0.3833, 0.1700, 0.0553, 0.3914; Rmat = 1.0000, 8.9199, 0.7503, 0.7503, 4.2441; gamma shape parameter = 0.4938; proportion of invariable sites = 0.2420). ML and BAY analyses showed similar tree topologies, but because the ML phylogeny was less resolved at deeper nodes, the BAY tree is presented (Figs. 1, 2).

### 3.1 Monophyly, paraphyly, and polyphyly of the infraorder Caridea

Our results can be interpreted to support monophyly of the infraorder Caridea as presently constituted, but at the same time they offer support for treatment of the family Procarididae as a separate infraorder (Fig. 1). While the basally positioned procaridids grouped more closely to carideans than to any other represented infraorder of pleocyemates, branch length between the procaridids and carideans was comparable to branch lengths between different infraorders of outgroup taxa, rather than those between other families of carideans. Furthermore, in unrooted trees (not shown here) the procaridids were positioned as a distinct lineage, separated from the remaining carideans.

There was no overwhelming support for the monophyly of the currently proposed superfamilies (those containing > 1 family). However, our analyses strongly suggested (bootstrap values > 0.9, $pP = 1.0$) three major multi-familial clades within the infraorder Caridea (Clades II, III, VIII, Figs. 1, 2). Additionally, there was weaker support ($pP \geq 0.88$) for the formation of two additional assemblages composed of two or more families (Clades I, IV, Fig. 1). Our analysis provides some evidence for a relationship between the families Agostocarididae, Oplophoridae, Nematocarcinidae, Pasiphaeidae, Psalidopodidae, and Alvinocarididae (Clade I, $pP = 0.92$). There is significant support for Clade II, which includes all families within Palaemonoidea, excluding Typhlocarididae, and there is no support for the inclusion of the typhlocaridids within the Palaemonoidea, as presently classified. The Ogyrididae is resolved as a sister clade to the Alpheidae (Clade III), and Atyidae + Xiphocarididae (Clade VIII) form a monophyletic assemblage with high support. Clade IV, uniting Crangonidae, Processidae, Thalassocarididae, and Glyphocrangonidae, has low support ($pP = .88$), but the subclade grouping Processidae and Thalassocarididae is marginally significantly supported with posterior probabilities ($pP = 0.94$). The remaining clades (V–VII) represent single families; two are weakly supported (Clade V: $pP = 0.70$, Clade VI: $pP = 0.90$) and one is strongly supported (Clade VII: bootstrap values = 1.0, $pP = 1.0$). The Hippolytidae, as currently defined, is split between clades V and VII, and Clade VI is limited to the Pandalidae.

Although superfamilial support is missing or low, our analyses suggest that many families form monophyletic units. Approximately 8 of 16 proposed superfamilies within the Caridea each contain a single family. Our present observations are limited to those families that have multiple genera represented in our tree, and thus we cannot comment on the monophyly of families represented by a single genus (i.e., Stylodactylidae, Rhynchocinetidae, Bathypalaemonellidae, Agostocarididae,

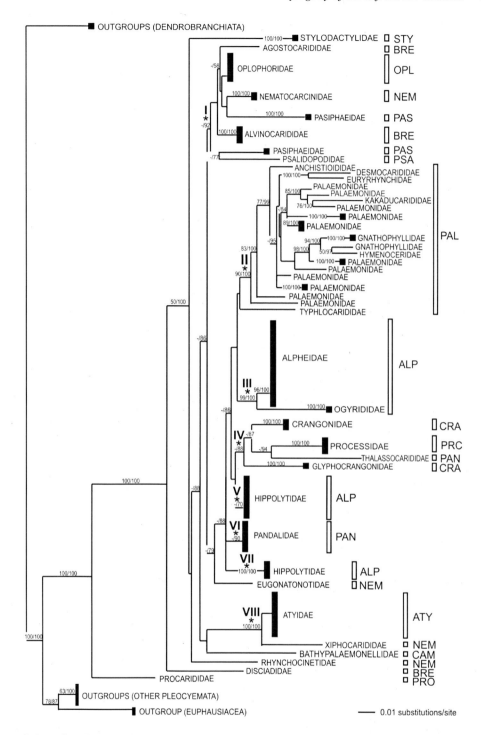

**Figure 2.** Bayesian (BAY) phylogram for the infraorder Caridea and selected outgroups based on 18S (rDNA) and 16S (rDNA) concatenated dataset. ML bootstrap values and BAY posterior probabilities are noted above the branches (ML/BAY). Values < 50% are not shown. For ease of interpretation, branches are collapsed to show caridean families (solid bars), superfamilies (open bars), and outgroup taxa (solid bars). I–VIII indicate the 8 major clades within the Caridea. * = node for each clade. STY = Stylodactyloidea, NEM = Nematocarcinoidea, CAM = Campylonotoidea, BRE = Bresilioidea, OPL = Oplophoroidea, PAS = Pasiphaeoidea, PSA = Psalidopodoidea, PAL = Palaemonoidea, ALP = Alpheoidea, CRA = Crangonoidea, PRC = Processoidea, PAN = Pandaloidea, ATY = Atyoidea, PRO = Procaridoidea.

Nematocarcinidae, Psalidopodidae, Anchistioididae, Hymenoceridae, Desmocarididae, Kakaducarididae, Euryrhynchidae, Typhlocarididae, Ogyrididae, Thalassocarididae, Eugonatonotidae, Disciadidae, Procarididae, and Glyphocrangonidae). Results are congruent with hypotheses of monophyly within the families Alvinocarididae, Alpheidae, Crangonidae, and Processidae. The monophyly of the Pandalidae is only marginally supported with posterior probabilities. Our findings suggest polyphyletic relationships among genera within the families Pasiphaeidae, Oplophoridae, Hippolytidae, and Palaemonidae (both Palaemoninae and Pontoniinae) and paraphyletic relationships within Gnathophyllidae and Atyidae (Figs. 1, 2).

Systematic placement of Typhlocarididae and Eugonatonotidae is unclear considering there is little support for their position in relation to other families within the tree's topology. The families Procarididae, Disciadidae, Rhynchocinetidae, Stylodactylidae, Bathypalaemonellidae, Atyidae, and Xiphocarididae represent basal (less derived) lineages, which we address in the discussion.

## 4 DISCUSSION

Aside from the phylogenetic discussions that follow, it does not escape our attention that euphausiaceans are positioned as a sister clade to the non-caridean pleocyemate outgroups included in the analysis. This is not entirely unexpected, because we did not enforce rooting to only the Euphausiacea as in a previous analysis by colleagues (Porter et al. 2005). While it is not our primary interest to resolve phylogenetic positioning of this group, it is noteworthy that other recent molecular studies have also yielded enigmatic placements for this putative sister group of the decapods. While sometimes at low support values, positioning in trees based on protein-coding genes can place euphausiaceans as an immediate sister group to the decapods or outside the eucarids altogether as a sister group to stomatopods (Podsiadlowski & Bartolomaeus 2006). Somewhat controversially, euphausiaceans, on the basis of 28S rDNA sequences, have been allied more closely to the mysidaceans than to dendrobranchiate decapods, but no pleocyemate decapods were included in that analysis (Jarman et al. 2000). Recent ontogenetic studies do not support a closer phylogenetic relationship to mysids than to dendrobranchiate decapods (Casanova et al. 2002).

### 4.1 *Procaridoidea + Caridea clade?*

Ever since the discovery of the anchialine shrimp *Procaris ascensionis* Chace & Manning, 1972, there has been a debate as to its systematic position in relationship to other shrimp-like decapods. Initially, procaridids were placed within their own family (Procarididae) and superfamily (Procaridoidea) within the infraorder Caridea (Chace & Manning 1972). Over the years, many studies have retained procaridids within the carideans (Chace & Manning 1972; Holthuis 1973; Abele & Felgenhauer 1986; Kensley & Williams 1986; Kim & Abele 1990). Kensley & Williams (1986) described a new genus and species of procaridid shrimp, *Vetericaris chaceorum*, and based on a suite of morphological characters agreed with the phylogenetic placement proposed by Chace & Manning some years earlier. Moreover, a phenetic and cladistic analysis suggested the procaridids be placed within the carideans on the basis of a single shared morphological character, the 2nd abdominal pleura overlapping the 1st and 3rd somites without the 1st being reduced (Abele & Felgenhauer 1986). In 1988, Felgenhauer & Abele discovered that *Procaris ascensionis* carried its eggs attached to the pleopods and secured the group's placement within the Pleocyemata. Molecular evidence presented by Kim & Abele (1990) again suggested a close affinity between the carideans and procaridids. However, this study lacked robust representation of caridean groups (n = 2), mandating a more thorough molecular investigation. While many studies position procaridids basally within the Caridea, there is some morphological evidence for the separation of the two groups (Felgenhauer & Abele 1983, 1985, 1989; Schram 1986). In foregut morphology, procaridids appear to be more like dendrobranchiates than carideans (Felgenhauer & Abele 1983, 1985, 1989), and after review of several morphological characters (e.g., gills, protocephalic, and foregut) Felgenhauer & Abele (1983) concluded that the procaridids be elevated to infraordinal level. Other characters potentially supporting separation of procaridids and carideans include distinct cephalic and thoracopodal anatomy (Fransen & De Grave this volume; Schram 1986).

Present results strongly separate (long branch length) procaridid shrimp basally as a sister group to all other putative carideans. The group is separated, along with carideans *sensu stricto*, from all other pleocyemate infraorders. This could be interpreted as support for treatment of the Procaridoidea at the infraordinal level within the Pleocyemata, especially if substantiated by analysis of additional genes and a more robust representation of pleocyemate taxa.

## 4.2 Superfamily Palaemonoidea

The superfamily Palaemonoidea is an extremely diverse group, currently composed of eight families, including Anchistioididae, Gnathophyllidae, Hymenoceridae, Palaemonidae, Desmocarididae, Kakaducarididae, Euryrhynchidae, and Typhlocarididae. Representatives from all the aforementioned families are presented in our analysis, and, with the exclusion of Typhlocarididae, Palaemonoidea is strongly supported.

Throughout the years, the systematic position of the freshwater troglobitic family, Typhlocarididae, has been controversial. Until recently, the typhlocaridids were thought to be close relatives of the euryrhynchids on the basis of overall mouthpart similarity (Chace 1992, 1993; Holthuis 1993). However, a recent review of morphological characters identifies a suite of fundamental differences between the two families and confirms that similarity in mouthpart structure is shared amongst many genera within Palaemonidae (De Grave 2007). Our analyses reject a close relationship between Euryrhynchidae and Typhlocarididae and question the systematic position of Typhlocarididae within Palaemonoidea, as defined by Chace (1992). Instead, our results strongly suggest Desmocarididae as the sister clade to Euryrhynchidae. Both families inhabit freshwater in South America (Euryrhynchidae) and West Africa (Euryrhynchidae, Desmocarididae) (De Grave et al. 2008) and share the presence of cuspidate setae on their appendix masculina in addition to other morphological features (De Grave 2007).

*Leptopalaemon gagadjui*, an Australian freshwater representative of the family Kakaducarididae, forms a strong affinity with the freshwater genera *Macrobrachium* and *Cryphiops*, which agrees with a recent molecular study (Page et al. 2008b). Although the placement of the Kakaducarididae in relation to these genera appears unclear in our analyses, Page et al. (2008) demonstrate how the use of many genes (16S/18S/28S/H3) help clarify the monophyletic position of this family.

The radiantly beautiful coral reef families, Gnathophyllidae and Hymenoceridae, had long been recognized as a single family (Gnathophyllidae) until Chace (1992) once again separated the two on the basis of the 3rd maxilliped. They both share morphological characteristics such as a broadened 3rd maxilliped and similarity in mandible structure (Holthuis 1993). Our analyses strongly support an affinity between Gnathophyllidae and Hymenoceridae, which is in accordance with results found by Mitsuhashi et al. (2007). However, our study includes the genus *Gnathophylloides*, which was lacking in the former study. This inclusion identifies Gnathophyllidae to be a paraphyletic assemblage with the genus *Gnathophyllum* more closely related to *Hymenocera* than to *Gnathophylloides*. Mitsuhashi et al. (2007) grouped the Gnathophyllidae + Hymenoceridae clade within the subfamily Pontoniinae, while providing evidence for the paraphyly of the Pontoniinae. Larval morphology corroborates the close relationship among the three aforementioned taxa (Bruce 1986, 1988; Yang & Ko 2002). Our analyses show an obvious association between Hymenoceridae, Gnathophyllidae, and the genus *Pontonia*, but we do not find strong support for the inclusion of the other pontoniine taxa (*Kemponia*, *Coralliocaris*, *Periclimenaeus*). This may be due to the limited number of pontoniine taxa in our analysis (n = 4 genera).

Our results suggest a polyphyletic Palaemonidae, which is not unexpected due to the high degree of morphological diversity found within this family. However, definitive conclusions about phylogenetic relationships cannot be drawn until a broader representation of taxa is included in the analysis, especially of the Pontoniinae. Undoubtedly, this group is ripe for multiple systematic and taxonomic revisions in the future.

### 4.3 Superfamily Alpheoidea

Currently, the superfamily Alpheoidea contains the families Alpheidae, Ogyrididae, Hippolytidae, and Barbouriidae. Our tree contains representatives from all families except Barbouriidae, and results reject the monophyly of Alpheoidea. It is evident the family Hippolytidae represents a polyphyletic assemblage that qualifies for partitioning into several families as formerly suggested (Kemp 1914; Gurney 1942; Christoffersen 1987, 1990; Chace 1997; Posada et al. 2002). Our tree infers a strong relationship between the genera *Thoralus* and *Latreutes*, while *Hippolyte*, *Tozeuma*, and *Trachycaris* fall out as a supported single unit. Moreover, the genus *Lysmata* forms a distinct clade, clearly separated from the remaining hippolytids. In the past, Christoffersen (1987, 1990) placed *Lysmata* with other related genera within the family Lysmatidae Dana, 1952, and our analysis supports this division. Since then, several studies have recognized unique morphological and reproductive traits (Bauer 2000; Lin & Zhang 2001; Bauer 2004) of these shrimp.

Results support Ogyrididae as a sister clade to Alpheidae, confirming proposals of previous workers (Banner & Banner 1982; Christoffersen 1987; Anker et al. 2006). Recently, Anker et al. (2006) performed a cladistic analysis on the family Alpheidae, examining the phylogenetic relationships among genera. Our results suggest some congruence with their morphological analysis such as the basal position of *Yagerocaris cozumel* and close associations between *Fenneralpheus* and *Leptalpheus*. However, our analysis does not place *Synalpheus* (including some representatives assigned to *Zuzalpheus* (Rios & Duffy 2007)) as sister taxon to *Alpheus*, as Anker et al. (2006) previously concluded. While the snapping claw, which is thought to have facilitated rich diversification found within *Alpheus* and *Synalpheus*, is concluded by morphological analyses to have evolved only once within the Alpheidae, our molecular evidence suggests this key innovation may have arisen more than one time.

### 4.4 Atyidae + Xiphocarididae clade

The genus *Xiphocaris* was formerly considered a primitive atyid by Bouvier (1925), and morphological studies have placed the xiphocaridids as a subfamily within the Atyidae (Christoffersen 1986). These taxa inhabit freshwater and possess a dactylar grooming comb on the 5th pereopod. However, other caridean families have dactylar grooming combs (e.g., palaemonids and campylonotids) and xiphocaridids lack the unique cheliped setal brushes used in filter feeding, a diagnostic character used to define membership in the family Atyidae. In 1992, Chace grouped xiphocaridids within the superfamily Nematocarcinoidea, because they shared large epipods on the anterior pereopods and similar mouthparts. Recently, a molecular analysis of atyid shrimp questioned the relationships between selected genera and revisited the issue of possible relationships between xiphocaridids and atyids (Page et al. 2008a). Due to the phylogenetic resolution of the genes used in that study (16S, COI), the position of Xiphocarididae remained unclear, and the authors recommended "the addition of more highly conserved nuclear genes ... to resolve the deeper nodes fully" (Page et al. 2008a). Our analysis clearly places the xiphocaridids as close relatives of the atyids, with *Xiphocaris* being positioned as the basal lineage of the group or nested within the Atyidae in many of our reconstructions.

With the exclusion of the enigmatic position of *Xiphocaris elongata*, the division of the genera concurs with the findings of Page et al. (2008a). While delimitation of subfamilies within the Atyidae is yet to be taxonomically resolved, two clades are strongly supported in our topology, one representing the subfamily Atyinae and the other containing members of the other three subfamilies within the Atyidae.

## 4.5 Crangonidae + Processidae + Thalassocarididae subclade

Our analysis suggests a weak affinity among the families Crangonidae, Processidae, and Thalassocarididae, and similar arrangements have been suggested in the past. The first proposed classification for the Caridea (Dana 1852) placed the processids with the crangonids, along with other selected taxa, in the family Crangonidae. More recently, in a cladistic analysis based on morphological characters, Christoffersen (1987) noted a relationship between the two groups and transferred the family Processidae from the Alpheoidea into the Crangonoidea. Christoffersen (1990) again treated the crangonids and processids within the superfamily Crangonoidea, uniting the taxa on the basis of the length of pereopod 2. Molecular evidence lends some support for a relationship between Crangonidae and Processidae. However, our subclade includes the family Thalassocarididae, a group traditionally assumed related to Pandalidae on the basis of mouthparts (Chace 1985). Other workers have suggested a close affinity between Thalassocarididae and Oplophoridae on the basis of larval morphology (Menon & Williamson 1971). The undivided carpus of the 2nd pereopod within some thalassocaridids (exception seen in *Chlorotocoides*) may suggest remote evolutionary ties with crangonids, and molecular evidence supports this grouping. Nevertheless, systematic placement of thalassocaridids remains controversial, and a more robust examination of this family is required.

## 4.6 Basal lineages

Felgenhauer & Abele (1989) suggested that morphological attributes of the foregut may provide insights into the evolutionary relationships among the carideans. They argued the armament of the foregut to be a conserved trait, more related to the phylogenetic history of the group than to feeding behavior and diet. In comparisons to the putatively ancestral state in the Dendrobranchiata, the least derived foregut among the carideans was thought to be a complete set of ossicles and a well-developed gastric mill. Any progressive reduction of chitinized structures was thus considered a derived feature. Felgenhauer & Abele (1983, 1985, 1989) reported primitive states of caridean foreguts to occur in the families Atyidae, Nematocarcinidae, Stylodactylidae, and Rhynchocinetidae, with the least derived state found within the Procarididae. In our analysis, each of these families, and to a lesser extent the Nematocarcinidae, represents a basal lineage in the phylogeny. Furthermore, this morphological observation concurs with molecular results that imply separation of the procaridids from the infraorder Caridea. To our knowledge the foreguts in the other basally positioned lineages such as *Discias* and *Bathypalaemonella* have not been examined, but it would appear worthwhile to determine if they follow the same trends. Derived foreguts were reported from families such as Alpheidae, Crangonidae, Palaemonidae, Hippolytidae, Gnathophyllidae, and Oplophoridae (Felgenhauer & Abele 1983, 1985, 1989). With the exception of the oplophorids, all these families can be considered derived within our phylogeny.

Perhaps more intriguing are observations Felgenhauer & Abele (1989) noted within the Pasiphaeidae. While the genus *Leptochela* was reported to have a primitive well-developed foregut, the foregut within *Pasiphaea* appeared less chitinized and thus more derived. Our analysis suggests the Pasiphaeidae to be polyphyletic, despite the striking similarities in mouthparts and pectinate nature of the anterior chelipeds (Holthuis 1993). This result is in congruence with the findings of Felgenhauer and Abele (1989) and appears to argue for the separation of this family.

Our findings argue that foregut morphology should be thoroughly revisited and considered as a potentially informative character in morphological cladistic analyses. Concordance between earlier reported trends in foregut morphology and our present molecular phylogenetic tree appears to be more than coincidental.

### 4.7 Testing morphological hypotheses with molecular data

Although our phylogeny is not in complete congruence with the classifications and/or relationships proposed by Thompson (1967), Christoffersen (1990), or Chace (1992), the current molecular analysis provides fresh insights on long-debated issues related to the evolution of caridean morphological characters and can also be used to formulate new testable hypotheses bearing on caridean phylogeny. For example, Thompson (1967), among others, believed an oplophorid-like ancestor gave rise to many lineages within the carideans. Our analyses show the Oplophoridae nested within a larger clade and do not support this hypothesis. In fact, we find the oplophorids to be a polyphyletic group that requires more examination. Other hypotheses have suggested the superfamilial grouping of Crangonidae and Glyphocrangonidae on the basis of the subchelate 1st pair of pereopods. Our results would argue against the aforementioned superfamily classification and position us to test for convergent evolution among those groups. Finally, there is widely held consensus that subdivision of the 2nd pereopod (polycarpidean lineage) occurred only once in the evolution of caridean families (Christoffersen 1990). Our tree suggests this trait arose multiple times throughout caridean history, a finding that agrees with Thompson's work (1967). Should these and other findings hold up to more exhaustive phylogenetic scrutiny, we are challenged, on a case-by-case basis, to find explanations in biology and evolutionary history, as well as to reflect them in taxonomic revisions.

## 5 CONCLUSIONS

Our study presents the most comprehensive treatment to date of caridean phylogeny. Results suggest the monophyly of the Caridea but also propose that this group may represent two separate infraorders. We find little congruence with present hypotheses of higher-level relationships among caridean families. There is no support for the current superfamily classification, and only the Alpheidae, Alvinocarididae, Crangonidae, and Processidae are retained as strongly supported monophyletic assemblages. Morphology has long suggested the procaridids may represent a distinct lineage separate from the remaining carideans, and molecular data provide evidence to justify this division.

Our phylogeny is not expected to resolve all debates currently surrounding classification of the group but, rather, should be treated as a milepost in our ongoing studies. It is intended to provide initial insights on a molecular genetic basis and lay groundwork for further testing. Our findings add validity to some current phylogenetic hypotheses while calling others into question, and in several cases suggest phylogenies that are difficult to rectify with morphological evidence and assumed biogeographic history. However, apparent polyphyletic and paraphyletic compositions of some caridean superfamilies and families are not surprising and have been suggested by previous morphological and molecular systematists.

## ACKNOWLEDGEMENTS

We thank F. Alvarez, G. Boxshall, T.-Y. Chan, R. Collin, A. Covich, J. Felder, S. Fredericq, E. Garcia, W. Klotz, R. Lemaitre, E. Palacios-Theil, L. M. Meja-Ortz, T. Page, V. Paul, R. Robles, K. Ruetzler, and A. Windsor for providing museum materials and/or helping us to otherwise obtain collections for this analyses. Staff and scientists at the National Museum of Natural History—Smithsonian Institution, Universidad Nacional Autónoma de México, Smithsonian Tropical Marine Station-Bocas del Torro (R/V Urraca) and Oxford University Museum of Natural History kindly assisted with research cruises, shipping of loans, or hosting our visits. We are grateful to R. Bauer, C. Chlan, B. Felgenhauer, S. France, and B. Thoma for valuable comments and advice on this manuscript. This study is part of the Assembling the Tree of Life—Decapoda and was supported under funding by U.S. National Science Foundation grants NSF/BS&I DEB-0315995 and NSF/AToL

EF-0531603 to D. Felder. Additional small travel grants were provided by the Smithsonian Marine Station, Ft. Pierce, Florida, and the Smithsonian Caribbean Coral Reef Ecosystems Program, Belize. This is University of Louisiana Laboratory for Crustacean Research contribution no. 124 and Smithsonian Marine Station Contribution no. 736.

REFERENCES

Abele, L.G. 1991. Comparison of morphological and molecular phylogeny of the Decapoda. *Mem. Qld. Mus.* 31: 101–108.

Abele, L.G. & Felgenhauer, B.E. 1982. Eucarida. In: Parker, S. (ed.), *Eucarida*: 294–325. New York: McGraw-Hill Book Company, Inc.

Abele, L.G. & Felgenhauer, B.E. 1986. Phylogenetic and phenetic relationships among the lower Decapoda. *J. Crust. Biol.* 6: 385–400.

Ahyong, S. & O'Meally, D. 2004. Phylogeny of the Decapoda Reptantia: resolution using three molecular loci and morphology. *Raff. Bull. Zool.* 52: 673–693.

Alvarez, F., Iliffe, T.M. & Villalobos, J.L. 2005. New species of the genus *Typhlatya* (Decapoda: Atyidae) from anchialine caves in Mexico, the Bahamas, and Honduras. *J. Crust. Biol.* 25: 81–94.

Anker, A., Ahyong, S.T., Nol, P.Y. & Palmer, A.R. 2006. Morphological phylogeny of alpheid shrimps: parallel preadaptation and the origin of a key morphological innovation, the snapping claw. *Evolution* 60: 2507–2528.

Anker, A. & Iliffe, T.M. 2000. Description of *Bermudacaris harti*, a new genus and species (Crustacea: Deacapoda: Alpheidae) from anchialine caves of Bermuda. *Proc. Biol. Soc. Wash.* 113: 761–775.

Apakupakul, K., Siddall, M.E. & Burreson, E.M. 1999. Higher level relationships of leeches (Annelida: Clitellata: Euhirudinea) based on morphology and gene sequences. *Mol. Phylogenet. Evol.* 12: 350–359.

Banner, D.M. & Banner, A.H. 1982. The alpheid shrimp of Australia. Part III: The remaining alpheids, principally the genus *Alpheus*, and the family Ogyridae. *Rec. Aust. Mus.* 34: 1–357.

Bauer, R.T. 2000. Simultaneous hermaphroditism in caridean shrimps: a unique and puzzling sexual system in the Decapoda. *J. Crust. Biol.* 20: 116–128.

Bauer, R.T. 2004. *Remarkable Shrimp: Adaptations and Natural History of the Carideans*. University of Oklahoma Press, Norman.

Bouvier, E.L. 1925. Recherches sur la morphologie, les variations, la distribution géographique des crevettes des la famille des Atyides. *Encyclopédie entomologique, sér. A*, 4: 1–370.

Bruce, A.J. 1986. Observations on the family Gnathophyllidae Dana, 1852 (Crustacea, Decapoda). *J. Crust. Biol.* 6: 463–470.

Bruce, A.J. 1988. A note on the first zoeal stage larva of *Hymenocera picta* Dana (Crustacea: Decapoda: Palaemonidae). *The Beagle* 5: 119–124.

Bull, J.J., Huelsenbeck, J.P., Cunningham, C.W., Swofford, D.L. & Waddell, P.J. 1993. Partitioning and combining data in phylogenetic analysis. *Syst. Biol.* 42: 384–397.

Burkenroad, M.D. 1963. The evolution of the Eucarida (Crustacea, Eumalacostraca), in relation to the fossil record. *Tulane Stud. Bio.* 2: 1–18.

Burkenroad, M.D. 1981. The higher taxonomy and evolution of Decapoda (Crustacea). *Trans. San Diego Soc. of Nat. Hist.* 19: 251–268.

Cai, Y. & Anker, A. 2004. On a collection of freshwater shrimps (Crustacea: Decapoda: Caridea) from the Philippines, with descriptions of five new species. *Trop. Zool.* 17: 233–266.

Casanova, B., Jong, L. & Moreau, X. 2002. Carapace and mandibles ontogeny in the Dendrobranchiata (Decapod), Euphausiacea, and Mysidea (Crustacea): a phylogenetic interest. *Can. J. Zoolog.* 80: 296–306.

Castresana, J. 2000. Selection of conserved blocks from multiple alignments for their use in phylogenetic analysis. *Mol. Biol. Evol.* 17: 540–552.

Chace, F.A., Jr. & Bruce, A.J., 1993. The caridean shrimps (Crustacea: Decapoda) of the Albatross Philippine expedition 1907–1910, part 6, Superfamily Palaemonoidea. *Smithsonian Contributions to Zoology* 543: 1–152.

Chace, F.A., Jr. 1985. The caridean shrimps (Crustacea: Decapoda) of the *Albatross* Philippine expedition, 1907–1910, part 3: families Thalassocarididae and Pandalidae. *Smith. Contr. Zool.* 411: 1–143.

Chace, F.A., Jr. 1992. On the classification of the Caridea (Decapoda). *Crustaceana* 63: 70–80.

Chace, F.A., Jr. 1997. The Caridean Shrimps (Crustacea: Decapoda) of the *Albatross* Philippine Expedition, 1907–1910, part 7: families Atyidae, Eugonatonotidae, Rhynchocinetidae, Bathypalaemonellidae, Processidae, and Hippolytidae. *Smith. Contr. Zool.* 587: 1–106.

Chace, F.A., Jr. & Manning, R.B. 1972. Two new caridean shrimps, on representing a new family, from marine pools on Ascension Island (Crustacea: Decapoda: Natantia). *Smith. Contr. Zool.* 131: 1–18.

Christoffersen, M.L. 1986. Phylogenetic relationships between Oplophoridae, Atyidae, Pasiphaeidae, Alvinocarididae fam. n., Bresiliidae, Psalidopodidae and Disciadidae (Crustacea Caridea Atyoidea). *Boletim de Zoologia*, Universidade de São Paulo 10: 273–281.

Christoffersen, M.L. 1987. Phylogenetic relationships of hippolytid genera, with an assignment of new families for the Crangonoidea and Alpheoidea (Crustacea, Decapoda, Caridea). *Cladistics* 3: 348–362.

Christoffersen, M.L. 1988a. Phylogenetic systematics of the Eucarida (Crustacea, Malacostraca). *Rev. Bras. Zool.* 5: 325–351.

Christoffersen, M.L. 1988b. Genealogy and phylogenetic classification of the world Crangonidae (Crustacea, Caridea), with a new species and new records for the South Western Atlantic. *Rev. Nord. Biol.* 6: 43–59.

Christoffersen, M.L. 1989. Phylogeny and classification of the Pandaloidea (Crustacea, Caridea). *Cladistics* 5: 259–274.

Christoffersen, M.L. 1990. A new superfamily classification of the Caridea (Crustacea: Pleocyemata) based on phylogenetic pattern. *Z. Zool. Syst. Evol. forsch.* 28: 94–106.

Crandall, K.A. & Fitzpatrick, J.F. 1996. Crayfish molecular systematics: using a combination of procedures to estimate phylogeny. *Syst. Biol.* 45: 1–26.

Dana, J.D. 1852. *United States Exploring Expedition. During the Years 1838, 1839, 1840, 1841, 1842. Under the Command of Charles Wilkes, U. S. N., Vol. XIII. Crustacea. Part I.* C. Sherman, Philadelphia.

De Grave, S. 2007. A new species of *Euryrhynchus* Miers, with a discussion of the systematic position of the Euryrhynchidae Holthuis (Crustacea, Decapoda). *Zool. Anz.* 246: 193–203.

De Grave, S., Cai, Y. & Anker, A. 2008. Global diversity of shrimps (Crustacea: Decapoda: Caridea) in freshwater. *Hydrobiologia* 595: 287–293.

De Grave, S. & Moosa, M.K. 2004. A new species of the engimatic shrimp genus *Pseudocheles* (Decapoda: Bresiliidae) from Sulawesi (Indonesia), with the designation of a new family Pseudochelidae. *Crustac. Res.* 33: 1–9.

Duffy, E.D. 2002. The ecology and evolution of eusociality in sponge dwelling shrimp. In: Kikuchi, T. (ed.), *Genes, Behavior and Evolution in Social Insects*: 1–38. Sapporo, Japan: University of Hokkaido Press.

Edgar, R.C. 2004. MUSCLE: multiple sequence alignment with high accuracy and high throughput. *Nucleic Acids Res.* 32: 1792–1797.

Felgenhauer, B.E. & Abele, L.G. 1983. Phylogenetic relationships among shrimp-like decapods In: *Crustacean Phylogeny*: 291–311. Rotterdam: A. A. Balkema.

Felgenhauer, B.E. & Abele, L.G. 1985. Feeding structures of 2 atyid shrimps, with comments on caridean phylogeny. *J. Crust. Biol.* 5: 397–419.

Felgenhauer, B.E. & Abele, L.G. 1989. Evolution of the foregut in the lower Decapoda. In: Felgenhauer, B.E., Watling, L. & Thistle, A.B. (eds.), *Evolution of the Foregut in the Lower Decapoda*: 205–219. Rotterdam: A.A. Balkema.

Felsenstein, J. 1985. Confidence-limits on phylogenies with a molecular clock. *Syst. Zool.* 34: 152–161.

Fransen, C.H. & De Grave, S. (this volume). Evolution and radiation of shrimp-like decapods: an overview. In: Martin, J.W., Crandall, K.A. & Felder, D.L. (eds.), *Crustacean Issues: Decapod Crustacean Phylogenetics*. Boca Raton, Florida: Taylor & Francis/CRC Press.

Giribet, G., Carranza, S., Baguna, J., Riutort, M. & Ribera, C. 1996. First molecular evidence for the existence of a Tardigrada plus Arthropoda clade. *Mol. Biol. Evol.* 13: 76–84.

Gurney, R. 1942. *Larvae of decapod Crustacea*. London: Ray Society.

Hayashi, K. 2002. A new species of the genus *Athanas* (Decapoda, Caridea, Alpheidae) living in the burrows of mantis shrimp. *Raff. Bull. Zool.* 48: 249–256.

Holthuis, L.B. 1955. The recent genera of the caridean and stenopodidean shrimps (class Crustacea, order Decapoda, supersection Natantia) with keys for their determination. *Zool. Verh.* Leiden 26: 1–157.

Holthuis, L.B. 1973. Caridean shrimps found in land-locked saltwater pools at four Indo-west Pacific localities (Sinai Peninsula, Funafuti Atoll, Maui and Hawaii Islands), with the description of one new genus and four new species. *Zool. Verh.* Leiden 128: 1–48.

Holthuis, L.B. 1993. *The recent genera of the caridean and stenopodidean shrimps (Crustacea, Decapoda): with an appendix on the order Amphionidacea*. Nationaal Natuurhistorisch Museum, Leiden.

Huelsenbeck, J.P. & Ronquist, F. 2001. MRBAYES: Bayesian inference of phylogeny. *Biometrics* 17: 754–755.

Jarman, S.N., Nicol, S., Elliott, N.G. & McMinn, A. 2000. 28S rDNA evolution in the Eumalacostraca and the phylogenetic position of Krill. *Mol. Phylogenet. Evol.* 17: 26–36.

Kemp, S. 1914. Notes on Crustacea Decapoda in the Indian Museum. V.- Hippolytidae. *Rec. Ind. Mus.* 10: 81–129.

Kensley, B. & Williams, D. 1986. New shrimps (families Procarididae and Atyidae) from a submerged lava tube on Hawaii. *J. Crust. Biol.* 6: 417–437.

Khan, R.N., Becker, J.H.A., Crowther, A.L. & Lawn, I.D. 2003. Sea anemone host selection by the symbiotic saddled cleaner shrimp *Periclimenes holthuisi*. *Mar. Freshwater Res.* 54: 653–656.

Kim, W. & Abele, L.G. 1990. Molecular phylogeny of selected decapod crustaceans based on 18S ribosomal–RNA nucleotide sequences. *J. Crust. Biol.* 10: 1–13.

Knowlton, N. 1980. Sexual selection and dimorphism in two demes of a symbiotic, pair-bonding snapping shrimp. *Evolution* 34: 161–173.

Knowlton, N. & Keller, B.D. 1983. A new, sibling species of snapping shrimp associated with the Caribbean sea anemone *Bartholomea annulata*. *Bul. Marine Sci.* 33: 1–17, figs. 1–5.

Komai, T., Giere, O. & Segonzac, M. 2007. New record of alvinocaridid shrimps (Crustacea: Decapoda: Caridea) from hydrothermal vent fields on the Southern Mid-Atlantic Ridge, including a new species of the genus *Opaepele*. *Spec. Div.* 12: 237–253.

Komai, T. & Segonzac, M. 2003. Review of the hydrothermal vent shrimp genus, *Mirocaris*, redescription of *M. fortunanta* and reassessment of the taxonomic status of the family Alvinocarididae (Crustacea: Decapoda: Caridea). *Cah. Biol. Mar.* 44: 199–215.

Lin, J. & Zhang, D. 2001. Reproduction in a simultaneous hermaphroditic shrimp, *Lysmata wurdemanni*: Any two will do? *Mar. Biol.* 139: 1155–1158.

Macdonald, K.S., Rios, R. & Duffy, J.E. 2006. Biodiversity, host specificity, and dominance by eusocial species among sponge-dwelling alpheid shrimp on the Belize Barrier Reef. *Divers. Distrib.* 12: 165–178.

Mantelatto, F.L., Robles, R. & Felder, D.L. 2007. Molecular phylogeny of the western Atlantic species of the genus *Portunus* (Crustacea, Brachyura, Portunidae). *Zool. J. Linn. Soc., Lond.* 150: 211–220.

Mantelatto, F.L.M., Robles, R., Biagi, R. & Felder, D.L. 2006. Molecular analysis of the taxonomic and distributional status for the hermit crab genera *Loxopagurus* Forest, 1964 and *Isocheles* Stimpson, 1858 (Decapoda, Anomura, Diogenidae). *Zoosystema* 28: 495–506.

Marin, I.N., Anker, A., Britayev, T.A. & Palmer, A.R. 2005. Symbiosis between the alpheid shrimp, *Athanas ornithorhynchus* Banner and Banner, 1973 (Crustacea: Decapoda), and the brittle star, *Macrophiothrix longipeda* (Lamarck, 1816) (Echinodermata: Ophiuroidea). *Zool. Stud.* 44: 234–241.

Martin, J.W. & Davis, G.E. 2001. An updated classification of the Recent Crustacea. *Nat. Hist. Mus. Los Angeles County, Sci. Series* 39: 1–124.

Martin, J.W. & Wicksten, M.K. 2004. Review and rediscription of the freshwater atyid shrimp genus *Syncaris* Holmes, 1900, in California. *J. Crust. Biol.* 24: 447–462.

Medlin, L.K., Elwood, H.J., Stickel, S. & Sogin, M.L. 1988. The characterization of enzymatically amplified eukaryotic Ids-like rRNA coding regions. *Gene* 71: 491–499.

Menon, P. & Williamson, D.I. 1971. Decapod Crustacea from the International Indian Ocean expedition. The species of *Thalassocaris* (Caridea) and their larvae. *J. Zool., Lond* 165: 27–51.

Mitsuhashi, M., Sin, Y.W., Lei, H.C., Chan, T.Y. & Chu, K.H. 2007. Systematic status of the caridean families Gnathophyllidae Dana and Hymenoceridae Ortmann (Crustacea: Decapoda): a preliminary examination based on nuclear rDNA sequences. *Invertebr. Syst.* 21: 613–622.

Page, T.J., Cook, B.D., von Rintelen, T., von Rintelen, K. & Hughes, J.M. 2008a. Evolutionary relationships of atyid shrimps imply both ancient Caribbean radiations and common marine dispersals. *J. N. Am. Benthol. Soc.* 27: 68–83.

Page, T.J., Short, J.W., Humphrey, C.L., Hillyer, M.J. & Hughes, J.M. 2008b. Molecular systematics of the Kakaducarididae (Crustacea: Decapoda: Caridea). *Mol. Phylogenet. Evol.* 46: 1003–1014.

Page, T.J., von Rintelen, K. & Hughes, J.M. 2007. An island in the stream: Australia's place in the cosmopolitan world of Indo-West Pacific freshwater shrimp (Decapoda: Atyidae: Caridina). *Mol. Phylogenet. Evol.* 43: 645–659.

Palumbi, S., Martin, A., Romano, S., McMillan, W.O., Stice, L. & Grabowski, G. 1991. *The Simple Fool's Guide to PCR*. Department of Zoology and Kewalo Marine Laboratory, University of Hawaii, Honolulu.

Podsiadlowski, L. & Bartolomaeus, T. 2006. Major rearrangements characterize the mitochondrial genome of the isopod *Idotea baltica* (Crustacea: Peracarida). *Mol. Phylogenet. Evol.* 40: 893–899.

Porter, M.L., Perez-Losada, M. & Crandall, K.A. 2005. Model-based multi-locus estimation of decapod phylogeny and divergence times. *Mol. Phylogenet. Evol.* 37: 355–369.

Posada, D., Crandall, K. & Holmes, E.C. 2002. Recombination in evolutionary genomics. *Annu. Rev. Genet.* 36: 75–97.

Posada, D. & Crandall, K.A. 1998. MODELTEST: testing the model of DNA substitution. *Bioinformatics* 14: 817–818.

Pratchett, M. 2001. Influence of coral symbionts on feeding preferences of crown of thorns starfish *Acanthaster planci* in the Western Pacific. *Mar. Ecol. Prog. Ser.* 214: 111–119.

Richardson, A.J. & Cook, R.A. 2006. Habitat use by caridean shrimps in lowland rivers. *Mar. Freshwater Res.* 57: 695–701.

Rios, R. & Duffy, J.E. 2007. A review of the sponge-dwelling snapping shrimp from Carrie Bow Cay, Belize, with description of *Zuzalpheus*, new genus, and six new species (Crustacea: Decapoda: Alpheidae). *Zootaxa*: 3–89.

Robles, R., Schubart, C.D., Conde, J.E., Carmona-Suarez, C., Alvarez, F., Villalobos, J.L. & Felder, D.L. 2007. Molecular phylogeny of the American *Callinectes* Stimpson, 1860 (Brachyura: Portunidae), based on two partial mitochondrial genes. *Mar. Biol.* 150: 1265–1274.

Scholtz, G. & Richter, S. 1995. Phylogenetic systematics of the Reptantian Decapoda (Crustacea, Malacostraca). *Zool. J. Linn. Soc.* 113: 289–328.

Schram, F.R. 1986. *Crustacea*. Oxford University Press, New York.

Schubart, C.D., Cuesta, J.A. & Felder, D.L. 2002. Glyptograpsidae, a new brachyuran family from Central America: larval and adult morphology, and a molecular phylogeny of the Grapsoidea. *J. Crustac. Biol.* 22: 28–44.

Schubart, C.D., Neigel, J.E. & Felder, D.L. 2000. Use of the mitochondrial 16S rRNA gene for phylogenetic and population studies of Crustacea. In: von Vaupel Klein, J.C. & Schram, F.R. (eds.), *Crustacean Issues 12, The Biodiversity Crisis and Crustacea*: 817–830. Rotterdam: Balkema.

Shank, T., Black, M., Halanych, K., Lutz, R. & Vrijenhoek, R. 1999. Miocene radiation of deep-sea hydrothermal vent shrimp (Caridea: Bresiliidae): evidence from mitochondrial cytochrome oxidase subunit 1. *Mol. Phylogenet. Evol.* 13: 244–254.

Silliman, B.R., Layman, C.A. & Altieri, A.H. 2003. Symbiosis between an alpheid shrimp and a xanthoid crab in salt marshes of mid-Atlantic states, USA. *J. Crust. Biol.* 23: 876–879.

Spears, T., Abele, L.G. & Applegate, M.A. 1994. Phylogenetic study of cirripedes and selected relatives (Thecostraca) based on 18S rDNA sequence analysis. *J. Crust. Biol.* 14: 641–656.

Spears, T., Abele, L.G. & Kim, W. 1992. The monophyly of brachyuran crabs: A phylogenetic study based on 18S rRNA. *Syst. Biol.* 41: 446–461.

Stamatakis, A., Ludwig, T. & Meier, H. 2005. RAxML-III: A fast program for maximum likelihood-based inference of large phylogenetic trees. *Bioinformatics* 21: 456–463.

Stillman, J.H. & Reeb, C.A. 2001. Molecular phylogeny of eastern Pacific porcelain crabs, genera *Petrolisthes* and *Pachycheles*, based on the mtDNA 16S rDNA sequence: phylogeographic and systematic implications. *Mol. Phylogenet. Evol.* 19: 236–245.

Swofford, D.L. 2003. *Phylogenetic Analysis Using Parsimony (*and Other Methods). Ver. 4*. Sinauer Associates: Sunderland, MA.

Thompson, J.R. 1967. Comments on phylogeny of section Caridea (Decapoda: Natantia) and the phylogenetic importance of the Oplophoroidea. *Proceedings of Symposium on Crustacea–Part 1*: 314–326.

Tudge, C.C. & Cunningham, C.W. 2002. Molecular phylogeny of the mud lobsters and mud shrimps (Crustacea: Decapoda: Thalassinidea) using nuclear 18S rDNA and mitochondrial 16S rDNA. *Invertebr. Syst.* 16: 839–847.

Xu, Y., Song, L.-s. & Li, X.-z. 2005. The molecular phylogeny of infraoder Caridea based on 16S rDNA sequences. *Marine Sciences* (Beijing) 29: 36–41.

Yang, H.J. & Ko, H.S. 2002. First zoea of *Palaemon ortmanni* (Decapoda, Caridea, Palaemonidae) hatched in the laboratory, with notes on the larval morphology of the Palaemonidae. *Korean J. Syst. Zool.* 18: 181–189.

# III ADVANCES IN OUR KNOWLEDGE OF THE THALASSINIDEAN AND LOBSTER-LIKE GROUPS

# Molecular Phylogeny of the Thalassinidea Based on Nuclear and Mitochondrial Genes

RAFAEL ROBLES[1], CHRISTOPHER C. TUDGE[2], PETER C. DWORSCHAK[3], GARY C.B. POORE[4] & DARRYL L. FELDER[1]

[1] *Department of Biology, University of Louisiana, Lafayette, Louisiana, U.S.A.*
[2] *Biology Department, American University, Washington, D.C., U.S.A.*
[3] *Dritte Zoologische Abteilung, Naturhistorisches Museum, Wien, Austria*
[4] *Department of Natural Sciences, Museum of Victoria, Abbotsford, Victoria, Australia*

## ABSTRACT

We conducted a molecularly based phylogenetic analysis with representatives of the thalassinidean families Axianassidae, Axiidae, Callianassidae, Callianideidae, Calocarididae, Ctenochelidae, Laomediidae, Micheleidae, Strahlaxiidae, Thalassinidae, Thomassiniidae, and Upogebiidae, along with decapod outgroup taxa representing the infraorders Anomura, Astacidea, Brachyura, Caridea, and Achelata. Analyses were based on two datasets, one corresponding to a partial fragment of the 16S mitochondrial gene and a second to a partial fragment of the 18S nuclear gene, representing roughly 1,800 nuclear and 550 mitochondrial characters. We incorporated 34 genera and 50 species in the analysis upon which our molecular phylogenetic trees were based and compared outcomes to morphologically based phylogenies. Our analysis finds the infraorder Thalassinidea to be paraphyletic, as presently comprised. We also find no support for monophyly in either the superfamily Axioidea or the superfamily Callianassoidea. Two large clades into which the infraorder is divided instead recall arrangements that were based upon larvae by Gurney and subsequently supported in some early taxonomic revisions. We conclude that these clades deserve separate infraordinal status, and we draw upon the work of de Saint Laurent for the name of each. One we refer to the infraorder Gebiidea, encompassing representatives of Upogebiidae, Laomediidae, Thalassinidae, and Axianassidae. The other we refer to Axiidea, encompassing Callianassidae, Ctenochelidae, Strahlaxiidae, Micheleidae, Callianideidae, Thomassiniidae, Axiidae, and Calocaridae. We accept previous evidence merging Eiconaxiidae with the Axiidae, and we suggest the Calocarididae should be likewise merged. We also present evidence to support merging of Thomassiniidae back into Callianideidae.

## 1 INTRODUCTION

The infraorder Thalassinidea encompasses a group of burrowing decapods that is almost global in distribution, with the northernmost record at 71° N and the southernmost at 55° S. Resembling hermit crabs in some features and lobsters in others (Borradaile 1903), they are known to populate sediments in depths from 0 to >2000 m (Dworschak 2005). Thalassinidean genera are in varied ways adapted morphologically to a fossorial existence, and many show evidence of a functional *linea thalassinica*, a hinge-line that to various degrees allows flexure of the carapacial branchiostegites for gill ventilation or cleaning while within a burrow. This character was invoked by some early workers to define membership in this group, but others discounted its systematic importance, as noted by Barnard (1950).

Thalassinideans often play major roles in mechanical bioturbation of sediments and mobilization of nutrients entrained in sediments or sedimentary pore-waters, with impacts on water chemistries as well as associated marine microbial, plant, and animal assemblages (Bird 2000, 2004; Dworschak 2000; Felder 2001; Atkinson & Taylor 2004; Coelho 2004; Dworschak et al. 2006; Klerks et al. 2007; Pillay et al. 2007). Larval life histories vary greatly within the group (Felder et al. 1985; Nates et al. 1997; Strasser & Felder 2000, 2005), as do burrow shapes, physiology, and trophic dependencies, which can also be phylogenetically informative (Felder 2001; Coelho 2004; Dworschak & Ott 1993). While classification of the thalassinideans has focused primarily on adult morphology, characters ranging from larval setation to fecal pellets at one time or another have been suggested as evidence for group relationships (Gurney 1942).

Recent accounts of thalassinidean diversity have usually recognized 11 families, 94 genera, and 556 species (Dworschak 2000, 2005). However, newly recognized species and genera can be added to these counts (bringing the count of genera to 99 and species to 600), and recognition of the family Axianassidae now appears to be justifiable on the bases of molecular (Tudge & Cunningham 2002) and comparative larval studies (Strasser & Felder 2005). The subfamily Gourretiinae was also raised to family rank (Sakai 2004), but in this case without supporting analyses and in clear contradiction to the cladistic evidence of Tudge et al. (2000), wherein members of Gourretiinae were shown to belong to Ctenochelidae. Also, the monogeneric family Eiconaxiidae has been proposed (Sakai & Ohta 2005) for *Eiconaxius*, but we continue to regard this group as a member of the monophyletic Axiidae in the absence of convincing morphological evidence that it is not just a specialized member of this family.

Phylogeny of the order Decapoda overall has been extensively debated at both higher and lower levels of classification but remains largely unresolved after a century of study (see de Saint Laurent 1973, 1979a, b; Felgenhauer & Abele 1983; McLaughlin & Holthuis 1985; Abele & Felgenhauer 1986; Kim & Abele 1990; Poore 1994; Scholtz & Richter 1995; Martin & Davis 2001; Schram 2001; Morrison et al. 2002; Tudge & Cunningham 2002; Dixon et al. 2003; Porter et al. 2005). Thalassinidean decapods were originally brought together by Borradaile (1903) into four families: Axiidae Huxley, 1879, Laomediidae Borradaile, 1903, Thalassinidae Dana, 1852, and Callianassidae Dana, 1852, with the callianassids subdivided to accommodate the subfamilies Callianassinae and Upogebiinae. While widely applied (de Man 1928; Bouvier 1940; Zariquiey Alvarez 1968), this classification did not conform to relationships deduced from larval morphology by Gurney (1938) who, lacking comparative materials of the Axianassidae and Thalassinidae, found larval similarities to group at least Callianassidae with Axiidae, and Upogebiidae with Laomediidae (see also Gurney 1942). This provided possible insight to phylogeny within the overall group, and suggested paraphyly within "Callianassidae" as it had been previously conceived, prompting at least some workers (Barnard 1950) to adopt Gurney's scheme. Following publication of a short paper in the early 1970s (de Saint Laurent 1973), which adopted Gurney's separation of the Upogebiidae and Callianassidae, there appeared several subsequent works applying revisions based upon adult morphology (Le Loeuff & Intès 1974; de Saint Laurent 1979a, b; de Saint Laurent & Le Loeuff 1979). In the following two decades, a host of morphologically based revisions impacted family and subfamily ranks among varied subgroups of the thalassinideans (Kensley 1989; Sakai & de Saint Laurent 1989; Manning & Felder 1991; Sakai 1992, 1999; Poore 1994).

Among recent workers to address the thalassinideans overall, some have proposed the group to be monophyletic (Poore 1994; Scholtz & Richter 1995; Schram 2001; Dixon et al. 2003; Ahyong & O'Meally 2004; Tsang et al. 2008b) and others paraphyletic or polyphyletic (de Saint Laurent 1973; Tudge 1997; Tudge & Cunningham 2002; Morrison et al. 2002; Tsang et al. 2008a). The group was morphologically rediagnosed less than 15 years ago on the basis of a single synapomorphy, the presence of a dense row of evenly spaced long setae along inferior margins of pereopod 2 (Poore 1994, 1997); it was also therewith reestablished that the *linea thalassinica* was a likely homolog of the *linea anomurica*, and that varied permutations of this character were thus not diagnostic.

However, monophyly of the group remains uncertain (see discussion in Martin & Davis 2001), as do evolutionary relationships among families assigned to the infraorder Thalassinidea, which makes for a problematic classification.

Based on morphological cladistic analyses, Poore (1994) distributed families among three superfamilies: Thalassinoidea (one family), Axioidea (four families), and Callianassoidea (six families). In a subsequent morphological cladistic analysis of the order Decapoda (Dixon et al. 2003) seven families of Thalassinidea were included. While the intention of the latter authors was not specifically to solve phylogenetic relationships within Thalassinidea, it is noteworthy that members of the superfamily Callianassoidea were found to be paraphyletic (Dixon et al. 2003: fig. 6), with *Jaxea* positioned basally instead of being clustered with *Callianassa*, *Upogebia*, and *Callianidea*. The latter grouping of three is also contrary to relationships suggested by larval evidence.

Some inconsistencies between views on the classification and systematics of Thalassinidea result from limited taxonomic representation. For example, Poore (1994) did not include *Axianassa*, only *Laomedia* (Axianassidae effectively excluded). The family Ctenochelidae (represented by four genera) appeared to be paraphyletic with respect to Callinassidae (one genus) in Poore's (1994) treatment, but in a more robust cladistic analysis involving six ctenochelid genera and numerous callianassid genera (Tudge et al. 2000), support was found for family status of both Callianassidae and Ctenochelidae. The latter analysis did not support all subfamilies proposed for membership within Callianassidae or Ctenochelidae.

Molecular genetic approaches also have been applied to understand evolutionary relationships within Thalassinidea. Tudge & Cunningham (2002) analyzed nuclear 18S and mitochondrial (mt) 16S sequence data from fourteen species representing seven of the twelve families of Thalassinidea. They found low support for monophyly of Thalassinidea, discovering instead two clades, one including Strahlaxiidae and Callianassidae (seven species) and the other Upogebiidae (two species), Axianassidae, Laomediidae (two species), and Thalassinidae. Porter et al. (2005) probed evolutionary relationships of the order Decapoda with the aid of four DNA fragments but included only members of Callianassidae in their analysis.

Our own molecular studies of Thalassinidea have been under way since 2002 (Felder et al. 2003; Felder & Robles 2004; Robles & Felder 2004). Recently, concurrent studies have come to our attention, bearing on many of the same questions we address (Tsang et al. 2008a, b). These studies differ from our own in terms of thalassinidean and outgroup taxa included and in outcomes. We take this opportunity to present our independent findings and compare them with those of other recent molecular phylogenetic studies. Principal objectives of our study are to resolve questions of monophyly of the Thalassinidea as a whole, but also to address monophyly and diagnostic characters of its constituent families and subfamilies. In a separate analysis (Felder & Robles this volume), other taxa are brought into an analysis of specifically the family Callianassidae.

## 2 MATERIALS AND METHODS

### 2.1 *Taxa included*

Our sample consisted of 55 organisms representing 12 currently accepted families of Thalassinidea (Table 1) and three commonly recognized superfamilies (*sensu* Martin & Davis 2001). To represent the superfamily Callianassoidea, we included representatives of Axianassidae, Callianassidae, Callianideidae, Ctenocheleidae, Laomediidae, Thomassiniidae, and Upogebiidae. For the superfamily Axioidea we included representatives of Axiidae, Calocarididae, Micheleidae, and Strahlaxiidae. We were unable to include Eiconaxiidae, a monogeneric family proposed by Sakai & Ohta (2005), which we regard as a highly specialized axiid. To represent the superfamily Thalassinoidea, we included a species of the genus *Thalassina*, the only genus in the family Thalassinidae.

To serve as outgroups, we included sequence data for 20 species representing as many genera, from infraorders (and listed families) as follow: Anomura (Galatheidae, Hippidae, Lithodidae), Astacidea (Astacidae, Cambaridae, Enoplometopidae, Nephropidae, Parastacidae), Brachyura (Cancridae, Portunidae), Caridea (Atyidae, Hippolytidae, Palaemonidae, Pandalidae), and Achelata (Palinuridae, Scyllaridae), to test for monophyly of Thalassinidea.

## 2.2 DNA extraction, PCR, and sequencing

DNA was extracted from muscle tissues excised from the abdomen or pleopods following standard protocols (Robles et al. 2007). Standard PCR amplification and automated sequencing protocols were used to sequence a fragment of approximately 550 bp of the 16S rDNA and 1,800 bp of the 18S rDNA genes. Both strands were sequenced. Primers used for PCR were 16ar (5-CGC CTG TTT ATC AAA AAC AT-3), 16br (5-CCG GTC TGA ACT CAG ATC ACG T-3) (Palumbi et al. 1991), 1472 (5-AGA TAG AAA CCA ACC TGG-3) (Crandall & Fitzpatrick 1996), and 16L2 (5-TGC CTG TTT ATC AAA AAC AT-3) (Schubart et al. 2002). Primers used for the 18S fragment were 18S-A (5'-AAC CTG GTT GAT CCT GCC AGT-3'), 18S-B (5'-TGA TCC TTC CGC AGG TTC ACC T-3') (Medlin et al. 1988), 18S-L (5'-CCA ACT ACG AGC TTT TTA ACT G-3'), 18S-C (5'-CGG TAA TTC CAG CTC CAA TAG-3'), 18S-Y (5'-CAG ACA AAT CGC TCC ACC AAC-3'), 18S-O (5'-AAG GGC ACC ACC AGG AGT GGA G-3') (Apakupakul et al. 1999).

## 2.3 Phylogenetic analyses

Consensus of complementary sequences was obtained with the Sequencher software program (ver 4.7, Genecodes, Ann Arbor, MI). Multiple sequence aligning was performed with the aid of BioEdit v.7.08.0 (Hall 1999) with the following settings: 6-2/6-2 penalty (opening-gap extension, pairwise/multiple alignment, respectively) following a profile alignment strategy. Base composition, pattern of substitution for pairwise comparison, and analysis of variability along both fragments of the 16S mtDNA and the 18S nDNA were performed as implemented in PAUP 4.0 beta 10 (Swofford 1998). Homogeneity of nucleotide frequency among taxa was also assessed for each gene with a $\chi^2$ test as implemented in PAUP. Previous to the analysis of the combined data, we performed an incongruence length difference (ILD) test or partition homogeneity test (Bull et al. 1993), as implemented in PAUP, to determine whether the 16S and 18S genes could be considered samples of the same underlying phylogeny.

Phylogenetic analyses were conducted using MRBAYES for Bayesian analysis (BAY) and PAUP 4.0 beta 10 for both maximum parsimony (MP) and neighbor joining (NJ) analyses; maximum likelihood (ML) analysis was conducted with RAxML v.7.0.4 (Stamatakis 2006) using the online version at the Cyber Infrastructure for Phylogenetic Research (CIPRES) website (Stamatakis et al. 2008). Prior to conducting the BAY and NJ analyses, the model of evolution that best fit the data was determined with the software MODELTEST (Posada & Crandall 1998). ML was performed with the default parameters for RAxML for the GTR model of evolution. BAY analysis was performed sampling one tree every 1,000 generations for 2,000,000 generations, starting with a random tree, thus obtaining 2,001 trees. A preliminary analysis showed that stasis was reached at approximately 30,000 generations. Thus, we discarded 51 trees corresponding to the first 50,000 generations and obtained a 50% majority rule consensus tree from the remaining 1,950 saved trees. NJ analysis was carried out with a distance correction set with the parameters obtained from MODELTEST (Posada & Crandall 1998). MP analysis was performed as a heuristic search with gaps treated as a fifth character, multistate characters interpreted as uncertain, and all characters considered as unordered. The search was conducted with a random sequence addition and 1,000 replicates, including tree bisection and reconnection (TBR) as a branch swapping option; branch swapping was performed on the best trees only.

To determine confidence values for the resulting trees, we ran 2,000 bootstrap pseudo-replicates for NJ and MP analysis, based on the same parameters as above. For ML analysis, we selected the option to automatically determine the number of bootstraps to be run in RAxML. Thus, 250 bootstrap pseudo-replicates were run. On the molecular trees, confidence values >50% were reported for ML, MP, and NJ analyses (bootstraps), while for the BAY analysis values were reported for posterior probabilities of the respective nodes among all the saved trees. Sequences as well as alignments have been submitted to GenBank as a Popset.

## 3 RESULTS

Unrooted trees (not shown here) yielded well-defined separations of Brachyura, Caridea, and Achelata, but not Thalassinidea. As Caridea was by this method shown to be the most distinct infraorder from all other infraorders, we used this clade thereafter to root our tree. Our final alignment included 2,094 bp, 1,729 for the 18S nuclear gene and 365 bp for the 16S mt gene (excluding primer regions, saturated and ambiguous fragments of both genes). Of these, 1,363 were invariable, 699 were variable but not parsimony informative, and 534 were parsimony informative characters. The ILD test showed no significant incongruence (P = 0.578). Thus we used the combined 16S and 18S fragments for our analysis. The nucleotide composition of this dataset can be considered homogeneous ($\chi^2$ = 65.96, df = 186, P = 1.00), with a slightly larger percentage of A-T (26.0%; 26.2 %).

The best-fitting model of substitution, selected with the Akaike information criterion (AIC, Akaike 1974) as implemented in MODELTEST (Posada & Crandall 1998), was the general time-reversible model, with invariable sites and a gamma distribution GTR+$\Gamma$+$\delta$ (Tavaré 1986) and with the following parameters: assumed nucleotide frequencies: A = 0.2677, C = 0.2066, G = 0.2592, T = 0.2665; substitution rates A-C = 1.6548, A-G = 5.2680, A-T = 2.7285, C-G = 1.1068, C-T = 6.5936, G-T = 1.0000; proportion of invariable sites $\Gamma$ = 0.5407; variable sites followed a gamma distribution with shape parameter $\delta$ = 0.5144. These values were used to obtain both BAY and NJ trees. All four phylogenetic methods yielded almost identical tree topologies with high support values (Fig. 1). Differences found between the methods were limited primarily to a few of the internal/terminal clades.

### 3.1 Testing for monophyly of the Thalassinidea

Our analyses showed Thalassinidea to be a distinctly paraphyletic group (Fig. 1). Members of the infraorder were separated into two well-supported clades. "Clade-A" grouped representatives of the families Upogebiidae, Laomediidae, Thalassinidae, and Axianassidae, thus encompassing our sole representative of the superfamily Thalassinoidea together with several families that are typically included in the superfamily Callianassoidea. "Clade-B" grouped representatives of the families Axiidae, Callianassidae, Calocarididae, Ctenochelidae, Micheleidae, Strahlaxiidae, and Thomassiniidae, thus encompassing remaining members of the superfamily Callianassoidea along with all members of the Axioidea, but clearly showing the latter superfamily to be polyphyletic. As rooted, our analysis positions Clade-B (hereafter called Axiidea) as a sister taxon of the other decapod infraorders (outgroup Caridea excepted), not of Clade-A (hereafter called Gebiidea) (Fig. 1).

### 3.2 The families of "Gebiidea"

One highly supported node shows a monophyletic family Upogebiidae while another well-supported node groups all representatives of Laomediidae, Thalassinidae, and Axianassidae. Structure within the Upogebiidae itself shows two sister clades, one of them moderately supported, that also suggest paraphyly in the genus *Upogebia* as presently applied. The companion clade includes Axianassidae positioned as a sister clade to a monophyletic Laomediidae, albeit at low support values.

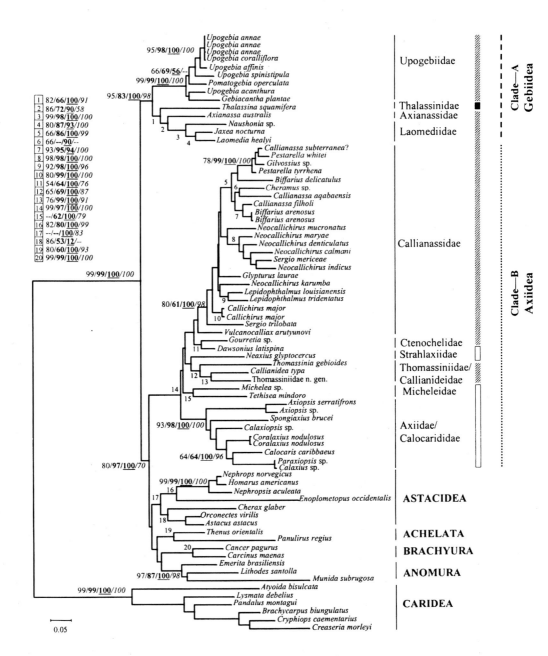

**Figure 1.** Evolutionary relationships among 12 families of Thalassinidea (*sensu* Martin & Davis 2001) inferred from a Bayesian analysis of 16S and 18S rDNA data. Support values shown from left to right are for NJ, MP, BAY, and ML respectively; "–" represents value equal to or lower than 50%; "?" indicates questioned identity of a sequence from GenBank. Vertical bar indicates assignments to herewith-rejected superfamilies Axioidea (open), Thalassinoidea (solid), and Callianassoidea (cross-hatched). We question identity of "*Callianassa subterranea*" in this tree, ostensibly representing the type species of that genus. It is included here on the basis of sequence data from GenBank (Table 1), originally used in Porter et al. (2005) and thereafter by Tsang et al. (2008b). Our own 16S sequence data for relatively topotypic specimens (morphologically confirmed as *C. subterranea*) do not match those in GenBank (DQ079706).

**Table 1.** List of specimens used for molecular analysis, as commonly classified (*sensu* Martin & Davis 2001). Letter abbreviations preceding catalog numbers indicate collections as follow: MV = Museum Victoria; NHMW = Naturhistorisches Museum, Wien; NMCR = National Museum of the Philippines, Manila; ULLZ = University of Louisiana—Lafayette Zoological Collection; USNM = National Museum of Natural History; ZRC = Zoological Reference Collection of the Raffles Museum of Biodiversity Research, National University of Singapore, Singapore; KC, MLP, and KAC = voucher IDs as reported in corresponding publication. Where two catalog numbers appear for the same sample, tissue was donated to the University of Louisiana at Lafayette and archived there under a ULLZ number, while original voucher retains number at the respective museum. Sequences obtained from GenBank are shown by accession number (Acc. No.) for the respective gene; the source where first published (S) is as follows: 1 = Porter et al. 2005; 2 = Bracken et al., this volume; 3 = Tudge & Cunningham 2002; 4 = Ahyong & O'Meally 2004; 5 = Pérez-Losada et al. 2002a; 6 = Pérez-Losada et al. 2002b; 7 = Pérez-Losada et al. 2004; 8 = Crandall et al. 2000; 9 = Giribet et al. 2001; 10 = Morrison et al. 2002. "?" following *Callianassa subterranea* indicates questionable identity of the sequence in GenBank.

| Taxon Name | Catalog No. | Acc. No. 18S | Acc. No. 16S | S |
|---|---|---|---|---|
| **OUTGROUP** | | | | |
| **Anomura** | | | | |
| Galatheidae | | | | |
| *Munida subrugosa* (White, 1847) | KACmusu | AF439382 | AY050075 | 6/5 |
| Hippidae | | | | |
| *Emerita brasiliensis* Schmitt, 1935 | KACembr | AF439384 | DQ079712 | 6/1 |
| Lithodidae | | | | |
| *Lithodes santolla* (Molina, 1782) | LAClisa | AF439385 | AY595927 | 6/7 |
| **Astacidea** | | | | |
| Astacidae | | | | |
| *Astacus astacus* (Linnaeus, 1758) | JF134 | AF235959 | AF235983 | 8 |
| Cambaridae | | | | |
| *Orconectes virilis* (Hagen, 1870) | JC897 | AF235965 | AF235989 | 8 |
| Enoplometopidae | | | | |
| *Enoplometopus occidentalis* (Randall, 1840) | | AY583966 | AY583892 | 4 |
| Nephropidae | | | | |
| *Homarus americanus* H. Milne Edwards, 1837 | KAChoam | AF235971 | AF370876 | 8/9 |
| *Nephrops norvegicus* (Linnaeus, 1758) | KC2163 | DQ079762 | DQ079726 | 1 |
| *Nephropsis aculeata* Smith, 1881 | KC2117 | DQ079761 | DQ079727 | 1 |
| Parastacidae | | | | |
| *Cherax glaber* Rieck, 1967 | KACchgl | DQ079745 | AF135978 | 1 |
| **Brachyura** | | | | |
| Cancridae | | | | |
| *Cancer pagurus* Linnaeus, 1758 | KC2158 | DQ079743 | DQ079708 | 1 |
| Portunidae | | | | |
| *Carcinus maenas* (Linnaeus, 1758) | KACcama | DQ079744 | DQ079709 | 1 |
| **Caridea** | | | | |
| Atyidae | | | | |
| *Atyoida bisulcata* (Randall, 1840) | KC2138 | DQ079747 | DQ079704 | 1 |
| Hippolytidae | | | | |
| *Lysmata debelius* Bruce, 1983 | MLP121 | DQ079752 | DQ079718 | 1 |
| Palaemonidae | | | | |
| *Creaseria morleyi* (Creaser, 1936) | MLP102 | DQ079746 | DQ079710 | 1 |
| *Cryphiops caementarius* (Molina, 1782) | JC1219 | DQ079747 | DQ079711 | 1 |
| *Brachycarpus biunguiculatus* (Lucas, 1846) | ULLZ 7430 | EU868779 | EU868685 | 2 |

Table 1. continued.

| Taxon Name | Catalog No. | Acc. No. 18S | Acc. No. 16S | S |
|---|---|---|---|---|
| **Pandalidae** | | | | |
| *Pandalus montagui* Leach, 1814 | ULLZ 6966 | EU868792 | EU868698 | 2 |
| **Achelata** | | | | |
| **Palinuridae** | | | | |
| *Panulirus regius* De Brito Capello, 1846 | KC2167 | DQ079765 | DQ079730 | 1 |
| **Scyllaridae** | | | | |
| *Thenus orientalis* (Lund, 1793) | NONE | EU875001 | EU874951 | 3 |
| **INGROUP** | | | | |
| **Thalassinidea** | | | | |
| **Axioidea** | | | | |
| **Axiidae** | | | | |
| *Axiopsis* | ULLZ 7750 | EU874970 | EU874920 | |
| *Axiopsis serratifrons* (A. Milne-Edwards, 1873) | ULLZ 8996 | EU874992 | EU874942 | |
| *Calaxius* sp. | ULLZ 7041 | EU874960 | EU874910 | |
| *Coralaxius nodulosus* (Meinert, 1877) | ULLZ 7011 | EU874959 | EU874909 | |
| *Coralaxius nodulosus* (Meinert, 1877) | ULLZ 7329 | EU874963 | EU874913 | |
| *Paraxiopsis* sp. | ULLZ 7559 | EU874967 | EU874917 | |
| *Spongiaxius brucei* (Sakai, 1986) | ULLZ 8937 MV J55585 | EU874991 | EU874941 | |
| **Calocarididae** | | | | |
| *Calaxiopsis* sp. | ULLZ 8918 MV J55576 | EU874988 | EU874938 | |
| *Calocaris ~caribbaeus* Kensley, 1996 | ULLZ 8285 | EU874979 | EU874929 | |
| **Micheleidae** | | | | |
| *Michelea* sp. | ULLZ 8920 MV J55702 | EU874990 | EU874940 | |
| *Tethisea mindoro* Poore, 1997 | ULLZ 8919 MV J55703 | EU874989 | EU874939 | |
| **Strahlaxiidae** | | | | |
| *Neaxius glyptocercus* von Martens, 1868 | MV J39643 | EU874994 | EU874944 | 3 |
| **Callianassoidea** | | | | |
| **Axianassidae** | | | | |
| *Axianassa australis* Rodrigues & Shimizu, 1992 | MV J44613 | EU874998 | EU874948 | 3 |
| **Callianassidae** | | | | |
| **Callianassinae** | | | | |
| *Biffarius arenosus* (Poore, 1975) | BaV3 | DQ079739 | DQ079705 | 1 |
| *Biffarius arenosus* (Poore, 1975) | MV J40669 | EU874995 | EU874945 | 3 |
| *Biffarius delicatulus* Rodrigues & Manning, 1992 | USNM 309754 | EU875003 | EU874953 | 3 |
| *Callianassa aqabaensis* Dworschak, 2003 | ULLZ 7924 | EU874975 | EU874925 | |
| *Callianassa filholi* A. Milne-Edwards, 1878 | MV J44818 | EU874999 | EU874949 | 3 |
| *Callianassa subterranea*? (Montagu, 1808) | KACcasu | DQ079740 | DQ079706 | 1 |
| *Gilvossius* sp. | ULLZ 7919 | EU874974 | EU874924 | |
| *Pestarella tyrrhena* (Petagna, 1792) | ULLZ 7931 | EU874977 | EU874927 | |
| *Pestarella whitei* (Sakai, 1999) | ULLZ 7932 NHMW 21948 | EU874978 | EU874928 | |

| Taxon Name | Catalog No. | Acc. No. 18S | Acc. No. 16S | S |
|---|---|---|---|---|
| **Callichirinae** | | | | |
| *Callichirus major* (Say, 1818) | MV J39044 | AF436002 | AF436041 | 10 |
| *Callichirus major* (Say, 1818) | KAC 1864 | DQ079741 | DQ079707 | 1 |
| *Glypturus laurae* (de Saint Laurent, 1984) | ULLZ 8446 NHMW 21939 | EU874985 | EU874935 | |
| *Lepidophthalmus louisianensis* (Schmitt, 1935) | ULLZ 7918 | EU874973 | EU874923 | |
| *Lepidophthalmus tridentatus* (von Martens, 1868) | ULLZ 7928 NMCR 27007 | EU874976 | EU874926 | |
| *Neocallichirus calmani* (Nobili, 1904) | ULLZ 8439 NHMW 21943 | EU874982 | EU874932 | |
| *Neocallichirus denticulatus* Ngoc-Ho, 1994 | ULLZ 8441 NHMW 21945 | EU874984 | EU874934 | |
| *Neocallichirus indicus* (de Man, 1905) | ULLZ 8437 NHMW 21942 | EU874981 | EU874931 | |
| *Neocallichirus karumba* (Poore & Griffin, 1979) | ULLZ 8435 ZRC 2002-0274 | EU874980 | EU874930 | |
| *Neocallichirus mucronatus* (Strahl, 1861) | ULLZ 8440 NHMW 21944 | EU874983 | EU874933 | |
| *Neocallichirus maryae* (Schmitt, 1935) | USNM 309751 | EU875002 | EU874952 | 3 |
| *Sergio mericae* Manning & Felder, 1995 | USNM 309755 | EU875004 | EU874954 | 3 |
| *Sergio trilobata* (Biffar, 1970) | ULLZ 7916 | EU874972 | EU874922 | |
| **Cheraminae** | | | | |
| *Cheramus* sp. | ULLZ 7313 | EU874962 | EU874912 | |
| **Vulcanocalliacinae** | | | | |
| *Vulcanocalliax arutyunovi* Dworschak & Cunha, 2007 | ULLZ 7620 NHMW 21927 | EU874969 | EU874919 | |
| **Callianideidae** | | | | |
| *Callianidea typa* H. Milne Edwards, 1837 | ULLZ 9179 | EU874993 | EU874943 | |
| **Ctenochelidae** | | | | |
| *Gourretia* sp. | ULLZ 7370 | EU874965 | EU874915 | |
| *Dawsonius latispina* (Dawson, 1967) | ULLZ 7306 | EU874961 | EU874911 | |
| **Laomediidae** | | | | |
| *Jaxea nocturna* Nardo, 1847 | MV J39045 | AF436006 | AF436046 | 10 |
| *Laomedia healyi* Yaldwyn & Wear, 1970 | MV J40697 | EU874996 | EU874946 | 3 |
| *Naushonia* sp. | ULLZ 8915 | EU874987 | EU874937 | |
| **Thomassiniidae** | | | | |
| *Thomassinia gebioides* de Saint Laurent, 1979 | ULLZ 8903 | EU874986 | EU874936 | |
| Thomassiniidae [unnamed genus] | ULLZ 7752 | EU874971 | EU874921 | |
| **Upogebiidae** | | | | |
| *Gebiacantha plantae* (Sakai, 1982) | MV J44914 | EU875000 | EU874950 | 3 |
| *Pomatogebia operculata* (Schmitt, 1924) | ULLZ 6905 | EU874957 | EU874907 | |
| *Upogebia acanthura* (Coelho, 1973) | ULLZ 7593 | EU874968 | EU874918 | |
| *Upogebia affinis* (Say, 1818) | MV J40668 | AF436007 | AF436047 | 10 |
| *Upogebia annae* Thistle, 1973 | ULLZ 6757 | EU874955 | EU874905 | |
| *Upogebia annae* Thistle, 1973 | ULLZ 7009 | EU874958 | EU874908 | |

**Table 1.** continued.

| Taxon Name | Catalog No. | Acc. No. 18S | Acc. No. 16S | S |
|---|---|---|---|---|
| *Upogebia annae* Thistle, 1973 | ULLZ 7522 | EU874966 | EU874916 | |
| *Upogebia coralliflora* Williams & Scott, 1989 | ULLZ 6765 | EU874956 | EU874906 | |
| *Upogebia spinistipula* Williams & Heard, 1991 | ULLZ 7360 | EU874964 | EU874914 | |
| **Thalassinoidea** | | | | |
| **Thalassinidae** | | | | |
| *Thalassina squamifera* de Man, 1915 | MV J41662 | EU874997 | EU874947 | 3 |

*Naushonia* is not isolated from the other two laomediid genera at high support values, while Axianassidae + Laomediidae form a sister group to Thalassinidae.

### 3.3 *The families of "Axiidea"*

Within this large clade, there is high support for grouping together members of Axiidae and Calocarididae into an internal clade, separated from representatives of all other axioid families as well as from Callianassidae and Ctenochelidae. Branch lengths are short for some of these separations, but support values are generally high. The two calocaridid genera included in this study, *Calaxiopsis* (already listed by Sakai & Ohta 2005 as an axiid) and *Calocaris*, were placed separately within Axiidae, casting doubt on the monophyly of Calocarididae (although it must be remembered that only two of six calocaridid genera and five of 21 axiid genera were included).

While clearly separated from the axiid and calocaridid genera, other axioid families were positioned immediately basal to the callianassids and ctenochelids, but without majority rule support. Although represented by only two species each, there is no evidence to contradict monophyly of either the Micheleidae or the Ctenochelidae (noting that we treat both *Dawsonius* and *Gourretia* within the Ctenochelidae, rather than in the Gourretiidae of Sakai 1999). Sister-group positioning of the Strahlaxiidae to a clade encompassing representatives of the Callianideidae and Thomassiniidae appears atypical at first glance, but Poore (1994) found *Strahlaxius* closer to Micheleidae than to Axiidae. Incorporation of *Callianidea* in a clade including *Thomassinia* and a thomassiniid-like species raises questions about the distinctiveness of these families. Within Callianassidae, there is some evidence to support current subfamilial groupings as well as some evidence of polyphyly among representative taxa, especially of the genera *Biffarius* and *Sergio*. These and other generic level issues are independently addressed in an expanded analysis of the Callianassidae (Felder & Robles, this volume).

## 4 DISCUSSION

### 4.1 *Monophyly or paraphyly*

While current schemes of classification treat Thalassinidea as an infraorder, issues such as its monophyly and its phylogenetic position, as well as the phylogenetic relationships among its constituent families, remain under debate. We have presented here a combined analysis based on two molecular datasets, one mitochondrial and one nuclear, and it does not support a monophyletic Thalassinidea.

When de Saint Laurent (1973) raised the subfamily Upogebiinae to family rank, she did so after concluding that its morphological differences were too striking to maintain the group within Callianassidae. In doing so, she commented on the family's affinities and suggested Upogebiidae was more closely related to Laomediidae and Thalassinidae than to Callianassidae and Axiidae. She relied on differences in larval morphology as justification, citing by footnote "Gurney ... 1940," in obvious reference to Gurney (1942).

Later, de Saint Laurent (1979a) cited differences in the union between the epistome and the carapace, in the number and kind of chelate legs, in larval development, in the appendix interna and in other undefined features, while discussing the difficulty in precisely defining what she called "Thalassinacea." Larval morphology had long suggested that Thalassinidea was composed of two distinct groups (Gurney 1938). One, the Callianassidae and Axiidae, was concluded to have a "homarine" zoea somewhat resembling that of Nephropidae, and the other, Upogebiidae and Laomediidae, an "anomuran" zoea (see also Gurney 1942; Felder et al. 1985). On the basis of this evidence, de Saint Laurent (1979a) suggested two groups, which she termed "sections": "Gebiidea" (Upogebiidae, Laomediidae s. l., and Thalassinidae) and "Axiidea" (Axiidae and Callianassidae). She illustrated these as two of ten distinct lines in a "radiation Triasique" of Reptantia (de Saint Laurent 1979a: Fig, 1). Nevertheless, she described tentative links between Gebiidea and "Dromiacea," "Anomala" and Brachyura as "sans doute articifielle." Subsequently, de Saint Laurent (1979b) followed this with more detailed diagnoses of the superfamily Axioidea and its families, Axiidae, Callianideidae, and Callianassidae, though it is unclear whether she believed the group to be other than monophyletic.

Poore (1994) conducted a morphologically based analysis of 22 genera of Thalassinidea, concluding that monophyly of the infraorder Thalassinidea was supported by the presence of a marginal setal fringe on pereopod 2 of all members. The monophyly view has been supported by some recent morphological and molecular studies. Morphological analyses of Dixon et al. (2003) found Thalassinidea to be monophyletic, with three characters to support that view: the curved articulation between the ischium and merus in pereopod 1; the presence of a row of setae on pereopod 2 (same as Poore 1994); and an enlarged and lobate seventh thoracic sternite (observed first by Scholtz & Richter 1995). A more recent analysis of Decapoda, based on a combination of morphological and molecular data, also supported monophyly of Thalassinidea (Ayhong & O'Meally 2004). Their study included sequences of the 16S, 18S, and 28S genes as well as morphological characters in what was called a "total evidence" analysis. These authors found the five families of Thalassinidea included in their parsimony analysis to be monophyletic. In a molecular analysis of 16S and 18S data for 13 thalassinidean genera, Tudge & Cunningham (2002) previously had shown only weak support for monophyly of Thalassinidea on the basis of 18S sequences, and no support for monophyly on the basis of 16S sequences. Interestingly, their composite tree showed the clade including Upogebiidae, Axianassidae, Thalassinidae, and Laomediidae positioned as a sister clade to five decapod outgroups, though at low support values. The molecular analysis of Porter et al. (2005) also infers thalassinideans to be monophyletic, but this analysis included representatives of only one family (Callianassidae), which we also find to be monophyletic, so no conclusion can be drawn for thalassinideans overall.

On the other hand, the molecular phylogenetic analyses of Morrison et al. (2002) presented evidence for polyphyly of Thalassinidea. Their analyses, based on sequences of the 16S, 18S, COII, and 28S genes, showed *Jaxea* and *Upogebia* (representing the families Laomediidae and Upogebiidae, respectively) allied with *Panulirus* (infraorder Palinura or Achelata) in a separate clade from *Neotrypaea* and *Callichirus* (representing the family Callianassidae). These results were used to show that Thalassinidea does not belong among the true Anomura, but explanation for the two separated clades of Thalassinidea was appropriately not addressed, given the few constituent taxa represented. It is noteworthy that Morrison et al. (2002), using 16S, 18S, 28S, and one additional gene, found thalassinideans to be paraphyletic. This different result from that of Ayhong & O'Meally (2004) could have resulted from inclusion of the COII gene by Morrison et al. and/or inclusion of the morphological database by Ahyong & O'Meally.

This debate continues, published results being difficult to compare between analyses because of differences in taxa chosen, data used, and phylogenetic methods. Sakai (2005) and Sakai & Sawada (2006) found thalassinideans to be "diphyletic" on the basis of pyloric ossicle structure, and they proposed superfamily or infraordinal separations on this basis, though without discussing group relationships. Very recent work on the basis of protein-coding genes (Tsang et al. 2008a) has shown evidence for at least paraphyly among the six included representatives of thalassindeans, the evidence for polyphyly having only weak support. The four axiids and calocaridids representing the Axiidea at the very least form a monophyletic clade. Their single thalassinid and single upogebiid did not group together as representatives of Gebiidea, but poor internodal support makes their positioning questionable.

Our molecular analysis argues against monophyly of the infraorder Thalassinidea, thus supporting conclusions of de Saint Laurent (1979a, b), Tudge et al. (2002), and Sakai & Sawada (2006), though not for the same reasons. Rooted to the Caridea, we find that the thalassinideans are distributed among two clades for which the rank of infraorder is more appropriate than superfamily, as the latter could imply membership in the same infraorder. One of these clades, first referred to as Gebiidea by de Saint Laurent (1979a), includes Upogebiidae, Thalassinidae, Axianassidae, and Laomediidae (Fig. 1: Clade-A). We reject the unnecessary replacement of this name by a restricted Thalassinidea (*sensu* Sakai & Sawada 2006) or redefined superfamily Thalassinoidea (*sensu* Sakai 2005; Tsang et al. 2008b).

The second clade we refer to as infraorder Axiidea, again using the term that de Saint Laurent (1979b) originally applied (Fig. 1: Clade-B). This is a monophyletic grouping of Axiidae, Calocarididae, Micheleidae, Thomassiniidae, Callianideidae, Strahlaxiidae, Ctenochelidae, and Callianassidae that is with strong support allied more closely to other decapod infraorders (outgroup taxa) than to the Gebiidea (Clade-A). We prefer Axiidea over the synonymous infraorder Callianassidea (*sensu* Sakai & Sawada 2006) or superfamily Callianassoidea (*sensu* Sakai 2005; Tsang et al. 2008b).

Our results differed somewhat from those of Tsang et al. (2008b: Fig. 1), even though we used the same 16S and 18S genetic markers. Among possible explanations are the following: 1) Our set of thalassinidean taxa was significantly different (55 thalassinidean specimens representing an additional family, more genera, and more species than in their sample of 27); 2) the two efforts may have differed slightly in parameters used to obtain alignments and in the way saturated fragments of genes were discarded (though unlikely as the efforts defined similar large clades); and 3) their selection of outgroups and of analyses was admittedly not designed to address the issue of thalassinidean monophyly. In addition, one could question our rooting of the tree to the Caridea even though, as noted in Results above, we selected this group in a preliminary unrooted analysis. To ascertain the impact of this selection on our analysis, we conducted an independent phylogenetic analysis excluding the Caridea but including all other outgroups otherwise used in Figure 1. That tree (not shown) showed no support for a monophyletic Thalassinidea and produced the same general groupings as in Figure 1.

Regardless of the rank ultimately assigned to our Clade-A and Clade-B, morphological characters summarized by other authors can be applied to diagnoses. The separation is supported by consistent group differences in larval morphology (Gurney 1938, 1942), possibly gastric mill morphology (Sakai 2005; Sakai & Sawada 2006), and the degree of chela development on the second pereopod (de Saint Laurent 1979a, b), even though questions remain as to whether all these shared character states represent synapomorphies. For example, while the second pereopod is never fully chelate in our Clade-B, as opposed to Clade-A, Poore (1994) has argued that this feature may have arisen multiple times among Decapoda. Our Clade-A is additionally supported by its members all lacking appendices internae on the pleopods, while they are present (with few exceptions among the axiids) in Clade-B.

## 4.2 *Previously applied superfamilies*

The most widely used current classification of the present infraorder Thalassinidea distributes all of its member families into three superfamilies, Axioidea, Thalassinoidea, and Callianassoidea (see Poore 1994; Martin & Davis 2001). Neither our analyses nor those of Tsang et al. (2008b), Sakai (2005), or Sakai & Sawada (2006) supported the monophyly of these superfamilies. One of our two major clades, Gebiidea, clustered representatives of the families Upogebiidae, Laomediidae, Axianassidae, and Thalassinidae. Poore's (1994) scheme would have the first three of these members of Callianassoidea and the last one a member of the Thalassinoidea (Fig. 1: Clade-A). Our second major clade, Axiidea, mixes members of Axioidea and Callianassoidea (Fig. 1: Clade-B).

In a morphologically based analysis, Dixon et al. (2003) supported Poore's (1994) superfamilies, but with some hesitation. Their only representative of Laomediidae (*Jaxea*) was positioned at the base of the clade for Thalassinidea instead of being clustered with Callianassidae, Callianideidae, and Upogebiidae (Dixon et al. 2003: Fig. 6). However, their goal was not to resolve internal relationships within Thalassinidea (their Thalassinida) but to suggest a new classification for the order Decapoda. We conclude that their support for the current superfamilies was overstated, since they included only one representative from each of five families of Thalassinidea and two specimens for another two families. In their analysis of the Decapoda, Ahyong & O'Meally (2004) also included five families of Thalassinidea. While having already noted our disagreement with their finding of monophyly for the group overall, we do agree to large extent with the interfamilial relationships they reported. They grouped *Upogebia*, *Jaxea*, and *Thalassina* in a single clade similar to our Clade-A. They also found *Biffarius*, *Callichirus*, and *Neaxius* in a second clade that resembles our Clade-B.

Sakai (2005) compared gastric mills among representatives of some thalassinidean families. He concluded that Thalassinidea should be divided into two superfamilies, Callianassoidea and Thalassinoidea, very similar to the clades we distinguish molecularly, acknowledging that his revision was being suggested on the basis of a single character and without comprehensive study of group representatives. In a second paper Sakai & Sawada (2006) elaborated on these observations and elevated the superfamilies to the infraorders Thalassinidea and the new name Callianassidea, effectively replacing de Saint Laurent's names, Gebiidea and Axiidea. They diagnosed their infraorders only in terms of pyloric ossicle shape and sought no supportive evidence from any other characters.

## 4.3 *Infraorder composition and internal family relationships*

Within our Clade-A, Gebiidea, family proximities are very similar to those reported in the recent molecular studies of Tsang et al. (2008b). As in Tudge & Cunningham (2002), members of the family Upogebiidae are grouped independently from the other three families, Thalassinidae, Axianassidae, and Laomediidae. Our support for separation of the family Axianassidae is weaker than that of Tudge & Cunningham (2002), but we judge neither our analysis nor that of Tsang et al. (2008b) to justify abandonment of this family. Topological placement appears to be external to the monophyletic Laomediidae, and a more robust coverage of axianassid species should be undertaken in subsequent analyses.

Recent work by Batang & Suzuki (2003) has examined the potential phylogenetic significance of gill-cleaning adaptations, as reviewed by Tsang et al. (2008b), calling attention to the striking dissimilarity of those in Upogebiibae from arrangements in the other three families that we place into the Gebiidea. Under our scenario, reported similarities of these structures in the Upogebiidae to those in the Callianassidae and Ctenochelidae must be regarded as convergent character states, likely in adaptation to similar sedimentary environments.

Within our Clade-B, Axiidea, we observe short branch lengths separating several of the primary clades, much as found by Tsang et al. (2008b). Separation of the Axiidae as the most basally positioned family is moderately to well supported, even at such short branch lengths. We also found that our molecular data did not support separation of monophyletic Calocarididae from a monophyletic Axiidae. Our calocaridid examples were unambiguously embedded in two separate subclades of the Axiidae. While our analysis did not include a representative of Eiconaxiidae, one was included in the analysis by Tsang et al. (2008b) and was clearly positioned among other clades of Axiidae and Calocarididae. Their evidence argues against retaining Eiconaxiidae as a separate family.

In as far as our two representatives tell us, Micheleidae are monophyletic and basal to the non-axiid lineage of Axiidea. Callianideidae (one species) appears embedded within Thomassiniidae

(*Thomassinia gebioides* plus a yet-to-be-named genus of thomassiniid). Tsang et al. (2008b) found a highly supported sister relationship between Micheleidae and Callianideidae but included no examples of Thomassiniidae.

Strahlaxiidae is in turn positioned topologically as a sister group to Callianideidae + Thomassiniidae, but without support, and the branch separating this entire group from the Callianassidae + Ctenochelidae clade lacks support. Given these poor resolutions, we must forego further interpretations.

Our analysis supports a monophyletic family Callianassidae but offers only modest support for positioning of the family Ctenochelidae as its sister group, a placement suggested on the basis of morphology (Poore 1994; Tudge et al. 2000). Without support, it was similarly positioned in the analyses of Tsang et al. (2008b), where the family was represented by the genus *Ctenocheles*. At moderate levels of support, the family Ctenochelidae appears to be monophyletic on the basis of the genera *Gourretia* and *Dawsonius* in our analyses. While our topology reflects some expected group relationships within the family Callianassidae, that issue is addressed more comprehensively in separate coverage of callianassid taxa (Felder & Robles this volume).

## 5 CONCLUSIONS

Our analysis shows paraphyly for what is presently referred to as the infraorder Thalassinidea and does not support its presently assigned taxa being redistributed among two constituent superfamilies or other subdivisons. There is no support for the superfamilies Axioidea, Thalassinoidea, and Callianassoidea (Poore 1994; Martin & Davis 2001). Rather, we support establishment of two separate infraorders that we label in accord with names introduced by de Saint Laurent (1979a): infraorder Gebiidea, composed of families Upogebiidae, Thalassinidae, Axianassidae, and Laomediidae; and infraoder Axiidea, composed of Axiidae, Calocarididae, Micheleidae, Thomassiniidae, Callianideidae, Strahlaxiidae, Ctenochelidae, and Callianassidae.

Our analysis supports family status for Axianassidae, Axiidae, Callianassidae, Ctenochelidae, Micheleidae, and Upogebiidae. While the limited support and sampling in our present analysis cannot confirm validity of the family Strahlaxiidae, there is no basis upon which to merge it with another family. On the other hand, its close relatives, Thomassiniidae and Callianideidae, appear to not represent distinct families. Similarly, highly supported clades in our own work and that of Tsang et al. (2008) show the families Eiconaxiidae and Calocaridiidae to be embedded within the Axiidae, rather than deserving independent family rank.

We do not suggest that our present analysis closes this debate, as sampling of genetic diversity in this group remains low. Rather, our continuing efforts are focused on adding representative taxa for molecular analyses, accumulating sequence data for additional genes, and preparing of a more thorough reappraisal of morphological characters. Our hope is that a reconciliation of molecular and morphological analyses will lead to a more stable classification.

## ACKNOWLEDGEMENTS

We thank C. Cunningham for making available some of the sequences used in the present analysis. Among many colleagues who in varied ways contributed to our specimen holdings for this analysis, we especially thank F. Alvarez, A. Anker, R. Atkinson, F. Bird, P. Clark, R. Collin, P. Collins, S. Fredericq, S. De Grave, G. Hernandez, R. Lemaitre, the late R. Manning, V. Paul, M. Rice, A. Tamaki, J.L. Villalobos, and M. Wicksten. This study was supported under funding from the U.S. National Science Foundation to D. Felder (BS&I grant no. DEB-0315995 & DEB/AToL grant no. EF-0531603), along with small travel grants from the Smithsonian Marine Station, Fort Pierce, Florida, and cruise support from the Smithsonian Tropical Research Institute, Panama. This is contribution number 738 from the Smithsonian Marine Station, Fort Pierce, Florida, and number 125 from the University of Louisiana Laboratory for Crustacean Research.

# REFERENCES

Abele, L.G. & Felgenhauer, B.E. 1986. Phylogenetic and phenetic relationships among the lower Decapoda. *J. Crust. Biol.* 6: 385–400.

Ahyong, S.T. & O'Meally, D. 2004. Phylogeny of the Decapoda Reptantia: resolution using three molecular loci and morphology. *Raffles Bull. Zool.* 52: 673–693.

Akaike, H. 1974. A new look at the statistical model identification. *IEEE Trans. Automat. Contr.* 19: 719–723.

Apakupakul, K., Siddall, M.E. & Burreson, E.M. 1999. Higher level relationships of leeches (Annelida: Clitellata: Euhirudinea) based on morphology and gene sequences. *Mol. Phylogenet. Evol.* 12: 350–359.

Atkinson, R.J.A. & Taylor, A.C. 2004. Aspects of the biology and ecophysiology of thalassinidean shrimps in relation to their burrow environment. In: Tamaki, A. (ed.), *Proceedings of the Symposium on "Ecology of large bioturbators in tidal flats and shallow sublittoral sediments–From individual behavior to their role as ecosystem engineers"*: 45–51. Nagasaki, Japan: Nagasaki Univ. Press.

Barnard, K.H. 1950. Descriptive catalog of South African decapod Crustacea (Crabs and Shrimps). *Ann. S. Afr. Mus.* 38: 1–815.

Batang, Z.B. & Suzuki, H. 2003. Gill-cleaning mechanisms of the burrowing thalassinidean shrimps *Nihonotrypaea japonica* and *Upogebia major* (Crustacea: Decapoda). *J. Zool. (Lond.)* 261: 69–77.

Bird, F. 2000. Physicochemical and microbial properties of burrows of the deposit-feeding thalassinidean ghost shrimp *Biffarius arenosus* (Decapoda: Callianassidae). *Estuar. Coast. Shelf. Sci.* 51: 279–292.

Bird, F. 2004. The interaction between ghost shrimp activity and seagrass restoration. In: Tamaki, A. (ed.), *Proceedings of the Symposium on "Ecology of large bioturbators in tidal flats and shallow sublittoral sediments—From individual behavior to their role as ecosystem engineers"*: 71–75. Nagasaki, Japan: Nagasaki Univ. Press.

Borradaile, L.A. 1903. On the classification of the Thalassinidea. *Ann. Mag. Nat. Hist.* 7: 534–551.

Bouvier, E.L. 1940. *Décapodes Marcheurs. Faune de France 37*. Paris: Paul Lechevalier et Fils.

Bracken, H.D., De Grave, S. & Felder, D.L. (this volume). Phylogeny of the infraorder Caridea based on mitochondrial and nuclear genes (Crustacea: Decapoda). In: Martin, J.W., Crandall, K.A. & Felder, D.L. (eds.), *Crustacean Issues: Decapod Crustacean Phylogenetics*. Boca Raton, Florida: Taylor & Francis/CRC Press.

Bull, J.J., Huelsenbeck, J.P., Cunningham, C.W., Swofford D.L. & Waddell, P.J. 1993. Partitioning and combining data in phylogenetic analysis. *Syst. Biol.* 42: 384–397.

Coelho, V.R. 2004. Feeding behavior, morphological adaptations and burrowing in thalassinidean crustaceans. In: Tamaki, A. (ed.), *Proceedings of the Symposium on "Ecology of Large Bioturbators in Tidal Flats and Shallow Sublittoral Sediments–From Individual Behavior to Their Role as Ecosystem Engineers"*: 1–6. Nagasaki, Japan: Nagasaki Univ. Press.

Crandall, K.A. & Fitzpatrick, J.F. 1996. Crayfish molecular systematics: using a combination of procedures to estimate phylogeny. *Syst. Biol.* 45: 1–26.

Crandall, K.A., Harris, D.J. & Fetzner, J.W. Jr. 2000. The monophyletic origin of freshwater crayfish estimated from nuclear and mitochondrial DNA sequences. *Proc. R. Soc. Lond., B.* 267: 1679–1686.

de Man, J.G. 1928. The Decapoda of the Siboga-Expedition. Part VII. The Thalassinidae and Callianassidae collected by the Siboga-Expedition with some remarks on the Laomediidae. *Siboga-Expedite, 39 a 6, Leiden*: 1–187, pls. 1–20.

de Saint Laurent, M. 1973. Sur la systématique et la phylogénie des Thalassinidea: definition des familles des Callianassidae et des Upogebiidae et diagnose de cinq genres nouveaux (Crustacea Decapoda). *C. R. Hebd. Acad. Sci.* 277: 513–516.

de Saint Laurent, M. & Le Loeuff, P. 1979. Campagnes de la *Calypso* au large des côtes Atlantiques Africaines (1956 et 1959) (suite) 22. Crustacés Décapodes Thalassinidea. I. Upogebiidae et Callianassidae. *Résultats Scientifiques des Campagnes de la Calypso* 11: 29–101.

de Saint Laurent, M. 1979a. Vers une nouvelle classification des Crustacés Décapodes Reptantia. *Bulletin de l'Office National des Pêches République Tunisienne, Ministere de L'Agriculture* 3: 15–31.

de Saint Laurent, M. 1979b. Sur la classification et la phylogénie des Thalassinides: définitions de la superfamille des Axioidea, de la sous-famille des Thomassiniinae et de deux genres nouveaux (Crustacea Decapoda). *C. R. Hebd. Acad. Sci.* 288: 1395–1397.

Dixon, C.J., Ahyong, S. & Schram, F.R. 2003. A new hypothesis of decapod phylogeny. *Crustaceana* 76: 935–975.

Dworschak, P.C. 2000. Global diversity in the Thalassinidea (Decapoda). *J. Crust. Biol.* 20 (Special Issue 2): 238–245.

Dworschak, P.C. 2005. Global diversity in the Thalassinidea (Decapoda): an update (1998–2004). *Nauplius.* 13: 57–63.

Dworschak, P.C., Koller, H. & Abed-Navandi, D. 2006. Burrow structure and feeding behaviour of *Corallianassa longiventris* and *Pestarella tyrrhena* (Crustacea, Thalassinidea, Callianassidae). *Mar. Biol.* 148: 1369–1382.

Dworschak, P.C. & Ott, J.A. 1993. Decapod burrows in mangrove-channel and back-reef environments at the Atlantic Barrier Reef, Belize. *Ichnos* 2: 277–290.

Felder, D.L. 2001. Diversity and ecological significance of deep-burrowing macrocrustaceans in coastal tropical waters of the Americas (Decapoda: Thalassinidea). *Interciencia* 26: 2–12.

Felder, D.L., Martin, J.W. & Goy, J.W. 1985. Patterns in early postlarval development of decapods. In: Wenner, A.M. (ed.), *Crustacean Issues 2, Larval Growth*: 163–225. Rotterdam: Balkema.

Felder, D.L., Nates, S.F. & Robles, R. 2003. Hurricane Mitch: Impacts of Bioturbating Crustaceans in Shrimp Ponds and Adjacent Estuaries of Coastal Nicaragua. *USGS Open File Report* 03–179: 1–47.

Felder, D.L. & Robles, R. 2004. Insights of molecular analyses in phylogeny and ecology of American callianassids (Decapoda: Thalassinidea). *Program and Abstracts, 3rd Brazilian Congress and TCS Meeting, Florianópolis, Brazil*, Abstract No. 183: 101.

Felder, D.L. & Robles, R. (this volume). Molecular phylogeny of the family Callianassidae based on preliminary analyses of two mitochondrial genes. In: Martin, J.W., Crandall, K.A. & Felder, D.L. (eds.), *Crustacean Issues: Decapod Crustacean Phylogenetics*. Boca Raton, Florida: Taylor & Francis/CRC Press.

Felgenhauer, B.E. & Abele, L.G. 1983. Phylogenetic relationships among shrimp-like decapods. In: Schram, F.R. (ed.), *Crustacean Issues 1, Crustacean Phylogeny*: 291–311. Rotterdam: Balkema.

Giribet, G., Edgecombe, G.D. & Wheeler, W.C. 2001. Arthropod phylogeny based on eight molecular loci and morphology. *Nature* 413: 121–122.

Gurney, R., 1938. Larvae of decapod Crustacea. Part 5. Nephropsidea and Thalassinidea. *Discovery Reports* 17: 291–344.

Gurney, R. 1942. *Larvae of Decapod Crustacea*. London: Ray Society.

Hall, T.A. 1999. BioEdit: a user-friendly biological sequence alignment editor and analysis program for Windows 95/98/NT. *Nucl. Acids Symp. Ser.* 41: 95–98.

Kensley, B. 1989. New genera in the thalassinidean families Calocarididae and Axiidae (Crustacea: Decapoda). *Proc. Biol. Soc. Wash.* 102: 960–967.

Kim, W. & Abele, L.G. 1990. Molecular phylogeny of selected decapod crustaceans based on 18S rRNA nucleotide sequences. *J. Crust. Biol.* 10: 1–13.

Klerks, P.L., Felder, D.L., Strasser, K.M. & Swarzenski, P.W. 2007. Effects of ghost shrimp on zinc and cadmium in sediments from Tampa Bay, FL. *Mar. Chem.* 104: 17–26.

Le Loeuff, P. & Intès, A. 1974. Les Thalassinidea (Crustacea, Decapoda) du Golfe de Guinée systématique-écologie. *Cahiers de l'Office de Recherches Scientifiques et Techniques Outre-Mer, série Océanographique* 12: 17–69.

Manning, R.B. & Felder, D.L. 1991. Revision of the American Callianassidae (Crustacea: Decapoda: Thalassinidea). *Proc. Biol. Soc. Wash.* 104: 764-792.

Martin, J.W. & Davis, G.E. 2001. An updated classification of the Recent Crustacea. *Nat. Hist. Mus. Los Angeles County, Science Series* 39: 1–124.

McLaughlin, P.A. & Holthuis, L.B. 1985. Anomura versus Anomala. *Crustaceana* 49: 204–209.

Medlin, L.K., Elwood, H.J., Stickel, S. & Sogin, M.L. 1988. The characterization of enzymatically amplified eukaryotic Ids-like rRNAcoding regions. *Gene* 71: 491–499.

Morrison, C.L., Harvey, A.W., Lavery, S., Tieu, K., Huang, Y. & Cunningham, C.W. 2002. Mitochondrial gene rearrangements confirm the parallel evolution of the crab-like form. *Proc. R. Soc. London, B.* 269(1489): 345–350.

Nates, S.F., Felder, D.L. & Lemaitre, R. 1997. Comparative larval development in two species of the burrowing ghost shrimp genus *Lepidophthalmus* (Crustacea, Decapoda, Callianassidae). *J. Crust. Biol.* 17: 497–519.

Palumbi, S., Martin, A., Romano, S., McMillan, W.O., Stice, L. & Grabowski, G. 1991. *The Simple Fool's Guide to PCR*. Honolulu: Department of Zoology and Kewalo Marine Laboratory, University of Hawaii.

Pérez-Losada, M., Bond-Buckup, G., Jara, C.G. & Crandall, K.A. 2004. Molecular systematics and biogeography of the southern South American freshwater 'crabs' *Aegla* (Decapoda: Anomura: Aeglidae) using multiple heuristic tree search approaches. *Syst. Biol.* 53: 767–780.

Pérez-Losada, M., Jara, C.G., Bond-Buckup, G. & Crandall, K.A. 2002a. Phylogenetic relationships among the species of *Aegla* (Anomura: Aeglidae) freshwater crabs from Chile. *J. Crust. Biol.* 22: 304–313.

Pérez-Losada, M., Jara, C.G., Bond-Buckup, G., Porter, M.L. & Crandall, K.A. 2002b. Phylogenetic position of the freshwater anomuran family Aeglidae. *J. Crust. Biol.* 22: 670–676.

Pillay, D., Branch, G.M. & Forbes, A.T. 2007. Experimental evidence for the effects of thalassinidean sandprawn *Callianassa kraussi* on macrobenthic communities. *Mar. Biol.* 52: 611–618.

Poore, G.C.B. 1994. A phylogeny of the families of Thalassinidea (Crustacea: Decapoda) with keys to families and genera. *Mem. Qld. Mus.* 54: 79–120.

Poore, G.C.B. 1997. A review of the thalassinidean families Callianideidae Kossman, Micheleidae Sakai, and Thomassiniidae de Saint Laurent (Crustacea, Decapoda) with descriptions of fifteen new species. *Zoosystema* 19: 345–420.

Porter, M.L., Pérez-Losada, M. & Crandall, K.A. 2005. Model-based multi-locus estimation of decapod phylogeny and divergence times. *Mol. Phylogenet. Evol.* 37: 355–369.

Posada, D. & Crandall, K.A. 1998. MODELTEST: testing the model of DNA substitution. *Bioinformatics* 14: 817–818.

Robles, R. & Felder, D.L. 2004. Phylogenetic analysis of the genus *Lepidophthalmus* based on the 16S mitochondrial gene. *Program and Abstracts, 3rd Brazilian Crustacean Congress and TCS Meeting, Florianópolis, Brazil*, Abstract No. 195: 107.

Robles, R., Schubart, C.D., Conde, J.E., Carmona-Suárez, C., Alvarez, F., Villalobos, J.L. & Felder, D.L. 2007. Molecular phylogeny of the American *Callinectes* Stimpson, 1860 (Brachyura: Portunidae), based on two partial mitochondrial genes. *Mar. Biol.* 150: 1265–1274.

Sakai, K. 1992. The families Callianideidae and Thalassinidae, with the description of two new subfamilies, one new genus and two new species. *Naturalists, Publications of Tokushima Biological Laboratory, Shikoku University* 4: 1–33.

Sakai, K. 1999. Synopsis of the family Callianassidae, with keys to subfamilies, genera and species, and the description of new taxa (Crustacea: Decapoda: Thalassinidea). *Zool. Verh.* 326: 1–152.

Sakai, K. 2004. Dr. R. Plante's collection of the families Callianassidae and Gourretiidae (Decapoda, Thalassinidea) from Madagascar, with the description of two new genera and one new species of the Gourretiidae Sakai, 1999 (new status) and two new species of the Callianassidae Dana, 1852. *Crustaceana* 77: 553–601.

Sakai, K. 2005. The diphyletic nature of the infraorder Thalassinidea (Decapoda, Pleocyemata) as derived from the morphology of the gastric mill. *Crustaceana* 77: 1117–1129.

Sakai, K. & Ohta, S. 2005. Some thalassinid collections by R/V "Hakuhou-maru" and R/V "Tensei-maru", University of Tokyo, in the Sulu Sea, Philippines, and in Sagami Bay and Suruga Bay, Japan, including two new species, one new genus, and one new family (Decapoda, Thalassinidea). *Crustaceana* 78: 67–93.

Sakai, K. & Sawada, T. 2006. The taxa of the infraorders Astacidea, Thalassinidea, Palinura, and Anomura (Decapoda, Pleocyemata) classified by the form of the prepyloric ossicle. *Crustaceana* 78: 1353–1368.

Sakai, K. & de Saint Laurent, M. 1989. A check list of Axiidae (Decapoda, Crustacea, Thalassinidea, Anomura), with remarks and in addition descriptions of one new subfamily, eleven new genera and two new species. *Naturalists, Publications of Tokushima Biological Laboratory, Shikoku University* 3: 1–104.

Scholtz, G. & Richter, S. 1995. Phylogenetic systematics of the reptantian Decapoda (Crustacea, Malacostraca). *Zool. J. Linn. Soc.* 113: 289–328.

Schram, F.R. 2001. Phylogeny of decapods: moving toward a consensus. *Hydrobiologia* 449: 1–20.

Schubart, C.D., Cuesta, J.A. & Felder, D.L. 2002. Glyptograpsidae, a new brachyuran family from Central America: larval and adult morphology, and a molecular phylogeny of the Grapsoidea. *J. Crust. Biol.* 22: 28–44.

Stamatakis, A. 2006. RAxML-VI-HPC: Maximum likelihood-based phylogenetic analyses with thousands of taxa and mixed models. *Bioinformatics* 22: 2688–2690.

Stamatakis, A, Hoover P. & Rougemont J. 2008. A rapid bootstrap algorithm for the RAxML web-Servers. *Syst. Biol.* 57: 758–771.

Strasser, K.M. & Felder, D.L. 2000. Larval development of the ghost shrimp *Callichirus islagrande* (Schmitt) (Decapoda: Thalassinidea). *J. Crust. Biol.* 20: 100–117.

Strasser, K.M. & Felder, D.L. 2005. Larval development of the mud shrimp *Axianassa australis* (Decapoda: Thalassinidea) under laboratory conditions. *J. Nat. Hist.* 39: 2289–2306.

Swofford, D.L. 1998. *PAUP\* Phylogenetic analysis using parsimony and other methods*. Sunderland, Massachusetts: Sinauer Assoc.

Tavaré, S. 1986. Some probabilistic and statistical problems in the analysis of DNA sequences. In: Miura, R.M. (ed.), *Some Mathematical Questions in Biology-DNA Sequence Analysis*: 57–86. Providence, Rhode Island: Amer. Math. Society.

Tsang, L.M., Ma, K.Y., Ahyong, S.T., Chan, T.-Y. & Chu, K. H. 2008a. Phylogeny of Decapoda using two nuclear protein-coding genes: Origin and evolution of the Reptantia. *Mol. Phylogenet. Evol.* 48: 359–368.

Tsang, L.M., Lin, F.-J., Chu, K.H. & Chan, T.-Y. 2008b. Phylogeny of Thalassinidea (Crustacea, Decapoda) inferred from three rDNA sequences: implications for morphological evolution and superfamily classification. *J. Zool. Syst. Evol. Res.* 46: 216–223.

Tudge, C.C. 1997. Phylogeny of the Anomura (Decapoda, Crustacea): Spermatozoa and spermatophor morphological evidence. *Contrib. Zool.* 67: 125–141.

Tudge, C.C. & Cunningham, C.W. 2002. Molecular phylogeny of the mud lobsters and mud shrimps (Crustacea: Decapoda: Thalassinidea) using nuclear 18s rDNA and mitochondrial 16s rDNA. *Invertebr. Syst.* 16: 839–847.

Tudge, C.C., Poore, G.C.B. & Lemaitre, R. 2000. Preliminary phylogenetic analysis of generic relationships within the Callianassidae and Ctenochelidae (Decapoda: Thalassinidea: Callianassoidea). *J. Crust. Biol.* 20 (Special Issue 2): 129–149.

Zariquiey Alvarez, R. 1968. Crustáceos Decápodos ibéricos. *Inv. Pesq., Barcelona* 32: i-xi, 1–510, figs. 1–164.

# Molecular Phylogeny of the Family Callianassidae Based on Preliminary Analyses of Two Mitochondrial Genes

DARRYL L. FELDER & RAFAEL ROBLES

*Department of Biology, University of Louisiana, Lafayette, Louisiana, U.S.A.*

## ABSTRACT

Recent revisions in callianassid subfamilies and genera are questionable and appear to be incongruous with relationships evident in morphologically based phylogenetic reconstructions. We generated molecular phylogenetic trees for the closely related families Callianassidae and Ctenochelidae as well as for outgroup representatives of the family Axiidae. Fragments of the 16S and 12S rDNA mitochondrial genes were sequenced for a total of 46 species, representing 18 genera of Callianassidae, two genera of Ctenochelidae, and five genera of Axiidae. Of approximately 1000 potential mitochondrial basepair characters, 903 were used in final alignments. Resolution in our phylogenetic tree was limited at some basal nodes of the topology, as might be expected with the genes chosen for this analysis. Callianassinae formed a well-supported monophyletic group, but Cheraminae was included within it. Support was found for continued recognition of many separate genera in this group and for the naming of additional ones, as opposed to their wholesale reassignment to the clearly separated genus *Callianassa*. Groupings within Callichirinae were not well resolved, though the subfamily appears to be paraphyletic at low support values. Genera of this group were monophyletic except for *Sergio*, which is paraphyletic and of questioned validity. Eucalliacinae appears to be paraphyletic at low to medium support, suggesting that the genus *Calliaxina* may share common lineage with the Ctenochelidae.

## 1 INTRODUCTION

Recent attempts by Sakai (1999a, b, 2002, 2004, 2005) to comprehensively review and revise systematics of the family Callianassidae and its closest relatives (collectively known as ghost shrimps) have brought together a diffuse taxonomic literature but do not offer objective assessments toward a natural classification. Sakai's major revisions at the level of subfamilies and genera remain questionable (Dworschak 2007), especially in that many appear to be incongruous with relationships evident in phylogenetic reconstructions based upon morphological character analysis (Poore 1994; Tudge et al. 2000). This applies to numerous cases in which previously erected genera of the subfamily Callianassinae were recently synonymized by Sakai (2005), who put them into a very broadly defined genus *Callianassa* Leach, 1814. This action dismissed a restricted definition of the genus previously made by Manning & Felder (1991), while imposing a retrograde taxonomy that potentially masked diversity within the group. Similarly, within the subfamily Callichirinae, Sakai synonymized *Corallianassa* Manning, 1987, with *Glypturus* Stimpson, 1866, on a questionable basis (Dworschak 2007). In the subfamily Eucalliacinae, both *Eucalliax* Manning & Felder, 1991, and *Calliaxina* Ngoc-Ho, 2003, were placed into questionable synonymy with *Calliax* de Saint Laurent, 1973. At a somewhat higher level, membership of the family Ctenochelidae was restricted (Sakai 1999a), and the family Gourretiidae was established to receive *Gourretia* de Saint Laurent, 1973, and *Dawsonius* Manning & Felder, 1991.

The present effort addresses some of the above issues by molecular genetic methods. A previous paper of this volume (Robles et al.) used a combination of 16S mitochondrial and 18S nuclear gene sequences to examine overall phylogenetics of thalassinidean taxa, and a review of previous analyses bearing on all of its member groups was undertaken there. Some of the callianassid taxa that appear in Robles et al. (this volume) are included in the present work, as are yet others treated in earlier brief reports (Felder & Robles 2004; Robles & Felder 2004). With these, we here incorporate additional taxa to potentially enable more robust interpretations at the generic and subfamily levels. Our combined analysis is based strictly upon 16S and 12S mitochondrial gene sequences, rather than on genes more suited to resolution at higher taxonomic levels. The present analysis is considered preliminary in that it is somewhat biased to American materials, along with some available western Pacific and European specimens, the latter including *Callianassa subterranea*, type species of that genus.

## 2 METHODS

### 2.1 *Specimens included*

Ghost shrimps were collected in Belize, Brazil, Colombia, Costa Rica, Ghana, Greece, Jamaica, Japan, Mexico, Nicaragua, Panama, Scotland, Spain, USA, and Venezuela, with some of these being obtained as gifts or loans from museums (Table 1). When possible, specimens were initially frozen in seawater or glycerine at −70°C or −20°C. In other cases, or after tissue was extracted for DNA analysis, they were placed directly into 70% ethyl alcohol. Our sample consisted of 74 specimens representing 46 species in 25 genera of the families Ctenochelidae, Callianassidae, and Axiidae, the latter family serving as the outgroup. Outgroup selection was based upon findings of Robles et al. (this volume), which placed Axiidae in a sister clade to that of the aforementioned families within the infraorder Axiidea. Where utilized following a taxon, *s.l.* = *sensu lato* and *s.s.* = *sensu stricto*.

### 2.2 *DNA extraction, PCR, and sequencing*

DNA was extracted from muscle tissues excised from the abdomen or pleopods following standard protocols (Robles et al. 2007). Standard PCR amplification and automated sequencing protocols were used to sequence a fragment of approximately 550 bp (basepairs) of the 16S and 450 bp of the 12S rDNA mitochondrial genes. Both strands were sequenced. Primers used for PCR were 16ar (5-CGC CTG TTT ATC AAA AAC AT-3), 16br (5-CCG GTC TGA ACT CAG ATC ACG T-3) (Palumbi et al. 1991), 1472 (5-AGA TAG AAA CCA ACC TGG-3) (Crandall & Fitzpatrick 1996), and 16L2 (5-TGC CTG TTT ATC AAA AAC AT-3) (Schubart et al. 2002). Primers used for the 12S fragment were 12Sai (5'-AAA CTA GCA TTA GAT ACC CTA TT AT-3') (Palumbi et al. 1991) and 12H2 (5'-ATG CAC TTT CCA GTA CAT CTA C-3') (Colbourne & Hebert 1996).

### 2.3 *Phylogenetic analyses*

Consensus of complementary sequences was obtained with the Sequencher software program (ver. 4.7, Genecodes, Ann Arbor, MI). Multiple sequence alignment was conducted with the aid of BioEdit v.7.08.0 (Hall 1999) at the following settings: 6-2/6-2 penalty (opening-gap extension, pairwise/multiple alignment respectively). Saturated parts of the alignment were removed with the web-accessible program Gblocks v. 0.91b (Castresana 2000, Talavera & Castresana 2007). Base composition, pattern of substitution for pair-wise comparison, and analysis of variability along both fragments of the 16S and 12S mtDNA were performed as implemented in PAUP 4.0 beta 10 (Swofford 1998). Homogeneity of nucleotide frequency among taxa was also assessed for each gene

**Table 1.** List of specimens used for molecular analysis. Letter abbreviations following species names refer to collection sites; these are sometimes sequentially numbered to indicate specimens identified as the same species. Catalog numbers refer to the following collections: CNCR = Colección Nacional de Crustáceos, UNAM; ULLZ = University of Louisiana at Lafayette, Zoological Collection. Asterisk (*) indicates sequences also used in Robles et al. (this volume). If a second catalog number is reported for a sample, tissue was donated to the University of Louisiana at Lafayette and a ULLZ catalog number was assigned to it, while the second number belongs to the original voucher that remains in the indicated museum. GenBank accession number (Acc. No.) is listed for each gene.

| Family Taxon Name | Collection Site | Catalog No. | Acc. No. (16S) | Acc. No. (12S) |
|---|---|---|---|---|
| **Outgroup** | | | | |
| **Axiidae** | | | | |
| *Axiopsis serratifrons* (A. Milne-Edwards, 1873) BEL | Caribbean, Belize | ULLZ-5827 | EU882909 | EU875019 |
| *Axiopsis* sp. PCR | Pacific, Costa Rica | ULLZ 7750 | EU874920* | EU875012 |
| *Calaxius* sp. GMX | Gulf of Mexico, Mexico | ULLZ 7041 | EU874910* | EU875007 |
| *Calocaris caribbaeus* Kensley, 1996 GMX-1 | Gulf of Mexico, Louisiana, USA | ULLZ 7877 | EU882902 | EU875014 |
| *Calocaris caribbaeus* Kensley, 1996 GMX-2 | Gulf of Mexico, Louisiana, USA | ULLZ 8285 | EU874929* | EU875016 |
| *Coralaxius nodulosus* (Meinert, 1877) GMX | Gulf of Mexico, Mexico | ULLZ 7329 | EU874913* | EU875010 |
| *Paraxiopsis* sp. GMX | Gulf of Mexico, Mexico | ULLZ 7559 | EU874917* | EU875011 |
| **Ingroup** | | | | |
| **Callianassidae** | | | | |
| **Callianassinae** | | | | |
| *Biffarius biformis* (Biffar, 1971) AFL | Atlantic, Florida, USA | ULLZ 6540 | EU882910 | EU875020 |
| *Biffarius fragilis* (Biffar, 1970) AFL | Atlantic, Florida, USA | ULLZ 6406 | EU882911 | EU875021 |
| *Biffarius fragilis* (Biffar, 1970) CMX-1 | Caribbean, Mexico | CNCR 8997 | EU882906 | EU875017 |
| *Biffarius fragilis* (Biffar, 1970) CMX-2 | Caribbean, Mexico | CNCR 8997 | EU882907 | EU875018 |
| *Biffarius fragilis* (Biffar, 1970) JAM | Jamaica | ULLZ 6532 | EU882912 | EU875022 |
| *Callianassa?* sp. GMX-1 | Gulf of Mexico, Louisiana, USA | ULLZ 8279 | EU882903 | EU875015 |
| *Callianassa?* sp. GMX-2 | Gulf of Mexico, Louisiana, USA | ULLZ 6058 | EU882915 | EU875025 |
| *Callianassa subterranea* (Montagu, 1808) SCO | Atlantic, Scotland | ULLZ 6368 | EU882924 | EU875034 |
| *Gilvossius setimanus* (De Kay, 1844) GFL-1 | Gulf of Mexico, Florida, USA | ULLZ 4500 | EU882934 | EU875044 |
| *Gilvossius setimanus* (De Kay, 1844) GFL-2 | Gulf of Mexico, Florida, USA | ULLZ 4500 | EU882935 | EU875045 |
| *Gilvossius setimanus* (De Kay, 1844) GFL-3 | Gulf of Mexico, Florida, USA | ULLZ 4500 | EU882936 | EU875046 |

Table 1. continued.

| Family<br>Taxon Name | Collection Site | Catalog No. | Acc. No.<br>(16S) | Acc. No.<br>(12S) |
|---|---|---|---|---|
| *Neotrypaea*? sp. JAP | Pacific, Hydrocarbon vents, Japan | ULLZ 9414 | EU882908 | EU875050 |
| *Neotrypaea californiensis* (Dana, 1854) USA | Pacific, Washington, USA | ULLZ 6405 | EU882947 | EU875058 |
| *Neotrypaea gigas* (Dana, 1852) PMX-1 | Pacific, Baja California, Mexico | ULLZ 4121 | EU882948 | EU875059 |
| *Neotrypaea gigas* (Dana, 1852) PMX-2 | Pacific, Baja California, Mexico | ULLZ 4121 | EU882949 | EU875060 |
| *Neotrypaea gigas* (Dana, 1852) PMX-3 | Pacific, Baja California, Mexico | ULLZ 4121 | EU882950 | EU875061 |
| *Neotrypaea gigas* (Dana, 1852) PMX-4 | Pacific, Baja California, Mexico | ULLZ 5176 | EU882943 | EU875054 |
| *Neotrypaea gigas* (Dana, 1852) PMX-5 | Pacific, Baja California, Mexico | ULLZ 5176 | EU882944 | EU875055 |
| *Neotrypaea gigas* (Dana, 1852) PMX-6 | Pacific, Baja California, Mexico | ULLZ 5176 | EU882945 | EU875056 |
| *Nihonotrypaea harmandi* (Bouvier, 1901) JAP | Pacific, Japan | ULLZ 5468 | EU882952 | EU875063 |
| *Nihonotrypaea japonica* (Ortmann, 1891) JAP | Pacific, Japan | ULLZ 5470 | EU882953 | EU875064 |
| *Paratrypaea*? sp. HWI | Pacific, Hawaii, USA | ULLZ 7080 | EU882919 | EU875029 |
| *Paratrypaea bouvieri* (Nobili, 1904) JAP-1 | Pacific, Japan | ULLZ 6367 | EU882913 | EU875023 |
| *Paratrypaea bouvieri* (Nobili, 1904) JAP-2 | Pacific, Japan | ULLZ 6367 | EU882914 | EU875024 |
| *Pestarella tyrrhena* (Petagna, 1792) SPN | Mediterranean, Spain | ULLZ 6366 | EU882965 | EU875078 |
| *Pestarella tyrrhena* (Petagna, 1792) GRE-1 | Mediterranean, Greece | ULLZ 6360 | EU882899 | EU875005 |
| *Pestarella tyrrhena* (Petagna, 1792) GRE-2 | Mediterranean, Greece | ULLZ 6360 | EU882900 | EU875006 |
| **Callichirinae** | | | | |
| *Callichirus major* (Say, 1818) BRA-1 | Atlantic, São Paulo, Brazil | ULLZ 6055 | EU882917 | EU875027 |
| *Callichirus major* (Say, 1818) BRA-2 | Atlantic, São Paulo, Brazil | ULLZ 6056 | EU882918 | EU875028 |
| *Callichirus islagrande* (Schmitt, 1935) GMX | Gulf of Mexico, Mississippi, USA | ULLZ 6052 | EU882916 | EU875026 |
| *Callichirus seilacheri* (Bott, 1955) PNI | Pacific, Nicaragua | ULLZ 6053 | EU882921 | EU875031 |
| *Callichirus seilacheri* (Bott, 1955) PMX | Pacific, Baja California, Mexico | ULLZ 6054 | EU882920 | EU875030 |
| *Callichirus* sp. PMX | Pacific, Baja California, Mexico | ULLZ 4163 | EU882922 | EU875032 |
| *Corallianassa* sp. JAM | Caribbean, Jamaica | ULLZ 6530 | EU882923 | EU875033 |
| *Glypturus acanthochirus* Stimpson, 1866 VEN | Caribbean, Isla Margarita, Venezuela | ULLZ 5642 | EU882928 | EU875038 |

Table 1. continued.

| Family Taxon Name | Collection Site | Catalog No. | Acc. No. (16S) | Acc. No. (12S) |
|---|---|---|---|---|
| *Glypturus acanthochirus* Stimpson, 1866 JAM | Caribbean, Montego Bay, Jamaica | ULLZ 6528 | EU882929 | EU875039 |
| *Glypturus acanthochirus* Stimpson, 1866 CPA | Caribbean, Panama | ULLZ 6488 | EU882930 | EU875040 |
| *Glypturus* sp. GMX-1 | Gulf of Mexico, Louisiana, USA | ULLZ 4659 | EU882932 | EU875042 |
| *Glypturus* sp. GMX-2 | Gulf of Mexico, Louisiana, USA | ULLZ 4659 | EU882933 | EU875043 |
| *Grynaminna tamakii* Poore, 2000 JAP-1 | Pacific, Japan | ULLZ 5474 | EU882937 | EU875047 |
| *Grynaminna tamakii* Poore, 2000 JAP-2 | Pacific, Japan | ULLZ 5475 | EU882938 | EU875048 |
| *Grynaminna tamakii* Poore, 2000 JAP-3 | Pacific, Japan | ULLZ 5476 | EU882939 | EU875049 |
| *Lepidophthalmus jamaicense* (Schmitt, 1935) JAM | Caribbean, Jamaica | ULLZ 5189 | EU882941 | EU875052 |
| *Lepidophthalmus louisianensis* (Schmitt, 1935) USA | Gulf of Mexico, Louisiana, USA | ULLZ 5617 | EU882940 | EU875051 |
| *Lepidophthalmus turneranus* (White, 1861) GHA | Atlantic, Ghana, Africa | ULLZ 4737 | EU882942 | EU875053 |
| *Neocallichirus cacahuate* Felder & Manning, 1995 GFL | Atlantic, Florida, USA | ULLZ 3552 USNM 374706 | EU882946 | EU875057 |
| *Neocallichirus grandimana* (Gibbes, 1850) AFL | Atlantic, Florida, USA | ULLZ 6491 | EU882951 | EU875062 |
| *Neocallichirus maryae* (Schmitt, 1935) AFL | Atlantic, Florida, USA | ULLZ 6492 | EU882954 | EU875065 |
| *Neocallichirus variabilis* (Edmondson, 1944) USA-1 | Pacific, Hawaii, USA | ULLZ 6043 | EU882955 | EU875066 |
| *Neocallichirus variabilis* (Edmondson, 1944) USA-2 | Pacific, Hawaii, USA | ULLZ 6039 | EU882957 | EU875068 |
| *Neocallichirus variabilis* (Edmondson, 1944) USA-3 | Pacific, Hawaii, USA | ULLZ 6045 | EU882956 | EU875067 |
| *Neocallichirus variabilis* (Edmondson, 1944) USA-4 | Pacific, Hawaii, USA | ULLZ 6047 | EU882958 | EU875069 |
| *Neocallichirus* sp. 1 PNI | Pacific, Nicaragua | ULLZ 4838 | EU882959 | EU875072 |
| *Neocallichirus* sp. 2 PNI | Pacific, Nicaragua | ULLZ 6536 | EU882961 | EU875074 |
| *Sergio mericeae* Manning & Felder, 1995 AFL | Atlantic, Florida, USA | ULLZ 6493 | EU882960 | EU875073 |
| *Sergio trilobata* (Biffar, 1970) GFL-1 | Gulf of Mexico, Florida, USA | ULLZ 4501 | EU882962 | EU875075 |
| *Sergio trilobata* (Biffar, 1970) GFL-2 | Gulf of Mexico, Florida, USA | ULLZ 4501 | EU882963 | EU875076 |
| *Sergio trilobata* (Biffar, 1970) GFL-3 | Gulf of Mexico, Florida, USA | ULLZ 4501 | EU882964 | EU875077 |

Table 1. continued.

| Family<br>Taxon Name | Collection Site | Catalog No. | Acc. No.<br>(16S) | Acc. No.<br>(12S) |
|---|---|---|---|---|
| **Cheraminae** | | | | |
| *Cheramus* sp. PCR | Pacific, Costa Rica | ULLZ 7751 | EU882901 | EU875013 |
| *Cheramus marginata* (Rathbun, 1901) GMX | Gulf of Mexico, Louisiana, USA | ULLZ 7313 | EU874912* | EU875009 |
| **Eucalliacinae** | | | | |
| *Calliaxina sakaii* (de Saint Laurent, 1979) JAP-1 | Pacific, Japan | ULLZ 8894 | EU882904 | EU875070 |
| *Calliaxina sakaii* (de Saint Laurent, 1979) JAP-2 | Pacific, Japan | ULLZ 8894 | EU882905 | EU875071 |
| *Eucalliax* sp. COL | Caribbean, Rosario Islands, Colombia | ULLZ 6543 | EU882926 | EU875036 |
| *Eucalliax* sp. JAM | Caribbean, Montego Bay, Jamaica | ULLZ 6531 | EU882927 | EU875037 |
| **Ctenochelidae** | | | | |
| *Dawsonius latispina* (Dawson, 1967) GMX | Gulf of Mexico, Mexico | ULLZ 7306 | EU874911* | EU875008 |
| *Gourretia* sp. GMX | Gulf of Mexico, Louisiana, USA | ULLZ 4673 | EU882925 | EU875035 |
| *Gourretia biffari* Blanco & Arana, 1994 CPA | Caribbean, Panama | ULLZ 5757 | EU882931 | EU875041 |

with a $\chi^2$ test as implemented in PAUP. Previous to the analysis of the combined data, we performed an incongruence length difference (ILD) test or partition homogeneity test (Bull et al. 1993), as implemented in PAUP, to determine whether the 16S and 12S genes could be considered samples of the same underlying phylogeny.

Phylogenetic analyses were conducted using MRBAYES for Bayesian analysis (BAY) and PAUP 4.0 beta 10 (Swofford 1998) for both maximum parsimony (MP) and neighbor joining (NJ) analyses; maximum likelihood (ML) analysis was conducted with RAxML v.7.0.4 (Stamatakis 2006) using the online version at the Cyberinfrastructure for Phylogenetic Research (CIPRES) website (Stamatakis et al. 2008). Prior to conducting the BAY and NJ analyses, the model of evolution that best fit the data was determined with the software MODELTEST (Posada & Crandall 1998). Maximum likelihood analysis was conducted with the default parameters for RAxML for the GTR model of evolution. Bayesian analysis was conducted by sampling one tree every 1,000 generations for 2,000,000 generations, starting with a random tree, thus obtaining 2,001 trees. A preliminary analysis showed that stasis was reached at approximately 75,000 generations. Thus, we discarded 101 trees corresponding to the first 100,000 generations and obtained a 50% majority rule consensus tree from the remaining 1,900 saved trees. NJ analysis was carried out with a distance correction set with the parameters obtained from MODELTEST (Posada & Crandall 1998). MP analysis was performed as a heuristic search with gaps treated as a fifth character, multistate characters interpreted as uncertain, and all characters considered as unordered. The search was conducted with a random sequence addition and 1,000 replicates, including tree bisection and reconnection (TBR) as a branch-swapping option; branch swapping was performed on the best trees only. To determine confidence values for the resulting trees, we ran 2,000 bootstrap pseudo-replicates for NJ and MP analysis, based on the same parameters as above. For ML analysis, we selected the option to automatically determine the number of bootstraps to be run in RAxML. Thus, 200 bootstrap pseudo-replicates were run. On the molecular trees, confidence values >50% were reported for ML, MP, and NJ analyses (bootstraps), while for the BAY analysis values were reported for posterior probabilities of the respective nodes among all the saved trees. Sequences as well as alignments have been submitted to GenBank as a Popset.

## 3 RESULTS

### 3.1 Description of datasets and model selection

We obtained sequences for 37 species of Callianassidae belonging to 18 genera. Our final alignment included 903 bp, 520 for the 16S and 383 bp for the 12S sequence data (excluding primer regions, saturated and ambiguous fragments of both genes). From these, 386 characters were found to be constant, 62 were variable but parsimony-uninformative, and 455 were parsimony-informative. The ILD test showed no significant incongruence (P = 0.110). Thus we used the combined 16S and 12S dataset for our phylogenetic analysis. The nucleotide composition of this dataset can be considered homogeneous ($\chi^2$ = 180.21, df = 219, p = 0.97), with a larger percentage of A-T (33.34%–34.54%, respectively). The best fitting model of substitution, selected with the Akaike information criterion (AIC, Akaike 1974) as implemented in MODELTEST (Posada & Crandall 1998), was the transversional model with invariable sites and a gamma distribution (TVM+$\Gamma$+$\delta$) (Rodríguez et al. 1990) and with the following parameters: assumed nucleotide frequencies: A = 0.3716, C = 0.1258, G = 0.1317, T = 0.3710; substitution rates A-C = 1.1541, A-G = 8.3551, A-T = 1.5835, C-G = 0.5502, C-T = 8.3551, G-T = 1.0000; proportion of invariable sites $\Gamma$ = 0.3104; variable sites followed a gamma distribution with shape parameter $\delta$ = 0.5690. These values were used for both NJ and BAY analyses.

### 3.2 Tree topologies, relations to Ctenochelidae, and basally positioned groups

All four phylogenetic methods produced similar tree topologies (Fig. 1). We illustrated one of two equally parsimonious trees of length 3013, CI = 0.326, and RI = 0.713, noting that both MP trees produced the same topology. Within the family Callianassidae, representatives of the four subfamilies included in our analysis were not uniformly monophyletic. The subfamily Eucalliacinae was not only paraphyletic (partitioned between Clades A and B, Fig. 1) but also more basally positioned than traditional classification would predict. Members of the genus *Calliaxina* were unexpectedly placed as a sister clade to members of Ctenochelidae, albeit only at low to moderate support levels. Regardless of their topological placement in our tree, three species representing two genera of Ctenochelidae formed a well-supported monophyletic group.

### 3.3 The Callichirinae

Clade C (Fig. 1) included all sampled genera presently assignable to the subfamily Callichirinae, except for *Lepidophthalmus*. *Lepidophthalmus* was instead positioned in clade D immediately basal to the Callianassinae, but at low support in ML and BAY analyses and without support in the MP and NJ analyses (75/–/–/59). Thus, *Lepidophthalmus* is here regarded as a monophyletic clade of unresolved subfamily assignment in our molecular analysis. Grouping of the Callichirinae was not well-resolved, but present topology suggests it is paraphyletic, though some clades are presently positioned at low support values. While clade C topologically grouped all members of the Callichirinae other than *Lepidophthalmus*, this node was not supported. Furthermore, genera assigned to the Callichirinae were not well-resolved in terms of intergeneric relationships, but with one exception were separated from one another with strong support. Only a single representative of *Corallianassa* was included, but multiple specimens were grouped for each of the genera *Callichirus* Stimpson, 1866, *Glypturus*, *Grynaminna* Poore, 2000, and *Neocallichirus* Sakai, 1988. Those for *Grynaminna* were all *a priori* assignable to *G. tamakii*, but all of the other three included multiple species, even when species names could not be assigned. Clearly grouped as a genus, species of *Callichirus* included at least one new species to be named from the eastern Pacific. Likewise, *Glypturus* included a long-recognized but unnamed species from the Gulf of Mexico, and *Neocallichirus* included two unnamed species from the Pacific coast of Nicaragua. An exception to monophyly was seen in branch positioning for two of the species presently assigned to *Sergio* Manning & Lemaitre, 1994, as *S. mericeae* and *S. trilobata* were positioned paraphyletically. It was also evident that *S. mericeae*, the species closest to *S. guassutinga* (Rodrigues, 1971) (type species of the genus), was placed unambiguously within what is otherwise a monophyletic grouping of species assignable to *Neocallichirus*. This raises a question as to the validity of the genus and, regardless of that issue, argues for generic reassignment of *S. trilobata*.

### 3.4 The Cheraminae and Callianassinae

Clade D (Fig. 1) included representatives of seven genera usually assigned to the subfamily Callianassinae and one assigned to the Cheraminae, in addition to *Lepidophthalmus*, which, as noted above, was questionably positioned as a basal branch with low support. Callianassinae formed a well-supported monophyletic group, but Cheraminae was included within it, also with strong support. While the two species representing the Cheraminae were clearly assignable to the genus *Cheramus* Bate, 1888, only one was assignable to a known species, given the need for further comparative studies and formal descriptions of several new congeners. Support was found for continued recognition of many separate genera in the Callianassinae, including *Pestarella* Ngoc-Ho, 2003, *Gilvossius* Manning & Felder, 1992, *Biffarius* Manning & Felder, 1991, *Neotrypaea*

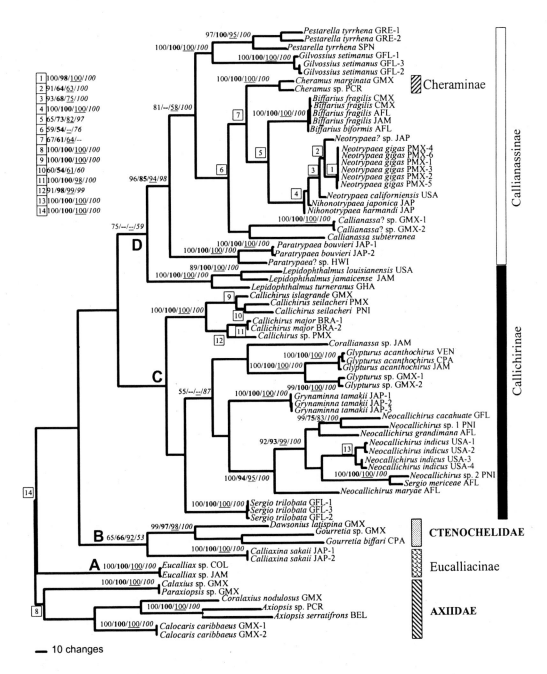

**Figure 1.** Evolutionary relationships of 18 genera of Callianassidae, two genera of Ctenochelidae, and five outgroup genera of Axiidae, inferred from an MP analysis of 16S and 12S rDNA data. Letters A–D adjacent to major nodes define clades that are referred to in Results. Support values shown from left to right are for ML, **MP**, NJ, and *BAY* respectively; – represents value equal to or lower than 50%.

Manning & Felder, 1991, and *Callianassa* s. s., rather than for their wholesale reassignment to the genus *Callianassa*. It is important to note that sequence data we have here identified with *C. subterranea*, type species of the genus, do not represent the same specimen and species for which sequence data are presently archived as GenBank Acc. No. DQ079706, originally reported in Porter et al. (2005). We propose that this previously published sequence possibly represents a source specimen assignable to *Pestarella*, as was also noted by Robles et al. (this volume). Extractions of *C. subterranea* for our present analyses were instead made from a more recently collected specimen taken in Scotland, for which we have carefully confirmed identification by morphological examination (Table 1). Thus identified, this type of the genus *Callianassa* defines a distinctly separate branch among other major clades of the subfamily, regardless of their present generic assignments. Support values are inadequate to confidently place an undescribed species listed as *Callianassa*? sp. from the Gulf of Mexico (GMX-1 and GMX-2) into this genus, despite its positioning in an immediate sister clade (albeit at long branch lengths). However, there is clear evidence to support the recent removal of *Paratyrpaea bouvieri* from *Callianassa* by Komai & Tachikawa (2008), while also suggesting that an apparently undescribed species from Hawaii is its likely congener.

Our samples of *Pestarella tyrrhena* and *Gilvossius setimanus* sort into a sister clade relationship at high support values. While samples represent multiple populations of both species, those of *P. tyrrhena* suggest, at very least, evidence of conspicuous population genetic structure. While all three specimens generally fit the present diagnosis for *P. tyrrhena*, the specimen from Spain appears to have a slightly broader telson and other features somewhat like those of *P. convexa* de Saint Laurent & LeLoeuff, 1979, from western Africa. Samples of the species we included to represent *Biffarius* reflect, in contrast, little measured genetic divergence between two species that separate readily on the basis of morphology. Finally, the representatives of *Neotrypaea* grouped together with those of *Nihonotrypaea* in a strongly supported monophyletic clade, encompassing somewhat less supported subclades that do not clearly resolve the status of the genus *Nihonotrypaea*.

The only present conclusion we draw for the two included species of *Nihonotrypaea* is that both are placed basally in the topology of this lineage, one without support. It is noteworthy that an undescribed species "*Neotrypaea*?", tentatively assigned by us to this genus on the basis of morphology, did indeed group among the other two, *N. californiensis* and *N. gigas*.

## 4 DISCUSSION

### 4.1 *Relationships of the family Callianassidae*

As the subfamily Eucalliacinae is placed within the Callianassidae by most recent authors (Manning & Felder 1991; Tudge et al. 2000; Ngoc-Ho 2003; Sakai 2005), present molecular phylogenetic placements for both of its clades are problematic, especially as one appears allied to the family Ctenochelidae. We can interpret that either the family Ctenochelidae is, undeserving of present rank, embedded within an otherwise monophyletic Callianassidae, or that the family Callianassidae is paraphyletic in present composition. The latter interpretation would infer that the subfamily Eucalliacinae is an unnatural grouping that encompasses at least one genus, *Calliaxina*, of ctenochelid affinities, and another, *Eucalliax*, which perhaps represents a yet-to-be-recognized family. We note that no such affinities were evident for a species of *Calliaxina* previously included in a morphological analysis under its earlier generic assignment, *Calliax punica*, by Tudge et al. (2000). However, relative development and positioning of the appendix masculina and appendix interna on the male pleopod 2 in members of Eucalliacinae is more like that seen in ctenochelids than in most callianassids (Felder & Manning 1994).

Given the low to medium support values that group *Calliaxina* with two other ctenochelid genera, we are not yet committed to family or subfamily level revisions reflecting this in taxonomy. Rather, we await inclusion of additional taxa in our analysis. Ideally, inclusion of *Calliax s.s.* and *Paraglypturus* Türkay & Sakai, 1995, would more comprehensively represent Eucalliacinae in this analysis, along with perhaps *Ctenocheles* Kishinouye, 1926, *Callianopsis* de Saint Laurent, 1973, *Anacalliax* de Saint Laurent, 1973, and *Paracalliax* de Saint Laurent, 1973, to represent likely members of the Ctenochelidae (*sensu* Manning & Felder 1991; Poore 1994; rather than that of Sakai 2005).

Sakai (2005) treated both *Calliaxina* and *Eucalliax* as junior synonyms of the genus *Calliax* de Saint Laurent, 1973. Clearly, molecular phylogenetic placement of at least *Calliaxina* corrects an error of that synonymy, but we do not yet have a molecular basis upon which to judge the other synonymy. In other revisions, Sakai (2004, 2005) removed both *Gourretia* and *Dawsonius* (see Sakai 2005: 245) from Ctenochelidae, placing them into separate subfamilies of a new family, Gourretiidae. Lacking representation of *Ctenocheles*, which Sakai left as the only genus assigned to Ctenochelidae, we cannot yet speak to the merits of this separation. However, the highly supported present grouping of *Dawsonius* and *Gourretia* raises doubt as to their warranting separation at the level of subfamily. These genera were also supported as a monophyletic group in a combined 18S and 16S molecular genetic analysis of higher-level thalassinidean relationships (Robles et al. this volume), where in the absence of eucalliacine representatives, Ctenochelidae was positioned immediately outside the Callianassidae. Similarly, where represented by a single species of *Ctenocheles* and a smaller group of callianassid taxa (Tsang et al. 2008), analysis of the same two genes placed the Ctenochelidae immediately outside the Callianassidae.

## 4.2 *Relationships within the subfamily Callichirinae*

No support was found for continued treatment of the genus *Lepidophthalmus* as a member of the subfamily Callichirinae, despite its previous placement among members of that group and wide separation from the Callianassinae in the morphological analysis of Tudge et al. (2000). We found weak support for its sharing a basal relationship with the subfamily Callianassinae but no evidence to contradict this topological placement on the basis of combined 16S and 18S sequence analyses (Robles et al. this volume; Tsang et al. 2008). In combined analysis of 16S, 18S, and 28S rDNA sequences (Tsang et al. 2008), there is in fact support for its separation from *Sergio* and *Callichirus*, the only other callianassid genera included, both of which are members of Callichirinae, though support for definition of that family, as traditionally defined, was lacking in our analysis. *Lepidophthalmus* was clearly monophyletic in our analysis, as in the morphological analysis of Tudge et al. (2000). In terms of habitat, physiology, and larval development, the genus is unique among the callianassids (Nates et al. 1997; Felder 2001, 2003), being highly adapted to muddy euryhaline coastlines and estuaries.

By contrast, members of the genus *Callichirus* are adapted to generally quartzite sandy sediments of high energy beaches and differ markedly from *Lepidophthalmus* and known members of the Callianassinae in varied aspects of larval morphology and life history (Strasser & Felder 1999, 2000; Felder 2001). The representatives in our analysis reflect a few of many remaining taxonomic problems at the species level and also a sister-clade relationship between members with eyes that end in long terminal spines (*C. islagrande* and *C. seilacheri*) and members with eyes that end in short terminal spines or blunt angles (*C. major* and relatives). Eastern Pacific populations of *C. seilacheri* obviously are separated into two populations, one of which may be identifiable with *C. garthi* Retamal, 1975. The latter species was placed into synonymy with *C. seilacheri* by Sakai (1999b) but without apparent study of its type or topotypic materials. Similarly, though our present tree represents only topotypic materials of *C. islagrande*, a sister lineage of *C. islagrande* is known to occur in the western Gulf of Mexico and may also warrant separate taxonomic treatment (Bilodeau et al. 2005). While only Brazilian populations (provisionally assigned to *C. major*) and yet another unnamed eastern Pacific species are included in the alternative major clade of this genus, it should also be noted that this group encompasses several divergent western Atlantic populations that potentially warrant further taxonomic revisions, and not all are represented in the present work (Staton & Felder 1995; Strasser & Felder 1999).

The highly supported grouping for two species of *Glypturus* included the widespread Caribbean species, *G. acanthochirus*, along with a Gulf of Mexico species that lacks a valid species name (without fixation of a holotype, see Dworschak 2007). While Sakai (2005) placed *Corallianassa* into synonymy with *Glypturus*, these genera were well separated in the morphological analyses of Tudge et al. (2000). There was also no support in our own analyses for placing of these genera into close relationship. However, our present analysis is based upon only one species of *Corallianassa* and two closely related species of *Glypturus*. Inclusion of additional members of these groups is needed to definitively resolve their generic status.

We have for now retained use of the genus *Grynaminna* for the species *G. tamakii*, instead of placing the genus into the synonymy of *Podocallichirus* Sakai, 1999, as called for by Sakai (2005) on rather subjective bases. As the genus *Podocallichirus* was derived by Sakai from subdivision of the genus *Callichirus*, it is of interest that *Grynaminna* was, with limited support, placed in a separate lineage from *Callichirus*. However, support is again low, and typical representatives of the genus *Podocallichirus* were not available for inclusion in our analysis.

As in the morphological analysis of Tudge et al. (2000), members of the genus *Neocallichirus* constituted a monophyletic group in our analysis, with the exception that *Sergio mericeae* was included among its subclades. The only other species of *Sergio* in our analysis, *S. trilobata*, was positioned independently, showing this genus to be paraphyletic, as was also evident in a combined analysis of 16S and 18S sequence data (Robles et al. this volume). This separation of *S. trilobata* from supposed congeners (including the type of the genus) was likewise the case in the previous morphological analysis of Tudge et al., where multiple species assigned to this genus were distributed among several clades. We continue to regard *S. mericeae* as a very close sibling species of *S. guassutinga*, type species of the genus, rather than placing it in synonymy with the latter species as advocated by Sakai (1999b). However, they are admittedly close, and thus we regard the clade including *S. mericeae* in our analysis to conservatively represent membership of the genus *Sergio*. If we hereafter treat these most typical members of *Sergio* to be *Neocallichirus*, as did Sakai (1999b), present congeners like *S. trilobata* must be assigned to one or more new genera. Thus, while we find no reason to disagree with Sakai (1999b) in placement of *Sergio s.s.* in synonymy with *Neocallichirus*, we cannot agree that such reassignment is justified for all members of *Sergio s.l.*

### 4.3 Relationships within the subfamily Callianassinae

In the course of deriving what has been termed a "controversial and retrograde classification" (Dworschak 2007), Sakai (1999b, 2005) merged a previously erected 12 genera of callianassids into synonymy with one large genus, *Callianassa*. Conceived as such, *Callianassa* in our analysis could be rationalized as monophyletic, but only provided one merged (from our analysis alone) eight monophyletic clades into it, thus giving high support at the same basal node for the genus that in our analysis defines a full subfamily. Were this to be adopted, a host of well-supported monophyletic genera evident in our phylogeny and that of Tudge et al. (2000) would be merged, serving to obfuscate evolutionary relationships and informative synapomorphies rather than to reflect them in classification and taxonomy. Virtually all nodes defining the represented generic membership of the Callianassinae prior to revisions by Sakai (1999b, 2005) are highly supported in our analysis. In addition, a basally positioned branch apparently defines *Paratrypaea*, recently separated from *Callianassa* on the basis of morphology (Komai & Tachikawa 2008).

While our continued recognition of these and perhaps other callianssine genera is in distinct disagreement with the recent works of Sakai, we submit that insight to reasonable generic groupings is best gained from overall study of tree topologies, branch lengths, and support values—based upon both morphological and molecular data when possible. Even so, outcomes of molecular and morphological analyses do not always agree in full and should not be expected to do so, given varied character sets and inconsistent taxonomic coverage among alternative studies. While the species set represented in our analysis produced strong evidence of monophyly for callianassine genera and supports the need for naming of generic-level monophyletic clades like that for *Paratrypaea*, inclusion of more species is certain to even further complicate this picture. For example, studies including other species of *Biffarius* analyzed with a different combination of genes do not definitively show monophyly among the represented species (Tsang et al. 2008; Robles et al. this volume). These could resolve differently in expanded analyses with additional genes or more likely become segregated into additional monophyletic clades supported by synapomorphies. We agree with Tudge et al. (2000:142) in that generic names are needed for these additional small groups of species, but those erected to date "should stand for the time being."

We do not support relegation of *Cheramus* to the synonymy of *Callianassa* as proposed by Sakai (1999b), but we cannot disagree with his conclusion that it belongs among the Callianassinae, rather than in its own subfamily. We thus advocate abandoning of the Cheraminae. Our analysis included only two species of the genus (one apparently unnamed), but they formed a well-supported monoplyletic group that was unambiguously positioned in topology, quite differently from the findings of Tudge et al. (2000).

A well-defined understanding of *Callianassa s.s.* was deemed essential to our analysis, so we made a concerted effort to ensure accurate representation of *C. subterranea*, the type species of the genus, in our analysis. Thus, the topological positioning for *C. subterranea* in the present work differs significantly from that for the currently available GenBank sequence of "*C. subterranea*" as depicted in Robles et al. (this volume), ostensibly for reasons already stated above in Results. The clade to which the specimen of *C. subterranea* is assigned in our analysis is not strongly supported and reflects a long-branch pairing with undescribed materials from hydrocarbon vent habitats of the Gulf of Mexico, provisionally assigned by us to this genus (*Callianassa?* sp. GMX-1, 2). While incomplete, our morphological studies suggest these materials may warrant treatment under a separate genus.

### 4.4 Pending analyses

Currently in progress, a molecular genetic analysis of all available species of *Lepidophthalmus* and its closest putative relatives should soon provide a somewhat more robust look at relationships of that genus. Likewise, a separate analysis targeted to the relationships of *Neotrypaea*, *Trypaea*, and *Nihonotrypaea* will address the unresolved status of the latter genus. In addition, collaborative work is currently under way to build the broadest overall taxonomic representation we can for a combined morphological and molecular analysis of not only the family Callianassidae but also other families in its infraoder, the Axiidea (*sensu* Robles et al. this volume).

### ACKNOWLEDGEMENTS

We thank F. Alvarez, R. Atkinson, R. Collin, J. Cuesta, S. De Grave, S. Nates, C. Schubart, and A. Tamaki for collecting or making available some of the specimens used in the present analysis. We also thank P. Dworschak, J. Goy, R. Lemaitre, J. Martin, G. Poore, and C. Tudge for providing useful comments on an earlier version of this manuscript. This study was supported under funding from the U.S. National Science Foundation to D. Felder (BS&I grant no. DEB-0315995 & DEB/AToL grant no. EF-0531603), along with small travel grants from the Smithsonian Marine Station–Fort Pierce, Florida and cruise support from the Smithsonian Tropical Research Institute, Panama. This is contribution number 739 from the Smithsonian Marine Station—Fort Pierce, Florida and number 131 from the University of Louisiana Laboratory for Crustacean Research.

## REFERENCES

Akaike, H. 1974. A new look at the statistical model identification. *IEEE Trans. Automat. Contr.* 19: 719–723.

Bilodeau, A.L., Felder, D.L. & Neigel, J.E. 2005. Population structure at two geographic scales in the burrowing crustacean *Callichirus islagrande* (Decapoda, Thalassinidea): historical and contemporary barriers to planktonic dispersal. *Evolution* 59: 2125–2138.

Bull, J.J., Huelsenbeck, J.P., Cunningham, C.W., Swofford D.L. & Waddell, P.J. 1993. Partitioning and combining data in phylogenetic analysis. *Syst. Biol.* 42: 384–397.

Castresana, J. 2000. Selection of conserved blocks from multiple alignments for their use in phylogenetic analysis. *Mol. Biol. Evol.* 17: 540–552.

Colbourne, J.K., & Hebert, P.D.N. 1996. The systematics of North American *Daphnia* (Crustacea: Anomopoda): a molecular phylogenetic approach. *Philos. Trans. R. Soc. B-Biol. Sci.* 351 (1337): 349–360.

Crandall, K.A. & Fitzpatrick, J.F. 1996. Crayfish molecular systematics: using a combination of procedures to estimate phylogeny. *Syst. Biol.* 45: 1–26.

Dworschak, P.C. 2007. Book review: Callianassoidea of the world (Decapoda, Thalassinidea). 2005. K. Sakai. Crustaceana Monographs 4. Koninklijke Brill, NV, Leiden, The Netherlands. *J. Crust. Biol.* 27:158–160.

Felder, D.L. 2001. Diversity and ecological significance of deep-burrowing macrocrustaceans in coastal tropical waters of the Americas (Decapoda: Thalassinidea). *Interciencia* 26: 440–449.

Felder, D.L. 2003. Ventrally sclerotized members of *Lepidophthalmus* (Crustacea: Decapoda: Callianassidae) from the Eastern Pacific. *Ann. Naturhist. Mus. Wien* 104 B: 429–442.

Felder, D.L. & Manning, R.B. 1994. Description of the ghost shrimp *Eucalliax mcilhennyi*, new species, from South Florida, with reexamination of its known congeners (Crustacea: Decapoda: Callianassidae). *Proc. Biol. Soc. Washington* 107: 340–353.

Felder, D. L. & Robles, R. 2004. Insights of molecular analyses in phylogeny and ecology of American callianassids (Decapoda: Thalassinidea). *Program and Abstracts, 3rd Brazilian Crustacean Congress and TCS Meeting, Florianópolis, Brazil*, Abstract No. 183: 101.

Hall, T.A. 1999. BioEdit: a user-friendly biological sequence alignment editor and analysis program for Windows 95/98/NT. *Nucl. Acids Symp. Ser.* 41: 95–98.

Komai, T. & Tachikawa, H. 2008. Thalassinidean shrimps (Crustacea: Decapoda) from the Ogasawara Islands, Japan. *Nat. Hist. Res.* 10: 19–52.

Manning, R.B. & Felder, D.L. 1991. Revision of the American Callianassidae (Crustacea: Decapoda: Thalassinidea). *Proc. Biol. Soc. Wash.* 104: 764–792.

Nates, S.F., Felder, D.L. & Lemaitre, R. 1997. Comparative larval development in two species of the burrowing shrimp genus *Lepidophthalmus* (Decapoda: Callianassidae). *J. Crust. Biol.* 17: 497–519.

Ngoc-Ho, N. 2003. European and Mediterranean Thalassinidea (Crustacea, Decapoda). *Zoosystema* 25: 439–555.

Palumbi, S., Martin, A., Romano, S., McMillan, W.O., Stice, L. & Grabowski, G. 1991. *The Simple Fool's Guide to PCR*. Honolulu: Department of Zoology and Kewalo Marine Laboratory, University of Hawaii.

Poore, G.C.B. 1994. A phylogeny of the families of Thalassinidea (Crustacea: Decapoda) with keys to families and genera. *Mem. Qld. Mus.* 54: 79–120.

Porter, M.L., Pérez-Losada, M. & Crandall, K.A. 2005. Model-based multi-locus estimation of decapod phylogeny and divergence times. *Mol. Phylogenet. Evol.* 37: 355–369.

Posada, D. & Crandall, K.A. 1998. MODELTEST: testing the model of DNA substitution. *Bioinformatics* 14:817–818.
Robles, R. & Felder, D.L. 2004. Phylogenetic analysis of the genus *Lepidophthalmus* based on the 16S mitochondrial gene. *Program and Abstracts, 3rd Brazilian Crustacean Congress and TCS Meeting, Florianópolis, Brazil*, Abstract No. 195: 107.
Robles, R., Schubart, C.D., Conde, J.E., Carmona-Suárez, C., Alvarez, F., Villalobos, J.L. & Felder, D.L. 2007. Molecular phylogeny of the American *Callinectes* Stimpson, 1860 (Brachyura: Portunidae), based on two partial mitochondrial genes. *Mar. Biol.* 150: 1265–1274.
Robles, R, Tudge, C.C., Dworschak, P.C., Poore, G.C.B. & Felder, D.L. (this volume). Molecular phylogeny of the Thalassinidea based on nuclear and mitochondrial genes. In: Martin, J.W., Crandall, K.A. & Felder, D.L. (eds.), *Crustacean Issues: Decapod Crustacean Phylogenetics*. Boca Raton, Florida: Taylor & Francis/CRC Press.
Rodríguez, F., Oliver, J.L., Marín, A. & Medina, J.R. 1990. The general stochastic model of nucleotide substitution. *J. Theor. Biol.* 142: 485–501.
Sakai, K. 1999a. Redescription of *Ctenocheles balssi* Kishinouye, 1926, with comments on its systematic position and establishment of a new subfamily Gourretiinae (Decapoda, Callianassidae). *Crustaceana* 72: 85–97.
Sakai, K. 1999b. Synopsis of the family Callianassidae, with keys to subfamilies, genera and species, and the description of new taxa (Crustacea: Decapoda: Thalassinidea). *Zool. Verh.* 326: 1–152.
Sakai, K. 2002. Callianassidae (Decapoda, Thalassinidea) in the Andaman Sea, Thailand, In: Bruce, N.L., Berggren, M. & Bussawarit, S. (eds.), *Biodiversity of Crustacea of the Andaman Sea. Proceedings of the International Workshop on Crustacea in the Andaman Sea. Phuket, 29 November — 20 December 1998*: 461–532. Phuket, Phuket Marine Biology Center.
Sakai, K. 2004. Dr. R. Plante's collection of the families Callianassidae and Gourretiidae (Decapoda, Thalassinidea) from Madagascar, with the description of two new genera and one new species of the Gourretiidae Sakai, 1999 (new status) and two new species of the Callianassidae Dana, 1852. *Crustaceana* 77: 553–601.
Sakai, K. 2005. Callianassoidea of the world (Decapoda, Thalassinidea). *Crustaceana Monographs* 4: 1–285.
Schubart, C.D., Cuesta, J.A. & Felder, D.L. 2002. Glyptograpsidae, a new brachyuran family from Central America: larval and adult morphology, and a molecular phylogeny of the Grapsoidea. *J. Crust. Biol.* 22: 28–44.
Stamatakis, A. 2006. RAxML-VI-HPC: Maximum likelihood-based phylogenetic analyses with thousands of taxa and mixed models. *Bioinformatics* 22: 2688–2690.
Stamatakis, A., Hoover, P. & Rougemont, J. 2008. A rapid bootstrap algorithm for the RAxML web-servers. *Syst. Biol.* 57: 758–771.
Staton, J.L. & Felder, D.L. 1995. Genetic variation in populations of the ghost shrimp genus *Callichirus* (Crustacea: Decapoda: Thalassinoidea) in the western Atlantic and Gulf of Mexico. *Bull. Mar. Sci.* 56: 523–536.
Strasser, K.M. & Felder, D.L. 1999. Larval development in two populations of the ghost shrimp *Callichirus major* (Decapoda: Thalassinidea) under laboratory conditions. *J. Crust. Biol.* 19: 844–878.
Strasser, K.M. & Felder, D.L. 2000. Larval development of the ghost shrimp *Callichirus islagrande* (Schmitt) (Decapoda: Thalassinidea). *J. Crust. Biol.* 20: 100–117.
Swofford D.L. 1998. *PAUP* Phylogenetic analysis using parsimony and other methods*. Sunderland, Massachusetts: Sinauer Assoc.
Talavera, G. & Castresana, J. 2007. Improvement of phylogenies after removing divergent and ambiguously aligned blocks from protein sequence alignments. *Syst. Biol* 56: 564–577.

Tsang, L.M., Lin, F.-J., Chu, K.H. & Chan, T.-Y. 2008. Phylogeny of Thalassinidea (Crustacea, Decapoda) inferred from three rDNA sequences: implications for morphological evolution and superfamily classification. *J. Zool. Syst. Evol. Res.* 46: 216–223.

Tudge, C.C., Poore, G.C.B. & Lemaitre, R. 2000. Preliminary phylogenetic analysis of generic relationships within the Callianassidae and Ctenochelidae (Decapoda: Thalassinidea: Callianassoidea). *J. Crust. Biol.* 20 (Special Issue 2): 129–149.

# The Timing of the Diversification of the Freshwater Crayfishes

JESSE BREINHOLT[1], MARCOS PÉREZ-LOSADA[2] & KEITH A. CRANDALL[1]

[1] *Department of Biology, Brigham Young University, Provo, Utah, U.S.A.*
[2] *CIBIO, Universidade do Porto, Vairão, Portugal*

## ABSTRACT

Freshwater crayfish (Astacidea) serve as model organisms for many diverse disciplines, from neurology to toxicology, and have been the focus of many physiological, ecological, and molecular-based studies. Although much of the recent work has focused on the evolutionary history, phylogeography, and conservation biology of freshwater crayfishes, estimations of their divergence times and radiations have never been made. Recently, divergence time estimations for decapods provided the first proposed molecular-timing hypothesis involving freshwater crayfish. In this study we focus specifically on estimating divergence among Astacidea. We employ a Bayesian method implemented in multidivtime for timing estimation, calibrated with multiple fossils including a Parastacoidea fossil newly discovered in Australia. With our narrow taxonomic focus, we increase the accuracy and provide divergence estimations more specific to freshwater crayfish. Our molecular time estimation supports a late Permian to early Triassic divergence from Nephropoidea with radiation and dispersal before the breakup of Pangaea, as well as subsequent speciation and radiation prior to or directly associated with Gondwana and Laurasia disassembly. The breakup of Gondwana and Laurasia resulted in the separation of Parastacoidea and Astacoidea during the Jurassic. The hypothesized divergence and radiation of these two superfamilies are also supported by our molecular time estimations. For the three families of crayfish, we estimate the Astacidae radiation at ∼153 million years ago (MYA), the Cambaridae radiation at ∼90 MYA, and diversification of Parastacidae at ∼161 MYA.

## 1 INTRODUCTION

Freshwater crayfish have a worldwide distribution, occurring on all continents except Antarctica and Africa excluding Madagascar. They are placed in the infraorder Astacidea, which includes three superfamilies: 1) Astacoidea—Northern Hemisphere crayfish, 2) Parastacoidea—Southern Hemisphere crayfish, and 3) Nephropoidea—the clawed lobsters. The crayfish form a monophyletic group (Crandall et al. 2000b) and have ∼640 described species (Crandall et al. 2008) with Nephropoidea, the clawed lobsters, hypothesized as their sister group (Crandall et al. 2000a). Parastacoidea contains one family, Parastacidae, with 15 genera (*Astacoides, Astacopsis, Cherax, Engaeus, Engaewa, Euastacus, Geocharax, Gramastacus, Ombrastacoides, Paranephrops, Parastacus, Samastacus, Spinastacoides, Tennuibranchiurus,* and *Virilastacus*) and 176 species. Astacoidea contains two families, Astacidae and Cambaridae. Astacidae has three genera (*Pacifastacus, Astacus, Austropotamobius*) (Hobbs 1974) to six genera (Starobogatov 1995), depending on whose taxonomy one prefers, and 16–39 species. Cambaridae has 2 subfamilies (Cambarellinae and Cambarinae) containing 11 genera (*Barbicambarus, Bouchardina, Cambarellus, Cambarus, Distocambarus, Fallicambarus, Faxonella, Hobbseus, Orconectes, Procambarus, Troglocambarus*), plus a distinct genus *Cambaroides* that appears to be more distantly related to these two subfamilies; Cambaridae has a total of 445 species (see Crandall & Buhay 2008 for a recent summary).

Freshwater crayfish relationships at higher taxonomic levels are well understood. The two superfamilies are monophyletic sister clades, and Parastacidae and Astacidae are monophyletic (Crandall et al. 2000b; Rode & Babcock 2003). Cambaridae is paraphyletic, as one of its genera, *Cambaroides*, is in a basal lineage to Astacidae and the rest of the Cambaridae genera (Braband et al. 2006; Crandall et al. 2000b). Most of the taxonomic relationships within Cambaridae are currently best explained by Hobbs' (1989) taxonomic revision. The following taxonomic groups within *Cambarinae* have been evaluated since Hobbs' (1989) revision: the genus *Orconectes* (Taylor and Knouft 2006); subgenus *Crockerinus* within *Orconectes* (Taylor and Hardman 2002); the subgenus *Scapulicambarus* within *Procambarus* (Busack 1989); and the subgenus *Aviticambarus* within *Cambarus* (Buhay et al. 2007). Within Astacidae, the taxonomy within *Astacus* and *Pacifastacus* is based on Hobbs' (1989) morphological taxonomic revision. The taxonomy within *Austropotamobius* was recently examined by Grandjean et al. (2000), Zaccara et al. (2004), and Fratini et al. (2004), all of whom reported multiple cryptic subspecies. However, Starobogatov (1995) provided a comprehensive overview of the Astacidae that resulted in 6 genera and 36 species, but his proposed taxonomy has not yet taken hold in the literature. The Astacidae in general is in need of a detailed examination to unify the diversity of ideas concerning its taxonomy.

The first comprehensive phylogenetic hypothesis of the Parastacoidea was morphologically based on male genitalia, cephalothorax, chelae, and body shape (Riek 1969). Studies that followed addressed the relations within this family using morphological, protein, and molecular data (Austin 1995; Crandall et al. 1995; Patak & Baldwin 1984; Patak et al. 1989; Riek 1972). These studies included limited sampling of genera and had conflicting results. The study by Crandall et al. (2000a) established well-supported relations within this family by analyzing 13 of the then 14 genera using mitochondrial DNA. Out of the now 15 genera in Parastacoidea, eight have been recently evaluated taxonomically and/or phylogenetically: *Engaewa* (by Horwitz and Adams 2000), *Cherax* (by Austin 1996), *Euastacus* (by Schull et al. 2005), two new genera *Spinastacoides* and *Ombrastacoides* (by Hansen and Richardson 2006), and *Engaeus*, *Geocharax*, and *Gramastacus* (by Schultz et al. 2007).

Through these recent studies, the problems of determining relationships among the freshwater crayfish become very apparent. Studies have not been fully comprehensive and have been limited in taxonomic sampling, due in part to the large number of freshwater crayfish taxa and their global distribution. The genetic and protein studies have shown high morphological and habitat variation within species and have demonstrated that convergent evolution is common (Braband et al. 2006; Crandall & Fitzpatrick 1996; Taylor & Hardman 2002). Additionally, these studies have revealed multiple cases of paraphyly, discovery of cryptic species, and even some unsupported described species (e.g., Austin 1996; Grandjean et al. 2000; Hansen & Richardson 2006; Schull et al. 2005; Schultz et al. 2007; Crandall et al. 2008). As a result, Sinclair et al. (2004) proposed the completion of a worldwide phylogeny based on multiple mitochondrial and nuclear genes. Because of the group's extensive convergent evolutionary history, only through molecular analysis and full taxonomic coverage will it be possible to infer the relationships within this group. While this goal is yet to be achieved, here we report on a phylogenetic status of the major genera of freshwater crayfish and the associated divergence times to put such a phylogeny into a temporal perspective.

Recently, Porter et al. (2005) published a phylogeny and associated divergence time estimates for the decapods as a whole. This study was the first molecular-based time hypothesis that included the freshwater crayfish. The goal of that study was to estimate decapod divergences; hence, only two of their fossil calibrations came from within the infraorder Astacidea. Multiple studies have shown that the most important factor affecting molecular divergence time estimation is the number and distribution of the calibration points throughout the tree (Lee 1999; Porter et al. 2005; Thorne & Kishino 2002; Yang & Yoder 2003; Yoder & Yang 2000). In this study we focus specifically on estimating divergence among Astacidea. By including multiple fossil calibrations and a specific taxonomic focus we increase the accuracy and can provide divergence estimations more specific to freshwater crayfish events. The use of molecular-based divergence time estimates has improved the understanding of the timing of evolutionary processes and events. A molecular time estimate for crayfish is particularly interesting because the current hypotheses of the divergence times correlates with estimates of the timing of the breakup of Pangaea and disassembly of Gondwana and Laurasia (Ahyong & O'Meally 2004; Crandall et al. 2000b; Porter et al. 2005; Rode & Babcock 2003). We test the hypotheses that freshwater crayfish diverged from *Nephropoidea* (clawed lobsters) during the Permian or Triassic, and that Parastacoidea (Southern Hemisphere) and Astacoidea (Northern Hemisphere) divergence occurred during the Jurassic (Ahyong & O'Meally 2004; Crandall et al. 2000b; Porter et al. 2005; Rode & Babcock 2003), using a comprehensive phylogeny at the genus level for the major lineages of freshwater crayfish.

## 2 METHODS

### 2.1 Taxon sampling, DNA extraction, PCR, and sequencing

Crayfish species were chosen to represent major crayfish lineages in order to date the divergence times of these major groups (Table 1). Multiple sequences were obtained from GenBank, and the remaining sequences were generated by Toon et al. (in prep.), as indicated by an asterisk in Table 1. Although specifics can be obtained from Toon et al. (this volume), crayfish collection, preservation, DNA extraction, and amplification were completed following protocols and methods described in Crandall & Buhay (2004) and Crandall & Fitzpatrick (1996) for 16S rDNA ($\sim$500 bp; Crandall & Fitzpatrick, 1996), 12S rDNA ($\sim$400; Mokady et al. 1999) and COI ($\sim$700 bp; Folmer et al. 1994), and two nuclear genes: 18S ($\sim$2,000 bp; Whiting et al. 1997) and 28S ($\sim$3,000 bp; Whiting et al. 1997).

### 2.2 Phylogenetic analyses

Astacoidea and Parastacoidea were aligned separately using MAFFT (Katoh et al. 2002; Katoh et al. 2005) implementing the G-INS-I alignment algorithm and then combined using the MAFFT profile alignment option with default parameters for each gene. *Homarus americanus* and *Sergio mericeae* were then aligned to the ingroup using MAFFT profile alignment for each gene. This multiple sequence alignment program has been shown to provide quick and accurate results by Notredame et al. (2000) and Katoh et al. (2005). The iterative algorithms used by MAFFT allow for repeatability of alignment. GBlocks 0.91b (Castresana 2000) was used to objectively trim sections of the alignment with questionable homology using the default parameter with the exception of the allowed gap positions parameter. The latter was set to allow gaps that are present in at least half of the sequences (Talavera & Castresana 2007). Models of evolution for each alignment were estimated in ModelTest (Posada & Crandall 1998) using the AIC criteria (Akaike 1973) to compare and choose best-fit models for the different gene partitions.

Phylogenies were estimated using maximum likelihood (ML) and Bayesian optimality criteria, with RAxML (Stamatakis 2006) and MrBayes (Ronquist & Huelsenbeck 2003), respectively (see Palero & Crandall, this volume, for a general description of these approaches). RAxML is a unique ML program in that it allows the use of multiple models, therefore giving better ML estimates. We partitioned the data set by gene and applied the model GTR+I+G to each gene allowing independent parameters to be estimated during analysis. We selected the tree with the best ML score after multiple independent runs with random starting positions and assessed confidence in nodal support through 1000 bootstrap pseudoreplications. Bayesian analysis was performed in MrBayes, in which four independent runs starting from random trees were run using the default flat priors for $5 \times 10^6$ generations sampling every 100 generations. We also ran two independent MrBayes runs with the same settings using the best RAxML tree as a start tree. The negative log likelihood posterior distribution was checked for convergence and length needed for burn-in using the program Tracer

**Table 1.** Taxa and GenBank accession numbers associated with each sample. Asterisks (*) indicate sequences from Toon et al. (submitted).

| Taxon | 12S | 16S | Gene 18S | 28S | CO1 |
|---|---|---|---|---|---|
| **Astacidea** Latreille 1802 | | | | | |
| Astacoidea Latreille 1802 | | | | | |
| *Astacus astacus* (Linnaeus 1758) | EU920881* | AF235983 | AF235959 | DQ079773 | AF517104 |
| *Cambarellus shufeldtii* (Faxon 1884) | EU921117* | AF235986 | AF235962 | DQ079778 | EU921149* |
| *Cambaroides japonicus* (de Haan 1841) | EU921118* | AF235987 | DQ079742 | DQ079779 | no seq |
| *Cambarus maculates* (Hobbs & Pflieger 1988) | EU921119* | AF235988 | AF235964 | DQ079780 | no seq |
| *Orconectes virilis* (Hagen 1870) | EU920900* | AF235989 | AF235965 | DQ079804 | AF474365 |
| *Pacifastacus leniusculus* (Dana 1852) | EU921116* | AF235985 | AF235961 | DQ079806 | EU921148* |
| *Procambarus clarkii* (Girard 1852) | EU920901* | AF235990 | EU920952* | EU920970* | AY701195 |
| Parastacoidea (Huxley 1879) | | | | | |
| *Astacoides betsileoensis* (Petit 1923) | EU920882* | EU920912* | EU920955* | EU920992* | EU921146* |
| *Astacoides crosnieri* (Hobbs 1987) | EU921112* | EU921122* | EU921129* | EU921136* | EU921147* |
| *Astacopsis tricornis* (Clark 1936) | DQ006419 | DQ006548 | EU921123* | EU921135* | DQ006290 |
| *Cherax cairnsensis* (Riek 1969) | EU921113* | EU921120* | EU921124* | EU921132* | EU921113* |
| *Cherax quadricarinatus* (von Martens 1868) | DQ006423 | DQ006552 | EU921125* | EU921139* | DQ006294 |
| *Engaeus fossor* (Erichson 1846) | EU921114* | EU921121* | EU921126* | EU921134* | EU921144* |
| *Euastacus sulcatus* (Riek 1951) | DQ006525 | DQ006651 | EU921127* | EU921133* | DQ006396 |
| *Geocharax gracilis* (Clark 1936) | EU921115* | AF235992 | AF235968 | EU921140* | EU921145* |
| *Paranephrops planifrons* (White 1842) | DQ006544 | AF135995 | EU921128* | EU921141* | DQ006415 |
| *Ombrastacoides huonensis* (Hansen & Richardson 2006) | EU920905* | AF135997 | EU920956* | EU920995* | EU921143* |
| *Parastacus brasiliensis* (von Martens 1869) | EF599134 | AF175244 | EU921130* | EU921138* | EF599158 |
| *Samastacus spinifrons* (Phillipi 1882) | EF599136 | AF175241 | EU921131 | EU921137 | EF599159 |
| Nephropoidea (Dana, 1852) | | | | | |
| *Homarus americanus* (H. Milne-Edwards 1837) | DQ298427 | HAU11238 | AF235971 | DQ079788 | DQ889104 |
| Outgroup | | | | | |
| Thalassinidea | | | | | |
| Callianassoidea (Dana 1852) | | | | | |
| *Sergio mericeae* (Manning & Felder 1995) | EU920909* | DQ079733 | DQ079768 | DQ079811 | no seq |

v1.4 (Rambaut & Drummond 2007) across all Bayesian runs. Converging MrBayes runs were combined after independent analysis and deletion of burn-in. Nodal confidence for the Bayesian trees was assessed using posterior probabilities compiled from the set of trees post-burn-in. We compared the support indices from our RAxML and MrBayes hypothesis and chose the phylogeny with the highest number of well-supported nodes considering bootstrap values $\geq 70$ and Bayesian posterior probabilities $\geq 95$ as high support for use in our molecular clock estimation.

### 2.3 Fossil calibrations

The fossil record is being continually updated, and relationships based on it are constantly being reanalyzed. The recent discovery of a new Australian fossil *Palaeoechinastacus australianus* (Martin et al. 2008) doubles the previously recorded geological time range of the family *Parastacidae* (Hasiotis 2002; Rode & Babcock 2003; Sokol 1987, 1988). Because fossil calibrations are a major source of error in molecular timing estimation, it is imperative to use multiple calibrations to get the best possible estimation, thus reducing the inherent amount of error associated with the fossil record (Table 2). Along with fossil calibrations, many studies have incorporated time estimations of vicariate events associated with the split in major land masses such as Pangaea, Laurasia, and Gondwana (Bocxlare et al. 2006; Porter et al. 2005). Our choice of Bayesian molecular time

**Table 2.** Fossil calibrations used for divergence time estimations, with the node referring to placement of the fossil on the crayfish chronogram.

| Taxonomy | Species | Reference | Geologic (MYA) | Node |
|---|---|---|---|---|
| **Infraorder** Astacidea | | | | |
| **Family** Chimaerastacidae | *Chimaerastacus pacifluvialis* | Amati et al. 2004 | Mid Triassic (Upper Ladinian) 227–234 | C1 |
| **Family** Parastacidae | *Palaeoechinastacus australianus* | Martin et al. 2008 | Early Cretaceous 106 | C3 |
| | *Paranephrops fordycei* | Feldmann & Pole 1994 | early middle Miocene (Otaian-Lillburnian) 21.7–12.7 | C4 |
| **Family** Astacidae | *Astacus licenti* | Van Straelen 1928 | Late Jurassic 144–159 | C5 |
| | *Astacus spinirostris* | Imaizumi 1938 | Late Jurassic 144–159 | C5 |
| **Family** Cambaridae | *Procambarus primaevus* | Feldmann et al. 1981 | Late early Eocene 52.6–53.4 | C6 |

Calibration C2 is 185 MYA, based on the splitting of Pangaea used as an upper limit

estimation requires that we have an estimation of at least one upper time limit (i.e., maximum age). Following Porter et al. (2005), we used the split of Pangaea at 185 MYA as an upper limit calibration for the divergence of the superfamilies Astacoidea and Parastacoidea (Crandall et al. 2000b). All other calibrations are estimated as the mean date of the fossil and set as the lower limit calibration indicating the absolute minimum age of the calibrated group (Porter et al. 2005). Additionally, we incorporated fossil calibrations for the origin of the family Astacidae and the split between Astacidea and Thalassinidea as the root node for our phylogenetic and molecular time estimation (Amati et al. 2004; Imaizumi 1938; Van Straelen 1928). Finally, we included three additional fossil calibrations: one to calibrate the genus *Procambarus* in Cambaridae and two to represent the family Parastacidae (Feldmann 2003; Feldmann et al. 1998; Martin et al. 2008). We agree with Porter et al. (2005) and others that trace fossil burrows are difficult to associate with crayfish with any amount of certainty (Babcock et al. 1998; Hasiotis 2002). Therefore, we chose to include only fossil records from descriptions of preserved animals. Our choice not to use trace fossils and to set each fossil calibration as the lower limit makes our estimate more conservative, while still allowing us to account for the fossil species existing for an undetermined amount of time before the actual fossilization event.

## 2.4 Divergence time estimation

Freshwater crayfish divergence times were estimated using the multi-locus Bayesian method of Thorne and Kishino (2002) as implemented in the Multidivtime package (http://statgen.ncsu.edu/thorne/multidivtime.html). This approach was built on the continual improvements of molecular clock theory and applications (Kishino et al. 2001; Thorne et al. 1998). This method allows the use of multiple genes while not requiring a full taxa set for all genes included, does not assume a molecular clock in branch estimation, and allows for multiple calibrations. The use of multiple genetic loci and multiple fossil calibrations improves divergence times and rate estimations (Pérez-Losada et al. 2004; Porter et al. 2005; Thorne & Kishino 2002; Yang 2004; Yang & Yoder 2003; Yoder & Yang 2000). Multidivtime estimates times and rates by minimizing the discrepancies in branch lengths and by minimizing rate changes over branches. This Bayesian method employs the rate evolution model of Thorne et al. (1998) and Kishino et al. (2001), which averages rates using a Markov chain Monte Carlo (MCMC) process.

We used three different parameter settings for Multidivtime. First, rttm and rttmsd (distribution of time separating the ingroup root from the present and the standard deviation, respectively) were set to 2.5 (250 MYA), and rtrate and rtratesd (prior evolutionary rate and standard deviation, respectively) were set to 0.0136 substitutions per million years. Second, rttm and rttmsd were set at 2.38 (238 MYA), and rtrate and rtratesd were set to 0.015 substitutions per million years, to see the effect of placing it closer to the age of the fossil calibration. Third, the rttm and rttmsd were set at 3.5 (350 MYA), and the rtrate and rtratesd were set to 0.0102 substitutions per million years to explore the effects of perturbations to the rttm setting. For each parameter setting, we applied two different burn-in period settings, $10^4$ and $10^6$ steps, combined with $5 \times 10^5$ samples collected at every 100th cycle. The default settings were used for the rest of the required parameters. A total of 12 runs were completed with three independent random starts for each parameter and burn-in period setting. The three runs for each burn-in and parameter setting were checked, and the set with the most consistent estimations was chosen for our time estimation.

## 3 RESULTS

### 3.1 Phylogenetics

All models selected by ModelTest were nst=6 with gamma and invariable sites (16S, 18S, and CO1=TVM+I+G ; 28S=TrN+I+G; and 12S=GTR+I+G). There are a limited number of models in RAxML and MrBayes; therefore, the GTR+G+I model was chosen for each partition, allowing the respective programs to estimate the parameters during phylogenetic estimation. The RAxML best tree likelihood score was -24658.608503. Our RAxML tree compared to our Bayesian tree resulted in fewer nodes with high bootstrap support ($\geq 70$) and Bayesian posterior probabilities ($\geq 95$). Therefore, the Bayesian tree was used for the molecular divergent time estimation (Fig. 1). The relationships within Astacidea are concordant with recent studies placing the genus *Cambariodes* basal to both Astacidae and Cambaridae. Although *Astacus* and *Pacifastacus* fall out independently, they both fall between the paraphyletic Cambaridae. Parastacids reflect the same relationships as in Crandall et al. (2000b), the most extensive study of the entire family to date.

### 3.2 Divergence time estimations

Changing the rttm parameter, defined as the distribution of time separating the ingroup root from the present, to 2.386 and 3.5 hardly affected the results, with the largest difference in estimations being $3 \times 10^5$ years (Table 3). Pérez-Losada et al. (2004) and Porter et al. (2005) found similar results using even larger perturbations and also reported a minimal effect on the overall time estimation. The burn-in period setting of $10^6$ steps produced three nearly identical independent time estimations. From these three estimates, we chose the estimation with the smallest 95% posterior intervals for the chronogram (Fig. 1 & Table 3).

Our divergence time estimates between the crayfish lineages (Astacoidea and Parastacoidea) and Nephropoidea is ~239 MYA (node 38). The divergence time estimates for the Northern Hemisphere families resulted in Astacidae divergence ~153 MYA (node 25) being significantly older than Cambaridae divergence at ~90 MYA (node 22). Parastacidae (the Southern Hemisphere crayfish) divergence time is estimated at ~161 MYA (node 36) with the genera having much older divergence times than Northern Hemisphere crayfish.

Timing of the Diversification of Freshwater Crayfishes 349

**Figure 1.** Crayfish divergence time chronogram estimated with a Bayesian tree topology. Bolded branches indicate posterior probability of 1. Nodes labeled C1–C6 indicate locations of fossil calibration (Table 2). Node number refers to the estimated time and 95% posterior interval (Table 3).

**Table 3.** Node time estimations referring to crayfish chronogram (Fig. 1). Time is represented in MYA with 95% interval, standard deviation, and well-supported ML bootstrap and Bayesian posterior probability.

| Node | Time MYA | 95% Posterior Interval Lower | 95% Posterior Interval Upper | Standard Deviation | ML Bootstraps | Bayesian Posterior Probability |
|---|---|---|---|---|---|---|
| 20 | 67.342 | 53.461 | 96.797 | 11.820 | 97 | 1 |
| 21 | 77.593 | 56.790 | 109.350 | 13.966 | 100 | 1 |
| 22 | 90.413 | 63.279 | 125.150 | 16.161 | 100 | 1 |
| 23 | 132.263 | 100.796 | 150.774 | 13.184 | 82 | 1 |
| 24 | 143.006 | 117.570 | 154.648 | 9.769 | - | 1 |
| 25 | 153.367 | 151.552 | 157.798 | 1.698 | 100 | 1 |
| 26 | 144.531 | 128.907 | 157.830 | 7.363 | 99 | 1 |
| 27 | 37.916 | 6.370 | 73.685 | 17.888 | 100 | 1 |
| 28 | 25.915 | 12.882 | 45.609 | 8.481 | 100 | 1 |
| 29 | 147.774 | 130.894 | 161.587 | 7.834 | - | - |
| 30 | 78.473 | 40.3 | 109.408 | 17.520 | 100 | 1 |
| 31 | 127.486 | 102.616 | 149.049 | 11.846 | - | - |
| 32 | 138.331 | 115.897 | 156.189 | 10.326 | 87 | 1 |
| 33 | 135.304 | 111.904 | 153.525 | 10.688 | 87 | - |
| 34 | 144.026 | 123.144 | 160.854 | 9.653 | 80 | 1 |
| 35 | 158.120 | 143.756 | 169.560 | 6.56 | - | 1 |
| 36 | 161.875 | 150.093 | 171.880 | 5.542 | 100 | - |
| 37 | 183.459 | 179.650 | 184.957 | 1.446 | 100 | 1 |
| 38 | 239.345 | 230.789 | 258.697 | 7.587 | | |

## 4 DISCUSSION

### 4.1 *Phylogeny and divergence time estimations*

The phylogenetic results were consistent with the most recent molecular studies for freshwater crayfishes (Crandall et al. 2000b; Porter et al. 2005; Rode & Babcock 2003). The tree is generally well supported with the monophyly of the freshwater crayfish being recovered in 100% of the Bayesian posterior distributions. Most lineages within the Parastacidea are similarly supported, with a few of the deeper nodes having low support values. Our divergence time estimations support the divergent time hypotheses of Crandall et al. (2000b), Rode and Babcock (2003), Ahyong & O'Meally (2004), and Porter et al. (2005). In the most current divergence hypothesis, Porter et al. (2005) estimated the divergence between the crayfish lineages Astacoidea and Parastacoidea from Nephropoidea at ∼278 MYA. Our estimation of ∼239 MYA (node 38) differs probably because of the calibration of the node prior to this estimation in each study. Although both studies used *Chimaerastacus pacifluvialis* (C1) as a lower limit, we additionally used it as a guideline to estimate the time from the root to the tip, setting it at 250 MYA. Our estimation falls between their two estimations when they calibrated the previous node as a lower limit and when it was calibrated as an upper and lower bound. We estimate the Astacidae radiation at ∼153 MYA (node 25), fitting within the range of the fossils used for calibration. We include *Cambaroides japonicus* in this estimation due to consistent placement of this genus within the Astacidae (Braband et al. 2006). Therefore, our estimate is significantly older than the Astacidae radiation estimate of Porter et al. (2005). Although their actual estimation is not reported, a visual inspection of the chronogram of Porter et al. (2005) reveals a similar estimation when including *Cambaroides japonicus*. The Cambaridae radiation was estimated at ∼90 MYA (node 22), which coincides with Porter et al. (2005). These divergence estimates support the idea that Astacoidea diversified and was widespread before the split of Laurasia during the late Cretaceous (Owen 1976) ∼65 MYA.

The diversification of Parastacidae was calibrated with a new fossil dated to 106 MYA (Martin et al. 2008), which resulted in our estimated divergence time of ∼161 MYA. This divergence time suggests that older Parastacidae fossils are likely to be found in Australia. The first stages of Gondwana separation are estimated to have begun ∼150 MYA with the separation of South America and Africa from Antarctica-India-Madagascar-Australia-New Zealand (Wit et al. 1999). Veevers (2006) estimates a later separation of Africa-India from Australia-Antarctica-South America at ∼132 MYA. Regardless of the specific Gondwana breakup theory ascribed, the divergence time estimates between South America and Australia-New Zealand crayfish (node 32) and the Madagascar and Australian crayfish (node 29) can be explained by vicariance associated with the disassembly of Gondwana. The split between *Ombrastacoides* (Australia) and *Paranephrops* (New Zealand) (node 32) ∼127 MYA is also consistent in that vicariance may have happened before or in sync with this separation, which is commonly estimated at ∼90 MYA, but rifting may have begun as early as ∼110-115 (Stevens 1980, 1985).

### 4.2 *Interpreting results*

Molecular time estimations are prone to multiple errors, partially due to complete reliance on fossil calibration, in which there is an inherent amount of error, including incorrect assignment of fossils, error in chronological and date assignment, and introduced topological errors in the phylogenetic estimation (Graur & Martin 2004). With the amount of possible error, it is encouraging to get results that are consistent with the current fossil record and/or that are supported by theories of distribution and divergence. Although most time estimations were discussed as point estimation (the expected estimate of posterior distribution), readers should be aware of, and consider, the 95% posterior interval for all estimations. The Bayesian method employed is one of the few methods that allows the user to set minimum age fossil calibrations, but in doing so it results in a larger variance, increasing the size of the posterior age interval. By setting fossil calibration intervals instead of minimum age estimates, you can effectively reduce the amount of variance resulting in a reduced size of the posterior age distribution. In the future, molecular clock estimates may consider using *Astacus licenti* and *Astacus spinirostris* fossil calibrations (C5) for Astacidae as an interval calibration instead of minimum age for two reasons. First, it is supported by two independent fossils. Second, our point estimation fits within the fossil estimated time interval. Including more upper limit calibrations or employing calibration intervals reduces the size of posterior interval estimates.

## 5 CONCLUSIONS

Our molecular clock estimation supports a late Permian to early Triassic divergence of freshwater crayfishes from Nephropoidea with radiation and dispersal before the breakup of Pangaea. Subsequent speciation and radiation prior to, or directly associated with, Gondwanan and Laurasian breakup resulted in the separation of the superfamilies Parastacoidea and Astacoidea during the Jurassic, thus supporting current divergent time estimations (Ahyong & O'Meally 2004; Crandall et al. 2000b; Porter et al. 2005; Rode & Babcock 2003). The hypothesized divergences and radiation of the two superfamilies attributed to the breakup of Laurasia and Gondwana are supported by our molecular time estimations. We do not expect this to be the last molecular divergence estimation for freshwater crayfishes, and we expect future estimates to improve in accuracy with the discovery of new fossils and new molecular dating techniques.

## ACKNOWLEDGEMENTS

We thank the wide variety of friends and colleagues who have helped us collect freshwater crayfish from around the world over the past 15 years. Likewise, this study was made possible by the exceptional undergraduates from Brigham Young University who have labored to collect DNA sequence data from freshwater crayfish. We thank Alicia Toon and two anonymous reviewers for helpful comments on this manuscript. Our work was supported by a grant from the US NSF EF-0531762 awarded to KAC.

## REFERENCES

Ahyong, S.T. & O'Meally, D. 2004. Phylogeny of the Decapoda Reptantia: Resolution using three molecular loci and morphology. *Raffl. Bull. Zool.* 52: 673–693.

Akaike, H. 1973. Information theory as an extension of the maximum-likelihood principle. In: Petrov, B. & Csake, F. (eds.), *Second International Symposium on Information Theory*. Budapest: Akademiai Kiado.

Amati, L., Feldmann, R.M. & Zonneveld, J.P. 2004. A new family of Triassic lobsters (Decapoda: Astacidea) from British Columbia and its phylogenetic context. *J. Paleo.* 78: 150–168.

Austin, C.M. 1995. Evolution in the genus *Cherax* (Decapoda: Parastacidae) in Australia: A numerical cladistic analysis of allozyme and morphological data. *Freshwater Crayfish* 8: 32–50.

Austin, C.M. 1996. Systematics of the freshwater crayfish genus *Cherax* erichson (Decapoda: Parastacidae) in northern and eastern Australia: Electrophoretic and morphological variation. *Aust. J. Zool.* 44: 259–296.

Babcock, L.E., Miller, M.F., Isbel, J.L., Collinson, J.W. & Hasiotis, S.T. 1998. Paleozoic-Mesozoic crayfish from Antarctica: Earliest evidence of freshwater Decapod crustaceans. *J. Geol.* 26: 539–542.

Bocxlare, I.V., Roelants, K., Biju, S., Nagaraju, J. & Bossuyt, F. 2006. Late Cretaceous vicariance in Gondwanan amphibians. *PLoS ONE* 1: e74.

Braband, A., Kawai, T. & Scholtz, G. 2006. The phylogenetic position of the east Asian freshwater crayfish *Cambaroides* within the northern hemisphere Astacoidea (Crustacea, Decapoda, Astacida) based on molecular data. *J. Zool. Syst. Evol. Res.* 44: 17–24.

Buhay, J.E., Moni, G., Mann, N. & Crandall, K.A. 2007. Molecular taxonomy in the dark: Evolutionary history, phylogeography, and diversity of cave crayfish in the subgenus *Aviticambarus*, genus *Cambarus*. *Mol. Phylogenet. Evol.* 42: 435–488.

Busack, C.A. 1989. Biochemical systematics of crayfishes of the genus *Procambarus*, subgenus *Scapulicambarus* (Decapoda: Cambaridae). *J. N. Amer. Benthol. Soc.* 8: 180–186.

Castresana, J. 2000. Selection of conserved blocks from multiple alignments for their use in phylogenetic analysis. *Mol. Biol. Evol.* 17: 540–552.

Crandall, K.A. & Buhay, J.E. 2004. Genomic databases and the tree of life. *Science* 306: 1144–1145.

Crandall, K.A. & Buhay, J.E. 2008. Global diversity of crayfish (Astacidae, Cambaridae, and Parastacidae-Decapoda) in freshwater. *Hydrobiologia* 595: 295–301.

Crandall, K.A., Fetzner, J.W., Jr., Jara, C.G. & Buckup, L. 2000a. On the phylogenetic positioning of the South American freshwater crayfish genera (Decapoda: Parastacidae). *J. Crust. Biol.* 20: 530–540.

Crandall, K.A. & Fitzpatrick, J.F., Jr. 1996. Crayfish molecular systematics: Using a combination of procedures to estimate phylogeny. *Syst. Biol.* 45: 1–26.

Crandall, K.A., Harris, D.J. & Fetzner, J.W. 2000b. The monophyletic origin of freshwater crayfishes estimated from nuclear and mitochondrial DNA sequences. *Proc. Roy. Soc. Lond. B. Biol. Sci.* 267: 1679–1686.

Crandall, K.A., Lawler, S.H. & Austin, C. 1995. A preliminary examination of the molecular phylogenetic relationships of the crayfish genera of Australia (Decapoda: Parastacidae). *Freshwater Crayfish* 10: 18–30.

Crandall, K.A., Robinson, H.W. & Buhay, J.E. 2008. Avoidance of extinction through nonexistence: The use of museum specimens and molecular genetics to determine the taxonomic status of an endangered freshwater crayfish. *Conservat. Genet.*: 10.1007/s10592-008-9546-9.

Feldmann, R.M. 2003. The Decapoda: New initiatives and novel approaches. *J. Paleo.* 77: 1021–1038.

Feldmann, R.M., Grande, L., Birkhimer, C.P., Hannibal, J.T. & McCoy, D.L. 1981. Decapod fauna of the Green River formation (Eocene) of Wyoming. *J. Paleo.* 55: 788–799.

Feldmann, R.M. & Pole, M. 1994. A new species of *Paranephrops* White, 1842: A fossil freshwater crayfish (Decapoda: Parastacidae) from the Manuherikia group (Miocene), central Otago, New Zealand. *New. Zeal. J. Geol. Geophys.* 37: 163–167.

Feldmann, R.M., Vega, F.J., Applegate, S.P. & Bishop, G.A. 1998. Early Cretaceous arthropods from the Tlayua formation at Tepexi de Rodriguez, Puebla, Mexico. *J. Paleo.* 72: 79–90.

Folmer, O., Black, M., Hoeh, W., Lutz, R. & Vrijenhoek, R. 1994. DNA primers for amplification of mitochondrial cytochrome c oxidase subunit i from diverse metazoan invertebrates. *Mol. Mar. Biol. Biotechnol.* 3: 294–299.

Fratini, S., Zaccara, S., Barbaresi, S., Grandjean, F., Souty-Grosset, C., Crosa, G. & Gherardi, F. 2004. Phylogeography of the threatened crayfish (genus Austropotamobius) in Italy: Implications for its taxonomy and conservation. *Heredity* 94: 108–118.

Grandjean, F., Harris, D.J., Souty-Grosset, C. & Crandall, K.A. 2000. Systematics of the European endangered crayfish species, *Austropotamobius pallipes* (Decapoda: Astacidae). *J. Crust. Biol.* 20: 522–529.

Graur, D. & Martin, W. 2004. Reading the entrails of chickens: Molecular timescales of evolution and the illusion of precision. *Trends in Genetics* 20: 80–86.

Hansen, B. & Richardson, A.M.M. 2006. A revision of the Taasmanian endemic freshwater crayfish genus *Parastacoides* (Crustacea: Decapoda: Parastacidae). *Invertebrate Systematics* 20: 713–769.

Hasiotis, S.T. 2002. Where is the fossil evidence for Gondwanan crayfish? *Gondwana Res.* 5: 872–878.

Hobbs, H.H., Jr. 1974. Synopsis of the families and genera of crayfishes (Crustacea:Decapoda). *Smithson. Contrib. Zoo.* 164: 1–32.

Hobbs, H.H., Jr. 1989. An illustrated checklist of the American crayfishes (Decapoda: Astacidae, Cambaridae, and Parastacidae). *Smithson. Contrib. Zoo.* 480: 1–236.

Horwitz, P. & Adams, M. 2000. The systematics, biogeography and conservation status of species in the freshwater crayfish genus *Engaewa* Riek (Decapoda: Parastacidae) from south-western Australia. *Invertebr. Taxon.* 14: 655–680.

Imaizumi, R. 1938. Fossil crayfishes from Jehol. *Sci. Rep. Tokyo Imperial Univer., Sendal, Japan, Second Series* 19: 173–179.

Katoh, K., Kuma, K., Toh, H. & Miyata, T. 2002. Mafft: A novel method for rapid multiple sequence alignment based on fast Fourier transform. *Nucleic Acids Res.* 30: 3059–3066.

Katoh, K., Kuma, K., Toh, H. & Miyata, T. 2005. Mafft version 5: Improvement in accuracy of multiple sequence alignment. *Nucleic Acids Res.* 33: 511–518.

Kishino, H., Thorne, J.L. & Bruno, W.J. 2001. Performance of a divergence time estimation method under a probabilistic model of rate evolution. *Mol. Biol. Evol.* 18: 352–361.

Lee, M.S. 1999. Molecular clock calibrations and metazoan diversity. *J. Mol. Evol.* 49: 385–391.

Martin, A.J., Rich, T.H., Poore, G.C.B., Schultz, M.B., Austin, C.M., Kool, L. & Vickers-Rich, P. 2008. Fossil evidence in Australia for oldest known freshwater crayfish of Gondwana. *Gondwana Res.* 14: 287–296.

Mokady, M., Loya, Y., Achituv, Y., Geffen, E., Graur, D., Rozenblatt, S. & Brickner, I. 1999. Speciation versus phenotypic plasticity in coral inhabiting barnacles: Darwin's observations in an ecological context. *J. Mol. Evol.* 49: 367–375.

Notredame, C., Higgins, D. & Heringa, J. 2000. T-coffee: A novel method for multiple sequence alignments. *J. Mol. Biol.* 302: 205–217.

Owen, H.G. 1976. Continental displacement and expansion of the earth during the Mesozoic and Cenozoic. *Phil. Trans. Roy. Soc. Lond.* 281: 223–291.

Palero, F. & Crandall, K.A. (this volume). Phylogenetic inference using molecular data. In: Martin, J.W., Crandall, K.A. & Felder, D.L. (eds.), *Crustacean Issues: Decapod Crustacean Phylogenetics*. Boca Raton, Florida: Taylor & Francis/CRC Press.

Patak, A. & Baldwin, J. 1984. Electrophoretic and immunochemical comparisons of haemocyanins from Australian fresh-water crayfish (family parastacidae): Phylogenetic implications. *J. Crust. Biol.* 4: 528–535.

Patak, A., Baldwin, J. & Lake, P.S. 1989. Immunochemical comparisons of haemocyanins of Australasian freshwater crayfish: Phylogenetic implications. *Biochem Systemat. Ecol.* 17: 249–252.

Pérez-Losada, M., Høeg, J.T. & Crandall, K.A. 2004. Unraveling the evolutionary radiation of the thoracican barnacles using molecular and morphological evidence: A comparison of several divergence time esimation approaches. *Syst. Biol.* 53: 244–264.

Porter, M.L., Pérez-Losada, M. & Crandall, K.A. 2005. Model based multi-locus estimation of decapod phylogeny and divergence times. *Mol. Phylogenet. Evol.* 37: 355–369.

Posada, D. & Crandall, K.A. 1998. Modeltest: Testing the model of DNA substitution. *Bioinformatics* 14: 817–818.

Rambaut, A. & Drummond, A. 2007. Tracer v1.4. Available from *http:slash/beast.bio.ed.ac.uk/Tracer*

Riek, E.F. 1969. The Australian freshwater crayfish (Crustacea: Decapoda: Parastacidae), with descriptions of new species. *Aust. J. Zool.* 17: 855–918.

Riek, E.F. 1972. The phylogeny of the Parastacidae (Crustacea: Astacoidea), and description of a new genus of Australian freshwater crayfishes. *Austral. J. Zool.* 20: 369–389.

Rode, A.L. & Babcock, L.E. 2003. Phylogeny of fossil and extant freshwater crayfish and some closely related nephropid lobsters. *J. Crust. Biol.* 23: 418–435.

Ronquist, F. & Huelsenbeck, J.P. 2003. MrBayes 3: Bayesian phylogenetic inference under mixed models. *Bioinformatics* 19: 1572–1574.

Schull, H.C., Pérez-Losada, M., Blair, D., Sewell, K., Sinclair, E.A., Lawler, S., Ponniah, M. & Crandall, K.A. 2005. Phylogeny and biogeography of the freshwater crayfish *Euastacus* (Decapoda: Parastacidae) based on nuclear and mitochondrial DNA. *Mol. Phylogenet. Evol.* 37: 249–263.

Schultz, M.B., Smith, S.A., Richardson, A.M.M., Horwitz, P., Crandall, K.A. & Austin, C.M. 2007. Cryptic diversity in *Engaeus* Erichson, *Geocharax* Clark and *Gramastacus* Riek (Decapoda: Parastacidae) revealed by mitochondrial 16S rDNA sequences. *Invertebrate Systematics* 21: 1–19.

Sinclair, E.A., Fetzner, J.W., Jr., Buhay, J. & Crandall, K.A. 2004. Proposal to complete a phylogenetic taxonomy and systematic revision for freshwater crayfish (Astacida). *Freshwater Crayfish* 14: 1–9.

Sokol, A. 1987. A note on the existence of Pre-Pleistocene fossils of parastacid crayfish. *Victorian. Nat.* 104: 81–82.

Sokol, A. 1988. Morphological variation in relation to the taxonomy of the *Destructor* group of the genus *Cherax. Invertebr. Taxon.* 2: 55–79.

Stamatakis, A. 2006. Raxml-vi-hpc: Maximum likelihood-based phylogenetic analyses with thousands of taxa and mixed models. *Bioinformatics* 22: 2688–2690.

Starobogatov, Y. 1995. Taxonomy and geographical distribution of crayfishes of Asia and East Europe (Crustacea Decapoda Astacoidei). *Arthropoda Selecta* 4: 3–25.

Stevens, G.R. 1980. *New Zealand Adrift: The Theory of Continental Drift in a New Zealand Setting.* Wellington: A.H. & A.W. Reed.

Stevens, G.R. 1985. *Lands in Collision: Discovering New Zealand's Past Geography.* Wellington: Science Information Publishing Centre, DSIR.

Talavera, G. & Castresana, J. 2007. Improvement of phylogenies after removing divergent and ambiguously aligned blocks from protein sequence alignments. *Syst. Biol.* 56: 564–577.

Taylor, C. & Hardman, M. 2002. Phylogenetics of the crayfish subgenus Crockerinus, genus, *Orconectes* (Decapoda: Cambaridae), based on cytochrome oxidase I. *J. Crust. Biol.* 22: 874–881.

Taylor, C.A. & Knouft, J.H. 2006. Historical influences on genital morphology among sympatric species: Gonopod evolution and reproductive isolation in the crayfish genus *Orconectes* (Cambaridae). *Biol. J. Linn. Soc.* 89: 1–12.

Thorne, J.L. & Kishino, H. 2002. Divergence time and evolutionary rate estimation with multilocus data. *Syst. Biol.* 51: 689–702.

Thorne, J.L., Kishino, H. & Painter, I.S. 1998. Estimating the rate of evolution of the rate of molecular evolution. *Mol. Biol. Evol.* 15: 1647–1657.

Toon, A., Finley, M., Staples, J. & Crandall, K.A. (this volume). Decapod phylogenetics and molecular evolution. In: Martin, J.W., Crandall, K.A. & Felder, D.L. (eds.), *Crustacean Issues: Decapod Crustacean Phylogenetics.* Boca Raton, Florida: Taylor & Francis/CRC Press.

Van Straelen, V. 1928. On a fossil freshwater crayfish from eastern Mongolia. *Bulletin of the Geological Society of China* 7: 173–178.

Veevers, J.J. 2006. Updated Gondwana (Permian–Cretaceous) earth history of Australia. *Gondwana Res.* 9: 231–260.

Whiting, M.F., Carpenter, J.C., Wheeler, Q.D. & Wheeler, W.C. 1997. The Strepsiptera problem: Phylogeny of the holometabolous insect orders inferred from 18S and 28S ribosomal DNA sequences and morphology. *Syst. Biol.* 46: 1–68.

Wit, M.D., Jeffrey, M., Bergh, H. & Nicolaysen, L. 1999. Gondawana reconstruction and dispersion. *Search and Discovery* Article #30001, http://www.searchanddiscovery.net/documents/97019/index.htm.

Yang, Z. 2004. A heuristic rate smoothing procedure for maximum likelihood estimation of species divergence times. *Acta Zoolog. Sin.* 50: 645–656.

Yang, Z. & Yoder, A.D. 2003. Comparison of likelihood and Bayesian methods for estimating divergence times using multiple gene loci and calibration points, with application to a radiation of cute-looking mouse lemur species. *Syst. Biol.* 52: 705–716.

Yoder, A.D. & Yang, Z. 2000. Estimation of primate speciation dates using local molecular clocks. *Mol. Biol. Evol.* 17: 1081–1090.

Zaccara, S., Stefani, F., Nardi, P.A. & Crosa, G. 2004. Taxonomic implications in conservation management of white-clawed crayfish *Austropotamobius pallipes* (Decapoda, Astacoidea). *Biol. Conservat.* 120: 1–10.

# Phylogeny of Marine Clawed Lobster Families Nephropidae Dana, 1852, and Thaumastochelidae Bate, 1888, Based on Mitochondrial Genes

DALE TSHUDY[1], RAFAEL ROBLES[2], TIN-YAM CHAN[3], KA CHAI HO[4], KA HOU CHU[4], SHANE T. AHYONG[5] & DARRYL L. FELDER[2]

[1] *Department of Geosciences, Edinboro University of Pennsylvania, Edinboro, Pennsylvania, U.S.A.*
[2] *Department of Biology, University of Louisiana at Lafayette, Lafayette, Louisiana, U.S.A.*
[3] *Institute of Marine Biology, National Taiwan Ocean University, Keelung, Taiwan*
[4] *Department of Biology, The Chinese University of Hong Kong, Hong Kong, China*
[5] *Marine Biodiversity and Biosecurity, National Institute of Water and Atmospheric Research, Kilbirnie, Wellington, New Zealand*

## ABSTRACT

Phylogenetic relationships of extant marine clawed lobsters of the families Nephropidae and Thaumastochelidae were analyzed based on partial sequences of the 12S and 16S mitochondrial rRNA genes. The ingroup sample consisted of 17 species and ten genera of the Nephropidae as well as two species and two genera of the Thaumastochelidae. The family Enoplometopidae was used as an outgroup. A total of 875 base pairs, with 241 parsimony informative sites, was analyzed. Bayesian (MRBAYES) and maximum likelihood (PAUP) analyses produced similar topologies. The ML tree was well supported at most nodes. Generic monophyly was confirmed for all five genera represented by two or more species. *Acanthacaris* is the least derived among genera included in the analysis. It was resolved as a sister taxon to all other nephropids (including thaumastochelids). The thaumastochelids are monophyletic but nested within Nephropidae; thus, family-level status for thaumastochelids was not supported. Some nephropid genera, previously regarded as close relatives on a morphological basis (e.g., *Homarus* and *Homarinus*, or *Nephrops* and *Metanephrops*), instead appear to be cases of morphological convergence.

## 1 INTRODUCTION

### 1.1 General

Marine clawed lobsters include the families Erymidae van Straelen, 1924 (Lower Triassic–Upper Cretaceous), Chimerastacidae Amatie et al., 2004 (Middle Triassic), Chilenophoberidae Tshudy & Babcock, 1997 (Middle Jurassic–Lower Cretaceous), Nephropidae Dana, 1852 (Lower Cretaceous–Recent), Thaumastochelidae Bate, 1888 (Upper Cretaceous–Recent), and Enoplometopidae de Saint Laurent, 1988 (Recent). The family Nephropidae is the most diverse, consisting of 14 genera (11 extant [*Acanthacaris* Bate, 1888; *Eunephrops* Smith, 1885; *Homarinus* Kornfield et al., 1995; *Homarus* Weber, 1795; *Metanephrops* Jenkins, 1972; *Nephropides* Manning, 1969; *Nephrops* Leach, 1814; *Nephropsis* Wood-Mason, 1873; *Thymopides* Burukovsky & Averin, 1976; *Thymops* Holthuis, 1974; *Thymopsis* Holthuis, 1974] and three extinct [*Hoploparia* McCoy, 1849; *Jagtia* Tshudy &

Sorhannus, 2000; *Palaeonephrops* Mertin, 1941]). The present study investigates phylogenetic relationships of the clawed lobster genera of the families Nephropidae and Thaumastochelidae.

Phylogeny of the clawed lobsters is of interest for more than their intrinsic generic relationships. It potentially provides insights into questions of general biological and paleontological interest such as rates of morphological and molecular evolution, or the frequency and distribution of molecular or morphologic homoplasy. Likewise, of general interest is the comparison of phylogenies produced by different methods, including traditional intuitive schemes versus cladistic analyses, and morphology- versus DNA-based cladistic analyses. Clawed lobsters, by virtue of their complex morphology, long range in the fossil record, wide geographic range, and ecological diversity, are a group well suited for such investigations.

### 1.2 *Previous work, morphological and molecular*

A number of workers have conducted morphology-based cladistic analyses on clawed lobsters (Tshudy 1993 [20 genera]; Williams 1995 [four genera]; Tshudy & Babcock 1997 [22 genera]; Tshudy & Sorhannus 2000a [19 genera], 2000b [13 genera]; Dixon et al. 2003 [four genera]; Rode & Babcock 2003 [nine genera]; Ahyong & O'Meally 2004 [five genera]; Amati et al. 2004 [seven genera]; Ahyong 2006 [26 genera]. Ahyong (2006) included all (14) nephropid and (three) thaumastochelid genera, fossil and extant, in the largest matrix published to date. Ahyong's (2006) character matrix is similar to earlier matrices of Tshudy (1993) and Tshudy & Babcock (1997), and thus does not constitute a robust test of those trees. Nonetheless, Ahyong (2006) added additional characters and included for the first time taxa such as *Neoglyphea* Forest & de Saint Laurent, 1975, *Enoplometopus* A. Milne-Edwards, 1862, and the Uncinidae Beurlen, 1928.

Few workers have conducted DNA-based cladistic analyses on the clawed lobsters. Tam & Kornfield (1998), using 16S mtDNA, produced a tree including five nephropid genera (*Homarus*, *Homarinus*, *Metanephrops*, *Nephrops*, *Nephropsis*). Ahyong & O'Meally (2004) used 16S mtDNA along with 18S and 28S nuclear DNA data (2,500 bp total) to evaluate reptant decapod phylogeny, including six lobster genera (*Enoplometopus*, *Homarus*, *Metanephrops*, *Neoglyphea*, *Nephropsis*, and *Thaumastochelopsis* Bruce, 1988). Porter et al. (2005) used 16S mtDNA along with 18S and 28S nuclear DNA data and the histone H3 gene (3,601 bp total) to evaluate decapod phylogeny (43 genera), including four lobster genera (*Acanthacaris*, *Homarus*, *Nephrops*, and *Nephropsis*). Chu et al. (2006) produced a 12S mtDNA-based tree for ten clawed lobster genera using *Neoglyphea* as an outgroup. The present study concerns the phylogenetic relationships of the Recent clawed lobster genera of the Nephropidae and Thaumastochelidae. Our analysis is based on partial sequences of mitochondrial 12S and 16S genes and includes 12 ingroup genera (adding *Homarinus*, *Thaumastochelopsis*, and *Thymops* to those analyzed by Chu et al. 2006).

## 2 MATERIALS AND METHODS

### 2.1 *Taxon sampling*

The ingroup (Table 1) consists of 21 terminals representing 17 species and ten genera of the Nephropidae as well as two species and two genera of the Thaumastochelidae. The family Thaumastochelidae was included in the analysis because family-level status has been debated and remains equivocal. In some studies, members of this family have been suggested to constitute their own family (Holthuis 1974; Tshudy & Sorhannus 2000a, b; Dixon et al. 2003; Schram & Dixon 2004; Ahyong & O'Meally 2004; Ahyong 2006), whereas other studies include them as part of Nephropidae (Tshudy & Babcock 1997; Chu et al. 2006).

The outgroup used in our study was the family Enoplometopidae, recently found to be the sister group to the Nephropidae + Thaumastochelidae in morphological (Ahyong & O'Meally 2004;

Table 1. List of specimens for which 16S mtDNA and 12S mtDNA were sequenced. CBM = Natural History Museum and Institute, Chiba; CNCR = Colección Nacional de Crustceos, Instituto de Biologa, UNAM; EUPG = Edinboro University of Pennsylvania; MNHN = Muséum National d'Histoire Naturelle, Paris; NTM = Museum of Art Gallery of the Northern Territory, Darwin; NTOU = National Taiwan Ocean University; USNM = National Museum of Natural History, Smithsonian Institution, Washington, D.C.; 1 = Aquarium shop, origin unknown; 2 = Supermarket, origin unknown.

| Species | Catalog No. | Locality | GenBank Accession No. 12S | GenBank Accession No. 16S |
|---|---|---|---|---|
| *Acanthacaris tenuimana* | MNHN-As639 | Solomon Islands | DQ298420 | EU882871 |
| *Enoplometopus crosnieri* | NTOU-M00602 | Taiwan | DQ298423 | EU882870 |
| *Enoplometopus daumi* | NTOU-M00171 | Singapore[1] | DQ298421 | EU882868 |
| *Enoplometopus debelius* | NTOU-00173 | Singapore[1] | DQ298422 | EU882869 |
| *Enoplometopus occidentalis* | NTOU-M00152 | Taiwan | DQ298424 | EU882871 |
| *Eunephrops cadenasi* | MNHN-As640 | Guadeloupe | DQ298425 | EU882873 |
| *Eunephrops manningi* | MNHN-As641 | Guadeloupe | DQ298426 | EU882874 |
| *Homarinus capensis* | USNM251453 | S. Africa | EU882895 | EU882887 |
| *Homarinus capensis* | USNM251454 | S. Africa | EU882896 | EU882888 |
| *Homarus americanus* | EUPGEO4001 | U.S.A. | DQ298427 | EU882875 |
| *Homarus gammarus* | NTOU-M00819 | France[2] | DQ298428 | EU882876 |
| *Metanephrops japonicus* | NTOU-M00521 | Japan | EU882897 | EU882889 |
| *Metanephrops rubellus* | NTOU-M00074 | Brazil | DQ298429 | EU882877 |
| *Metanephrops thomsoni* | NTOU-M00504 | Taiwan | DQ298430 | EU882878 |
| *Nephropides caribaeus* | MNHN-As642 | Guadeloupe | DQ298432 | EU882879 |
| *Nephrops norvegicus* | CBM-ZC7438 | France[2] | DQ298433 | EU882881 |
| *Nephropsis aculeata* | CNCR21650 | Mexico | EU882892 | EU882884 |
| *Nephropsis aculeata* | CNCR21660 | Mexico | EU882893 | EU882885 |
| *Nephropsis rosea* | CNCR21631 | Mexico | EU882894 | EU882886 |
| *Nephropsis serrata* | NTOU-M00157 | Taiwan | DQ298434 | EU882881 |
| *Nephropsis stewarti* | NTOU-M00505 | Taiwan | DQ298435 | EU882882 |
| *Thaumastocheles japonicus* | NTOU-M00168 | Taiwan | DQ298438 | EU882866 |
| *Thaumastochelopsis wardi* | NTM-Cr.004231 | Australia | EU882891 | EU882867 |
| *Thymopides grobovi* | MNHN-As181 | Kerguelen Island | DQ298436 | EU882883 |
| *Thymops birsteni* | USNM291290 | Chile | EU882898 | EU882890 |

Ahyong 2006) and molecular analyses (Ahyong & O'Meally 2004; Tsang et al. 2008; Chu et al. this volume). The monogeneric Enoplometopidae is represented in the analysis by four species: *Enoplometopus crosnieri* Chan & Yu, 1998, *E. daumi* Holthuis, 1983, *E. debelius* Holthuis, 1983, and *E. occidentalis* (Randall, 1840).

2.2 *Tissue sampling*

Tissue samples used in this study were derived from freshly collected specimens or, more often, from preserved museum collections (Table 1). On collection, specimens were either frozen on site and later transferred to 70% ethyl alcohol (ETOH) or directly preserved in 70% ETOH. Species identification was based on morphology (Holthuis 1974, 1991; Tshudy 1993).

## 2.3 DNA extraction

DNA extraction, amplification, and sequencing were conducted at both the University of Louisiana Lafayette and the Chinese University of Hong Kong. Total genomic DNA was extracted from fresh or ethanol-fixed tissue samples collected from the abdomen (ventral side) or pereiopods. Muscle was ground and then incubated for 1–12 h in 600 $\mu$l of lysis buffer (100 mM EDTA, 10 mM tris pH 7.5, 1% SDS) at 65°C; protein was separated by addition of 200 $\mu$l 7.5 M of ammonium acetate and subsequent centrifugation. DNA was precipitated by addition of 600 $\mu$l of cold isopropanol followed by centrifugation; the resulting pellet was rinsed in 70% ETOH, dried in a speed vacuum system (DNA110 Speed Vac), and resuspended in 10–20 $\mu$l of TE buffer (10 mM TRIS, 1 mM EDTA). For samples extracted at the Chinese University of Hong Kong, total DNA was obtained from pleopod muscles (10–15 mg) with the QIAamp DNA Mini Kit (QIAGEN) following manufacturer's instructions. DNA was eluted in 200 $\mu$l of distilled water.

## 2.4 DNA amplification and sequencing

Two mitochondrial ribosomal genes, the 12S and 16S rRNA, were selected because of their proven utility in resolving generic relationships for other decapods (Kornfield et al. 1995; Schubart et al. 2000; Robles et al. 2007; Chan et al. 2008). Standard PCR amplification and automated sequencing protocols were used to sequence a fragment of approximately 400 bp of the 12S mtDNA and 550 bp of the 16S mtDNA. Both strands were sequenced for each gene. In all cases, the 12S and 16S sequences were derived from the same specimen. When possible, more than one species of each genus was included in our analysis.

Primers used for the 12S fragment were 12Sai (5'-AAA CTA GCA TTA GAT ACC CCT ATT AT-3') (Palumbi et al. 1991) and 12H2 (5'-ATG CAC TTT CCA GTA CAT CTA C-3') (Colbourne & Hebert 1996). Primers used for the 16S fragment were 16ar (5'-CGC CTG TTT ATC AAA AAC AT-3'), 16br (5'- CCG GTC TGA ACT CAG ATC ACG T-3') (Palumbi et al. 1991), 1472 (5'-AGA TAG AAA CCA ACC TGG-3') (Crandall & Fitzpatrick 1996), and 16L2 (5'-TGC CTG TTT ATC AAA AAC AT-3') (Schubart et al. 2002). Reactions were performed in 25 $\mu$l volumes (200 M each dntp, 1X buffer, 0.5 $\mu$M each primer, 1 unit Taq polymerase, 1 $\mu$l extracted DNA). Thermal cycling was performed as follows: initial denaturation for 10 min at 94–95°C followed by 40–42 cycles of 1 min at 94–95°C, 1–1:30 min at 48°C and 1:30–2 min at 72°C, with a final extension of 10 min at 72°C. PCR products were purified using 100,000 MW filters (Microcon-100® Millipore Corp.) and sequenced with the ABI BigDye® terminator mix (Applied Biosystems). Both PCR and cycle sequence reactions were conducted on a Robocycler® 96 cycler. Sequencing products were run on either a 310 or 3100 Applied Biosystems® automated sequencer.

## 2.5 Sequence alignment and nucleotide composition

Consensus of complementary sequences of the gene was obtained with the Sequencher® software program (ver 4.1, Genecodes, Ann Arbor, MI). Alignment of consensus sequences was performed with Clustal W, as implemented in Bioedit (Hall 1999) with the following settings: 6-2/6-2 penalty (opening-gap extension, pairwise/multiple alignment respectively). Base composition, pattern of substitution for pairwise comparison, and analysis of variability along both fragments of the 12S and the 16S mtDNA were analyzed in PAUP 4.0 beta 10 (Swofford 1993). Homogeneity of nucleotide frequency among taxa was also assessed for each gene with a $\chi^2$ test as implemented in PAUP. The 12S and 16S data sets were combined for analysis. Partition homogeneity was assessed by the incongruence length difference (ILD) test as implemented in PAUP (Swofford 1993).

## 2.6 Phylogenetic analysis

Phylogenetic analyses were conducted using MRBAYES v.3.17 software for Bayesian analysis (BAY) and PAUP 4.0 beta 10 (Swofford 1993) for maximum likelihood (ML) analysis. Prior to conducting the BAY or ML analyses, the model of evolution that best fit the data was determined using MODELTEST v.3.7 (Posada & Crandall 1998). The Bayesian analysis was performed by sampling one tree every 100 generations for 1,000,000 generations, starting with a random tree, thus generating 10,001 trees. A preliminary analysis showed that stasis was reached at approximately 10,000 generations. Thus, we discarded 101 trees corresponding to those generations and obtained 50% majority rule consensus trees from the remaining 9,900 saved trees using PAUP. ML analysis was carried out with a distance correction set with the parameters obtained from MODELTEST (Posada & Crandall 1998). Analysis was performed as a heuristic search with gaps treated as missing data, multistate characters interpreted as uncertain, and all characters unordered. The search was conducted with a random sequence addition of taxa and tree bisection and reconnection as branch swapping option. Relative stability of clades under ML was determined from 100 bootstrap pseudoreplicates based on the same parameters as above. Bootstrap proportions >50% (for ML) and posterior probabilities (for BAY) are indicated in Figure 1.

## 3 RESULTS

### 3.1 Nucleotide composition

We produced 12S and 16S sequence data for 23 species (25 specimens) resulting in an alignment of 50 sequences. Sequences and alignments were submitted to GenBank as a PopSet. Our 12S alignment included a total of 407 bp of which 246 bp were constant, 33 were variable but not parsimony informative, and 128 characters were parsimony informative. The nucleotide composition of the database can be considered homogeneous ($\chi^2$ = 27.293, df = 72, P = 0.999) with a larger percentage of A–T (36.7%–37.0% respectively). Our 16S alignment included a total of 537 bp, of which 305 bp were constant, 65 were variable but parsimony uninformative, and 167 were parsimony informative. The nucleotide composition of the database can be considered homogeneous ($\chi^2$ = 31.636, df = 72, P = 0.999) with a larger percentage of A–T (32.8%–34.8% respectively). The combined alignment included 944 bp. We also excluded 69 saturated characters, 21 from the 12S fragment and 48 from the 16S fragment. From the remaining 875 characters, 544 were constant, 90 were variable but not parsimony informative, and 241 were parsimony informative. The ILD test showed no significant incongruence among gene segments (P = 0.462). Thus, all phylogenetic analyses were performed with a single data set including both genes.

### 3.2 Phylogenetic analyses

The best-fit model of nucleotide substitution, selected with the Akaike information criterion (AIC; Akaike 1974) as implemented in MODELTEST (Posada & Crandall 1998), was the HKY model (Hasegawa et al. 1985), with proportion of invariable sites ($\Gamma$) and a gamma distribution ($\delta$), with the following parameters: assumed nucleotide frequencies: A = 0.3518, C = 0.0890, G = 0.1804, T = 0.3788; with transition/transversion ratio = 3.967; proportion of invariable sites $\Gamma$ = 0.315; variable sites followed a gamma distribution with shape parameter $\delta$ = 0.498. These values were used for both ML and BAY analyses, which produced the same topology. We thus present a single tree obtained with ML analysis (ML score = 4986.170) that includes both ML bootstrap as well as Bayesian posterior probabilities (Fig. 1). In both analyses, monophyly of all five genera represented by two or more species received strong support values.

The ML tree based on the 12S and 16S genes is generally well supported at most, though not all, nodes (Fig. 1). Representative species of the putative family Thaumastochelidae were found

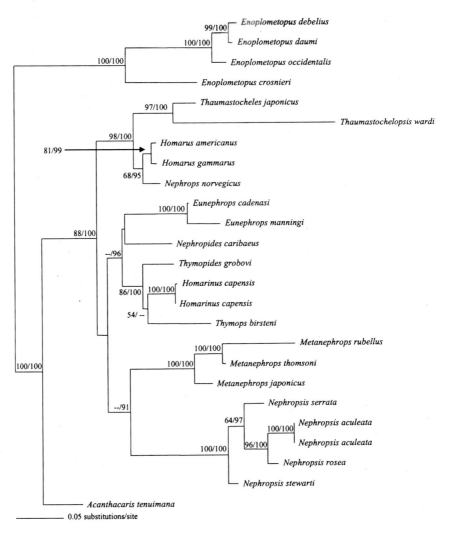

**Figure 1.** ML tree based on combined 16S and 12S sequences. ML bootstrap proportions (>50%) and Bayesian posterior probabilities indicated at nodes (ML/BAY).

to be monophyletic but nested within the Nephropidae (Fig. 1). *Acanthacaris* is the sister taxon to the remaining nephropids *sensu lato*. Among the latter, three clades were recovered. Relationships among these three clades cannot be considered resolved since they were not well supported by either bootstrap or Bayesian posterior probabilities. One clade included *Homarus*, *Nephrops*, *Thaumastocheles*, and *Thaumastochelopsis*. A second clade included the genera *Eunephrops*, *Nephropides*, *Thymopides*, *Homarinus*, and *Thymops*, although it was supported only by BAY. *Metanephrops* and *Nephropsis* formed a third clade, though it too was supported only by BAY.

## 4  DISCUSSION

To more fully understand relationships of the marine clawed lobsters, it is optimal to have a taxonomically comprehensive (all extant genera) molecular phylogenetic analysis based on multiple genes along with an equally comprehensive morphological study (all extant and extinct genera) based on a large data matrix. The present study analyzes two mitochondrial genes (12S and 16S) as a step toward this objective. While it would be ideal to root both the morphological and molecular trees to the same outgroup, that is so far impractical. The most appropriate outgroup for the present DNA analysis, Enoplometopidae, has no fossil record, although potentially Uncinidae may be an enoplometopid (Ahyong 2006). Fortunately, in the case of the marine clawed lobsters, our unpublished DNA data indicate that ingroup topology is insensitive to a range of potential outgroups such as freshwater crayfish (*Astacus, Parastacus, Cambarus*), glypheoids (*Neoglyphea*), or Enoplometopidae.

## 4.1 Comparison with previous works

Our results for *Acanthacaris* corroborate those of Porter et al. (2005), who found good support for *Acanthacaris* as a sister taxon to the three remaining nephropid genera in their 43-genus analysis of decapod phylogeny. Topology of the *Nephropsis* + *Nephrops* + *Homarus* clade in Porter et al. (2005) is also consistent with our results. Topology of the present 12S–16S tree (12 genera) is nearly identical to the 12S tree (ten genera) of Chu et al. (2006), despite their using *Neoglyphea* as the outgroup. The topology of our 12S–16S tree differs somewhat from that of the 16S–18S–28S tree of Ahyong & O'Meally (2004), who included five genera of clawed lobsters (*Enoplometopus, Homarus, Metanephrops, Nephropsis, Thaumastochelopsis*) in their analysis of 45 decapod genera. The disagreement in topologies is in the arrangement of three nephropid genera: *Homarinus, Metanephrops*, and *Nephropsis*. Ahyong & O'Meally (2004), analyzing three nephropid genera, found *Nephropsis* to be the sister to *Metanephops* + *Homarus*. Our analysis shows *Metanephrops* and *Nephropsis* are closer to each other than either is to *Homarus*. However, in addition to their analysis encompassing a taxonomically broader group of decapod genera, they used a species of *Stenopus* Latreille, 1819, representing the Stenopodidae Claus, 1872 (consistently identified as sister group to reptantian decapods by Ahyong & O'Meally 2004), as their outgroup. Tam & Kornfield (1998) analyzed five nephropid genera using mitochondrial 16S rRNA and produced trees that, while not well resolved, show either *Nephropsis* (via maximum parsimony) or *Metanephrops* (via neighbor joining) as sister to the remaining nephropid genera analyzed.

## 4.2 Acanthacaris

*Acanthacaris* is determined here (Fig. 1), as in the multi-locus analysis of Porter et al. (2005) and the 12S analysis by Chu et al. (2006), to be the sister taxon to the remaining nephropoids. Most previous morphological studies (Tshudy & Babcock 1997; Tshudy & Sorhannus 2000a, b; Ahyong 2006) found *Acanthacaris* to be deeply nested within Nephropidae rather than the sister taxon to the remaining genera. This disagreement between morphological and molecular topologies is marked and is largely due to the many autapomorphies of *Acanthacaris* and unstable rooting of the morphological trees. In comparison to other nephropoid genera, *Acanthacaris* has many distinctive autapomorphies including: 1) a laterally compressed rostrum; 2) a single row of dorsal rostral spines; 3) parallel submedian carinae on the telson; 4) an extremely large scaphocerite extending almost to the end of the antennal peduncle; and 5) delicately constructed, symmetrical claws, each with a narrow, cylindrical palm and fingers bearing acicular denticles. However, these features, being unique, are cladistically uninformative. Thus, very few character states remain to robustly position *Acanthacaris* (irrespective of whether they are convergent). In addition, the position of the root, and thus *Acanthacaris*, in the morphological analysis is sensitive to outgroup choice (Tshudy et al., unpublished data). Significantly, however, morphological analyses, using an identical group of taxa, recover an identical position for *Acanthacaris* as sister to the remaining nephropids (Tshudy et al., unpublished data). In terms of branch support, the molecular data provide strongest support for the "basal" position of *Acanthacaris*, using a range of outgroups, so we may be justified in favoring the molecular results. Future morphological studies should closely reconsider the apparently autapomorphous character states of *Acanthacaris* to determine whether, on closer inspection, they might be related to states in other taxa.

*Acanthacaris* is a blind, deep-sea (229–2161 m) lobster with no known close extant relatives and no known fossil relatives. *Palaeophoberus* Glaessner, 1932, previously thought to be related to *Acanthacaris* (Glaessner 1932, 1969; Mertin 1941; Burukovsky & Ckreko 1986), is now regarded as a chilenophoberid. At present, we cannot reliably infer whether the blind *Acanthacaris* evolved in the deep sea or, like the *Oncopareia-Thaumastocheles* lineage, lost its eyes through a migration from shallow, shelf depths into deeper, aphotic habitats.

### 4.3 *Status of Thaumastochelidae*

The family Thaumastochelidae is represented in this analysis by both of its Recent genera, *Thaumastocheles* and *Thaumastochelopsis*. These genera, along with the fossil (Late Cretaceous-Miocene) genus *Oncopareia* Bosquet, 1854, form a morphologically distinctive and cladistically cohesive group. The monophyly of the thaumastochelids has been supported by previous morphological studies (Tshudy & Babcock 1997; Tshudy & Sorhannus 2000a, b). Tshudy et al. (unpublished data), analyzing a 90-character morphology matrix, found the thaumastochelids united by three unambiguous synapomorphies: first pereiopod palm bulbous; telson wider than long; and uropodal endopod much smaller than exopod. Aside from these synapomorphies, all thaumastochelids have very distinctive abdominal pleura that are wider than long and quadrate, and even more distinctive first pereiopods with very long, slender fingers armed with acicular denticles. The close relationship among these three genera is undisputed, but their family-level status has been debated and has remained equivocal. Holthuis (1974) recognized the family, as did morphological cladistic analyses of Tshudy & Sorhannus (2000a, b), Dixon et al. (2003), Schram & Dixon (2004), Ahyong & O'Meally (2004), and Ahyong (2006). Molecular phylogenetic analyses support (Ahyong & O'Meally 2004) or dispute (Chu et al. 2006; Tsang et al. 2008) family level status for the thaumastochelids. In the DNA tree of Ahyong & O'Meally (2004), which did not include *Acanthacaris*, *Thaumastochelopsis* is the sister taxon to the three nephropid genera analyzed. Our molecular analysis supports monophyly of thaumastochelids, similar to all previous morphological studies. However, it does not support family level status for thaumastochelids because they are nested within Nephropidae *sensu stricto*. The paraphyly of this taxon is also evident in the decapod tree based on nuclear protein coding genes (Tsang et al. 2008; Chu et al. this volume). We thus regard thaumastochelids as members of the Nephropidae. As with the putative Thaumastochelidae, the nephropid subfamilies Nephropinae (*Eunephrops, Homarus, Metanephrops, Nephrops*) and Thymopinae (*Nephropides, Nephropsis, Thymops, Thymopsis*) of Holthuis (1974) are not recovered by present results.

### 4.4 *Morphological convergence*

The present and recently published DNA studies facilitate detailed comparison with morphology-based phylogenies of nephropid genera. These agree in some aspects, for example, the placement of *Acanthacaris* (as discussed above) and *Eunephrops* and *Nephropides* forming a clade in some morphological studies (Tshudy & Babcock 1997; not Ahyong 2006) and in DNA studies (this study; Chu et al. 2006 [*Eunephrops* is a sister taxon to *Nephropides* + *Thymopides*]). However, morphological and DNA studies disagree in other aspects of nephropid phylogeny (discussed below), and these differences seem largely attributable to morphological convergence.

#### 4.4.1 Homarus *and* Homarinus
A previous study based on 16S sequence data (Tam & Kornfield 1998; five nephropid genera) and also the present 12S–16S study position *Homarus* as the sister taxon to *Nephrops*, instead of *Homarinus*, as is common in morphological analyses (Tshudy & Babcock 1997; Ahyong 2006; Tshudy

et al. unpublished data). If these molecular results are interpreted to be more phylogenetically accurate than existing morphological studies (alpha-taxonomic and phylogenetic), then morphological similarities between *Homarus* and *Homarinus* are most parsimoniously explained as morphological convergence. *Homarus* and *Homarinus* are "plain-looking" nephropids that lack many of the distinguishing external features of other nephropid genera, features such as cephalothoracic carinae and spines, sculptured abdominal terga, and carinate claws. Ahyong (2006) found *Homarus* and *Homarinus* to be the most plesiomorphic of nephropids. Until recently, these two genera were considered congeneric. Kornfield et al. (1995) examined what were at that time three species of *Homarus* (*H. americanus*, *H. gammarus*, *H. capensis*) and removed *H. capensis* to a new genus, *Homarinus*, on the basis of DNA sequence comparisons and morphology. They reported 16S sequence (380 bp) divergence between *H. americanus* and *H. gammarus* at 1.3%, compared to average divergence between these and the "cape lobster" at 9.7% (Kornfield et al. 1995). Recent and present molecular analyses strongly support *Homarus* and *Homarinus* as having evolved in separate lineages, and both genera are "safely" nested in ornamented clades. Therefore, their morphologic similarities are interpreted as morphologic convergence.

### 4.4.2 *Nephrops* and *Metanephrops*

Similar to the *Homarus–Homarinus* example, *Nephrops* and *Metanephrops* are sister taxa in morphological analyses (Tshudy & Sorhannus 2000b; Tshudy et al., unpublished data) and are widely disparate in DNA-based trees (Chu et al. 2006; this study). In a morphological study parallel to this one (Tshudy et al., unpublished data), *Metanephrops* and *Nephrops* are the most derived nephropids and are sister taxa united by one unambiguous synapomorphy: the male pleopod 1 distal end is a posteriorly curving/terminating hook. There are several other obvious external similarities between these genera (ambiguous synapomorphies), which are apparently convergent. These similarities include their intermediate and lateral thoracic carinae, the complexly sculptured abdominal tergites, and their carinate and spiny claws. DNA analyses (Tam & Kornfield 1998 [16S]; present study [12S, 16S]) find *Nephrops* and *Metanephrops* well separated on the cladogram, indicating that the morphological similarities between these genera are the result of convergence.

## 5 CONCLUSIONS AND FUTURE WORK

This DNA analysis of clawed lobster genera facilitates detailed comparison with similarly comprehensive morphology-based topologies. There are major differences between the DNA and morphological results to date. *Eunephrops* and *Nephropides* form a sister group in some morphological studies and in DNA studies. Aside from that, there are conflicts at the level of family and genus.

*Acanthacaris* is determined to be the least derived of the genera in this analysis and is the sister group to all the nephropids, including the putative Thaumastochelidae. Published morphological studies have determined *Acanthacaris* to be more highly derived within the nephropids, and notably more so than the thaumastochelids.

Our molecular analysis supports monophyly of thaumastochelids, similar to all previous morphological studies. However, it does not support family level status for thaumastochelids, on the basis of their phylogenetic placement within Nephropidae. Thaumastochelidae should therefore be synonymized with Nephropidae.

*Homarus* and *Homarinus* form a clade in the morphological analyses, but our DNA analyses suggest they belong to different lineages, indicating that their similarities are the result of convergence. *Nephrops* and *Metanephrops*, likewise, form a clade in morphological analyses but are not closely related according to DNA analyses. Our molecular data suggest that *Homarus* and *Nephrops* are sister taxa, despite their being well separated in morphology-based trees.

Given the sensitivity of morphological analyses to taxon and character selection, which we interpret mainly to convergence, we should work toward further testing of DNA trees as guides to the phylogeny of extant and, ultimately, extinct lobsters. Thus far, sequences from four gene regions have been applied (12S, 16S, 18S, 28S), with as many as three in one analysis (Ahyong & O'Meally [2004] used 16S, 18S, and 28S). If the addition of new data (e.g., protein coding genes; see Tsang et al. 2008) stabilizes these trees, we could, through reverse extrapolation, infer which morphological characters are most phylogenetically reliable for analysis of extinct genera. Future work should also combine morphological and molecular data in a total evidence analysis.

ACKNOWLEDGEMENTS

Support for D. Tshudy was provided under U.S. National Science Foundation (NSF) RUI:ROA amendment to NSF-BS&I grant 0315995 to S. Fredericq and D.L. Felder at UL Lafayette and a grant from the Edinboro University Faculty Senate. Additional support was provided under NSF-AToL grant 0531603 to D.L. Felder and grant CUHK4419/04M from the Research Grants Council, Hong Kong Special Administrative Region, China, to K.H. Chu and T.Y. Chan. This work was partially supported by research grants from the National Science Council, Taiwan, R.O.C., and Center for Marine Bioscience and Biotechnology of the National Taiwan Ocean University to T.Y. Chan. STA gratefully acknowledges support from the New Zealand Foundation for Research, Science and Technology (C01X0502) and Biosecurity New Zealand (ZBS200524). Two anonymous reviewers provided detailed comments that significantly improved the manuscript. This is contribution number 126 from the UL Lafayette Laboratory for Crustacean Research.

REFERENCES

Ahyong, S.T. 2006. Phylogeny of the clawed lobsters (Crustacea: Decapoda: Homarida). *Zootaxa* 1109: 1–14.
Ahyong, S.T. & O'Meally, D. 2004. Phylogeny of the Decapoda Reptantia: resolution using three molecular loci and morphology. *Raffles Bull. Zool.* 52: 673–693.
Akaike, H. 1974. A new look at the statistical model identification. *IEEE Trans. Automat. Contr.* 19: 716–723.
Amati, L., Feldmann, R.M. & Zonneveld, J.-P. 2004. A new family of Triassic lobsters (Decapoda: Astacidea) from British Columbia and its phylogenetic context. *J. Paleontol.* 78: 150–168.
Burukovsky, R.N. & Ckreko, B.T. 1986. Archaic lobsters. *Nature, Moscow* 12: 93–95. [In Russian.]
Chan, T.Y., Tong, J.G., Tam, Y.K. & Chu, K.H. 2008. Phylogenetic relationships among the genera of the Penaeidae (Crustacea: Decapoda) revealed by mitochondrial 16S rRNA gene sequences. *Zootaxa* 1694: 38–50.
Chu, K.H., Li, C.P. & Qi, J. 2006. Ribosomal RNA as molecular barcodes: a simple correlation analysis without sequence alignment. *Bioinformatics* 22: 1690–1701.
Chu, K.H., Tsang, L.M., Ma, K.H., Chan, T.Y. & Ng, P.K.L. (this volume). Decapod phylogeny: what can protein-coding genes tell us? In: Martin, J.W., Crandall, K.A. & Felder, D.L. (eds.), *Crustacean Issues: Decapod Crustacean Phylogenetics*. Boca Raton, Florida: Taylor & Francis/CRC Press.
Colbourne, J.K. & Hebert, P.D.N. 1996. The systematics of North American *Daphnia* (Crustacea: Anomopoda): a molecular phylogenetic approach. *Philos. T. Roy. Soc.* B 351 (1337): 349–360.
Crandall, K.A. & Fitzpatrick, Jr., J.F. 1996. Crayfish molecular systematics: inferences using a combination of procedures to estimate phylogeny. *Syst. Biol.* 45: 1–26.
Dixon, C., Ahyong, S.T. & Schram, F.R. 2003. A new hypothesis of decapod phylogeny. *Crustaceana* 76: 935–975.
Glaessner, M.F. 1932. Zwei ungenügend bekannte mesozoische Dekapodenkrebse. *Pemphix sueuri* (Desm.) und *Palaeophoberus suevicus* (Quenstedt). *Paläontologische Zeitschrift* 14: 108–121.

Glaessner, M.F. 1969. Decapoda. In: Moore, R.C. (ed.), *Treatise on Invertebrate Paleontology, Part R, Arthropoda 4(2)*: R399-R651. Lawrence, Kansas: Geological Society of America and University of Kansas Press.

Hall, T.A. 1999. BioEdit: a user-friendly biological sequence alignment editor and analysis program for Windows 95/98/NT. *Nucl. Acids Symp. Ser.* 41: 95–98.

Hasegawa, M., Kishino, H. & Yano, T. 1985. Dating of human-ape splitting by a molecular clock of mitochondrial DNA. *J. Mol. Evol.* 22: 160–174.

Holthuis, L.B. 1974. The lobsters of the Superfamily Nephropoidea of the Atlantic Ocean (Crustacea: Decapoda). *Bull. Mar. Sci.* 24: 723–884.

Holthuis, L.B. 1991. *FAO Species Catalogue. Volume 13, Marine Lobsters of the World.—FAO Fisheries Synopsis* 125: 1–292.

Kornfield, I., Williams, A.B. & Steneck, R.S. 1995. Assignment of *Homarus capensis* (Herbst, 1792), the Cape lobster of South Africa, to the new genus *Homarinus* (Decapoda: Nephropidae). *Fish. B-NOAA* 93: 97–102.

Mertin, H. 1941. Decapode Krebse aus dem subhercynen und Braunschweiger Emscher und Untersenon sowie Bemerkungen über einige verwandte Formen in der Oberkreide. *Nova Acta Leopoldina* 10: 1–264.

Palumbi, S.R., Morton, A.P., Romano, S.L., McMillan, W.G., Stice, L. & Grabowski, G. 1991. *The Simple Fools Guide to PCR*. Honolulu: Best of Zoology, University of Hawaii.

Porter, M.L., Pérez-Losada, M. & Crandall, K.A. 2005. Model-based multi-locus estimation of decapod phylogeny and divergence times. *Mol. Phylogenet. Evol.* 37: 355–369.

Posada, D. & Crandall, K.A. 1998. MODELTEST: testing the model of DNA substitution. *Bioinformatics* 14: 817–818.

Robles, R., Schubart, C.D., Conde, J.E., Carmona-Suárez, C., Alvarez, F., Villalobos, J.L. & Felder, D.L. 2007. Molecular phylogeny of the American *Callinectes* Stimpson, 1860 (Brachyura: Portunidae), based on two partial mitochondrial genes. *Mar. Biol.* 150: 1265–1274.

Rode, A.L. & Babcock, L.E. 2003. Phylogeny of fossil and extant freshwater crayfish and some closely related nephropid lobsters. *J. Crust. Biol.* 23: 418–435.

Schram, F.R. & Dixon, C.J. 2004. Decapod phylogeny: addition of fossil evidence to a robust morphological cladistic data set. *Bull. Mizunami Fossil Mus.* 31: 119.

Schubart, C.D., Cuesta, J.A. & Felder, D.L. 2002. Glyptograpsidae, a new brachyuran family from Central America: larval and adult morphology, and a molecular phylogeny of the Grapsoidea. *J. Crust. Biol.* 22: 28–44.

Schubart C.D., Neigel, J.E. & Felder, D.L. 2000. Use of the mitochondrial 16S rRNA gene for phylogenetic and population studies of Crustacea. In: von Vaupel Klein, J.C. & Schram, F.R. (eds.), *Crustacean Issues 12, The Biodiversity Crisis and Crustacea*: 817–830. Rotterdam: Ballama.

Swofford, D.L. 1993. *PAUP* 4.0. Phylogenetic Analysis Using Parsimony (*and Other Methods)*. Sunderland, Massachusetts: Sinauer Associates.

Tam, Y.K. & Kornfield, I. 1998. Phylogenetic relationships of clawed lobster genera (Decapoda: Nephropidae) based on mitochondrial 16s rRNA gene sequences. *J. Crust. Biol.* 18: 138–146.

Tsang, L.M., Ma, K.Y., Ahyong, S.T., Chan, T.Y. & Chu, K.H. 2008. Phylogeny of Decapoda using two nuclear protein-coding genes: origin and evolution of the Reptantia. *Mol. Phylogenet. Evol.* 48: 359–368.

Tshudy, D. 1993. *Taxonomy and Evolution of the Clawed Lobster Families Chilenophoberidae and Nephropidae*. Unpublished Ph.D. Thesis, Kent State University, Kent, Ohio.

Tshudy, D. & Babcock, L. 1997. Morphology-based phylogenetic analysis of the clawed lobsters (family Nephropidae and the new family Chilenophoberidae). *J. Crust. Biol.* 17: 253–263.

Tshudy, D. & Sorhannus, U. 2000a. *Jagtia kunradensis*, a new genus and species of clawed lobster (Decapoda: Nephropidae) from the Upper Cretaceous (upper Maastrichtian) Maastricht Formation, The Netherlands. *J. Paleontol.* 74: 224–229.

Tshudy, D. & Sorhannus, U. 2000b. Pectinate claws in decapod crustaceans: convergence in four lineages. *J. Paleontol.* 74: 474–486.

Williams, A.B. 1995. Taxonomy and evolution. In: Factor, J.R. (ed.), *Biology of the Lobster Homarus americanus*: 13-21. New York: Academic Press.

# The Polychelidan Lobsters: Phylogeny and Systematics (Polychelida: Polychelidae)

SHANE T. AHYONG

*National Institute of Water and Atmospheric Research, Kilbirnie, Wellington, New Zealand*

ABSTRACT

Decapods of the infraorder Polychelida are unusual in having chelate pereopods 1–4 and reduced eyes in extant species. Polychelidans traditionally have been included with the achelate lobsters in the infraorder Palinura. Polychelida, however, is depicted as basal in the Reptantia by most recent studies. The polychelidan fossil record extends back to the Upper Triassic, with four families recognized to date, of which only Polychelidae is extant. Interrelationships of the fossil and living polychelidan lobsters were studied by cladistic analysis of morphology, with emphasis on Polychelidae. Coleiidae was found to be sister to Polychelidae, to the exclusion of *Palaeopentacheles*, previously placed in the latter. A new family, Palaeopentachelidae, is recognized for *Palaeopentacheles*. All other recognized polychelidan families are also diagnosed. An incomplete fossil taxon from the Upper Triassic attributed to Polychelidae, *Antarcticheles antarcticus*, is confirmed as a polychelid and is most closely related to the extant genus *Willemoesia*. The strong similarities between *Willemoesia* and *Antarcticheles* indicate that differentiation of the 'polychelid form' was well established by the late Jurassic. Among extant Polychelidae, *Willemoesia* is least derived, though the shallow dorsal orbits, regarded by some as plesiomorphic, are a derived condition. *Stereomastis* is removed from the synonymy of *Polycheles*. Six extant polychelid genera are recognized: *Cardus*, *Homeryon*, *Pentacheles*, *Polycheles*, *Stereomastis*, and *Willemoesia*. All extant polychelid genera are diagnosed, and keys to genera and species are provided. Phylogenetic trends within Polychelida include a general narrowing of the carapace and abdomen; shortening of the carapace front with respect to the anterolateral margins, leading to a shift in eye orientation from anterior to transverse; dorsal exposure of the base of the antennules and development of a stylocerite; and a shift in the form of the major chelipeds from relatively robust with short, triangular carpi to elongated and slender, with slender carpi. These trends within Polychelida appear to correspond to a shift from a shallow-water, epibenthic habit to the deep-water, fossorial lifestyle currently evident in Polychelidae. Phylogenetic trends within Polychelidae include a consistent reduction in length of the maxilliped 3 and pereopodal epipods. Epipod length is not known for any of the fossil taxa, but character polarization among extant taxa predicts that extinct taxa bore well-developed epipods.

## 1 INTRODUCTION

Among reptant decapods, polychelidans (Figs. 1, 2) are conspicuous in the possession of chelae on pereopods 1–4 and sometimes pereopod 5. Glaessner (1969) recognized four polychelidan families: Eryonidae, Coleiidae, Tetrachelidae, and Polychelidae. Polychelida was most morphologically diverse during the Mesozoic, with all known families then present. Only a single family, Polychelidae, survives to the present. Polychelids are often referred to as deep-sea blind lobsters because all extant forms live in deep water and have strongly reduced eyes. The well-developed eyes

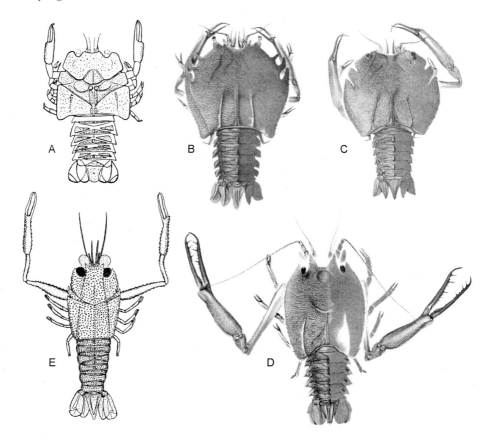

**Figure 1.** Fossil Polychelida. (A) *Tetrachela raiblana* (Tetrachelidae). (B) *Eryon arctiformis* (Eryonidae). (C) *Cycleron propinquus* (Eryonidae). (D) *Pentacheles roettenbacheri* (Palaeopentachelidae). (E) *Coleia longipes* (Coleiidae). A, from Glaessner (1969: fig. 272). B–D, from Garassino & Schweigert (2006: pl. 6, 7, 9). E, from Schweigert & Dietl (1999).

and palaeoecology of most extinct polychelidans, however, implies a shallow water origin for the group.

Polychelidae is thus the sole extant family of the infraorder Polychelida. The polychelids and achelate lobsters (Palinuroidea) have traditionally constituted Palinura (see Holthuis 1991), but recent morphological (Scholtz & Richter 1995; Schram 2001; Dixon et al. 2003) and molecular phylogenies (Ahyong & O'Meally 2004; Tsang et al. 2008) recognize independent status of both groups as separate infraorders: Achelata and Polychelida. Significantly, most of these analyses place the Polychelida as the sister group to all other reptants, apart from Tsang et al. (2008), which places Polychelida as sister to Achelata, though with low nodal support. Either way, all results recognize reciprocal monophyly of Polychelida and Achelata.

Internal relationships of Polychelidae have received scant attention aside from that implied by generic arrangements or from use of species exemplars in broader studies of decapod phylogeny (e.g., Dixon et al. 2003; Schram & Dixon 2004; Ahyong & O'Meally 2004). Unfortunately, the generic system of the Polychelidae has been in a constant state of confusion for more than a century. Over much of this period, four generic names have been applied to adult polychelids: *Polycheles* Heller, 1862 [type species *P. typhlops* Heller, 1862], *Pentacheles* Bate, 1878 [type species: *Pe. laevis* Bate, 1878], *Stereomastis* Bate, 1888 [type species: *S. suhmi* (Bate, 1878)], and

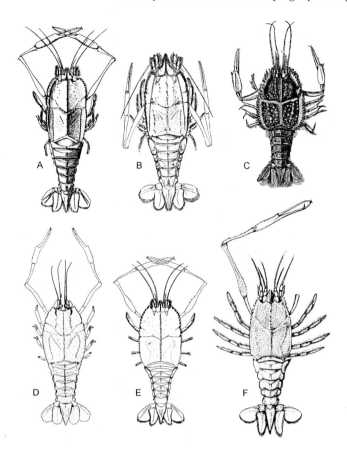

**Figure 2.** Extant Polychelidae. (A) *Polycheles typhlops*. (B) *Stereomastis sculpta*. (C) *Cardus crucifer*. (D) *Homeryon armarium*. (E) *Pentacheles laevis*. (F) *Willemoesia pacifica*. A, E, from Selbie (1914). B, from Smith (1882: pl. 3). C, from Bate (1888: fig. 31). F, from Kensley (1968: fig 4).

*Willemoesia* Grote, 1873 [type species: *W. leptodactyla* (Thomson, 1873)]. The status of *Willemoesia* has not been controversial owing to its distinctive shallow dorsal orbital concavities and the accessory spine on the pollex of the first cheliped. In contrast, the status of *Pentacheles*, *Polycheles*, and *Stereomastis* has been in constant flux. Much of the confusion has stemmed from inadequate original descriptions regarding the length of the epipod of the third maxilliped and the use of unreliable characters as diagnostic. This is particularly so in the case of *Pentacheles*, in which the original primary diagnostic character was the chelate or non-chelate condition of pereopod 5 (Bate 1878). The chelation of pereopod 5 was soon recognized to be subject to allometry and sexual dimorphism in species of *Pentacheles*, *Polycheles*, and *Stereomastis* (see Faxon 1895). Consequently, *Pentacheles* was treated as a synonym of *Polycheles* by most workers (Kemp & Sewell 1912; Selbie 1914; de Man 1916; Firth & Pequegnat 1971; Griffin & Stoddart 1995). Several workers have emphasized the reduced maxillipedal and pereopodal epipods as a defining character of *Stereomastis*, but characterization of *Polycheles* remained difficult because of variability in the length of the epipod of the third maxilliped in species then assigned to the genus (see Firth & Pequegnat 1971). Separation of *Stereomastis* from *Polycheles* has never been satisfactorily resolved, such that most workers could only distinguish the two genera based on a unitary difference in the number of lateral carapace spines — whether more or fewer than 20 — hardly a satisfactory situation. Further progress in separating polychelid genera was stalled until Galil (2000) comprehensively revised the world species of the Polychelidae, emphasizing the lengths of the epipod of the third maxilliped and excluding the pereopodal epipods. Galil (2000) recognized two new genera, *Cardus* and *Homeryon*, for several unusual species previously assigned to *Polycheles*. One of the most significant advances made by Galil (2000), however, was resurrection of *Pentacheles*, but under a significantly different generic concept from that originally proposed by Bate (1878). In removing *Pentacheles*, *Cardus*, and *Homeryon* from *Polycheles*, Galil (2000) also regarded *Stereomastis* as a synonym of *Polycheles*. *Polycheles sensu* Galil (2000) became a speciose, morphologically diverse genus united by a vestigial epipod on the third maxilliped.

The obvious relationship between the polychelids and the extinct eryonids was recognized early on (see Glaessner 1969). The phylogenetic position of several taxa has been speculated on, such as a basal or derived position of *Willemoesia* on the basis of its shallow dorsal orbits (Bouvier 1917), but relationships have never been comprehensively studied. Therefore, the present study examines the interrelationships of the Polychelida by cladistic analysis with a focus on the extant Polychelidae.

## 2 MATERIALS AND METHODS

### 2.1 *Terminal taxa*

All 37 recognized extant species of Polychelidae (Galil 2000; Ahyong & Brown 2002; Ahyong & Chan 2004; Ahyong & Galil 2006) are included as terminals. Character state scoring for each species is derived from examination of specimens and/or published accounts (see Appendix 1). Characters were polarized using *Tetrachela raiblana* (Tetrachelidae) as the outgroup. In addition, *Cycleryon propinquus*, *Eryon arctiformis* and *Knebelia bilobata* (all Eryonidae), *Palaeopentacheles roettenbacheri* (originally placed in Polychelidae), and *Coleia longipes* (Coleiidae) were included in the ingroup as exemplars of the extinct polychelidan families, in order to assess their phylogenetic positions and act as potential tests of polychelid monophyly. Each of the aforementioned fossil taxa was selected because of the availability of excellent reconstructions including details of cheliped morphology (Schweigert & Dietl 1999; Garassino & Schweigert 2006). The extinct *Antarcticheles antarcticus* is known only from the carapace and partial abdomen but is regarded as a polychelid (Aguirre-Urreta et al. 1990); it was included in a separate analysis (Analysis 2) to assess its phylogenetic position. Specimens are deposited in the following institutions: Australian Museum (AM); Muséum National d'Histoire Naturelle, Paris (MNHN); National Fisheries University, Shimonoseki, Japan (NFU); National Institute of Water and Atmospheric Research, Wellington, New Zealand (NIWA); National Taiwan Ocean University (NTOU); Raffles Museum of Biodiversity Research, National University of Singapore (NUS); South Australian Museum (SAM); Texas A & M University, Texas (TAMU); National Museum of Natural History, Smithsonian Institution (USNM); Western Australian Museum, Perth (WAM); and Zoological Museum, Berlin (ZMB).

### 2.2 *Morphological characters*

The 71 morphological characters used in the analysis are listed in Appendix 3, along with character states, brief descriptions (and references to Fig. 3), and selected definitions.

### 2.3 *Analytical methods*

The data matrix was constructed in MacClade 4.0 (Maddison & Maddison 2000) and includes 44 taxa and 71 characters (Appendix 2). Some characters are applicable only to some species and

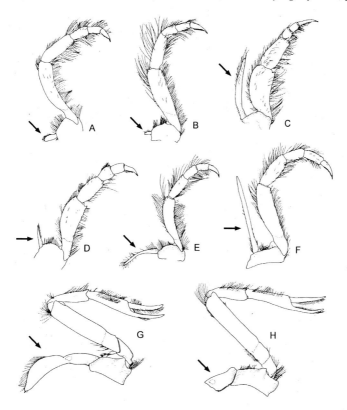

**Figure 3.** Extant Polychelidae. A–F, maxilliped 3. G–H, pereopod 3 (branchiae omitted). (A, G) *Polycheles enthrix*. (B, H) *Stereomastis suhmi*. (C) *Cardus crucifer*. (D) *Homeryon armarium*. (E) *Pentacheles laevis*. (F) *Willemoesia forceps*. Arrows indicate epipod.

cannot be meaningfully scored for the remaining taxa. Coding of inapplicable characters, either as a '?' or as a state called 'inapplicable,' has been shown to be problematic based on currently available computer algorithms (Maddison 1993). Although Platnick et al. (1991) suggested that the '?' coding can lead to implications of unlikely ancestral states, the alternative coding as a character may lead to branches being supported by the non-existent character state 'inapplicable.' Inapplicables were therefore scored '?' but are indicated as '-' in Appendix 2 to distinguish them from unknowns.

All characters were unordered (non-additive) and equally weighted, missing data were scored unknown, and polymorphisms were scored as such rather than assuming a plesiomorphic state. Characters were unordered, so the score given for each state (i.e., 0, 1, 2) implies nothing about order in a transformation series. Trees were generated in PAUP*4.0b10 (Swofford 2002) under the heuristic search (MULTREES, tree-bisection-reconnection, 500 replications with random input order). Relative stability of clades was assessed by parsimony jackknifing (Farris et al. 1996) with 500 pseudoreplicates and 30% character deletion as implemented in PAUP*.

## 3 RESULTS

Analysis 1 retrieved 10 minimal length trees of length 191, consistency index (CI) 0.4974, and retention index (RI) 0.8580 (Fig. 4A). Unambiguous character state changes for 1 of 10 most parsimonious topologies are listed in Appendix 3 and correspond to nodes numbered in Fig. 5. All

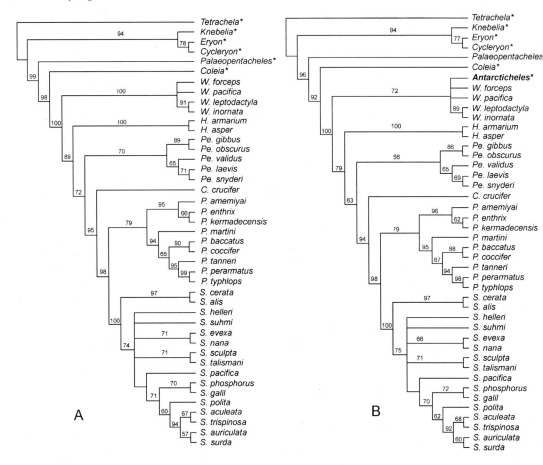

**Figure 4.** Phylogeny of the Polychelida. (A) Analysis 1, strict consensus of 10 most parsimonious topologies (TL = 191, CI = 0.4974, RI = 0.8580). (B) Analysis 2, strict consensus of 20 most parsimonious topologies (TL = 192. CI = 0.4948, RI = 0.8578). Jackknife proportions indicated at nodes. Generic names abbreviated as: *Cardus* (C.), *Homeryon* (H.), *Pentacheles* (Pe.), *Polycheles* (P.), *Stereomastis* (S.), *Willemoesia* (W.). Extinct taxa (*).

polychelid genera as recognized by Galil (2000) were recovered by the analysis. The most basal polychelid clade is *Willemoesia*, followed by *Homeryon* and *Pentacheles*. *Cardus* is sister to *Polycheles sensu* Galil (2000). *Polycheles sensu* Galil (2000) comprises two major clades corresponding to *Stereomastis* and *Polycheles sensu stricto*. Monophyly of crown-group Polychelidae received 100% jackknife support, suggesting a monophyletic origin for all extant forms. *Coleia* (Coleiidae), rather than *Palaeopentacheles*, was sister to crown-group polychelids, suggesting that the latter should be excluded from Polychelidae. The eryonid clade is sister to *Palaeopentacheles* + (*Coleia* + Polychelidae). Jackknife values for the genera are as follows: *Homeryon* (100%), *Stereomastis* (100%), *Pentacheles* (70%), *Polycheles* (79%), and *Willemoesia* (100%). The *Polycheles* + *Stereomastis* clade is robust to jackknifing (98%), but relationships between other genera received lower jackknife support (72–95%). Analysis 2 (including *Antarcticheles*) recovered 20 minimal-length trees of length 192, CI = 0.4948, RI = 0.8578 (Fig. 4B). The strict consensus reflected the strict consensus of Analysis 2, with *Antarcticheles* in a clade with *Willemoesia*. Jackknife proportions for most nodes in Analysis 2 were similar to those of Analysis 1.

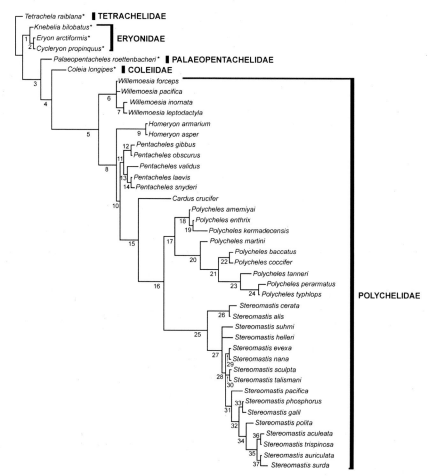

Figure 5. Phylogeny of the Polychelida. 1 of 10 most parsimonious topologies derived from Analysis 1 (TL = 191, CI = 0.4974, RI = 0.8580). Clade number indicated at nodes. Unambiguous character state changes for nodes are given in Appendix 4.

## 4 DISCUSSION

### 4.1 *The polychelid sister group and the position of* Palaeopentacheles

Coleiidae is sister to the Polychelidae (to the exclusion of *Palaeopentacheles*). Both share distinct cervical and postcervical incisions in the carapace margins, with a well-marked postcervical groove, distinct postorbital carinae, and the slender, elongate carpus of pereopod 1. Note, however, that the pereopod 1 carpus condition is not strictly uniform in coleids and polychelids: the carpus is short and stout in one polychelid (*Cardus crucifer*) and several coleids: *Proeryon hartmanni* (von Meyer, 1836) and several species of *Coleia* (see Teruzzi 1990; Schweigert 2000; Karasawa et al. 2003). Coleiidae otherwise differs from Polychelidae chiefly in the 2-segmented uropodal exopod, in having postorbital carinae (when present) that are aligned with the branchial carinae and a second abdominal pleuron that is similar to that of the third pleuron, rather than being distinctly larger. The unisegmental uropodal exopod of Polychelidae is not unique, being present in all polychelidans except Coleiidae and Tetrachelidae. The distinctly enlarged second pleuron that overlaps both the first and third pleura, however, is a synapomorphy of Polychelidae. As with other known fossil polychelidans, the eyes of coleids are well developed rather than reduced as in extant polychelids. Further synapomorphies of extant Polychelidae (unknown in the fossil *Antarcticheles*) are the reduced eyes and laterally expanded basal antennular segment with stylocerite.

The position of *Palaeopentacheles* as sister to Coleiidae + Polychelidae is significant. Though *Palaeopentacheles* has always been assigned to Polychelidae on account of its unisegmental uropodal exopod, well-marked cervical groove (only medially), and deep orbits, each of these features is plesiomorphic. *Palaeopentacheles* is excluded from the Coleiidae + crown-group polychelid clade by lacking postorbital carinae; in lacking an anterior median carina on the carapace; in the possession of sharp, angular, pleural terminations; and in lacking any trace of cervical and postcervical incisions on the lateral carapace margin. *Palaeopentacheles* is herein placed in a new family, Palaeopentachelidae, diagnosed below (section 5.1).

## 4.2 The genera of the Polychelidae

Galil (2000) synonymised *Stereomastis* with *Polycheles*, but present results indicate that both genera are monophyletic and readily distinguished. Both are recognized herein. *Stereomastis* and *Polycheles* differ from all other polychelids by the vestigial instead of well-developed epipod on maxilliped 3. *Stereomastis* is readily distinguished from *Polycheles* by the following synapomorphies: the reduced instead of long epipod on pereopods 1–5, deep; U-shaped instead of V-shaped dorsal orbital sinuses in the frontal margin of the carapace; the bilobed instead of unilobate eye; and the presence of a pleural spine on abdominal tergite 1 (except in *S. cerata* and *S. alis*; present in *Polycheles tanneri*). The aforementioned diagnostic characters of *Stereomastis* are far more 'satisfactory' than former distinctions that relied on lateral spine counts of the carapace, whether more than or fewer than 20 (Firth & Pequegnat 1971). Within *Stereomastis*, species allied to *S. phosphorus*, namely *S. aculeata*, *S. auriculata*, *S. galil*, *S. polita*, *S. surda*, and *S. trispinosa*, are united by the presence of spines on the coxae of pereopods 2–3. *Stereomastis alis* and *S. cerata* form a clade that is sister to the remaining species of the genus. Though *Polycheles* is monophyletic in the present analysis, its support is low, suggesting possible heterogeneity. Few unambiguous characters support monophyly of *Polycheles* (Clade 17), and, at present, the genus is most easily recognized by a combination of character states, most of which are plesiomorphies: the V-shaped dorsal orbital sinus (plesiomorphic), vestigial epipod of maxilliped 3 (plesiomorphic), rounded anterolateral margin of the basal antennular segment (plesiomorphic), and absence of an arthrobranch on maxilliped 3 (apomorphic). Although overall monophyly of *Polycheles* is not well supported, it consists of two well-supported clades (jackknife > 90%). One clade contains six species including the type species, and the other contains *P. enthrix*, *P. kermadecensis*, and *P. amemiyai*. The most important characters separating the second clade from the first are the chelate instead of simple pereopod 5 in males and the articulating instead of fused ischium and basis on pereopods 3–5. The pereopod 3–5 ischium and basis is fused in all other extant polychelids except *Homeryon*. Further study may justify removal of *P. enthrix* and allies to a separate genus.

Support for monophyly of *Pentacheles* is low, suggesting that it could be paraphyletic. Species of *Pentacheles* share similar general morphology, but most previously employed diagnostic characters, such as the well-developed epipod of the third maxilliped and angular anterolateral margin of the basal antennular segment, are plesiomorphies present also in *Homeryon* and *Willemoesia*. The single synapomorphy of *Pentacheles* identified here is the indistinct to absent branchial carina. In other polychelids, the branchial carina is well defined.

*Homeryon* is readily recognized by its strongly curved pereopod 2–4 dactyli, prominently angled carina laterally bordering the buccal cavity, and elongate pereopodal epipods. An unusual feature of *Homeryon* shared with *Polycheles amemiyai*, *P. enthrix*, and *P. kermadecensis* is the articulated rather than fused basis and ischiomerus, with a diagonal rather than transverse junction (Char. 69, 70). In other polychelids the basis and ischiomerus are fused, with a transverse junction (except in *Willemoesia*, with a diagonal junction).

*Cardus* is unique among extant polychelids for its ovate carapace, short pereopod 1 carpus, and small maximum size (reaching about 30 mm carapace length). The median spines on the abdominal terga are also unusual for their slenderness, being usually stout and triangular in other genera. In these respects, *Cardus* resembles the eryoneicus larva and as such may be neotenous.

## 4.3 *The position of* Willemoesia

Bouvier (1917) identified *Willemoesia* as the most 'primitive' of extant polychelids based on the eryonid-like shallow dorsal orbits and well-developed pereopodal epipods. Although *Willemoesia* (or *Willemoesia* + *Antarcticheles*) was found to be sister to remaining extant genera, present results suggest that the resemblance to eryonids is superficial. The eyes of eryonids are well developed and directed forwards. Conversely, the eyes of *Willemoesia* are poorly developed and the stalk is oriented transversely along the anterior wall of the carapace as in all other extant polychelids. In extant polychelids (other than *Willemoesia*), the base of the eyestalk is swollen and protrudes dorsally, occupying the dorsal orbital sinus, and the cornea protrudes laterally through the lateral orbital sinus. In *Willemoesia*, however, the eye is shorter than in other polychelids, not reaching the lateral carapace margins. The cornea is fused with the anterior wall of the carapace. Although the base of the eyestalk is reduced and does not protrude through the carapace, the homologous position and apparent outline of the dorsal orbital sinuses present in other polychelids are visible in most species of *Willemoesia* as a depressed, aspinulate area above the eyestalk bases. Thus, in *Willemoesia*, degeneration of the eyes possibly has been accompanied by closure of the dorsal orbits. Species of *Willemoesia* are the deepest living polychelids (exceeding 5000 m; Galil 2000), and it appears that vision is correspondingly degenerate. The shallow dorsal orbits of *Willemoesia* thus appear to be a derived feature, not homologous with those of eryonids. Moreover, the presence of deep dorsal orbital sinuses in the extinct palaeopentachelids and most coleiids, which are more closely related to the polychelids than are the eryonids, indicates that the orbital condition in *Willemoesia* is probably derived. Further study of the diverse coleiids, however, is required to assess the degree of the orbital variation and thus the likely stem condition in Polychelidae. Bouvier (1917) was incorrect to homologize the orbital condition of *Willemoesia* with that of eryonids, but the polarization of character 59 suggests that well-developed pereopodal epipods are plesiomorphic as supposed. Other plesiomorphies of *Willemoesia* placing it outside the remaining extant polychelids are the absence of a lateral orbital sinus, a bulbous rather than slender cornea, and an unarmed anterolateral margin of the basal antennular segment (Clade 8).

The sister relationship between *Willemoesia* and *Antarcticheles* recovered by Analysis 2 is noteworthy. Appendages, pereopods, and the tailfan are unknown in *Antarcticheles*, but discernable carapace characters are virtually identical to those of *Willemoesia*, with the full complement of carapace grooves and carinae that are present in extant polychelids. Aguirre-Urreta et al. (1990) interpreted the dorsal orbits of *Antarcticheles* as 'very deep,' but their fig. 2b appears to show broad, shallow dorsal orbits as in *Willemoesia*. The presence in *Antarcticheles* of carapace morphology resembling contemporary taxa suggests that differentiation of the 'polychelid form' was well established by the late Jurassic.

## 4.4 *Morphological trends*

Extant polychelids differ most obviously from extinct polychelidans in the degenerate instead of well-developed eyes and distinctly concave anterior carapace margin. The polarization of character 6 indicates that a general shortening of the frontal carapace margin has occurred in Polychelidae. In other polychelidans, especially *Palaeopentacheles* and coleids, the frontal margin is level with or advanced beyond the anterolateral carapace margins, concealing the bases of the antennae and antennules. This suggests that the projecting carapace front was probably a feature of at least some stem-lineage Polychelidae. In crown-group polychelids, the frontal margin does not extend anteriorly as far as the anterolateral carapace margins, exposing the bases of the antennae and antennules. In coleids and *Palaeopentacheles*, the eyes project laterally into wide dorsal orbital sinuses.

In Polychelidae, the shortening of the front is accompanied by a corresponding shortening and narrowing of the dorsal orbits. The eyes become positioned at the far anterior of the frontal region, lying parallel to the frontal margin. In extant Polychelidae (except *Willemoesia*), the bases of the eyes fill the dorsal orbits, and the cornea (or its remnants) is narrow and elongated, projecting laterally into the lateral orbits. In *Willemoesia*, the dorsal orbits are reduced to a shallow concavity and the remnants of the eyes are fused to the anterior wall of the carapace; the cornea is globular but does not project laterally as far as the lateral carapace margin as in other polychelids.

An additional characteristic feature of polychelids (but unknown in *Antarcticheles*) is the well-developed basal antennular segment with stylocerite. The degenerate eyes of polychelids are plausibly accounted for by their deep-water habitat. The structure of the stylocerite, however, bears little relationship to bathymetry, instead probably reflecting a fossorial habit. The stylocerites, when placed together, form what appears to be a respiratory canal enabling individuals to breathe whilst buried in the substrate (Gore 1984) in a similar fashion to penaeoid prawns.

The major chelipeds exhibit a general trend towards elongation within Polychelida. In tetrachelids, eryonids, and palaeopentachelids, the chelipeds are robust and the carpus is short, being, at most, little longer than high (Fig. 1A–D). In polychelids (except *Cardus*; unknown in *Antarcticheles*), the major chelipeds are long, slender, and considerably less robust than those of tetrachelids, eryonids, and palaeopentachelids, with the carpus slender and distinctly longer than high (Fig. 2). Interestingly, the coleids, which are phylogenetically intermediate between palaeopentachelids and polychelids, exhibit both robust and slender cheliped forms, though the latter condition is apparently more common (Teruzzi 1990; Schweigert & Dietl 1999). Coleiidae has a late Triassic to late Jurassic geologic range (Terruzi & Garassino 2007), and it is not inconceivable that coleids may be paraphyletic with respect to Polychelidae. If so, the shift from shallow to deep-water habitats may have commenced within the coleids, in which case the stem polychelids evolved in deep water. In this context, it is significant that the late Jurassic *Coleia longipes* has been attributed superposition eyes, suggesting adaptation to reduced light conditions (Schweigert & Dietl 1999).

Modern polychelids appear to be ambush predators, striking from a buried position with the chelipeds folded against the lateral margins of the carapace. In underwater footage, polychelids are typically buried in the substrate, as reported by Gore (1984) for species of *Willemoesia*. In contrast to extant polychelids, the unspecialized basal antennular segment and more robust major cheliped of extinct forms suggest that they may have actively foraged or were at least epibenthic. Another derivation in polychelids, including the Jurassic *Antarcticheles*, is the antrorse median spine or tooth on one of more of the abdominal tergites of most species, and the prominently enlarged second abdominal pleuron that overlaps the first and third pleura. Dorsal median spines, when present in other fossil families, are directed posteriorly instead of anteriorly as in modern forms.

Thus, general morphological trends within Polychelida include a shortening of the carapace front with respect to the anterolateral margins, leading to dorsal exposure of the base of the antennules and a shift in eye orientation from anterior to transverse; development of the basal antennular segment stylocerite to form a respiratory canal; and a shift in the form of the major chelipeds from relatively robust with short, triangular carpi to elongated and slender, with slender carpi. A further trend is toward narrowing of the body, marked by a reduction in carapace width, and stronger taper of the abdomen including enlargement of the second pleuron (compare Figs. 1, 2). The carapace in tetrachelids and eryonids distinctly overhangs the pereopods, covering much of the merus of pereopod 1. The posterior width of the carapace in tetrachelids and eryonids significantly exceeds the width of the anterior abdomen, which is itself relatively broad with little taper. In palaeopentachelids, the carapace is proportionally narrower than eryonids and tetrachelids, though distinctly wider than the anterior abdomen. In coleids and polychelids, the carapace is generally narrowed and 'boxlike' with little lateral overhang of pereopod 1. The posterior width of the carapace is similar to the anterior abdominal width so the dorsal outline of the carapace is confluent with that of the abdomen. The abdomen is tapered in coleids, but is even more so in polychelids, enabling more efficient burying. The lateral surfaces of the carapace of extant polychelids are near vertical, allowing individuals to fold the chelipeds against the carapace sides and strike prey from a buried position. These general morphological trends within Polychelida appear to correspond to a shift from a shallow-water, epibenthic habit to the deep-water, fossorial lifestyle, currently evident in Polychelidae.

Within Polychelidae, several topological trends are noteworthy. First, the length of the maxilliped 3 epipod shows a consistent reduction in living taxa. In *Cardus*, *Willemoesia*, and *Pentacheles*, the maxilliped 3 epipod is as long as or longer than the ischium, and in *Homeryon*, it is about one-third the ischium length. In *Polycheles* and *Stereomastis*, the maxilliped 3 epipod is vestigial. Though the maxilliped 3 epipod length is not known for any of the fossil taxa, the polarization of character 57 predicts that they bore well-developed epipods. Similarly, the reduced epipods of pereopods 1–5 in *Stereomastis* is a derived state, so the well-developed condition of other extant genera could be expected in the fossil taxa.

## 5 SYSTEMATICS

The focus of this study is extant Polychelidae, but appraisal of polychelid phylogeny has required assessment of the overall polychelidan system. Notably, *Palaeopentacheles*, formerly placed in Polychelidae, is demonstrated above to lie outside a Polychelidae + Coleiidae clade. Therefore, *Palaeopentacheles* is referred to a new family, Palaeopentachelidae, diagnosed below. Many fossil taxa are poorly known and require revision, but as basis for further research, the families of Polychelida are all diagnosed below. The stratigraphic ranges of the polychelidan families are illustrated in Fig. 6.

### 5.1 *Diagnoses of higher taxa*

Infraorder Polychelida de Haan, 1841

*Diagnosis.* Reptantia. Carapace dorsoventrally flattened; lateral margins cristate, well-defined. Antennal segments free. Pereopods 1–4 chelate. Pereopod 5 chelate in one or both sexes.

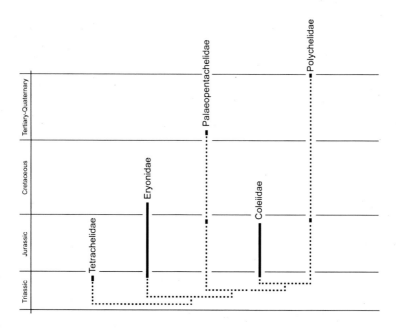

**Figure 6.** Stratigraphic range of Polychelidan families. Broken lines are inferred ranges.

*Remarks.* Polychelida presently includes five families of which Palaeopentachelidae is diagnosed as new. The chief synapomorphy uniting polychelidans is the chelate pereopods 1–4. Other features also unite the extant polychelidans, such as the unique abdominal–thoracic 'fastening' device (Scholtz & Richter 1995) and the dorsally directed aperture of the renal gland. These features remain to be confirmed in fossil forms.

Family Coleiidae Van Straelen, 1924

*Diagnosis.* Carapace with deeply incised, U-shaped dorsal orbits; eyes well-developed, directed laterally; with or without distinct median carina anterior to cervical groove; postorbital carinae (when present) aligned with branchial carinae; cervical and postcervical grooves distinct across carapace, indicated at lateral margins by shallow notches. Abdominal pleuron 2 similar to pleuron 3, not overlapping pleuron 1. Uropodal exopod with curved diaeresis. Telson triangular.

*Composition. Coleia* Broderip, 1835; *Hellerocaris* Van Straelen, 1925; *Proeryon* Beurlen, 1928; *Pseudocoleia* Garassino & Teruzzi, 1993; *Tropifer* Gould, 1857; *Willemoesiocaris* Van Straelen, 1925. *Stratigraphic range.* Late Triassic to late Jurassic (Teruzzi & Garassino 2007).

*Remarks. Willemoesiocaris* Van Straelen, 1925, from the mid-Jurassic of France, regarded as a polychelid by Glaessner (1969), is transferred to Coleiidae. *Willemoesiocaris* is known only from the carapace of its type species, *W. ovalis* (Van Straelen, 1923). According to Van Straelen (1923), *W. ovalis* lacks a median carina anterior to the cervical groove, the postorbital carinae is aligned with the branchial carina, and the carapace front extends anteriorly slightly beyond the anterolateral margins, features of Coleiidae.

Family Eryonidae de Haan, 1841

*Diagnosis.* Carapace with shallow orbits, located on anterior margin, eyes well-developed, directed anteriorly; usually without median carina anterior to cervical groove; cervical groove absent or indicated medially and at carapace margins, not extending across carapace; postcervical groove indicated only at carapace margins. Branchiocardiac grooves absent. Abdominal pleuron 2 similar to pleuron 3, not overlapping pleuron 1. Uropodal exopod entire, without diaeresis. Telson triangular or subrectangular. Pereopod 1 dactylus with triangular subdistal lobe, longer than pollex.

*Composition. Eryon* Desmarest, 1822; *Cycleryon* Glaessner, 1965; *Knebelia* Van Straelen, 1922; *Rosenfeldia* Garassino, Teruzzi, & Dalla Vecchia, 1996.

*Stratigraphic range.* Late Triassic to Lower Cretaceous (Glaessner 1969; Garassino et al. 1996).

Family Palaeopentachelidae, new family

*Diagnosis.* Carapace with dorsal orbits deeply incised, narrow, U-shaped; eyes well-developed, directed laterally; without median carina anterior to cervical groove; cervical groove indicated medially only, not extending to lateral carapace margins; branchiocardiac grooves absent. Posterior margin of carapace distinctly wider than anterior margin of abdomen. Abdominal pleuron 2 similar to pleuron 3, not overlapping pleuron 1. Uropodal exopod entire, without diaeresis. Telson triangular. Pereopod 1 dactylus tapering distally, as long as pollex; occlusal margins of dactylus and pollex lined with spines.

*Composition. Palaeopentacheles* von Knebel, 1907 (type genus).

*Stratigraphic range.* Upper Jurassic, possibly to the Oligocene (Schweitzer & Feldmann 2001).

*Remarks. Palaeopentacheles* was previously placed in Polychelidae, but results of the present study exclude it from Polychelidae *sensu stricto* by the incursion of Coleiidae. As sister to Coleiidae + Polychelidae, *Palaeopentacheles* cannot be accommodated within either Coleiidae or Polychelidae without subsuming all three taxa into an enlarged Polychelidae, significantly diluting the concept of the family. Thus, the new family Palaeopentachelidae is herein proposed for *Palaeopentacheles*. Moreover, Palaeopentachelidae differs from Coleiidae and Polychelidae by lacking postorbital carinae, in lacking an anterior median carina on the carapace, in the possession of sharp pleural terminations, in having entire lateral carapace margins without any trace of cervical or postcervical incisions, and in the multispinose occlusal margins of the pereopod 1 dactylus and pollex. *Palaeopentacheles* is presently known only from its type species, *P. roettenbacheri* (Upper Jurassic of Germany), and from *P. starri* Schweitzer & Feldmann, 2001 (Oligocene of North America). The holotype of *P. starri* is in poor condition, though, so its assignment to *Palaeopentacheles* was tentative, based on the spinose fingers of the major cheliped (Schweitzer & Feldmann 2001).

Family Polychelidae Wood-Mason, 1874

*Diagnosis*. Carapace with dorsal orbits shallow or deeply incised, U- or V-shaped; eyes reduced, fused to anterior margin of carapace, directed laterally; with distinct median carina anterior to cervical groove; postorbital carinae not aligned with branchial carinae but terminating distinctly mesial to branchial carinae; cervical and branchiocardiac grooves distinct across carapace, indicated at lateral margins by notches. Abdominal pleuron 2 distinctly larger than, and overlapping, pleura 1 and 3. Uropodal exopod entire, without diaeresis. Telson triangular. Pereopod 1 dactylus tapering distally, as long as pollex.

*Stratigraphic range*. Upper Jurassic to Recent (Aguirre-Urreta et al. 1990).

*Composition*. *Antarcticheles* Aguirre-Urreta, et al. 1990 (upper Jurassic); *Cardus* Galil, 2000; *Homeryon* Galil, 2000; *Pentacheles* Bate, 1878; *Polycheles* Heller, 1862; *Stereomastis* Bate, 1888; *Willemoesia* Grote, 1873.

*Remarks*. The Jurassic *Antarcticheles* is retained in Polychelidae on the basis of carapace characters: a median carina anterior to the cervical groove is present, and the cervical and postcervical grooves are distinct dorsally and marked laterally by notches in the carapace margins. Unfortunately, the pereopods and tailfan are not known in *Antarcticheles*. *Willemoesiocaris*, placed in Polychelidae by Glaessner (1969), is transferred above to Coleiidae.

Family Tetrachelidae Beurlen, 1930

*Diagnosis*. Carapace with shallow orbits located on anterior margin; eyes well-developed, apparently directed anteriorly; without median carina anterior to cervical groove; cervical and branchiocardiac grooves distinct across carapace, not meeting, indicated at lateral margins by notches. Abdominal pleuron 2 similar to pleuron 3, not overlapping pleuron 1. Uropodal exopod with straight diaeresis. Telson rounded distally. Pereopod 1 dactylus tapering distally, longer than pollex. (Based on Glaessner 1969.)

*Composition*. *Tetrachela* Reuss, 1858.

*Stratigraphic range*. Upper Triassic (Glaessner 1969).

## 5.2  *Diagnoses of Recent genera and keys to species of Polychelidae*

Key to Recent genera of Polychelidae

1. Carapace ovate, slightly long than wide .................................................... *Cardus*
   - Carapace distinctly longer than wide ............................................................ 2
2. Dorsal orbital sinuses forming a shallow concavity. Pollex of major chela with perpendicular spine on inner margin. Anterolateral margin of basal antennular segment unarmed .................................................................................... *Willemoesia*

- Dorsal orbital sinuses deep, slit-like, U- or V-shaped. Pollex of major chela without perpendicular spine on inner margin. Anterolateral margin of basal antennular segment with 1 or more spines .................................................................................................... 3
3. Dorsal orbital notch U-shaped. Epipod of pereopods 1–5 reduced, shorter than coxal width ........................................................................... *Stereomastis*
   - Dorsal orbital notch V-shaped or slit-like. Epipod of pereopods 1–5 well-developed, markedly longer than coxal width ................................................................ 4
4. Basal antennular segment with rounded anterolateral margin (though bearing 1 or 2 small spines). Maxilliped 3 epipod vestigial ........................................ *Polycheles*
   - Basal antennular segment with quadrate anterolateral margin. Maxilliped 3 epipod well-developed ....................................................................... 5
5. Dactylus and pollex of pereopods 2–4 strongly curved. Basal antennular segment with 2 anterolateral spines. Maxilliped 3 epipod one-third to half length of ischium ....... *Homeryon*
   - Dactylus and pollex of pereopods 3–4 relatively straight, weakly curved. Basal antennular segment with 1 outer spine. Maxilliped 3 epipod as long as or longer than ischium. ..................................................................... *Pentacheles*

## Genus *Cardus* Galil, 2000

*Diagnosis.* Carapace ovate, slightly longer than wide. Dorsal orbital sinus slit-like. Pollex of major chela without perpendicular spine on inner margin. Anterolateral margin of basal antennular segment with rounded outer margin and 1 or 2 anterolateral spines. Dactylus and pollex of pereopods 2–4 relatively straight. Maxilliped 3 epipod as long as ischium. Pereopods 1–5 epipod well-developed.

*Type species.* *Deidamia crucifer* Thomson, 1873, by original designation and monotypy.

*Composition.* *Cardus crucifer* (Thomson, 1873).

## Genus *Homeryon* Galil, 2000

*Diagnosis.* Carapace distinctly longer than wide. Dorsal orbital sinus V-shaped or slit-like. Pollex of major chela without perpendicular spine on inner margin. Anterolateral margin of basal antennular segment with quadrate outer margin and 1 anterolateral spine. Dactylus and pollex of pereopods 2–4 strongly curved. Maxilliped 3 epipod one-third to half length of ischium. Pereopods 1–5 epipod well-developed.

*Type species.* *Homeryon armarium* Galil, 2000, by original designation.

*Composition.* *H. armarium* Galil, 2000, *H. asper* (Rathbun, 1906).

### Key to species of *Homeryon*

1. Lateral margins of carapace posterior to postcervical incision cristate, serrulate. Median abdominal carinae blunt. Abdominal pleuron 2 cordiform. Uropods smooth ......... *H. asper*
   - Lateral margins of carapace posterior to postcervical incision rounded, bearing rows of antrorse spinules. Median abdominal carinae with distinct notch. Abdominal pleuron 2 reniform. Uropods granulate .................................................... *H. armarium*

## Genus *Pentacheles* Bate, 1878

*Diagnosis.* Carapace distinctly longer than wide. Dorsal orbital sinuses deep, V-shaped. Pollex of major chela without perpendicular spine on inner margin. Anterolateral margin of basal antennular segment with quadrate outer margin and 1 anterolateral spine. Dactylus and pollex of pereopods 3–4 relatively straight, weakly curved. Maxilliped 3 epipod as long as or longer than ischium. Pereopods 1–5 epipod well-developed.

*Type species. Pentacheles laevis* Bate, 1878, designated by Fowler (1912).

*Composition. Pe. Gibbus* Alcock, 1894; *Pe. Laevis* Bate, 1878; *Pe. Obscurus* Bate; 1878, *Pe. Snyderi* (Rathbun, 1906); *Pe. Validus* A. Milne-Edwards, 1880.

Key to species of *Pentacheles*

1. Inner angle of dorsal orbital sinus unarmed ............................................. 2
   - Inner angle of dorsal orbital sinus spinose .......................................... 3
2. Carapace depressed, flattened. Abdominal tergites and pleura nearly smooth ... *Pe. obscurus*
   - Carapace strongly convex in lateral profile. Abdominal tergites and pleura set with conical tubercles ................................................................. *Pe. gibbus*
3. Abdominal tergites 1–3 with distinct antrorse tooth ........................... *Pe. laevis*
   - Abdominal tergites 1–3 without antrorse tooth, at most with blunt rounded prominence .. 4
4. Outer angle of dorsal orbit unarmed or with at most 2 spines ................... *Pe. validus*
   - Anterior margin of carapace between outer orbital angle and anterolateral spine lined with 3 or 4 spines ................................................................. *Pe. snyderi*

Genus *Polycheles* Heller, 1862

*Diagnosis.* Carapace distinctly longer than wide. Dorsal orbital sinuses V-shaped. Pollex of major chela without perpendicular spine on inner margin. Anterolateral margin of basal antennular segment rounded, with anterolateral spines. Maxilliped 3 epipod vestigial. Pereopod 15 epipod well-developed. Dactylus and pollex of pereopods 34 relatively straight, weakly curved.

*Type species. Polycheles typhlops* Heller, 1862, by monotypy.

*Composition. P. amemiyai* Yokoya, 1933; *P. baccatus* Bate, 1878; *P. coccifer* Galil, 2000; *P. enthrix* Bate, 1878; *P. kermadecensis* Sund, 1920; *P. martini* Ahyong & Brown, 2002; *P. perarmatus* Holthuis, 1952; *P. tanneri* Faxon, 1893; *P. typhlops* Heller, 1862.

Key to species of *Polycheles*

1. One (rarely two) rostral spine. Inner basal margin of dorsal orbit spinose ................. 2
   - Two rostral spines. Inner basal margin of dorsal orbit unarmed ....................... 3
2. Abdominal pleuron 2 trianguloid anteriorly with rounded apex. Uropodal exopod ventrally bicarinate ................................................................. *P. typhlops*
   - Abdominal pleuron 2 semicircular anteriorly, evenly rounded. Uropodal exopod ventrally tricarinate ................................................................. *P. perarmatus*
3. Frontal submarginal tooth prominent, longer than separate rostral spines ............... 4
   - Frontal submarginal tooth shorter than rostrum, or rostrum bifid ..................... 5
4. Gastro-orbital region bispinose; median postrostral and postcervical carinae irregularly granulate. Abdominal pleuron 2 with broadly convex anteroventral margin. Dorsal margin of first chela prominently spinulose ................................................. *P. baccatus*
   - Gastro-orbital region quadrispinose; median postrostral and postcervical carinae set with antrorse tubercles. Abdominal pleuron 2 with concave anteroventral margin. Dorsal margin of first chela granulose ................................................. *P. coccifer*

5. Frontal margin of carapace with several spinules on either side of rostral spines .......... 6
   - Frontal margin of carapace on either side of rostral spines unarmed except for spine on inner angle of dorsal orbital sinus ............................................... *P. tanneri*
6. Median carina on abdominal tergites 2–5 notched or crenulate. Abdominal tergites 2–5 with distinct, oblique grooves. Dorsal surface of carapace strongly granulate .......... *P. martini*
   - Median carina on abdominal tergites 2–5 entire, without median notch. Abdominal tergites 2–5 relatively smooth, without distinct oblique grooves. Dorsal surface of carapace smooth or sparsely spinose but not strongly granulate ......................................... 7
7. Branchial carina indicated at most by low granules; branchial groove not flanked by row of spines; gastric region of carapace with 1 or 2 spines of similar size to spines of median carina; postcervical groove without antrorse spine on posterior margin between median carina and branchial carina ....................................................... 8
   - Branchial carina indicated by row of 4–6 spines; branchial groove flanked by row of 4 or 5 small spines; gastric region of carapace covered by numerous spines of similar size to spines of median carina; postcervical groove with antrorse spine on posterior margin between median carina and branchial carina .................................. *P. kermadecensis*
8. Frontal margin with 1 spine between rostral spines and spine of inner angle of dorsal orbit ............................................................... *P. amemiyai*
   - Frontal margin with 2 or more spines between rostral spines and spine of inner angle of dorsal orbit ............................................................ *P. enthrix*

Genus *Stereomastis* Bate, 1888

*Diagnosis.* Carapace distinctly longer than wide. Dorsal orbital sinuses U-shaped. Pollex of major chela without perpendicular spine on inner margin. Anterolateral margin of basal antennular segment rounded, with 1 or 2 anterolateral spines. Maxilliped 3 epipod vestigial. Pereopods 1–5 epipod vestigial. Dactylus and pollex of pereopods 3–4 relatively straight, weakly curved.

*Type species. Pentacheles suhmi* Bate, 1878, designated by Holthuis (1962).

*Composition. S. alis* (Ahyong & Galil, 2006) comb. nov.; *S. aculeata* (Galil, 2000) comb. nov.; *S. auriculata* (Bate, 1878) comb. nov.; *S. cerata* (Alcock, 1894) comb. nov.; *S. evexa* (Galil, 2000) comb. nov.; *S. galil* (Ahyong & Brown, 2002) comb. nov.; *S. helleri* (Bate, 1878) comb. nov.; *S. nana* (Smith, 1884) comb. nov.; *S. pacifica* (Faxon, 1893); *S. phosphorus* (Alcock, 1894) comb. nov.; *S. polita* (Galil, 2000) comb. nov.; *S. sculpta* (Smith, 1880) comb. nov.; *S. suhmi* (Bate, 1878), *S. surda* (Galil, 2000) comb. nov.; *S. talismani* (Bouvier, 1917) comb. nov.; *S. trispinosa* (de Man, 1905) comb. nov.

Key to species of *Stereomastis*

1. Outer proximal margin of basal antennular segment with 1 spine ...................... 2
   - Outer proximal margin of basal antennular segment with 2 spines ..................... 4
2. Median carina of abdominal tergites 1–5 with antrorse spine ...................... *S. galil*
   - Median carina of abdominal tergites 1–4 with antrorse spine ........................ 3
3. Dorsum of carapace between branchial and median postcervical carinae unarmed; branchial carina obsolescent; branchial groove unarmed ................................. *S. polita*
   - Dorsum of carapace between branchial and median postcervical carinae with antrorse spine; branchial carina indicated by row of spines; branchial groove with row of spines ............................................................... *S. phosphorus*
4. Inner angle of dorsal orbital sinus unarmed ........................................ 5
   - Inner angle of dorsal orbital sinus spinose ....................................... 7

5.  Branchial carina unarmed ................................................................. 6
    - Branchial carina spinose ................................................................. *S. helleri*
6.  Antrorse spine on abdominal tergite 5 large, overhanging anterior margin of tergite 4. Postorbital carina ill-defined, without spines ......................................... *S. cerata*
    - Antrorse spine on abdominal tergite 5 not overhanging anterior margin of tergite 4. Postorbital carina defined by arcuate row of spines .................................... *S. alis*
7.  Median carina on abdominal tergite 5 (usually also tergites 2–4) with short, upright posterior tooth in addition to strong antrorse spine. Ischium and merus of pereopod 2 articulated ................................................................. *S. suhmi*
    - Median carina on abdominal tergites 2–5 without short, upright posterior tooth. Ischium and merus of pereopod 2 fused ................................................. 8
8.  Branchial groove with 1 or more anterior spines ................................................. 9
    - Branchial groove unarmed ................................................................. 12
9.  Median carina on abdominal tergite 5 without antrorse spine ................................ 10
    - Median carina on abdominal tergite 5 with antrorse spine ................................. 11
10. Median carina on abdominal tergite 4 with strong antrorse spine. Region of carapace between branchial and median postcervical carinae unarmed posteriorly ............. *S. pacifica*
    - Median carina on abdominal tergite 3 bearing long antrorse spine; median carina on abdominal tergite 4 unarmed anteriorly. Region of carapace between branchial and median postcervical carinae posteriorly spinose. ............................... *S. trispinosa*
11. Antrorse spine on abdominal tergite 3 largest; lyre-shaped carina on abdominal tergite 6 prominently denticulate; basal tubercle on telson pointed ................... *S. nana*
    - Antrorse spine on abdominal tergite 5 largest; lyre-shaped carina on abdominal tergite 6 smooth; basal tubercle on telson blunt ................................. *S. evexa*
12. Median carina of abdominal tergite 5 with antrorse spine ..................................... 13
    - Median carina of abdominal tergite 5 without antrorse spine ............................... 14
13. Abdominal tergite 6 bearing denticulate, lyre-shaped, mesial carinae. Lateral margins of carapace posterior to postcervical incision, usually with 7–10 spines ............. *S. talismani*
    - Abdominal tergite 6 bearing parallel smooth carinae, confluent anteriorly and posteriorly. Lateral margins of carapace posterior to postcervical incision, usually with 6–8 spines ................................................................. *S. sculpta*
14. Posterior margin of cervical groove with single antrorse spine midway between median postcervical and branchial carinae. Frontal submarginal tooth prominent, visible in dorsal view ................................................................. *S. aculeata*
    - Posterior margin of cervical groove with 2–4 (usually 3 or 4) antrorse spines midway between median postcervical and branchial carinae. Frontal submarginal tooth small ....... 15
15. Lateral margins of carapace posterior to postcervical incision with 7 or 8 spines. Oblique grooves on abdominal tergites deeply marked; lyre-shaped carina on sixth tergite prominent ................................................................. *S. auriculata*
    - Lateral margins of carapace posterior to postcervical incision with 10–14 spines. Oblique grooves on abdominal tergites obsolescent; lyre-shaped carina on sixth tergite obsolescent ................................................................. *S. surda*

Genus *Willemoesia* Grote, 1873

*Diagnosis.* Carapace distinctly longer than wide. Dorsal orbital sinuses obsolete, indicated by shallow concavities. Pollex of major chela with perpendicular spine on inner margin. Anterolateral margin of basal antennular segment quadrate, without anterolateral spine. Maxilliped 3 epipod as long as ischium. Pereopods 1–5 epipod well-developed. Dactylus and pollex of pereopod 3 crossing; relatively straight, weakly curved on pereopod 4.

*Type species. Deidamia leptodactyla* Willemoes-Suhm, 1873, by monotypy.

*Composition. W. forceps* A. Milne-Edwards, 1880; *W. inornata* Faxon, 1893; *W. leptodactyla* (Willemoes-Suhm, 1873); *W. pacifica* Sund, 1920.

Key to species of *Willemoesia*

1. Abdominal tergite 6 sculptured .................................................... 2
   - Abdominal tergite 6 nearly smooth ............................................... 3
2. Lateral margins of carapace posterior to postcervical incision with 10 or fewer spines. Dorsal margin of chela of pereopod 1 with 2 rows of spines ......................... *W. inornata*
   - Lateral margins of carapace posterior to postcervical incision with 15 or more spines. Dorsal margin of chela of pereopod 1 with several rows of spines ................. *W. leptodactyla*
3. Lateral margins of carapace anterior to cervical incision with 15–19 spines. Abdominal tergites 2–5 with deep, oblique grooves. Telson with rounded apex ................ *W. forceps*
   - Lateral margins of carapace anterior to cervical incision with 6–10 spines. Abdominal tergites smooth, without deep, oblique grooves. Telson with sharp apex ........... *W. pacifica*

ACKNOWLEDGEMENTS

This study was supported by the New Zealand Foundation for Research, Science and Technology (BBBI093 and BBDC083), the NIWA capability fund, and Biosecurity New Zealand (ZBS200524). Most NIWA specimens were collected under the research programme "Seamounts: their importance to fisheries and marine ecosystems," funded by the New Zealand Foundation for Research, Science and Technology (contracts CO1X0028, CO1X0224). Tin-Yam Chan (NTOU), Regis Cleva and Alain Crosnier (MNHN), Oliver Coleman (ZMB), Ken-Ichi Hayashi (formerly NFU), Stephen Keable (AM), Thierry Laperousaz (SAM), Peter Ng and Swee Hee Tan (NUS), Kareen Schnabel (NIWA), Melissa Titelius (WAM), and Mary Wicksten (TAMU) are thanked for the opportunity to study polychelid collections under their care. Jody Martin and Bella Galil provided constructive comments on the manuscript. Michelle Yerman kindly prepared Fig. 3C.

NOTE

After this chapter went to press, the description of a new species, *Stereomastis panglao* (Ahyong & Chan 2008), was published. *Stereomastis panglao* is closest to *S. polita* and *S. galil*; it differs from *S. galil* by having unarmed branchial grooves, and from *S. polita* in having an antrorse spine on abdominal tergite 5.

APPENDIX 1

Terminal taxa and sources of character scoring. Extinct (*).

**TETRACHELIDAE***
*Tetrachela* **Reuss, 1858**
*T. raiblana* (Bronn, 1858): Glaessner (1969).

**ERYONIDAE***
*Eryon* **Desmarest, 1822**
*E. arctiformis* (Schlotheim, 1820): AM; Garassino & Schweigert (2006).

*Cycleryon* **Glaessner, 1965**
*C. propinquus* (Schlotheim, 1822): Garassino & Schweigert (2006).

*Knebelia* **Van Straelen, 1922**
*K. bilobatus* (Münster, 1839): Garassino & Schweigert (2006).

**PALAEOPENTACHELIDAE new family***
*Palaeopentacheles* **von Knebel, 1907**
*Pa. roettenbacheri* (von Münster, 1839): Garassino & Schweigert (2006).

**COLEIIDAE***
*Coleia* **Broderip, 1835**
*C. longipes* (O. Fraas, 1855): Schweigert & Dietl (1999).

**POLYCHELIDAE**
*Antarcticheles* **Aguirre-Urreta, Buatois, Chernoglasov & Medina, 1990***
*A. antarcticus* Aguirre-Urreta et al., 1990: Aguirre-Urreta et al. (1990).

*Cardus* **Galil, 2000**
*C. crucifer* (Thomson, 1873): TAMU, males and females.

*Homeryon* **Galil, 2000**
*H. armarium* Galil, 2000: NFU, females; Galil (2000). *H. asper* (Rathbun, 1906): Rathbun (1906); Galil (2000).

*Pentacheles* **Bate, 1878**
*Pe. gibbus* Alcock, 1894: Alcock (1894); Galil (2000). *Pe. laevis* Bate, 1878: AM, NIWA, males and females. *Pe. validus* A. Milne-Edwards, 1880: AM, NIWA, males and females. *Pe. snyderi* Rathbun, 1906: MNHN, female; Galil (2000).

*Polycheles* **Heller, 1862**
*P. amemiyai* Yokoya, 1933: NTOU, NUS, males and females. *P. baccatus* Bate, 1878: AM, males and females. *P. coccifer* Galil, 2000: NTOU, NUS, males and females. *P. enthrix* (Bate, 1878): AM, NIWA, males and females. *P. kermadecensis* (Sund, 1920): AM, males and females. *P. martini* Ahyong & Brown, 2002: AM, males and females. *P. perarmatus* Holthuis, 1952: USNM, MNHN, males and females. *P. tanneri* Faxon, 1893: ZMB, male; Galil (2000). *P. typhlops* Heller, 1862: AM, SAM, NTOU, males and females.

*Stereomastis* **Bate, 1888**
*S. aculeata* (Galil, 2000) comb. nov.: AM, MNHN, males and females. *S. alis* (Ahyong & Galil, 2006) comb. nov.: MNHN, female holotype. *S. auriculata* (Bate, 1878) comb. nov.: AM, MNHN, males and females. *S. cerata* (Alcock, 1894) comb. nov.: Alcock (1894); Galil (2000); de Man (1916). *S. evexa* (Galil, 2000) comb. nov.: Galil (2000); Faxon (1895) (as *P. nana*). *S. galil* (Ahyong & Brown, 2002) comb. nov.: WAM, AM, NTOU, males and females. *S. helleri* (Bate, 1878) comb. nov.: AM, NTOU, males and females. *S. nana* (Smith, 1884) comb. nov.: AM, NIWA, males and females. *S. pacifica* (Faxon, 1893) comb. nov.: AM, male. *S. phosphorus* (Alcock, 1894) comb. nov.: AM, SAM, males and females. *S. polita* (Galil, 2000) comb. nov.: MNHN, males and females. *S. suhmi* (Bate, 1878) comb. nov.: AM, NIWA, males and females. *S. surda* (Galil, 2000) comb. nov.: AM, NIWA, males and females. *S. sculpta* (Smith, 1880) comb. nov.: AM, NTOU, males and females.

*S. talismani* (Bouvier, 1917) comb. nov.: Galil (2000). *S. trispinosa* (de Man, 1905) comb. nov.: de Man (1905); Galil (2000).

### *Willemoesia* Grote, 1873

*W. forceps* A. Milne Edwards, 1880: MNHN, NTOU, males and females. *W. inornata* Faxon, 1893: Faxon (1893), Galil (2000). *W. pacifica* Sund, 1920: AM, NIWA, males and females. *W. leptodactyla* (Willemoes-Suhm, 1873): MNHN, NTOU, males and females.

## APPENDIX 2

Data matrix. Missing data indicated by question marks (?); inapplicable data by hyphens (-); and extinct data are marked with asterisks (*).

```
Tetrachela*        ???20000000000-0-00-?1000-1200??10???????00??002?3????????000???00??0
Eryon*             0??2000020000000-00-10100-1200??1000???000000002030???????0100???00?00
Cycleryon*         0002000020000000-00-10100-1200??1000???000000002030???0???010000000000
Knebelia*          ???2000020000000--0---0000-02-0??1000???000000002??????????0100???00?0
Coleia*            ???2000020000000--0--10000-1200001010???001000002030???22???0010???00???
Palaeopentacheles* ???2001021200000-00112200-1200001000???001000000203??000???00000??00??1
Antarcticheles*    ?????10?000?00-?011110000110000??10???00?????????????????0???????????
C.crucifer         0000101111000011011011000110110022110000111021012120102110000000000110
H.armarium         0010111111001010011110000110000001011000001110110101200111001000010001
H.asper            0010111111001010011110000110000001011000001110110101001200011?100??000100?1
Pe.gibbus          0010101110001100111100001000000101100000111?11011????02?10010???00??1
Pe.laevis          0010101111010010011110000100000010110000011101101002000211001100000111
Pe.obscurus        0010101111000010011110000100010010110000011101101102000?21001000000111
Pe.snyderi         0010111111010010011110000100000010110000011101101020002?1001100000111
Pe.validus         0010101111000010111110201100000010110000011101101002000211001100000111
P.baccatus         111211112010111211110000112000011111001011102110111100010010000000110
P.coccifer         111211112010111212110010112000011111001011102110111100010010000000110
P.amemiyai         101010111100011111110000010101002110020110211100111000100010000000001
P.enthrix          101010111100011111110000110110021110020110211001110001000000001
P.kermadecensis    101010111100001111111101111111110021110020110211100111000100010000000001
P.martini          101111111001011211110000120010010001110021110011110010000100000000110
P.perarmatus       11001111121110111111102111110012110011011021110011100?1001100000110
P.tanneri          1112111112011011111102101111010211101011021110011100010010000000110
P.typhlops         110211112111011111102101110012110011011021110011100010011000000110
S.aculeata         103210121100011011231010111012111011101111211101211001000010011111100111
S.alis             10311012100001101123101001010100110011011121110020000?0001100000?11?
S.auriculata       10311012110001101123102010111101111011011121110121100100011111100111
S.cerata           10311012100001100120101001101010011001101111?21110????00?00011??00???
S.evexa            103010121100011011231011111111111211001101111211100211?00?00001111??00111
S.galil            10311012110001101123101111111011221100111111121110001100100011111000111
S.helleri          10301012100001101123101111111011221001011111211100211000011100000111
S.nana             10301012110001101123101111111111211001011112110021100010011100000111
S.pacifica         10301012110001101133101111111011111100110111121110000001001110000111
S.phosphorus       10311012110001101123101111111011111100111111121110001100100011111000111
S.polita           10311012110001101123101011010111110110101112111000011001000011110000111
S.sculpta          10301012110001101133101101111011211001011111211100211000011100000111
S.suhmi            01301012110000101123102111111011211011011112110021001001001000111
S.surda            10311012110001101123102101111011211001011111211101211001000011111100111
S.talismani        10301012110001101133101011011012110010111112111002110021100?0011100000111
S.trispinosa       10321012110001101123101101111011111101110111?21110012100?0011?1100??1
W.forceps          00001000010000100111100000110000010110000001100110000200021101100000111102
W.inornata         00001000010000100111100000110010012110000111001110000200021101100001102
W.leptodactyla     000010000100001001111000011000012110000111001100002100211101100001102
W.pacifica         00001000010000100111100001100000201100000110011000020002111011000001102
```

APPENDIX 3

Morphological characters used in analysis.

1. Carapace, buccal carina: absent (0); present (1). The buccal carina, unique to *Homeryon*, is a prominent, angular projection along the lateral margins of the buccal cavity (Galil 2000).
2. Carapace, sublateral carina: indistinct, indicated by setae or granules (0); distinct, spinose (1). The sublateral carina is present on the lateral surface of the carapace starting behind the lateral orbit and is indicated by rows of setae or granules or by a row of spines. In most taxa, the sublateral carina reaches to almost the posterior margin, though in species of *Polycheles* related to *P. typhlops* and *P. baccatus*, the carina reaches posteriorly only to about the carapace midlength.
3. Carapace, sublateral carina, length: long, almost reaching posterior margin of carapace (0); short, reaching to about midlength of carapace (1).
4. Carapace, rostral spines: one (0); two (1); absent (2); two, basally fused (3).
5. Carapace, frontal submarginal tooth: absent or obsolete (0); small, rounded (1); prominent, conical (2).
6. Carapace, frontal margin, position: reaching or projecting anteriorly beyond anterolateral carapace margin (0); distinctly behind anterolateral carapace margin (1). A synapomorphy of the Polychelidae is the position of the frontal margin of the carapace, being distinctly behind the level of the anterolateral margins.
7. Carapace, anterior margin between outer orbital margin and anterolateral spine: unarmed (0); spinose (1).
8. Carapace, lateral orbital sinus: absent (0); present (1). In extant polychelids, the eyes are aligned transversely along the frontal margin of the carapace. The bases of the eyes are exposed dorsally via the dorsal orbital sinus, but the remnants of the cornea are directed laterally through the lateral margin of the carapace, forming the lateral orbital sinus. A lateral orbit is present only in those species whose eyes project laterally beyond the dorsal orbit.
9. Carapace, dorsal orbit, shape: broadly concave (0); V-shaped (1); U-shaped (2); slit (3). The dorsal orbits range in shape from broadly concave in *Willemoesia* and *Tetrachela* to U-shaped in eryonids, *Stereomastis*, *Coleia*, and *Palaeopentacheles*; V-shaped in *Pentacheles*; and a narrow slit in *Homeryon* and *Cardus*.
10. Carapace, dorsal orbit, length: distinctly shorter than wide (0); as long as or longer than wide (1). In tetrachelids, eryonids, *Willemoesia*, and *Antarcticheles*, the dorsal orbital length is very short, distinctly shorter than wide. In other taxa, the dorsal orbit is as long as or longer than wide.
11. Carapace, inner angle of dorsal orbit: rounded (0); spinous (1); triangular (2).
12. Carapace, inner margin of dorsal orbit: smooth (0); spinous (1).
13. Carapace, outer orbital spine: absent (0); present (1).
14. Carapace, outer orbital margins: smooth (0); spinose (1).
15. Carapace, lateral spine spacing: evenly spaced (0); spacing becoming wider posteriorly (1). In most polychelidans with lateral spines on the carapace, the spines are evenly spaced. In *Stereomastis*, however, the lateral spines become more widely spaced posteriorly.
16. Carapace, postorbital carina: indistinct or absent (0); distinct (1). The position of the postorbital carina is usually indicated by slight surface swelling and a row of spines or granules.
17. Carapace, postorbital carina, orientation: arcuate, divergent anteriorly (0); subparallel or slightly convergent anteriorly (1). The postorbital carina is present in the fossil *Antarcticheles*, but its orientation cannot be satisfactorily interpreted from Aguirre-Urreta et al.'s (1990) account, so it is scored as unknown.
18. Carapace, postorbital carina, ornamentation: unarmed (0); spined (1); tuberculate (2).

19. Carapace, anterior median carina: absent or indistinct (0); present, well-developed (1).
20. Carapace, anterior median carina, ornamentation: unarmed (0); irregularly spinous or tuberculate (1); with spine formula 1:1:2:1, 1:2:1 (2). In *Stereomastis*, the median carina of the carapace is armed with spines in the arrangement 1:1:2:1 anterior to the cervical groove, and 1:2:1 posterior to the cervical groove.
21. Carapace, posterior median carina, ornamentation: unarmed (0); irregularly spinous (1); unarmed at midlength (2); paired spines at midlength (3).
22. Carapace, cervical groove: absent or only faintly indicated (0); distinct across dorsum (1); indicated medially only (2). The cervical groove is distinct across the dorsum in polychelids, coleids, and tetrachelids. The cervical groove is indicated only medially in *Palaeopentacheles* and is faintly indicated or absent in eryonids.
23. Carapace, cervical and postcervical groove, lateral notches: shallow (0); deeply incised (1); absent (2). The cervical and postcervical grooves are indicated by shallow notches in the carapace margins in most polychelidans. *Palaeopentacheles* lacks any trace of cervical and postcervical notches in the carapace margins. In *Eryon*, *Cycleryon*, and *Cardus*, the cervical and postcervical notches are deeply incised.
24. Carapace, cervical groove, midpoint spines: absent (0); one spine (1); two or more spines (2).
25. Carapace, spine on cervical groove near junction with postcervical groove: absent (0); present (1).
26. Carapace, branchial groove, ornamentation: unarmed (0); spined (1); tuberculate (2); absent (3).
27. Carapace, branchial groove, orientation: absent or indistinct (0); divergent (1); parallel (2).
28. Carapace, branchial carina: indistinct (0); distinct (1); absent (2).
29. Carapace, branchial carina, ornamentation: unarmed (0); spined (1); tuberculate (2).
30. Carapace, posterior margin with median spines: absent (0); present (1).
31. Carapace, posterior margin, ornamentation on either side of midline: unarmed (0); with row of spines (1).
32. Abdominal tergite 1, sublateral spine: absent (0); present (1). The sublateral spine is present on the anterior margin of abdominal tergite 1, slightly dorsal to the pleuron. It is present in all species of *Stereomastis* and in *Polycheles martini*, *P. kermedecensis*, *P. enthrix* and *P. amemiyai*.
33. Abdominal tergite 1, anterior pleural spine: absent (0); present (1). The spine is present in *Polycheles tanneri* and most species of *Stereomastis*.
34. Abdominal tergites 2–5, submedian groove: absent (0); distinct (1); indistinct (2).
35. Abdominal tergites 4–5, antrorse spine: absent on AS4–5 (0); absent on AS5 (1); present on AS5 (2). A feature of most polychelids is the presence of an anterodorsally directed spine (termed 'antrorse') on one or more of the abdominal tergites.
36. Abdominal pleural terminations: sharp, angular (0); rounded (1). The pleural terminations in coleids and polychelids are rounded; they are sharp and angular in other taxa. The pleura of *Stereomastis suhmi* are ventrally rounded, but with a small spine present; it is scored as state 1.
37. Abdominal tergite 2, pleuron size: similar to that of pleuron 3 (0); distinctly larger than pleuron 3 (1). The second abdominal pleuron is distinctly enlarged in all extant polychelids, unknown in *Antarcticheles*, and similar to pleuron 3, in other taxa.
38. Abdominal tergite 2, pleuron shape: ovate (0); triangular (1).
39. Abdominal tergite 2, pleuron, anterior spine: absent (0); present (1).
40. Abdominal tergite 2, pleuron, surface carina: absent (0); crescent shaped (1).
41. Abdominal tergite 6, surface, double carina: absent (0); present (1); partial (2).
42. Abdominal tergite 6, surface: uniform or slightly irregular (0); sculptured (1). This character distinguishes species of *Willemoesia* in which two species have a distinctly sculptured surface of abdominal tergite 6. In other polychelidans, the surface of tergite 6 is uniform or slightly irregular.

43. Eye orientation: directed anteriorly (0); transverse, directed laterally (1). The eyes are directed laterally in polychelids, coleids, and palaeopentachelids, and anteriorly in eryonids. The eyes of tetrachelids are not known, but the anterior position of the orbits, as in eryonids, suggests an anterior orientation.
44. Eye articulaton: free (0); fused to anterior margin of carapace (1). The eyes of extant polychelids are fused to the anterior margin of the carapace; the condition is unknown in *Antarticheles*. The eyes of other polychelidans are articulated.
45. Cornea shape: globular (0); slender (1). The cornea is globular in extinct taxa and *Willemoesia* and is tapering in other extant polychelids.
46. Apex of eye: simple (0); bilobed (1). In *Stereomastis*, the apex of the eye is distally widened and somewhat T-shaped or bilobed.
47. Basal antennular segment, anterolateral margin: obsolete, not expanded (0); expanded, quadrate (1); expanded, round (2). The basal antennular segment in non-Polychelidae is unspecialized and similar to the following segment. In extant Polychelidae, the basal antennular segment is expanded anterolaterally, and the stylocerite is strongly produced anteromedially to form a spiniform or triangular projection. The antennules are not known in *Antarticheles*.
48. Basal antennular segment, stylocerite: absent (0); present (1).
49. Basal antennular segment, stylocerite length: not extending beyond peduncle, upturned medially (0); as long as or longer than peduncle (1); obsolete (2).
50. Basal antennular segment, anterolateral spines: absent (0); one or two (2).
51. Basal antennular segment, stylocerite form: triangular (0); foliaceous (1); spinular (2); obsolete (3). The stylocerite is triangular in most Polychelidae, but is spinular in *Cardus*, and foliaceous in *Pentacheles gibbus* and *Pe. obscurus*.
52. Antennular peduncle, segment 1, inner spine: absent (0); present (1).
53. Antennal protopod, segment 1, inner spine: absent (0); large, prominent (1); small (2).
54. Antennal peduncle, segment 1, inner spine or tooth: absent (0); present (1).
55. Antennal peduncle, segment 2, inner spine or tooth: absent (0); present (1).
56. Antennal scale shape: lanceolate (0); convex outer margin (1); circular (2).
57. Maxilliped 3, epipod: vestigial (0); about one-third ischium length (1); as long as or longer than ischium (2). The maxilliped 3 epipod is as long as or longer than the ischium in *Cardus*, *Willemoesia*, and *Pentacheles* (Fig. 3C, E, F); about one-third the ischium length in *Homeryon* (Fig. 3D); and vestigial in *Polycheles* and *Stereomastis* (Fig. 3A, B).
58. Maxilliped 3, arthrobranch: absent (0); present (1).
59. Pereopods 1-5 epipod: reduced (0); well-developed (1). The epipods of pereopods 1-5 are very short and reduced in *Stereomastis* (Fig. 3H) and well-developed in other extant polychelids (Fig. 3G). Bate (1888) used the length of the pereopods 1-5 epipods to distinguish *Stereomastis* from *Pentacheles*, but his concept of *Pentacheles* included species now assigned to *Polycheles*, which have vestigial rather than well-developed maxilliped 3 epipods. The epipod length is not known in any extinct taxa.
60. Pereopod 1, pollex accessory spine: absent (0); present (1). The pereopod 1 pollex accessory spine is unique to *Willemoesia* (Fig. 2F).
61. Pereopod 1, dactylus: distally evenly tapering (0); with small, triangular subdistal lobe. Distally tapering pereopod dactyli are present in all taxa except eryonids, in which the dactylus terminates in a small, triangular subdistal lobe.
62. Pereopod 1, carpus length: very short, triangular (0); elongate, slender (1). The short, triangular carpus is a feature of tetrachelids, eryonids, palaeopentachelids, and the extant *Cardus*. In other taxa, where known, the carpus is elongate and slender.
63. Pereopod 1, carpus, upper distal spine: absent (0); present (1).
64. Pereopod 2, ischium-merus: articulating (0); fused (1). The pereopod 2 ischium and merus are fused in most species of *Stereomastis* and articulated in other polychelids. The condition in fossil taxa is not known except for *Cycleryon*, in which the ischium and merus are articulated.

65. Pereopod 2, coxal spines: absent (0); present (1).
66. Pereopod 3, coxal spines: absent (0); present (1).
67. Pereopods 2–4, dactyli curvature: weak (0); strong (1). Strongly curved pereopods 2–4 dactyli are a synapomorphy of *Homeryon*. In other taxa, the pereopods 2–4 dactyli are only weakly curved.
68. Pereopod 3, cheliped fingers: apices not crossing (0); apices crossing (1). State 1 is unique to *Willemoesia*.
69. Pereopods 3–5, basis-ischium-merus fusion: articulating (0); fused (1). Scholtz & Richter (1995) proposed that a fused basis-ischium-merus of pereopods 3–5 is a synapomorphy of Polychelidae. Although the basis-ischium-merus are fused in most extant polychelids, the basis and ischiomerus segments are articulated in *Homeryon* and *P. amemiyai*, *P. enthrix*, and *P. kermadecensis*. The condition in fossil taxa is not known except for *Cycleryon*, in which the basis and ischiomerus are articulated.
70. Pereopods 3–5, basis-ischium-merus junction: diagonal (0); perpendicular (1). The basis-ischium-merus junction of pereopods 3–5 is perpendicular to the segment axis in extant polychelids except for *Willemoesia*, *Homeryon*, *P. amemiyai*, *P. enthrix*, and *P. kermadecensis*, in which the junction is diagonal to the segment axis. The condition in fossil taxa is not known except for *Cycleryon* and *Eryon*, in which the basis-ischium-merus junction is also diagonal to the segment axis.
71. Pereopod 5, dactylus in adult males: simple (0); partially chelate, dactylus distinctly longer than pollex (1); fully chelate, dactylus as long as pollex (2).

APPENDIX 4

Unambiguous character state changes for 1 of 10 most parsimonious topologies derived from Analysis 1 shown in Fig. 5. Clade numbers correspond to those indicated in Fig. 5.

**Clade 1.** 22: 1→0, 61: 0→1. **Clade 2.** 23: 0→1. **Clade 3.** 10: 0→1, 11: 0→2, 43: 0→1, 71: 0→1. **Clade 4.** 16: 0→1, 32: 0→1, 62: 0→1. **Clade 5.** 6: 0→1, 11: 2→1, 19: 0→1, 27: 0→1, 29: 2→0, 37: 0→1, 44: 0→1, 47: 0→1, 48: 0→1, 49: 2→0, 51: 3→0. **Clade 6.** 10: 1→0, 60: 0→1, 63: 0→1, 68: 0→1, 71: 1→2. **Clade 7.** 35: 0→2, 42: 0→1. **Clade 8.** 8: 0→1, 45: 0→1, 50: 0→1. **Clade 9.** 1: 0→1, 7: 0→1, 14: 0→1, 52: 0→1, 56: 0→1, 57: 2→1, 67: 0→1. **Clade 10.** 70: 0→1. **Clade 11.** 28: 1→0. **Clade 12.** 11: 1→0, 51: 0→1. **Clade 13.** 63: 0→1. **Clade 14.** 13: 0→1. **Clade 15.** 30: 0→1, 35: 0→2, 47: 1→2, 55: 0→1. **Clade 16.** 2: 0→1, 18: 0→1, 32: 0→1, 41: 0→1, 49: 0→1, 57: 2→0. **Clade 17.** 53: 2→1; 58: 1→0. **Clade 18.** 34: 1→0, 41: 1→2, 69: 1→0, 70: 1→0. **Clade 19.** 31: 0→1. **Clade 20.** 7: 0→1, 13: 0→1, 29: 0→2, 71: 1→0. **Clade 21.** 3: 0→1, 11: 1→2, 32: 1→0. **Clade 22.** 15: 0→1, 35: 2→1, 52: 0→1. **Clade 23.** 14: 0→1, 24: 0→2, 29: 2→1, 31: 0→1. **Clade 24.** 4: 0→1, 12: 0→1, 40: 0→1, 63: 0→1. **Clade 25.** 4: 1→3, 9: 1→2; 15: 0→1, 20: 1→2, 21: 1→3, 24: 0→1, 40: 0→1, 46: 0→1, 59: 1→0, 63: 0→1. **Clade 26.** 5: 0→1, 11: 1→0, 34: 1→0, 35: 2→1. **Clade 27.** 25: 0→1, 29: 0→1, 33: 0→1. **Clade 28.** 64: 0→1. **Clade 29.** 31: 0→1. **Clade 30.** 20: 2→3. **Clade 31.** 35: 2→1, 53: 2→0. **Clade 32.** 5: 0→1, 65: 0→1. **Clade 33.** 42: 0→1. **Clade 34.** 26: 1→0, 39: 0→1. **Clade 35.** 52: 0→1, 53: 0→2, 66: 0→1. **Clade 36.** 5: 1→2. **Clade 37.** 24: 1→2.


REFERENCES

Aguirre-Urreta, M.B., Buatois, L.A., Chernoglasov, G.C.B. & Medina, F.A. 1990. First Polychelidae (Crustacea, Palinura) from the Jurassic of Antarctica. *Ant. Sci.* 2: 157–162.

Ahyong, S.T. & Brown, D.E. 2002. New species and new records of Polychelidae from Australia (Decapoda: Crustacea). *Raff. Bull. Zool.* 50: 53–79.

Ahyong, S.T. & Chan, T.-Y. 2004. Polychelid lobsters of Taiwan (Decapoda: Polychelidae). *Raff. Bull. Zool.* 52:171–182.

Ahyong, S.T. & Chan, T.-Y. 2008. Polychelidae from the Bohol and Sulu seas collected by "PANGLAO 2005" (Crustacea: Decapoda: Polychelida). *Raff. Bull. Zool.* Supp. 19: 63–70.

Ahyong, S.T. & Galil, B.S. 2006. Polychelidae from the southern and western Pacific (Decapoda, Polychelida). *Zoosystema* 28: 757–767.

Ahyong, S.T. & O'Meally, D. 2004. Phylogeny of the Decapoda Reptantia: resolution using three molecular loci and morphology. *Raff. Bull. Zool.* 52: 673–693.

Alcock, A. 1894. Natural history notes from H. M. Indian marine survey steamer *Investigator*, Commander R. F. Hoskyn, R. N., commanding. Series II, number 1. On the results of deep-sea dredging during the season 1890–91. *Ann. Mag. Nat. Hist.* (6) 13: 225–245.

Bate, C.S. 1878. XXXII. On the *Willemoesia* group of Crustacea. *Ann. Mag. Nat. Hist.* 5: 273–283, pl. 13.

Bate, C.S. 1888. Report on the Crustacea Macrura dredged by H.M.S. *Challenger* during the years 1873–1876. *Rep. Sci. Res. H.M.S. Challenger 1873–76, Zool.* 24: 1–942, 154 pls.

Beurlen, K. 1928. Die Decapoden des schwäbischen Jura mit Ausnahme der aus den oberjurassichen Plattenkalken stammenden. *Palaeontographica* 70: 115–278, pl. 6–8.

Beurlen, K. 1930. Nachträge zur Decapodenfauna des Schwäbischen Jura. *Neues Jahrbuch für Mineralogie, Geologie und Paläontologie, Abteilung B, Beilagen-Band*, 64: 219–234.

Bouvier, E.L. 1917. Crustacés décapodes (Macrours marcheurs) provenant des campages des yachts Hirondelle et Princesse-Alice (1885–1915). *Rés. Camp. Sci. Prince de Monaco* 50: 1–140.

Broderip, W.J. 1835. Description of some fossil Crustacea and Radiata. *Proc. Geol. Soc. Lond.* 2: 201–202.

Desmarest, A.G. 1822. Les Crustacés proprement dits. In: Brongniart, A. & Desmarest, A.G. (eds.), *Histoire Naturelle des Crustacés fossiles, sous les rapports zoologiques et géologiques*: 67–154, pl. 5–11. Paris: F.G. Levraut.

de Haan, W. 1833–1850. Crustacea. In: von Siebold, Ph.F. (ed.), *Fauna Japonica sive descriptio animalium, quae in itinere per Japoniam, jusse et auspiciis superiorum, qui summum in India Batavia Imperium tenent, suscepto, annis 1823–1830 collegit, notis observationibus et adumbrationibus illustravit*: 1–243. Lugdunum Batavorum: A. Arnz.

de Man, J.G. 1905. Diagnoses of new species of macrurous decapod Crustacea from the "Siboga Expedition." I. *Tijds. Neder. dier. Vereen.* (2) 9: 587–614.

de Man, J.G. 1916. The Decapoda of the Siboga Expedition. Part III. Families Eryonidae, Palinuridae, Scyllaridae and Nephropsidae. *Siboga-Exped.* 39 A2: 1–122.

Dixon, C.J., Ahyong, S.T. & Schram, F.R. 2003. A new hypothesis of decapod phylogeny. *Crustaceana* 76: 935–975.

Farris, J.S., Albert, V.A., Källersjo, M., Lipscomb, D. & Kluge, A.G. 1996. Parsimony jackknifing outperforms neighbor-joining. *Cladistics* 12: 99–124.

Faxon, W. 1893. No. 7. Reports on the dredging operations off the West Coast of Central America to the Galapagos by the *Albatross*. VI. Preliminary descriptions of new species of Crustacea. *Bull. Mus. Comp. Zool. Harvard* 24: 149–220.

Faxon, W. 1895. XV. The stalk-eyed Crustacea. Reports on an exploration off the west coasts of Mexico, Central and South America, and off the Galapagos Islands, in charge of Alexander Agassiz, by the U.S. Fish Commission Steamer "Albatross," during 1891, Lieut.-Commander Z.L. Tanner U.S.N., Commanding. *Mem. Mus. Comp. Zool. Harvard* 18: 1–292, pls. 1–56.

Firth, R.W. & Pequegnat, W.E. 1971. *Deep Sea Lobsters of the Families Polychelidae and Nephropidae (Crustacea, Decapoda) in the Gulf of Mexico and Caribbean Sea*. 106 pp. College Station: Texas A & M University.

Fowler, H.W. 1912. The Crustacea of New Jersey. *Ann. Rep. New Jersey State Mus.* 1911: 29–650, pl. 1–150.

Fraas, O. 1855. Beiträge zum obersten weissen Jura in Schwaben. *Jh. Ver. vaterl. Naturkde. Württemberg* 11: 76–107.

Galil, B.S. 2000. Crustacea decapoda: review of the genera and species of the family Polychelidae Wood-Mason, 1874. In: Crosnier, A. (ed.), *Résultats des Campagnes MUSORSTOM, Volume 21*. 285–387. Paris: Mém. Mus. nat. Hist. naturelle.

Garassino, A. & Schweigert, G. 2006. The Upper Jurassic Solnhofen decapods Crustacean fauna: review of the types from old descriptions. Part I. Infraorders Astacidea, Thalassinidea, Palinura. *Mem. Soc. Ital. Sci. Nat. Museo civ. Stor. Nat Milano* 34: 1–64.

Garassino, A. & Teruzzi, G. 1993. A new decapod crustacean assemblage from the upper Trassic of Lombardy (N. Italy). *Paleo. Lomb.* 1: 1–27.

Garassino, A., Teruzzi, G. & Dalla Vecchia, F.M. 1996. The macruran decapod crustaceans of the Dolomia di Forni (Norian, Upper Triassic) of Carnia (Udine, NE Italy). *Atti Soc. Ital. Sci. Nat. Museo civ. Stor. Nat. Milano* 136(1): 15–60.

Glaessner, M.F. 1965. Vorkommen fossiler Dekapoden (Crustacea) in Fisch-Schiefern. *Senck. Leth.* 46a: 111–122.

Glaessner, M.F. 1969. Decapoda. In: Moore, R.C. (ed.), *Arthropoda 4. Part R, vol. 2. Treatise on Invertebrate Paleontology*: 399–533. Lawrence: Geological Society of America and University of Kansas Press.

Gore, R.H. 1984. Abyssal lobsers, genus *Willemoesia* (Palinura, Polychelidae), from the Venezuela Basin, Caribbean Sea. *Proc. Acad. Nat. Sci. Phil.* 136: 1–11.

Gould, C. 1857. On a new fossil crustacean (*Tropifer laevis* C. Gould) from the Lias bone bed. *Quart. J. Geol. Soc. Lond.* 13: 360–363.

Griffin, D.J.G. & Stoddart, H.E. 1995. Deepwater decapod crustacea from eastern Australia: lobsters of the families Nephropidae, Palinuridae, Polychelidae and Scyllaridae. *Rec. Aust. Mus.* 47: 231–263.

Grote, A.R. 1873. Deidamia. *Nature* 8: 485.

Heller, C. 1862. Beiträge zur näheren Kenntnis der Macrouren. *Sitzung. Akad. Wiss. Wien math.-physik. Klasse* 45: 389–426, 2 pls.

Holthuis, L.B. 1952. Crustacés Décapodes Macrures. *Rés. Sci. Expéd. Ocean. Belge Eaux Côt. Afr. Atl. Sud (1948–1949)* 3: 1–88.

Holthuis, L.B. 1962. *Stereomastis* Bate, 1888 (Crustacea, Decapoda), proposed validation under the plenary powers. Z.N.(S.) 1497. *Bull. Zool. Nomen.* 19: 182–183.

Holthuis, L.B. 1991. Marine lobsters of the world. *FAO Fisheries Synopsis* 125(13): 1–292.

Karasawa, H., Takahashi, F., Doi, E. & Ishida, H. 2003. First notice of the family Coleiidae Van Straelen (Crustacea: Decapoda: Eryonoidea) from the upper Triassic of Japan. *Paleo. Res.* 7: 357–362.

Kemp, S.W. & Sewell, R.B.S. 1912. Notes on Decapoda in the Indian Museum. III. The species obtained by R.I.M.S.S. 'Investigator' during the survey season 1910–11. *Rec. Ind. Mus.* 7:15–32, plate 1.

Kensley, B.F. 1968. Deep sea decapod Crustacea from west of Cape Point, South Africa. *Ann. S. Afr. Mus.* 50: 283–323.

Maddison, W.P. 1993. Missing data versus missing characters in phylogenetic analysis. *Syst. Biol.* 42: 576–581.

Maddison, D.R. & Maddison, W.P. 2000. *MacClade. Analysis of Phylogeny and Character Evolution. Version 4.0.* Sunderland, Massachusetts: Sinauer Associates.

Milne-Edwards, A. 1880. No.1. Reports on the results of dredging under the supervision of Alexander Agassiz, in the Gulf of Mexico, and in the Caribbean Sea, 1877, 78, 79, by the MS coast survey steamer *Blake*. VIII. 'Etudes préliminaries sur les Crustacés. *Bull. Mus. Comp. Zool. Harvard* 8: 1–68, pls 1–2.

Platnick, N.I., Griswold, C.E. & Coddington, J. A. 1991. On missing entries in cladistic analysis. *Cladistics* 7: 337–343.

Rathbun, M.J. 1906. The Brachyura and Macrura of the Hawaiian Islands. *Bull. U.S. Fish Com.* 23: 827–930, pls. 1–24.

Reuss, A. 1858. Über fossile Krebse aus den Raibler Schichten. *Beitr. Paläo. Oest.* 1: 1–6.

Scholtz, G. & Richter, S. 1995. Phylogenetic systematics of the reptantian Decapoda (Crustacea, Malacostraca). *Zool. J. Linn. Soc.* 113: 289–328.

Schram, F. R. 2001. Phylogeny of decapods: moving towards a consensus. *Hydrobiologia* 449: 1–20.

Schram, F.R. & Dixon, C.J. 2004. Decapod phylogeny: addition of fossil evidence to a robust morphological cladistic data set. *Bull. Mizunami Foss. Mus.* 31: 1–19.

Schweigert, G. 2000. News about Jurassic eryonid decapods (Coleiidae, Eryonidae) from southern Germany. *Stud. Ricerche – Ass. Amici Mus. Civ. "G. Zannato," Montechio Maggiore (Vicenza)* 2000: 63–65.

Schweigert, G. & Dietl, G. 1999. Neubeschreibung von "*Eryon longipes* O. Fraas" (Crustacea, Decapoda, Eryonidea) aus dem Nusplinger Plattenkalk (Ober-Kimmeridgium, Schwäbische Alb). *Stuttgarter Beiträge zur Naturkunde Serie B (Geologie und Paläontologie)* 274: 1–19.

Schweitzer, C. & Feldmann, R. 2001. New Cretaceous and Tertiary decapod crustaceans from western North America. *Bull. Mizunami Foss. Mus.* 28: 173–210.

Selbie, C.M. 1914. The Decapoda Reptantia of the coasts of Ireland. Part 1. Palinura, Astacura and Anomura (except Paguridea). *Fish. Ireland Sci. Inv.* 1: 1–116, pl. 1–15.

Smith, S.I. 1880. Notice of a new species of the "*Willemoesia* Group of Crustacea," Recent Eryontidae. *Proc. U.S. Nat. Mus.* 2: 345–353, pl. 7.

Smith, S.I. 1882. XVII. Report on the Crustacea. I. Decapoda. Reports on the results of dredging, under the supervision of Alexander Agassiz, on the east coast of the United States, during the summer of 1880, by the U.S. Coast Survey Steamer "Blake," Commander J.R. Bartlett, U.S.N., Commanding. *Bull. Mus. Comp. Zool. Harvard* 10: 1–108, pl. 1–15.

Smith, S.I. 1884. XV. Report on the Decapod Crustacea of the *Albatross* dredgings off the east-coast of the United States in 1883. *Rep. U.S. Fish Com.* 10(1882): 345–426, pls. 1–10.

Sund, O. 1920. The *Challenger* Eryonidea (Crustacea). *Ann. Mag. Nat. Hist.* (9), 6: 220–226.

Swofford, D.L. 2002. *PAUP\*. Phylogenetic Analysis Using Parsimony (\* and Other Methods). Version 4.0b10*. Sunderland, Massachusetts: Sinauer Associates.

Teruzzi, G. 1990. The genus *Coleia* Broderip, 1835 (Crustacea, Decapoda) in the Sinemurian of Osteno in Lombardy. *Atti Soc. Ital. Sci. Nat. Museo Civ. Stor. Nat. Milano* 131: 85–104.

Terruzi, G. & Garassino, A. 2007. *Coleia* Broderip, 1835 (Crustacea, Decapoda, Coleiidae) from the Mesozoic of Italy: an update. *Mem. Soc. Ital. Sci. Nat. Museo civ. Stor. Nat Milano*: 95–96.

Tsang, L.M., Ma, K.Y., Ahyong, S.T., Chan, T.-Y. & Chu, K.H. 2008. Phylogeny of Decapoda using two nuclear protein-coding genes: origin and evolution of the Reptantia. *Mol. Phylogenet. Evol.* 48: 359–368.

Thomson, C.W. 1873. Notes from the "Challenger." *Nature* 8: 246–249, 266–267.

Van Straelen, V. 1922. Les Crustacés décapodes du Callovien de La Voulte-sur-Rhône (Ardèche). *C. Rend. Acad. Sci. Paris* 175: 982–984.

Van Straelen, V. 1923. Description de Crustacés décapodes macroures nouveaux des terrain secondaires. *Ann. Soc. Roy. Zool. Malacol. Belg. ser.* 2, 7: 1–462

Van Straelen, V. 1924–1925. Contribution a l'étude des crustacés décapodes de la Période Jurassique. *Mém. Acad. Roy. Belg. Pt. 4, ser.* 2, 7: 1–462.

von. Knebel, W. 1907. Die Eryoniden des oberen Weissen Jura von Süddeutschland. *Arch. Biont.* 2: 193–233.

von Meyer, H. 1836. Beiträge zu *Eryon*, einem Geschlechte fossilier langschwänziger Krebse. *Nova Acta Acad. Caesar. Leopold.-Carol. German. Nat. Curios.* 18/1: 263–283.

von Münster, G. 1839. Decapoda Macrura. Abbildung und Beschreibung der fossilen langschwänzigen Krebse in den kalkschiefern von Bayern. *Beitr. Pefrefactenkd.* 2: 1–88.

von Schlotheim, E.F. 1820. *Die Petrefactenkunde auf ihrem jetzigen Standpunkte, durch die Beschreibung seiner Sammlung versteinerter und fossiler Überreste des Thier- und Pflanzenreichs der Vorwelt erläutert*. Gotha: Becker.

von Schlotheim, E.F. 1822. Nachträge zur Petrefactenkunde, 2: 1–88. Gotha: Becker.

von Willemoes-Suhm, R. 1873. *Deidamia leptodactyla*. In: Thomson, C.W. Notes from the "Challenger." *Nature* 8: 51–53.

von Willemoes-Suhm, R. 1875. Von der *Challenger* Expedition. *Zeit. Wiss. Zool.* 25: xxv–xlvi.

Wood-Mason, J. 1874. On blind crustaceans. *Proc. Asiatic Soc. Bengal* 1874: 180–181.

Yokoya, Y. 1933. On the distribution of decapod crustaceans inhabiting the continental shelf around Japan, chiefly based upon the materials collected by the S.S. Sôyô-Maru, during the year 1923–1930. *J. Coll. Agr., Imp. Uni. Tokyo* 12: 1–226.

# IV ADVANCES IN OUR KNOWLEDGE OF THE ANOMURA

# Anomuran Phylogeny: New Insights from Molecular Data

SHANE T. AHYONG, KAREEN E. SCHNABEL & ELIZABETH W. MAAS

*National Institute of Water and Atmospheric Research, Private Bag 14901, Kilbirnie, Wellington, New Zealand*

ABSTRACT

High-level classifications of Anomura typically recognize three major clades: Galatheoidea (squat lobsters and porcelain crabs), Paguroidea (hermit and king crabs), and Hippoidea (mole crabs). The general stability of this classification, however, has masked the vigorous debate over internal relationships. Phylogenetic relationships of the Anomura are analyzed based on sequences from three molecular loci (mitochondrial 16S; nuclear 18S and 28S), with multiple exemplars representing 16 of 17 extant families. The dataset assembled is the largest analyzed to date for Anomura. Analyses under maximum parsimony and Bayesian inference recognize a basal position for Hippoidea, corroborating several recent studies, but point to significant polyphyly in the two largest superfamilies, Galatheoidea and Paguroidea. Three independent carcinization events are identified (in Lithodidae, Porcellanidae, and Lomisidae). The polyphyletic origin of asymmetrical hermit crabs is a radical departure from previous studies and suggests independent derivations of asymmetry in three separate clades: Paguridae, Coenobitidae + Diogenidae, and Parapaguridae. Such a scenario may seem unlikely owing to the complex characters involved, but if carcinization has multiple, independent origins, then adaptation to dextral shell habitation may also be plausible. Polyphyly of Galatheoidea, however, while unexpected, is morphologically tenable—characters traditionally used to unify Galatheoidea are plesiomorphies. Chirostylid squat lobsters are more closely related to an assemblage including aegloids, lomisoids, and parapagurids than to other galatheoids. Galatheidae may be paraphyletic on the basis of an internally nested Porcellanidae, and a similar situation may obtain for Chirostylidae with respect to Kiwaidae. Present topologies are not sufficiently robust to justify significant changes to the classification, but they point to fruitful lines for further research.

## 1 INTRODUCTION

Few major decapod groups have had as unstable a taxonomic history as the Anomura. Historically, the composition of Anomura has been significantly fluid, with inclusion or exclusion of the major groups such as the thalassinidean shrimps and the dromiacean crabs (reviewed by Martin & Davis 2001; McLaughlin et al. 2007). Even the name has not been universally accepted, with some authors favouring Anomala over Anomura (see McLaughlin & Holthuis 1985). Most classifications recognize three major anomuran groups: Galatheoidea (squat lobsters and porcelain crabs), Paguroidea (hermit and king crabs), and Hippoidea (mole crabs). The general anomuran classification has been relatively stable for the last two to three decades, but this stability has masked the vigorous and ongoing debate over their internal relationships.

Nevertheless, advances have been made. The monophyly of Anomura is now well established. The relationship between thalassinideans and anomurans has long been ambiguous, leading workers to variously recognize independent status for each group or a single, expanded Anomura (e.g., Henderson 1888; Borradaile 1907; Balss 1957; Burkenroad 1963, 1981; Glaessner 1969;

McLaughlin 1983b). McLaughlin & Holthuis (1985) excluded thalassinideans from Anomura, and this has been corroborated by numerous phylogenetic analyses (e.g., Martin & Abele 1986; Poore 1994; Scholtz & Richter 1995; Ahyong & O'Meally 2004; Tsang et al. 2008). The dromiacean crabs, which were variously regarded as anomuran or brachyuran based largely on plesiomorphic larval features, are confirmed as Brachyura (the 'true' crabs) (see Spears et al 1992; Ahyong et al. 2007). Moreover, the sister group to Anomura is now widely accepted as Brachyura, the two clades constituting Meiura (Scholtz & Richter 1995; Schram 2001; Dixon et al. 2003; Ahyong & O'Meally 2004; Tsang et al. 2008). The ingroup for analysis is thus well circumscribed in terms of composition and monophyly.

Anomura presently includes 7 superfamilies, 17 families, almost 200 genera, and about 1500 species. Although less speciose than its sister clade by more than one-quarter, recovering the pattern of anomuran evolution is no less challenging. Anomura presents a morphological array that spans the generalized squat lobsters, symmetrical and asymmetrical hermit crabs, the brachyuran-like king and porcelain crabs, and fossorial mole crabs. Overlying this diversity is the phenomenon of carcinization (Borradaile 1916), the evolution of a crab-like form, which has occurred independently in multiple anomuran lineages. Anomurans may thus prove to be a particularly fruitful group for investigating evolution of form. Were one so inclined, the meiuran morphospace might even be viewed as an evolutionary 'testing ground' for different ground-plans, out of which the Brachyura was singularly most successful (at least numerically) and most effectively carcinized. Consequently, although highly diverse, brachyurans still exhibit a greater degree of morphological uniformity than does Anomura. Anomurans, on the other hand, emerge with a much wider array of forms, exhibiting considerably greater morphological disparity than the 'true' crabs. Discovering the connections between these morphologically disparate clades, however, presents significant challenges to phylogenetic reconstruction, not least because their conditions of existence presumably exert considerable influence on the expression of form.

The advent of cladistic analysis has seen a steady rise in efforts to understand anomuran evolution and interrelationships (Fig. 1). In addition to the increasing application of cladistic methods, mostly based on somatic morphology, new sources of data have become increasingly accessible, the most significant being DNA sequences. Most phylogenetic studies of anomurans are based on morphology, most recently McLaughlin et al. (2007); few have explored molecular data to any great extent. Thus, to reconstruct phylogenetic interrelationships of the Anomura, we assembled existing and newly generated sequence data from three molecular loci (mitochondrial 16S; nuclear 18S and 28S) encompassing 16 of 17 recognized anomuran families in the largest anomuran dataset to date.

## 2 MATERIALS AND METHODS

### 2.1 *Taxon sampling*

Representatives of all anomuran families, *sensu* McLaughlin et al. (2007) (except Pylojacquesidae), were included as terminals, with emphasis on the Galatheoidea (Table 1). Representatives of all three galatheid subfamilies were included, representing 11 of 34 recognized genera. Porcellanidae was represented by three exemplars and Chirostylidae was represented by five of six recognized genera. Tissue samples were derived from specimens in the collections of the Muséum National d'Histoire Naturelle, Paris (MNHN); National Institute of Water and Atmospheric Research, Wellington, New Zealand (NIWA); and National Taiwan Ocean University, Keelung, Taiwan (NTOU). The 28S sequence of *Shinkaia* was amplified from genomic DNA generously provided by K. H. Chu (Chinese University of Hong Kong), who also shared unpublished 16S and 18S *Shinkaia* sequences. Brachyura is the sister group to Anomura (Scholtz & Richter 1995; Ahyong & O'Meally 2004;

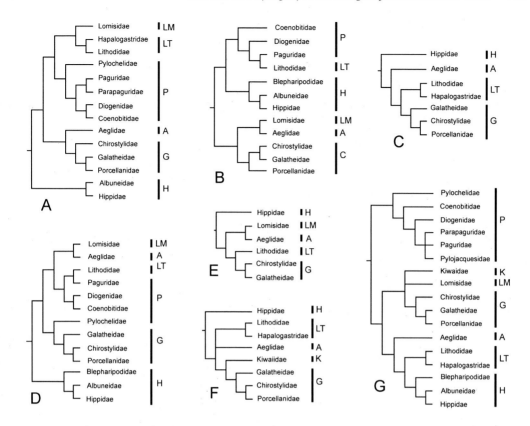

**Figure 1.** Selected hypotheses of anomuran relationships. (A) based on Martin & Abele (1986); (B) based on Morrison et al. (2002); (C) based on Pérez-Losada et al. (2002); (D) based on Ahyong & O'Meally (2004); (E) based on Porter et al. (2005); (F) based on Macpherson et al. (2005); (G) based on McLaughlin et al. (2007). Superfamilies as recognized by McLaughlin et al. (2007) abbreviated as follows: Aegloidea – A; Kiwaoidea – K; Galatheoidea – G; Hippoidea – H; Lithodoidea – LT; Lomisoidea – LM; Paguroidea – P.

Tsang et al. 2008), so the analysis was rooted to two brachyuran exemplars, *Lauridromia dehaani* and *Paromola japonica*.

## 2.2 Molecular data

Two nuclear ribosomal genes (18S rRNA and the D1 region of 28S rRNA) and one mitochondrial ribosomal gene (16S rRNA) were selected for their utility in resolving phylogenetic history at different taxonomic levels (Crandall et al. 2000; Ahyong & O'Meally 2004). We collected new sequence data for 19 species, resulting in 53 new sequences (see Table 1). Other sequences were available in GenBank. For the *Pagurus* terminal, 16S and 28S sequences were derived from *P. bernhardus* and the 18S sequence from *P. longicarpus*.

**Table 1.** Classification of terminal taxa with GenBank accession numbers for gene sequences. New sequences are indicated (*). *Shinkaia* 16S and 18S sequence provided by K. H. Chu (KHC, Chinese University of Hong Kong). For convenience, the high-level classification follows McLaughlin et al. (2007). Location of voucher specimens for new sequences: MNHN (Muséum National d'Histoire Naturelle, Paris), NIWA (National Institute of Water and Atmospheric Research, Wellington, New Zealand), NTOU (National Taiwan Ocean University, Keelung, Taiwan).

| | 16S | 18S | 28S | Voucher |
|---|---|---|---|---|
| **ANOMURA** | | | | |
| AEGLOIDEA | | | | |
|   AEGLIDAE | | | | |
|     *Aegla uruguyana* Schmitt, 1942 (*Aegla 1*) | AF436051 | AF436012 | AF435992 | |
|     *Aegla violacea* Bond-Buckup & Buckup, 1994 (*Aegla 2*) | AY595880 | AY595799 | AY596051 | |
| HIPPOIDEA | | | | |
|   ALBUNEIDAE | | | | |
|     *Lepidopa californica* Efford, 1971 | AF436054 | AF436015 | AF435996 | |
|   BLEPHARIPODIDAE | | | | |
|     *Blepharipoda occidentalis* Randall, 1840 | AF436053 | AF436014 | AF435994 | |
|   HIPPIDAE | | | | |
|     *Emerita emeritus* (Linnaeus, 1767) | AY583898 | AY583971 | AY583990 | |
| KIWAOIDEA | | | | |
|   KIWAIDAE | | | | |
|     *Kiwa hirsuta* Macpherson, Jones & Segonzac, 2005 | *EU831284 | DQ219316 | *EU831286 | MNHN |
| PAGUROIDEA | | | | |
|   COENOBITIDAE | | | | |
|     *Coenobita compressus* H. Milne Edwards, 1837 | AF436059 | AF436023 | AF435999 | |
|   DIOGENIDAE | | | | |
|     *Calcinus obscurus* Stimpson, 1859 | AF436058 | AF436022 | AF435998 | |
|     *Clibanarius albidigitatus* Nobili, 1901 | AF425323 | AF438751 | AF425362 | |
|     *Isocheles pilosus* (Holmes, 1900) | AF436057 | AF436021 | – | |
|   PAGURIDAE | | | | |
|     *Bythiopagurus macroculus* McLaughlin, 2003 | *EU821532 | *EU821548 | *EU821565 | NIWA |
|     *Discorsopagurus schmitti* (Stevens, 1925) | AF436055 | AF436017 | – | |
|     *Pagurus bernhardus* (Linnaeus, 1758) | AF425335 | – | AF425354 | |
|     *Pagurus longicarpus* Say, 1817 | – | AF436018 | – | |
|   PARAPAGURIDAE | | | | |
|     *Parapagurus latimanus* Henderson, 1888 | *EU821534 | *EU821550 | *EU821567 | NIWA |
|     *Sympagurus dimorphus* (Studer, 1883) | *EU821533 | *EU821549 | *EU821566 | NIWA |
|   PYLOCHELIDAE | | | | |
|     *Pylocheles macrops* Forest, 1987 | AY583897 | AY583970 | AY583989 | |
|     *Trizocheles spinosus* (Henderson, 1888) | *EU821535 | *EU821551 | *EU821568 | NIWA |
| LITHODOIDEA | | | | |
|   LITHODIDAE | | | | |
|     *Lithodes santolla* (Molina, 1782) | AF595927 | AF439385 | AF596100 | |
|   HAPALOGASTRIDAE | | | | |
|     *Oedignathus inermis* (Stimpson, 1860) | AF425334 | Z104062 | AF425353 | |

**Table 1.** continued.

|  | 16S | 18S | 28S | Voucher |
|---|---|---|---|---|
| **LOMISOIDEA** | | | | |
| **LOMISIDAE** | | | | |
| *Lomis hirta* (Lamarck, 1818) | AF436052 | AF436013 | AF435993 | |
| **GALATHEOIDEA** | | | | |
| **CHIROSTYLIDAE** | | | | |
| *Chirostylus novaecaledoniae* Baba, 1991 | *EU821539 | *EU821555 | *EU821572 | MNHN |
| *Eumunida sternomaculata* Saint Laurent & Poupin, 1996 | AY351063 | AF436011 | AF435991 | |
| *Gastroptychus novaezelandiae* Baba, 1974 | *EU821538 | *EU821554 | *EU821571 | NIWA |
| *Pseudomunida fragilis* Haig, 1979 | *EU821536 | *EU821552 | *EU821569 | MNHN |
| *Uroptychus nitidus* (A. Milne-Edwards, 1880)(*Uroptychus 1*) | AY595925 | AF439387 | AY596096 | |
| *Uroptychus scambus* Benedict, 1902 (*Uroptychus 2*) | *EU831282 | *EU821553 | *EU831283 | NIWA |
| **GALATHEIDAE** | | | | |
| Galatheinae | | | | |
| *Agononida longipes* (A. Milne-Edwards, 1880) (*Agononida 1*) | – | AF439381 | – | |
| *Agononida procera* Ahyong & Poore, 2004 (*Agononida 2*) | *EU821540 | *EU821556 | *EU821573 | NIWA |
| *Allogalathea elegans* (Adams & White, 1848) | *EU821543 | *EU821560 | *EU821577 | MNHN |
| *Cervimunida johni* (Porter, 1903) | *EU821546 | *EU821563 | *EU821580 | NIWA |
| *Galathea* sp. | *EU821544 | *EU821561 | *EU821578 | NIWA |
| *Leiogalathea laevirostris* (Balss, 1913) | *EU821541 | *EU821557 | *EU821574 | NIWA |
| *Munida quadrispina* Benedict, 1902 (*Munida 1*) | AF436050 | AF436010 | AF435990 | |
| *Munida gregaria* (Fabricius, 1793) (*Munida 2*) | AY050075 | AF439382 | AY596099 | |
| *Pleuroncodes monodon* (H. Milne Edwards, 1837) | *EU821545 | *EU821562 | *EU821579 | NIWA |
| *Sadayoshia* sp. | *EU821547 | *EU821564 | *EU821571 | MNHN |
| Munidopsinae | | | | |
| *Galacantha rostrata* (A. Milne-Edwards, 1880) | – | *EU821559 | *EU821576 | NIWA |
| *Munidopsis bairdii* (Smith, 1884) | *EU821542 | *EU821558 | *EU821575 | NIWA |
| Shinkaiinae | | | | |
| *Shinkaia crosnieri* Baba & Williams, 1998 | KHC | KHC | *EU831285 | NTOU |
| **PORCELLANIDAE** | | | | |
| *Pachycheles rudis* Stimpson, 1859 | AF260598 | AF436048 | AF435988 | |
| *Petrolisthes armatus* (Gibbes, 1850) | AF436049 | AF436009 | AF435989 | |
| *Porcellanella triloba* White, 1851 | *EU834069 | – | – | |

## 2.3 DNA extraction and analysis

Genomic DNA was either directly extracted from fresh or ethanol-fixed tissue samples that were soaked 24 hours in a buffer containing 500 mM Tris-HCL (pH 9.0), 20mM EDTA, and 10 mM NaCl. Extraction followed the standard protocol of the QIAGEN DNeasy Blood & Tissue Kit and subsequent quantification of DNA concentration using PicoGreen TM (Molecular Probes Inc., USA). For problematic taxa, a linear acrylamide precipitation was used overnight to increase concentration of DNA. Sequences of two nuclear (the nearly complete sequence of 18S and the 28S D1 expansion region) and one mitochondrial (16S) ribosomal RNA genes were obtained. Primers used are indicated in Table 2. Polymerase chain reactions (PCR) were conducted in 25-$\mu$L volumes with 1–5 $\mu$L of genomic DNA and using Invitrogen Platinum PCR SuperMix containing 22 mM Tris-HCL, 55 mM KCl, 1.65 mM $MgCl_2$, and 220 $\mu$M dNTP. Conditions for 18S and 28S amplification were an initial denaturation at 94°C for two minutes, then 30 cycles of 94°C for one minute, annealing for 1 minute at 50°C, extension at 72°C for two minutes, and a final extension at 72°C for seven minutes. Conditions for 16S amplification were an initial denaturation at 94°C for 5 minutes followed by 30 cycles of 94°C for 30 seconds, annealing for 30 seconds at 50°C, extension at 72°C for one and a half minutes, and a final extension at 72°C for seven minutes. PCRs were checked by running 5 $\mu$L of the reaction on a 1% agarose gel.

In most cases, a single band was obtained and purified using the Qiagen MinElute PCR Purification kit. In the event of multiple bands, the correct-sized fragment was excised from a 2% agarose gel over UV light and purified using QIAquick PCR purification spin columns. Forward and reverse strands were sequenced using sequencing services of Macrogen Inc., Korea (BigDyeTM terminator and ABI Sequencer 3730x, www.macrogen.com). Forward and reverse sequences were combined and checked for errors using ChromasPro Version 1.34 (Technelysium Pty Ltd). Final sequences were aligned in Clustal W using default parameters and adjusted by eye. Regions of ambiguous alignment were excluded and gaps were treated as missing.

## 2.4 Phylogenetic analysis

Following the principle of 'total evidence' (e.g., Prendini et al. 2003), the 16S, 18S, and 28S sequences were analyzed simultaneously. The combined sequences contained about 2.6 kilobases of nucleotide data. Maximum parsimony analyses (MP) were conducted in PAUP* 4.0b10 (Swofford 2002) (heuristic search, TBR, random addition sequence, 500 replicates). Initial analyses were conducted under equal character weights. Topological robustness was assessed using parsimony jackknifing (Farris et al. 1996). Jackknife frequencies were calculated in PAUP* using 1000 pseudoreplicates under a heuristic search with 30% character deletion.

Analyses using Bayesian inference (BI) were conducted in MrBayes Version 3.1.2 (Huelsenbeck & Ronquist 2001). Metropolis coupled Monte Carlo Markov Chains were run for 2,000,000 generations. Four differentially heated chains were run in each of two simultaneous runs. Topologies were sampled every 100 generations. Likelihood settings were determined during the run. Base frequencies were estimated, as were the rates of the six substitution types (nst = 6). A discrete gamma distribution was assumed for variation in the rate of substitution between nucleotide positions in the alignment, and the shape parameter of this distribution was estimated. After inspection of the likelihoods of the sampled trees, the first 50,000 generations were discarded as 'burn in.' All remaining topologies had likelihoods within 0.1% of the long-term asymptote in each run, suggesting that these were sampled after the Markov Chain's convergence to a stable posterior probability distribution. The standard deviation of split frequencies converged to a value of 0.004946. All trees remaining after discarding 'burn in' were used to calculate posterior probabilities using a majority rule consensus.

**Table 2.** Sequencing primers used.

| Primer name | Sequence | | Source |
|---|---|---|---|
| 18S-F07 | 5' – CTG GTT GAT CCT GCC AG – 3' | 18S PCR primer | Medlin et al. (1998) |
| 18S-R1514 | 5' – TGA TCC TTY GCA GGT TCA C – 3' | 18S PCR primer | Sogin (1990) |
| 18S-R651 | 5' – CGA GGT CCT ATT CCA TTA TTC C – 3' | 18S Sequencing primer | Newly designed herein |
| 18S-F551 | 5' – GGT AAT TCG AGC TCC RRT AGC G – 3' | 18S Sequencing primer | Newly designed herein |
| 18S-F1053 | 5' – GAT TCT ATG GGT GGT GGT – 3' | 18S Sequencing primer | Newly designed herein |
| 28S-F216 | 5' – CTG AAT TTA AGC ATA TTA ATT AGK GSA GG – 3' | 28S PCR & sequencing primer | Newly designed herein |
| 28S-R443 | 5' – CCT CAC GGT ACT TGT TCG CTA TCG G – 3' | 28S PCR & sequencing primer | Newly designed herein |
| LR-N-13398 | 5' – CGC CTG TTT AAC AAA AAC AT – 3' | 16S forward PCR & sequencing primer | Morrison et al. (2002) |
| LR-J-12887 | 5' – CCG GTC TGA ACT CAG ATC ACG T – 3' | 16S reverse PCR & sequencing primer | Morrison et al. (2002) |

## 3 RESULTS

### 3.1 *Sequence data*

We collected 54 new sequences from 19 species (18 for 16S, 17 for 18S, and 19 for 28S) (GenBank accession numbers: EU821536, EU821532–821536, E821571–821581, EU831282–831286, EU834069). The aligned combined dataset contained 44 taxa and 2627, characters of which 795 are parsimony informative. The aligned 16S rRNA dataset contained 422 characters, of which 297 are variable (70%) and 216 are parsimony informative (51%). The aligned 18S rRNA dataset contained 1913 characters with 693 variable sites (36%), of which 450 are parsimony informative sites (24%). The aligned 28S rRNA dataset contained 292 characters, of which 170 are variable (58%) and 129 parsimony informative (44%). The 16S fragment is relatively AT rich compared to the other two fragments. Departures from base homogeneity, according to $\chi^2$ tests of nucleotide composition for each gene fragment, were significant for 16S and insignificant for 18S and 28S (16S, df = 132, $P$ = 0.55; 18S, df = 132, $P$ = 1.00; 28S, df = 132, $P$ = 1.00).

### 3.2 *Analyses: maximum parsimony and Bayesian inference*

MP analysis under equal weights retrieved a single, fully resolved topology of length (TL) 3836, consistency index (CI) 0.4726, retention index (RI) 0.6184 (Fig. 2). Hippoidea, containing *Emerita*, *Lepidopa*, and *Blepharipoda*, representing Hippidae, Lepidopidae, and Blepharipodidae, respectively, was monophyletic and sister to the remaining anomurans, corroborating Martin & Abele (1986), Pérez-Losada et al. (2002), Ahyong & O'Meally (2004), Porter et al. (2005), Macpherson et al. (2005), and Tsang et al. (2008). Galatheoidea and Paguroidea, however, are significantly polyphyletic. Three clades of paguroids, corresponding respectively to Diogenidae + Coenobitidae, Parapaguridae + *Trizocheles*, and Paguridae + *Pylocheles*, are widely dispersed. Notably, the two pylochelid terminals, *Pylocheles* and *Trizocheles*, are never in close proximity, instead being associated with Paguridae and Parapaguridae, respectively. *Lithodes* + *Oedignathus* (representing

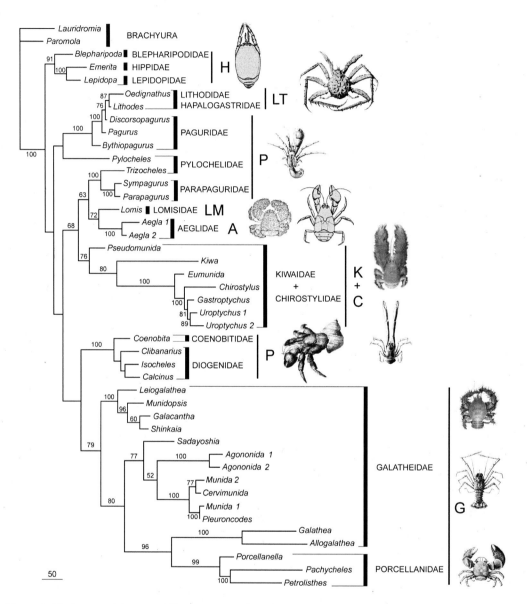

**Figure 2.** Phylogeny of Anomura. Single most parsimonious topology derived from MP analysis under equal weights (TL = 3836, CI = 0.4726, RI = 0.6184). Jackknife proportions indicated at nodes. Superfamilies as recognized by McLaughlin et al. (2007) abbreviated as follows: Aegloidea – A; Chirostylidae – C; Kiwaoidea – K; Galatheoidea – G; Hippoidea – H; Lithodoidea – LT; Lomisoidea – LM; Paguroidea – P.

Lithodidae + Hapalogastridae) is nested within Paguridae. The Paguridae + *Pylocheles* clade is sister to the major clade containing the remaining paguroids and galatheoids *sensu lato*. Aeglidae and Lomisidae are sister taxa, which together are sister to Parapaguridae + *Trizocheles*. The parapagurid-aeglid-lomisid clade is sister to a monophyletic Chirostylidae (with the inclusion of *Kiwa*). Diogenidae is sister to Galatheidae + Porcellanidae. *Shinkaia* (representing Shinkaiinae), *Munidopsis* and *Galacantha* (representing Munidopsinae), and *Leiogalathea* (Galatheiinae) together form a clade that is sister to the remaining galatheids/porcellanids. Within this larger galatheid/porcellanid clade, Porcellanidae is deeply nested, rendering Galatheidae paraphyletic. Jackknife support for 'backbone' nodes was generally low, though clades corresponding to currently recognized families were usually strongly supported (Fig. 2).

Results of BI (Fig. 3) were compatible with, but 'basally' less resolved than, MP results. A hippoid clade, diogenid clade, galatheid + porcellanid clade, pagurid clade, and chirostylid-kiwaid-parapagurid-lomisid-aeglid clade were all recovered with strong support (posterior probability 0.98 or higher). Notably, each of the paguroid clades was dispersed, as were the major galatheoid clades. As in MP results, the two pylochelid terminals were never associated and a monophyletic Porcellanidae nests within a paraphyletic Galatheidae. Under both MP and BI, the Galatheidae and Chirostylidae are not closely related to each other.

## 4 DISCUSSION

### 4.1 Polyphyly of Paguroidea and Galatheoidea

The most striking aspect of the present results is the radical polyphyly of Paguroidea and Galatheoidea. Despite ongoing controversy over internal interrelationships, general consensus has recognized three major clades corresponding to Hippoidea, Galatheoidea, and Paguroidea, irrespective of debate over the positions of one or other constituent groups (e.g., Lomisidae: McLaughlin 1983a; Aeglidae: Pérez-Losada et al. 2002, Ahyong & O'Meally 2004; and, more recently, Pylochelidae: Ahyong & O'Meally 2004). Present results retrieve well-supported clades of paguroids corresponding to Paguridae, Parapaguridae, and Diogenidae + Coenobitidae, respectively. Pylochelidae, however, represented by *Pylocheles* and *Trizocheles*, is not supported as monophyletic. Most significantly, a monophyletic Paguroidea is never recovered. MacDonald et al. (1957) questioned the monophyly of the paguroids based on larval characters, and Tudge (1997), using spermatozoal morphology, found Paguroidea not to be strictly monophyletic owing to incursion of galatheoids. Others, however, have cogently defended paguroid monophyly (McLaughlin 1983b; Richter & Scholtz 1994). Under BI, the positions of major clades of paguroids are either unresolved or dispersed to the proximity of the chirostylids-kiwaids-lomisids-aeglids. Under MP, however, topologies are fully resolved: one paguroid clade (Diogenidae) aligns with the galatheid + porcellanid clade; another (Parapaguridae + *Trizocheles*) forms a clade together with aeglids, lomisids, and chirostylids; and a third clade (Paguridae + Lithodidae + Hapalogastridae) is distant from both Galatheidae and Chirostylidae. Several of the nodes that are unresolved under BI are recovered by MP, but with low jackknife support. Exclusion of parapagurids + *Trizocheleles* from other paguroids is well supported, but the relationship among other paguroid clades is less clear. The pattern of paguroid polyphyly is thus difficult to interpret, though analyses are unequivocal in challenging a strictly monophyletic origin of the hermit crabs. That a monophyletic Pylochelidae is not recovered is perhaps not surprising — likely paraphyly has already been recognized (e.g., Richter & Scholtz 1994; McLaughlin et al. 2007). However, polyphyly of the asymmetrical hermit crabs is difficult to reconcile with somatic morphology. *A priori*, the suite of associated modifications required for gastropod shell habitation, present in all asymmetrical paguroids, is compelling evidence of monophyly. Significant convergence is implied if the hermit crabs are polyphyletic, with independent derivations of asymmetry in Paguridae, Coenobitidae + Diogenidae, and Parapaguridae. Such a scenario seems unlikely, though perhaps plausible, given the discovery that development of abdominal asymmetry

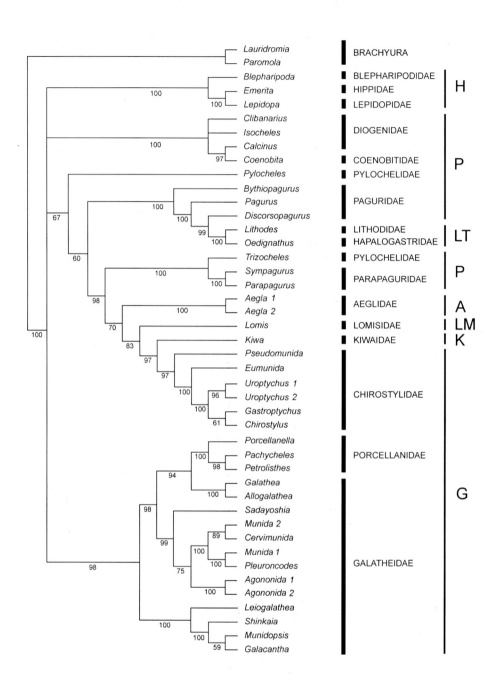

**Figure 3.** Phylogeny of Anomura. Bayesian topology; posterior probabilities indicated on branches as percentages. Superfamilies as recognized by McLaughlin et al. (2007) abbreviated as follows: Aegloidea – A; Kiwaoidea – K; Galatheoidea – G; Hippoidea – H; Lithodoidea – LT; Lomisoidea – LM; Paguroidea – P.

is mediated, at least in part, by environmental factors (Przibam 1907; Harvey 1998). It is also perhaps of more than passing interest that the asymmetrical hermit crab exemplars align basally with different paguroid clades, respectively (under MP: *Trizocheles* with Parapaguridae; *Pylocheles* with Paguridae). Our molecular data strongly corroborate monophyly of the three major paguroid clades (i.e., family level taxa), so the absence of molecular support for overall paguroid monophyly is significant. It should be noted, however, that important phylogenetic information could be contained in hypervariable regions that presently defy alignment and were excluded from the analysis. Also, taxon sampling within speciose families is limited, so a more extensive taxon set may influence topologies.

Galatheoidea, universally recognized to at least include the squat lobsters (Galatheidae and Chirostylidae) and porcelain crabs (Porcellanidae), is not supported as monophyletic. The chirostylids are well removed from the galatheids and porcellanids, being more closely related to an assemblage including aeglids, kiwaids, lomisids, and some hermit crabs. This wide phylogenetic separation, while unexpected, is not counterintuitive. As with Aeglidae, which was formerly assigned to Galatheoidea (e.g., Martin & Davis 2001), the remaining galatheoids have been thought related on the basis of overall habitus, having the generally elongated cephalothorax and 'long tail.' These features, however, are plesiomorphies, and little otherwise unites the galatheoid families. Indeed, McLaughlin et al. (2007) reported only a single unifying synapomorphy of Galatheoidea: the progressive development of the orbits. The orbital structure in galatheids, chirostylids, and porcellanids, though similar, appears to be linked to the well-developed rostrum, which is a plesiomorphy. Thus, given the absence of robust synapomorphies, the polyphyly of Galatheoidea is not surprising.

The 'hairy crab,' *Kiwa hirsuta* (Kiwaidae), was originally posited as sister to the Galatheidae + (Chirostylidae + Porcellanidae) clade with strongest morphological similarities to aeglids and chirostylids (Macpherson et al. 2005). These observations are consistent with present results in the close molecular relationship between chirostylids, aeglids, and kiwaids. Indeed, under MP, *Kiwa* is nested within Chirostylidae, albeit with moderate jackknife support, raising questions about the validity of Kiwaidae. *Kiwa* and chirostylids uniquely share the complete loss of the last thoracic sternite, which was initially regarded as a parallelism (Macpherson et al. 2005; McLaughlin et al. 2007) but is now more parsimoniously interpreted as a synapomorphy. That the chirostylids may be closer to non-galatheoids than galatheids or porcellanids is consistent with observations of other workers. Larval characters of *Chirostylus* are markedly dissimilar to larval *Galathea* (see Clark & Ng 2008), and chirostylid sperm morphology is more similar to that of hermit crabs than to other galatheoids (Tudge 1995, 1997).

Although aeglids are usually classified as galatheoids on the basis of general habitus, their affinities have been widely debated, notably with regards to paguroid affinities (Dana 1852; Martin & Abele 1988, 1986). Similarly, lomisoids have been variously treated as porcellanids, paguroids, or as independent (Pilgrim 1965; McLaughlin 1983a). The *Lomis* + *Aegla* clade recovered here under MP corroborates other recent studies based on mitochondrial gene rearrangements (Morrison et al. 2002), somatic morphology and molecular data (Ahyong & O'Meally 2004; Porter et el. 2005), and spermatozoal morphology (Tudge & Scheltinga 2002). Only very recently were aeglids formally removed to their own superfamily (McLaughlin et al. 2007).

Three subfamilies of Galatheidae are currently recognized (Baba & Williams 1998): Galatheinae, Munidopsinae, and Shinkaiinae. Representatives of the munidopsines (*Munidopsis* and *Galacantha*) and shinkaiines (*Shinkaia*) together with the galatheine, *Leiogalathea*, form a well-supported clade that is sister to the remaining galatheids/porcellanids. The position of *Leiogalathea* is unexpected, because it closely resembles other galatheines such as *Allogalathea* and *Galathea*. *Leiogalathea* thus warrants further scrutiny for morphological corroboration of molecular patterns. The close relationship between Galatheidae and Porcellanidae is widely recognized (e.g., McLaughlin et al. 2007), but the possibility that porcellanids are derived from within the galatheids is novel. The crab-like form of porcellanids, an example of carcinization within the Anomura, is derived.

However, the chief characters separating galatheids from porcellanids, namely the well-developed rostrum; deeper, more elongate cephalothorax; more muscular and more elongate abdomen; and anteriorly directed chelipeds, are plesiomorphic. Thus, derivation of Porcellanidae from within Galatheidae is morphologically plausible. Further studies with larger suites of both families are required to test the reciprocal monophyly implied by the current classification. *Munida* is not monophyletic under either BI or MP; the two exemplars are more closely related to *Cervimunida* or *Pleuroncodes*, respectively. With almost 250 known species of *Munida*, this result must be considered indicative only, though recent studies already suggest that *Munida* requires further division (e.g., Machordom & Macpherson 2004; Cabezas et al. 2008).

## 4.2 Carcinization

Borradaile (1916) first coined the term carcinization for evolution of the crab-like form, with the best known example being the derivation of king crabs (Lithodoidea: Lithodidae and Hapalogastridae) from within the asymmetrical hermit crabs. Derivation of the king crabs from within the paguroids has been widely supported by both molecular and morphological studies (e.g., Boas 1880; Bouvier 1894a–c, 1895 a, b; Cunningham et al. 1992; Richter & Scholtz 1994; McLaughlin et al. 1997; Morrison et al. 2002; Ahyong & O'Meally 2004; Tsang et al. 2008), though several recent studies dispute pagurid derivation of lithodids on the basis of apparently implausible transformation pathways (e.g., McLaughlin & Lemaitre 1997; McLaughlin et al. 2004, 2007). The 'hermit to king' hypothesis, however, is unequivocally corroborated here: Lithodidae + Hapalogastridae is nested within Paguridae. Independent carcinization events are also identified in the Porcellanidae and Lomisidae.

## 4.3 Implications for anomuran classification

The phylogenetic patterns recovered here are not compatible with recent anomuran classifications, either the four-superfamily system of Martin & Davis (2001) or the seven-superfamily system of McLaughlin et al. (2007). At the family level, few major problems are identified: polyphyly of Pylochelidae, paraphyly of Galatheidae with respect to Porcellanidae, and possible inclusion of Kiwaidae within Chirostylidae. The most significant and far-reaching challenges are in the likely polyphyly of the two largest superfamilies, Paguroidea and Galatheoidea. Of the superfamilies collectively recognized by Martin & Davis (2001) and McLaughlin et al. (2007), only Aegloidea, Hippoidea, and Lomisoidea remain uncontroversial from a nomenclatural perspective. Kiwaoidea and Lithodoidea are not compatible with present results. Lithodidae and Hapalogastridae are nested within Paguridae, rendering recognition of Lithodoidea problematical. *Kiwa* may be nested within Chirostylidae, which would preclude separate familial or superfamilial status for the former. Moreover, Chirostylidae itself is excluded from Galatheoidea and would warrant its own superfamily. Similarly, among the asymmetrical hermit crabs, Parapaguridae appears to be independent of the other major paguroid clades, also warranting superfamilial status. For the remaining major hermit crab clades, recognition of either one or two superfamilies is more ambiguous. The pagurid and diogenid + coenobitid clades are independent under MP, but nodal support for their separation is equivocal, so these potentially could constitute a monophylum. The current classification will require either abandonment of superfamilies or recognition of several more.

## 5 CONCLUSIONS

The internal phylogenetic relationships of the Anomura remain contentious, and consensus is still far off. The diversity of phylogenetic hypotheses proposed, even in the last two decades, highlights the complexity of the issue. The present analyses, based on the largest molecular dataset for the Anomura analyzed to date, offer new perspectives on the issue. Results corroborate several previous studies in the basal position of Hippoidea (Martin & Abele 1986; Pérez-Losada et al. 2002; Ahyong & O'Meally 2004; Macpherson et al. 2005) but point to significant polyphyly in the two largest superfamilies, Galatheoidea and Paguroidea. Whereas previous cladistic analyses have identified anomalous positions for one or other galatheoid or paguroid taxa, all have recovered major clades that substantially correspond to Paguroidea, Galatheoidea, and Hippoidea (e.g., Martin & Abele 1986; Morrison et al. 2002; Ahyong & O'Meally 2004; McLaughlin et al. 2007). Thus, present results are a significant departure from predecessors in suggesting that the asymmetrical hermit crabs have a strongly polyphyletic origin. Similarly, the chirostylids are derived independently of galatheids/porcellanids. Although it would be premature to change the classification at this stage, the phylogenetic patterns recovered suggest significant changes will be required.

Some patterns recovered herein, while unexpected, are not counterintuitive — namely, polyphyly of Galatheoidea. Reconsideration of the unifying characters of Galatheoidea shows that the group lacks synapomorphies. To date, Galatheoidea has been recognized on the basis of plesiomorphies, so it is hardly surprising that it collapses under phylogenetic analysis. Likewise, at a lower taxonomic level, Galatheidae may be paraphyletic on the basis of an internally nested Porcellanidae, and a similar situation may be obtained for Chirostylidae with respect to Kiwaidae. The close relationship between aegloids, lomisoids, and parapagurids to chirostylids and kiwaids recovered here has precedence to various degrees in other studies and is an obvious focus of further research.

Other patterns recovered herein are both unexpected and counterintuitive — namely, polyphyly of the asymmetrical hermit crabs. Morphological synapomorphies unifying the Paguroidea are often complex and related to the almost universal habit of occupying gastropod shells. The apparent polyphyly of the paguroids suggests independent derivations of asymmetry in three separate clades: Paguridae, Coenobitidae + Diogenidae, and Parapaguridae. Such a result, however, should not be automatically dismissed. If carcinization can have multiple, independent origins (e.g., in Lithodoidea, Porcellanidae, Lomisoidea) (Morrison et al. 2002), then why not adaptation to dextral shell habitation? Much of the recent debate in anomuran phylogenetics is over the reality of carcinization and revolves around the position of lithodids with respect to the hermit crabs. However, present results pose even more fundamental questions about whether the Paguroidea is even a natural group.

Clearly, further research is required using more taxa and more data; available data sources, both morphological and molecular, are certainly far from exhausted. To this end, further investigations are currently underway, combined with morphological data and an expanded taxon set focused on the galatheoids. Nevertheless, the phylogenetic patterns suggested here ought to stimulate closer scrutiny of morphology, especially for unrecognized synapomorphies that could corroborate (or further challenge) unexpected molecular results. Ultimately, morphological plausibility is the criterion by which molecular phylogenetic hypotheses are evaluated, though that is not to say that morphology is yet fully understood.

## ACKNOWLEDGEMENTS

We are most grateful to Enrique Macpherson, Régis Cleva, Chia-Wei Lin, and Tin-Yam Chan for assistance with tissue samples of selected species. Special thanks go to K.H. Chu and L.M. Tsang for sharing genomic DNA of *Kiwa hirsuta* and *Shinkaia crosnieri* along with unpublished 16S and 18S sequences of the latter. We acknowledge support through the NIWA program "Seamounts: their importance to fisheries and marine ecosystems," from the New Zealand Foundation for Research, Science and Technology (BBBI091 and BBDC083) and the NIWA capability fund.

## REFERENCES

Ahyong, S.T, Lai, J.C.Y., Sharkey, D., Colgan, D.J. & Ng, P.K.L. 2007. Phylogenetics of the brachyuran crabs (Crustacea: Decapoda) part 1: the status of Podotremata based on small subunit nuclear ribosomal RNA. *Mol. Phylogenet. Evol.* 45: 576–586.

Ahyong, S.T. & O'Meally, D. 2004. Phylogeny of the Decapoda Reptantia: resolution using three molecular loci and morphology. *Raff. Bull. Zool.* 52: 673–693.

Baba, K. & Williams, A.B. 1998. New Galatheoidea (Crustacea, Decapoda, Anomura) from hydrothermal systems in the West Pacific Ocean: Bismark Archipelago and Okinawa Trough. *Zoosystema* 20: 143–156.

Balss, H. 1957. Decapoda. VIII. Systematik. In: Bronns, H.G. (ed.), *Klassen und Ordnungen des Tierreichs.* Funfter Band, 1. Abteilung 7, Buch 12: 1505–1672. Leipzig and Heidelberg: Winter.

Boas, J.E.V. 1880. II. Wissenschaftliche Mittheilungen. I. *Lithodes und Pagurus. Zool. Anz.* 3: 349–352.

Borradaile, L.A. 1907. On the classification of the decapods crustaceans. *Ann. Mag. Nat. Hist.* 19: 457–486.

Borradaile, L.A. 1916. Crustacea. Part II. *Porcellanopagurus*: an instance of carcinization. *Brit. Ant. (Terra Nova) Exped., 1910. Nat. Hist. Rep. Zool.* 3: 111–126.

Bouvier, E.-L. 1894a. Sur les caractres et l'évolution des Lomisinés, nouveau groupe de crustacés anomures. *Comptes Rend. Séances Acad. Sci.* 118: 1353–1355.

Bouvier, E.-L. 1894b. Sur la signication de *Hapalogaster* dans l'évolution des Lithodinés. *Compte Rend. Som. Séances Soc. Philomath. Paris* 18: 1–5.

Bouvier, E.-L. 1894c. Sur la transformation des Paguriens en crabes anomures de la sous-famille des Lithodinés. *Comptes Rend. Hebdomad. Séances. Acad. Sci.* 119: 350–352.

Bouvier, E.-L. 1895a. Sur les Lithodinés hapalogastriques *Hapalogaster* et *Dermaturus*. *Compte Rend. Som. Séances Soc. Philomath. Paris* 18: 56.

Bouvier, E.-L. 1895b. Recherches sur les afnités des *Lithodes* & des *Lomis* avec les Pagurides. *Ann. Sci. Nat. Zool. Pal.* (7)18: 157–213.

Burkenroad, M.D. 1963. The evolution of the Eucarida (Crustacea, Eumalacostraca) in relation to the fossil record. *Tul. Stud. Geol.* 2: 3–16.

Burkenroad, M.D. 1981. The higher taxonomy and evolution of Decapoda (Crustacea). *Trans. San Diego Soc. Nat. Hist.* 19: 251–268.

Cabezas, P., Macpherson, E. & Machordom, A. 2008. A new genus of squat lobster (Decapoda: Anomura: Galatheidae) from the South West Pacific and Indian Ocean inferred from morphological and molecular evidence. *J. Crust. Biol.* 28: 68–75.

Clark, P.F. & Ng, P.K.L. 2008. The lecithotrophic zoea of *Chirostylus ortmanni* Miyake & Baba, 1968 (Crustacea: Anomura: Galatheoidea: Chirostylidae) described from laboratory hatched material. *Raff. Bull. Zool.* 56: 85–94.

Crandall, K.A., Harris D.J. & Fetzner Jr., J.W. 2000. The monophyletic origin of freshwater crayfish estimated from nuclear and mitochondrial DNA sequences. *Proc. Roy. Soc. London*, B, 267: 1679–1686.

Cunningham, C.W., Blackstone, N.W. & Buss, L.W. 1992. Evolution of king crabs from hermit crab ancestors. *Nature* 355: 539–542.

Dana, J.D. 1852. *Crustacea, Part 1. United States Exploring Expedition during the years 1838, 1839, 1840, 1841, 1842, under the command of Charles Wilkes, U.S.N.* 13: 1–685. Phildelphia: C. Sherman.

Dixon, C.J., Ahyong, S.T. & Schram, F.R. 2003. A new hypothesis of decapod phylogeny. *Crustaceana* 76: 935–975.

Farris, J.S., Albert, V.A., Källersjö, M., Lipscomb, D. & Kluge, A.G. 1996. Parsimony jackknifing outperforms neighbour-joining. *Cladistics* 12: 99–124.

Glaessner, M.F. 1969. Decapoda. In: Moore, R.C. (ed.), *Arthropoda 4. Part R, vol. 2. Treatise on Invertebrate Paleontology*: 399–533. Lawrence: Geological Society of America and University of Kansas Press.

Harvey, A.W. 1998. Genes for asymmetry overruled. *Nature* 392: 345–346.

Henderson, J.R. 1888. Report on the Anomura collected by H.M.S. Challenger during the years 1873–76. *Report on the Scientific Results of the Voyage of H.M.S. Challenger during the years 1873–76. Zoology* 27: 1–221, 221 pls.

Huelsenbeck, J.P. & Ronquist, F.R. 2001. MRBAYES: Bayesian inference of phylogeny. *Bioinformatics* 17: 754–755.

Machordom, A. & Macpherson, E. 2004. Rapid radiation and cryptic speciation in galatheid crabs of the genus *Munida* and related genera in the South West Pacific: molecular and morphological evidence. *Mol. Phylogenet. Evol.* 33: 259–279.

Macpherson, E., Jones, W.J. & Segonzac, M. 2005. A new squat lobster family of Galatheoidea (Crustacea, Decapoda, Anomura) from the hydrothermal vents of the Pacific-Antarctic Ridge. *Zoosystema* 27: 709–723.

Martin, J.W. & Abele, L.G. 1986. Phylogenetic relationships of the genus *Aegla* (Decapoda: Anomura: Aeglidae), with comments on anomuran phylogeny. *J. Crust. Biol.* 6: 576–616.

Martin, J.W. & Abele, L.G. 1988. External morphology of the genus *Aegla* (Crustacea: Anomura: Aeglidae). *Smith. Contr. Zool.* 453: 1–46.

Martin, J.W. & Davis, G.E. 2001. An updated classification of the Recent Crustacea. *Natural History Museum of Los Angeles County Science Series* 39: 1–124.

MacDonald, J.D., Pike, R.B. & Williamson, D.I. 1957. Larvae of the British species of *Diogenes*, *Pagurus*, *Anapagurus* and *Lithodes* (Crustacea, Decapoda). *Proc. Zool. Soc. London* 128: 209–257.

McLaughlin, P.A. 1983a. A review of the phylogenetic position of the Lomidae (Crustacea: Decapoda: Anomala). *J. Crust. Biol.* 3: 431–437.

McLaughlin, P.A. 1983b. Hermit crabs — are they really polyphyletic? *J. Crust. Biol.* 3: 608–621.

McLaughlin, P.A. & Holthuis, L.B., 1985. Anomura versu Anomala. *Crustaceana* 49: 204–209.

McLaughlin, P.A. & Lemaitre, R. 1997. Carcinization in the Anomura — fact or fiction? I. Evidence from adult morphology. *Contr. Zool.* 67: 79–123.

McLaughlin, P.A., Lemaitre, R. & Sorhannus, U. 2007. Hermit crab phylogeny: a reappraisal and its "fall out." *J. Crust. Biol.* 21: 97–115.

McLaughlin, P.A. & Lemaitre, R. & Tudge, C. 1997. Carcinization in the Anomura — fact or fiction? II. Evidence from larval, megalopal and early juvenile morphology. *Contr. Zool.* 73: 165–205.

McLaughlin, P.A., Lemaitre, R. & Tudge, C.C. 2004. Carcinization in the Anomura — fact or fiction? II. Evidence from larval, megalopal and early juvenile morphology. *Contr. Zool.* 73: 165–205.

Medlin, L., Elwood, H.J., Stickel, S. & Sogin, M.L. 1998. The characterization of enzymatically amplified eukaryotic 16S-like rRNA-coding regions. *Gene* 71: 491–499.

Morrison, C.L., Harvey, A.W., Lavery, S., Tieu, K., Huang, Y. & Cunningham, C.W. 2002. Mitochondrial gene rearrangements confirm the parallel evolution of the crab-like form. *Proc. Roy. Soc. London*, B, 269: 345–350.

Pérez-Losada, M., Jara, C.G., Bond-Buckup, G., Porter, M.L. & Crandall, K.A. 2002. Phylogenetic position of the freshwater anomuran family Aeglidae. *J. Crust. Biol.* 22: 670–676.

Poore, G.C.B. 1994. A phylogeny of the families of Thalassinidea (Crustacea: Decapoda) with keys to the families and genera. *Mem. Mus. Vic.* 54: 79–120.

Porter, M.L., Pérez-Losada, M. & Crandall, K.A. 2005. Model-based multi-locus estimation of decapod phylogeny and divergence times. *Mol. Phylogenet. Evol.* 37: 355–369.

Pilgrim, R.L.C. 1965. Some features in the morphology of *Lomis hirta* (Lamarck) (Crustacea: Decapoda) with a discussion of its systematic position and phylogeny. *Aust. J. Zool.* 13: 545–557.

Prendini, L., Crowe, T.M. & Wheeler, W.C. 2003. Systematics and biogeography of the family Scorpionidae (Chelicerata: Scorpiones), with a discussion on phylogenetic methods. *Invert. Syst.* 17: 185–259.

Przibam, H. 1907. Differenzierung des Abdomens enthäuster Einsiedlerkrebse. *Arch. Entw. Mech.* 23: 245–261.

Richter, S. & Scholtz, G. 1994. Morphological evidence for a hermit crab ancestry of lithodids (Crustacea, Decapoda, Anomala, Paguroidea). *Zool. Anz.* 233: 187–210.

Scholtz, G. & Richter, S. 1995. Phylogenetic systematics of the reptantian Decapoda (Crustacea, Malacostraca). *Zool. J. Linn. Soc.* 113: 289–328.

Schram, F.R. 2001. Phylogeny of decapods: moving towards a consensus. *Hydrobiologia* 449: 1–20.

Sogin, M.L. 1990. Amplification of ribosomal RNA genes for molecular evolution studies. In: Innis, M., Gelfand, D., Sninsky, J. & White, T. (eds.), *PCR Protocols: A Guide to Methods and Applications*: 307–314. Amsterdam: Elsevier Scientific.

Spears, T., Abele, L.G. & Kim, W. 1992. The monophyly of brachyuran crabs: a phylogenetic study based on 18S rRNA. *Syst. Biol.* 41: 446–461.

Swofford, D.L. 2002. *PAUP*. Phylogenetic Analysis Using Parsimony (* and Other Methods), Version 4.0b10*. Sunderland, Massachusetts: Sinauer Assoc.

Tsang, L.M., Ma, K.Y., Ahyong, S.T., Chan, T.-Y. & Chu, K.H. 2008. Phylogeny of Decapoda using two nuclear protein-coding genes: origin and evolution of the Reptantia. *Mol. Phylogenet. Evol.* 48: 359–368.

Tudge, C.C. 1995. Ultrastructure and phylogeny of the spermatozoa of the infraorders Thalassinidea and Anomura (Decapoda, Crustacea). In: Jamieson, B.G.M., Ausio, J. & Justine, J.-L. (eds.), Advances in Spermatozoal Phylogeny and Taxonomy. *Mém. Mus. Nat. Hist. Nat. Paris* 166: 251–263.

Tudge, C.C. 1997. Phylogeny of the Anomura (Decapoda, Crustacea): spermatozoa and spermatophore morphological evidence. *Contr. Zool.* 67: 125–141.

Tudge, C.C. & Scheltinga, D.M. 2002. Spermatozoal morphology of the freshwater anomuran *Aegla longirostris* Bond-Buckup & Buckup, 1994 (Crustacea: Decapoda: Aeglidae) from South America. *Proc. Biol. Soc. Wash.* 115: 118–128.

# V ADVANCES IN OUR KNOWLEDGE OF THE BRACHYURA

# Is the Brachyura Podotremata a Monophyletic Group?

GERHARD SCHOLTZ[1] & COLIN L. MCLAY[2]

[1] *Humboldt-Universität zu Berlin, Institut für Biologie/Vergleichende Zoologie, Berlin, Germany*
[2] *University of Canterbury, School of Biological Sciences, Christchurch, New Zealand*

ABSTRACT

We undertook a morphological analysis to test whether the Podotremata or primitive crabs including Dromiacea, Homoloidea, Raninoidea, and Cyclodorippoidea form a monophyletic group. We can show that the podotrematan subgroups are all monophyletic. Furthermore, our data clearly suggest that Cyclodorippoidea is the sister group to Eubrachyura, that the Raninoidea is the sister group to both, that the Homoloidea is the sister group to this clade, and that all of them are the sister group to Dromiacea ((((Eubrachyura, Cyclodorippoidea), Raninoidea), Homoloidea), Dromiacea). Hence the Podotremata is a paraphyletic assemblage. With this result we corroborate recent molecular studies.

## 1 INTRODUCTION

With almost 7000 species the Brachyura or true crabs form the largest and most diverse decapod group (Ng et al. 2008). Brachyura are found in the deep sea, at thermal vents, and in freshwater and terrestrial habitats. Based on a number of morphological and molecular analyses, there is now a growing consensus that the sister group of Brachyura is the Anomala or Anomura, with both groups together forming the Meiura (Scholtz & Richter 1995; Schram 2001; Dixon et al. 2003; Ahyong & O'Meally 2004; Miller & Austin 2006; Ahyong et al. 2007; Tsang et al. 2008). However, brachyuran internal phylogenetic relationships are far from clear, and even their monophyly has been doubted (e.g., Gordon 1963; Williamson 1974; Rice 1980; Spears et al. 1992). This relates in particular to the brachyuran taxa whose representatives do not show the characters that are considered to make a true brachyuran crab. These taxa, the Dromiacea, Homoloidea, Raninoidea, and the Cyclodorippoidea, are often either seen as primitive brachyuran crabs or their brachyuran status is doubted. For instance, H. Milne Edwards (1837) excluded Raninoidea and Dromiacea (including Homoloidea) from Brachyura, Gordon (1963) proposed the exclusion of all podotreme crabs, Ortmann (1896) excluded the Dromiacea (including Homoloidea), and Williamson (1974) and Rice (1980, 1981b, 1983) excluded the Dromiacea. Even a relatively recent molecular phylogenetic analysis suggested the exclusion of dromiaceans from the Brachyura (Spears et al. 1992). Since the seminal work on Brachyura systematics by Guinot in the 1970s, these "primitive" crabs have been unified in a taxon called Podotremata as opposed to the sternitreme crabs or Eubrachyura containing the brachyuran crabs *sensu stricto*. According to de Saint Laurent (1980), the monophyly of Eubrachyura is well supported by the apomorphic sternal position of the female gonopores in combination with a seminal receptacle connected to the oviduct, which leads to internal fertilization. The problem is that Guinot (1977, 1978, 1979a) erected the group Podotremata based on the coxal position of the gonopores. However, coxal genital openings are found in all other decapods and in most malacostracans, and this is a clearly plesiomorphic character. Since then the Podotremata has remained problematic. Several authors, using sperm characters and other morphological data, argued for a monophyletic Podotremata, although an unambiguous apomorphy for this group has not been established (Guinot 1978, 1979a; Jamieson 1994; Jamieson et al. 1995). Guinot & Tavares (2001), Tavares (2003), and Guinot & Quenette (2005) discuss the spermathecal invagination at the sternal boundary between the 7th and 8th thoracic segment as an apomorphy supporting the Podotremata. And indeed, this complex character involving two sternites is restricted to podotrematan representatives, but it suffers from a problematic polarization because nothing comparable exists in other reptant groups. However, we must note that the seminal receptacle and spermathecae may not be homologous structures, so the derivation of one from the other (see Hartnoll 1979) is difficult. Accordingly, several authors suggested a paraphyletic Podotremata (e.g., Scholtz & Richter 1995; Martin & Davis 2001; Dixon et al. 2003; Brösing et al. 2007), and an older (Spears et al. 1992) and a recent (Ahyong et al. 2007) molecular analysis support this view. In addition to the general question of podotrematan monophyly versus paraphyly, the internal relationships between the major podotrematan groups are a continuous matter of debate. For instance, some authors include Homoloidea within Dromiacea (e.g., Boas 1880; Borradaile 1907), while other authors (e.g., Guinot 1978) separate them. Števčić (1995) even synonymizes Dromiacea with Podotremata. Furthermore, Guinot (1978) erected a group Archaeobrachyura that includes Homoloidea, Cyclodorippoidea, and Raninoidea, although later she excluded the Homoloidea from the Archaeobrachyura (Guinot & Tavares 2001).

Here we test whether morphological data contribute to the question of podotrematan monophyly or paraphyly and whether the Archaeobrachyura is a valid taxon. We investigate a comprehensive number of different characters. Our analysis indicates that podotrematan Brachyura are a paraphyletic assemblage. Our results are largely congruent with those of a recent analysis based on a molecular data set (Ahyong et al. 2007).

## 2 MATERIALS AND METHODS

### 2.1 *Animals*

We examined the following brachyuran species from our personal collections: Homolodromiidae: *Dicranodromia karubar* Guinot, 1993; Dromiidae: *Moreiradromia sarraburei* (Rathbun, 1910), *Hypoconcha arcuata* Stimpson, 1858; Dynomenidae: *Dynomene pilumnoides* Alcock, 1900; Homolidae: *Dagnaudus petterdi* (Grant, 1905), *Homola barbata* (Fabricius, 1793); Latreilliidae: *Eplumula australiensis* (Henderson, 1888); Raninidae: *Lyreidus tridentatus* de Haan, 1841, *Ranina ranina* (Linnaeus, 1758); Cyclodorippidae: *Krangalangia spinosa* (Zarenkov, 1970); Cymonomidae: *Cymonomus aequilonius* Dell 1971; Cyclodorippidae: *Tymolus brucei* Tavares, 1991; Majidae: *Prismatopus filholi* (A. Milne Edwards, 1876); Dorippidae: *Medorippe lanata* (Linnaeus, 1767); Xanthidae *Xantho poressa* (Olivi, 1792); Portunidae: *Nectocarcinus antarcticus* (Hombron & Jacquinot, 1846), *Ovalipes catharus* (White in White & Doubleday, 1843); Varunidae *Eriocheir sinensis* H. Milne Edwards, 1853, *Hemigrapsus crenulatus* (H. Milne Edwards, 1837). For outgroup comparison we used the following species: Anomala: *Petrolisthes elongatus* (H. Milne Edwards, 1837), *Galathea strigosa* (Linnaeus, 1767); Astacida: *Paranephrops zealandicus* (White, 1847), *Procambarus clarkii* (Girard, 1852). In addition, we considered data from the literature.

### 2.2 *Microscopy*

The morphological investigations were done with the aid of a dissecting microscope and a scanning electron microscope (SEM) (Leica). Some dissected specimens were boiled with 5% KOH to remove the soft parts. Alizarin-red stain was used to highlight calcified parts of the skeleton and appendages (for detail see Brösing et al. 2002). The specimens prepared for SEM were transferred to an ethanol series up to pure ethanol for dehydration and then dried at critical point, mounted on stubs, and sputter-coated with gold.

## 2.3 Analysis

In this analysis we reconstruct the phylogenetic tree "by hand" and brain following a Hennigian approach (Hennig 1966). In the first step we provide evidence that the brachyuran subgroups under consideration are monophyletic, and in a second step we reconstruct their phylogenetic relationships following a top-down approach starting with the Eubrachyura and looking for its sister taxon, then looking for the sister taxon to this unified clade, etc. (see below).

# 3 RESULTS

## 3.1 The monophyly of the brachyuran subtaxa

### 3.1.1 Dromiacea

The Dromiacea *sensu* Guinot (1978, 1979a) consist of the Homolodromiidae, the Dynomenidae, and the Dromiidae (see McLay 1999). The Homoloidea, which in older concepts were part of the Dromiacea, are excluded. The clade Dromiacea *sensu* Guinot is well supported by a number of apomorphies (character set 1):

The renal opening in the coxal segment of the 2nd antennae is surrounded by upper and lower projections in a beak-like manner (Fig. 1). A corresponding structure is not found in any other decapod taxon (see below). We find this character in all investigated species of the Homolodromiidae, Dynomenidae, and Dromiidae, including *Hypoconcha*. In the relevant literature we see no exception.

The fingers of the chelae are hollow and serrated, and the serrate tips of the fingers engage (Fig. 2). Plesiomorphically, the fingers are compact and show pointed tips. As with the previous character, this is seen in all investigated dromiacean species and also found in the literature (McLay 1993, 1999; Guinot 1995; Guinot & Tavares 2003).

The 2nd pleopod of the male is flagellate with a needle-like tip and a multi-segmented basal part. The plesiomorphic condition is a stout 2nd pleopod (see McLay 1993, 1999; Guinot 1995).

In addition, the shape of the flattened acrosome of the sperm (Jamieson 1994) and the set of foregut ossicles (Brösing et al. 2002, 2007) corroborate dromiacean monophyly.

### 3.1.2 Homoloidea

The Homoloidea include the Homolidae, the Latreilliidae, and the Poupiniidae (Guinot & Richer de Forges 1995). All these subgroups share the following apomorphies (character set 2):

The telson projects between the bases of the third maxillipeds (Fig. 3). In most other cases, the telson ends posterior to the maxilliped segments. Only some leucosiids are slightly similar in this respect, but a detailed analysis reveals the fundamental difference (see Guinot 1979a). The representatives of Latreilliidae and Homolidae studied by us all showed the same pattern. For Poupiniidae, we find a corresponding character state in the publication of Guinot (1991).

The retention of the pleon is achieved by two devices, namely paired projections on the 3rd thoracic sternite and little protrusions of the basal parts of the 3rd maxillipeds. All other brachyurans show a different pattern of pleon retention structures (see below and Guinot & Bouchard 1998).

These are not many apomorphies, but as far as we know there are no exceptions found within the Homoloidea. Jamieson (1994) and Jamieson et al. (1995) mentioned several sperm characters such as numerous radial extensions of the operculum and a spiked wheel form of the anterior expansion of the perforatorium supporting the Homoloidea clade. Furthermore, larval features are interpreted as homolid apomorphies (Rice 1980).

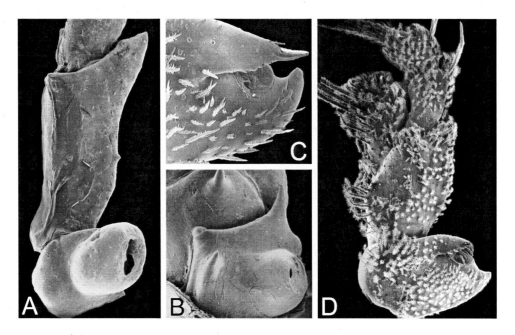

Figure 1. Renal openings, I. The renal opening of a homolid (*Dagnaudus petterdi*) (A) and an astacid (*Paranephrops zealandicus*) (B) showing the plesiomorphic condition of a tube positioned on the proximal part of the 2nd antenna. The beak-like structure around the renal opening is exemplified in a dromiid (*Moreiradromia sarraburei*) (C) and a dynomenid (D) (*Dynomene pilumnoides*) apomorphic for Dromiacea.

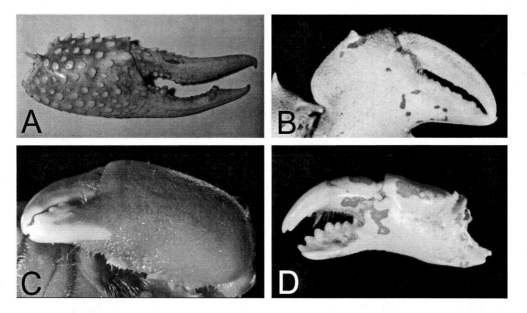

Figure 2. Chelae. (A) Chela of an astacid (*Procambarus clarkii*) and (B) of the raninoid crab *Lyreidus tridentatus* showing the pointed tips of the dactylus and propodus. (C, D): The chelae of a dynomenid (*Dynomene pilumnoides*) (C) and a homolodromiid (*Dicranodromia karubar*) (D) with hollow fingers and serrated margins that show interlocking teeth.

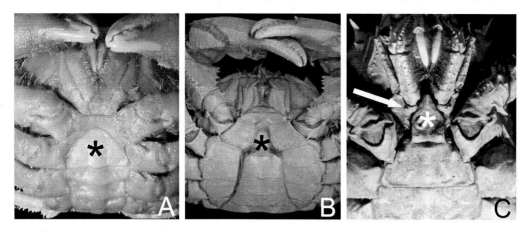

**Figure 3.** Telson position. (A) The telson of a dynomenid (*Dynomene pilumnoides*), (B) of a eubrachyuran (*Eriocheir sinensis*), and (C) of the homoloid species (*Dagnaudus petterdi*). Telsons marked with (*). The telson in *Dagnaudus* reaches apomorphically between the basal parts of the 3rd maxilliped, which possesses a coxal process as a pleon retention device (arrow).

### 3.1.3 *Raninoidea*

The Raninoidea is a very uniform and easy to identify group of crabs. Accordingly, there are a number of clear apomorphies supporting this clade (character set 3):

The exopod of the 1st maxilliped is flattened, lacks a flagellum, and is involved in the exhalant water current channel (see also Bourne 1922) (Fig. 4). The plesiomorphic state is a more or less round exopod equipped with a flagellum.

The paired spermathecal openings lead into an unpaired median atrium. This is associated with the 7th thoracic sternite (see also Gordon 1963; Guinot 1993). In the other podotrematan crabs the spermathecal openings are separate and positioned between the 7th and 8th thoracic sternites.

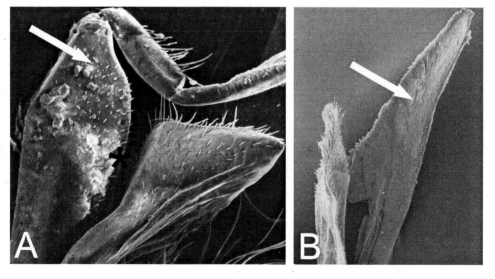

**Figure 4.** Exopod of the 1st maxilliped. (A) The flat and flagellate exopod of the 1st maxilliped (arrow) of a eubrachyuran (*Prismatopus filholi*) representing the plesiomorphic condition. (B) The apomorphic aflagellate and widened exopod (arrow) in *Lyreidus tridentatus*, a raninoid species.

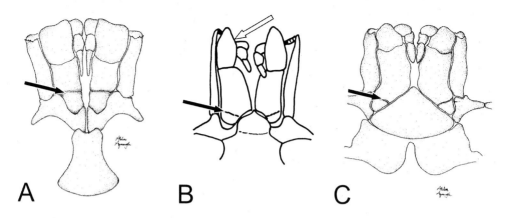

Figure 5. The 3rd maxillipeds of (A) a dromiacean (*Moreiradromia sarraburei*), (B) a cyclodorippoid (*Cymonomus aequilonius*), and (C) a eubrachyuran (*Nectocarcinus antarcticus*). The black arrows point to the basis-ischium boundary showing that there is a characteristic pattern apomorphically shared by cyclodorippoids and eubrachyurans. The white arrow in (B) points to the apomorphically posteriorly situated endopodal palp of the cyclodorippoid 3rd maxilliped.

The sternum is narrowed posterior to the 4th or 5th sternites (see Bourne 1922; Guinot 1993) (see Fig. 9). Plesiomorphically, the posterior part of the sternum is much wider.

Additional data from sperm morphology and the foregut ossicles also support a monophyletic Raninoidea (Jamieson 1994; Brösing et al. 2007).

### 3.1.4 *Cyclodorippoidea*

The Cyclodorippoidea are subdivided into the Cyclodorippidae, Cymonomidae, and Phyllotymolinidae (Tavares 1998). We found relatively few putative apomorphies, and thus the status of the group is debatable (character set 4):

The palp of the 3rd maxilliped is in a very sub-distal position (Fig. 5). The plesiomorphic condition is a more distal position. This character can be seen in *Tymolus*, *Cymonomus*, and *Krangalangia* (see also Tavares 1993).

The first three pleon segments are visible dorsally when the crab is in a horizontal position. In other crabs either no segments or at most two segments are seen in the dorsal aspect.

The tip of the telson reaches only to the segment of the 3rd pereopods. In most other crabs it extends more anteriorly, with the notable exception of some raninoids (see Fig. 3).

Further morphological evidence for a Cyclodorippoidea clade comes from sperm data (Jamieson et al. 1995).

### 3.1.5 *Eubrachyura*

The Eubrachyura *sensu* de Saint Laurent (1980) or sternitreme crabs (Balss 1940; Gordon 1963; Guinot 1978, 1979a) are composed of the Heterotremata and Thoracotremata (Guinot 1978). It was not the task of the present study to investigate the internal relationships of the Eubrachyura and to test the monophyly of Heterotremata and Thoracotremata (Guinot 1978). Here we discuss only the putative apomorphies of this taxon (character set 5):

The position of female gonopores is on the 6th thoracic sternite. The plesiomorphic condition is a coxal position of female gonopores. This is without exception the case in the specimens studied by us.

The seminal receptacle is part of the oviduct. Plesiomorphically, all sperm receptacles (if present) in other decapods, including podotrematan crabs, are not connected to oviducts, but are instead part of the external thoracic surface.

The fertilization is internal. In all other reptants there is external fertilization.

The epistome encircles the base of the 2nd antenna. This can even lead to the complete fusion and fixation of the base of the 2nd antenna in some groups (e.g., majids and parthenopids). Plesiomorphically, the base of the 2nd antenna is free.

Subsequent to Guinot's papers, the validity of this group has rarely been doubted. Only Brösing et al. (2007) found some evidence in foregut ossicle patterns for the resurrection of a taxon Oxystomata, which would include the raninoids, cyclodorippoids, and some basal heterotreme groups.

## 3.2 The phylogenetic relationships among brachyuran subtaxa

Below we reconstruct, in stepwise fashion, the phylogenetic relationships of Brachyura, starting with the sister group to Eubrachyura.

### 3.2.1 Synapomorphies of Eubrachyura and Cyclodorippoidea (character set 6)

The 3rd thoracic sternite is wide, separating the basis and ischium of the 3rd maxilliped in a characteristic manner (Fig. 5). The plesiomorphic state is a narrow sternite, with the basis and the ischium of the 3rd maxilliped lying in an adjacent position. This character is found in all Eubrachyura without exception and in the cyclodorippoidean species investigated by us.

The coxal segment of the 2nd antenna is scale-like and conceals the renal opening (Fig. 6). The epistome forms a counterpart. This pattern is not found in any other brachyuran or other decapod group. The beak-like structure of Dromiacea is exclusively formed by the coxa, and in other groups there is a simple tube-like projection. The pattern is in detail slightly different in some Eubrachyura. For instance, in Majidae the coxa is completely fused to the epistome and is thus immobile.

The epipodite of the 1st maxilliped is elongated and strengthened with a calcified rod (dorsal gill cleaner and flabellum) (Fig. 7). The epipod is triangular and relatively short and lacks the calcified rod in the other Brachyura. This character seems to occur in all eubrachyuran species studied by

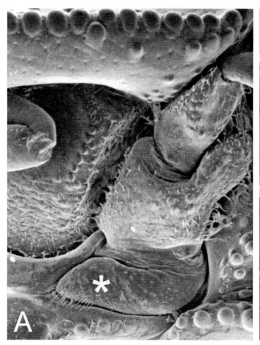 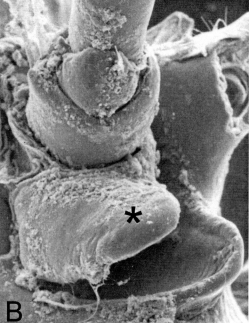

**Figure 6.** Renal openings, II. The scale-like cover (*) of the renal opening in the eubrachyuran *Hemigrapsus crenulatus* (A) and in the cyclodorippoid *Krangalangia spinosa* (B). Compare to Figure 1.

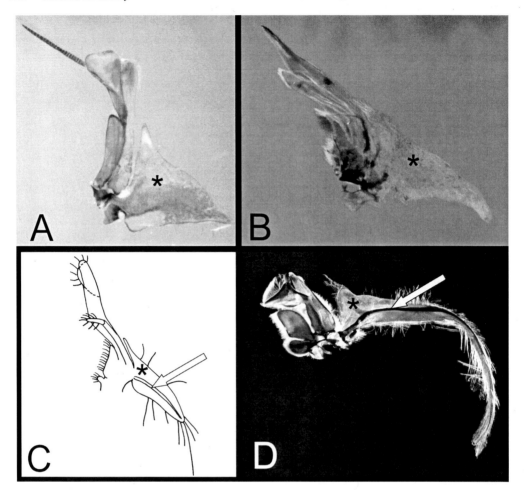

**Figure 7.** The epipods of the 1st maxillipeds. The 1st maxillipeds of (A) the dromiacean *Dynomene pilumnoides*, (B) the raninoid *Lyreidus tridentatus*, (C) the cyclodorippoid *Tymolus brucei*, and (D) the eubrachyuran *Ovalipes catharus*. The epipod (*) forms a triangular lobe that is elongated and supported by a calcified rod (arrows) in cyclodorippoids and eubrachyurans. At least in the latter two clades, the epipod serves as a gill cleaning brush (flabellum).

us and described in the literature. However, the database is not very large, and further studies are necessary.

A sterno-pleonic cavity is present (see also Guinot & Bouchard 1998) (see Fig. 9D). Plesiomorphically, there is a more or less flat sternum that lacks a corresponding cavity. Again we found no exception, only different degrees of the sharpness of the boundaries of the cavities (see Tavares 1993).

The cladistic analysis of brachyuran relationships based on ossicle patterns of the foregut by Brösing et al. (2007) does not resolve a eubrachyuran–cyclodorippoidean sister group relationship, but a certain affinity of these two taxa plus the Raninoidea, to the exclusion of the Dromiacea and Homoloidea, is also shown.

3.2.2 *Synapomorphies of Eubrachyura-Cyclodorippoidea and Raninoidea (character set 7)*
The palp of the 3rd maxilliped is inserted and articulates in the plane of the operculum, i.e., it moves in a medial-lateral direction (Fig. 8). In the plesiomorphic condition the palp moves dorso-ventrally, as is seen in all outgroup representatives.

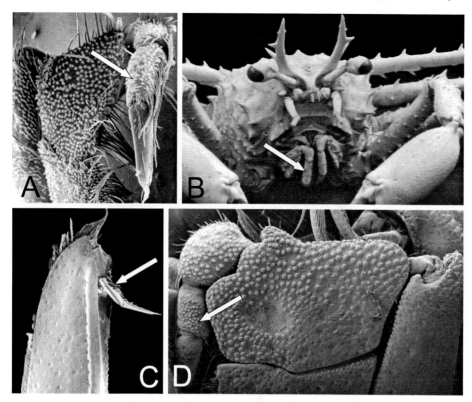

**Figure 8.** The orientation of the palps (arrows) of the 3rd maxillipeds in (A) the dromiacean *Dynomene pilumnoides*, (B) the homoloid *Dagnaudus petterdi*, (C) the raninoid *Lyreidus tridentatus*, and (D) the eubrachyuran *Xantho poressa*. In C and D the palps lie in one plane with the rest of the maxilliped, whereas in (A) and (B) they are situated at an angle that implies a different plane of movement. This more pediform appearance is the plesiomorphic condition.

The *crista dentata* on the inner margin of the basis-ischium is a plesiomorphic reptant character that is present in the homolodromiids, dromiids, dynomenids, and homolids (except latreilliids), but it has been lost in the ancestor of the cyclodorippids, cymonomids, phyllotymolinids, and raninids, as well as in the Eubrachyura (and independently in latreilliids).

The 3rd maxilliped is truly operculiform. This means that all elements lie in one plane tightly covering the buccal field. The plesiomorphic condition is a pediform third maxilliped. Compared to the condition in crayfish, the 3rd maxilliped of all crabs, including homolodromiids and homoloideans, is slightly flattened (see Scholtz & Richter 1995), and in dromiids and dynomenids it is flattened even more so, resulting in a convergent operculum-like structure. But this is not the same as forming a completely flat and closed field. The condition found in the anomalan porcelain crab *Petrolisthes* and in some thalassinids is only superficially similar, as indicated by the position of the *crista dentata* (see Balss 1940; Scholtz & Richter 1995).

All elements of the sternum form a flat plane, including the episternites (Fig. 9). The plesiomorphic state is that the episternites lie in a dorsal position and the pereopod coxae are withdrawn dorsally.

The coxae of the pereopods are narrow and triangular in ventral view, lacking an anterior lobe (Fig. 9). Homoloidea and Dromiacea as well as the outgroup representatives have a differently shaped coxa.

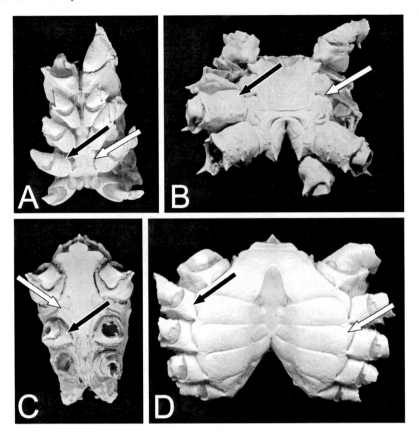

Figure 9. Sternal elements (white arrows) and coxae (black arrows) of (A) the crayfish *Paranephrops zealandicus*, (B) the homoloid *Dagnaudus petterdi*, (C) the raninoid *Lyreidus tridentatus*, and (D) the eubrachyuran *Hemigrapsus crenulatus*. The white arrows point to the lateral elements of the sternal complex, which plesiomorphically are situated in a different level compared to the sternites bearing the sterno-coxal joints (A and B). Apomorphically, all elements lie in the same plane. The coxae are plesiomorphically relatively wide. In the apomorphic condition they are narrow and triangular in ventral view and are pointed to the sterno-coxal joints (C and D).

A vertical notch is formed in the epimeral walls of the P1 and P2 segments. A corresponding structure is absent in all other investigated taxa.

An anterior tooth forms a clip for attachment of the carapace to the epimeral wall. A corresponding structure is absent in all other investigated taxa.

The facets of the compound eyes are hexagonal (Fig. 10). This character is found in the Eubrachyura genera *Cancer*, *Ovalipes*, *Nectocarcinus*, and *Hemigrapsus* and appears to be a general feature of eubrachyuran crabs indicating apposition and parabolic superposition eye types (see also Fincham 1980; Nilsson 1983, 1988; Gaten 1998; Richter 2002), the Cyclodorippoidea *Krangalanga* and *Tymolus*, and in the Raninoidea *Lyreidus* and *Ranina* (in contrast to the findings of Gaten 1998, but see Fincham 1980). The cyclodorippid *Cymonomus* has reduced eyes. All representatives of Homoloidea and Dromiacea have square facets, which occur in reflecting superposition eyes. This is apparently the plesiomorphic condition for reptant Decapoda since it occurs in crayfish and lobsters and plesiomorphically in Anomala as is seen in *Petrolisthes* and *Galathea* studied by us (see Fincham 1980; Gaten 1998; Richter 2002; but see also Porter & Cronin this volume).

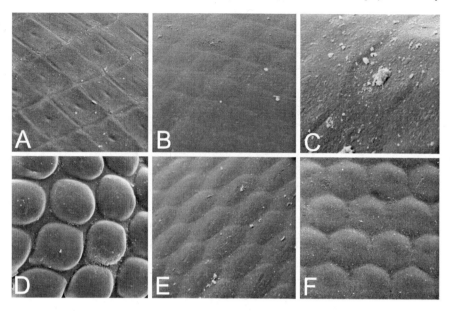

**Figure 10.** Facets of compound eyes. The crayfish *Paranephrops zealandicus* (A) and the dromiacean *Dynomene pilumnoides* (B) show squared facets, plesiomorphic for reptants, whereas the raninoid *Lyreidus tridentatus* (C), the cyclodorippoid *Krangalangia spinosa* (D), and the eubrachyurans *Nectocarcinus antarcticus* (E) and *Hemigrapsus crenulatus* (F) possess apomorphic round/hexangular facets.

### 3.2.3 *Synapomorphies of Eubrachyura-Cyclodorippoidea-Raninoidea and Homoloidea (character set 8)*

The arthrophragmal skeleton of the last thoracic segment is elongated, completely fused in the midline, and forming two anterior wings, i.e. "*sella turcica*" *sensu stricto* (Fig. 11). In the brachyuran literature the term "*sella turcica*" is used in many ways. Some authors consider a "*sella turcica*" as an apomorphy of all Brachyura (e.g., Jamieson et al. 1995; Števčić 1995). In contrast to this, Secretan (1998) restricts the word "*sella turcica*" to the situation found in Eubrachyura. We see no fundamental difference between the condition of homoloids, raninoids, and eubrachyurans. In contrast to this, we recognize a distinct difference between the condition found in Dromiacea and in the other brachyuran crabs. This relates to the fact that the fusion of the arthrophragm in dromiaceans is incomplete, leaving a hole in the center (see below). This hole is plesiomorphic because, in the outgroups, the corresponding endoskeletal parts are not medially fused at all (Fig. 11). In several crab lineages the "*sella turcica*" is reduced.

The pleonal retention mechanism involves a pair of cavities (ball-and-socket principle, "*bouton-pression*") at the posterior margin of the 6th pleon segment (Fig. 12). No uropods are involved. In raninoids this character is present only in the genus *Lyreidus* (Guinot & Bouchard 1998; our study). We consider the presence of this mechanism as plesiomorphic within the Raninoidea, and the absence (loss) is correlated to a more posterior position of the tip of the telson. This seems also the case in Cyclodorippoidea, which lack the ball-and-socket principle. Guinot & Bouchard (1998) discuss the origin of the cavities in the 6th pleon segment from uropods, but this needs confirmation by developmental data.

Uropod vestiges are completely absent. Dromiacea possess small articulated plates at the posterior margin of the 6th pleomere (Guinot & Bouchard 1998; McLay 1999). These are generally interpreted as vestigial uropods. No corresponding structures exist in Homoloidea, Cyclodorippoidea, and Eubrachyura. Hence, the existence of uropods (also vestigial) is the plesiomorphic condition.

The gills are of the phyllobranchiate type (Fig. 13). The plesiomorphic condition is trichobranchiate gills, as seen in crayfish, lobsters, and Anomala/Anomura (Balss 1940). (*Petrolisthes*

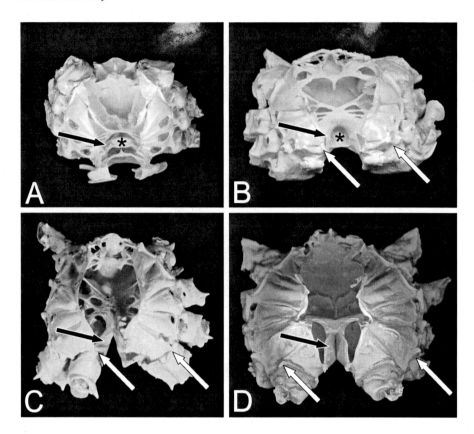

**Figure 11.** The endoskeleton. (A) The anomalan *Petrolisthes elongatus*. (B) The dromiacean *Dynomene pilumnoides*. (C) The homoloidean *Dagnaudus petterdi*. (D) The eubrachyuran *Ovalipes catharus*. The black arrows point to the arthrophragm of the last thoracic segment. In (A) they form small dorsally projecting lobes. In (B) to (D) they project anteriorly and fuse with more anterior endosternal elements. The asterisk (*) marks the open area between the two arthrophragm lobes. This hole is still present in the Dromiacea (B), but closed in the Homoloidea (C) and in all other Brachyura. The white arrows mark the little process at the epimeral walls of the 4th and 5th pereopodal segments that form a clip-on mechanism with the carapace margin.

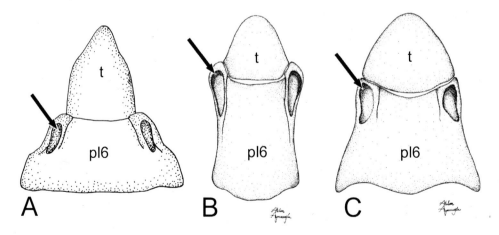

**Figure 12.** Pleon retention structures. The 6th pleomere is equipped with sockets at the posterior margin in representatives of homoloids (*Dagnaudus petterdi*) (A), raninoids (*Lyreidus tridentatus*) (B), and eubrachyurans (*Medorippe lanata*) (C).

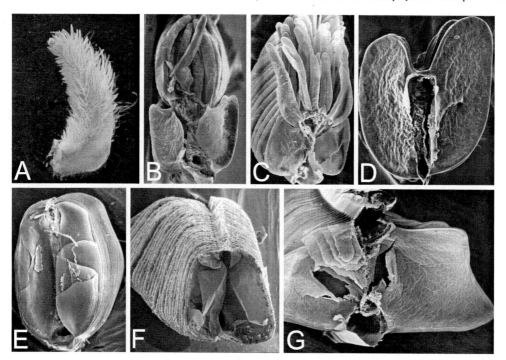

**Figure 13.** Gill structures. The plesiomorphic trichobranchiate gills of a freshwater crayfish (A) and of two species of dromiaceans, a homolodromiid (*Dicranodromia karubar*) (B) and a dynomenid (*Dynomene pilumnoides*) (C), the latter with a kind of intermediate gill type between trichobranchiate and phyllobranchiate gills (cross-section). (D) The heart-shaped special type of phyllobranchiate gills that evolved within Dromiacea (*Hypoconcha arcuata*). (E–G): Phyllobranchiate gills of the homoloid *Dagnaudus petterdi* (E), the raninoid *Lyreidus tridentatus* (F), and the eubrachyuran *Hemigrapsus crenulatus* (G).

and *Galathea* are examples of convergent evolution towards phyllobranchiate gills in anomalans). Interestingly enough, dromiaceans show patterns of transition between trichobranchiate and phyllobranchiate gills (see Bouvier 1896) (Figs. 13B–D). The latter occur, in particular, in the Dromiidae. These are differently shaped from the phyllobranchiate gills of the remainder of the crabs (Homoloidea, Cyclodorippoidea, Eubrachyura) (Figs. 13E–G) and are a clear case of convergence.

### 3.2.4 Synapomorphies of Eubrachyura-Cyclodorippoidea-Raninoidea-Homoloidea and Dromiacea = apomorphies of Brachyura (character set 9)

The endopod of the 1st maxilliped is characteristically shaped with a rectangular bend to form the bottom of a tunnel for the breathing current (Fig. 14). The endopods of the 1st maxilliped in other reptants are flat.

The carapace is locked posteriorly by projections of the epimeral walls of the segments of pereopods 4 and 5 (Fig. 11). Corresponding structures were not found in outgroup species, not even in the very crab-like *Petrolisthes* (Fig. 11A).

The arthrophragms of the last thoracic segment are elongated, incompletely fused medially, and forming two anterior wings (primitive "*sella turcica*" with hole) (see Fig. 9). The outgroups show short and separated arthrophragms of the last thoracic segment.

There are a number of other morphological characters indicating the monophyly of the Brachyura (see Scholtz & Richter 1995; Jamieson et al. 1995; Števčić 1995; Schram 2001; Dixon et al. 2003; Brösing et al. 2007).

Fig. 15 presents an overview of the phylogenetic relationships of Brachyura resulting from our morphological analysis. The numbers refer to the character sets mentioned in the text.

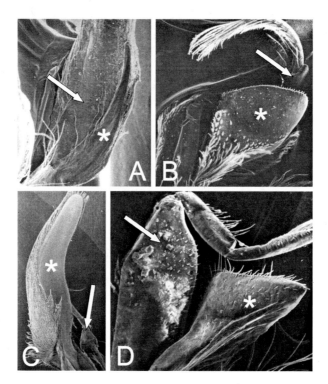

**Figure 14.** The endopods of the 1st maxillipeds (*) of the crayfish *Paranephrops zealandicus* (A), the dromiacean *Dynomene pilumnoides* (B), and the eubrachyurans *Medorippe lanata* (C) and *Prismatopus filholi* (D). In all brachyuran crabs the endopod shows a characteristic bend, which is absent in the flat crayfish endopod. The arrows mark the exopods.

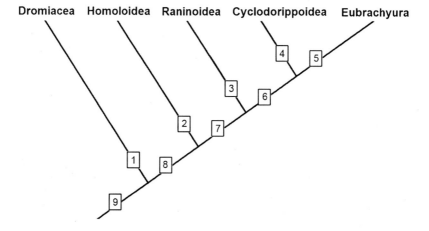

**Figure 15.** The cladogram of Brachyura resulting from our morphological analysis. Each branch is supported by at least one apomorphy. The numbers refer to the apomorphic character sets mentioned in the text.

## 4 DISCUSSION

### 4.1 Paraphyly of Podotremata

When Guinot erected the taxon Podotremata in the late 1970s, she used the coxal gonopores of both sexes as the constituting character for this group (Guinot 1977, 1978, 1979a). This was part of a comprehensive approach to a new subdivision of the entire Brachyura based on the position and differentiation of gonopores and the associated organs such as the spermathecae. Gordon had already proposed a similar approach in 1963, but she suggested excluding all peditreme representatives from the Brachyura, proposing that only sternitreme groups should constitute the true crabs. The major part of crabs, the Eubrachyura (*sensu* de Saint Laurent 1980), is convincingly supported by an apomorphic sternal position of the genital openings in females in combination with a spermatheca connected to the oviduct and internal fertilization. In contrast to this, the coxal position of gonopores of the Podotremata is a clear plesiomorphy since a corresponding condition is found in all other decapods and in the vast majority of Malacostraca to which the Decapoda and thus the Brachyura belong. The absence of an apomorphic character does not necessarily disprove monophyly of the group under consideration, but it at least casts doubt about its validity. Accordingly, Guinot herself discusses this issue critically (1979b). Cladistic studies mainly based on sperm ultrastructure and on some other characters seemingly support the monophyly of Podotremata (Jamieson 1994; Jamieson et al. 1995). Moreover, Tavares (2003) and Guinot & Quenette (2005) discuss the type of external sperm receptacles (here we follow the terminology of Guinot & Quenette 2005, who discriminate between a seminal receptacle as seen in eubrachyurans and the spermathecae as seen in podotrematans) occurring in a characteristic pattern in podotrematan crabs as a putative apomorphy. However, the sperm data are not very convincing. The only three sperm characters in favor of Podotremata are (i) a depressed acrosome, (ii) a predominantly horizontal zonation of the acrosome, and (iii) a bilaterally symmetrical capitate perforatorial head (Jamieson 1994; Jamieson et al. 1995). The first two characters are probably not independent of each other, and whether the conditions seen in raninoids and cyclodorippoids have to be scored as depressed and horizontally zoned is at least disputable (see the figures in Jamieson 1994; Jamieson et al. 1995). The third character occurs only in some species of the dromiaceans, and even Jamieson et al. (1995) doubt its relevance. The polarization of the spermathecal character is problematic because comparable structures do not occur in anomalans or astacids, and the eubrachyuran condition might be derived from that found in podotrematan groups. In contrast to these investigations, two molecular studies dealing with this topic have so far resolved podotrematans as paraphyletic or even polyphyletic with respect to the Eubrachyura (Spears et al. 1992; Ahyong et al. 2007). This is also suggested in a recent study using the ossicle pattern of the foregut of brachyuran crabs (Brösing et al. 2007). The molecular study by Tsang et al. (2008) is somewhat ambiguous. The only depicted tree (Tsang et al. 2008: fig 2) based on sequence data of two nuclear protein coding genes resolves Podotremata as monophyletic, but in the discussion the authors state that a tree based on just one gene shows paraphyletic podotrematans. Furthermore, their taxon sampling did not include Cyclodorippoidea, the putative sister group of Eubrachyura, which might have led to a different result.

The major podotrematan groups Dromiacea, Homoloidea, Raninoidea, and Cyclodorippoidea are all monophyletic in our analysis. However, not all groups are equally well supported. In particular, for the Homoloidea and Cyclodorippoidea more characters are needed to unambiguously support these clades. The Dromiacea do not include the Homoloidea as some authors suggest (Boas 1880; Borradaile 1907). Thus, they form the Dromiacea *sensu stricto* of Guinot (1978, 1979a). There are no apomorphies to support the separate Homolodromioidea superfamily proposed by Ng et al. (2008). A proposed group composed of the homoloids, raninoids, and cyclodorippoids, the Archaeobrachyura (Guinot 1978), finds no support from our data. We can clearly show that the Podotremata is a paraphyletic assemblage. This is revealed not only by the result that the Cyclodorippoidea is the sister group to the Eubrachyura, but also by the general topology and character distribution found by us. For example, the fact that some characters of the Homoloidea and Raninioidea are shared with the rest of the crabs, but not with the dromiaceans, renders the Podotremata paraphyletic. Our suggestion of internal brachyuran relationships is also supported by larval data. Williamson (1974) and, in particular, Rice (1980, 1981a, 1983) stress the similarities of homolid and raninoid zoea and megalopa larvae to those of eubrachyurans to the exclusion of dromiaceans. Moreover, several characteristics of raninoid zoeae (e.g., the overall appearance, the ventrally directed rostrum, and the dorsal and paired lateral spines on the carapace) and megalopae (reduced uropods) indicate a closer relationship to Eubrachyura than to homoloids (Rice 1980, 1981a, 1981b, 1983). Little is known about the larval development of Cyclodorippoidea, but the description of megalopa larvae lacking uropods, as is the case in Eubrachyura, corroborates our conclusion of a sister group relationship between Eubrachyura and Cyclodorippoidea (Rice 1981b).

Our tree is largely congruent with that of the most recent study of brachyuran phylogeny by Ahyong et al. (2007). The only difference is that these authors found a close relationship between dromiids, dynomenids, and homoloids, which all form a common clade, the Dromiacea *sensu lato*. Morphologically, we did not observe any character supporting such a group, and it is also not resolved in other molecular studies on Brachyura phylogeny (Tsang et al. 2008).

### 4.2 *Brachyuran monophyly*

Although a number of carcinologists suggested that the Brachyura form a natural group or monophyletic taxon (e.g., Boas 1880; Borradaile 1907; Guinot 1978), the monophyly has been doubted by several authors based on different levels of evidence such as adult morphology, larval characters, or molecular data (Milne Edwards 1837; Gordon 1963; Williamson 1974; Rice 1980, 1981a, 1983; Spears et al. 1992). In particular, the Raninoidea and the Dromiacea have been excluded from brachyurans due to their adult morphology and the anomuran-like larvae. However, in phylogenetic systematics the exclusion of taxa is only relevant if they can be related to other taxa based on shared apomorphies. In their molecular phylogeny of the Brachyura, Spears et al. (1992) found that the dromiacean representative *Hypoconcha arcuata* clusters with hermit crabs. Accordingly, these authors suggested that dromiaceans should be excluded from Brachyura. In contrast to this view, Scholtz & Richter (1995) and Jamieson et al. (1995) listed a number of characters supporting a monophyletic Brachyura. Here we found several additional characters supporting the Brachyura as monophyletic. These characters include the shape of the endopod of the first maxilliped and the fusion of the arthrodial membranes of the last thoracic segments forming anteriorly directed wings. What is more, our reinvestigation of *Hypoconcha arcuata* reveals that in addition to brachyuran characters, this species shows all apomorphies of the Dromiacea. These apomorphies are nested within the brachyuran characters. Hence, there is no doubt that *Hypoconcha* is a brachyuran and, in particular, a dromiacean. Our results concur with those of the molecular analysis of Ahyong et al. (2007) and the morphological analyses of Jamieson et al. (1995) and Brösing et al. (2007).

### ACKNOWLEDGEMENTS

This study was made possible through an Erskine Fellowship of the University of Canterbury, Christchurch, which allowed GS to work in New Zealand and to collaborate with CLM. We thank Jan McKenzie for her help in using the scanning electron microscope and Atilim Aynacioglu for the drawings in Figs. 5 and 12. We are grateful to the organizers of the symposium "Advances in Decapod Crustacean Phylogenetics" at the SICB meeting in San Antonio 2008 for the opportunity to present our thoughts and results.

# REFERENCES

Ahyong, S.T. & O'Meally, D. 2004. Phylogeny of the Decapoda Reptantia: resolution using three molecular loci and morphology. *Raffl. Bull. Zool.* 52: 673–693.

Ahyong, S.T., Lai, J.C.Y., Sharkey, D., Colgan, D.J. & Ng, P.K.L. 2007. Phylogenetics of the brachyuran crabs (Crustacea: Decapoda): the status of Podotremata based on small subunit nuclear ribosomal RNA. *Mol. Phylogenet. Evol.* 45: 576–586.

Balss, H. 1940. 5. Band, 1. Abteilung, 7. Buch Decapoda, 1. Lieferung. In: Schellenberg, A. (ed.), *Dr. H.G. Bronns Klassen und Ordnungen des Tierreichs 2. Auflage*: 1–165. Leipzig: Akademische Verlagsgesellschaft Becker & Erler.

Boas, J.E.V. 1880. Studier over Decapodernes Slaegtskabsforhold. *K. Danske Vidensk. Selsk. Skr.* 6: 163–207.

Borradaile, L.A. 1907. On the classification of the decapod crustaceans. *An. Mag. Nat. Hist.* 19: 457–486.

Bourne, G.C. 1922. The Raninidae: a study in carcinology. *J. Linn. Soc. (Zool.)* 35: 25–79.

Bouvier, M.E.-L. 1896. Sur l'origine Homarienne des crabs: étude comparative des dromiacés vivants et fossiles. *Bull. de la Soc. Philomath. Paris*, ser. 8, 8: 34–77.

Brösing, A., Richter, S. & Scholtz, G. 2002. The foregut-ossicle system of *Dromia wilsoni*, *Dromia personata* and *Lauridromia intermedia* (Decapoda, Brachyura, Dromiidae), studied with a new staining method. *Arth. Struc. & Dev.* 30: 329–338.

Brösing, A., Richter, S. & Scholtz, G. 2007. Phylogenetic analysis of the Brachyura (Crustacea, Decapoda) based on characters of the foregut with establishment of a new taxon. *J. Zool. Syst. Evol. Res.* 45: 20–32.

de Saint Laurent, M. 1980. Sur la classification et la phylogènie des Crustacès Dècapodes Brachyoures. Podotremata Guinot, 1977, et Eubrachyura sect. nov. *C.R. Acad. Sc. Paris* 290: 1265–1268.

Dixon, C.J., Ahyong, S.T. & Schram, F.R. 2003. A new hypothesis of decapod phylogeny. *Crustaceana* 76: 935–975.

Fincham, A.A. 1980. Eyes and classification of malacostracan crustaceans. *Nature* 287: 729–731.

Gaten, E. 1998. Optics and phylogeny: is there any insight? The evolution of superposition eyes in the Decapoda (Crustacea). *Contrib. Zool.* 67: 223–236.

Gordon, I. 1963. On the relationship of Dromiacea, Tymolidae and Raninidae to the Brachyura. In: Whittington, H.B. & Rolfe, W.D.I. (eds.), *Phylogeny and Evolution of Crustacea*: 51–57. Cambridge: Museum of Comparative Zoology.

Guinot, D. 1977. Propositions pour une nouvelle classification des Crustacé Décapodes Brachyoures. *C. R. Hebd. Séances Acad. Sci. Sér. D* 285: 1049–1052.

Guinot, D. 1978. Principes d'une classification èvolutive des crustacès dècapodes brachyoures (Principles of a new classification of the Crustacea Decapoda Brachyura). *Bull. Biol. France et Belgique, N. S.* 112: 211–293.

Guinot, D. 1979a. Donnèes Nouvelles sur la Morphologie, la Phylogenese et la Taxonomie des Crustacès Dècapodes Brachyoures. *Mèm. Mus. Natn. Hist. Nat. Sèrie A, Zoologie* 112: 1–354.

Guinot, D. 1979b. Problmes pratiques d'une classification cladistique des Crustacés Dècapodes Brachyoures. *Bull. Off. Nat. Pech. Tunisie* 3: 33–46.

Guinot, D. 1991. Ètablissement de la famille des Poupiniidae pour *Poupinia hirsuta* gen. nov., sp. nov. de Polynésie (Crustacea Decapoda Brachyura Homoloidea). *Bull. Mus. natn. Hist. nat. Paris, 4e s'er.*, 12, sect. A: 577–605.

Guinot, D. 1993. Données nouvelles sur les Raninoidea de Haan, 1841 (Crustacea Decapoda Brachyura Podotremata). *C. R. Acad. Sci. Paris, Science de la vie* 316: 1324–1331.

Guinot, D. 1995. Crustacea Decapoda Brachyura: Révision de la famille des Homolodromiidae Alcock, 1900. *Mém. Mus. natn. Hist. nat.* 163: 283–511.

Guinot, D. & Bouchard, J.-M. 1998. Evolution of the abdominal holding systems of brachyuran crabs (Crustacea, Decapoda, Brachyura). *Zoosystema* 20: 613–694.

Guinot, D. & Quenette, G. 2005. The spermatheca in podotreme crabs (Crustacea, Decapoda, Brachyura, Podotremata) and its phylogenetic implications. *Zoosystema* 27: 267–342.

Guinot, D. & Richer de Forges, B. 1995. Crustacea Decapoda Brachyura: Révision de la famille des Homolidae de Haan, 1839. *Mém. Mus. Natn. Hist. Nat.* 163: 155–282.

Guinot, D. & Tavares, M. 2001. Un nouvelle famille de Crabes du Crétacé, et la notion de Podotremata Guinot, 1977 (Crustacea, Decapoda, Brachyura). *Zoosystema* 23: 507–546.

Guinot, D. & Tavares, M. 2003. A new subfamilial arrangement for the Dromiidae de Haan, 1833, with diagnoses and descriptions of new genera and species (Crustacea, Decapoda, Brachyura). *Zoosystema* 25: 43–129.

Hartnoll, R.G. 1979. The phyletic implications of spermathecal structure in the Raninidae (Decapoda, Brachyura). *J. Zool. Lond.* 187: 75–83.

Hennig, W. 1966. *Phylogenetic Systematics*. Urbana: University of Illinois Press.

Jamieson, B.G.M. 1994. Phylogeny of the Brachyura with particular reference to the Podotremata: evidence from a review of spermatozoal ultrastructure (Crustacea, Decapoda). *Phil. Trans. R. Soc. Lond. B.* 345: 372–393.

Jamieson, B.G.M., Guinot, D. & Richer de Forges, B. 1995. Phylogeny of the Brachyura (Crustacea, Decapoda): evidence from spermatozoal ultrastructure. *Mém. Mus. Natn. Hist. Nat.* 166: 265–283.

Martin, J.W. & Davis, G.E. 2001. An updated classification of the Recent Crustacea. *Natural History Museum of Los Angeles County, Science Series* 39: 1–124.

McLay, C.L. 1993. Crustacea Decapoda: The sponge crabs (Dromiidae) of New Caledonia and Philippines with a review of the genera. *Mém. Mus. Natn. Hist. Nat.* 156: 111–251.

McLay, C.L. 1999. Revision of the family Dynomenidae. *Mém. Mus. Natn. Hist. Nat.* 180: 427–569.

Miller, A.D. & Austin, C.M. 2006. The complete mitochondrial genome of the mantid shrimp *Harpiosquilla harpax*, and a phylogenetic investigation of the Decapoda using mitochondrial sequences. *Mol. Phylogenet. Evol.* 38: 565–574.

Milne Edwards, H. 1837. Histoire naturelle des Crustacès, comprenant l'anatomie, la physiologie et la classification de ces animaux. *Librairie Encyclopedique de Roret*, Paris, 1: 1–532.

Ng, P.K.L., Guinot, D. & Davie, P.J. 2008. Systema Brachyurorum: Part I. An annotated checklist of extant brachyuran crabs of the world. *Raffl. Bull. Zool.* 17: 1–286.

Nilsson, D.-E. 1983. Evolutionary links between apposition and superposition optics in crustaceans. *Nature* 302: 818–821.

Nilsson, D.-E. 1988. A new type of imaging optics in compound eyes. *Nature* 332: 76–78.

Ortmann, A.E.1896. Das System der Decapoden Krebse. *Zool. Jb. Syst.* 9: 409–453.

Porter, M.L. & Cronin, T.W. (this volume). A shrimp's eye view of evolution: how useful are visual characters in decapod phylogenetics? In: Martin, J.W., Crandall, K.A. & Felder, D.L. (eds.), *Crustacean Issues: Decapod Crustacean Phylogenetics*. Boca Raton, Florida: Taylor & Francis/CRC Press.

Rice, A.L. 1980. Crab zoeal morphology and its bearing on the classification of the Brachyura. *Trans. Zool. Soc. Lond.* 35: 271–424.

Rice, A.L. 1981a. Crab zoeae and brachyuran classification: a re-appraisal. *Bull. Br. Mus. nat. Hist. (Zool.)* 40: 287–296.

Rice, A.L. 1981b. The megalopa stage in brachyuran crabs. The Podotremata Guinot. *J. Nat Hist.* 15: 1003–1011.

Rice, A.L. 1983. Zoeal evidence for brachyuran phylogeny. In: Schram, F.R. (ed.), *Crustacean Issues 1, Crustacean Phylogeny*: 313–329. Rotterdam: Balkema.

Richter, S. 2002. Evolution of optical design in the Malacostraca (Crustacea). In: Wiese, K. (ed.), *The Crustacean Nervous System*: 512–524. Berlin: Springer Verlag.

Scholtz, G. & Richter, S. 1995. Phylogenetic systematics of the reptantian Decapoda (Crustacea, Malacostraca). *Zool. J. Linn. Soc.* 113: 289–328.

Schram, F.R. 2001. Phylogeny of decapods: moving towards a consensus. *Hydrobiologia* 449: 1–20.

Secretan, S. 1998. The sella turcica of crabs and the endophragmal system of decapods. *J. Nat. Hist.* 32: 1753–1767.

Spears, T., Abele, L.G. & Kim, W. 1992. The monophyly of brachyuran crabs: a phylogenetic study based on 18S rRNA. *Syst. Biol.* 41: 446–461.

Števčić, Z. 1995. Brachyuran systematics and the position of the family Raninidae reconsidered. *Arthr. Select.* 4: 27–36.

Tavares, M. 1993. Crustacea Decapoda: Les Cyclodorippidae et Cymonomidae de l'Indo-ouest-Pacifique à l'exclusion du genre Cymonomus. *Mém. Mus. Natn. Hist. Nat.* 156: 253–313.

Tavares, M. 1998. Phyllotymolonidae, nouvelle famille de Brachyoures Podotremata (Crustacea, Decapoda). *Zoosytema* 20: 109–122.

Tavares, M. 2003. A new theoretical approach for the study of monophyly of the Brachyura (Crustacea: Decapoda) and its impact on the Anomura. *Mem. Mus.Victoria* 60: 145–149.

Tsang, L.M., Ma, K.Y., Ahyong, S.T., Chan, T.-Y. & Chu, K.H. 2008. Phylogeny of Decapoda using two nuclear protein-coding genes: origin and evolution of the Reptantia. *Mol. Phylogenet. Evol.* 48: 359–368.

Williamson, D.I. 1974. Larval characters and the origin of crabs (Crustacea, Decapoda, Brachyura). *Thalass. Jugosl.* 10: 401–414.

# Assessing the Contribution of Molecular and Larval Morphological Characters in a Combined Phylogenetic Analysis of the Superfamily Majoidea

KRISTIN M. HULTGREN [1], GUILLERMO GUERAO[2], FERNANDO P.L. MARQUES[3] & FERRAN P. PALERO[4]

[1] *Smithsonian Tropical Research Institute, Apartado 0843-03092, Balboa, Ancon, Panamá, Republica de Panamá*
[2] *IRTA, 42540 Sant Carles de la Ràpita, Tarragona, Spain*
[3] *Departamento de Zoologia, Instituto de Biociências, Universidade de São Paulo, Brazil*
[4] *Dept. Genetica, Fac. Biologica, University of Barcelona, Spain*

## ABSTRACT

Although the crab superfamily Majoidea is well recognized as a distinct grouping within the Brachyura, resolving the classification of and relationships between different majoid families has been more difficult. In this study, we combine molecular and larval morphology data in a total evidence approach to the phylogeny of the Majoidea, using sequence data from three different loci and 53 larval morphology characters from 14 genera representing 7 majoid families. We examine the relative contribution of morphological and molecular characters in resolving relationships within the superfamily Majoidea and how different alignment and tree construction methods affect tree topology. Using maximum parsimony analyses and partitioned Bremer support, we show that molecular and larval morphology partitions are congruent in combined analyses and that both types of characters contribute positively to resolution of the tree and support for major nodes. Both Bayesian analysis and direct optimization of nucleotide sequences under parsimony supported some similar relationships, including a monophyletic Oregoniidae branching at the base of the majoid tree. However, Bayesian and direct optimization trees differed in their resolution of some relationships, namely in placement of inachid and tychid species relative to the remaining majoids. Neither Bayesian nor direct optimization trees of the combined dataset supported monophyly of the majority of majoid families proposed in recent taxonomic revisions of the group, suggesting the adult morphological characters used to classify majoids into families may be incongruent with larval characters and molecular data used in this study.

## 1 INTRODUCTION

The crab superfamily Majoidea Samouelle, 1819, is one of the most species-rich groups of the Brachyura and is estimated to contain more than 800 species (Rice 1988) assembled into >170 different genera (Ng et al. 2008). Majoids occupy a diverse range of marine habitats worldwide (Rathbun 1925; Rice 1988), and are commonly known as "spider crabs" or "decorator crabs" because of their characteristically long legs and their distinctive behavior of attaching materials from their environment to hooked setae on their carapace to camouflage themselves against predators (Wicksten 1993). As a group, the majoids are typically thought to be one of the earliest brachyuran lineages, based on evidence from spermatozoal ultrastructure (Jamieson 1994), larval characters (Rice 1980, 1981, 1988), and molecular characters (Spears et al. 1992; Porter et al. 2005). Exact estimates of the age of this group vary; studies using model-based methods estimated that the majoids diverged from the rest of the Brachyura ~254 MYA (Porter et al. 2005), although the earliest unequivocal majoid fossils are from the Eocene (Spears et al. 1992). The monophyly of the superfamily Majoidea is often assumed based on adult and larval morphological synapomorphies: all majoids have a terminal molt upon maturity (in contrast to other brachyurans) and only two zoeal stages (Rice 1980, 1981, 1983, 1988; but see Clark & Ng 2004). However, no study thus far has rigorously tested the monophyly of this group, and some workers have suggested inclusion of the Hymensomatidae based on affinities between hymenosomatids and inachoids (Guinot & Richer de Forges 1997; Ng et al. 2008). However, hymenosomatids differ from the majoids as they typically possess three zoeal stages and no true megalopa (Guinot & Richer de Forges 1997), and placement of the Hymensomatidae in the Majoidea is still provisional (Ng et al. 2008).

Formerly known as the family Majidae, the Majoidea were recently reclassified as a superfamily (Hendrickx 1995; Martin & Davis 2001; McLaughlin et al. 2005; Števčić 2005). Diversity of the former family Majidae is very high, and recognition or treatment of the majoids as a superfamily was suggested by many early workers (Guinot 1978; Drach & Guinot 1983; Clark & Webber 1991; Števčić 1994). Nevertheless, many difficulties exist in establishing different families within the Majoidea. Clark & Webber (1991) proposed recognition of family Macrocheiridae based on a reevaluation of the larval features of the genus *Macrocheira* and suggested that extant majoids be partitioned among four families: Oregoniidae, Macrocheiridae, Majidae, and Inachidae. Števčić (1994) recognized six traditional families (Majinae, Mithracinae, Tychinae, Pisinae, Epialtinae, and Inachinae) and also included the Pliosominae, Planotergiinae, Micromajinae, and Eurynolambrinae within Majidae. McLaughlin et al. (2005), following Griffin & Tranter (1986), recognized eight families (Epialtidae, Inachidae, Inachoididae, Majidae, Mithracidae, Pisidae, Tychidae, and Oregoniidae), the first seven of which were recognized by Martin & Davis (2001) in their recent reorganization of the Crustacea. Števčić (2005) partitioned the traditionally recognized majoids into two families, the Majidae and the Inachoididae, and proposed inclusion of the families Lambrachaeidae Števčić, 1994, and Paratymolidae Haswell, 1882. Most recently, Ng et al. (2008) included the hymenosomatids in the superfamily, and recognized six majoid families: Epialtidae, Hymenosomatidae, Inachidae, Inachoididae, Majidae and Oregoniidae. Here we use the traditional classification of majoids as a superfamily, split into eight recognized majoid families (Epialtidae, Inachidae, Inachoididae, Majidae, Mithracidae, Pisidae, Tychidae, and Oregoniidae; Griffin & Tranter 1986; Martin & Davis 2001; McLaughlin et al. 2005) and use molecular and morphological data to review the monophyly of (and relationships among) these groups. The majority of these familial associations follow from elevation of formerly recognized majoid subfamilies to familial status in recent taxonomic monographs (Hendrickx 1995; Martin & Davis 2001).

Several workers have examined relationships among the major groups using larval characters, primarily spination, presence and segmentation of appendages, and setation on the zoeal and megalopal stages (Kurata 1969; Rice 1980, 1988; Clark & Webber 1991; Marques & Pohle 1998; Pohle & Marques 2000; Marques & Pohle 2003). Despite differences in the conceptual framework of assessing homology in these studies (e.g., the identity of the "ancestral" and "derived" forms of majoids), they agree on some points. Kurata (1969) assumed reduction of spination and setation in larval majoids was the derived condition, and he proposed six parallel, heterogeneous lineages of majoids preceded by four different "ancestral" majoids: *Camposcia* (Inachidae), *Schizophrys* (Majidae), *Maia* (Majidae), and *Pleistacantha* (Inachidae). Although he also assumed that reduction of spination and setation was the derived condition, Rice (1980, 1988) hypothesized that the Oregoniidae family retained the "ancestral" majoid larvae, and he proposed two additional lines of majoids: 1) the Inachidae, and 2) a line including the Majidae and another clade of the Pisidae and Epialtidae (formerly the Pisinae and Acanthonychinae subfamilies). Although the family Mithracidae was not considered, Rice (1988) concluded using megalopal characters that the Mithracidae was closely related to the Pisidae and Epialtidae. Phylogenies constructed from larval characters concur on some of these relationships, including a monophyletic Oregoniidae clade branching at the base of the majoid tree (Clark & Webber 1991; Marques & Pohle 1998), and close phylogenetic relationships between the Epialtidae, Pisidae, and Mithracidae families (Pohle & Marques 2000; Marques & Pohle 2003). Marques and Pohle (2003) evaluated support for the monophyly of majoid families and found that while most majoid families were paraphyletic (with the exception of the Oregoniidae), tree lengths in which families were constrained to be monophyletic were not significantly longer than unconstrained topologies, and they concluded that larval characters could not definitively reject monophyly of majoid families. However, support for monophyly varied among different families; for example, the Oregoniidae and the Inachidae + Inachoididae groups (with the exception of *Macrocheira*) formed a clade in unconstrained analyses, while the family Pisidae never formed a clade, and tree lengths of topologies where this group was constrained to be monophyletic were significantly longer than unconstrained trees.

More recently, a molecular phylogeny of this group based on partial sequences of 16S, COI, and 28S genes has corroborated some relationships proposed from phylogenies based on larval morphology (Hultgren & Stachowicz in press). These include: 1) strong support for a monophyletic Oregoniidae; 2) poor support for monophyly of most other majoid families; and 3) close phylogenetic relationships among the families Mithracidae, Pisidae, and Epialtidae. However, molecular data could not resolve key relationships at the base of the majoid tree, namely which of three family groupings—the Inachidae, Oregoniidae, or Majidae—represented the most basally branching majoid group. This may have been due in part to difficulties with aligning portions of the DNA dataset, in particular portions of the 28S locus, suggesting it may be useful to explore if branching patterns at the base of the tree are sensitive to different alignment methods.

Prior to this study, there has been no systematic work addressing the results of simultaneous analyses of molecular and larval morphology characters to examine phylogenetic relationships in the Majoidea, despite intriguing similarities between molecular and morphological phylogenies of this group (Marques & Pohle 1998; Pohle & Marques 2003; Hultgren & Stachowicz in press) and the demonstrated utility of combining multiple sources of data in many phylogenetic studies (Baker et al. 1998; Ahyong & O'Meally 2004). In this study, we combine molecular and larval morphological data in a 'total-evidence' approach to the phylogeny of the superfamily Majoidea, using ~1450 bp of sequence data from 3 loci (16S, COI, and 28S) and 53 larval morphology characters from 14 genera (representing 7 majoid families) to provide a more robust phylogenetic hypothesis for selected members of the Majoidea. We evaluate the relative contribution of morphological and molecular characters and explore how different alignment (static homology and dynamic homology) and tree construction methods (Bayesian and direct optimization using parsimony) affect tree topology in the superfamily Majoidea.

## 2 MATERIALS AND METHODS

### 2.1 *Larval morphology*

To assemble the larval morphology character database, we expanded the data matrix of Marques & Pohle (2003) by adding additional larval characters (for a total of 53) and additional taxa using species-specific descriptions of majoid larval stages. We analyzed the larval characters and codified characters of species with available DNA sequences (summarized in Appendix 1). These included *Acanthonyx petiverii* (Hiyodo et al. 1994), *Menaethius monoceros* (Gohar & Al-Kholy 1957), *Pugettia quadridens* (Kornienko & Korn 2004), *Taliepus dentatus* (Fagetti & Campodonico 1971), *Stenorhynchus seticornis* (Yang 1976; Paula & Cartaxana 1991), *Maja brachydactyla* (Clark 1986), *Micippa thalia* (Kurata 1969), *Micippa platipes* (Siddiqui 1996), *Chionoecetes japonicus* (Motoh 1976), *Hyas coarctatus alutaceus* (Christiansen 1973; Pohle 1991), *Hyas araneus* (Christiansen 1973; Pohle 1991), *Libinia dubia* (Sandifer & Van Engel 1971), *Libinia emarginata* (Johns & Lang 1977), *Pitho lherminieri* (F.P.L. Marques, unpublished data), *Herbstia condyliata* (Guerao et al. 2008), *Mithraculus sculptus* (Rhyne et al. 2006), *Mithraculus forceps* (Wilson et al.

1979), and *Microphrys bicornatus* (Gore et al. 1982). Although this represents a small taxon sample relative to the number of described majoid species, we were limited to taxa (primarily Atlantic species) for which both molecular and morphological data were available. Descriptions of character states are summarized in Appendix 2. Phylogenetic trees constructed from an earlier version of this matrix (Marques & Pohle 1998), using a non-majoid outgroup, found strong evidence for a monophyletic Oregoniidae branching at the base of the tree, similar to trees constructed from molecular data (Hultgren & Stachowicz in press). However, as larval characters coded from non-majoid crabs with >2 zoeal stages may not be homologous to characters coded from majoid crabs (which have only 2 zoeal stages), subsequent phylogenetic analyses based on larval morphology used oregoniid species as the rooting point to the remaining majoids (Marques & Pohle 2003). As larval morphology data for megalopal stages of *Heterocrypta occidentalis* were not available, we coded morphological data for this outgroup species as missing (< 5% of the total dataset for the outgroup).

## 2.2 Molecular data

We used sequence data from the 18 species for which we had morphological data, in addition to 7 additional congeners of those species for which we had only molecular data; in the latter case, morphological data were coded as missing (Table 1). Sampling, extraction, amplification, and sequencing methods have been described previously (Hultgren & Stachowicz in press). Briefly, we used partial sequence data from 3 loci: nuclear 28S ribosomal RNA (~600 bp), mitochondrial 16S ribosomal RNA (~430 bp), and the mitochondrial protein-coding gene cytochrome oxidase I (~580 bp, hereafter COI). Although approximately 25% of the species in the molecular data set were sequenced for only 2 out of the 3 loci, we chose to include terminals (taxa) with missing loci, as simulation studies suggested that the addition of taxa with some missing data (generally <50%) increased accuracy of the final tree (Wiens 2005, 2006). For the molecular dataset, we additionally included sequences from one outgroup species, the parthenopid crab *Heterocrypta occidentalis*.

Molecular data were initially aligned using the program MUSCLE v. 3.6 (Edgar 2004), using default parameters to align nucleotide sequences from each individual locus. Hyper variable regions were excluded from further analysis due to the ambiguity of the alignment, using the program GBlocks v.091b (Castresana 2000) and allowing all gap positions. In total, GBlocks excluded 21% of the 16S alignment, 17% of the COI alignment, and 24% of the 28S alignment. The final combined (and trimmed) molecular dataset consisted of 1478 total base pairs (BP) of sequence data. This alignment was used to test incongruence between molecular and morphological data in all analyses examining the relative contribution of molecular vs. morphological data and in Bayesian analyses of the combined molecular + morphology dataset.

## 2.3 Comparisons of molecular and morphological data partitions

To test whether there were significant incongruities between molecular and morphological datasets, we excluded all additional species from a genus that were not explicitly described in the larval morphology studies. Using the program PAUP ver. 4.0b10 (Swofford 2002) and the molecular alignment described above, we used the incongruence length difference (ILD) test (Farris et al. 1994) implemented under maximum parsimony (MP) to test whether molecular and morphological data were congruent.

Because molecular data often comprise a much higher proportion of characters in combined datasets relative to morphological data and may overwhelm the phylogenetic signal from morphological data (Baker et al. 1998; Wahlberg et al. 2005), we examined the relative contribution of both datasets. Using taxa with both morphology and molecular data, we examined the relative contribution of molecular and morphological characters in the combined dataset by calculating the number

Table 1. Familial associations, molecular data, and larval morphology references for different species used in the study. Familial associations are given according to the classifications of McLaughlin et al. (2005) and Ng et al. (2008).

| Species | Family (McLaughlin et al. 2005) | Family (Ng et al. 2008) | GenBank Accession Nos. 16S | COI | 28S | Larval morphology reference |
|---|---|---|---|---|---|---|
| *Chionoecetes bairdi* (Rathbun, 1924) | Oregoniidae | Oregoniidae | AY227446 | AB21159 | – | – |
| *Chionoecetes japonicus* (Rathbun, 1924) | | | AB188685 | AB211611 | – | Motoh 1976 |
| *Chionoecetes opilio* (Fabricius, 1788) | | | EU682768 | EU682832 | EU682875 | – |
| *Hyas araneus* (Linnaeus, 1758) | | | EU682771 | EU682834 | EU682878 | Christiansen 1973, Pohle 1991 |
| [a]*Hyas coarctatus alutaceus* Brandt, 1851 | | | EU682774 | EU682835 | – | Christiansen 1973, Pohle 1991 |
| [b]*Stenorhynchus* | Inachidae | Inachidae | unpublished | unpublished | – | Yang 1976, Paula & Cartaxana 1991 |
| [c]*Maja brachydactyla* (Balss, 1922) | Majidae | Majidae (sf. Majinae) | DQ079723 | EU000832 | DQ079799 | Clark 1986 |
| *Micippa thalia* (Herbst, 1803) | Mithracidae | Majidae (sf. Mithracidae) | EU682780 | EU682844 | EU682883 | Kurata 1969 |
| *Micippa platipes* (Ruppell 1830) | | | EU682779 | – | EU682884 | Siddiqui 1996 |
| *Microphrys bicornatus* (Latreille, 1825) | | | EU682781 | EU682843 | EU682885 | Gore et al. 1982 |
| *Mithraculus forceps* (Milne-Edwards, 1875) | | | EU682782 | EU682840 | EU682886 | Wilson et al. 1979 |
| *Mithraculus sculptus* (Lamarck, 1818) | | | EU682784 | EU682841 | EU682887 | Rhyne et al. 2006 |
| *Pitho lherminieri* (Schramm, 1867) | Tychidae | Epialtidae (sf. Tychinae) | EU682789 | EU682839 | EU682891 | Marques et al. unpublished data |
| *Acanthonyx petiverii* (Milne-Edwards, 1834) | Epialtidae | Epialtidae (sf. Epialtinae) | EU682803 | EU682855 | EU682903 | Hiyodo et al. 1994 |
| *Menaethius monoceros* (Latreille, 1825) | | | EU682805 | EU682857 | EU682904 | Gohar & Al-Kholy 1957 |
| *Pugettia dalli* (Rathbun, 1893) | | | EU682810 | EU682860 | EU682907 | – |
| *Pugettia gracilis* (Dana, 1851) | | | EU682813 | EU682863 | EU682909 | – |
| *Pugettia minor* (Ortmann, 1893) | | | EU682815 | – | EU682910 | – |
| *Pugettia producta* (Randall, 1840) | | | EU682817 | EU682865 | EU682912 | – |

Table 1. continued.

| Species | Family | | GenBank Accession Nos. | | | Larval morphology reference |
| --- | --- | --- | --- | --- | --- | --- |
| | (McLaughlin et al. 2005) | (Ng et al. 2008) | 16S | COI | 28S | |
| Pugettia quadridens (deHaan, 1850) | | | EU682824 | EU682869 | EU682916 | Kornienko & Korn 2004 |
| Pugettia richii (Dana, 1851) | | | EU682826 | EU682871 | EU682917 | – |
| Taliepus dentatus (Milne-Edwards) | | | EU682827 | EU682872 | EU682918 | Fagetti & Campodonico 1971 |
| Herbstia condyliata (Fabricius, 1787) | Pisidae | Epialtidae (sf. Pisinae) | EU682790 | EU682845 | – | Guerao et al. 2008 |
| Libinia dubia (H. Milne Edwards, 1834) | | | EU682794 | EU682847 | EU682894 | Sandifer & Van Engel 1971 |
| Libinia emarginata (Leach, 1815) | | | EU682796 | EU682849 | EU682896 | Johns & Lang 1977 |
| Libinia mexicana (Rathbun, 1892) | | | EU682797 | – | EU682897 | – |
| Heterocrypta occidentalis (Dana, 1854) | Parthenopidae | | EU682767 | EU682829 | EU682874 | – |

[a] Molecular data from *Hyas coarctatus* (Leach, 1815), morphological data from *Hyas coarctatus alutaceus* (Brandt, 1851).
[b] Molecular data came from *Stenorhynchus lanceolatus* (Brullé, 1837) (16S) and *Stenorhynchus seticornis* (Herbst, 1788) (28S); morphological data from *Stenorhynchus seticornis*.
[c] Molecular data for 16S and 28S came from GenBank *Maja squinado* specimen (Porter et al. 2005); subsequent revisions of this genus and comparison of sequence data with several *Maja* species (Sotelo et al. 2008) indicate the GenBank specimen is likely *Maja brachydactyla*.

of phylogenetically informative characters (PI) for each partition using PAUP*. We also calculated partitioned Bremer support (PBS) (Baker & Desalle 1997; Baker et al. 1998) for each data partition at each node using the program TreeRot v.2 (Sorenson 1999).

## 2.4 Bayesian phylogenetic analysis

Bayesian trees were run using the combined molecular + morphological dataset (with the molecular alignment produced by MUSCLE and GBlocks as described above) using the program MrBayes v3.1.2. (Ronquist & Hulsenbeck 2003). Prior to Bayesian analyses, we used the program Modeltest v.3.7 (Posada & Crandall 1998) to select the appropriate model of molecular evolution for each of the individual molecular loci (i.e., the model that best fit the data) using the Akaike Information Criterion (Posada & Buckley 2004) and allowing MrBayes to estimate parameters for each partition substitution model. Bayesian posterior probabilities (BPP) were obtained for different clades by performing three independent runs with four Markov chains (consisting of 2,000,000 generations sampled every 100 generations). When the log-likelihood scores were found to stabilize, we calculated a majority rule consensus tree after omitting the first 25% of the trees as burn-in.

## 2.5 Direct optimization analysis (dynamic homology)

The direct optimization method was first proposed by Wheeler (1996) as an algorithm to process unaligned nucleotide sequences alone or in conjunction with morphological and aligned molecular data to search optimal topologies using maximum parsimony. Cladogram length during tree search is calculated by the sum of the costs for all hypothesized substitutions and insertion/deletion events (INDELs) via simultaneous evaluation of nucleic acid sequence homologies and cladograms (Wheeler et al. 2006). Throughout the analysis, for each examined topology, potentially unique schemes of positional homologies are dynamically postulated, tested by character congruence, and selected based on the overall minimal cost of character transformations. Detailed properties of the method and its relative advantages in comparison to conventional phylogenetic analysis have been extensively discussed elsewhere and will not be explored here (Wheeler 1996; Wheeler & Hayashi 1998; Phillips et al. 2000; Wheeler et al. 2001; but see Kjer et al. 2006, 2007).

Phylogenetic inference based on nucleotide sequences requires the assignment of specific numerical values for alignment and analysis parameters that define cost regimes for INDELs and transformations (e.g., transversion and transition costs), which can be expressed as cost ratios such as gap: transversion: transition. Because utilization of a single cost regime (traditionally used for phylogenetic studies based on molecular data) does not allow evaluation of how sensitive tree topologies are to any specific set of cost regimes, Wheeler (1995) suggested the selection of a number of parameter sets consisting of the combination of different values for each component of the cost regime (i.e., gap: transversion: transition) within his concept of sensitivity analysis. Sensitivity analysis identifies robust clades, which would be considered those present under most or all parameter sets, from more "unstable" clades resulting from one or a few cost regimes. Since different cost regimes can often generate conflicting topologies, character congruence among different data partitions can be used as an external criterion to choose among parameter sets (Wheeler 1995; Wheeler & Hayashi 1998; Schulmeister et al. 2002; Aagesen et al. 2005). Using this criterion, the combination of parameter values that maximize character congruence (and hence minimize homoplasy inherent in a combined analysis) can be calculated with the incongruence length difference (ILD; Mickevich & Farris 1981; Farris et al. 1994).

Within this framework, in the present study, we submitted all data partitions to a simultaneous cladistic analysis using direct optimization as implemented in POY ver. 4.0 (Varon et al. 2007). We performed tree search using 7200 random addition sequences followed by branch swapping with simulated annealing algorithm (Kirkpatrick et al. 1983), keeping one best tree for each starting tree, on a 24 x 3.2 GHz AMD64 CPU cluster. We used an array of 9 parameter sets to examine the stability of the phylogenetic hypotheses in relation to cost regimes for INDELs (gaps), transversions, and transitions. These parameters considered ranges of costs of 1 to 8 for gaps and 1 to 4 for transformations, resulting in the following cost ratios for gap: transversion: transition: 111, 112, 121, 211, 212, 221, 411, 412, and 421. To compute ILD values (= $\text{Length}_{combined} - (\text{Length}_{MORPH} + \text{Length}_{DNA})/\text{Length}_{combined}$), we submitted the molecular partition to the same search protocol as described above, and analyzed the morphological matrix in TNT version 1.1 (Goloboff & Giannini 2008) with 1000 random additions and branch swapping by alternate SPR and TBR algorithms.

## 3 RESULTS

### 3.1 Comparisons of molecular and morphological data partitions

ILD tests indicated that morphological and molecular datasets were strongly congruent (p = 0.99). The majority of nodes had positive PBS values for both molecular (86% of nodes > 0) and morphological (73% of nodes > 0) data partitions, indicating both sets of characters contributed positively to resolution of the tree in the combined analysis. Relative to the molecular data, morphological data also had a greater percentage of phylogenetically informative (PI) characters (56% of morphological characters, versus 30% of the molecular character set). We calculated the relative support provided by molecular and morphological data partitions by summing the PBS values of all nodes ($\text{PBS}_{DNA}$ = 134.6, $\text{PBS}_{MORPH}$ = 11.3) and examining information content relative to the number of phylogenetically informative characters for each partition (e.g., Baker et al. 1998). Although morphological characters represented <4% of the total character matrix, they had higher overall PBS values relative to the number of phylogenetically informative (PI) sites ($\text{PBS}_{MORPH}/\text{PBS}_{DNA} > \text{PI}_{MORPH}/\text{PI}_{DNA}$), suggesting the morphological data provided more support for nodes in the tree relative to the size of its character set.

### 3.2 Bayesian analysis

The Bayesian combined-analysis tree resolved several major groupings of taxa (Fig. 1). A clade including the Oregoniidae and the mithracid genus *Micippa* branched first (BPP = 81), and then a clade (with the majid species *Maja* branching at the base) consisting of the Epialtidae, Mithracidae, Pisidae, and the inachid genus *Stenorhynchus*. Within this latter grouping, there were well-supported clades of mithracid and tychid genera (*Pitho*, *Microphrys*, and *Mithraculus*; BPP=100); two epialtid genera (*Acanthonyx* and *Menaethius*; BPP = 99); and a clade of epialtid and pisid taxa (*Taliepus*, *Pugettia*, *Herbstia*, and *Libinia*; BPP = 91). Members of Oregoniidae (*Chionoecetes* + *Hyas*, BPP = 100) and the family Pisidae (*Libinia* + *Herbstia*, BPP = 100) both formed monophyletic groups, but there was otherwise no support for monophyly of majoid families recognized by Ng et al. (2008), McLaughlin et al. (2005), or Clark & Webber (1991).

### 3.3 Direct optimization analysis

For direct optimization analyses, the set of alignment cost parameters that minimized homoplasy between datasets (i.e., had the lowest ILD value) corresponded to the 1:1:1 cost weighting scheme (gaps: transversions: transitions; ILD values not shown). To evaluate support for different nodes in this topology given different sets of cost parameters, we used the sensitivity plot to indicate the proportion of parameter sets supporting a given node. In this topology (Fig. 2), the Oregoniidae formed a monophyletic group branching at the base of the majoids, followed by the majid genus *Maja* (similar to the Bayesian tree). The mithracid genus *Micippa* branched at the base of the remaining majoids. In contrast to the Bayesian tree (where it grouped with the mithracid genera *Mithraculus* and *Microphrys*), the tychid species *Pitho lherminieri* formed an idiosyncratic clade with the inachid

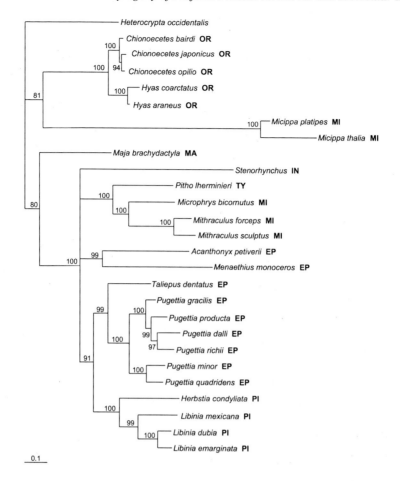

**Figure 1.** Bayesian tree of the Majoidea based on combined molecular and morphological partitions. Numbers by each node indicate Bayesian posterior probability values for that node. Abbreviations in bold after each species indicate family affiliations (after McLaughlin et al. 2005; OR = Oregoniidae, MI = Mithracidae, MA = Majidae, IN = Inachidae, TY = Tychidae, EP = Epialtidae, PI = Pisidae).

*Stenorhynchus* and the epialtid species *Menaethius monoceros* (Fig. 2). Remaining epialtid species formed a clade with the Pisidae. As in the Bayesian tree, there was support for monophyly only for the Oregoniidae and Pisidae families.

## 4 DISCUSSION

In this study, we found that molecular and larval morphology data were strongly congruent, with both partitions independently contributing positively to the support of most relationships. Given the increasing availability of DNA sequence data, the utility of morphological data in phylogenetic inference is often debated (Scotland et al. 2003; Jenner 2004; Lee 2004), in part because many combined-analysis studies show significant incongruence between relationships inferred from morphological and molecular character sets and/or an insignificant contribution of morphological data to tree topology (Baker et al. 1998; Wortley & Scotland 2006). Indeed, previous studies have shown relationships among the majoids inferred from molecular data (Hultgren & Stachowicz in press) are incongruent with familial relationships inferred from adult morphology, even with the most recent reclassifications of majoid families (e.g., Ng et al. 2008). The high levels of congruence between

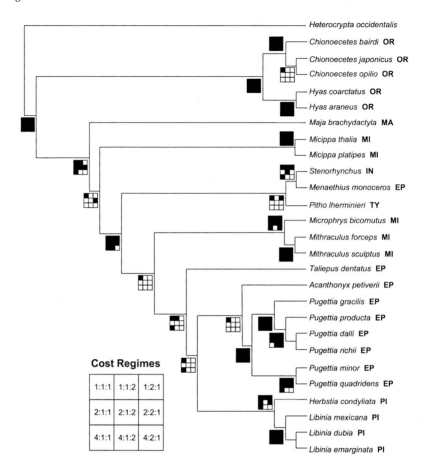

**Figure 2.** Most congruent phylogenetic hypothesis based on direct optimization of molecular and morphological data for Majoidea assuming cost ratios of 1:1:1 (gap: transversion: transition ratios). Sets of boxes below each node indicate the sensitivity plots for which dark fields indicate those parameter sets in which the respective group came out as monophyletic. The order of parameter sets is represented in the box at the bottom left of the figure. Abbreviations in bold after each species indicate family affiliations (abbreviations as in Figure 1).

molecular and larval morphology datasets in this study suggest that for the majoids, molecular and larval characters may provide more phylogenetic information than the adult morphological characters used to place majoids into families, although no phylogeny based on adult morphology has been published to date. That one source of morphological data should be more congruent than another with regards to relationships proposed by molecular data supports earlier observations made by decapod workers that adult morphological characters are often more convergent than larval characters (Williamson 1982). This result also suggests that any decisions to include additional morphological data in a particular study should involve investigation of whether characters in question are under strong selection that might obscure branching patterns (e.g., convergence of similar adult body morphologies due to selection patterns rather than homology). The difficulty of defining morphological characters and making accurate assessments of primary homology (*sensu* de Pinna 1991) often limits the number of characters in these datasets, relative to obtaining sequence data (Baker et al. 1998; Scotland et al. 2003; Wahlberg et al. 2005; Wortley & Scotland 2006). However, morphological characters will always represent a unique set of characters that is independent of sequence data, unlike, for example, a "multi-locus" dataset consisting of two different mitochondrial loci.

Additionally, morphological characters often exhibit less homoplasy and a higher proportion of phylogenetically informative characters than molecular data (Lee 2004) and can often resolve different (but complementary) portions of the tree from molecular data (Jenner 2004), suggesting that combining these multiple types of data may contribute positively to phylogenetic reconstruction (Baker et al. 1998; Ahyong & O'Meally 2004; Wahlberg et al. 2005).

Although trees constructed with direct optimization vs. Bayesian methods reconstructed similar relationships at many of the apical nodes in our study, branching patterns of deeper nodes appear to be sensitive to sequence alignment and inclusion or exclusion of insertion/deletion events (INDELs). For example, the idiosyncratic mithracid genus *Micippa* grouped with the Oregoniidae at the base of the Bayesian tree but branched in a different region in the direct optimization tree. This pattern may not be surprising, given that > 60% of the molecular data consisted of ribosomal gene sequences (16S and 28S) in which INDELs may make multiple alignment problematic. However, it is difficult to compare the effects of different alignment methods and INDEL inclusion independently of differences in phylogenetic inference methods, e.g., model-based methods (utilized in the Bayesian tree) versus maximum parsimony (utilized in the direct-optimization tree). Additionally, support for certain clades in the majoid combined analyses is difficult to directly compare between the topology produced by direct optimization, in which clade stability was assessed using sensitivity plots for a particular node, and trees produced by Bayesian analysis, in which support for a certain clade was assessed by posterior probability.

Despite differences in deep branching patterns due to differences in alignment, inclusion or exclusion of INDELs, and optimality criteria, some groupings were supported in multiple forms of analysis. One such grouping was a monophyletic Oregoniidae branching at the base of the tree. Although previous molecular phylogenies also supported a monophyletic Oregoniidae, they did not conclusively resolve the position of this clade relative to the remaining majoids (Hultgren & Stachowicz in press). Utilization of a combined molecular and morphological dataset in this study strongly supports the Oregoniidae as the most basally branching majoid family, as has been proposed in earlier studies of this group (Rice 1983, 1988; Clark & Webber 1991; Marques & Pohle 1998). Unlike the majority of majoid families, which contain species distributed worldwide, all members of the Oregoniidae are primarily limited to boreal regions (Griffin & Tranter 1986), and similarity in geographic range and/or habitat may help explain why this family is the only group unambiguously resolved in analyses of larval morphology, molecular data, and adult morphology. Although the two pisid genera represented in this study (*Herbstia* and *Libinia*) were monophyletic, molecular and morphological studies with higher taxon sampling (Marques & Pohle 2003; Hultgren & Stachowicz in press) find no support for the monophyly of the Pisidae. The Mithracidae were paraphyletic in all trees in this study, primarily because the genus *Micippa* never grouped with the remaining mithracids. Placement of *Micippa* relative to the remaining majoids was generally unstable (as has been noted in other studies, e.g., Hultgren & Stachowicz in press) and sensitive to different alignment and tree construction methods (Figs. 1, 2). There was likewise no support for the Majidae family *sensu* Ng et al. (2008) (Mithracinae + Majinae). The family Epialtidae was paraphyletic in this study, though in both Bayesian and direct optimization trees there was a close phylogenetic alliance between selected members of the Epialtidae and Pisidae. In this case, the recent Ng et al. (2008) reclassification of the Epialtinae and Pisinae (i.e., Epialtidae and Pisidae) as subfamilies within a larger family (Epialtidae *sensu* Ng et al. 2008) is supported; close relationships between the Pisidae and Epialtidae also were noted in some of the earliest systematic investigations of majoid relationships and larval morphology (Rice 1980, 1988).

The difficulty of using adult morphological characters to establish different family groupings within the Majoidea is reflected in frequent reclassification of majoid families (Griffin & Tranter 1986; Clark & Webber 1991; Martin & Davis 2001; McLaughlin et al. 2005; Ng et al. 2008) and in the failure of subsequent molecular and larval morphology phylogenies to support monophyly of most of these families. However, molecular and larval morphology data in this study both supported a few key taxonomic groupings in combined-analysis Bayesian and direct optimization trees. Both trees supported a monophyletic Oregoniidae branching near the base of the tree, confirming earlier studies suggesting this group represents one of the oldest majoid lineages (Rice 1980, 1988; Clark & Webber 1991; Marques & Pohle 1998). Our study also suggests at least two distinct groupings of the Mithracidae, namely one (*Mithraculus* + *Microphrys*) that may be related to the tychid species *Pitho lherminieri* and one (the mithracid genus *Micippa*) more distantly related to the remaining mithracids. Sampling molecular and morphological characters from additional taxa, especially from hyper diverse regions underrepresented in our study (such as the Indo-Pacific), is warranted to further examine these hypothesized groupings.

We would like to emphasize that the relationships suggested herein represent tentative hypotheses based on the data at hand, namely, <10% of the 170+ majoid genera in the world. Additional focus on the Inachidae, Majidae, and Inachoididae (the latter of which was not sampled in this study) is crucial to further resolve branching patterns at the base of the majoid tree. Rigorous testing of the monophyly of the Majoidea—namely, whether it includes the Lambrachaeidae, Paratymolidae, and Hymenosomatidae (Guinot & Richer de Forges 1997; Števčić 2005; Ng et al. 2008)—is also important in order to properly describe the higher-level systematics of this group. However, the positive contribution of both molecules and morphology to resolution of relationships within the majoids suggests that combining these different sources of data may hold strong potential for researchers to establish a more stable classification of majoid families in the future.

ACKNOWLEDGEMENTS

We would like to thank Jody (Joel) Martin and the other organizers for the opportunity to participate in the symposium *Advances in Decapod Crustacean Phylogenetics*. We would also like to thank two anonymous reviewers for constructive comments on earlier drafts of the manuscript.

# APPENDIX 1

Larval morphology character matrix for taxa in the study. A "?" indicates missing data for that character; parentheses surround characters ambiguous for two states.

| Family | Species | Character matrix | Reference |
|---|---|---|---|
| Epialtidae | *Acanthonyx petiverii* (Milne-Edwards, 1834) | 01001212010212110110200110011002100012??01?434127???? | Hiyodo et al. 1994 |
| | *Menaethius monoceros* (Latreille, 1825) | 011012110102121111102011101110121000127???????????????? | Gohar & Al-Kholy 1957 |
| | *Pugettia quadridens* (de Haan, 1850) | 01001202010112111102001200110021000101001?2430211370 | Kornienko & Korn 2004 |
| | *Taliepus dentatus* (Milne-Edwards) | 010012011102122111102000100110021000127?7112313127??? | Fagetti & Campodonico 1971 |
| Inachidae | *Stenorhynchus seticornis* (Herbst, 1788) | 1100121201012121111020002111110210111020147225202?701? | Yang 1976, Paula & Cartaxana 1991 |
| Majidae | *Maja brachydactyla* (Balss, 1922) | 00001100011012111110200011111100010011110127243102010000 | Clark 1986 |
| Mithracidae | *Micippa thalia* (Herbst, 1803) | 0110120011212010110200010111100010001?????????????? | Kurata 1969 |
| | *Micippa platipes* (Ruppell, 1830) | 0110120011212010110200010111100010001?????????????? | Siddiqui 1996 |
| | *Microphrys bicornutus* (Latreille, 1825) | 01001200010212110110011011110021000101011?67301010110 | Gore et al. 1982 |
| | *Mithraculus forceps* (Milne-Edwards, 1875) | 01001200010(01)2121101101001101110021000100011062310011001 | Wilson et al. 1979 |
| | *Mithraculus sculptus* (Lamarck, 1818) | 01001200010(01)2121101101001101110021000100011062310011001 | Rhyne et al. 2006 |
| Oregoniidae | *Chionoecetes japonicus* (Rathbun, 1924) | 0000000010(12)5010100010002000000100000000037677117?7? | Motoh 1976 |
| | *Hyas coarctatus alutaceus* Brandt, 1851 | 00000000101501010000100020000001000000007241212???? | Christiansen 1973, Pohle 1991 |
| | *Hyas araenus* (Linnaeus, 1758) | 00000000101501010000100020000001000000007241212???? | Christiansen 1973, Pohle 1991 |
| Pisidae | *Herbstia condyliata* (Fabricius, 1787) | 01001200110212110110200000010021000101017413010200 | Guerao et al. 2008 |
| | *Libinia dubia* (H. Milne Edwards, 1834) | 0100120111011011101110200110011002100010001174(01)30011110 | Sandifer & Van Engel 1971 |
| | *Libinia emarginata* (Leach, 1815) | 0100120111011011101110200172771100210001001741300117171? | Johns & Lang 1977 |
| Tychidae | *Pitho lherminieri* (Schramm, 1867) | ?100120111021210170100110111100210001177?????????????? | Marques et al. unpublished data |

## APPENDIX 2

Morphological characters of majoid larvae used in the analyses.

1. Zoeal rostral spine: present (0), absent (1).
2. Zoeal lateral spines: present (0), absent (1).
3. Zoeal dorsal spine: present (0), absent (1).
4. Zoeal carapace serrulation: ornamentation absent (0), ornamentation present (1).
5. Zoea II subterminal setation on the antennule: present (0), absent (1).
6. Zoeal exopod morphology of the antenna: terminal spine minute, less than half length of apical setae (0), terminal spine half or more length of apical setae but not extending beyond tip of setae (1), terminal spine extending beyond tip of setae, latter inserted distally to proximal half of shaft (2), terminal spine extending much beyond setae, latter inserted on proximal half of shaft (3).
7. Proximal segment of the zoeal maxillulary endopod: seta present (0), seta absent (1).
8. Distal segment of the zoeal maxillulary endopod: six setae (0), 5 setae (1), 4 setae (2), 3 setae (3).
9. Ontogenetic setal transformation of the maxillulary coxa from zoae I to zoea II: stasis at 7 additional 8th seta (0), additional 9th seta (1).
10. Zoeal proximal setation of maxillulary basis: plumodenticulate (0), pappose (1).
11. Ontogenetic setal transformation of the maxillulary basis from ZI to ZII: 7 to 10 (0), 7 to 9 (1), 7 to 8 (2).
12. Ontogenetic setal transformation of the proximal lobe of the maxillary coxa from ZI to ZII: stasis at 3 (0), stasis at 4 (1), statis at 5 (2), 3 to 4 (3), 4 to 5 (4), stasis at 4 (5).
13. Ontogenetic setal transformation of the proximal lobe of the maxillary basis from ZI to ZII: 5 to 6 (0), stasis at 5 (1).
14. Ontogenetic setal transformation of the distal lobe of the maxillary basis from ZI to ZII: stasis at 4 (0), 5 to 6 (1), 4 to 5 (2).
15. Zoeal setation of the maxillary endopod: 6 setae (0), 5 setae (1), 4 setae (2), 3 setae (3).
16. Lobes of the zoeal maxillary endopod: bilobed (0), single lobed (1).
17. Setation on the zoeal basis maxilliped 1: 10 setae (0), 9 setae (1), 11 setae (2).
18. Setation on the zoeal basis of maxilliped 2: 4 setae (0), 3 setae (1), 2 setae (2), 1 setae (3), absent (4).
19. Setation on the proximal zoeal endopod segment of maxilliped 2: seta present (0), seta absent (1).
20. Setation on the penultimate segment of the zoeal endopod of maxilliped 2: seta present (0), seta absent (1).
21. Setation on the distal segment of the zoeal endopod of maxilliped 2: 6 setae (0), 5 setae (1), 4 setae (2), 3 setae (3).
22. Relative length of terminal setae on the distal segment of the zoeal endopod of maxilliped 2: one shorter (0), same length (1).
23. Spine on the distal segment of the zoeal endopod of maxilliped 2: present (1), absent (0).
24. Dorsal lateral process on the third zoeal abdominal somite: present (0), absent (1).
25. Middorsal setae on the first abdominal somite in zoea II: 5 setae (0), 3 setae (1), 2 setae (2), absent (3).
26. Middorsal setae on the second abdominal somite in zoea II: present (0), absent (1).
27. Middorsal setae on the third abdominal somite in zoea II: present (0), absent (1).
28. Middorsal setae on the fourth abdominal somite in zoea II: present (0), absent (1).
29. Middorsal setae on the fifth abdominal somite in zoea II: present (0), absent (1).
30. Zoeal acicular process on the second abdominal somite: present (1), absent (0).
31. 6th somite in zoae II: differentiated (0), not differentiated (1).
32. Zoeal telson furcal spination: 3 spines (0), 2 spines (1), 1 spine (2), no spine (3).
33. Zoeal II telson furcal arch setation: 8 setae (0), 6 setae (1).
34. Megalopa uropods (pleopods on the 6th abdominal somite): present (0), absent (1).
35. Pronounced antennal exopod process in megalopa: present (1), absent (0).
36. Fusion of megalopa antennal flagellar articles 2+3: present (1), absent (0).
37. Fusion of megalopa antennal flagellar articles 4+5: present (1), absent (0).
38. Seta on the first segment of the peduncle of the antennule: present (1), absent (0).
39. Seta on the second segment of the peduncle of the antennule: 2 setae (0), 1 seta (1), absent (2).
40. Seta on the third segment of the peduncle of the antennule: 1 seta (0), 2 setae (1).
41. Setae on the distal segment of the antenna: 4 setae (0), 3 setae (1).
42. Setation of the palp of the mandible: 8 setae (0), 5 setae (1), 4 setae (2), 11 setae (3), 6 setae (4), 1 seta (5).

43. Epipod setae on the maxillule: present (1), absent (0).
44. Setation on the endopod of the maxillule: 6 setae (0), 5 setae (1), 4 setae (2), 3 setae (3), 2 setae (4), 1 seta (5), seta absent (6).
45. Ontogenetic change from zoea II to megalopa on the coxal endite of the maxillule: 8 to 11 (0), 8 to 10 (1), 7 to 10 (2), 7 to 11 (3), 7 to 9 (4), 7 to 8 (5), stasis to 7 (6).
46. Ontogenetic change from zoea II to megalopa in the distal lobe of the coxal endite of the maxilla: 4 to 6 (0), 4 to 5 (1), stasis at 4 (2), stasis at 3 (3).
47. Seta on the proximal segment of the exopod on the third maxilliped: present (1), absent (0).
48. Setation on the distal segment on the exopod of the third maxilliped: 6 setae (0), 5 setae (1), 4 setae (2).
49. Setation on the second abdominal somite: 8 setae (0), 6 setae (1): 2 setae (3).
50. Setation on the third abdominal somite: 8 setae (0), 6 setae (1), 2 setae (3).
51. Setation on the fourth abdominal somite: 8 setae (0), 6 setae (1), 10 setae (2), 4 setae (3).
52. Setation on the fifth abdominal somite: 8 setae (0), 6 setae (1).
53. Setation on the sixth abdominal somite: 2 setae (0), none (1).


REFERENCES

Aagesen, L., Peterson, G. & Seberg, O. 2005. Sequence length variation, indel costs, and congruence in sensitivity analysis. *Cladistics* 21: 15–30.

Ahyong, S.T. & O'Meally, D. 2004. Phylogeny of the Decapoda Reptantia: resolution using three molecular loci and morphology. *Raff. Bull. Zool.* 52: 673–693.

Baker, R.H. & Desalle, R. 1997. Multiple sources of character information and the phylogeny of Hawaiian drosophilids. *Syst. Biol.* 46: 654–673.

Baker, R.H., Yu, X. & Desalle, R. 1998. Assessing the relative contribution of molecular and morphological characters in simultaneous analysis trees. *Mol. Phylogenet. Evol.* 9: 427–436.

Castresana, J. 2000. Selection of conserved blocks from multiple alignments for their use in phylogenetic analysis. *Mol. Biol. Evol.* 17: 540–552.

Christiansen, M.E. 1973. The complete larval development of *Hyas araneus* (Linnaeus) and *Hyas coarctatus* (Leach) (Decapoda, Brachyura, Majidae) reared in the laboratory. *Norw. J. Zool.* 21: 63–89.

Clark, P.F. 1986. The larval stages of *Maja squinado* (Herbst, 1788) (Crustacea: Brachyura: Majidae) reared in the laboratory. *J. Nat. Hist.* 20: 825–836.

Clark, P.F. & Ng, P.K.L. 2004. Two zoeal stages and the megalop of *Pilumnus sluiteri* De Man, 1892 [Crustacea: Brachyura: Xanthoidea: Pilumnidae] described from laboratory-reared material. *Invertebr. Reprod. Dev.* 45: 205–219.

Clark, P.F. & Webber, W.R. 1991. A redescription of Macrocheira-Kaempferi (Temminck, 1836) zoeas with a discussion of the classification of the majoidea Samouelle, 1819 (Crustacea, Brachyura). *J. Nat. Hist.* 25: 1259–1279.

de Pinna, M.G.G. 1991. Concepts and tests of homology in the cladistic paradigm. *Cladistics* 7: 67–394.

Drach, P. & Guinot, D. 1983. The Inachoididae Dana, a family of Majoidea characterized by a new type of connexions between carapace, pleurites, sternites and pleon (Crustacea, Decapoda). *Comptes rendus des séances de l'Académie des sciences. Série 3, Sciences de la vie* 297: 37–42.

Edgar, G.J. 2004. MUSCLE: multiple sequence alignment with high accuracy and high throughput. *Nucleic Acids Res.* 32: 1792–1797.

Fagetti, E.Y. & Campodonico, I. 1971. Desarrollo larval en el laboratorio de *Taliepus dentatus* (Milne-Edwards) (Crustacea: Brachyura: Majidae, Acanthonychinae). *Rev. Biol. Mar. Valpariso.* 14: 1–14.

Farris, J.S., Kallersjo, M., Kluge, A.G. & Bult, C. 1994. Testing significance of incongruence. *Cladistics* 10: 315–319.

Gohar, H.A.F. & Al-Kholy, A.A. 1957. The larvae of four decapod Crustacea from the Red Sea. *Publ. Mari. Bio. Stn. Al-Ghardaqua* 9: 177–202.

Goloboff, P. & Giannini, N. 2008. An optimization-based method to estimate and test character correlation for continuous characters. *Cladistics* 24: 91–91.

Gore, R.H., Scotto, L.E. & Yang, W.T. 1982. *Microphrys bicornutus* (Latreille, 1825): the complete larval development under laboratory conditions with notes on other Mithracine larvae (Decapoda, Brachyura, Majidae). *J. Crust. Biol.* 2: 514–534.

Griffin, D.J.G. & Tranter, H.A. 1986. *The Decapoda Brachyura of the Siboga Expedition, Part VIII. Majidae.* Leiden: E.J. Brill.

Guerao, G., Abello, P. & Hispano, C. 2008. Morphology of the larval stages of the spider crab *Herbstia condyliata* (Fabricius, 1787) obtained in laboratory conditions (Brachyura: Majoidea: Pisidae). *Zootaxa* (in press).

Guinot, D. 1978. Principes d'une classification evolutive des crustaces decapodes brachyoures. *Bull. Biol. France Belgique* 112: 211–292.

Guinot, D. & Richer de Forges, B. 1997. Affinités entre les Hymenosomatidae MacLeay, 1838 et les Inachoididae Dana, 1851 (Crustacea, Decapoda, Brachyura). *Zoosystema* 19: 453–502.

Hendrickx, M.E. 1995. Checklist of brachyuran crabs (Crustacea: Decapoda) from the eastern tropical Pacific. *Med. K. Belg. Inst. Nat. Wet.* 65: 125–150.

Hiyodo, C.M., Fransozo, A. & Fransozo, M.L.N. 1994. Larval development of the spider crab *Acanthonyx petiverii* H. Milne Edwards, 1834 (Decapoda, Majidae) in the laboratory. *Crustaceana* 66: 53–66.

Hultgren, K.M. & Stachowicz, J.J. in press. Molecular phylogeny of the brachyuran crab superfamily Majoidea indicates close congruence with trees based on larval morphology. *Mol. Phylogenet. Evol.*

Jamieson, B.G.M. 1994. Phylogeny of the Brachyura with particular reference to the Podotremata: evidence from a review of spermatozoal ultrastructure (Crustacea, Decapoda). *Phil. Trans. Roy. Soc. Lond. B.* 345: 373–393.

Jenner, R.A. 2004. Accepting partnership by submission? Morphological phylogenetics in a molecular millennium. *Syst. Biol.* 53: 333–342.

Johns, D.M. & Lang, W.H. 1977. Larval development of the spider crab, *Libinia emarginata* (Majidae). *Fish. Bull.* 75: 831–841.

Kirkpatrick, S., Gelatt, C.D. & Vecchi, M.P. 1983. Optimization by simulated annealing. *Science* 220: 671–680.

Kjer, K.M., Gillespie, J.J. & Ober, K.A. 2006. Structural homology in ribosomal RNA, and a deliberation on POY. *Arthr. Syst. Phy.* 64: 159–164.

Kjer, K.M., Gillespie, J.J. & Ober, K.A. 2007. Opinions on multiple sequence alignment, and an empirical comparison of repeatability and accuracy between POY and structural alignment. *Syst. Biol.* 56: 133–146.

Kornienko, E.S. & Korn, O.M. 2004. Morphological features of the larvae of spider crab *Pugettia quadridens* (Decapoda: Majidae) from the northwestern Sea of Japan. *Russian J. Mar. Biol.* 30: 402–413.

Kurata, H. 1969. Larvae of Decapoda Brachyura of Arasaki, Sagami Bay-IV. Majidae. *Bull. Tokai. Reg. Fish. Res. Lab.* 57: 81–127.

Lee, M.S.Y. 2004. Molecular and morphological datasets have similar numbers of relevant phylogenetic characters. *Taxon* 53: 1019–1022.

Marques, F. & Pohle, G. 1998. The use of structural reduction in phylogenetic reconstruction of decapods and a phylogenetic hypothesis for 15 genera of Majidae: testing previous larval hypotheses and assumptions. *Invertebr. Reprod. Dev.* 22: 241–262.

Marques, F.P.L. & Pohle, G. 2003. Searching for larval support for majoid families (Crustacea: Brachyura) with particular reference to Inachoididae Dana, 1851. *Invertebr. Reprod. Dev.* 43: 71–82.

Martin, J.W. & Davis, G.E. 2001. An updated classification of the Recent Crustacea. *Nat. Hist. Mus. Los Angeles County, Science Series* 39: 1–124.

McLaughlin, P.A., Camp, D.K., Angel, M.V., Bousfied, E.L., Brunel, P., Brusca, R.C., Cadien, D., Cohen, A.C., Conlan, K., Eldredge, L.G., Felder, D.L., Goy, J.W., Haney, T.A., Hann, B., Heard, R.W., Hendrycks, E.A., Hobbs, H.H., Holsinger, J., Kensley, B., Laubitz, D.R., LeCroy, S.E., R., L., Maddocks, R.F., Martin, J.W., Mikkelsen, P., Nelson, E., Newman, W.A., Overstret, R.M., Poly, W.J., Price, W.W., Reid, J.W., Robertson, A., Rogers, D.C., Ross, A., Schotte, M., Schram, F.R., Shih, C.-T., Watling, L., Wilson, G.D.F. & Turgeon, D.D. 2005. Common and scientific names of aquatic invertebrates from the United States and Canada: crustaceans. *Am. Fish. Soc. Sp. Publ.* 31: 1–545.

Mickevich, M.L. & Farris, J.S. 1981. The implications of congruence in Menidia. *Syst. Zool.* 30: 351–370.

Motoh, H. 1976. The larval stages of Benizuwai-gani, *Chionoecetes japonicus* Rathbun reared in the laboratory. *Bull. Jap. Soc. Sci. Fish.* 42: 533–542.

Ng, P.K.L., Guinot, D. & Davie, P.J.F. 2008. Systema Brachyurorum: Part I. An annotated checklist of extant Brachyuran crabs of the world. *Raff. Bull. Zool.* 17: 1–286.

Paula, J. & Cartaxana, A. 1991. Complete larval development of the spider crab *Stenorhynchus-Lanceolatus* (Brulle, 1837) (Decapoda, Brachyura, Majidae), reared in the laboratory. *Crustaceana* 60: 113–122.

Phillips, A., Janies, D. & Wheeler, W.C. 2000. Multiple sequence alignment in phylogenetic analysis. *Mol. Phylogenet. Evol.* 16: 317–330.

Pohle, G. & Marques, F. 2000. Larval stages of *Paradasygyius depressus* (Bell, 1835) (Crustacea: Decapoda: Brachyura: Majidae) and a phylogenetic analysis for 21 genera of Majidae. *Proc. Biol. Soc. Wash.* 113: 739–760.

Pohle, G. & Marques, F.P.L. 2003. Zoeal stages and megalopa of *Leucippa pentagona* H. Milne Edwards, 1833 (Decapoda: Brachyura: Majoidea: Epialtidae) obtained from laboratory culture and a comparison with other epialtid and majoid larvae. *Invertebr. Reprod. Dev.* 43: 55–70.

Pohle, G.W. 1991. Larval development of Canadian Atlantic oregoniid crabs (Brachyura, Majidae), with emphasis on *Hyas-Coarctatus-Alutaceus* Brandt, 1851, and a comparison with Atlantic and Pacific conspecifics. *Can. J. Zool.* 69: 2717–2737.

Porter, M.L., Pérez-Losada, M. & Crandall, K.A. 2005. Model-based multi-locus estimation of decapod phylogeny and divergence times. *Mol. Phylogenet. Evol.* 37: 355–369.

Posada, D. & Buckley, T.R. 2004. Model selection and model averaging in phylogenetics: advantages of the AIC and Bayesian approaches over likelihood ratio tests. *Syst. Biol.* 53: 793–808.

Posada, D. & Crandall, K.A. 1998. MODELTEST: testing the model of DNA substitution. *Bioinformatics* 14: 817–818.

Rathbun, M.J. 1925. *The Spider Crabs of America*. Washington, D.C.: Smithsonian Institution, United States National Museum Bulletin.

Rhyne, A.L., Fujita, Y. & Calado, R. 2006. Larval development and first crab of *Mithraculus sculptus* (Decapoda: Brachyura: Majoidea: Mithracidae) described from laboratory-reared material. *J. Mar. Biol. Assoc. UK* 86: 1133–1147.

Rice, A.L. 1980. Crab zoeal morphology and its bearing on the classification of the Brachyura. *Trans. Zool. Soc. Lond.* 35: 271–424.

Rice, A.L. 1981. The megalopa stage in brachyuran crabs — the Podotremata Guinot. *J. Nat. Hist.* 15: 1003–1011.

Rice, A.L. 1983. Zoeal evidence for brachyuran phylogeny. In: Schram, F.R., (ed.), *Crustacean Issues Volume 1, Crustacean Phylogeny*: 313–399. Rotterdam: Balkema.

Rice, A.L. 1988. The megalopa stage in majid crabs, with a review of spider crab relationships based on larval characters. *Symp. Zool. Soc. Lond.* 59: 27–46.

Ronquist, F. & Hulsenbeck, J.P. 2003. MRBAYES 3: Bayesian phylogenetic inference under mixed models. *Bioinformatics* 20: 407–415.

Sandifer, P.A. & Van Engel, W.A. 1971. Larval development of the spider crab, *Libinia dubia* H. Milne Edwards (Brachyura, Majidae, Pisinae), reared in laboratory culture. *Ches. Sci.* 12: 18–25.

Schulmeister, S., Wheeler, W.C. & Carpenter, J.M. 2002. Simultaneous analysis of the basal lineages of Hymenoptera (Insecta) using sensitivity analysis. *Cladistics* 18: 455–484.

Scotland, R.W., Olmstead, R.G. & Bennett, J.R. 2003. Phylogeny reconstruction: the role of morphology. *Syst. Biol.* 52: 539–548.

Siddiqui, F.A. 1996. Larval development of *Micippa platipes* Ruppell, 1830 reared under laboratory conditions (Crustacea, Decapoda, Majidae). *J. Mar. Sci.* 5: 155–160.

Sorenson, M.D. 1999. *TreeRot, version 2.* Boston, Massachusetts: Boston University.

Sotelo, G., Morán, P. & Posada, D. 2008. Genetic identification of the northeastern Atlantic spiny spider crab as *Maja brachydactyla* Balss, 1922. *Journal of Crust Ocean Biology* 28(1): 76–81.

Spears, T., Abele, L.G. & Kim, W. 1992. The monophyly of brachyuran crabs: a phylogenetic study based on 18S rRNA. *Syst. Biol.* 41: 446–461.

Števčić, Z. 1994. Contribution to the re-classification of the family Majidae. *Per. Biol.* 96: 419–420.

Števčić, Z. 2005. The reclassification of brachyuran crabs (Crustacea: Decapoda: Brachyura). *Nat. Croat.* 14 (Suppl. 1): 1–159.

Swofford, D.L. 2002. *PAUP*. Phylogenetic Analysis Using Parsimony (*and other methods), Version 4.* Sunderland, Massachusetts: Sinauer Associates.

Varon, A., Vinh, L.S., Bomash, I. & Wheeler, W.C. 2007. *POY 4.0 Beta 2635.* http://research.amnh.org/scicomp/projects/poy.php. American Museum of Natural History.

Wahlberg, N., Braby, M.F., Brower, A.V.Z., de Jong, R., Lee, M.M., Nylin, S., Pierce, N.E., Sperling, F.A.H., Vila, R., Warren, A.D. & Zakharov, E. 2005. Synergistic effects of combining morphological and molecular data in resolving the phylogeny of butterflies and skippers. *Proc. R. Soc. Lond. B Biol. Sci.* 272: 1577–1586.

Wheeler, W.C. 1995. Sequence alignment, parameter sensitivity and the phylogenetic analysis molecular-data. *Syst. Biol.* 44: 321–331.

Wheeler, W.C. 1996. Optimization alignment: the end of multiple sequence alignment in phylogenetics? *Cladistics* 12: 1–9.

Wheeler, W.C., Aagesen, L., Arango, C.P., Faivoich, J., Grant, T., D'Haese, C., Janies, D., Smith, W.L., Varon, A. & Giribet, G. 2006. *Dynamic Homology and Systematics: A Unified Approach.* New York: American Museum of Natural History.

Wheeler, W.C. & Hayashi, C.Y. 1998. The phylogeny of the extant chelicerate orders. *Cladistics* 14: 173–192.

Wheeler, W.C., Whiting, M., Wheeler, Q.D. & Carpenter, J.M. 2001. The phylogeny of the extant hexapod orders. *Cladistics* 17: 113–169.

Wicksten, M.K. 1993. A review and a model of decorating behavior in spider crabs (Decapoda, Brachyura, Majidae). *Crustaceana* 64: 314–325.

Wiens, J.J. 2005. Can incomplete taxa rescue phylogenetic analyses from long-branch attraction? *Syst. Biol.* 54: 731–742.

Wiens, J.J. 2006. Missing data and the design of phylogenetic analyses. *J. Biomed. Inform.* 39: 34–42.

Williamson, D.I. 1982. Larval morphology and diversity. In: Abele, L.G. (ed.), *Embryology, Morphology, and Genetics*: 43–110. New York: Academic Press.

Wilson, K.A., Scotto, L.E. & Gore, R.H. 1979. Studies on decapod Crustacea from the Indian River region of Florida. XIII. Larval development under laboratory conditions of the spider crab *Mithrax forceps* (A. Milne-Edwards, 1875) (Brachyura, Majidae). *Proc. Biol. Soc. Wash.* 92: 307–327.

Wortley, A.H. & Scotland, R.W. 2006. The effect of combining molecular and morphological data in published phylogenetic analyses. *Syst. Biol.* 55: 677–685.

Yang, W.T. 1976. Studies on the western Atlantic arrow crab genus *Stenorhynchus* (Decapoda, Brachyura, Majidae) I. Larval characters of two species and comparison with other larvae of Inachinae. *Crustaceana* 31: 157–177.

# Molecular Genetic Re-Examination of Subfamilies and Polyphyly in the Family Pinnotheridae (Crustacea: Decapoda)

EMMA PALACIOS-THEIL[1], JOSÉ A. CUESTA[2], ERNESTO CAMPOS[3] & DARRYL L. FELDER[1]

[1] University of Louisiana at Lafayette, Department of Biology and Laboratory for Crustacean Research, Lafayette, Louisiana, U.S.A.

[2] Instituto de Ciencias Marinas de Andalucía, Consejo Superior de Investigaciones Científicas (ICMAN-CSIC), Avenida República Saharaui, Puerto Real (Cádiz), Spain

[3] Universidad Autónoma de Baja California, Facultad de Ciencias, Ensenada, Baja California, México

## ABSTRACT

The family Pinnotheridae de Haan, 1833 is a highly adapted group of largely symbiotic species distributed among 49–56 genera, some of debatable status. Many species remain to be described, a task complicated by the confused state of systematics in the group. Despite a massive taxonomic literature base, illustrations of morphology are of limited scope and quality, hampering morphologically based phylogenetic comparisons. Striking post-planktonic changes in ontogeny, related to unique life histories, can occur among subadults, and different stages of the same species have occasionally been named independently. Polyphyly of the Pinnotheridae has been previously suggested in our own preliminary analyses that combined findings from adult and larval morphology with molecular genetic data. While some issues of polyphyly center at the generic level, questions also remain as to how family and subfamily ranks should be applied to reflect monophyletic clades. The present molecular analysis was based on combined sequence data for the partial mitochondrial large subunit 16S rRNA gene, the tRNA-Leu gene, and the partial mitochondrial gene for NADH1, primarily to examine generic assignments. The results of mitochondrial gene analyses are relatively unambiguous, with strong support values for transfer of Xenophthalminae and Asthenognathinae out of Pinnotheroidea. The family Pinnotheridae is partitioned between two primary clades representing the subfamilies Pinnothereliinae and Pinnotherinae, and smaller clades may justify one or more additional subfamilies. Members of several genera within these subfamilies require taxonomic revision. Analyses based upon the 18S nuclear gene, while supporting morphologically and mitochondrial gene-based definition of the Pinnothereliinae, did not clarify relationships between most other pinnotherid genera and were thus not incorporated into our analysis.

## 1 INTRODUCTION

Crabs of the family Pinnotheridae de Haan, 1833, the pea crabs, are typically symbiotic crustaceans found with ascidians, annelids, other crustaceans, echiurans, echinoderms, or molluscs (Schmitt et al. 1973) and are rarely free living. Their adaptation to this variety of host organisms likely accounts for their diversity. By the most commonly used current taxonomy, there are about 313 described species, or 287 if excluding Asthenognathinae and Xenophthalminae (Ng et al. 2008). These are distributed among a maximum of 56 genera (49 according to Ng et al. 2008), and some are of debatable generic assignment (Zmarzly 1992; Manning 1993a; Campos 1996a). The largest genera are *Pinnotheres* (71 spp.), *Pinnixa* (56 spp.), and *Arcotheres* (20 spp.), while the other genera contain fewer than 10 species each, and 23 of those are monotypic. Since description of the first pinnotherid, *Nepinnotheres pinnotheres* (described as *Cancer pinnotheres* Linnaeus, 1758), discovery and description of new species have continued almost unabated. From the Gulf of Mexico alone, we estimate our present holdings to include no fewer than 20 undescribed species. In addition to increasing numbers of species and genera, taxonomy has become very unstable over recent decades. Some genera and species have been excluded from the family, species have been reassigned from one genus to another, and many synonymies have been recommended (e.g., Campos 1989; Manning 1993b; Ahyong & Ng 2007; Ng et al. 2008).

Complicating the taxonomic problems even further, post-planktonic development in pinnotherids can involve more complex metamorphoses than in most other brachyurans, often involving several morphologically distinct subadult stages during the postlarval ontogeny of a single species. Changes can involve carapace shape, abdominal morphology, and development of the pleopods, many of these altering characters used for morphological diagnoses of genera. As noted by Campos (1989), taxonomists have on some occasions assigned separate names to two different stages of the same species.

Classification of the pinnotherids has been the object of multiple revisions, especially since the late 1980s (Griffith 1987; Manning & Felder 1989; Campos 1996a, b; Coelho 1997; Campos 2006; Ahyong & Ng 2007; Ng et al. 2008). Most of these were partial revisions, limited to a certain subfamily or genus, or confined to a limited geographic region. However, even when only partial revisions, they often defined species and genera that remain of uncertain phylogenetic placement in the group.

Polyphyly of the Pinnotheridae in its present composition (sensu Schmitt et al. 1973) has already been supported in several studies based upon morphological analyses (Marques & Pohle 1995; Campos 1996b, 1999; Števčić 1996; Pohle & Marques 1998; Campos & Manning 2000), as well as in preliminary molecular analyses (Cuesta et al. 2001). Very recently, new arrangements at family and subfamily levels have been proposed (Cuesta et al. 2005; Števčić 2005; Ng et al. 2008; Campos 2009).

We herewith provide molecular phylogenetic analyses that bear on recently proposed revisions. In so doing, we evaluate clade relationships in a tree based upon the partial 16S rRNA gene, the tRNA-Leu gene, and the partial NADH1 gene from the mitochondrial genome. We also attempt phylogenetic analyses based upon the nuclear 18S rRNA for potential clarification of relationships at the subfamily and family levels.

## 2 MATERIALS AND METHODS

### 2.1 *Specimens used in analyses*

We attempted to include as many pinnotherid genera as possible, but especially those representing diverse morphologies or taxa that have been questionably placed in the past. Specimens represented the four putative subfamilies Pinnothereliinae, Pinnotherinae, Xenophthalminae, and Asthenognathinae, thus excluding only the monospecific Anomalifrontinae previously included in the family by Schmitt et al. (1973). Sequences were obtained from our own extractions, supplemented by some from GenBank (Table 1). For outgroups, we chose species from other brachyuran families of putative close or distant relationship to pinnotherids for which comparable 16S or 18S sequences were available (Table 2). In mitochondrial sequence analyses, we included a single member of Xenophthalminae, two species of two genera assigned to Asthenognathinae, 21 species representing three genera assigned to Pinnothereliinae, and 19 species of 16 genera recognized by Schmitt et al. (1973) as members of Pinnotherinae. For *Clypeasterophilus stebbingi*, *Clypeasterophilus rugatus*, *Tunicotheres moseri*, and *Zaops ostreum*, we sequenced specimens from more than one geographic location. In addition, we included two undescribed species that are morphologically assignable to

**Table 1.** Species used in molecular phylogenetic analyses of the family Pinnotheridae (*sensu* Schmitt et al. 1973). For collection catalog numbers (Cat. No.), abbreviations are as follow: CBM-ZC = Natural History Museum and Institute, Zoology, Crustacea, Chiba, Japan; CBR-ICM = Colección Biológica de Referencia, Instituto de Ciencias del Mar, Barcelona, Spain; RMNH = Rijksmuseum van Natuurlijke Historie, Nationaal Naturhistorisch Museum, Leiden; SMF = Senckenberg Museum, Frankfurt a.M., Germany; ULLZ = University of Louisiana at Lafayette Zoological Collections; USNM = U.S. National Museum of Natural History, Smithsonian Institution, Washington, D.C.

| Species | Location | Cat. No. | GenBank Accession No. 16S | GenBank Accession No. 18S |
|---|---|---|---|---|
| **Family PINNOTHERIDAE de Haan, 1833** | | | | |
| Pinnotherid sp. 1 | Bahía de los Ángeles, México | ULLZ 9337 | EU934955 | EU934919 |
| Pinnotherid sp. 2 | Northern Gulf of Mexico | ULLZ 5582 | EU934991 | |
| **Subfamily XENOPHTHALMINAE Alcock, 1900** | | | | |
| *Xenopththalmus pinnotheroides* White, 1846 | Hiroshima Bay, Seto Is. Sea, Japan | CBM-ZC 7784 | EU934951 | EU934922 |
| **Subfamily ASTHENOGNATHINAE Stimpson, 1856** | | | | |
| *Asthenognathus atlanticus* Monod, 1933 | Mauritania, off Banc d'Arguin | RMNH 40008 | EU934952 | |
| *Tritodynamia horvathi* Nobili, 1905 | Aitsu Mar. Biol. St., Japan | ULLZ 5585 | EU934953 | EU934950 |
| **Subfamily PINNOTHERINAE de Haan, 1833** | | | | |
| *Austinotheres angelicus* (Lockington, 1877) | San Felipe, México | ULLZ 9601 | EU935002 | |
| *Calyptraeotheres granti* (Glassell, 1933) | San Felipe, México | ULLZ 9599 | EU934979 | |
| *Clypeasterophilus rugatus* (Bouvier, 1917) | Twin Keys, Belize | ULLZ 9511 | EU934981 | |
| *Clypeasterophilus rugatus* (Bouvier, 1917) | East Coast Florida, USA | ULLZ 5546 | EU934980 | EU934924 |
| *Clypeasterophilus stebbingi* (Rathbun, 1918) | Praia do Leste, Brazil | ULLZ 5543 | EU934984 | EU934941 |
| *Clypeasterophilus stebbingi* (Rathbun, 1918) | Is. Margarita, Venezuela | ULLZ 5545 | EU934983 | |
| *Dissodactylus crinitichelis* Moreira, 1901 | Praia do Sul, Isla Anchieta, Ubatuba, Brazil | ULLZ 5561 | EU934982 | EU934942 |
| *Dissodactylus latus* Griffith, 1987 | East Coast Florida, USA | ULLZ 5548 | EU934985 | |
| *Fabia subquadrata* Dana, 1851 | California, USA | ULLZ 5575 | EU935000 | EU934947 |
| *Limotheres* sp. | off southeastern USA | ULLZ 9176 | EU934996 | EU934923 |
| *Holothuriophilus pacificus* (Poeppig, 1836) | Bahía de Concepción, Cocholque, Chile | ULLZ 5569 | EU934997 | EU934948 |
| *Juxtafabia muliniarum* (Rathbun, 1918) | San Felipe, México | ULLZ 9600 | EU934990 | |
| *Nepinnotheres pinnotheres* (Linnaeus, 1758) | Bahía de Cádiz, Spain | CBR-ICM pending | EU935001 | |
| *Orthotheres barbatus* (Desbonne, 1867) | Los Roques, Venezuela | ULLZ 5559 | EU934999 | EU934921 |

**Table 1.** continued.

| Species | Location | Cat. No. | GenBank Accession No. 16S | GenBank Accession No. 18S |
|---|---|---|---|---|
| *Pinnaxodes chilensis* (H. Milne Edwards, 1837) | Caleta Coquimbo, Chile | ULLZ 5570 | EU934998 | EU934949 |
| *Pinnotheres pisum* (Linnaeus, 1767) | Regensburg, Germany (mussel import) | SMF 30947 | AM180694 | |
| *Scleroplax granulata* Rathbun, 1893 | Bodega Bay, California, USA | ULLZ 5576 | EU934972 | EU934930 |
| *Tumidotheres maculatus* (Say, 1818) | Praia do Lazaro, Ubatuba, Brazil | ULLZ 9512 | EU934986 | |
| *Tumidotheres maculatus* (Say, 1818) | Isla Coche, Venezuela | ULLZ 5534 | EU934945 | |
| *Tumidotheres margarita* (Smith, 1869) | Bahía Margarita, Baja California Sur, México | ULLZ 5533 | EU934987 | EU934946 |
| *Tunicotheres moseri* (Rathbun, 1918) | Tampa Bay, Florida, USA | ULLZ 4516 | EU934988 | EU934925 |
| *Tunicotheres moseri* (Rathbun, 1918) | Isla Margarita, Venezuela | ULLZ 5536 | EU934989 | EU934926 |
| *Zaops ostreum* (Say, 1817) | Fort Pierce, Florida, USA | ULLZ 5537 | EU934994 | EU934943 |
| *Zaops ostreum* (Say, 1817) | Isla Margarita, Venezuela | ULLZ 5535 | EU934995 | |
| **Subfamily PINNOTHERELIINAE Alcock, 1900** | | | | |
| *Austinixa aidae* (Righi, 1967) | Praia do Perequê Açú, Ubatuba, Brazil | ULLZ 5538 | EU934966 | EU934936 |
| *Austinixa behreae* (Manning & Felder, 1989) | Mustang Is., Texas, USA | ULLZ 5541 | EU934956 | EU934939 |
| *Austinixa chacei* (Wass, 1955) | Navarre, Florida, USA | ULLZ 4405 | EU934957 | EU934940 |
| *Austinixa cristata* (Rathbun, 1900) | Fort Pierce, Florida, USA | ULLZ 5556 | EU934967 | |
| *Austinixa felipensis* (Glassell, 1935) | San Felipe, Baja California Norte, México | ULLZ 5558 | EU934969 | EU934927 |
| *Austinixa gorei* (Manning & Felder, 1989) | Islas del Rosario, Colombia | ULLZ 5586 | EU934965 | EU934920 |
| *Austinixa hardyi* Heard & Manning, 1997 | Blood Bay, Tobago, Trinidad and Tobago | USNM 284177 | AF503185 | |
| *Austinixa patagoniensis* (Rathbun, 1918) | Praia do Araçá, São Sebastião, Brazil | ULLZ 5549 | EU934970 | EU934935 |
| *Pinnixa chaetopterana* Stimpson, 1860 | Fort Pierce, Florida, USA | ULLZ 5553 | EU934961 | EU934937 |
| *Pinnixa cylindrica* (Say, 1818) | Corpus Christi Bay, Texas, USA | ULLZ 5560 | EU934963 | EU934929 |
| *Pinnixa faba* (Dana, 1851) | State of Washington, USA | ULLZ 5571 | EU934976 | EU934933 |
| *Pinnixa franciscana* Rathbun, 1918 | Bodega Bay, California, USA | ULLZ 5624 | EU934974 | |
| *Pinnixa littoralis* Holmes, 1894 | Tahuya, Washington, USA | ULLZ 5572 | EU934975 | EU934932 |
| *Pinnixa monodactyla* (Say, 1818) | Fort Pierce, Florida, USA | ULLZ 8713 | EU934964 | |
| *Pinnixa pearcei* Wass, 1955 | Tampa Bay, Florida, USA | ULLZ 5557 | EU934971 | EU934934 |
| *Pinnixa rapax* Bouvier, 1917 | São Sebastião, Brazil | ULLZ 5568 | EU934959 | |
| *Pinnixa retinens* Rathbun, 1918 | Fort Pierce, Florida, USA | ULLZ 9347 | EU934992 | |

**Table 1.** continued.

| Species | Location | Cat. No. | GenBank Accession No. 16S | GenBank Accession No. 18S |
|---|---|---|---|---|
| *Pinnixa sayana* Stimpson, 1860 | Fort Pierce, Florida, USA | ULLZ 5620 | EU934962 | |
| *Pinnixa schmitti* Rathbun, 1918 | Japonski Is., Stika, Alaska, USA | ULLZ 5574 | EU934978 | EU934931 |
| *Pinnixa tomentosa* Lockington, 1877 | Brown's Beach, Baranof Is., Sitka, Alaska | ULLZ 5522 | EU934977 | |
| *Pinnixa tubicola* Holmes, 1894 | Brown's Beach, Baranof Is., Sitka, Alaska | ULLZ 5521 | EU934973 | |
| *Pinnixa valerii* Rathbun, 1931 | Estero Corrientes, Nicaragua | ULLZ 9336 | EU934993 | |
| *Pseudopinnixa carinata* Ortmann, 1894 | Moji, Fukuoka prefecture, Japan | ULLZ 5628 | EU934954 | EU934944 |
| *Austinixa* sp. 1 | Nagualapa, Nicaragua | ULLZ 5566 | EU934958 | EU934938 |
| *Austinixa* sp. 2 | Las Enramadas, Cosigüina, Nicaragua | ULLZ 5564 | EU934968 | EU934928 |
| *Pinnixa* sp. | Tampa Bay, Florida, USA | ULLZ 8126 | EU934960 | |

*Austinixa* (*Austinixa* sp. 1 and sp. 2) and two more undescribed species that are morphologically questionable as to placement among the Pinnotheridae (Pinnotherid sp. 1 and Pinnotherid sp. 2; Table 1).

For the 18S gene, we extracted DNA from a single species of each of the subfamilies Xenophthalminae and Asthenognathinae, 12 species of Pinnothereliinae (representing the genera *Austinixa*, *Pinnixa* and *Pseudopinnixa*), and 13 species of Pinnotherinae representing 11 genera, all of which were also included in the mitochondrial analyses. In this case we obtained sequences from different locations for two species (*Tumidotheres maculatus* and *Tunicotheres moseri*). The above-mentioned undescribed species of *Austinixa* again were used, and one of the questionably placed undescribed pinnotherid species was included (Pinnotherid sp. 1).

## 2.2 DNA extraction and PCR

Total genomic DNA was extracted from muscle tissue with a DNeasy Blood and Tissue Kit (QIAGEN, Valencia, CA) following the manufacturer's protocol or with the standard DNA extraction protocols (Robles et al. 2007). Polymerase chain reaction (PCR) was conducted to amplify a fragment of the mitochondrial genome that extends from the gene for the large ribosomal subunit 16S rRNA through the tRNA-Leu to and including part of the protein coding region of the mitochondrial nitrogen dehydrogenase subunit 1 (NADH1). For this fragment we used the primers 16SH2 (5'-AGA TAG AAA CCA ACC TGG-3') (Schubart et al. 2000, equivalent to the primer 1472 described in Crandall & Fitzpatrick 1996), 16SL2 (5'-TGC CTG TTT ATC AAA AAC AT-3'), 16SL6 (5'-TTG CGA CCT CGA TGT TGA AT-3') (developed by JAC and C. Schubart), and NADH1 (5'-TCC CTT ACG AAT TTG AAT ATA TCC-3'). We also used five internal primers designed specifically for pinnotherids, including PH1 (5'-CGC TGT TAT CCC TAA AGT AAC-3'), PH2 (5'-CCT GGC TCA CGC CGG TCT GAA-3'), PH3 (5'-AAT CCT TTC GTA CTA AAA-3'), PL1 (5'-AAC TTT TAA GTG AAA AGG CTT-3'), and PL2 (5'-TTA CTT TAG GGA TAA CAG CG-3').

For 18S rRNA the primers developed by Medlin et al. (1988) were used, including 18SC (5'-CGG TAA TTC CAG CTC CAA TAG-3'), 18SL (5'-AGT AAA AGC TCG TAG TGG-3'),

**Table 2.** Outgroup sequences from GenBank used in phylogenetic analyses based upon mitochondrial 16S rRNA and the nuclear 18S rRNA genes.

| | | | GenBank accession no. | |
|---|---|---|---|---|
| Superfamily | Family | Species | 18S | 16S |
| **Heterotremata** | | | | |
| Majoidea | Majidae | *Maja crispata* Risso, 1827 | | EU000852 |
| | | *Maja squinado* (Herbst, 1788) | DQ079758 | EU000851 |
| | Oregoniidae | *Chionoecetes opilio* (Fabricius, 1788) | | AB188684 |
| Portunoidea | Portunidae | *Carcinus maenas* (Linnaeus, 1758) | DQ079757 | |
| | | *Necora puber* (Linnaeus, 1767) | DQ079759 | |
| Potamoidea | Potamidae | *Geothelphusa* sp. | DQ079750 | |
| Xanthoidea | Panopeidae | *Panopeus herbstii* H. Milne Edwards, 1834 | | AJ130815 |
| | Xanthidae | *Xantho poressa* (Olivi, 1792) | | AM076937 |
| **Thoracotremata** | | | | |
| Grapsoidea | Gecarcinidae | *Cardisoma crassum* Smith, 1870 | | AJ130805 |
| | | *Gecarcinus lateralis* (Freminville, 1835) | | AJ130804 |
| | Grapsidae | *Pachygrapsus marmoratus* (Fabricius, 1787) | DQ079763 | |
| | | *Pachygrapsus transversus* (Gibbes, 1850) | | AJ250641 |
| | Plagusiidae | *Euchirograpsus americanus* A. Milne-Edwards, 1880 | | AJ250648 |
| | | *Plagusia dentipes* de Haan, 1835 | | AJ308421 |
| | Sesarmidae | *Sesarma reticulatum* (Say, 1817) | | AJ130799 |
| | Varunidae | *Cyrtograpsus altimanus* Rathbun, 1914 | | AJ487319 |
| | | *Gaetice depressus* (de Haan, 1835) | AY859577 | |
| | | *Helice tridens tientsinensis* Rathbun, 1931 | Z70526 | |
| | | *Varuna litterata* (Fabricius, 1798) | | AJ308419 |
| Ocypodoidea | Dotillidae | *Dotilla wichmani* De Man, 1892 | | AB002126 |
| | | *Scopimera globosa* (de Haan, 1835) | | AB002124 |
| | Macrophthalmidae | *Macrophthalmus banzai* Wada & K. Sakai, 1989 | | AB002132 |
| | | *Macrophthalmus japonicus* (de Haan, 1835) | EU284156 | |
| | | *Macrophthalmus latifrons* Haswell, 1882 | | Z79669 |
| | Ocypodidae | *Minuca minax* (LeConte, 1855) | | Z79670 |
| | | *Ocypode quadrata* (Fabricius, 1878) | AY743942 | Z79679 |

18SO (5'-AAG GGC ACC ACC AGG AGT GGA G-3'), 18SY (5'-GTT GGT GGA GCG ATT TGT CTG-3'), and 18SB (5'-AGG TGA ACC TGC GGA AGG ATC A-3'). Instead of primer 18SA indicated by Medlin et al. (1988), we used the slight variant 18SEF (5'-CTG GTT GAT CCT GCC AGT-3') (Hillis & Dixon 1991), which is three basepairs (bp) shorter at the 5' end.

## 2.3 *Phylogenetic analyses*

Sequences for each gene region were assembled and edited with Sequencher 4.7 (Genecodes, Ann Arbor, MI). Preliminary alignments were checked for accuracy with BioEdit 7.0.9.0 (Hall 1999) and then aligned with MUSCLE (Edgar 2004) on the website of the European Bioinformatics Institute (www.ebi.ac.uk). Outgroup sequences of 18S rRNA and 16S rRNA were obtained from GenBank (Table 2). Once all the sequences were added and aligned, regions where primers were located were trimmed to avoid artefacts. In addition, poorly aligned and gapped positions were removed after identification with Gblocks (v. 0.91b, Castresana 2000). The resulting sequence lengths were 786 bp for the combined mitochondrial sequences and 1625 bp for the 18S sequences.

The combined mitochondrial sequence data were tested for partition homogeneity (Bull et al. 1993), as implemented in PAUP* 4.0 beta 10 (Swofford 2002) with 1000 replicates. PAUP* was also used for determining base composition, pattern of substitution for pairwise comparison, and analysis of variability along the 16S rRNA and 18S rRNA fragments. The alignment file was submitted for processing with RAxML version 7.0.4 (Stamatakis et al. 2008) and with bootstrapping at the Cyberinfrastructure for Phylogenetic Research (CIPRES) Web Portal (www.phylo.org). We used this program for a maximum likelihood search (ML), selecting the option of automatically determining the number of necessary bootstrapping runs. Once we obtained the results, the trees were analyzed and edited with Mega 4 (Tamura et al. 2007). In addition to ML analysis, Bayesian (BAY) phylogenetic analyses were performed using MrBayes for the mitochondrial combined data. Before conducting BAY analysis, the model of evolution that best fit the data was estimated with the computer program MODELTEST (Posada & Crandall 1998).

The phylogenetic analysis was conducted sampling one tree every 500 generations for 1,000,000 generations, starting with a random tree. We obtained 2001 trees, of which we discarded 4%. In a previous analysis we could determine that stasis was reached after approximately 35,000 generations, so we discarded the first 40,000 generations, or, in other words, the 81 first trees sampled. With the remaining trees we obtained 50% majority rule consensus trees by means of PAUP* 4.0 (see above).

Support values for analyses based on the 18S nuclear gene were in general so low that phylogenetic trees based upon these sequence data were not reproduced in the present manuscript. Where the 18S analyses did support phylogenetic groupings based on the combined mitochondrial sequence data or morphology, mention is made in the following sections.

## 3 RESULTS

### 3.1 Utility of the combined mitochondrial analyses for the Pinnotheridae

The concatenated BAY analysis of mitochondrial genes resulted in a well-resolved consensus tree (Fig. 1). Topologies for the Pinnotheridae in the separate ML and BAY trees (not shown) were virtually the same, with only minor differences. While in the ML tree *Zaops* was grouped with low support values into Clade IIA, it was in the BAY tree grouped at low support values into Clade IIC. Also, while the ML tree shows Clade III to include Pinnotherid sp. 2 with weak support, it was placed external to this group in the BAY tree. Aside from these differences, both analyses define the same membership in Clades I, II, and III.

### 3.2 Restriction of the Pinnotheridae in the mitochondrial phylogenetic analyses

In our molecular phylogeny, *Xenophthalmus pinnotheroides*, *Asthenognathus atlanticus*, and *Tritodynamia horvathi* are by ML and BAY analysis positioned among outgroup families rather than among other putative pinnotherids (Fig. 1). *Asthenognathus atlanticus* and *T. horvathi* are placed in both analyses with high support values into a common clade with the two outgroup species of the family Varunidae. On the other hand, *X. pinnotheroides* is grouped with strong support with representatives of the family Dotillidae. With the exception of *Pseudopinnixa carinata*, all other putative pinnotherids that were included in these analyses are joined together into a well-supported single clade, which is in turn subdivided into two major and one minor clade. The enigmatic *Pseudopinnixa carinata* is positioned basally to all other putative pinnotherid groups, but in a poorly resolved polytomy. It is clearly excluded from a highly supported node that groups Clades I, II, and III of the Pinnotheridae in our ML and BAY analyses. Among these clades, Clade III is of most limited membership, grouping *Pinnixa valerii*, *P. retinens*, and, with modest support, an undescribed species

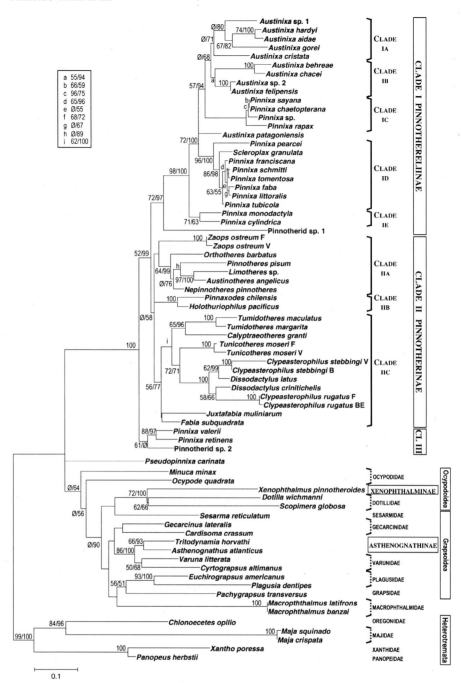

**Figure 1.** Phylogeny for species of the family Pinnotheridae, superimposed on classification of Schmitt et al. (1973), inferred from a maximum likelihood (ML) analysis of 786 bp of the mitochondrial genes for 16S rRNA (604 bp), tRNA-Leu, and NADH1 (together 182 bp). Bootstrap values for ML and Bayesian posterior probabilities are shown (ML bootstrap value first); ø means value < 50%. Where value is the same for both, only one number is shown; no number is shown if both values < 50%. Letters follow some species names to separate conspecific populations from Brazil (B), Belize (BE), Florida (F), and Venezuela (V).

(Pinnotherid sp. 2) of uncertain generic assignment. Given the polyphyletic stature of *Pinnixa* in the overall analysis, proper generic placement of all species grouped into this clade must be open to question.

### 3.3 Definition of pinnotherid subfamilies in the mitochondrial phylogenetic analyses

Two major groups of the putative pinnotherids included in our analyses are segregated in the molecular phylogenetic tree, and these are supported in both the ML and BAY analyses. The more strongly supported of these groups (Clade I) encompasses those pinnotherids that current taxonomy assigns to the subfamily Pinnothereliinae, thus including analyzed members of the genera *Austinixa* and *Pinnixa* but in this case also the species *Scleroplax granulata* (placed in Pinnotherinae instead of Pinnothereliinae by Schmitt et al. 1973). Clade I also includes a basally positioned undescribed species (Pinnotherid sp. 1) that is pending generic assignment. A less well-resolved second major group (Clade II) encompasses taxa currently assigned by most workers to the taxonomically diverse subfamily Pinnotherinae, thus including *Austinotheres, Limotheres, Orthotheres, Pinnotheres, Nepinnotheres, Zaops, Holothuriophilus, Pinnaxodes, Fabia, Juxtafabia, Calyptraeotheres, Tunicotheres, Tumidotheres, Clypeasterophilus*, and *Dissodactylus*.

### 3.4 Subfamily Pinnothereliinae

The Pinnothereliinae of Clade I are subdivided into five subgroups, two of which consist exclusively of species assignable to the genus *Austinixa*. The 8 species originally included in this genus (Heard & Manning 1997) were all represented in our analysis, in addition to two new species pending description. Additional congeners, *A. bragantina* and *A. leptodactyla*, placed in the genus by Coelho (1997, 2005), were not available for inclusion. As presently constituted, *Austinixa* appears to be polyphyletic. While 7 of the 8 named species, including the type species of the genus, *A. cristata*, share a common lineage (Clades IA–C), *Austinixa patagoniensis* is separated from this group in a poorly resolved polytomy.

Other members of *Austinixa* (Clade IA plus IB) are positioned as a sister clade to a grouping of four species (Clade IC) that are presently treated under *Pinnixa*, though these are not grouped in our analysis with the type species of that genus, *P. cylindrica*. With support only in BAY analysis, Clade IA includes *A. hardyi, A. aidae, A. gorei, Austinixa* sp. 1, and *A. cristata*, while Clade IB includes the closely related species *A. behreae* and *A. chacei* along with *A. felipensis* and the undescribed species *Austinixa* sp. 2 from eastern Pacific waters of Central America. Clade IC encompasses the very closely related sister species *P. chaetopterana* and *P. sayana*, along with *P. rapax* and an undetermined *Pinnixa* sp. from Tampa Bay, Florida. A fourth clade (ID) within the apparent Pinnothereliinae includes almost all remaining members of the genus *Pinnixa* that we analyzed (having previously excluded *Pinnixa valerii* and *P. retinens* from both subfamilies), along with *Scleroplax granulata*. However, *Pinnixa cylindrica*, type species of the genus *Pinnixa*, and *P. monodactyla* (Clade IE) are with strong support grouped separately from both Clade IC and ID, thus segregating them from all present congeners included in this analysis and underscoring polyphyly of this genus as presently recognized.

### 3.5 Subfamily Pinnotherinae

Clade II of our phylogeny (Fig. 1) includes a diverse set of genera that broadly represents the present subfamily Pinnotherinae, albeit without the previously affiliated genus *Scleroplax*, as noted above. While a number of its encompassed lower subclades are well supported, support for grouping of the subfamily overall is very limited and found only in the BAY analysis. Clades IIA–B are separated only as a polytomy. Without support, topology of our tree positions populations of *Zaops ostreum* basally in Clade IIA, which contains a well-supported grouping of *Orthotheres barbatus*, *Pinnotheres pisum*, *Limotheres* sp., *Austinotheres angelicus*, and *Nepinnotheres pinnotheroides*. A second clade (IIB) defines the highly supported grouping of *Holothuriophilus pacificus* and *Pinnaxodes chilensis*, while a third clade (IIC) groups our included species of *Fabia*, *Juxtafabia*, *Calyptraeotheres*, *Tumidotheres*, *Tunicotheres*, *Clypeasterophilus*, and *Dissodactylus*.

As noted above, the BAY tree (not shown) also groups *Zaops* here, rather than with Clade IIA, but only with low support. While *Clypeasterophilus* and *Dissodactylus* are expectedly grouped together with high support within Clade IIC, neither of these genera appears to be monophyletic in our analyses, their constituent species being in both cases distributed between alternative sister subclades.

## 4 DISCUSSION

### 4.1 *Exclusions from Pinnotheridae, and exceptional members of the group*

While long affiliated with Pinnotheridae by some workers (see Schmitt et al. 1973), Asthenognathinae and Xenophthalminae have been the subject of several recent re-examinations. The subfamily Asthenognathinae was proposed by Števčić (2005) for elevation to the family level and transfer to the superfamily Grapsoidea. On the other hand, the subfamily Xenophthalminae was elevated by the same author to family level, though he retained it within the Pinnotheroidea. Based in part on a preliminary report (Cuesta et al. 2005), Ng et al. (2008) have instead recently placed both of the asthenognathine species that we analyzed (*Asthenognathus atlanticus* and *Tritodynamia horvathi*) among the varunids, and our present findings clearly offer further support for this placement. Thus, Ng et al. moved some present members of the subfamily Asthenognathinae to the family Varunidae H. Milne Edwards, 1853, but concluded that the genus *Tritodynamia* is polyphyletic, to the point that some of its members may warrant assignment to separate families. In their opinion, *T. horvathi* appears related to the varunids, but its congeners are more closely related to macrophthalmids. They thus transferred most members of *Tritodynamia* to the Macrophthalmidae Dana, 1851, as Tritodynamiinae Števčić, 2005. Among the species presently assigned to *Tritodynamia*, only *T. horvathi* was available for inclusion in our analysis, and therefore we can provide no support for division of the genus *Tritodynamia* as suggested. Studies with more members of this genus are thus warranted to support the proposed new classification.

In the case of Xenophthalminae, Ng et al. (2008) recommended the elevation of this group to the family level as Xenophthalmidae Stimpson, 1858, with the two subfamilies Anomalifrontinae Rathbun, 1931, and Xenophthalminae Stimpson, 1858, placing them in the superfamily Ocypodoidea. At least from our analysis of *Xenophthalmus pinnotheroides*, we can support this revision, as the species clearly is not placed by molecular genetics among other members of the family Pinnotheridae; rather, our molecular data and larval morphology suggest the close relationship of *X. pinnotheroides* with the family Dotillidae. Future molecular analyses should ideally include another member of *Xenophthalmus* White, 1846 (*X. wolffi*), the two species of the genus *Neoxenophthalmus* Serène & Umali, 1972, and the only representative of the subfamily Anomalifrontinae.

*Pseudopinnixa carinata* is presently considered a pinnotherid belonging to the subfamily Pinnothereliinae (Schmitt et al. 1973; Ng et al. 2008). Our results show this monospecific genus to be excluded from the highly supported grouping of Pinnothereliinae (Clade I) and Pinnotherinae (Clade II), being affiliated with neither of these major clades nor our newly defined Clade III of the Pinnotheridae. *Pseudopinnixa* is left in a poorly resolved basal polytomy, but given the distance by which it is separated from other putatively pinnotherid groups, it may warrant eventual treatment as a separate family of the Pinnotheroidea. Further molecular analyses must examine the relationship of *Pseudopinnixa* Ortmann, 1894, to a full array of both heterotrematan and thoracotrematan families of the Brachyura. Larval morphology suggests relationships with Grapsoidea, especially the family Macrophthalmidae, a proposed sister family of

Varunidae (Cuesta et al. 2005), but our present analysis does not lend any clear support to this hypothesis.

Several other taxa also do not easily resolve as to their exact relationships with other included taxonomic groups of the pinnotherids, some because of morphology and others because of their placement in the present molecular analyses, though we confidently conclude they are members of the family on the basis of molecular characters. Pinnotherid sp. 1 (currently in description as a new genus by EC) exhibits unique morphological characters that could perhaps justify assignment to a unique subfamily. However, it is unambiguously placed as the most basal branch of Clade I (Pinnothereliinae) in our analysis. Two other taxa (*Pinnixa valerii* and *P. retinens*) and more questionably Pinnotherid sp. 2 (external to this grouping in the separate BAY tree) form a well-supported clade that may also deserve separate subfamilial treatment. Of these, detailed morphological study has been completed only for *P. valerii*, which is pending assignment to a new genus (DLF and JAC in description). There appear to be clear morphological similarities of *Pinnixa valerii* with both *Pinnixa retinens* Rathbun, 1918, and *Alarconia seaholmi* Glassell, 1938, along with some evidence that these three species share characters of the carapace, sternum, abdomen, and third maxilliped that are distinct from other members of the Pinnotheridae. Should further molecular and morphological study support this grouping, these three species and Pinnotherid sp. 2 may deserve assignment to the tribe Alarconiini Števčić, 2005, which in turn could be rediagnosed for elevation to subfamily level.

## 4.2 The Pinnotheridae restricted, two major subfamilies and more

Clade I corresponds remarkably well to generally accepted membership of the current subfamily Pinnothereliinae. With the exception of species already pending assignment to new genera, including those in our Clade III (see above), its molecular definition includes all species of *Pinnixa* for which specimens were available in our analyses and all available specimens of the genus *Austinixa*, but it surprisingly also included *Scleroplax granulata*. Members of the subfamily Pinnothereliinae are characterized by a third maxilliped with the ischium not fused to the merus, which is oriented longitudinally or is skewed toward a longitudinal orientation. The palp is comparatively large, occasionally as wide as the ischiomerus; the carapace is ovoid in outline, usually much wider than long; and the fifth walking leg is often reduced (Balss 1957). In contrast to other members of Clade I, *Scleroplax* has been assigned previously to the Pinnotherinae by Schmitt et al. (1973). However, this genus does share with the genera *Pinnixa* and *Austinixa* a wider than long carapace and a distinct lateral exopod lobe on the third maxilliped (Campos 2006), characters that may be of more significance than previously thought.

In Clade II we find representatives of a restricted subfamily Pinnotherinae. Morphological characters typical of this subfamily are a third maxilliped ischium that is not distinguishable from, or is at least rudimentarily fused with, the merus, which usually lies transversely or is skewed toward a transverse orientation. The palp is not as wide as the ischiomerus and the carapace usually does not have a clearly transverse rectangular shape (Balss 1957). For the most part, our results agree with the reorganization adopted by Ng et al. (2008), which leaves two subfamilies within the Pinnotheridae, namely Pinnothereliinae Alcock, 1900, and Pinnotherinae de Haan, 1833. However, contrary to their placements, the monotypic genus *Scleroplax* belongs instead among the Pinnothereliinae, supported both by our results and by morphological characters (Campos 2006). Also, it does not appear that either of these two subfamilies encompasses at least one other minor clade (Clade III) that is well-supported in our molecular phylogenetic analyses.

Clearly, our molecular phylogenetic analysis contradicts a close monophyletic grouping of the genera *Pinnixa*, *Fabia*, and *Juxtafabia* that was previously postulated on the basis of larval morphology (Marques & Pohle 1995), as these genera represent members of separate subfamilies that are divergent at a basal node. The molecular data suggest that ostensibly synapomorphic larval features of the abdominal somites are instead best regarded as convergences. Adult morphological differences between the *Pinnixa/Austinixa/Scleroplax* group and *Fabia/Juxtafabia* group would also support present molecular evidence indicating that these two groups of genera do not have a close sister relationship (see Campos 1993, 1996a, 2006).

4.3 *Constituents of the subfamily Pinnothereliinae*

Within Clade I, the subfamily Pinnothereliinae, five internal clades were distinguished. Clades IA and IB included most species of *Austinixa* in our analysis, with only *Austinixa patagoniensis* distinctly excluded from these groups. The character that differentiates members of this genus (formerly treated as the "*Pinnixa cristata* complex," Manning & Felder 1989) from others in the subfamily Pinnothereliinae is a complete (side to side) transverse ridge or carina across the cardiac region of the carapace (Heard & Manning 1997). In previous molecular genetic studies of species assigned to *Austinixa*, varied trees were based upon analyses of 16S and COI mitochondrial genes, and slight differences from our outcomes were evident in some (Harrison 2004). As in our present results, *A. aidae* and *A. hardyi* were resolved in at least some of those previous analyses as distinct but closely related species, forming a sister group to *A. gorei*. Placement of the undescribed *Austinixa* sp. 1 into this clade suggests yet other members of this grouping remain to be named. Our results also agree with the previous report of Harrison (2004) in placing *A. behreae* and *A. chacei* as sister species, and in both cases *A. cristata* is somewhat separated from the two aforementioned clades, being in our analysis basally positioned in Clade IA. Differences arise in that *A. patagoniensis* occupies a basal position within the genus *Austinixa* in the earlier analysis (Harrison 2004), but it must be noted that this earlier work included only two species of the subfamily Pinnothereliinae. Thus, the position of *A. patagoniensis* relative to varied members of the genus *Pinnixa* could not be robustly evaluated. But also of potential impact, *Scleroplax granulata* was used in this previous analysis as an outgroup, while present evidence suggests it is in fact a member of the subfamily Pinnothereliinae.

Our own phylogeny suggests that revisions may be justified for the genus *Austinixa*. Whether or not one were to split Clades IA and IB into separate genera, a separate genus does appear warranted for *A. patagoniensis*, which differs morphologically from all other present members of the genus in having branchial ridges that extend fully to the orbits (Manning & Felder 1989). The positions of *A. bragantina* and *A. leptodactyla* relative to the clades we have defined remain unknown, since these species were not available for inclusion. At the specific level, it has been recommended recently that *A. hardyi* and *A. aidae* be treated as synonyms (Harrison & Hanley 2005) on the bases of mitochondrial genetic and morphometric analyses. However, the genetic distances we observed for these two species had high support values and were not smaller than others shown by different species pairs, as, for example, between some of the species within Clade ID for *Pinnixa* or between *Pinnixa chaetopterana* and *P. sayana* (Clade IC). While Harrison & Hanley (2005) reported a genetic distance of only 0.28% within the COI region, and no differences at all for the 16S region they analyzed, we found a genetic distance of 1.53% for the 16S region we studied (8 mutations in the 16S region, since the genes for tRNA-Leu and NADH1 were not included in the sequence for *A. hardyi* we obtained from GenBank). Our differing outcomes are not readily explained, but we also find no ambiguity in applying diagnostic morphological characters (*sensu* Heard & Manning 1997) to the separation of these species. Clearly, additional analyses would be welcomed, but for now we must recommend treatment of *A. hardyi* and *A. aidae* as separate species.

A third internal clade (IC) of the subfamily Pinnothereliinae included four species of the genus *Pinnixa* (*P. rapax, P. chaetopterana, P. sayana,* and the undescribed *Pinnixa* sp.), while a fourth clade (ID, dominated by northeastern Pacific species of *Pinnixa*) was also formed along with the northeastern Pacific *Scleroplax granulata* as previously discussed. With good support for most branches among species of *Pinnixa* in our analyses, the topology strongly suggests that this genus is polyphyletic and requires revision. However, our present representation of this largest genus of the Pinnotheridae includes but a fraction of its almost 60 presently named species. Furthermore, only one other species of the genus aligned closely with the type species, *Pinnixa cylindrica*, which was basally positioned within the subfamily Pinnothereliinae; this suggests that most species presently assigned to the genus would better be treated under some other generic name. In addition, no readily apparent morphological character sets have been found to support most of the branch groupings among species of *Pinnixa* that were here defined by molecular methods. Morphological and further molecular analyses of *Pinnixa* sensu lato are in progress, and revision of the genus must follow.

Finally, it is imperative that *Pinnotherelia*, type genus of Pinnothereliinae, eventually be included in molecular phylogenetic analyses. This genus is morphologically very different from all putative members of the Pinnothereliinae included in our present analysis, and may require restricted application of this subfamily name. The genera we have treated do indeed form a morphologically and molecularly defined group, but one that may instead warrant recognition as a separate subfamily, perhaps equivalent to the tribe Pinnixini of Števčić (2005).

## 4.4 Constituents of the subfamily Pinnotherinae

Within Clade II, the subfamily Pinnotherinae, three internal clades were recognized, with one of them (IIA) questionably including *Zaops* with a well-defined grouping of the genera *Orthotheres*, *Pinnotheres*, *Limotheres*, *Austinotheres*, and *Nepinnotheres*. The composition of this clade is particularly of interest in that it lends provisional support to a revised classification recently proposed by one of us (Campos 2009) on the basis of adult and larval morphological characters. Under this pending revision, 25 genera (8 tentatively) are proposed to constitute a restricted, monophyletic subfamily Pinnotherinae in which all members share a soft, thin carapace and a unique protuberance on the basal antennal article. Of the 25 genera so grouped, to date we have been able to represent only the aforementioned six in our molecular analyses, but they may indeed be definable as in a single clade. To this end, additional analyses with more representative genera will be essential, especially to resolve the questionable placement of genera like *Zaops*.

The remaining genera that were treated as Pinnotherinae in the Schmitt et al. (1973) classification (excepting *Scleroplax*, as earlier noted) but excluded from the subfamily by Campos (2009) are grouped into at least two other clades (IIB and IIC), which again generally conform with Campos' revised grouping of subfamilies. Separated as Clade IIB, under strong support values, are the genera *Pinnaxodes* and *Holothuriophilus*, which have long been regarded as close relatives, with species having been transferred back and forth between them and remaining debate as to the proper assignment of species for each (see Manning 1993b; Ng & Manning 2003). Members of both these genera use holothurians as hosts and exhibit very similar morphology in the third maxilliped (Ng & Manning 2003).

Clade IIC, by contrast, encompasses a more complex topology, with some internal subgroupings that appear to reflect morphological similarities. Considering that *Clypeasterophilus* was originally erected to receive some members of *Dissodactylus* by Campos & Griffith (1990), it is not surprising to see these genera positioned closely in our phylogeny, given that they share adaptive synapomorphies such as bifid walking leg dactyls and a similar fusion of abdominal somites. However, it is also evident that our present molecular phylogenetic analysis does not support monophyly in either of these genera. Both *Clypeasterophilus* and *Dissodactylus* may warrant further subdivision and/or revisionary reassignments in membership.

A sister clade to the *Clypeasterophilus/Dissodactylus* group is formed by *Tunicotheres*, while *Tumidotheres* and *Calyptraeotheres* are strongly grouped as a more basal branch. At least some support for these groupings may be found in morphology, though it is not entirely congruent with proximities suggested by molecular phylogenetics. Some species of the *Clypeasterophilus/Dissodactylus* group share a two-segmented third maxilliped palp with *Tunicotheres*, though shape of the palp articles in the latter genus differs. Morphology in the former genera appears nearer that of *Calyptraeotheres*, which contains species with very similar third maxillipeds (and other features), even though they may bear a two- or three-segmented palp. It is noteworthy that members of the genus

*Tumidotheres* most resemble *Fabia* in this character (Campos 1996a, b). Zoeal morphology of the *Clypeasterophils/Dissodactylus* group and of at least the type species of *Calyptraeotheres* is very similar, even though it has not been formally described (but see Marques & Pohle 1995). On the other hand, *Tumidotheres*, *Calyptraeotheres*, and *Tunicotheres* are morphologically and ecologically very different from one another. The only shared feature presently apparent among them is the dactylus of the walking leg 4 (pereopod 5), which is larger than the others, a character that develops in the adult female. Thus, we cannot at present offer a set of morphological features that uniquely groups all of these genera to support the genetically defined Clade IIC. Present knowledge of larval and adult morphology would suggest a closer relationship of the *Clypeasterophilus/Dissodactylus* group to *Calyptraeotheres* than to other genera of Clade IIC.

Finally, we note a highly supported separation between the included populations of *Clypeasterophilus stebbingi* from Brazil and Venezuela, respectively. Distances between these two populations suggest they likely represent separate species.

### 4.5 Limited utility of the nuclear 18S rRNA in phylogenetic analysis of the Pinnotheridae

The nuclear gene for the large ribosomal subunit 18S rRNA has been used previously for phylogenetic studies of many crustacean groups at varied phylogenetic levels, including studies of decapods at the level of family and above (e.g., Kim & Abele 1990; Crandall et al. 2000; Oakley 2005; Porter et al. 2005). Initially, our analyses of this gene looked promising for study of pinnotherid genera, as the genetic variation that we found among the first set of genera that we analyzed appeared to be larger than that reported previously among genera of other decapod families (Crandall et al. 2000). However, while 18S rRNA sequences served to differentiate among pinnotherid genera, and in some cases even species, it does not allow us to infer a well-supported phylogeny within the family. While the overall topology of the pinnotherids and their putative relatives by ML (not shown) approximated the phylogeny based upon our mitochondrial sequences, bootstrap values generally did not exceed the 50% majority consensus rule. Nonetheless, it provided a definition of the subfamily Pinnothereliinae that grouped the included species of *Austinixa*, *Pinnixa* (*P. valerii* not included in analysis), and *Scleroplax granulata*, as inferred from the combined genes 16S rRNA, tRNA-Leu, and NADH1, albeit with somewhat different internal topology. This adds evidence for reassignment of *Scleroplax* to this subfamily. It is also of interest that *Pinnixa cylindrica* is separated in the 18S ML analysis at high support values from the other included members of *Pinnixa* (*P. monodactyla* not available for inclusion).

Membership of the subfamily Pinnotherinae (sensu Schmitt et al. 1973) is not resolved by the 18S analyses. Some taxa like *Xenophthalmus pinnotheroides* were peculiarly placed among the pinnotherine genera, perhaps because of long-branch attraction. While positioned external to the pinnotherids among representatives of the outgroup families as in our mitochondrially based analysis, the asthenognathine genus *Tritodynamia* is not definitively affiliated to any one grapsoid family in the 18S analysis; this should be expected, as there was no strong support for separation of these families from one another in the 18S analysis, at least based on our presently limited sampling. Yet, as in our mitochondrially based phylogeny, *Zaops* and *Limotheres* were grouped, and *Dissodactylus*, *Clypeasterophilus*, and *Tumidotheres* were grouped, in both cases at moderate levels of support. *Pinnaxodes* and *Holothuriophilus* were also grouped together, and *Pseudopinnixa* was positioned basally, both as in the 16S analysis, but in both cases at low support values.

We must conclude that genetic variability in the 18S rRNA gene within the members of the family Pinnotheridae is not high enough to allow general resolution of the relationships among most of its constituent genera or thus a bootstrap-supported topology of its subfamilies. Indicative of this is the difference between the overall mean distance for the mitochondrial pinnotherid sequences (0.17) and those for 18S (0.013). Limitations of 18S analyses have been previously noted (Hillis & Dixon 1991; Aleshin & Petrov 1999). While this gene can be informative, its utility is apparently defined not only by the phylogenetic level at which it is applied but also by unique evolutionary histories of the taxonomic group under investigation.

4.6 *Perspectives for the future*

While present results from our analyses of mitochondrial genes allow a number of conclusions, work is under way to confirm and refine these results. On one front, we will integrate additional sequence data into our analyses, including at least the nuclear 28S rRNA gene and two more mitochondrial genes, the cytochrome oxidase subunit I (COI) and the 12S rRNA gene. We are also expanding taxonomic coverage in these analyses, seeking to more comprehensively represent a greater diversity of named and pending pinnotherid genera. We are also continuing to add coverage at the species level in our analyses, especially in large genera like *Pinnixa*, to undertake taxonomic revisions that appear to be warranted, and to define ecologically informative clades. At the other extreme, we seek to integrate all of these data into a comprehensive analysis of phylogeny of brachyuran decapods that will provide improved resolutions at the family and superfamily level. As possible, we are integrating further efforts in our respective labs to draw upon multiple genes in our molecular phylogenies as well as adult and larval characters in morphological analyses.

ACKNOWLEDGEMENTS

We thank A. Anker, A. Baeza, A. Baldwin, J. Bolaños, R. Collin, A.R. de Campos, E.G. Farley, E. Felix-Pico, E. García, Y. Henmi, G. Hernández, G. Jensen, R. Heard, Y. Katakura, R. King, I. López de la Rosa, F. Mantelatto, K. Matsuo, S. Morgan, S. Nates, J. Neigel, R. Robles, I.T. Rodríguez, C. Schubart, M. Takahashi, M. Thiel, and B. Thoma for assistance in obtaining crucially important specimens or with various aspects of data collection, analysis, and manuscript preparation. We are also grateful to T. Mokai for providing loans from the Natural History Museum and Institute, Chiba, Japan; C. Fransen for loans from the Rijksmuseum van Natuurlijke Historie, Nationaal Naturhistorisch Museum, Leiden, The Netherlands; and R. Lemaitre for loans from the National Museum of Natural History, Smithsonian Institution, Washington D.C., USA. This study was supported by U.S. National Science Foundation grants NSF/BS&I DEB-0315995 and NSF/AToL EF-0531603 to D.L. Felder, as well as several small travel grants from the Smithsonian Marine Station, Ft. Pierce, Florida. Additional support for J.A. Cuesta was provided through a postdoctoral research fellowship funded by the "Ministerio de Educación, Cultura y Deportes," Spain. E. Campos received scholarship support (2005–2008) from the "Consejo Nacional de Ciencia y Tecnología" (CONACyT), México, and grant support from CONACyT S52903-Q "Sistemática, relaciones filogenéticas y evolución de los géneros Americanos de la subfamilia Pinnotherinae (Crustacea, Brachyura, Pinnotheridae)". This is University of Louisiana Laboratory for Crustacean Research contribution no. 128 and Smithsonian Marine Station (Ft. Pierce) contribution no. 741.

REFERENCES

Ahyong, S. & Ng, P.K.L. 2007. The pinnotherid type material of Semper (1880), Nauck (1880) and Bürger (1895) (Crustacea: Decapoda: Brachyura). *Raff. Bull. Zool.* Supplement 16: 191–226.

Aleshin, V.V. & Petrov, N.B. 1999. Implicaciones del gen 18S ARNr en la evolución y filogenia de los Arthropoda. In: Evolución y Filogenia de Arthropoda. Sección II: Los artrópodos en el árbol de la vida. *Bol. S.E.A.* 26: 177–196.

Balss, H. 1957. Decapoda. In: Bronns, H.G. (ed.), *Klassen und Ordnungen des Tierreiches. Fünfter Band*, Abteilung 1: 1505–1672.

Bull, J.J., Huelsenbeck, J.P., Cunningham, C.W., Swofford, D.L. & Waddell, P.J. 1993. Partitioning and combining data in phylogenetic analysis. *Syst. Biol.* 42: 384–397.

Campos, E. 1989. *Tumidotheres*, a new genus for *Pinnotheres margarita* Smith, 1869, and *Pinnotheres maculatus* Say, 1818 (Brachyura: Pinnotheridae). *J. Crust. Biol.* 9: 672–679.

Campos, E. 1993. Systematics and taxonomic remarks on *Pinnotheres muliniarum* Rathbun, 1918 (Crustacea: Brachyura: Pinnotheridae). *Proc. Biol. Soc. Wash.* 106: 92–101.

Campos, E. 1996a. Partial revision of the genus *Fabia* Dana, 1851 (Crustacea: Brachyura: Pinnotheridae). *J. Nat. Hist.* 30: 1157–1178.

Campos, E. 1996b. Partial revision of pinnotherid crab genera with a two-segmented palp on the third maxilliped (Decapoda: Brachyura). *J. Crust. Biol.* 16: 556–563.

Campos, E. 1999. Inclusion of the austral species *Pinnotheres politus* (Smith, 1869) and *Pinnotheres garthi* Fennuci, 1975 within the genus *Calyptraeotheres* Campos, 1990 (Crustacea: Brachyura: Pinnotheridae). *Proc. Biol. Soc. Wash.* 112: 536–540.

Campos, E. 2006. Systematics of the genus *Scleroplax* Rathbun, 1893 (Crustacea: Brachyura: Pinnotheridae). *Zootaxa* 1344: 33–41.

Campos, E. 2009. A new species and two new genera of pinnotherid crabs from the northeastern Pacific Ocean, with a reappraisal of the subfamily Pinnotherinae de Haan, 1833 (Crustacea: Brachyura: Pinnotheridae). *Zootaxa*, 2022: 29–44.

Campos, E. & Griffith, H. 1990. *Clypeasterophilus*, a new genus to receive the small-palped species of the *Dissodactylus* complex (Brachyura: Pinnotheridae). *J. Crust. Biol.* 10: 550–553.

Campos, E. & Manning, R.B. 2000. The identities of *Pinnotheres nudus* Holmes, 1895 and *P. nudus sensu* Weymouth, 1910 (Crustacea: Decapoda: Pinnotheridae). *Proc. Biol. Soc. Wash.* 113: 799–805.

Castresana, J. 2000. Selection of conserved blocks from multiple alignments for their use in phylogenetic analysis. *Mol. Biol. Evol.* 17: 540–552.

Coelho, P.A. 1997. Revisão do gênero *Pinnixa* White, 1846, no Brasil (Crustacea, Decapoda, Pinnotheridae). *Trab. Oceanog. Univ. Fed. PE, Recife* 25: 163–193.

Coelho, P.A. 2005. Descriçao de *Austinixa bragantina* sp. nov. (Crustacea: Decapoda: Pinnotheridae) do litoral do Pará, Brasil. *Rev. Brasil. Zool.* 22: 552–555.

Crandall, K.A. & Fitzpatrick Jr., J.F. 1996. Crayfish molecular systematics: using a combination of procedures to estimate phylogeny. *Syst. Biol.* 45: 1–26.

Crandall, K.A., Harris, D.J. & Fetzner Jr, J.W. 2000. The monophyletic origin of freshwater crayfish estimated from nuclear and mitochondrial DNA sequences. *Proc. R. Soc. Lond. B* 267: 1679–1686.

Cuesta, J.A., Schubart, C.D. & Felder, D.L. 2001. Larval morphology and preliminary molecular systematics for the family Pinnotheridae de Haan, 1833, as evidence for a revised classification. *[Abstracts of the] Fifth International Crustacean Congress, Melbourne*: 56.

Cuesta, J.A., Schubart, C.D. & Felder, D.L. 2005. Systematic position of Asthenognathidae Stimpson, 1858 and *Pseudopinnixa carinata* Ortman (Decapoda, Brachyura): new findings from larval and DNA comparisons. *[Abstracts of the] Sixth International Crustacean Congress, Glasgow*: 127.

Edgar, R.C. 2004. MUSCLE: a multiple sequence alignment method with reduced time and space complexity. *BMC Bioinformatics* 5: 113.

Griffith, H. 1987. Taxonomy of the genus *Dissodactylus* (Crustacea: Brachyura: Pinnotheridae) with descriptions of three new species. *Bull. Mar. Sci.* 40: 397–422.

Hall, T.A. 1999. BioEdit: a user-friendly biological sequence alignment editor and analysis program for Windows 95/98/NT. *Nucl. Acids Symp. Ser.* 41: 95–98.

Harrison, J.S. 2004. Evolution, biogeography, and the utility of mitochondrial 16S and COI genes in phylogenetic analysis of the crab genus *Austinixa* (Decapoda: Pinnotheridae) *Mol. Phylogen. Evol.* 30: 743–754.

Harrison, J.S. & Hanley, P.W. 2005. *Austinixa aidae* Righi, 1967 and *A. hardyi* Heard and Manning, 1997 (Decapoda: Brachyura: Pinnotheridae) synonymized, with comments on molecular and morphometric methods in crustacean taxonomy. *J. Nat. Hist.* 39: 3649–3662.

Heard, R.W. & Manning, R.B. 1997. *Austinixa*, a new genus of pinnotherid crab (Crustacea: Decapoda: Brachyura), with the description of *A. hardyi*, a new species from Tobago, West Indies. *Proc. Biol. Soc. Wash.* 110: 393–398.

Hillis, D.M. & Dixon, M.T. 1991. Ribosomal DNA: molecular evolution and phylogenetic inference. *Quart. Rev. Biol.* 66: 411–453.

Kim, W. & Abele, L.G. 1990. Molecular phylogeny of selected decapod crustaceans based on 18S rRNA nucleotide sequences. *J. Crust. Biol.* 10: 1–13.

Manning, R.B. 1993a. West African pinnotherid crabs, subfamily Pinnotherinae (Crustacea, Decapoda, Brachyura). *Bull. Mus. natl. Hist. Nat., Paris 4$^e$ sér., 15, section A, n$^o$s 1–4*: 125–177.

Manning, R.B. 1993b. Three genera removed from the synonymy of *Pinnotheres* Bosc, 1802 (Brachyura: Pinnotheridae). *Proc. Biol. Soc. Wash.* 106: 523–531.

Manning, R.B. & Felder, D.L. 1989. The *Pinnixa cristata* complex in the Western Atlantic, with a description of two new species (Crustacea: Decapoda: Pinnotheridae). *Smith. Contr. Zool.* 473: 1–26.

Marques, F. & Pohle, G. 1995. Phylogenetic analysis of the Pinnotheridae (Crustacea, Brachyura) based on the larval morphology, with emphasis on the *Dissodactylus* species complex. *Zool. Scripta* 24: 347–364.

Medlin, L., Hille, J.E., Shawn, S. & Sogin, M.L. 1988. The characterization of enzymatically amplified eukaryotic 16S-like rRNA-coding regions. *Gene* 71: 491–499.

Ng, P.K.L. & Manning, R.B. 2003. On two new genera of pea crabs parasitic in holothurians (Crustacea: Decapoda: Brachyura: Pinnotheridae) from the Indo-West Pacific, with notes on allied genera. *Proc. Biol. Soc. Wash.* 116: 901–919.

Ng, P.K.L., Guinot, D. & Davie, P.J.F. 2008. Systema Brachyurorum: Part I. An annotated checklist of extant brachyuran crabs of the world. *Raff. Bull. Zool.* 17: 1–286.

Oakley, T.H. 2005. Mydocopa (Crustacea: Ostracoda) as models for evolutionary studies of light and vision: multiple origins of bioluminescence and extreme sexual dimorphism. *Hydrobiologia* 538: 179–192.

Pohle, G. & Marques, F. 1998. Phylogeny of the Pinnotheridae: larval and adult evidence, with emphasis on the evolution of gills. *Invert. Repr. Dev.* 33: 229–239.

Porter, M.L., Pérez-Losada, M. & Crandall, K.A. 2005. Model-based multi-locus estimation of decapod phylogeny and divergente times. *Mol. Phylog. Evol.* 37: 355–369.

Posada, D. & Crandall, K.A. 1998. MODELTEST: testing the model of DNA substitution. *Bioinformatics* 14: 817–818.

Rathbun, M.J. 1918. The grapsoid crabs of America. Smithsonian Institution, *US Nat. Mus. Bull.* 97: 1–461.

Robles, R., Schubart, C.D., Conde, J.E., Carmona-Suárez, C., Álvarez, F., Villalobos, J. & Felder, D.L. 2007. Molecular phylogeny of the American *Callinectes* Stimpson, 1860 (Brachyura: Portunidae), based on two partial mitochondrial genes. *Mar. Biol.* 150: 1265–1274.

Schmitt, W.L., McCain, J.C. & Davidson, E.S. 1973. Decapoda I, Brachyura I, family Pinnotheridae. In: Gruner, H.E. & Holthuis, L.B. (eds.), *Crustaceorum Catalogus*. Den Haag, The Netherlands: Dr. W. Junk BV.

Schubart, C.D., Neigel, J.E. & Felder, D.L. 2000. Use of the mitochondrial 16S rRNA gene for phylogenetic and population studies of Crustacea. In: von Vaupel Klein, J.C. & Schram, F.R. (eds.), *Crustacean Issues 12, The Biodiversity Crisis and Crustacea*: 817–830. Rotterdam: Balkema.

Stamatakis, A., Hoover, P. & Rougemont, J. 2008. A rapid bootstrap algorithm for the RAxML web servers. *Syst. Biol.* 57: 758–771.

Števčić, Z. 1996. Preliminary revision of the family Pinnotheridae. *[Abstracts of the] Colloquium Crustacea Decapoda Mediterranea, Florencia*: 87.

Števčić, Z. 2005. The reclassification of brachyuran crabs (Crustacea: Decapoda: Brachyura). *Nat. Croat.* 14 (suppl. 1): 1–159.

Swofford, D.L. 2002. PAUP*: Phylogenetic Analysis Using Parsimony (*and other methods), 4.0 Beta. Sunderland, Massachusetts: Sinauer Associates.

Tamura, K., Dudley, J., Nei, M. & Kumar, S. 2007. MEGA4: Molecular Evolutionary Genetics Analysis (MEGA) software version 4.0. *Mol. Biol. Evol.* 24: 1596–1599.

Zmarzly, D. 1992. Taxonomic review of pea crabs in the genus *Pinnixa* (Decapoda: Brachyura: Pinnotheridae) occurring on the Californian shelf, with descriptions of two new species. *J. Crust. Biol.* 12: 677–713.

# Evolutionary Origin of the Gall Crabs (Family Cryptochiridae) Based on 16S rDNA Sequence Data

REGINA WETZER[1], JOEL W. MARTIN[1] & SARAH L. BOYCE[2]

[1] *Natural History Museum of Los Angeles County, 900 Exposition Boulevard, Los Angeles, CA 90007*
[2] *Harvard University, Cambridge, MA 02138*

ABSTRACT

Gall crabs (family Cryptochiridae) are small brachyuran crabs living on or in depressions formed in scleractinian corals. Their adaptation to this unusual habitat has led to specializations, including mucous feeding, small body size, and relatively short appendages. Currently, gall crabs are treated as constituting a distinct superfamily (Cryptochiroidea) that contains the sole family Cryptochiridae. There has never been an attempt to elucidate the relationships of the gall crabs to other brachyurans. The group is therefore an ideal candidate for employing molecular data to deduce phylogenetic relationships. We sequenced a 545-bp fragment of the 16S mitochondrial gene from specimens of a widespread species of cryptochirid (*Hapalacarcinus marsupialis*) from Mexico and French Polynesia and compared these to other crab sequences available in GenBank. Our preliminary analyses confirm the placement of the cryptochirids in the Brachyura subsection Thoracotremata. Our results also indicate that cryptochirids are members of the superfamily Grapsoidea and are probably closely allied with the family Grapsidae. The Grapsoidea as presently defined is considered a paraphyletic assemblage.

1 INTRODUCTION

Crabs of the family Cryptochiridae Paul'son, 1875, are among the most unusual of all groups of decapod crustaceans. From what little we know about their biology and natural history, it appears that young crabs settle on scleractinian corals, and most species somehow induce the coral to grow over and around the crab. For some cryptochirids, the result is merely a protective indentation or crevice within the coral, and there appears to be little modification of the host. Females, and in some cases males, live in open pits or tunnels in the corals, or on the surface of the corals. Some species (notably *Hapalocarcinus marsupialis* and *Pseudohapalocarcinus ransoni*) live within the protective confines of a coral "gall" that completely or partially (in the case of *Pseudohapalocarcinus*) encompasses and protects the crab, where it remains for the remainder of its life (see Kropp 1986, 1988; Abelson et al. 1991; Carricart-Ganivet et al. 2004 for reviews of species-specific life histories). Males, which are far smaller than females, and about which less is generally known, are also sometimes found in pits or depressions on the same coral (e.g., the crab genus *Fungicola*, which inhabits fungiid corals) or are not directly associated with the coral as far as is known. Currently, the family includes 46 extant species (there are no known fossil species) partitioned among 20 genera (Table 1; see also Ng et al. 2008: 212). Cryptochirids are probably found wherever scleractinian coral reefs occur worldwide, although some reef systems have yet to be rigorously sampled for them. There are also species associated with deep-water, ahermatypic corals found far from reefs. Although roughly circumtropical in distribution, the group is most diverse in the Indo-West Pacific. Table 1 is the first compilation

Table 1. Comprehensive list of described genera (in bold) and species of the family Cryptochiridae, with a summary of the coral families and genera that the crabs inhabit, general biogeographic distributions of the crab genera, and depth records. Depth applies to the entire geographic range.

| Genus and Species<br>Known Coral Hosts | General Distribution (of crab) | Primary References |
|---|---|---|
| ***Cecidocarcinus* Kropp & Manning, 1987**<br>Dendrophylliidae: *Dendrophyllia, Enallopsammia* | Atlantic: Valdivia Ridge (southeastern Atlantic, off Namibia); depth 512 m | Kropp & Manning 1987 |
| *Cecidocarcinus brychius* Kropp & Manning, 1987 | | |
| *Cecidocarcinus zibrowii* Manning, 1991 | | |
| ***Cryptochirus* Heller, 1861**<br>Faviidae: *Cyphastrea, Barabatoia, Favia, Favites, Goniastrea, Leptoria, Montastrea, Platygyra*<br>Oculinidae: *Cyathelia* | Red Sea<br>Pacific: Vietnam, Japan, Micronesia (Palau, Guam, Pohnpei); depth <1 to 30 m | Kropp 1990<br>Wei et al. 2006 |
| *Cryptochirus coralliodytes* Heller, 1861 | | |
| *Cryptochirus planus* (Takeda & Tamura, 1983) | | |
| *Cryptochirus rubrilineatus* Fize & Serène, 1957 | | |
| ***Dacryomaia* Kropp, 1990**<br>Siderastreidae: *Psammocora* | Pacific: Vietnam, Japan (Isu Islands, Ogasawara Islands, Ryukyu Islands), Micronesia (Palau, Guam); depth <1 to 8 m | Kropp 1990<br>Wei et al. 2006 |
| *Dacryomaia edmondsoni* (Fize & Serène, 1956a) | | |
| *Dacryomaia japonica* (Takeda & Tamura, 1981b) | | |
| *Dacryomaia* sp. 1 | Pacific: Micronesia (Guam) | Paulay et al. 2003 |
| *Dacryomaia* sp. 2 | Pacific: Micronesia (Guam) | Paulay et al. 2003 |
| ***Detocarcinus* Kropp & Manning, 1987**<br>Caryophylliidae: *Asterosimilia, Caryophyllia*<br>Dendrophylliidae: *Dendrophyllia* (questionable)<br>Oculinidae: *Schizoculina*<br>Rhizangiidae: *Phyllangia* | Atlantic: off Ghana | Kropp & Manning 1987 |
| *Detocarcinus balssi* (Monod, 1956) | | |

Table 1. continued.

| Genus and Species<br>Known Coral Hosts | General Distribution (of crab) | Primary References |
|---|---|---|
| **Fizesereneia Takeda & Tamura, 1980b** | | |
| Mussidae: *Acanthastrea, Lobophyllia, Symphyllia* | Pacific: Vietnam, Indonesia, Japan (Izu Islands, Ryukyu Islands), Australia, Micronesia (Palau, Guam, Pohnpei); depth 1 to 15 m | Kropp 1990 |
| *Fizesereneia heimi* (Fize & Serène, 1956a) | | |
| *Fizesereneia ishikawai* (Takeda & Tamura, 1980b) | | |
| *Fizesereneia latisella* Kropp, 1994 | | |
| *Fizesereneia stimpsoni* (Fize & Serène, 1956b) | | |
| *Fizesereneia tholia* Kropp, 1994 | | |
| **Fungicola Serène, 1966** | | |
| Fungiidae: *Fungia, Podobacia, Sandalolitha* | Pacific: Vietnam, Indonesia, Japan (Ryukyu Islands), Micronesia (Palau, Guam); depth 1 to 15 m | Kropp 1990 |
| *Fungicola fagei* (Fize & Serène, 1956a) | | |
| *Fungicola utinomii* (Fize & Serène, 1956a) | | |
| **Hapalocarcinus Stimpson, 1859** | | |
| Pocilloporidae: *Pocillopora, Seriatopora, Stylophora* | Pacific: Indo-West Pacific to Eastern Pacific (Colombia) Red Sea; depth 1 to 27 m | Kropp 1990<br>Wei et al. 2006 |
| *Hapalocarcinus marsupialis* Stimpson, 1859 | | |
| **Hiroia Takeda & Tamura, 1981a** | | |
| Faviidae: *Cyphastrea, Hydnophora*<br>Merulinidae: *Merulina* | Pacific: Vietnam, Japan (Izu Islands, Ryukyu Islands), Micronesia (Palau, Guam); depth 1 to 19 m | Kropp 1990<br>Wei et al. 2006 |

Table 1. continued.

| Genus and Species Known Coral Hosts | General Distribution (of crab) | Primary References |
|---|---|---|
| *Hiroia krempfi* (Fize & Serène, 1956a) | | |
| **Lithoscaptus Milne Edwards, 1862** | Pacific: Réunion, Vietnam, Japan (Izu Islands, Kushimoto, Ogasawara Islands, Ryukyu Islands), Micronesia (Palau, Guam, Pohnpei), Palmyra Island, Teraina; depth <1 to 12 m | Kropp 1990 Wei et al. 2006 |
| Favidae: *Cyphastrea, Echinopora, Favia, Favites, Hydnophora, Goniastrea, Leptastrea, Platygyra, Plesiastrea* | | |
| Merulinidae: *Merulina* | | |
| *Lithoscaptus grandis* (Takeda & Tamura, 1983) | | |
| *Lithoscaptus helleri* (Fize & Serène, 1957) | | |
| *Lithoscaptus nami* (Fize & Serène, 1957) | | |
| *Lithoscaptus* (?) *pacificus* (Edmondson, 1933)[1] | | |
| *Lithoscaptus paradoxus* Milne Edwards, 1862 | | |
| *Lithoscaptus pardalotus* Kropp, 1995 | | |
| *Lithoscaptus prionotus* Kropp, 1994 | | |
| *Lithoscaptus tri* (Fize & Serène, 1956b) | | |
| **Luciades Kropp & Manning, 1996** | Pacific: Micronesia (Guam); depth 128 to 137 m | Kropp & Manning 1996 |
| Pavonidae: *Leptoseris* | | |
| *Luciades agana* Kropp & Manning, 1996 | | |
| **Neotroglocarcinus Takeda & Tamura, 1980a** | Pacific: Vietnam, Japan (Izu Islands, Ryukyu Islands), Micronesia (Palau, Guam, Pohnpei), Enewetak, Hong Kong; depth <1 to 13 m | Kropp 1990 Wei et al. 2006 |
| Dendrophyllidae: *Turbinaria* | | |
| *Neotroglocarcinus hongkongensis* (Shen, 1936) | | |
| *Neotroglocarcinus dawydoffi* (Fize & Serène, 1956a) | | |

Table 1. continued.

| Genus and Species Known Coral Hosts | General Distribution (of crab) | Primary References |
|---|---|---|
| **Opecarcinus Kropp & Manning, 1987** Agariciidae: *Agaricia*, *Gardineroseris*, *Leptoseris*, *Pavona* Siderasteriidae: *Coscinaraea*, *Siderastrea* | Pacific: Vietnam, Japan, to west coast of Mexico Indian Ocean: Christmas Island Atlantic Ocean: Ascension Island and western Atlantic (Caribbean, Gulf of Mexico south to Brazil); depth <1 to 82 m | Kropp & Manning 1987 Kropp 1990 Wei et al. 2006 |
| *Opecarcinus aurantius* Kropp, 1989 | | |
| *Opecarcinus crescentus* (Edmondson, 1925) | | |
| *Opecarcinus granulatus* (Shen, 1936) | | |
| *Opecarcinus hypostegus* (Shaw & Hopkins, 1977) | | |
| *Opecarcinus lobifrons* Kropp, 1989 | | |
| *Opecarcinus peliops* Kropp, 1989 | | |
| *Opecarcinus pholeter* Kropp, 1989 | | |
| *Opecarcinus sierra* Kropp, 1989 | | |
| **Pelycomaia Kropp, 1990** Faviidae: *Cyphastrea*, *Leptastrea* | Pacific: Vietnam, Micronesia (Guam), Hawaii; depth < 2 m | Kropp 1990 |
| *Pelycomaia minuta* (Edmondson, 1933) | | |
| **Pseudocryptochirus Hiro, 1938** Dendrophyllidae: *Turbinaria* | Pacific: Vietnam, Indonesia, Japan (Isu Islands), Micronesia (Palau, Guam, Pohnpei); depth 1 to 6 m | Kropp 1990 Wei et al. 2006 |
| *Pseudocryptochirus viridis* Hiro, 1938 | | |

Table 1. continued.

| Genus and Species Known Coral Hosts | General Distribution (of crab) | Primary References |
|---|---|---|
| **Pseudohapalocarcinus Fize & Serène, 1956a** | | |
| Agariciidae: *Pavona* | Pacific: Vietnam, Japan (Ryukyu Islands), Micronesia (Palau, Guam, Pohnpei); depth <1 to 21 m | Kropp 1990 |
| *Pseudohapalocarcinus ransoni* Fize & Serène, 1956a | | |
| **Sphenomaia Kropp, 1990** | Central Pacific (Teraina); depth not recorded | Kropp 1990 |
| *Sphenomaia pyriforma* (Edmondson, 1933) | | |
| **Troglocarcinus Verrill, 1908** | Atlantic: Bermuda, Florida, Caribbean south to Brazil, Ascension Island, eastern Atlantic; depth <1 to 75 m | Kropp & Manning 1987 |
| Astrocoeniidae: *Stephanocoenia* | | |
| Caryophylliidae: *Polychathu* | | |
| Faviidae: *Diploria, Manicina* | | |
| Meandrinidae: *Dichocoenia* | | |
| Mussidae: *Isophyllia, Mussa, Mussimilia, Mycetophyllia, Scolymia* | | |
| Oculinidae: *Oculina* | | |
| Siderastreidae: *Siderastrea* | | |
| *Troglocarcinus corallicola* (Fize & Serène, 1956a) | | Carricart-Ganivet et al. 2004 |
| **Utinomiella Kropp & Takeda, 1988** | | |
| Pocilloporidae: *Pocillopora, Stylophora* | Pacific: Japan (Ryukyu Islands), Micronesia (Palau, Guam, Pohnpei), Hawaii Indian Ocean: Andaman Islands; depth 1 to 29 m | Kropp 1990 Wei et al. 2006 |
| *Utinomiella dimorpha* (Henderson, 1906) | | |

Table 1. continued.

| Genus and Species / Known Coral Hosts | General Distribution (of crab) | Primary References |
|---|---|---|
| **Xynomaia Kropp, 1990** | | |
| Faviidae: *Favia, Goniastrea, Montastrea, Oulophllia, Platygyra* Merulinidae: *Merulina* Pectiniidae: *Pectinia* | Pacific: Vietnam, Sumatra, Japan (Izu Islands, Kushimoto), Micronesia (Palau, Guam); depth 1 to 15 m | Kropp 1990 |
| *Xynomaia boissoni* (Fize & Serène, 1956a) | | |
| *Xynomaia sheni* (Fize & Serène, 1956b) | | |
| *Xynomaia verrilli* (Fize & Serène, 1957) | | |
| **Zibrovia Kropp & Manning, 1996** | | |
| Phyllangiidae: *Phyllangia* | Pacific: Philippines Indian Ocean: Madagascar; depth 81 to 100 m | Kropp & Manning 1996 |
| *Zibrovia galea* Kropp & Manning, 1996 | | |

[1] The question mark after the genus name in *Lithoscaptus pacificus* refers to the fact that, because of the poor condition of the type of *Cryptochirus pacificus* Edmondson, Kropp (1990) placed the species in the genus *Lithoscaptus* only tentatively.

that includes all genera and species of the family, the host scleractinian coral genus from which they have been reported, and the general distribution patterns of each cryptochirid genus.

Presumably as an adaptation to their environment (their close association with corals), the cryptochirids have evolved a small, squat, and distinctive body that, although perhaps superficially similar to crabs of the family Pinnotheridae in some species, is unlike that of other crab families, even those that also live as obligate commensals of corals (e.g., trapeziids and domeciids). Based on their morphology, in the most current (and indeed in all other) classifications, the gall crabs are placed in their own family (Cryptochiridae) and superfamily (Cryptochiroidea). There is some (unpublished) information indicating that the family is probably monophyletic (Kropp 1988), but little beyond that. Even placement of the superfamily within the Eubrachyura (higher crabs) has been historically uncertain. For example, Martin & Davis (2001) placed the gall crabs within the subsection Heterotremata, whereas the most recent treatment of the Brachyura (Ng et al. 2008) places the superfamily Cryptochiroidea in the subsection Thoracotremata. It would seem, therefore, that the question of the origin and evolutionary relationships of the cryptochirid crabs is a question perfectly suited to investigation with molecular systematic techniques. We address for the first time the evolutionary relationships of gall crabs to other brachyuran families using molecular sequence data. This study must be considered preliminary in that only two populations of a single species (the widespread *Hapalocarcinus marsupialis* Stimpson, 1859) were included, but the results seem sufficiently robust to suggest affinities of the gall crabs at the superfamily and possibly family level.

## 2 MATERIALS AND METHODS

We sequenced a ~545-bp fragment of the 16S mitochondrial gene from Mexican and French Polynesian specimens of the cryptochirid *Hapalocarcinus marsupialis* Stimpson, 1895. The Mexican material was extracted from crabs removed from corals that had been in the collections of the Natural History Museum of Los Angeles County. The Polynesian material was collected in 2001 and was preserved in ethanol. Locality and collection details as well as GenBank numbers are included in Table 2. Muscle tissue was taken from the fifth pereopod and was extracted with a QIAGEN DNeasy Kit (Qiagen, Valencia, CA). The manufacturer's protocol was followed for extraction, and tissue was macerated in a PCR tube with a pestle and then incubated in a 55°C incubator overnight on a shaking table set to medium speed. Polymerase chain reaction (PCR, Sakai et al. 1988) was carried out with standard PCR conditions (2.5 $\mu$l of 10x PCR buffer, 1.5 $\mu$l of 50 mM $MgCl_2$, 4 $\mu$l of 10 mM dNTPs, 2.5 $\mu$l each of two 10 pmol primers, 0.15 Platinum *Taq* (5 units/$\mu$l), 9.6 $\mu$l double distilled water, and 1 $\mu$l template) and thermal cycling as follows: an initial denaturation at 96°C for 3 minutes followed by 40 cycles of 95°C for 1 minute, 46°C for 1 minute, and 72°C for 10 minutes. 16SrDNA was amplified in both directions with universal 16Sar and 16Sbr primers (Palumbi et al. 1991). PCR products were visualized by agarose (1.2%) gel electrophorsis with Sybr Gold (Invitrogen, Carlsbad, CA), PCR product was purified with Sephadex (Sigma Chemical, St. Louis, MO) on millipore multiscreen filter plates, and DNA was cycle sequenced with ABI Big-dye ready-reaction kit and following the standard cycle sequencing protocol with one quarter of the suggested reaction volume.

Sequences were edited and assembled in Sequencher (Gene Codes Corporation); 16S rDNA was aligned using MAFFT (Multiple Alignment Program for amino acid or nucleotide sequences, Katoh et al. 2002; Katoh et al. 2005) and manually adjusted where mismatches were made. All three LINS, EINS, and GINS alignment protocols were reviewed. Phylogenetic trees were estimated with maximum likelihood (GARLI, Genetic Algorithm for Rapid Likelihood Inference, Zwickl 2006). GARLI phylogenetic searches on aligned nucleotide datasets begin with an assumed model of nucleotide substitutions (GTR), with gamma distributed rate heterogeneity and an estimated proportion of invariable sites. The implementation of this model is exactly equivalent to that in PAUP*, making the log likelihood (lnL) scores obtained directly comparable. All model parameters were estimated, including the equilibrium base frequencies. The gamma model of rate heterogeneity

**Table 2.** Cryptochirids sequenced and GenBank sequences used in analyses.

| Subsection Superfamily | Family | Genus/species | GenBank No. |
|---|---|---|---|
| Cryptochiroidea | | | |
| | Cryptochiridae | *Hapalocarcinus marsupialis* | EU743929 |

Mexico, Baja California Sur, Palmas Bay, Rancho Buena Vista, *Pocillopora* with barnacles, 4.57 m. Original fixative unknown, specimen in 70% ethanol. 15 Sep. 1962. AHF, 1963-13, lot 13, cat. no. 530, JM-2005-003. Coll. Edmond Hobsen. RW05.301.1154.

EU743930

Pacific, Society Islands, French Polynesia, Moorea, 6 km south of airport, site 9, ~17.533°S ~149.783°W, *Pocillopora* with barnacles, snorkel to motu, very close to outer reef, original fixative rum 50% ethanol, subsequently transfered to 95% ethanol. 25 Jul. 2001. JM-2005-004, ST01.055. Coll. Sandy Trautwein. RW05.302.1155.

| Subsection Superfamily | Family | Genus/species | GenBank No. |
|---|---|---|---|
| Heterotremata | | | |
| Potamoidea | Gecarcinucidae | *Sartoriana spinigera* | AM234649 |
| | Potamidae | *Geothelphusa pingtung* | AB266168 |
| Thoracotremata | | | |
| Grapsoidea | Gecarcinidae | *Cardisoma carnifex* | AM180687 |
| | | *Gecarcinus lateralis* | AJ130804 |
| | | *Gecarcoidea lalandii* | AM180684 |
| | Glyptograpsidae | *Glyptograpsus impressus* | AJ250646 |
| | | *Platychirograpsus spectabilis* | AJ250645 |
| | Grapsidae | *Geograpsus lividus* | AJ250651 |
| | | *Goniopsis cruentata* | AJ250652 |
| | | *Grapsus grapsus* | AJ250650 |
| | | *Leptograpsus variegatus* | AJ250654 |
| | | *Metopograpsus latifrons* | AJ784028 |
| | | *Metopograpsus quadridentatus* | DQ062732 |
| | | *Metopograpsus thukuhar* | AJ784027 |
| | | *Pachygrapsus crassipes* | AB197814 |
| | | *Pachygrapsus marmoratus* | DQ079728 |
| | | *Pachygrapsus minutus* | AB057808 |
| | | *Pachygrapsus transversus* | AJ250641 |
| | | *Planes minutus* | AJ250653 |
| | Plagusiidae | *Euchirograpsus americanus* | AJ250648 |
| | | *Percnon gibbesi* | AJ130803 |
| | | *Plagusia squamosa* | AJ311796 |
| | Sesarmidae | *Armases elegans* | AJ784011 |
| | | *Sarmatium striaticarpus* | AM180680 |
| | | *Sesarma meridies* | AJ621819 |
| | | *Sesarma windsor* | AJ621824 |
| | | *Sesarmoides longipes* | AJ784026 |
| | Varunidae | *Austrohelice crassa* | AJ308416 |
| | | *Brachynotus atlanticus* | AJ278831 |
| | | *Cyrtograpsus affinis* | AJ130801 |
| | | *Eriocheir sinensis* | AJ250642 |
| | | *Gaetice americanus* | AJ250643 |
| | | *Helograpsus haswellianus* | AJ308417 |
| | | *Hemigrapsus oregonensis* | AJ250644 |

Table 2. continued.

| Subsection Superfamily | Family | Genus/species | GenBank No. |
|---|---|---|---|
| | | Hemigrapsus sanguineus | AJ493053 |
| | | Paragrapsus laevis | AJ308418 |
| | | Varuna litterata | AJ308419 |
| Ocypodoidea | Camptandriidae | Baruna trigranulum | AB002129 |
| | | Paracleistostoma depressum | AB002128 |
| | Mictyridae | Mictyris brevidactylus | AB002133 |
| | Ocypodidae | Dotilla wichmanni | AB002126 |
| | | Ilyoplax deschampsi | AB002117 |
| | | Scopimera globosa | AB002125 |
| | | Tmethypocoelis ceratophora | AB002127 |
| | Palicidae | Crossotonotus spinipes | AJ130807 |
| | | Palicus caronii | AM180692 |
| Pinnotheroidea | Pinnotheridae | Austinixa hardyi | AF503185 |
| | | Austinixa patagoniensis | AF503186 |
| | | Pinnotheres pisum | AM180694 |

assumes four rate categories. GARLI uses a genetic algorithm approach to simultaneously find the topology, branch lengths, and model parameters that maximize the lnL (Zwickl 2006).

The phylogeny was also estimated with Mr. Bayes 3.0b4 (Ronquist & Hulsenbeck 2003) using Bayesian inferences coupled with Markov chain Monte Carlo techniques. Four Markov–Monte Carlo chains were run for ten million generations, and a sample tree was saved every 1000 generations. Trees chosen from the first one million generations were discarded as "burn in." Trees that were chosen once likelihood scores converged on a stable value were used to construct a 50% majority rule consensus tree in PAUP*.

A ~1860-bp double-stranded fragment of 18SrDNA was also sequenced but not used due to a lack of sequence variation (GenBank numbers EU743931 and EU743932). Taxon selection for the analyses was repeatedly refined, as it was determined that Cryptochiridae are members of Thoracotremata and the Grapsoidea and are nested within the Grapsidae. This realization changed our approach from focusing on 18S rDNA to the more appropriate 16S rDNA for this analysis. Taxa selected for the 16S dataset included broad, but not exhaustive, sampling of Varunidae, Grapsidae, Plagusiidae, Sesarmidae, Camptandridae, Gecarcinidae, Pinnotheridae, and Mictridae, with the goal of associating the Cryptochiridae with its closest relatives.

## 3 RESULTS

Analyses of our cryptochirids from Mexico and Polynesia revealed that despite their geographic separation, both samples were the same species, the widespread and relatively common *Hapalocarcinus marsupialis* Stimpson, 1859. In all of our analyses, the cryptochirids are nested within a group of crabs considered by most workers to constitute the Thoracotremata. More specifically, the genus *Hapalocarcinus* falls within a clade that includes the familiar grapsid genera *Grapsus*, *Geograpsus*, *Goniopsis*, *Leptograpsus*, *Planes*, and *Pachygrapsus* (Fig. 1). Branch lengths for the two *Haplocarcinus* sequences are long, as is the branch length of the *Mictyris* sequence (not shown). Interestingly, however, *Hapalocarcinus* was not close to some of the grapsoids that are common reef inhabitants, such as the genera *Percnon* and *Plagusia*, both of which were at one time considered members of the family Plagusiidae (but see below). Beyond our observations on the gall crabs (based on this single species), our results also indicate that the genus *Pachygrapsus* is not monophyletic, with *P. marmoratus* not clustering with the other four *Pachygrapsus* species.

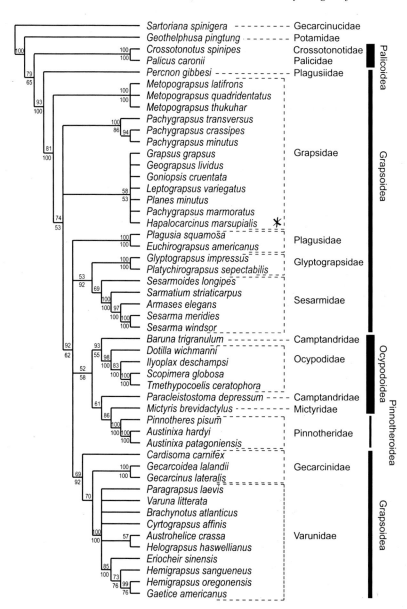

**Figure 1.** Phylogenetic placement of the Cryptochiridae, represented by the genus *Hapalocarcinus* (*), and relationships of Ocypodoidea, Grapsoidea, Pinnotheroidea, and Palicoidea based on 16S mtDNA sequences of 51 taxa, 589 characters, nucleotide frequences: f(A) = 0.24387, f(C) = 0.24433, f(G) = 0.27220, f(T) = 0.23960. This tree is rooted in Gecarcinidae and Potamidae. Topology derived from Bayesian inference 50% majority rule consensus of 18,000 trees. Significance values are posterior probabilities >50% above the branches. GARLI maximum likelihood ln score = -8935.92, 50% majority rule consensus of 74 trees; bootstrap values are below the branches.

Maximum likelihood and Bayesian analyses converged on the same topology. All of our analyses recognize *Glyptograpsus* and *Platychirograpsus* as sister taxa, confirming their placement in the family Glyptograpsidae. The species of *Pinnotheres* and *Austinixa* selected for this analysis constitute a monophyletic clade (the Pinnotheridae). The Varunidae (*Austrohelice*, *Brachynotus*,

*Cyrtograpsus, Eriocheir, Gaetice, Helograpsus, Hemigrapsus, Paragrapsus*, and *Varuna*) is a well-supported monophyletic clade. Gecarcinidae are basal to the Varunidae (posterior probability 69% and bootstrap support 92%). As alluded to above, the plagusiid genera *Plagusia* and *Euchirograpsus* are sister taxa, but they are not at all closely related to the genus *Percnon*, previously included in the Plagusiidae.

At the superfamily level, Pinnotheroidea appears monophyletic, although only three taxa were used in our analysis. The Palicoidea appears as monophyletic and basal to the "grapsoids" in our phylogeny. In our analysis, the superfamilies Ocypodoidea and Grapsoidea are not monophyletic clades.

## 4 DISCUSSION

As noted earlier, in all of our analyses, which must be considered preliminary because of the single species used to represent the gall crabs, the cryptochirids are nested within a group of crabs considered by most workers to constitute the Thoracotremata. This group is defined primarily by having the location of the opening of the vas deferens through the sternum rather than through the coxa of the fifth pereopod (Ng et al. 2008: 8). This placement agrees with the most recent compilation and classification of crabs by Ng et al. 2008 and not with the classification suggested by Martin & Davis (2001), in which the cryptochirids were treated as members of the more diverse Heterotremata. The Ng et al. (2008) classification treats the Thoracotremata as being composed of 17 extant families distributed among four superfamilies: Cryptochiroidea, Grapsoidea, Ocypodoidea, and Pinnotheroidea.

Within the Thoracotremata, our best tree places the gall crab genus *Hapalocarcinus* within a clade that includes the familiar grapsid genera *Grapsus, Geograpsus, Goniopsis, Leptograpsus, Planes*, and *Pachygrapsus*. Since only a single species was sampled in the family, the long branch length of *Haplocarcinus* precludes more accurate placement within the grapsids in this analysis. The association of *Haplocarcinus* with grapsid genera is a somewhat surprising result, in part because there are other groups of crabs that are closely associated with reefs (e.g., trapeziids, domeciids, and some other coral-associated taxa). Also surprising to us was that, even among grapsoids, there are genera more typically associated with reef-dwelling than those with which *Hapalocarcinus* clusters, such as *Percnon* and *Plagusia*; these were not close to the gall crabs in our results. The transition from a coral-obligate commensal group of crabs (such as the trapeziids, tetraliids, or domeciids) to a more heavily coral-dependent group such as the gall crabs would have been, in some ways, easier to understand. However, no such coral-obligates are seen among the crabs that appear closest to *Hapalocarcinus* in our analysis. We should also point out that adaptation to a coral-associated lifestyle does not always result in similar modifications, even among decapods (e.g., consider the morphological differences between trapeziids and domeciids such as *Maldivia*, or between the shrimp genera *Paratypton* and *Alpheus*) despite similar lifestyles and diets.

Some traditional groupings, such as the families Varunidae, Pinnotheridae, Ocypodidae, Sesarmidae, and Glyptograpsidae, are supported in this analysis. However, other traditionally recognized families, such as the Camptandriidae and Plagusiidae, are not supported (see also Schubart et al. 2002; Schubart et al. 2006). Although a case could be made for recognition of the superfamily Pinnotheroidea, and possibly the Ocypodoidea (with the exception of the genera *Paracleistostoma* and *Mictyris*), there is no support for the superfamilies Cryptochiroidea, Grapsoidea, and Ocypodoidea as previously defined (Fig. 1). This perhaps is not surprising in light of the rather weak and likely convergent morphological characters that have been used to define these superfamilies in the past (such as the "rectangular" carapace shape of the grapsoids and the long eyestalks of many ocypodoids).

The pinnotherids, all of which are highly modified (most having extremely short and wide bodies) for a commensal existence, appear to be monophyletic and are not closely related to cryptochirids despite an apparently superficial resemblance (see Introduction), although this result is based on only three representatives of that family. The former family Palicidae (*Crossotonotus* + *Palicus*) (now treated as two families, Crossotonotidae and Palicidae, within a superfamily Palicoidea; Ng et al. 2008) appears basal to the other (non-outgroup) crabs in our study. Palicids are morphologically very unusual in that they have greatly reduced fifth pereopods (see Castro 2000).

Our results are in general agreement with the findings of Schubart et al. (2002, 2006) in their studies of the Glyptograpsidae and of the relationships within the Grapsoidea, respectively. As in the conclusion of Schubart et al. (2006), our results cast doubt on the usefulness of the superfamily categories Grapsoidea and Ocypodoidea, and confirm that *Percnon* is not allied to *Plagusia* and *Euchirograpsus*, such that the family Plagusiidae cannot be recognized as monophyletic.

For the gall crabs, the superfamily status of the Cryptochiroidea is now difficult to justify, as, based on our admittedly small dataset, the gall crabs appear to be highly modified grapsids. For practical reasons, and until more cryptochirid sequences from a broader family sampling are included in future analyses, we suggest maintaining the family status of the Cryptochiridae but treating it as one of many separate "grapsoid" families. We recommend dropping the superfamily category (Cryptochiroidea), while at the same time recognizing that the Grapsoidea, as previously defined, is itself an artificial assemblage. The rather wide geographical range of the gall crabs, summarized in Table 1, and the fact that, despite the geographical distance between the populations sampled in this study (Mexico and French Polynesia), our sequences came from a single species, also are reasons to suspend making any higher-level classificatory changes, as it is possible that convergence to a coral-dwelling habitat has occurred more than once.

## ACKNOWLEDGEMENTS

This work was supported in part by grant number DEB 0531616 from the National Science Foundation's "Assembling the Tree of Life" program to J. W. Martin, in conjunction with collaborative awards to Keith Crandall and Nikki Hannegan (Brigham Young University), Darryl Felder (University of Louisiana Lafayette), and Rodney Feldmann and Carrie Schweitzer (Kent State University). The symposium during which these results were first presented was funded by NSF grant DEB 0721146, with additional support from the American Microscopical Society, the Crustacean Society, the Society of Systematic Biologists, the Society of Integrative and Comparative Biology (SICB), and the SICB Divisions of Invertebrate Zoology and Systematics & Evolutionary Biology. We sincerely thank Roy Kropp and Gustav Paulay for providing information and literature on gall crab biology and for helpful reviews of the manuscript. The participation of J. Martin was also made possible by funding from the U.S. National Oceanographic and Atmospheric Administration NOAA for systematic work on crabs of the Hawaiian Islands. Additional support (for R. Wetzer) was provided by NSF grant DEB-0129317. We especially thank Keith Crandall and the members of his laboratory, especially Rebecca Scholl, Katharina Dittmar, Marcos Pérez-Losada, and Megan Porter, for their hospitality, assistance, and mentoring at the bench as well for their help with the analyses during R. Wetzer's working visits to Brigham Young University in 2006–2007.

## REFERENCES

Abelson, A., Galil, B.S. & Loya, Y. 1991. Skeletal modifications in stony corals caused by indwelling crabs: hydrodynamic advantages for crab feeding. *Symbiosis* 10: 233–248.

Carricart-Ganivet, J.P., Carrera-Parra, L.F., Quan-Young, L.I. & Garia-Madrigal, M.S. 2004. Ecological note on *Troglocarcinus corallicola* (Brachyura: Cryptochiridae) living in symbiosis with *Manicina areolata* (Cnidaria: Scleractina) in the Mexican Caribbean. *Coral Reefs* 23: 215–217.

Castro, P. 2000. Crustacea Decapoda: A revision of the Indo-West Pacific species of palicid crabs (Brachyura Palicidae). In: Crosnier, A. (ed.), *Résultats des Campagnes Musorstom* 21: 437–610.

Edmondson, C.H. 1925. Marine zoology of tropical central Pacific. Crustacea. *Bull. Bernice Bishop Mus.* 27: 3–62.

Edmondson, C.H. 1933. *Cryptochirus* of the central Pacific. *Bernice Bishop Mus. Occas. Papers* 10: 1–23.

Fize, A. & Seréne, R. 1957. Les Hapalocarcinidés du Việt-Nam. *Mém. l'Institut Océanograph. Nhatrang* 10: 1–202.

Fize, A. & Serène, R. 1956a. Note prèliminaire sur huit especes nouvelles, dont une d'un genre nouveau, d'Hapalocarcinidés. *Bull. Soc. Zool. France* 80: 375–378.

Fize, A. & Serène, R. 1956b. Note préliminaire sur quatre espèces nouvelles d'Hapalocarcinids avec quelques remarques au sujet du *Cryptochirus rugosus* Edmonson. *Bull. Soc. Zool. France* 80: 379–382.

Heller, C. 1861. Synopsis der im rothen Meere vorkommenden Crustaceen. *Verhandl. Zoolog.-Botan. Gesellsch. Wien* 11: 3–32.

Henderson, J.R. 1906. On a new species of coral-infesting crab taken by the R.I.M.S. 'Investigator' at the Andaman Islands. *Ann. Mag. Nat. Hist. Ser. 7* 18: 211–219.

Hiro, F. 1938. A new coral-inhabiting crab, *Pseudocryptochirus viridis* gen. et. sp. nov. (Hapalocarcinidae, Brachyura). *Zool. Magazine, Tokyo* 50: 149–151.

Katoh, K., Kuma, K., Toh, H. & Miyata, T. 2005. MAFFT verson 5: improvement in accuracy of multiple sequence alignment. *Nucleic Acids Res.* 33: 511–518.

Katoh, K., Misawa, K., Kuma, K., & Miyata, T. 2002. MAFFT: a novel method for rapid multiple sequence alignment based on fast Fourier transform. *Nucleic Acids Res.* 30: 3059–3066.

Kropp, R.K. 1986. Feeding biology and mouthpart morphology of three species of coral gall crabs (Decapoda: Cryptochiridae). *J. Crust. Biology* 6: 377–384.

Kropp, R.K. 1988. Biology and systematics of coral gall crabs (Crustacea: Cryptochiridae). Ph.D. dissertation, University of Maryland, College Park. 354 pp.

Kropp, R.K. 1989. A revision of the Pacific species of gall crabs, genus *Opecarcinus* (Crustacea: Cryptochiridae). *Bull. Mar. Sci.* 45: 98–129.

Kropp, R.K. 1990. Revision of the genera of gall crabs (Crustacea: Cryptochiridae) occurring in the Pacific Ocean. *Pacific Sci.* 44: 417–448.

Kropp, R.K. 1994. The gall crabs (Crustacea: Decapoda: Brachyura: Cryptochiridae) of the Rumphius Expeditions revisited, with descriptions of three new species. *Raffles Bull. Zool.* 42: 521–538.

Kropp, R.K. 1995. *Lithoscaptus pardalotus*, a new species of coral-dwelling gall crab (Crustacea: Brachyura: Cryptochiridae) from Belau. *Proc. Biol. Soc. Washington* 108: 637–642.

Kropp, R.K. & Manning, R.B. 1987. The Atlantic gall crabs, family Cryptochiridae (Crustacea: Decapoda: Brachyura). *Smithson. Contrib. Zool.* 462: 1–21.

Kropp, R.K. & Manning, R.B. 1996. Crustacea Decapoda: two new genera and species of deep water gall crabs from the Indo-west Pacific (Cryptochiridae). In: Crosnier, A. (ed.), Résultats des Campagnes MUSORSTOM, vol. 15. *Mém. Mus. Nat. His. Natur., Paris* 168: 531–539.

Kropp, R.K. & Takeda, M. 1988. *Utinomiella*, a replacement name for *Utinomia* Takeda et Tamura, 1981 (Crustacea, Decapoda), *non* Tomlinson, 1963 (Crustacea, Acrothoracica). *Bull. Biogeog. Soc. Japan* 43: 29.

Manning, R.B. 1991. Crustacea Decapoda: *Cecidocarcinus zibrowii*, a new deep water gall crab (Cryptochiridae) from new Caledonia. In: Crosnier, A. (ed.), Résultats des Campagnes MUSORSTOM, vol. 9. *Mém. Mus. Nat. His. Natur., Paris* 152: 515–520.

Martin, J.W. & Davis, G.E. 2001. An updated classification of the Recent Crustacea. *Natural History Museum of Los Angeles County, Science Series* 39: 1–124.

Milne Edwards, A. 1862. (Annexe F) Faune carcinologique de l'île de las Réunion. In: Maillard, L. (ed.), *Notes sur l'île de la Réunion*. Paris. 1–16.

Monod, T. 1956. Hippidea et Brachyura ouest-africains. *Mém. Institut Franç. Afrique Noire* 45: 1–674.

Ng, P.K.L., Guinot, D. & Davie, P.J.F. 2008. Systema Brachyurorum: Part I. An annotated checklist of extant brachyuran crabs of the world. *Raffles Bull. Zool.* 17: 1–286.

Palumbi, S.R., Martin, A., Romano, S., McMillan, W.O., Stice, L. & Grabowski, G. 1991. *The Simple Fool's Guide to PCR, version 2*. Department of Zoology and Kewalo Marine Laboratory, University of Hawaii, Honolulu, Hawaii. 43 pp.

Paul'son, O.M. 1875. Izsledovaniya rakoobraznykh krasnago morya s zametkami otnositel'no rakoobraznykh drugikh morei. Chast' 1. Podophthalmata i Edriophthalmata (Cumacea). S.V. Kul'zhenko, Kiev. 164 pp.

Paulay, G., Kropp, R., Ng, P.K.L. & Eldredge, L.G. 2003. The crustaceans and pycnogonids of the Mariana Islands. *Micronesica* 35–36: 456–513.

Ronquist, F. & Huelsenbeck, J.P. 2003. MrBayes 3: Bayesian phylogenetic inference under mixed models. *Bioinformatics* 19: 1572–1574.

Sakai, R., Gelfand, D.J., Srofell, S., Scharf, S.J., Higuchi, R., Horn, G.T., Mullis, K.B. & Erlich, H.A. 1988. Primer-directed enzymatic amplification of DNA with a termostable DNA polymerase. *Science* 239: 487–491.

Schubart, C.D., Cannicci, S., Vannini, M. & Fratini, S. 2006. Molecular phylogeny of grapsoid crabs (Decapoda, Brachyura) and allies based on two mitochondrial genes and a proposal for refraining from current superfamily classification. *J. Zool. System. Evolut. Res.* 44: 193–199.

Schubart, C.D., Cuesta, J.A. & Felder, D.L. 2002. Glyptograpsidae, a new brachyuran family from Central America: larval and adult morphology, and a molecular phylogeny of the Grapsoidea. *J. Crust. Biol.* 22: 28–44.

Serène, R. 1966. Sur deux espéces nouvelles de Brachyoures (Crustacés Décapodes) et sur une troisiéme peu connue, récoltées dans la région malaise. *Bull. Mus. Nat. Hist. Nat., Paris* 38: 817–827.

Shaw, K. & Hopkins, T.S. 1977. The distribution of the family Hapalocarcinidae (Decapoda, Brachyura) on the Florida Middle Ground with a description of *Pseudocryptochirus hypostegus* new species. *Proc. 3rd Internat. Coral Reef Symp., Miami* 1: 177–183.

Shen, C.J. 1936. Notes on the family Hapalocarcinidae (coral-infesting crabs) with description of two new species. *Hong Kong Nat. Suppl.* 5: 21–26.

Stimpson, W. 1859. [Communication, *Hapalocarcinus marsupialis*]. *Proc. Boston Soc. Nat. Hist.* 6: 412–413.

Takeda, M. & Tamura, Y. 1980a. Coral-inhabiting crabs of the family Hapalocarcinidae from Japan. *Bull. Nat. Sci. Mus., Tokyo, Series A (Zoology)* 3: 147–151.

Takeda, M. & Tamura, Y. 1980b. Coral-inhabiting crabs of the family Hapalocarcinidae from Japan, III. New genus *Fizesereneia*. *Bull. Nat. Sci. Mus., Tokyo, Series A (Zoology)* 6: 137–146.

Takeda, M., & Tamura, Y. 1981a. Coral-inhabiting crabs of the family Hapalocarcinidae from Japan. VIII. Genus *Pseudocryptochirus* and two new genera. *Bull. Biogeog. Soc. Japan* 36: 13–27.

Takeda, M. & Tamura, Y. 1981b. Coral-inhabiting crabs of the family Haplocarcinidae from Japan. VII. Genus *Favicola*. *Res. on Crustacea, Carcin. Soc. Japan* 11: 41–50.

Takeda, M. & Tamura, Y. 1983. Coral-inhabiting crabs of the family Hapalocarcinidae from Japan. IX. A small collection made at Kushimoto and Koza, Kii Peninsula. *Bull. Nat. Sci. Mus., Tokyo, Ser. A (Zoology)* 9: 1–12.

Verrill, A.E. 1908. Decapod Crustacea of Bermuda; 1. Brachyura and Anomura. Their distribution, variations, and habits. *Trans. Connecticut Acad. Arts Sci.* 13: 299–473.

Wei, T.-P., Hwang, J.-S., Tsai, M.-L. & Fang, L.-S. 2006. New records of gall crabs (Decapoda, Cryptochiridae) from Orchid Island, Tawain, Northwestern Pacfic. *Crustaceana* 78: 1063–1077.

Zwickl, D.J. 2006. Genetic algorithm approaches for the phylogenetic analysis of large biological sequence datasets under the maximum likelihood criterion. Ph.D. dissertation, The University of Texas at Austin, Austin, TX.

# Systematics, Evolution, and Biogeography of Freshwater Crabs

NEIL CUMBERLIDGE[1] & PETER K.L. NG[2]

[1] *Department of Biology, Northern Michigan University, Marquette, Michigan, U.S.A.*
[2] *Department of Biological Sciences, National University of Singapore, Singapore, Republic of Singapore*

## ABSTRACT

Freshwater crabs are a large group of aquatic animals, with more than 1,280 described species worldwide found in freshwater ecosystems throughout the warmer parts of the Neotropical, Afrotropical, Palaearctic, Oriental, and Australasian zoogeographical regions. We report here on the changes in the understanding of the higher systematics of these decapods over the past 25 years associated with attempts to put freshwater crab taxonomy into a phylogenetic framework. The distributional patterns of the freshwater crabs on continents and islands are interpreted in terms of their dispersal abilities and barriers to their distribution. Theories on freshwater crab origins are discussed in the light of their phylogeny and present-day distributions. Adaptations to a permanent existence in freshwater and the adaptive radiation of freshwater crabs into such ecosystems worldwide are discussed.

## 1 DIVERSITY

The term 'freshwater crab' is most commonly used to refer to the large and diverse group of brachyurans found worldwide throughout freshwater ecosystems of inland waters of the continents in the tropics and subtropics (here called the 'true' freshwater crabs). However, the term 'freshwater crab' also has been applied commonly by different workers to such different groups of decapod crustaceans as the exclusively freshwater anomurans (Aeglidae) (Bond-Buckup et al. 2008) and even to species of predominantly marine brachyuran families (Sesarmidae, Varunidae, Hymenosomatidae) that spend time in freshwater (Ng 1988, 2004; Schubart & Koller 2005), making it necessary to distinguish here between the vernacular use of terms to refer to these very different groups of freshwater decapods. True freshwater crabs are defined here as heterotreme brachyurans that are found exclusively in freshwater habitats (never in brackish or marine environments) and that all reproduce exclusively by direct development (never with larval stages). The recent surge in taxonomic interest in this group has led to the realization that the biodiversity of freshwater crabs is not only much higher than previously thought (Martin & Davis 2001) but that they, in fact, constitute the largest natural group (18.8%) within a vastly expanded and reorganized Brachyura (Ng et al. 2008). The number of species of freshwater crabs has grown tremendously in the past 25 years, with more than 50% of all species described since 1980.

## 2 PHYLOGENY AND HIGHER TAXONOMY

Our understanding of freshwater crab relationships has been boosted by recent morphological and molecular studies, and the relationships of these decapods at the family, genus, and species levels are now becoming much clearer (e.g., Daniels et al. 2006; Klaus et al. 2006; Cumberlidge et al. 2008),

Table 1. Freshwater crab diversity by zoogeographical region and family.

| Family | Region | No. Genera | No. Species |
|---|---|---|---|
| TRICHODACTYLIDAE | Neotropical | 15 | 47 |
| PSEUDOTHELPHUSIDAE | Neotropical | 40 | 251 |
| POTAMONAUTIDAE | Afrotropical | 18 | 132 |
| POTAMIDAE | Afrotropical, Palaearctic, Oriental | 90 | 505 |
| GECARCINUCIDAE | Oriental, Australasian | 57 | 345 |
| **Total:** | | **220** | **1,280** |

although molecular studies on the Neotropical crabs are still not available. The most recent evaluations of freshwater crab biodiversity (Yeo et al. 2008; Ng et al. 2008) recognized more than 1,280 species of freshwater crabs worldwide (Table 1, Fig. 1).

Changes in our understanding of freshwater crab higher taxonomy in recent years (Table 2) has also meant that the number of families has been significantly reduced from the high point of 12 families recognized by Bott (1969, 1970a, b, 1972) and Cumberlidge (1999) and the eight families of Martin & Davis (2001). Recently, Cumberlidge et al. (2008) and Ng et al. (2008) assigned the freshwater crabs to only six families (Pseudothelphusidae, Potamonautidae, Potamidae, Gecarcinucidae, Parathelphusidae, and Trichodactylidae). Six other freshwater crab families, Potamocarcinidae, Deckeniidae, Platythelphusidae, Sundathelphusidae, Isolapotamidae, and Sinopotamidae, have been synonymized. The six valid families of freshwater crabs are separated into two main monophyletic lineages, each assumed to have a different (unknown) marine crab sister group (Sternberg et al. 1999). One of these lineages includes five families (Pseudothelphusidae, Potamonautidae, Potamidae, Gecarcinucidae, and Parathelphusidae), and the other includes only a single family (Trichodactylidae). Klaus et al. (2006) recently argued that the Gecarcinucidae and Parathelphusidae should be regarded as synonymous (the former having priority), supported by Klaus et al. (this

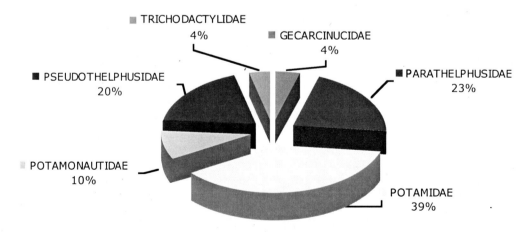

Figure 1. (**See Color Figure 2 in the Color Insert at the end of the book.**) Freshwater crab diversity (Table 2C).

**Table 2.** Recent changes in the higher taxonomy of the true freshwater crabs. (A) Freshwater crab higher taxonomy (Bott 1970b, Cumberlidge 1999). (B) Freshwater crab higher taxonomy (Martin & Davis 2001). (C) Freshwater crab higher taxonomy (Cumberlidge et al. 2008; Ng et al. 2008). (D) Freshwater crab higher taxonomy (present work).

### A. Freshwater crab higher taxonomy (Bott 1970b; Cumberlidge 1999)

Pseudothelphusoidea Ortmann, 1853
    Pseudothelphusidae Rathbun, 1893
    Potamocarcinidae
Potamoidea Ortmann, 1896
    Potamidae Ortmann, 1896
    Potamonautidae Bott, 1970b
    Deckeniidae Ortmann, 1897
    Platythelphusidae Colosi, 1920
    Sinopotamidae Bott, 1970a
    Isolapotamidae Bott, 1970a
Gecarcinucoidea Rathbun, 1904
    Gecarcinucidae Rathbun. 1904
    Parathelphusidae Alcock, 1910
    Sundathelphusidae Bott, 1969
Portunoidea Rafinesque, 1815
    Trichodactylidae H. Milne Edwards, 1853

### B. Freshwater crab higher taxonomy (Martin & Davis 2001)

Pseudothelphusoidea Ortmann, 1853
    Pseudothelphusidae Rathbun, 1893
Potamoidea Ortmann, 1896
    Potamidae Ortmann, 1896
    Potamonautidae Bott, 1970b
    Deckeniidae Ortmann, 1897
    Platythelphusidae Colosi, 1920
Gecarcinucoidea Rathbun, 1904
    Gecarcinucidae Rathbun. 1904
    Parathelphusidae Alcock, 1910
Portunoidea Rafinesque, 1815
    Trichodactylidae H. Milne Edwards, 1853

### C. Freshwater crab higher taxonomy (Cumberlidge et al. 2008; Ng et al. 2008)

Pseudothelphusoidea Ortmann, 1853
    Pseudothelphusidae Rathbun, 1893
Potamoidea Ortmann, 1896
    Potamidae Ortmann, 1896
    Potamonautidae Bott, 1970b
Gecarcinucoidea Rathbun, 1904
    Gecarcinucidae Rathbun, 1904
    Parathelphusidae Alcock, 1910
Trichodactyloidea H. Milne Edwards, 1853
    Trichodactylidae H. Milne Edwards, 1853

**Table 2.** continued.

### D. Freshwater crab higher taxonomy (present work)

Potamoidea Ortmann, 1896
    Pseudothelphusidae Rathbun, 1893
    Potamidae Ortmann, 1896
    Potamonautidae Bott, 1970b
    Gecarcinucidae Rathbun. 1904
Trichodactyloidea H. Milne Edwards, 1853
    Trichodactylidae H. Milne Edwards, 1853

---

volume). There have been only a few phylogenetic studies on freshwater crab family-level relationships, but those that are available indicate that the lineage that includes the five families shares common ancestry, and this warrants their assignment to a single higher taxonomic unit above the family level (Sternberg et al. 1999). We consider that the most appropriate choice would be at the superfamily level, thereby keeping this group of heterotremes consistent with other groups of families elsewhere in the Brachyura (Ng et al. 2008).

This contrasts with the traditional taxonomy that assigned the 12 freshwater crab families to three different superfamilies (Bott 1969, 1970a, b, 1972): the Pseudothelphusoidea (for Pseudothelphusidae and Potamocarcinidae), the Potamoidea (for Potamidae, Potamonautidae, Deckeniidae, Platythelphusidae, Sinopotamidae, and Isolapotamidae), and the Gecarcinucoidea (for Gecarcinucidae, Parathelphusidae, and Sundathelphusidae). Bott (1970a) left the Trichodactylidae without a superfamily assignment, although Banarescu (1990) referred it to a new superfamily, the Trichodactyloidea. Ng et al. (2008) adopted a conservative approach to the higher taxonomy of the freshwater crabs and placed them in four superfamilies: Pseudothelphusoidea (with Pseudothelphusidae), Potamoidea (with Potamidae and Potamonautidae), Gecarcinucoidea (with Gecarcinucidae and Parathelphusidae), and Trichodactyloidea (with Trichodactylidae).

In view of the existing evidence, we propose to provisionally recognize here a single superfamily, the Potamoidea, for the lineage of four families of freshwater crabs (Pseudothelphusidae, Potamonautidae, Potamidae, and Gecarcinucidae). The Potamoidea as defined here is a group with a global distribution and includes species of freshwater crabs from both the New World (Pseudothelphusidae) and the Palaeotropics (Potamonautidae, Potamidae, and Gecarcinucidae). This monophyletic potamoid superfamily, however, excludes the 47 species of Neotropical river crabs assigned to the Trichodactylidae, given that the latter group of species forms a separate clade (Sternberg et al. 1999; Martin & Davis 2001; Schubart & Reuschel this volume).

### 2.1 *Evolution of mandibular palp characters in the potamoid freshwater crabs*

Freshwater crabs traditionally have been assigned to families and superfamilies using characters of the mandibular palp, gonopods, and frontal median triangle (Bott 1970b; Ng 1988). However, these characters may not be as reliable as previously thought. In recent years, phylogenetic character mapping of mandibular palp characters in the five potamoid freshwater crab families onto a consensus phylogeny based on morphological and molecular studies (Fig. 2) has raised doubts. Although mandibular palp characters (such as the number of segments and the form of the terminal article) are invariant in the Pseudothelphusidae, Potamidae, and Gecarcinucidae, this is not true for the Potamonautidae, where the form of the terminal article of the mandibular palp is highly variable across taxa. For example, *Seychellum* Ng, Števčić & Pretzmann, 1995, from the Seychelles

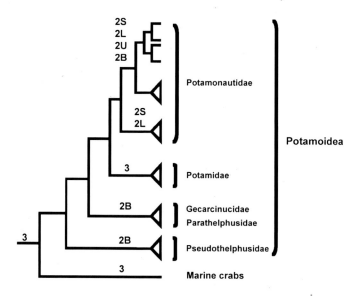

**Figure 2.** Phylogenetic character mapping of characters of the mandibular palp of potamoid freshwater crabs plotted onto a simplified consensus phylogeny (based on mitochondrial and nuclear DNA sequences as well as morphological data) for the freshwater crabs excluding the Trichodactylidae (after Sternberg et al. 1999; Daniels et al. 2006; and Klaus et al. 2006). The mandibular palp characters are consistent at the family level except for the Potamonautidae. 3 = 3-segmented mandibular palp, simple terminal segment; 2S = 2-segmented, simple terminal segment; 2L = 2-segmented, terminal segment with ledge; 2U = 2-segmented, terminal segment unequal bilobed; 2B = 2-segmented, terminal segment subequal bilobed.

and *Deckenia* Hilgendorf, 1869, from East Africa have a strongly supported sister group relationship and both have a 2-segmented mandibular palp. However, the mandibular palp of *Seychellum* has a bilobed terminal segment, whereas that of *Deckenia* has a simple terminal segment. Traditional taxonomic thinking placed these two genera in different families and superfamilies (Ng et al. 1995; Števčić, 2005), whereas these taxa are now both included in the Potamonautidae (Daniels et al. 2006; Cumberlidge et al. 2008; Klaus et al. this volume). Clearly, mandibular palp characters on their own are unreliable for assigning species of Afrotropical freshwater crabs to a family, and this undermines confidence in their use as a high-weight character for the assignment of specimens to other potamoid families. In practice, mandibular palp characters remain useful for family-level placement of species of Pseudothelphusidae and Gecarcinucidae (there are no known exceptions), but there is reason to believe that these characters may be homoplastic, and as such they may not be reliable indicators of higher phylogenetic relationships.

A three-segmented mandibular palp with a simple terminal segment is generally agreed to be the common brachyuran condition and is found in nearly all heterotrematous marine crab families (unpublished data). However, only one potamoid family (Potamidae) has this form of mandibular palp, and (perhaps surprisingly) this is not positioned most basally on the phylogenetic tree (Fig. 2). Instead, it is the Pseudothelphusidae that appears as the most basal family, suggesting perhaps that the 2-segmented bilobed mandibular palp evolved early in this group (Fig. 2) and this is invariant in the family Pseudothelphusidae. The next branch in the tree is a division into two Palaeotropical lineages: (1) a branch with the Gecarcinucidae for specimens with a 2-segmented bilobed mandibular palp, and (2) a branch grouping the Potamidae and Potamonautidae together for specimens with either a 3-segmented mandibular palp (Potamidae) or a 2-segmented mandibular palp (Potamonautidae). Members of the Potamidae all have a 3-segmented mandibular palp with a simple terminal segment, while members of the Potamonautidae have a 2-segmented palp with a terminal segment that exhibits a variety of forms (either simple, or with a small ledge, or bilobed) in different species and genera (Cumberlidge et al. 2008). The only possible explanation for the 3-segmented mandibular palp with a simple terminal segment that is diagnostic of the Potamidae at the moment is that it appears to be a reversal of the apomorphic 2-segmented palp back to the pleisiomorphic condition.

## 2.2 Evolution of gonopod 1 characters in the potamoid freshwater crabs

The gonopod 1 (G1) morphology of potamoid and trichodactylid freshwater crabs has become modified in the course of life in freshwater, and it is now distinctly different from each other and from the G1 of most marine crabs in both of these freshwater crab lineages. Because the marine sister group of the freshwater crabs is unknown, it is difficult to theorize about the original form of the gonopods seen in freshwater crabs. This search is hampered by the specialized and highly derived G1 seen in many families of marine heterotremes and thoracotremes (see Guinot 1977, 1978, 1979). Nevertheless, a stout columnar 3-segmented G1 is typical of many marine crabs, and no marine crabs have the four-part (4-segmented) G1 as seen in Palaeotropical freshwater crabs. Within the five families of potamoid freshwater crabs, the Pseudothelphusidae are the most basal, indicating that the stout 3-segmented G1 (Rodriguez 1982) may be closest to the marine crab ancestral form. However, the G1 of pseudothelphusids is distinguished from the superficially similar G1 of marine crabs such as panopeids (see Martin & Abele 1986) and pseudorhombilids (see Ng et al. 2008) by the highly ornamented and lobed distal end, the degree of complexity of which is not found in any other brachyurans. In contrast, the three Palaeotropical families of freshwater crabs (Potamidae, Potamonautidae, and Gecarcinucidae) usually share a similar 4-part G1 — the three segments seen in all brachyurans, plus a distinct terminal article (Bott 1970b). Within the Palaeotropical freshwater crab families, G1 characters are by no means uniform, and differences in G1 morphology are sufficient to distinguish between families in most cases. For example, both the Potamidae and Potamonautidae have a G1 with a long symmetrical tapered terminal article that may possess complex folds and lobes (Ng & Naiyanetr 1993; Cumberlidge 1999; Dai 1999; Ng 2004). Members of the Gecarcinucidae may have a simple terminal article but often also lack one, in which case the distal part sometimes displays a variety of different forms (Bott 1970b; Ng & Naiyanetr 1993; Ng 1988, 2004; Dai 1999).

## 2.3 The Potamoidea

The Potamoidea, as redefined here, is now a freshwater crab superfamily with a very wide distributional range that stretches from the tropical and subtropical parts of the Americas across to Africa, Eurasia, Indonesia, and Australia. Possible alternatives to characters of the mandibular palp and G1 as indicators of family and superfamily level groupings of the Potamoidea include the following suite of synapomorphic characters of the carapace, mouthparts, sternum, and pereiopods (Sternberg et al. 1999). The anterolateral margin of the carapace has a distinct exorbital tooth and a distinct epibranchial tooth; the margin behind the epibranchial tooth is well defined, convex, and lined with numerous small teeth or tubercles (which in some species may be secondarily lost); there is a vertical (= cervical) sulcus on the carapace sidewall dividing the suborbital from the subhepatic regions, beginning just posterior to the epibranchial tooth and extending inferiorly to meet the longitudinal (= pleural) sulcus on the sidewall of the carapace. The antennae are short and are only half the length of the eyestalk. The third maxillipeds are broad and fill the entire buccal field; the medial margins of the ischium and merus of the third maxillipeds are vertical and touch along their entire length; there is a distinct, triangular epistomial tooth on the lower margin of the epistome, and the epistomial tooth is flanked by incisions. The median septum of the endophragmal system is interrupted between interosternites 4/5, 5/6, 6/7, but interosternite 7/8 is complete and not medially erased. The anterior–inferior margin of the merus of pereiopod 1 (cheliped) has distinct, irregular teeth; the dorsal surface of the merus of pereiopods 1–5 are rugose (either vague or distinct); and the dactyli of pereiopods 2–5 have at least four longitudinal rows of distinct corneous spines.

## 2.4 The Trichodactylidae

The Trichodactylidae is a freshwater crab family found primarily in the drainage basins of the Amazon, Orinoco, and Paraguay-Parana rivers in South America, with a small number of taxa distributed in Mexico and Trinidad (Rodriguez 1992; Magalhães & Türkay 1996a, b, c). The Trichodactylidae are morphologically unusual crabs that form a well-defined monophyletic group that is sharply isolated systematically. Other than direct development and a strict freshwater habitat, the trichodactylids have little in common morphologically with the Potamoidea as defined here. We list below the likely synapomorphies for the Trichodactylidae that include characters of the carapace, mouthparts, sternum, pereiopods, and G1 based on Magalhães (2003) and Sternberg & Cumberlidge (2003). The medial margins of the third maxillipeds meet along the midline, and the meri are slim and do not fill the entire buccal frame when closed; the endopod of maxilliped 1 has a distinct portunoid lobe (Rodriguez 1992). The antennae are long, either equal to or longer than the length of the eyestalk. The dactyli of the walking legs (P2–P5) have fields of dense soft setae rather than corneous spines. The median septum of the endophragmal system is dorsoventrally reduced, and interosternite 7/8 is extensively interrupted medially (Sternberg & Cumberlidge 2003). The male abdomen is broadly triangular with segments a3–a5 often fused. G1 is in three parts and is tubular (Sternberg 1998).

## 2.5 Marine crab sister group of the Potamoidea and Trichodactylidae

Other decapods such as crayfish (Astacoidea and Parastacoidea) that live exclusively in freshwater have identifiable (extant) marine lobster-like relatives (e.g., the Nephropoidea: Nephropsidae). The exclusively freshwater Aeglidae are included in the same anomuran superfamily (Galatheoidea) as the Galatheidae and other marine anomurans (Crandall 2007). However, the marine sister group of the Potamoidea (as defined here) has proven difficult to identify, and the identity of the closest living relatives of the potamoid freshwater crabs is still the subject of much active discussion (Sternberg et al. 1999; Sternberg & Cumberlidge 2003). This knowledge is necessary to both understand the evolutionary history of the freshwater crabs and to establish the proper placement of the group within the Brachyura.

According to several morphological studies (Sternberg et al. 1999, Sternberg & Cumberlidge 2001a, b) and preliminary molecular evidence (T. Spears pers. comm.) a possible candidate for the marine sister group of the potamoids would be an unspecified basal member of the Grapsoidea (which may now be extinct). In support of this hypothesis, Sternberg et al. (1999) listed a number of apomorphic characters that are shared by grapsoids (thoracotremes) and potamoids (heterotremes). These include a pair of epigastric crests on the anterior carapace, a pair of postorbital crests on the anterior carapace, clear exorbital and epigastric teeth on the anterolateral margins of the carapace, a posterior carina (a long raised line) running parallel with the posterolateral margin of the carapace, fields of carinae (short raised lines) on the posterolateral surfaces of the carapace, fields of carinae on the carapace sidewalls, a vertical sulcus on the carapace sidewall, a distinct triangular epistomial tooth, a notch flanking the epistomial tooth, a pereiopod 2–5 merus with a triangular cross-section, an anterior trough (groove) running parallel to the superior margin, and fields of carinae on the sides.

If the grapsoid sister group hypothesis of Sternberg et al. (1999) were to be supported by further studies, then the common ancestor (a heterotreme marine crab) gave rise to two monophyletic lineages, one that is exclusively freshwater that resulted in the several heterotreme potamoid crab families extant today, and the other, mostly marine, that produced a number of thoracotreme families. Interestingly, the predominantly marine families Sesarmidae and Grapsidae (all thoracotremes) resemble the true freshwater crabs in that both are mainly tropical and subtropical groups with a circumglobal distribution. In addition, a number of inland and coastal species of sesarmids spend large parts of their life cycle in freshwater habitats, and some have large eggs and abbreviated development (Hartnoll 1964; Soh 1969; Schubart & Cuesta 1998; Cuesta et al. 1999; Ng 2004). One species of sesarmid (*Geosesarma notophorum*) even has direct development similar to that seen in all true freshwater crabs, and this species never needs to return to the sea to complete its life cycle (Ng & Tan 1995). Interestingly, there are no known species of Grapsidae *sensu stricto* (see Ng et al. 2008) that are freshwater, although there are several terrestrial species that live near coasts. None have abbreviated or direct development. Similarly, although many members of the Varunidae (another major grapsoid group) live in freshwater, all need to return to the sea to release their small eggs. The predominantly marine brachyuran family Hymenosomatidae (false spider crabs, Majoidea) also has a few exclusively freshwater species, some of which reproduce by direct development (Ng & Chuang 1996).

A lack of knowledge also surrounds the identity of the marine sister group of the Trichodactylidae. A basal (possibly extinct) member of the Portunoidea has been suggested based on morphological evidence (Rodriguez 1992, Sternberg & Cumberlidge 2003), but this persuasive idea is not supported by preliminary molecular studies of selected species of modern portunids (Schubart et al. this volume). However, that study was also unable to shed light on the possible identity of the marine sister group of the trichodactylids (and therefore on the proper placement of this family within the Brachyura), and to date, this remains unknown.

## 3 DISTRIBUTION

The massive increase in our knowledge of the taxonomy of freshwater crabs worldwide has led to a refinement of the understanding of the distribution patterns of families, genera, and species, which are now the most resolved they have ever been. It is clear that freshwater crabs have a circumglobal distribution that is restricted to tropical and subtropical freshwater ecosystems. Cold temperatures, arid lands, deserts, high mountains, and large tracts of oceans are all barriers to the dispersal of true freshwater crabs, and these decapods are never found naturally in aquatic ecosystems that have even low levels of salt water. These warm-water decapods are represented in the Neotropical, Afrotropical, Oriental, Palaearctic, and Australasian zoogeographic regions and are absent from the Nearctic and Antarctic regions and from the cooler temperate zones of the Palaearctic, Neotropical, and Australasian regions (including New Zealand). Elsewhere in the tropics, freshwater crabs are completely absent from all remote oceanic islands in the Pacific (such as the Galapagos islands, the Hawaiian archipelago, the Society Islands) and from the remote oceanic islands in the Atlantic and Indian oceans.

Some families of freshwater crabs (e.g., the Pseudothelphusidae and Trichodactylidae) are restricted to the Neotropical zoogeographical region, and no species of Palaeotropical crabs are found in that region naturally. The same family-level endemism is largely true for the freshwater crabs found in the Afrotropical region: all belong to the Potamonautidae, except for three species of potamids on the island of Socotra. However, family-level endemism at the continental/zoogeographical region level is not seen in the Palaearctic, Oriental, and Australasian regions, where the parathelphusids are found in all three regions, and the potamids and gecarcinucids are found only in the Palaearctic and Oriental regions (and are both absent from the Australasian region).

### 3.1 *The Neotropical region*

Freshwater crabs are found throughout the Neotropical region in Central America (from Mexico to Panama and several Caribbean islands) and South America (from Colombia to Argentina). This region hosts two phylogenetically unrelated monophyletic lineages (families) of freshwater crabs — the Pseudothelphusidae (with 251 species) and the Trichodactylidae (with 47 species). Each of these families has representatives throughout the warmer parts of Central and South

America (from Mexico to northern Argentina), including the islands in the Caribbean and Pacific, and both families are absent from the cooler parts of the region (Chile, southern Argentina). Our present knowledge of the Pseudothelphusidae comes in large part from the landmark monograph by Rodriguez (1982) that brought together a literature that is widely scattered across time and in many different journals. Since then there have been a number of important contributions dealing with aspects of this family from specialists working in Central America (Alvarez 1989; Alvarez et al. 1996; Alvarez & Villalobos 1997, 1998, 1990, 1991, 1994, 1998), the Colombian Andes (Campos 2005 and publications therein), Venezuela and the Caribbean (Rodriguez 1992), and the Amazon (Rodriguez & Magalhães 2005). As for the Trichodactylidae, our present knowledge is based largely on the monographs of Rodriguez (1992), Magalhães & Türkay (1996a, b, c), and Magalhães (2003). The rate of description of new species of trichodactylids is now slowing compared to the past (Yeo et al. 2008), and this may indicate that we are close to knowing the true diversity of this family.

## 3.2 The Afrotropical region

The Afrotropical region is dominated by the endemic family Potamonautidae (with 132 species), which is distributed throughout the African continent and its associated islands in the Atlantic and western Indian Ocean (except for Socotra Island, where there are three endemic species with affinities to the Palaearctic-Oriental Potamidae). The first authors to treat the freshwater crab fauna of the Afrotropical region as a whole were Rathbun (1904, 1905, 1906), Chace (1942), and Bott (1955, 1965), and these works are still used by many as the standard taxonomic references for this group. Elsewhere in Africa recent taxonomic revisions are available for the freshwater crab faunas of West Africa (Cumberlidge 1999), Tanzania (Reed & Cumberlidge 2006), Lake Tanganyika (Cumberlidge et al. 1999; Marijnissen et al. 2004), Angola (Cumberlidge & Tavares 2006), southern Africa (Cumberlidge & Daniels 2008), and the Nile basin (Cumberlidge 2008), but large geographic areas such as Central Africa and East Africa are still in need of taxonomic revision. Recent works by Daniels et al. (2006), Cumberlidge et al. (2008), Yeo et al. (2008), and Cumberlidge (2008) have all advanced our knowledge of the phylogeny, higher classification, and biodiversity of the freshwater crabs of the Afrotropical region.

## 3.3 The Palaearctic region

In the vast Palaearctic region, freshwater crabs (Potamidae) are found only on its warmer southern margins stretching from North Africa to northern China and northern Japan, but these are not endemic to the region because they are also found in the Oriental region. The Palaearctic region is dominated by species of the family Potamidae, and potamonautids and gecarcinucids (Table 2D) are largely absent. The Potamidae is divided into two subfamilies (the western Palaearctic Potaminae and the eastern Palaearctic and Oriental Potamiscinae) whose distributional ranges overlap in northeast India and Myanmar (Yeo & Ng 2003). Freshwater crabs occur in the warmer freshwater habitats bordering the Mediterranean, the Middle East, the Himalayas, China, and Japan, and are not found in the colder, more northerly parts of the region. For example, freshwater crabs are absent from the Palaearctic region in Asia north of the Himalayas, Tibet, northern China, and the Korean peninsula, with the exception of a few species of potamids (subfamily Potamiscinae) found on the main islands of Japan (Dai 1999). In contrast, the southern islands of Japan (the Ryukyu Islands including Okinawa) and Taiwan lie in the Oriental region, and these have a rich freshwater crab fauna (mainly potamids). Potamid freshwater crabs of the subfamily Potaminae are found in Myanmar, the Himalayan states of north India, and Nepal, Pakistan, Afghanistan, the Middle East, southeastern Europe, and North Africa, which represents a wide distribution that (except for Myanmar) lies in the Palaearctic region (Brandis et al. 2000). Most of Europe lacks freshwater crabs except for a few species found in Italy, Greece, the Balkans, and the Black Sea region (Brandis et al. 2000). In North Africa, which is dominated by the Sahara desert, a single species of potamid is found along the

Mediterranean side of the Atlas Mountains in Morocco, Algeria, and Tunisia, but Libya completely lacks freshwater crabs. Most of Egypt also lacks freshwater crabs, except for the Sinai Peninsula, which has a single species of potamid (Potaminae), and the Nile valley, which has two species of Afrotropical potamonautid crabs (Bott 1970b; Williams 1976; Cumberlidge 2008).

### 3.4 *The Oriental region*

The Oriental region is home to three phylogenetically distinct monophyletic lineages of freshwater crabs recognized here as natural families — the Potamidae and Gecarcinucidae (including Parathelphusidae). Each of these families has representatives throughout the warmer parts of this region, both on the mainland and on most of the nearby islands. Crabs of the vastly diverse and widely distributed family Potamidae are found throughout the Oriental region as well as being well represented in the Palaearctic region. Potamids are completely absent from peninsular India south of the Ganges. The Potamidae reaches its greatest diversity in the Oriental region (which hosts about 450 out of the more than 500 species) (Dai 1999; Yeo & Ng 2007; Yeo et al. 2008). The southern boundary of the distributional range of the Potamidae is marked by Wallace's Line, whereby the islands of the Sunda Shelf (Sumatra, Java [only the western part], Borneo) and the southern Philippines have potamids, but the islands to the east of this (from Lombok to Sulawesi and eastwards to Australasia) all lack them. Besides mainland Asia, the Potamidae in the Oriental region (subfamily Potamiscinae) has representatives on many of the smaller islands in the South China Sea, the Yellow Sea, and the East China Sea. Smaller numbers of potamids (subfamily Potamiscinae) are found in the Palaearctic region in northern China and Japan, and there are several species (subfamily Potaminae) found in the Himalayas, the Middle East, southern Europe, and North Africa (Brandis et al. 2000). Interestingly, there are three species of potamids found in the Afrotropical region on the island of Socotra (Apel & Brandis 2000; Cumberlidge & Wranik 2002). The newly defined Gecarcinucidae, including the Parathelphusidae of Ng et al. (2008) (*sensu* Klaus et al. this volume) has a total of 345 species and is very diverse in the Oriental region (Sri Lanka, northeast India, Myanmar, Indochina, Thailand, Malaysia, Indonesia, Taiwan, the Philippines) but is also well represented in the Australasian region as far east as northern Australia and the Solomon islands (Bott 1970b; Yeo & Ng 1999; Bahir & Yeo 2007).

### 3.5 *The Australasian region*

Wallace's Line marks the edge of the continental margin at the Sunda Shelf and divides the Australasian and Oriental regions. Bali, Borneo, and the Philippines lie on the western (Oriental) side and Lombok and Sulawesi lie on the eastern (Australasian) side. The Australasian zoogeographical region stretches from the Lesser Sunda Islands (Lombok, Flores, and Sambawa) and Sulawesi eastward to include the Moluccas and the Aru Islands, New Guinea and its neighboring islands, and Australia. The Australasian region is relatively poor in freshwater crab species compared to the neighboring Oriental region. All freshwater crabs found in Australasia belong to the family Gecarcinucidae, and potamids are completely absent from this region. Sulawesi and New Guinea are the largest islands in this region and have the highest diversity of freshwater crab species. It would appear that the gecarcinucid freshwater crabs found in these islands today are all derived from ancestral southeast Asian forms that dispersed east across the seawater barrier represented by Wallace's Line (see Klaus et al. this volume for discussion). The deep water of the Lombok Strait between the islands of Bali and Lombok and the Philippines and Sulawesi has always represented a significant seawater barrier, even when lower sea levels linked many of the now-separated islands in this region with the landmasses on either side. Freshwater crab diversity in Australasia is highest in Sulawesi, Moluccas, and New Guinea and declines towards Australia. In Australia, only seven species of freshwater crabs (all in the endemic genus *Austrothelphusa*) are found in the northern tropical and subtropical parts of the continent, although several more species remain undescribed (P.J.F. Davie, pers. comm.). They are absent in southern and western Australia, Tasmania, and New Zealand. This distribution pattern strongly suggests that crabs entered Australia relatively recently from New Guinea, presumably during periods of lowered sea level (corresponding to the Pleistocene Ice Ages) when Australia and New Guinea were connected across the Torres Strait. Presumably the ancestors of *Austrothelphusa* crossed to the forested Cape York Peninsula and from there dispersed over time throughout the river systems of northeastern Australia, spreading in all directions and eventually reaching most of inland and coastal Queensland, the Lake Eyre basin, and the Darling River drainage system in western New South Wales. The southern boundary of the distribution of *Austrothelphusa* was presumably established by the cooler, more temperate climates in the south and the lack of water in the west.

## 4 BIOGEOGRAPHY

The realization that five families of freshwater crabs may share common ancestry has revolutionized the way that these brachyurans are now viewed, because their worldwide distribution on continents and islands today includes vast tracts of ocean in between. These crabs are not only found on every continent in the tropics: these exclusively freshwater animals are also found on most of the large and small offshore islands associated with the continents. All around the world freshwater crab families include species found on offshore islands, and some species have a distribution that includes both the mainland and nearby islands. In many cases, the presence of freshwater crabs on islands near continental landmasses can best be explained by past sea level changes that created land bridges. However, there are a number of islands with established freshwater crab faunas that have never been connected to the mainland, even when sea levels were at historical lows. In the latter cases it is clear that freshwater crabs must have somehow crossed tracts of seawater to reach these islands, perhaps in a similar way to that proposed for amphibians on oceanic islands (Measey et al. 2007).

Single ancestry for the potamoid freshwater crabs has profound implications for biogeographical theories, as does a detailed knowledge of the global distribution and phylogenetic relationships within this group. However, an important piece of information — the age of origin of the freshwater crabs — is still not available. The oldest known freshwater crab fossil (see Feldmann et al. 2007) is still quite recent. Equally important is the lack of knowledge of the physiological abilities of the freshwater crabs to survive in seawater (it is widely assumed that they cannot survive for long). Single ancestry for the potamoid freshwater crabs could be explained by postulating a colonization event by a marine crab ancestor into the freshwaters of a single continent followed by a worldwide overland radiation (see Ng et al. 1995). This would require an ancient origin for the freshwater crabs because it would have to have taken place in the Jurassic (about 250 mya) when the continents were fused into a single landmass (Pangaea). In this scenario, crabs could have established a global distribution without crossing tracts of seawater, because they were carried to their present positions on fragmenting and drifting continents. However, there is no evidence that freshwater crabs, or even the Eubrachyura for that matter, are that old.

However, other explanations must be sought if freshwater crabs first evolved after the initial breakup of Pangaea into Laurasia and Gondwana (200 mya). In this case it is necessary to postulate at least two separate colonization events by marine crabs (one into Laurasian freshwaters and one into Gondwanan freshwaters). An even later origin of freshwater crabs after the further fragmentation of these two landmasses into smaller continental fragments (160—80 mya) would require either a separate colonization of each landmass by multiple marine crab ancestors or a single colonization event by a marine crab ancestor followed by overseas dispersal across oceanic barriers by its freshwater crab descendants to reach each of the widely separated continents. However, neither of the preceding scenarios is congruent with the phylogenetic relationships of the freshwater crab families found today in the Neotropics, Afrotropics, and the Indian subcontinent (i.e., on the continental plates that were once part of Gondwana), and they are not congruent with the sequence of continental breakup predicted by geological data. This argues against vicariance theories that postulate that freshwater crabs are an ancient group present on Gondwana before continental breakup that reached its present distribution when continents separated and moved. Vicariance theories of biogeography do not require the assumption that crabs crossed one or more seawater barriers (Rodriguez 1986; Ng et al. 1995; Ng & Rodriguez 1995).

Alternately, present-day distribution patterns could be explained by a separate colonization of the freshwaters of each continental landmass during the Cenozoic by members of a single widespread marine crab ancestral stock living in the circumtropical Tethys Sea from the Neotropics to the Pacific. Single ancestry and a recent origin for freshwater crabs require that explanations be sought for explaining present-day distributions on widely separated continents and islands with seawater barriers in between. Sternberg et al. (1999) theorized that potamoid freshwater crabs descended from a widespread freshwater-adapted marine crab ancestor that had a global distribution across the shallow tropical seas from tropical America to Southeast Asia. This was at a time before the closing of the Mediterranean Sea and before the collision of India with Asia, when the Atlantic, Indian, and Pacific oceans formed a continuous water body around the tropics. At this time, ancestral crabs living in different parts of the range of the same widespread species entered suitable freshwater ecosystems in the Neotropical, Afrotropical, Palaearctic, and Oriental regions. Once established in freshwater, these colonizers lost their ability to survive in seawater and effectively became isolated in freshwater habitats over time. Evolution in isolation in each of these regions led to their radiation, adaptation, and speciation to produce monophyletic groups in each of these continents. Freshwater crabs then spread slowly throughout continental freshwaters and also colonized many of the offshore islands. This process led to the development of morphologically distinct lineages of freshwater crabs in each of the zoogeographic regions that are separable at the family level. The founder effect on islands led to some freshwater crabs' becoming morphologically atypical, and in some cases this led taxonomists to recognize higher taxa or make family-level transfers for some of the more apomorphic species (e.g., in Madagascar, the Seychelles, and East Africa) (Bott 1960, 1965; Ng et al. 1995; Cumberlidge et al. 1999, Števčić 2005), all of which later proved to belong to the same family (Daniels et al. 2006; Cumberlidge et al. 2008).

4.1 *Colonization of freshwater*

Today, there are several species of catadromous marine crabs such as *Varuna litterata* (Varunidae) that have wide distributional ranges over tens of thousands of sq. km. of ocean, and that have the physiological ability to live both in the sea and in freshwater habitats for long periods of time. For example, *Varuna litterata* ranges from East Africa and Madagascar in the Indian Ocean to Japan and Polynesia in the Pacific Ocean. These catadromous brachyurans have free-living larval stages that require saltwater for development, and all need to return to the sea to breed, a strategy that has the advantage of achieving a wide dispersal range when developing larvae are carried long distances by ocean currents.

The first step in the colonization of freshwaters must have involved the development of the physiological ability to osmoregulate and gain some control over the movement of ions and water in and out of the body. The ability of marine crabs to osmoregulate in low-salinity environments encompasses adaptations ranging from short-term survival in brackish water to long-term colonization of freshwater. These were presumably the stages through which the ancestors of the true freshwater crabs passed on their way to becoming exclusively freshwater organisms. Once the ancestors of freshwater crabs had become fully adapted to freshwater, they would have lost their ability to survive for long in seawater. The best "analogy" in the modern crabs would probably be genera of Sesarmidae like *Geosesarma*, whose members are all freshwater or semiterrestrial and species have varying larval strategies, from eggs hatching into planktotrophic larvae, very advanced zoeae, megalopae or even direct development (see discussion in Ng et al. 2004).

Another important adaptation to life in freshwater was the ability of freshwater crabs to complete their life cycle without returning to the sea to release eggs and larvae. Like other freshwater decapods (such as crayfish and many species of palaemonid and atyid shrimps), the ancestors of freshwater crabs evolved direct development and could remain in freshwater habitats year round without having to return to the sea to release their larvae. The lack of dispersive planktonic larval stages restricted the dispersal abilities of freshwater crabs, and their distributional ranges in freshwater habitats necessarily became much smaller. Oceans now became barriers to their dispersal rather than facilitators. The result was that freshwater crab populations became reproductively isolated much more easily, and this isolation led to their adaptation, speciation, and diversification over time.

## 4.2 *Theories on origins*

There is some morphological, molecular, and fossil evidence (Sternberg et al. 1999; Daniels et al. 2006; Brösing 2008) that the evolution of freshwater crabs from a brachyuran heterotreme stock happened sometime in the Late Cetaceous/early Cenozoic. The study by Brösing (2008) provided a temporalized cladogram that estimated the divergence time of the potamoids (represented by a potamonautid terminal taxon) from marine crab stock just prior to the Cretaceous-Tertiary boundary. If this estimate of freshwater crab origins is supported by further studies, then the potamoids appeared well after major tectonic events such as the breakup of Pangaea that separated the northern continent (Laurasia) from the southern continent (Gondwana). Similarly, freshwater crabs were therefore not present on the continents when the Laurasian supercontinent broke up into the Nearctic from the Palaearctic landmasses. It also follows that freshwater crabs were not present when Gondwana began to fragment, first splitting off the South American landmass from the western coast of Africa, and then splitting off the Madagascar-Seychelles-India landmass from the eastern coast of Africa, followed by the eventual breakup of Madagascar-Seychelles-India.

A Late Cetaceous/early Cenozoic origin of the freshwater crabs would mean that these decapods colonized freshwaters at a time when the continental landmasses of North America, South America, Africa, and India were all islands, and when the southern margins of the Eurasian landmass were fragmented and constituted a series of small shifting plates. At this time the warm, shallow Tethys Sea formed a continuous marine connection between all of these landmasses around the equator from the Americas to Asia, joining the Atlantic with the Indian and Pacific oceans. This continuous marine connection was later broken when the Mediterranean Sea closed, separating the Atlantic and Indian oceans, and when peninsular India collided with Eurasia.

The collision of India with Asia had a big impact on the three families of freshwater crabs that are found today in Eurasia, India, and the Oriental region. It is likely that these three families were well established in these areas long before the Indian collision with Asia and the building of the Himalayas. For example, the Gecarcinucidae most likely evolved in isolation on peninsular India (where it is most diverse today) and was already present before this landmass collided with Asia. There probably were no gecarcinucids on the mainland of Asia before the contact with India. This is consistent with today's distribution pattern of this family, where there are now only a handful of gecarcinucid taxa to the east of peninsular India (in Myanmar, Thailand, Malaysia, and Sarawak) and where there is a similar tapering off of diversity to the west (in Pakistan, Afghanistan, and Iran). This pattern is most likely the result of the subsequent gradual spread of gecarcinucids out of India following the collision of India with Eurasia. It is significant that there are still no potamids in peninsular India today, an observation that is consistent with the fact that the potamids evolved on the mainland of Asia and were never able (for some reason) to disperse south into India despite the favorable habits for them there.

The present-day distribution pattern of the Potamidae indicates that it most likely evolved in the warmer eastern parts of the Palaearctic landmass (where it is most diverse today) and was widely distributed in the freshwater ecosystems along the southern shores of the Tethys Sea from Europe to southeast Asia before the collision of India with Asia. Potamids most likely evolved when India was still an island continent, which would explain their absence there to this day. The high numbers of potamid taxa found to the east of India (in Myanmar, Indochina, Malaysia, China, the Sunda Shelf Islands, the Philippines, Taiwan, and Japan) and the relatively few taxa found to the west of India (stretching from Pakistan to North Africa) are likely the result of the isolation of the eastern and western potamids after the collision of peninsular India with Eurasia. The building of the Himalayan mountains likely eliminated most of the potamids already living along the collision zone and became a barrier to subsequent potamid dispersal, after which the western potamids (Potaminae) evolved separately from the eastern potamids (Potamiscinae) (see Shih et al. in press).

With regard to the rest of Gecarcinucidae (the Parathelphusidae in Ng et al. 2008), it is possible they first evolved in Southeast Asia (Myanmar, Thailand, Indochina, Malaysia, southwest China) where it is most diverse today (see also Klaus et al. this volume). The number of gecarcinucid taxa declines eastwards from this center towards China and Taiwan and westwards (in northern India), but the family is well represented in the Philippines, most of the islands in the Sunda Shelf, the Indonesian islands east of Java, and in the chain of Indonesian islands as far east as New Guinea and northern Australia and the Solomons. The collision of India with Eurasia no doubt explains the decline and then absence of this family west of Myanmar, but the origin of the rich gecarcinucid fauna of Sri Lanka is difficult to explain (Ng & Tay 2001; Bossuyt et al. 2004). The southeasterly dispersal of gecarcinucids from southeast Asia to the Philippines and northern Australia is likely the result of their subsequent spread across marine barriers over time because the landmasses in this part of the world between the Sunda Shelf and Australia are greatly divided and dominated by islands (see also Klaus et al. this volume).

## REFERENCES

Álvarez, F. 1989. *Smalleyus tricristatus*, new genus, new species, and *Pseudothelphusa parabelliana*, new species (Brachyura: Pseudothelphusidae) from Los Tuxtlas, Veracruz, Mexico. *Proc. Biol. Soc. Wash.* 102: 45–49.

Álvarez, F., Villalobos, J.L. & Lira, E. 1996. Decápodos. In: Llorente, J., García-Aldrete, A.N. & González-Soriano, E. (eds.), *Biodiversidad, Taxonomía y Biogeografiá de Artropodos Mexicanos: Hacia una Síntesis de su Conocimiento*. Mexico: Instituto de Biología, Universidad Nacional Autónoma de México, CONABIO.

Álvarez, F. & Villalobos, J.L. 1997. *Pseudothelphusa ayutlaensis*, a new species of freshwater crab (Brachyura: Pseudothelphusidae) from Mexico. *Proc. Biol. Soc. Wash.* 110: 388–392.

Álvarez, F. & Villalobos, J.L. 1998. Six new species of freshwater crabs (Brachyura: Pseudothelphusidae) from Chiapas, Mexico. *J. Crust. Biol.* 18: 187–198.

Álvarez, F. & Villalobos, J.L. 1990. *Pseudothelphusa galloi*, a new species of freshwater crab (Crustacea: Brachyura: Pseudothelphusidae) from the Southwestern Mexico. *Proc. Biol. Soc. Wash.* 103: 103–105.

Álvarez, F. & Villalobos, J.L. 1991. A new genus and two new species of freshwater crabs from Mexico, *Odontothelphusa toninae* and *Stygothelphusa lopezformenti* (Crustacea: Brachyura: Pseudothelphusidae). *Proc. Biol. Soc. Wash.* 104: 288–294.

Álvarez, F. & Villalobos, J.L. 1994. Two new species and one new combination of freshwater crabs (Brachyura, Pseudothelphusidae) from Mexico. *Proc. Biol. Soc. Wash.* 107: 729–737.

Álvarez, F. & Villalobos, J.L. 1998. Six new species of fresh-water crabs (Brachyura: Pseudothelphusidae) from Chiapas, Mexico. *J. Crust. Biol.* 18: 187–198.

Apel, M. & Brandis, D. 2000. A new species of freshwater crab (Crustacea: Brachyura: Potamidae) from Socotra Island and description of *Socotrapotamon* n. gen. *Fauna Arabia* 18: 133–144.

Bahir, M.M. & Yeo, D.C.J. 2007. The gecarcinucid freshwater crabs of southern India (Crustacea: Decapoda: Brachyura). *Raffles Bull. Zool.* Supplement No. 16: 309–354.

Banarescu, P. 1990. *Zoogeography of Fresh Waters. Volume 1. General Distribution and Dispersal of Freshwater Animals*. Wiesbaden: AULA-Verlag.

Bond-Buckup, G., Jara, C.G., Pérez-Losada, M, Buckup, L. & Crandall, K.A. 2008. Global diversity of crabs (Aeglidae: Anomura: Decapoda) in freshwater. In: Balian, E.V., Lévequè, C., Segers, H. & Martens, M. (eds.), *Freshwater Animal Diversity Assessment. Hydrobiologia* 595: 267–273.

Bossuyt, F., Meegaskumbura, M., Beenaerts, N., Gower, D.J., Pethiyagoda, R., Roelants, K., Mannaert, A., Wilkinson, M., Bahir, M.M., Manamendra-Arachchid, K., Ng, P.K.L., Schneider, C.J., Oommen, O.V. & Milinkovitch, M.C. 2004. Local endemism within the western Ghats-Sri Lanka biodiversity hotspot. *Science* 306: 479–481.

Bott, R. 1955. Die Süßwasserkrabben von Afrika (Crust., Decap.) und ihre Stammesgeschichte. *Ann Mus. Congo Belge. (Tervuren, Belgique) C-Zoologie*, Séries III 3: 209–352.

Bott, R. 1960. Crustacea (Decapoda): Potamonidae. In: Hanstrom, B., Brinck, P. & Ruderbeck, G. (eds.), *South African Animal Life: Results of the Lund University Expedition in 1950-1951*, Vol. 7: 13–18. Uppsala: Almqvist and Wiksells.

Bott, R. 1965. Die Süßwasserkrabben von Madagaskar. *Bull. Mus. Nat. Hist. Nat.* Paris 37: 335–350.

Bott, R. 1969. Die Flußkrabben aus Asien und ihre Klassifikation. *Senck. Biol.* 50: 339–366.

Bott, R. 1970a. Betrachtungen über die Entwicklungsgeschichte und Verbreitung der Süßwasser-Krabben nach der Sammlung des Naturhistorischen Museums in Genf/Schweiz. *Rev. Suisse Zool.* 77: 327–344.

Bott, R. 1970b. Die Süßwasserkrabben von Europa, Asien, Australien und ihre Stammesgeschichte. *Abh. Sencken. Natur. Ges.* 526: 1–33.

Bott, R. 1972. Stammesgeschichte und geographische Verbreitung der Süßwasserkrabben. *Natur und Museum, Frankfurt* 102: 63–77.

Brandis, D., Storch, V. & Türkay, M. 2000. Taxonomy and zoogeography of the freshwater crabs of Europe, North Africa, and the Middle East (Crustacea, Decapoda, Potamidae). *Senck. Biol.* 80: 5–56.

Brösing, A. 2008. A reconstruction of an evolutionary scenario for the Brachyura (Decapoda) in the context of the Cretaceous-Tertiary boundary. *Crustaceana* 81: 271–287.

Campos, M.R. 2005. *Freshwater Crabs from Colombia. A Taxonomic and Distributional Study.* Academia Colombiana de Ciencia Exactas. Físicas 4 Naturales, Coleccion Jorge Álvarez Lleras, Bogota, 24: 1–363.

Chace, F.A. 1942. Scientific results of a fourth expedition to forested areas in eastern Africa, III: Decapod Crustacea. *Bull. Mus. Comp. Zool. Harvard College* 91: 185–233.

Colosi, G. 1920. I Potamonidi conservati del R. Museo Zoologico di Torino. *Boll. Mus. Zool. Anat. Comp. R. Universita di Torin* 35(734): 1–39.

Crandall, K.A. 2007. Anomura. Version 10, January 2007 (temporary). *http://tolweb.org/Anomura/6658/2007.01.10* in The Tree of Life Web Project, *http://tolweb.org*.

Cuesta, J.A., Schuh, M., Diesel, R. & Schubart, C.D. 1999. Abbreviated development of *Armases miersii* (Grapsidae: Sesarminae), a crab that breeds in supralittoral rock pools. *J. Crust. Biol.* 19: 26–41.

Cumberlidge, N. 2008. Insular species of Afrotropical freshwater crabs (Crustacea: Decapoda: Brachyura: Potamonautidae and Potamidae) with special reference to Madagascar and the Seychelles. *Contrib. Zool.* (in press).

Cumberlidge, N. & Daniels, S.R. 2008. A conservation assessment of the freshwater crabs of southern Africa (Brachyura: Potamonautidae). *Af. J. Ecol.* 46: 74–79.

Cumberlidge, N. & Tavares, M. 2006. Remarks on the freshwater crabs of Angola, Southwestern Africa, with the description of *Potamonautes kensleyi*, new species (Brachyura: Potamoidea: Potamonautidae). *J. Crust. Biol.* 26: 248–257.

Cumberlidge N. & Wranik, W. 2002. A new genus and new species of freshwater crab (Potamoidea: Potamidae) from Socotra Island, Yemen. *J. Nat. Hist.* 36: 51–64.

Cumberlidge, N., Sternberg, R.v., Bills, I.R. & Martin, H.A. 1999. A revision of the genus *Platythelphusa* A. Milne-Edwards, 1887 from Lake Tanganyika, East Africa (Decapoda: Potamoidea: Platythelphusidae). *J. Nat. Hist.* 33: 1487–1512.

Cumberlidge, N. 1986. Ventilation of the branchial chambers in the amphibious West African freshwater crab *Sudanonautes (Convexonautes) aubryi monodi* (Balss, 1929) (Brachyura, Potamonautidae). *Hydrobiologia* 134: 53–65.

Cumberlidge, N. 1999. *The Freshwater Crabs of West Africa. Family Potamonautidae*: 1-382. Faune et Flore Tropicales 35. Paris: ORSTOM.

Cumberlidge, N., Sternberg, R.v. & Daniels, S.R. 2008. A revision of the higher taxonomy of the Afrotropical freshwater crabs (Decapoda: Brachyura) with a discussion of their biogeography. *Biol. J. Linn. Soc.* 93: 399–413.

Dai, A.Y. 1999. *Fauna Sinica Arthropoda Crustacea Malacostraca Decapoda Parathelphusidae Potamidae*. Beijing, China: Science Press. [In Chinese with English.]

Daniels, S.D., Cumberlidge, N., Pérez-Losada, M., Marijnissen, S.A.E. & Crandall, K.A. 2006. Evolution of Afrotropical freshwater crab lineages obscured by morphological convergence. *M. Phylo. Evol.* 40: 225–235.

Feldmann, R.M., O'Connor, P.M., Stevens, N.J., Gottfried, M.D., Roberts, E.M., Ngasala, S., Rasmusson, E.L. & Kapilima, S. 2007. A new freshwater crab (Decapoda: Brachyura: Potamonautidae) from the Paleogene of Tanzania, Africa. *N. Jb. Geol. Paläont. Abh.* 244/1: 71–78.

Guinot, D. 1977. Propositions pour une nouvelle classification des Crustacés Décapodes Brachyoures. *Comptes rend. hebd. Séan. Acad. Sci.* 285 (D): 1049–1052.

Guinot, D. 1978. Principes d'une classification évolutive des Crustacés Décapodes Brachyoures. *Bull. Biol. France Belg.* n.s. 112 (3): 211–292, figs. 1-3, 1 table.

Guinot, D. 1979. Donnés nouvelles sur la morphologie, la phylogenèse et la taxonomie des Crustacés Décapodes Brachyoures. *Mém. Mus. Nat. Hist. nat.* 112 (A), 354 p., 70 figs., 27 pls., 5 tables.

Hartnoll, R.G. 1964. The freshwater grapsid crabs of Jamaica. *Proc. Linn. Soc., Lond.*, 175: 145–169.

Klaus, S., Schubart, C.D. & Brandis, D. 2006. Phylogeny, biogeography and a new taxonomy for the Gecarcinucoidea Rathbun, 1904 (Decapoda: Brachyura). *Org. Div. Evol.* 6: 199–217.

Klaus, S., Brandis, D., Ng, P.K.L., Yeo, D.C.J., Schubart, C.D. (this volume). Phylogeny and biogeography of Asian freshwater crabs of the family Gecarcinucidae (Brachyura: Potamoidea). In: Martin, J.W., Crandall, K.A. & Felder, D.L. (eds.), *Crustacean Issues: Decapod Crustacean Phylogenetics*. Boca Raton, Florida: Taylor & Francis/CRC Press.

Magalhães, C. 2003. Famílias Pseudothelphusidae e Trichodactylidae. In: de Melo, G.A.S. (ed.), *Manual de Identificação dos Crustacea Decapoda de Água Doce do Brasil*: 143–287. São Paulo: Editora Loyola.

Magalhães, C. & Türkay, M. 1996a. Taxonomy of the Neotropical freshwater crab family Trichodactylidae I. The generic system with description of some new genera (Crustacea: Decapoda: Brachyura). *Senck. Biol.* 75 (1/2): 63–95.

Magalhães, C. & Türkay, M. 1996b. Taxonomy of the Neotropical freshwater crab family Trichodactylidae II. The genera *Forsteria, Melocarcinus, Sylviocarcinus*, and *Zilchiopsis* (Crustacea: Decapoda: Brachyura). *Senck. Biol.* 75 (1/2): 97–130.

Magalhães, C. & Türkay, M. 1996c. Taxonomy of the Neotropical freshwater crab family Trichodactylidae II. The genera *Fredilocarcinus* and *Goyazana* (Crustacea: Decapoda: Brachyura). *Senck. Biol.* 75 (1/2): 131–142.

Marijnissen, S., Schram, F., Cumberlidge, N. & Michel, E. 2004. Two new species of *Platythelphusa* A. Milne-Edwards, 1887 (Decapoda, Potamoidea, Platythelphusidae) and comments on the taxonomic position of *P. denticulata* Capart, 1952 from Lake Tanganyika, East Africa. *Crustaceana* 77: 513–532.

Martin, J.W. & Abele, L.G. 1986. Notes on male pleopod morphology in the brachyuran crab family Panopeidae Ortmann, 1893, sensu Guinot (1978) (Decapoda). *Crustaceana* 50: 182–198.

Martin, J.W. & Davis, G.E. 2001. An updated classification of the Recent Crustacea. *Nat. Hist. Mus. Los Angeles, Science Series* 39: 1–124.

Measey, G.J., Vences, M., Drewes, R.C., Chiari, Y., Melo, M. & Bourles, B. 2007. Freshwater paths across the ocean: molecular phylogeny of the frog *Ptychadena newtoni* gives insights into amphibian colonization of oceanic islands. *J. Biogeog.* 34: 7–20.

Ng, P.K.L. 1988. *The Freshwater Crabs of Peninsular Malaysia and Singapore*. Dept. Zool., Natn. Univ. Singapore. Shinglee Press, Singapore.

Ng, P.K.L. 2004. Crustacea: Decapoda, Brachyura. In: Yule C.M. & Yong H.S. (eds.), *The Freshwater Invertebrates of the Malaysian Region*: 311–336. Kuala Lumpur: National Academy of Sciences.

Ng, P.K.L. & Chuang, C.T.N. 1996. The Hymenosomatidae (Crustacea: Decapoda: Brachyura) of Southeast Asia, with notes on other species. *Raffles Bull. Zool.* Suppl. 3, 82 pp.

Ng, P.K.L., Guinot, D. & Davie, P. 2008. Systema Brachyuorum: Part I. An annotated checklist of extant Brachyuran crabs of the world. *Raffles Bull. Zool.* Suppl. 17: 1–286.

Ng, P.K.L., Liu, H.-C. & Schubart, C.D. 2004. *Geosesarma hednon*, a new species of terrestrial crab (Crustacea: Decapoda: Brachyura: Sesarmidae) from Taiwan and the Philippines. *Raffles Bull. Zool.* 52: 239–249.

Ng, P.K.L. & Naiyanetr, P. 1993. New and recently described freshwater crabs (Crustacea: Decapoda: Brachyura: Potamidae, Gecarcinucidae and Parathelphusidae) from Thailand. *Zool. Verh.* 284: 1117, fig. 168.

Ng, P.K.L. & Rodriguez, G. 1995. Freshwater crabs as poor zoogeographical indicators: a critique of Banarescu (1990). *Crustaceana* 68: 636–645.

Ng, P.K.L. & Tay, W.M. 2001. The freshwater crabs of Sri Lanka. (Decapoda: Brachyura: Parathelphusidae). *Zeylanica* 6: 113–199.

Ng, P.K.L. & Tan, C.G.S. 1995. *Geosesarma notophorum* sp. nov. (Decapoda, Brachyura, Grapsidae, Sesarminae), a terrestrial crab from Sumatra, with novel brooding behaviour. *Crustaceana* 68: 390–395.

Ng, P.K.L., Števčić, Z. & Pretzmann, G. 1995. A revision of the family Deckeniidae Ortmann, 1897 (Crustacea, Decapoda, Brachyura, Potamoidea) from the Seychelles, Indian Ocean. *J. Nat. Hist.* 29: 581–600.

Ortmann, A. 1896. Das System der Decapoden-Krebse. *Zool. Jahrbuch. (Syst.)* 9: 409–453.

Rathbun, M.J. 1904. Les crabes d'eau douce (Potamonidae). *Nouv. Arch. Mus. Hist. Nat* 6: 255–312.

Rathbun, M.J. 1905. Les crabes d'eau douce (Potamonidae). *Nouv. Arch. Mus. Hist. Nat* 7: 159–322.

Rathbun, M.J. 1906. Les crabes d'eau douce (Potamonidae). *Nouv. Arch. Mus. Hist. Nat* (4) 8: 33–122, Figs. 106–124.

Reed, S.K. & Cumberlidge, N. 2006. Taxonomy and biogeography of the freshwater crabs of Tanzania, East Africa (Brachyura: Potamoidea: Potamonautidae, Platythelphusidae, Deckeniidae). *Zootaxa* 1262: 1–139.

Rodriguez, G. 1982. Les crabes d'eau douce d'Amerique Famille des Pseudothelphusidae. Paris: Faune Tropicale 22, Office de la Recherche Scientifique d'Outre Mer (Orstom).

Rodriguez, G. 1986. Centers of radiation of freshwater crabs in the Neotropics. In: Gore, R.H. & Heck, K.L. (eds.), *Crustacean Issues* 4, *Crustacean Biogeography*: 51–67. Rotterdam: Balkema.

Rodriguez, G. 1992. The freshwater crabs of America. Family Trichodactylidae. Paris: Office de la Recherche Scientifique d'Outre Mer (ORSTOM).

Rodriguez, G. & Magalhães, C. 2005. Recent advances in the biology of the Neotropical freshwater crab family Pseudothelphusidae (Crustacea, Decapoda, Brachyura). *Rev. Bras. Zool.* 22: 354–365.

Schubart, C.D. & Cuesta, J.A. 1998. The first zoeal stages of four *Sesarma* species from Panama, with identification keys and remarks on the American Sesarminae (Crustacea: Brachyura: Grapsidae). *J. Plank. Res.* 20: 61–84.

Schubart, C.D. & Koller, P. 2005. Genetic diversity of freshwater crabs (Brachyura: Sesarmidae) from central Jamaica with description of a new species. *J. Nat. Hist.* 39: 469–481.

Schubart, C.D. & Reuschel, S. (this volume). A proposal for a new classification of Portunoidea and Cancroidea (Brachyura: Heterotremata) based on two independent molecular phylogenies. In: Martin, J.W., Crandall, K.A. & Felder, D.L. (eds.), *Crustacean Issues: Decapod Phylogenetics*. Boca Raton, Florida: Taylor & Francis CRC Press.

Shih, H.-T., Yeo, D.C.J. & Ng, P.K.L. In press. Impact of the collision of the Indian Plate with Asia on the phylogeny of the freshwater crab family Potamidae (Crustacea: Decapoda: Brachyura) revealed by morphological and molecular evidence. *Mol. Ecol.*

Soh, C.L. 1969. Abbreviated development of a non-marine crab, *Sesarma (Geosesarma) perracae* (Brachyura; Grapsidae) from Singapore. *J. Zool.* 158: 357–370.

Sternberg, R.v. 1998. The sister group of the freshwater crab family Trichodactylidae (Crustacea: Decapoda: Eubrachyura). *J. Comp. Biol.* 3: 93–101.

Sternberg, R.v., Cumberlidge, N. & Rodriguez, G. 1999. On the marine sister groups of the freshwater crabs (Crustacea: Decapoda: Brachyura). *J. Zool. Syst. Evol. Res.* 37: 19–38.

Sternberg, R.v. & Cumberlidge, N. 2001a. Notes on the position of the true freshwater crabs within the Brachyrhynchan Eubrachyura (Crustacea: Decapoda: Brachyura). *Hydrobiologia* 449: 21–39.

Sternberg, R.v. & Cumberlidge, N. 2001b. On the heterotreme thoracotreme distinction in the Eubrachyura de Saint-Laurent, 1980. *Crustaceana* 74: 321–338.

Sternberg, R.v. & Cumberlidge, N. 2003. Autapomorphies of the endophragmal system in trichodactylid freshwater crabs (Crustacea: Decapoda: Eubrachyura). *J. Morph.* 256: 23–28.

Števčić, Z. 2005. The reclassification of Brachyuran crabs (Crustacea: Decapoda: Brachyura). *Fauna Croat.* 14: 1–159.

Williams, T.R. 1976. Freshwater crabs of the Nile system. In: Rzoska, J. (ed.), *The Nile Biology of an Ancient River*, The Hague: Dr. W. Junk Publishers.

Yeo, D.C.J. & Ng, P.K.L. 1999. The state of freshwater crab taxonomy in Indochina (Decapoda, Brachyura). In: Schram, F.R. & von Vaupel Klein, J.C. (eds.), *Crustaceana*, Special Volume, *Crustaceans and the Biodiversity Crisis*: 637–646. Leiden, The Netherlands.

Yeo, D.C.J. & Ng, P.K.L. 2003. Recognition of two subfamilies in the Potamidae Ortmann, 1896 (Brachyura, Potamidae), with a note on the genus Potamon Savigny. *Crustaceana* 76: 1219–1235.

Yeo, D.C.J. & Ng, P.K.L. 2007. On the genus *"Potamon"* and allies in Indochina (Crustacea: Decapoda: Brachyura: Potamidae). *Raffles Bull. Zool.* Suppl. 16: 273–308.

Yeo, D.C.J., Ng, P.K.L., Cumberlidge, N., Magalhães, C., Daniels, S.R. & Campos, M. 2008. A global assessment of freshwater crab diversity (Crustacea: Decapoda: Brachyura). In: Balian, E.V., Lévequè, C., Segers, H. & Martens, M. (eds.), Freshwater Animal Diversity Assessment. *Hydrobiologia* 595: 275–286.

# Phylogeny and Biogeography of Asian Freshwater Crabs of the Family Gecarcinucidae (Brachyura: Potamoidea)

SEBASTIAN KLAUS[1], DIRK BRANDIS[2], PETER K.L. NG[3], DARREN C.J. YEO[3] & CHRISTOPH D. SCHUBART[4]

[1] *Abteilung Ökologie & Evolution, Goethe-Universität Frankfurt, Siesmayerstr. 70-72, D-60054 Frankfurt am Main, Germany*

[2] *Zoologisches Museum, Universität Kiel, Hegewischstr. 3, 24105 Kiel, Germany*

[3] *Department of Biological Sciences, National University of Singapore, Science Drive 4, 117543 Singapore, Republic of Singapore*

[4] *Fakultät für Biologie 1 (Zoologie), Universität Regensburg, 93040 Regensburg, Germany*

## ABSTRACT

The phylogeny of the Asian freshwater crabs of the family Gecarcinucidae is investigated using the mitochondrial large subunit rRNA gene and the nuclear encoded histone 3 gene. The results confirm the monophyly of the Gecarcinucidae. A division into two families, Gecarcinucidae and Parathelphusidae, is not supported. Therefore, and in consideration of the unresolved family relationships, all Old World freshwater crabs are assigned to one superfamily, the Potamoidea. The evolution of structures of the second gonopod within the Gecarcinucidae is shown to involve convergent reduction of a complex-type groove to a simple-type groove or its complete absence. Gecarcinucids without a frontal triangle are shown to form a paraphyletic group. Thus, these morphological characters are of minor importance for clarifying phylogenetic relationships within the Gecarcinucidae. Genetically, the Gecarcinucidae can be differentiated and separated into seven monophyletic lineages and an assemblage of as yet unresolved Indian groups. We identify the Malay Peninsula and Borneo (particularly Sabah and Sarawak), where representatives of four of these lineages occur, as a hotspot of gecarcinucid diversity. In agreement with our phylogenetic results, an early radiation of the Gecarcinucidae on the Indian subcontinent is postulated along with several dispersal events from Sundaland into the Malesian (Malaysian) Archipelago.

## 1 INTRODUCTION

The Southeast Asian biota has been a constant focus of biogeography since the 19th century (e.g., Wallace 1869; Hall 2003). This interest is mainly because the region's biodiversity hotspots (Myers et al. 2000) coincide with a complex geography and geological history (Hall and Holloway 1998; Morley 2000). The phylogeny of the freshwater crab family Gecarcinucidae (*sensu* Klaus et al. 2006) appears to be well suited to reflect both the geography and history of Southeast Asia. In general, freshwater crabs are believed to have limited dispersal capabilities (Ng & Rodríguez 1995), and crabs within hydrographic drainage systems can be expected to be more closely related. This is of particular interest within Sundaland, consisting of the Malay Peninsula and the Greater Sunda Islands (Borneo, Sumatra, and Java), as these land masses, now separated by the sea, were connected by palaeoriver systems in times of lower sea level (Voris 2000).

The range of the Gecarcinucidae (*sensu* Klaus et al. 2006) covers both the Australian and Oriental zoogeographic regions, and it is the only freshwater crab family that crosses Wallace's Line.

With currently 345 described species in 57 genera, gecarcinucids make up about 35% of the total species diversity and 46% of the genus diversity of the Old World freshwater crabs (Ng et al. 2008). Important local species radiations, based on molecular markers, have been described for Sri Lanka (Bossuyt et al. 2004), Sulawesi (Schubart and Ng 2008), and Taiwan (Shih et al. 2007). Nevertheless, no phylogenetic analysis of the whole family has been conducted until now. Recent molecular phylogenies that included gecarcinucid species primarily addressed family and superfamily relationships with only a limited number of gecarcinucid representatives (Bossuyt et al. 2004: 40 specimens, 20 species, 10 genera; Daniels et al. 2006: 18 species, 10 genera; Klaus et al. 2006: 25 species, 19 genera). All previous systematic approaches to the Gecarcinucidae were based primarily on morphology, focusing on the mandibular palp (Alcock 1910), the frontal triangle (Bott 1970b), or second gonopod characters (Klaus et al. 2006).

Our aim is to identify major evolutionary lineages within the Gecarcinucidae. Our study includes 76 gecarcinucid species of 40 genera. These genera cover 70% of the gecarcinucid genus-level diversity and 85% of the known species. Several genera, especially among the Indian fauna (see Bahir and Yeo 2007), are not included. Nevertheless, the present data allow conclusions to be drawn on the historical biogeography of the Gecarcinucidae and provide a phylogenetic framework that sets the context for future locality or genus-based revisions. This study also contributes to a better understanding of the evolution of morphological characters previously used for taxonomic assignments.

## 2 HISTORICAL SYSTEMATIC APPROACHES TO THE GECARCINUCIDAE

Rathbun (1904) divided the Asian freshwater crabs (which were all included in the family Potamidae Ortmann, 1896) into two subfamilies: the Potaminae, containing most of the Asian freshwater crab fauna, and the monotypic Gecarcinucinae for the genus *Gecarcinucus*. This system was fundamentally altered by Alcock (1910). He assigned all Asian species with a bilobed terminal segment of the mandibular palp to the Gecarcinucinae, and retained species with a simple terminal segment within the Potaminae. Within this redefined Gecarcinucinae, Alcock (1910) recognized two genera: *Parathelphusa* and *Gecarcinucus*. Possibly because he doubted the validity of the genus *Gecarcinucus*, he introduced the name Parathelphusinae as a synonym for the Gecarcinucinae but kept the latter name throughout his work. Influenced by these ideas, Colosi (1920) established within the Gecarcinucinae the tribes Parathelphusini Alcock, 1910, and Hydrothelphusini Colosi, 1920, the latter to include the Madagascan genus *Hydrothelphusa* with a bilobed mandibular palp.

A major change to this taxonomy by Bott (1969, 1970a, 1970b) recognized a superfamily Parathelphusoidea Alcock, 1910 (later corrected to Gecarcinucoidea Rathbun, 1904, by Holthuis 1979), which included Alcock's Gecarcinucinae and several African genera with a bilobed mandibular palp. The Gecarcinucinae *sensu* Alcock (1910) was split into three families, applying diagnostic characters of the frontal triangle: the Gecarcinucidae Rathbun, 1904, with the subfamilies Gecarcinucinae Rathbun, 1904, and Liotelphusinae Bott, 1969; the Parathelphusidae Alcock, 1910, with the subfamilies Spiralothelphusinae Bott, 1968, the monogeneric Ceylonthelphusinae Bott, 1969, and the East– and Southeast Asian Somanniathelphusinae Bott, 1968; and as the third family the Sundathelphusidae Bott, 1969, from the Sunda islands, the Philippines, New Guinea, and Australia. The latter was not further divided into subfamilies. Bott recognized within the Gecarcinucoidea 31 genera with 98 species (115 including subspecies). Later, the Sundathelphusidae were synonymized with the Parathelphusidae (Ng and Sket 1996).

This system was adopted by Martin & Davis (2001) with the reservation that the African species should possibly be excluded from the Gecarcinucoidea. However, Bott's system of subfamilies was not generally adopted by other researchers, and there have been doubts about their validity (see Ng & Tay 2001; Ng 2004; Bahir & Yeo 2007). The distinction of the Gecarcinucidae and Parathelphusidae has been questioned by several workers (e.g., Holthuis 1979; Ng 1988, 2004; Yeo & Ng 1999; Daniels et al. 2006), but Klaus et al. (2006) formally recognized only one family of gecarcinucoid freshwater crabs in Asia, the Gecarcinucidae, on the basis of gonopod morphology and mtDNA phylogeny. All African members of the Gecarcinucidae were assigned to the Deckeniidae (the Deckeniinae within the Potamonautidae according to Cumberlidge et al. 2008). The Gecarcinucidae was divided into two subfamilies based on the morphology of the second gonopod (Klaus et al. 2006): the Indian-Sri Lankan Gecarcinucinae and the Parathelphusinae with their main distribution in East- and Southeast Asia. Cumberlidge et al. (2008), Ng et al. (2008), and Yeo et al. (2008), however, provisionally recognized both Gecarcinucidae and Parathelphusidae as separate families, although, like Klaus et al. (2006), they excluded all African freshwater crabs from the Gecarcinucidae.

## 3 MATERIALS AND METHODS

### 3.1 *Molecular analysis*

Samples for this study were obtained from different museum holdings, aquarists, and collections by the authors between 1999 and 2006 (Table 1). Some of the museum specimens, which include type material, were more than 100 years old and made amplification of longer DNA sequences impossible. Genomic DNA was extracted from the muscle tissue of walking legs using the Puregene kit (Gentra Systems). Selective amplification of an approximately 560 basepair (bp) fragment, excluding primers, from the mitochondrial large ribosomal subunit (16S rRNA) and of a 320-bp fragment of the nuclear histone 3 gene (H3) was carried out by polymerase chain reaction (PCR) under the following conditions: 40 cycles, with 45 sec denaturing at 94°C, 1 min annealing at 48°C, and 1 min extension at 72°C (with 4 min initial denaturation and 10 min final extension time). Especially for the H3 gene amplification, touchdown PCRs were performed to prevent unspecific binding of primers; denaturation and elongation times as well as the corresponding temperatures were identical to the previous PCR profile, but the annealing temperature in the first eight cycles was decreased from 52°C to 48°C (steps of 0.5°C), followed by 40 cycles with an annealing temperature of 48°C. Primers used were 16L29 (5'-YGCCTGTTT-ATCAAAAACAT-3', Schubart, this volume) and 16H37 (5'-CCGGTYTGAACTCAAATCATGT-3', Klaus et al. 2006) or 16H12 (5'-CTGTTATCCCTAAAGTAACTT-3', Schubart, this volume) for the 16S and H3AF (5'-ATGGCTCGTACCAAGCAGACVGC-3') in combination with H3AR (5'-ATATCCTTRGGCATRATRGTGAC-3', both Colgan et al. 1998) or the H3H2 (5'-GGCATRATGG-TGACRCGCTT-3') for the H3. PCR products were purified with the Sure Clean Kit (Bioline) and sequenced with the ABI BigDye terminator mix in an ABI Prism 310 Genetic Analyzer (Applied Biosystems, Foster City, USA). In addition to the sequences generated in this study, our phylogenetic analyses include previously published sequences corresponding to the same 16S and H3 gene regions from GenBank, originating from the studies of Bossuyt et al. (2004), Daniels et al. (2006), Klaus et al. (2006), and Shih et al. (2007).

Sequences were aligned manually with the software BioEdit 7.0.9.0 (Hall 1999) with alignment lengths of 557 bp for 16S RNA and 318 bp for H3. A partition homogeneity test as implemented in PAUP 4.0b was performed (100 replicates). As expected, this test showed significant differences between the genes, as the H3 sequences are much more conserved than the 16S rRNA gene. Thus within the phylogenetic analysis each gene supports different splits at different points in time. The data sets for both genes were combined in one alignment. *Epilobocera sinuatifrons* (Pseudothelphusidae) was designated as the outgroup taxon.

Bayesian analysis (MrBayes 3.1.2, Huelsenbeck and Ronquist 2001) was run with four MCMC chains for 20 million generations, until the average standard deviation of split frequencies decreased to 0.00248. A tree was saved every 1000 generations (with a corresponding output of 20,000 trees). Prior settings as suggested by MODELTEST 3.7 (Posada and Crandall 1998) following the Akaike information criterion were applied (the HKY+I+G model for the H3 and the TrN+I+G model for the 16S partition). The first 1,000,000 generations, i.e., 1000 trees ("burn-in phase"), were excluded

**Table 1.** Freshwater crab species used for DNA-sequencing and subsequent phylogeny reconstruction, including taxonomic authority, museum catalogue number, locality of collection, and genetic database (EMBL) accession numbers for the H3 and 16S sequences.

| Species | Cat. No. | Provenance | H3 | 16S rRNA |
|---|---|---|---|---|
| *Epilobocera sinuatifrons* (A. Milne Edwards, 1866) | R 199 | Puerto Rico, Guajataca | FM 178885 | AJ 130810 |
| *Johora singaporensis* Ng, 1986 | SMF 32717 | Singapore, Bukit Batok | FM 178886 | FM 180114 |
| *Malayopotamon* aff. *brevimarginatum* (De Man, 1392) | SMF 32718 | S-Sumatra, Danau Ranau, Gng Raya | FM 178887 | FM 180115 |
| *Potamon persicum* Pretzmann, 1962 | ZUTC | Zagros mountains, Iran | FM 178888 | FM 180116 |
| *Stoliczia bella* Ng & Ng, 1987 | SMF 32719 | Malaysia, Pulau Langkawi | FM 178889 | FM 180117 |
| *Deckenia mitis* Hilgendorf, 1898 | SAEM | Tanzania, Mwangombe near Tanga, site 23 | FM 178890 | FM 180118 |
| *Hydrothelphusa madagascariensis* (A. Milne Edwards, 1872) | SAEM | Madagascar, Ambolitsara; M. Vences coll. | FM 178891 | FM 180119 |
| *Madagapotamon humberti* Bott, 1965 | MNHN B 25562 | Madagascar | FM 178892 | AM 234641 |
| *Platythelphusa armata* A. Milne Edwards, 1887 | SAEM | Tanzania, Lake Tanganyika, Kigoma Bay | FM 178893 | FM 180120 |
| *Seychellum alluaudi* (A. Milne Edwards & Bouvier, 1893) | SMF 30157 | Seychelles, La Digue | FM 178894 | AM 234653 |
| *Arachnothelphusa rhadamanthysi* Ng & Goh, 1987 | ZRC 1990.443; type | Malaysia, Borneo, Sabah | FM 178895 | FM 180121 |
| *Austrothelphusa transversa* (Roux, 1911) | RMNH 31622 | Papua New Guinea | FM 178896 | FM 180122 |
| *Austrothelphusa* sp. | ZMB | Australia, 16°3'S, 129°11'E | FM 178897 | FM 180123 |
| *Bakousa sarawakensis* Ng, 1995 | ZRC 1995.235 | Malaysia, Borneo, Sarawak | FM 178899 | FM 180124 |
| *Balssiathelphusa cursor* Ng, 1986 | ZRC 1989.3036; type | Indonesia, Borneo, E-Kalimantan, Wanariset | FM 178900 | FM 180126 |
| *Balssiathelphusa natunaensis* Bott, 1970 | RMNH 29300; holotype | Indonesia, Natuna Island | – | FM 180125 |
| *Ceylonthelphusa kandambyi* Bahir, 1999 | uncatalogued | Sri Lanka | FM 178901 | FM 180127 |
| *Currothelphusa asserpes* Ng, 1990 | ZRC 1989.2156 | Indonesia, Moluccas, Halmahera | FM 178902 | FM 180128 |
| *Cylindrotelphusa* sp. | SMF 2754 | India, Malabar | FM 178903 | AM 234635 |
| *Gecarcinucus jacquemonti* H. Milne Edwards, 1844 | NHML 1895.11.8 | India, Bombay, Kaman River | FM 178904 | AM 234637 |
| *Geelvinkia holthuisi* Bott, 1974 | RMNH 29371; paratype | New Guinea, Tanah Merah | FM 178908 | FM 180129 |

Table 1. continued.

| Species | Cat. No. | Provenance | H3 | 16S rRNA |
|---|---|---|---|---|
| *Geithusa pulchra* Ng, 1989 | SMF 32720 | Malaysia, Pulau Redang | – | FM 180130 |
| *Heterothelphusa fatum* Ng, 1997 | SMF 32721 | Singapore, aquarist | FM 178905 | FM 180131 |
| *Holthuisana biroi* (Nobili, 1905) | SMF 7373 | New Guinea, Borowei, Lake Senkani | FM 178906 | FM 180132 |
| *Holthuisana festiva* (Roux, 1911) | SMF 4280 | Papua New Guinea | FM 178907 | FM 180133 |
| *Irmengardia johnsoni* Ng & Yang, 1985 | SMF 30158 | Singapore, Nee Soon swamp forest | FM 178908 | AM 234640 |
| *Lepidothelphusa cognetti* (Nobili, 1903) | ZRC | Malaysia, Borneo, Sarawak | FM 178909 | FM 180134 |
| *Liotelphusa gageii* (Alcock, 1909) | NHMB 1027 a | Bhutan, Kaeme | FM 178910 | FM 180135 |
| *Maydelliathelphusa edentula* (Alcock, 1909) | NHMB 1028 a | Bhutan, Samchi | FM 178911 | FM 180136 |
| *Maydelliathelphusa lugubris* (Wood-Mason, 1871) | NHMB 1025 | Bhutan | FM 178912 | FM 180137 |
| *Niasathelphusa wirzi* (Roux, 1930) | ZRC 1990.447-448 | Indonesia, Nias | FM 178913 | FM 180138 |
| *Oziothelphusa ceylonensis* (Fernando, 1960) | uncatalogued | Sri Lanka, aquarist | FM 178914 | FM 180139 |
| *Oziothelphusa* sp. | uncatalogued | South India, aquarist | FM 178915 | FM 180140 |
| *Parathelphusa convexa* (De Man, 1879) | RMNH 348; syntype | Indonesia, East Java, Besuki | FM 178916 | FM 180141 |
| *Parathelphusa maculata* De Man, 1879 | ZRC 1989.2472-75 | Malaysia, Pahang, Sg. Kinchin | FM 178917 | FM 180142 |
| *Parathelphusa oxygona* (Nobili, 1901) | ZRC 1998.547 | Malaysia, Sarawak, Sg. Sham Tomcu | FM 178918 | FM 180143 |
| *Parathelphusa pantherina* (Schenkel, 1902) | ZRC 2000.1705 | Indonesia, Sulawesi | FM 178919 | FM 180144 |
| *Parathelphusa sarawakensis* (Ng, 1986) | ZRC 1998.545 | Malaysia, Borneo, Sarawak, Sg. Kuhas | FM 178920 | FM 180145 |
| *Perithelphusa borneensis* (von Martens, 1868) | RMNH 33955 | Malaysia, Borneo, Sarawak, Gunung Jambusan | FM 178921 | FM 180146 |
| *Perithelphusa lehi* Ng, 1986 | ZRC 1989.2770 | Malaysia, Borneo, Sarawak | FM 178922 | FM 180147 |
| *Phricotelphusa amnicola* Ng, 1994 | ZRC 1997.315 | Malaysia, Kedah, Gunung Jerai | FM 178923 | FM 180148 |
| *Phricotelphusa gracilipes* Ng & Ng, 1987 | SMF 32722 | Malaysia, Pulau Langkawi | FM 178924 | FM 180149 |
| *Phricotelphusa hockpingi* Ng, 1986 | ZRC 7318-7346 | Malaysia, Taiping, Bukit Larut | FM 178925 | FM 180150 |
| *Phricotelphusa limula* (Hilgendorf, 1882) | ZRC 2000.1917 | Thailand, Phuket, Ton Sai Falls | FM 178926 | FM 180151 |
| *Phricotelphusa sirindhorn* Naiyanetr, 1989 | SMF 32726; paratype | Thailand, Ranong Prov., Amphoe Muang | FM 178927 | FM 180152 |
| *Salangathelphusa brevicarinata* (Hilgendorf, 1882) | SMF 32723 | Malaysia, Pulau Langkawi | FM 178928 | FM 180153 |

Table 1. continued.

| Species | Cat. No. | Provenance | H3 | 16S rRNA |
|---|---|---|---|---|
| *Sartoriana blandfordi* (Alcock, 1909) | SMF 5524 | Iran, Bam | FM 178929 | FM 180154 |
| *Sartoriana spinigera* (Wood-Mason, 1871) | SMF 9344 | India, West Bengal | 178930 | FM 180155 |
| *Sayamia sexpunctata* (Lanchester, 1906) | RMNH 38015 | Malaysia, Pulau Langkawi | FM 178932 | FM 180156 |
| *Sendleria gloriosa* (Balss, 1923) | SMF 4350 | New Britain, 35 km SE Cap Lambert | FM 178933 | FM 180157 |
| *Siamthelphusa improvvisa* (Lanchester, 1901) | SMF 32724 | Malaysia, Pulau Langkawi | FM 178934 | FM 180158 |
| *Siamthelphusa* sp. | uncatalogued | Thailand, aquarist | FM 178935 | FM 180159 |
| *Snaha escheri* (Roux, 1931) | NHMB 803 a; paratype | India, Palnis, Vandaravu | – | FM 180160 |
| *Sundathelphusa boex* Ng & Sket, 1996 | ZRC 2000.2088 | Philippines, Bohol, Anteguera | FM 178936 | FM 180161 |
| *Sundathelphusa cavernicola* Takeda, 1983 | ZRC 2000.2080 | Philippines, Bohol, Anteguera | FM 178937 | FM 180162 |
| *Sundathelphusa celer* (Ng, 1991) | RMNH 36577; type | Philippines, Luzon, Laguna de Bay | – | FM 180163 |
| *Sundathelphusa hades* Takeda & Ng, 2001 | ZRC 2001.1000; type | Philippines, Mindanao, Surigao del Sur | FM 178938 | FM 180164 |
| *Sundathelphusa halmaherensis* (von Martens, 1868) | SMF 4273; holotype | Indonesia, Moluccas, Halmahera | – | FM 180165 |
| *Sundathelphusa minahassae* (Schenkel, 1902) | ZRC 2000.1681 | Indonesia, Sulawesi, Tomohon | FM 178939 | AM 234651 |
| *Sundathelphusa picta* (von Martens, 1868) | RMNH 35242 | Philippines, Luzon, Cabrazan River | FM 178940 | FM 180166 |
| *Sundathelphusa rubra* (Schenkel, 1902) | ZRC 2000.1695 | Indonesia, Sulawesi, Kakaskasan | FM 178941 | FM 180167 |
| *Sundathelphusa sutteri* (Bott, 1970) | NHMB 35 a; holotype | Philippines, Luzon, Bagüis | – | FM 180168 |
| *Sundathelphusa tenebrosa* Holthuis, 1979 | RMNH 31972; type | Malaysia, Borneo, Sarawak, Gunung Mulu Nat. P. | FM 178942 | FM 180169 |
| *Sundathelphusa* sp. | ZRC 2000.1684 | Indonesia, Sulawesi, Mayoa | – | AM 292919 |
| *Stygothelphusa bidiensis* (Lanchester, 1900) | ZRC 1998.541 | Malaysia, Borneo, Sarawak, Gua serih | FM 178943 | FM 180170 |
| *Stygothelphusa* sp. | ZRC 1999.8.0690 | Malaysia, Borneo, Sarawak | FM 178944 | FM 180171 |
| *Terrathelphusa kuhli* (De Man, 1883) | SMF 32725 | Indonesia, Java, Cibodas | FM 178945 | FM 180172 |
| *Thaksinthelphusa yongchindaratae* (Ng & Naiyanetr, 1993) | ZRC 1991.1882-1884; type | Thailand, Bang Phrik waterfall, Takua Pa Distr., Phangna Prov. | FM 178946 | FM 180173 |
| *Thelphusula baramensis* (De Man, 1902) | ZRC 1997.804 | Brunei, Laba, Bukit Teraja | FM 178947 | FM 180174 |

Table 1. continued.

| Species | Cat. No. | Provenance | H3 | 16S rRNA |
|---|---|---|---|---|
| *Thelphusula hulu* Tan & Ng, 1997 | ZRC 1997.103 | Malaysia, Borneo, Sabah | FM 178948 | FM 180175 |
| *Thelphusula sabana* Tan & Ng, 1998 | ZRC 1997.808; type | Malaysia, Borneo, Sabah, Lahad Datu, Juraco | FM 178949 | FM 180176 |
| *Thelphusula tawauensis* Tan & Ng, 1998 | ZRC 1997.810; paratype | Malaysia, Borneo, Sabah, Tawau Hills Park | FM 178950 | FM 180177 |
| *Travancoriana pollicaris* (Alcock, 1909) | NHMB 799 a | India, Tandikudi, Palnis | – | FM 180179 |
| *Travancoriana schirnerae* Bott, 1969 | SMF 5086; paratype | India, Nilgiris, Coonor | – | FM 180178 |
| *Vanni malabarica* (Henderson, 1912) | NHMB 798 b | India, Naduar Riv., Anamalais | FM 178951 | FM 180180 |
| *Vanni nilgiriensis* (Roux, 1931) | NHMB 802 a; paratype | India, Ootacamund, Nilgiris | – | FM 180181 |

Abbreviations: MNHN: Muséum National d'Histoire Naturelle, Paris; NHML: Natural History Museum, London; NHMB: Naturhistorisches Museum Basel; R: Collection Rudolf Diesel; RMNH: Nationaal Natuurhistorisch Museum, Leiden; SAEM: Collection S.A.E. Marijnissen; SMF: Senckenberg Museum, Frankfurt am Main; ZRC: Zoological Reference Collection, Raffles Museum at the National University of Singapore; ZMB: Museum für Naturkunde, Berlin; ZUTC: Zoological Museum, University of Tehran.

**Table 2.** Freshwater crab species used for analysis of the second gonopod (G2), and the respective type of second gonopod groove. Histological data are new (in bold) or from Klaus et al. (2006).

| Species | Catalogue No. | Provenance | Type of G2 groove |
| --- | --- | --- | --- |
| *Austrothelphusa angustifrons* (A. Milne Edwards 1869) | SMF 4272 | Australia, Kimberley Res. Stat. | complex |
| *Ceylonthelphusa rugosa* (Kingsley 1880) | SMF 4378 | Sri Lanka | simple |
| *Ceylonthelphusa soror* (Zehntner 1880) | SMF 4394 | Sri Lanka | simple |
| *Deckenia imitatrix* Hilgendorf 1869 | SMF 2877 | East Africa | simple |
| *Gecarcinucus jacquemonti* A. Milne Edwards 1844 | SMF 1763 | India, Bombay | simple |
| **Geithusa pulchra** Ng 1989 | SMF 32720 | Malaysia, Pulau Redang | simple |
| *Holthuisana biroi* (Nobili 1905) | SMF 7373 | New Guinea, Borowai, Lake Sentani | complex |
| **Holthuisana subconvexa** (Roux 1927) | SMF 7373 | New Guinea, Borowai, Lake Sentani | complex |
| **Irmengardia pilosimana** (Roux 1936) | ZRC 1984.7288-7302 | Malaysia, Pahang, Bukit Chintamani | complex |
| *Oziothelphusa ceylonensis* (Fernando 1960) | uncatalogued | Sri Lanka | simple |
| *Oziothelphusa senex* (Fabricius 1798) | SMF 4368 | Sri Lanka, Kanniyat, near Trincomalee | simple |
| *Oziothelphusa* sp. | SMF 24914 | India, Kerala, Mavoor/Mapram | simple |
| **Oziothelphusa** sp. | uncatalogued | South India | simple |
| *Parathelphusa celebensis* Schenkel 1909 | SMF 1790 | Sulawesi, Mankoka | complex |
| *Parathelphusa bogorensis* Bott 1970 | SMF 2753 | Indonesia, Java, Bogor | complex |
| *Parathelphusa maculata* (De Man 1879) | SMF 2757 | Singapore, Mardai Road | complex |
| *Perbrinckia enodis* (Kingsley 1880) | SMF 4391 | Sri Lanka, Kandy | simple |
| *Potamonautes perlatus* (A. Milne Edwards 1837) | SMF 23255 | South Africa | tube |
| **Phricotelphusa gracilipes** Ng & Ng 1987 | SMF 32722 | Malaysia, Pulau Langkawi | complex |
| **Phricotelphusa hockpingi** Ng 1986 | uncatalogued | Malaysia, Bukit Larut | complex |
| *Platythelphusa armata* A. Milne Edwards 1887 | SMF 6882 | Tanzania, Lake Tanganjika, Gombe Nat. Park | tube |
| *Salangathelphusa brevicarinata* (Hilgendorf 1882) | SMF 12019 | Thailand | simple |
| *Sartoriana spinigera* (Wood-Mason 1871) | SMF 26 057 | India, Nagaland, market in Dimapur | complex |
| *Snaha escheri* (Roux 1931) | SMF 5140 | India, Shembaganur | complex |
| *Spiralothelphusa hydrodroma* (Herbst 1794) | SMF 2823 | Sri Lanka, Lake Mundale | simple |
| *Spiralothelphusa wuellerstorfi* (Heller 1862) | SMF 4406 | India, Nicobar islands | simple |
| **Stoliczia bella** Ng & Ng 1987 | SMF 32719 | Malaysia, Pulau Langkawi | tube |

Table 2. continued.

| Species | Catalogue No. | Provenance | Type of G2 groove |
|---|---|---|---|
| **Stygothelphusa bidiensis** (Lanchester 1900) | ZRC 1998.540 | Malaysia, Sarawak, Guah Serih | complex |
| **Sundathelphusa boex** Ng & Sket 1996 | ZRC 2000.2088 | Philippines, Bohol, Anteguera | simple |
| **Sundathelphusa cassiope** (De Man 1902) | SMF 1802 | Moluccas, Batjan | complex |
| **Sundathelphusa cavernicola** Takeda 1983 | ZRC 2000.2080 | Philippines, Bohol, Anteguera | simple |
| **Sundathelphusa rubra** (Schenkel 1902) | ZRC 2000.1695 | Indonesia, Sulawesi, Kakaskasen | simple |
| **Sundathelphusa tenebrosa** Holthuis 1979 | ZRC 2000.0064 | Malaysia, Sarawak, Niah | simple |
| **Thelphusula baramensis** (De Man 1902) | ZRC 1997.806 | Brunei, Kuala Belait district, Seria | groove absent |
| **Terrathelphusa kuhlii** (De Man 1883) | SMF 5088 | Indonesia, Java, Cibodas | complex |
| **Travancoriana schirnerae** Bott 1969 | SMF 5086 | South India, Nilgiris, Coono | complex |

Abbreviations: SMF: Senckenberg Museum, Frankfurt am Main; ZRC: Zoological Reference Collection, Raffles Museum at the National University of Singapore.

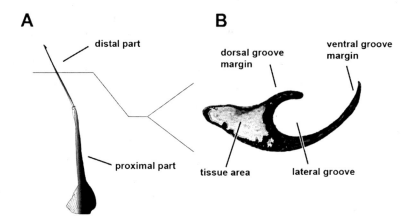

**Figure 1.** Terminology used for describing the second gonopod (G2) of freshwater crabs as proposed by Klaus et al. (2006). (A) Model of a G2. (B) Cross-section of the distal part of the G2 with a complex type of groove (*Parathelphusa bogorensis*). Scales are different.

from the analysis. Besides the combined analysis, the 16S partition was analyzed separately to show the contribution of each of the two genes to the final phylogenetic conclusions. The 87 sequences include additional sequences from GenBank (accession number indicated in the tree, see Fig. 2) and sequences of species for which we failed to amplify the orthologous H3 sequence (see Table 1). Bayesian analysis was run with four MCMC chains for 10 million generations (final average standard deviation of split frequencies = 0.00606) with the prior settings as suggested by MODELTEST 3.7 (HKY+I+G). The "burn-in" phase was of 1,000,000 generations and was excluded from the subsequent analysis.

### 3.2 Morphological analysis

Cross-sections of second gonopods (G2) available from the study of Klaus et al. (2006) and specimens additionally investigated for this study are listed in Table 2. Second gonopods were stored in 70% EtOH, decalcified in 5% trichloroacetic acid for 24 hours, dehydrated in a series of EtOH, and embedded in Spurrs resin or Durcupan® (Fluka AG, Buchs, Switzerland), respectively. Semi-thin sections of 2 $\mu$m thickness were cut using an ultramicrotome with a diamond-knife and stained with Richardsons blue. The terminology used for describing the different G2 morphologies is introduced in Fgure 1.

## 4 RESULTS

The combined H3–16S phylogenetic analysis (Fig. 2) and the 16S-only analysis (Fig. 3) strongly support the monophyly of the Gecarcinucidae *sensu* Klaus et al. (2006) and confirm the separation of the Gecarcinucidae from the Potamidae by the morphology of the mandibular palp as proposed by Alcock (1910) and by sperm morphology (Klaus et al. 2008). Yet the division of the Gecarcinucidae into Gecarcinucinae and Parathelphusinae is not reflected by the molecular phylogenies. In contrast, several major clades are recognizable.

In the 16S-only analysis all deeper splits within the Gecarcinucidae remain polytomous or are weakly supported. Primarily congeneric groups have maximum posterior probabilities. This indicates a much faster evolution of this mitochondrial gene compared to the nuclear encoded histone H3. Nevertheless, the 16S rRNA sequence contains valuable phylogenetic information that increases

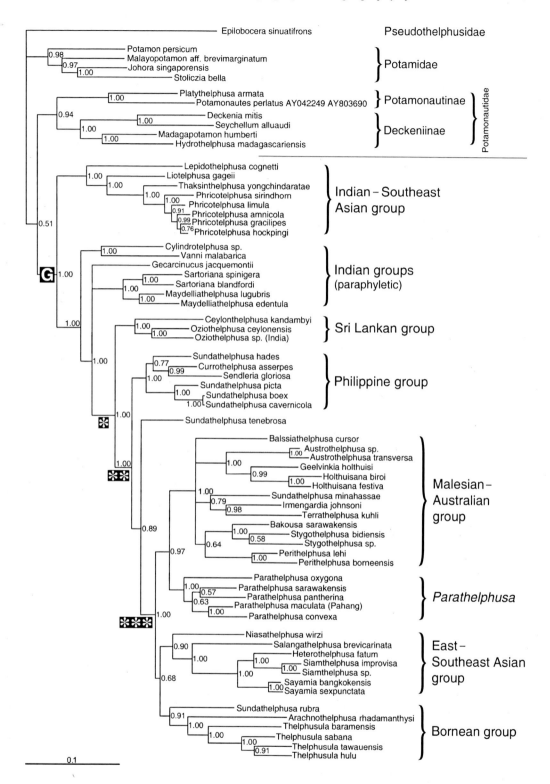

**Figure 2.** Bayesian analysis of the combined H3–16S rRNA data set, with the different lineages within the Gecarcinucidae (G). Indicated are: a clade similar to the "Parathelphusidae" of Bott (*); a monophyletic clade excluding all Indian species (**); and the sister clade to *Sundathelphusa tenebrosa* consisting of four gecarcinucid lineages (***).

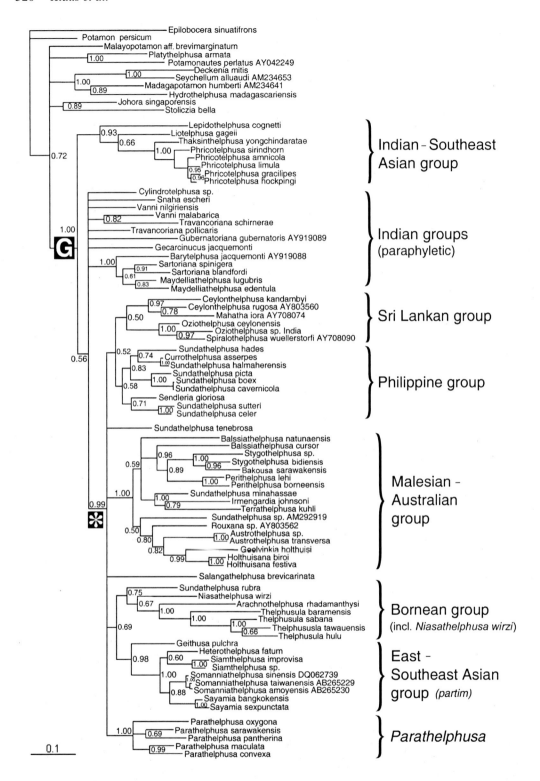

**Figure 3.** Bayesian analysis of the 16S rRNA data only, including sequences of species for which the amplification of the H3 fragment failed and sequences of further species from GenBank. Indicated are the Gecarcinucidae (G), the clade similar to Bott's "Parathelphusidae" (*), and the different gecarcinucid lineages.

the accuracy of the combined analysis. The following groups can be identified within the Gecarcinucidae (referring to the combined H3–16S rRNA analysis, if not indicated otherwise).

## 4.1   Indian–Southeast Asian group

This monophyletic clade branches off first in the Gecarcinucidae and is the basal sister group to all other gecarcinucids. It consists of the genera *Lepidothelphusa* (Borneo), *Liotelphusa* (India and the Himalayas), *Thaksinthelphusa* (Thailand), and *Phricotelphusa* (northern Burma to the Malay Peninsula).

The groove of the second gonopod of *Phricotelphusa gracilipes* and *P. hockpingi* is intermediate in morphology between the complex (where both ventral and dorsal groove margins are broadened, e.g., in *Travancoriana schirnerae*) and simple (where only the ventral groove margin is broadened, e.g., in *Gecarcinucus jacquemonti*) types of the G2 grooves. In *P. hockpingi* the groove is formed by a much thicker cuticle compared to the cuticle surrounding the tissue area, typical of the complex type of G2. However, a true dorsally broadened groove margin is absent. The dorsal margin is more prominent in *P. gracilipes* but is not solid and contains soft tissue.

## 4.2   A paraphyletic group of continental Indian species

Several Indian species included in this analysis dissociate in the combined analysis into several clades. In the first assemblage, the genera *Vanni* and *Cylindrotelphusa* cluster together and form the earliest split with respect to all other gecarcinucids listed below. Well supported is a clade that includes the genera *Sartoriana* and *Maydelliathelphusa* (and *Barytelphusa*, 16S-only). This clade occurs on the Indian subcontinent excluding Sri Lanka, and its range extends north into the Himalayas with *Maydelliathelphusa* and into Afghanistan and Iran in the west with *Sartoriana blandfordi*. The relationship of *Gecarcinucus jacquemonti*, representing the type genus of the Gecarcinucidae, to this clade and to all other gecarcinucids is unresolved. In the 16S-only analysis, all continental Indian species, even congeners, remain polytomous, except the clade that contains *Sartoriana*, *Maydelliathelphusa*, and *Barytelphusa*.

Different character states of the second gonopod occur within these Indian species. *Cylindrotelphusa* and *Maydelliathelphusa* have the distal part of the G2 completely reduced, while the continental Indian species of *Travancoriana*, *Sartoriana*, and *Snaha escheri* (*Gubernatoriana* in Klaus et al. 2006) possess the complex type of second gonopod groove. The specimen identified as *Travancoriana* sp. (see Klaus et al. 2006, SMF 24914), and showing the simple type of second gonopod, turned out to belong to *Oziothelphusa* after reexamination. *Gecarcinucus jacquemonti* is so far the only species of this set of Indian gecarcinucids with the ventral groove margin of the G2 broadened (simple type of G2).

## 4.3   Sri Lankan group

This clade from the Indian subcontinent is represented in the combined analysis by *Oziothelphusa* and *Ceylonthelphusa*. In the 16S-only analysis, *Oziothelphusa* and *Spiralothelphusa* cluster together but connect to the Sri Lankan genera *Ceylonthelphusa* and *Mahatha* with only weak support. The study of Bossuyt et al. (2004), based on mitochondrial sequence data, shows that two more genera of Sri Lanka that are not included here, *Pastilla* and *Perbrinckia*, also belong to this clade. The sister group relationship of the Sri Lankan group to the following lineages of East and Southeast Asian gecarcinucids is well supported (not in the 16S-only analysis). In all investigated species of this group, the simple type of G2 occurs (Klaus et al. 2006). Within the genus *Ceylonthelphusa*, the groove of the G2 is reduced and the distal part of the G2 forms a leaf-like structure.

All non-Indian gecarcinucids, excluding the genera *Lepidothelphusa*, *Thaksinthelphusa*, and *Phricotelphusa* from the Indian–Southeast Asian group, form a monophyletic clade (Fig. 2).

### 4.4 Philippine group

Branching off first within this clade is a group containing species from the Philippines and the Moluccas and reaching with the genus *Sendleria* to New Guinea and the Solomon Islands. *Sundathelphusa picta*, *S. boex*, and *S. cavernicola* from the Philippines cluster together in both the combined H3–16S and 16S-only analyses. *Currothelphusa asserpes* from Halmahera and *Sendleria gloriosa* from the Solomon Islands group together, while in the 16S-only analysis *Sundathelphusa halmaherensis* is sister species to *C. asserpes*, and *Sundathelphusa sutteri* and *S. celer* from Luzon form the sister group to *Sendleria gloriosa*. The G2 of *Sundathelphusa picta* and *S. boex* is of the simple type. Interestingly, *Sundathelphusa tenebrosa* from Borneo does not cluster with the previous clade but is the sister group to all remaining freshwater crabs from East and Southeast Asia. These in turn form a strongly supported monophyletic assemblage (Fig. 2). This set can be subdivided as outlined below.

### 4.5 East–Southeast Asian group

Within this group, *Siamthelphusa*, *Heterothelphusa*, and *Sayamia* cluster together with high support. *Salangathelphusa* separates at a more basal level, and *Niasathelphusa wirzi* appears as the sister group to all other species of this group.

In the 16S-only analysis *Salangathelphusa brevicarinata* and *Niasathelphusa wirzi* do not connect to this clade. The East Asian genus *Somanniathelphusa* appears as the sister group to the Southeast Asian species, while *Geithusa pulchra* (Redang Island, Malay Peninsula) appears as the sister taxon to all other species of the East–Southeast Asian group. Although having a very weak posterior probability, *Niasathelphusa wirzi* clusters in the 16S-only analysis within the Bornean assemblage. However, in the combined H3–16S analysis, its relationship to the East–Southeast Asian group is well supported.

The range of this group covers East Asia (China, Taiwan) and Southeast Asia down to the Malay Peninsula with the isolated occurrence of *Niasathelphusa wirzi* on Nias island west of Sumatra. In the species *Salangathelphusa brevicarinata* and *Geithusa pulchra* the simple type of G2 occurs, whereas all other species in this clade show a completely reduced distal part of the G2. This argues for the simple type of G2 being the plesiomorphic character state within this group, with complete reduction being an apomorphy.

### 4.6 Bornean group

In both analyses, this clade clusters with the East–Southeast Asian group, although this interrelationship is not supported by the very low posterior probabilities. The topology of the deeper splits is similar in both analyses, with *Sundathelphusa rubra* of Sulawesi diverging first, followed by *Arachnothelphusa rhadamanthysi* and then the species of the genus *Thelphusula*. As mentioned above, however, in the 16S-only analysis *Niasathelphusa wirzi* arises between *S. rubra* and *A. rhadamanthysi*. This is not supported by the posterior probabilities, but again this indicates the close relationship of the East–Southeast Asian group and the Bornean group. The G2 of *Sundathelphusa rubra* is of the simple type with a broad ventral groove margin. Although *Thelphusula baramensis* has a G2 with elongated distal part, it lacks any groove structures.

The Malesian–Australian group and the genus *Parathelphusa* cluster together in the combined H3–16S analysis as a monophyletic clade.

### 4.7 Malesian–Australian group

With *Austrothelphusa*, *Balssiathelphusa*, *Geelvinkia*, *Holthuisana*, *Irmengardia*, *Perithelphusa*, members of the genus *Sundathelphusa*, *Stygothelphusa*, *Rouxana*, and *Terrathelphusa*, this group contains a diverse set of genera. Its range covers most of the phytogeographic region of Malesia (ranging from the Isthmus of Kra on the Malay Peninsula to the Solomon Islands in the East) including northern Australia.

Within the Malesian–Australian freshwater crabs, there are two well-supported clades. One clade contains the New Guinean-Australian genera *Austrothelphusa*, *Geelvinkia*, *Holthuisana*, and *Rouxana* (16S-only), and the other clade contains the three species *Irmengardia johnsoni* (Malay Peninsula), *Terrathelphusa kuhli* (Java), and *Sundathelphusa minahassae* (Sulawesi). Of the Bornean genera belonging to the Malesian–Australian group, the genera *Bakousa* and *Stygothelphusa* cluster together. The phylogenetic relationships of these clades along with the Bornean genera *Balssiathelphusa* and *Perithelphusa* are not sufficiently resolved. In the Malesian–Australian group, a G2 with both groove margins broadened is present, although weaker developed in *Terrathelphusa kuhli* and *Irmengardia pilosimana*. *Sundathelphusa cassiope* from Halmahera (Moluccas), which has a complex type of G2 groove, probably also belongs to this lineage, and not, like *S. halmaherensis*, to the Philippine group.

## 4.8 *The genus* Parathelphusa

The five representatives of the speciose genus *Parathelphusa* form a monophyletic group with identical topologies in both analyses. In the combined H3–16S analysis, *Parathelphusa* is the sister group to the Malesian-Australian clade. Compared to the other Southeast Asian groups, rather short branches occur within *Parathelphusa*, even between species from the western (*P. maculata*, Malay Peninsula) and the eastern (*P. pantherina*, Sulawesi) margin of the range. *Parathelphusa oxygona* from Borneo is in a sister group relationship to the other species. All examined second gonopods of this genus have a complex type of groove.

## 5 DISCUSSION

### 5.1 *Monophyly of the Gecarcinucidae*

This study supports the monophyly of the Gecarcinucidae as previously defined by Klaus et al. (2006), corresponding to the Gecarcinucinae *sensu* Alcock (1910) and the Gecarcinucoidea *sensu* Cumberlidge et al. (2008) and Ng et al. (2008). The family relationships among the Gecarcinucidae, Potamidae, and Potamonautidae are not resolved. This is also the case in the molecular analyses of Daniels et al. (2006) and Klaus et al. (2006). Sperm morphology also provides no evidence on the familial relationships (Klaus et al. 2008). The only morphological character shared between Potamidae and Potamonautidae (Potamonautinae) is the distal part of the G2 forming a closed tube (Klaus et al. 2006). However, the Deckeniinae within the Potamonautidae have a G2 with a lateral open groove. If this simple character state is the plesiomorphic condition in the Potamonautidae, then the conformation of the G2 tube in the Potamidae and Potamonautinae are convergent developments. In fact, the potamid tube is formed by groove margins that are involuted, while in the Potamonautinae these margins broadly overlap (see Klaus et al. 2006).

There is therefore no phylogenetic evidence to unite Potamidae and Potamonautidae in a superfamily Potamoidea and on the other hand maintain a separate superfamily Gecarcinucoidea with the single family Gecarcinucidae. As already proposed by several authors (von Sternberg et al. 1999; von Sternberg & Cumberlidge 2001; Klaus et al. 2006; Klaus et al. 2008), we favor the recognition of only one superfamily of Old World freshwater crabs, the Potamoidea, that includes the Gecarcinucidae, Potamidae, and Potamonautidae.

### 5.2 *Gecarcinucid lineages and the morphology of the frontal triangle and the second gonopod*

The present analysis does not support the differentiation of the Gecarcinucidae into two or three families based on character states of the frontal triangle as introduced by Bott (1970a) and adopted by Martin & Davis (2001) and Cumberlidge et al. (2008). The use of the absence or presence of the frontal triangle as a diagnostic character for the two sister groups (Gecarcinucidae and Parathelphusidae) implies that one of the two groups might be paraphyletic, as one of the two character states must represent the plesiomorphic condition. This is confirmed by the present molecular phylogeny. Moreover, there are several genera (e.g., *Ceylonthelphusa* and *Perbrinckia*) for which it is difficult to separate the different character states, as they show intermediate morphologies. It appears that the plesiomorphic character state within the Gecarcinucidae is the complete absence of the frontal triangle, as indicated by its absence in the Indian and Indian–Southeast Asian groups, not to mention its absence in the Potamidae and Potamonautidae as comparative outgroups. The same criticism for the use of the frontal triangle can be applied for the two character states of the second gonopod (simple groove versus complex groove) that were used by Klaus et al. (2006) as diagnostic characters for the gecarcinucid sister groups Gecarcinucinae and Parathelphusinae. However, it is more difficult to identify the plesiomorphic state of the second gonopod. If the complex type of G2 groove of the genus *Phricotelphusa* and several Indian species is homologous, it would probably represent the plesiomorphic character state in the Gecarcinucidae. In the paraphyletic Indian group, both types of G2 groove occur. In the common ancestors of the Malesian–Australian group and the genus *Parathelphusa*, the complex type of G2 groove evolved, while the East–Southeast Asian and the Bornean groups retained a simple type of G2 groove, as it occurs in the Philippine group (Fig. 4).

The complete reduction of the distal part of the second gonopod occurs independently in several Indian genera and in the East–Southeast Asian group. Probably this correlates with a dramatic change in the mechanisms involved in sperm transfer. This is also evident from the absence of a flexible terminal joint in the first gonopod, the generally reduced length of the first gonopod, and in modifications of the female genital apparatus in species lacking the distal part of the second gonopod (unpublished data).

### 5.3 Similarities with the system of Bott

Superficially, the splitting of the Gecarcinucidae into several subclades resembles the taxonomic grouping of Bott (1970a), although his use of the frontal triangle as a diagnostic character and the resulting system of three different families (Gecarcinucidae, Parathelphusidae, and Sundathelphusidae) is strongly contradicted by this study. Most of Bott's subfamilies appear as para- or polyphyletic assemblages. In detail, groups with certain congruence to Bott's taxa are:

(1) The Indian–Southeast Asian group. This clade corresponds to Bott's Liotelphusinae with exclusion of *Sartoriana*, *Thelphusula*, and *Travancoriana*, while the position of *Adeleana* with representatives on Borneo and Sumatra still remains unknown. *Lepidothelphusa cognetti* of Borneo was previously suggested to be closely related to *Phricotelphusa* based on morphological characters (Bott 1970a).

(2) The Sri Lankan group. This group comprises, with *Oziothelphusa* and *Spiralothelphusa*, part of Bott's Spiralothelphusinae (excluding *Balssiathelphusa* and *Irmengardia*) and, with *Ceylonthelphusa*, his Ceylonthelphusinae.

(3) The East–Southeast Asian group. This monophyletic clade includes all the genera of Bott's subfamily Somanniathelphusinae (*Salangathelphusa*, *Somanniathelphusa*, and *Siamthelphusa*).

(4) The genus *Parathelphusa*. Bott's Parathelphusinae included the genera *Parathelphusa*, *Nautilothelphusa*, and *Palawanthelphusa*. The latter was synonymized with *Parathelphusa* (Ng & Goh 1987), while *Nautilothelphusa* seems to nest deeply within the genus *Parathelphusa* of Sulawesi (Schubart & Ng 2008), making the latter paraphyletic.

As this study includes only selected gecarcinucid representatives, it is likely that the phylogeny may change with a larger sample size. This might affect the placements of the Indian gecarcinucid taxa and relationships within the described groups. However, we are reasonably confident that many of the present ideas will be reinforced. Certainly, a clade of *Lepidothelphusa* and *Phricotelphusa*

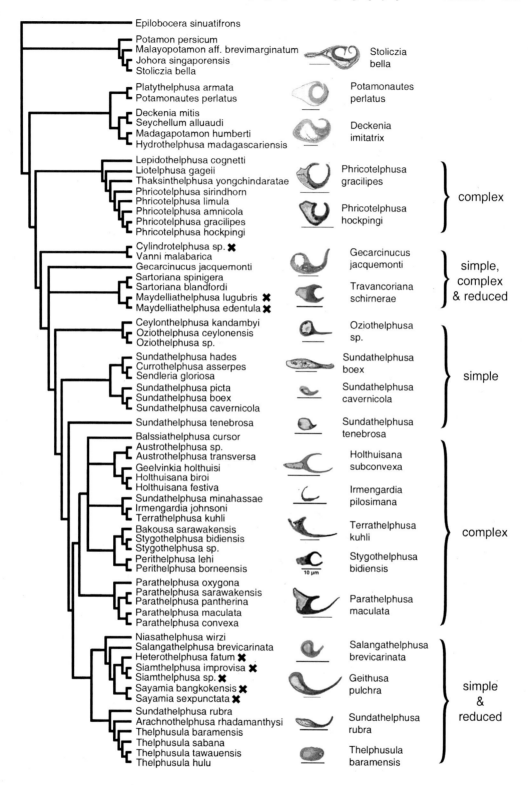

**Figure 4.** Different morphologies of the second gonopod (cross-sections of its distal part) correlated with the topology of the combined gecarcinucid H3–16S rRNA data (Fig. 2). Crosses (×) indicate complete reduction of the distal part of the G2. Scale bars = 50 μm if not indicated otherwise.

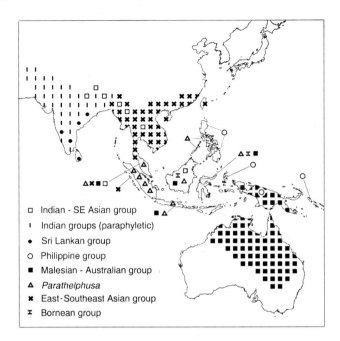

**Figure 5.** Approximate distribution of the different lineages of the Gecarcinucidae.

can also be justified through a suite of morphological characters (unpublished data). We refrain from recognizing formal taxonomic ranks here.

## 5.4 *The genus* Sundathelphusa *Bott, 1969*

The genus *Sundathelphusa* contains 27 species, of which 18 are described from the Philippines with several dozen more that need to be described (unpublished data). Together with *Parathelphusa* and *Somanniathelphusa*, it is one of the most speciose genera within the Gecarcinucidae.

It is evident that the current taxonomic definition of *Sundathelphusa* is flawed, as the species included here are distributed among at least three different lineages. *Sundathelphusa rubra* (Sulawesi) is sister to the other species of the Bornean lineage (Fig. 2). Within *Sundathelphusa* from the Philippines and Halmahera, *Currothelphusa* and *Sendleria* are nested, and *Sundathelphusa* sp. from Sulawesi clusters within the Malesian–Australian assemblage (Fig. 3). The same applies for *S. minahassae* from Sulawesi, described as a subspecies of *S. cassiope* by Bott (1970b). *Sundathelphusa cassiope* itself is the type species of *Sundathelphusa* and originates from Sulawesi. Therefore, the genus name will stay with the species from Sulawesi (excluding *S. rubra*). The genus *Sundathelphusa* needs to be revised (Chia and Ng 2006), and only more detailed morphological and molecular investigations will clarify relationships and taxonomy of this polyphyletic assemblage.

## 5.5 *Biogeography*

Remarkably, species distribution among the lineages is more or less equal (treating the poorly resolved Indian groups as one paraphyletic assemblage, see Figs. 5, 6). Only the Bornean group and the Indian–Southeast Asian group show comparably lower species numbers (Fig. 6). As expected, most of the gecarcinucid species occur in continental Asia. Nevertheless, there are remarkable radiations of gecarcinucid crabs on Sri Lanka and Borneo. New Guinea and Sulawesi also display

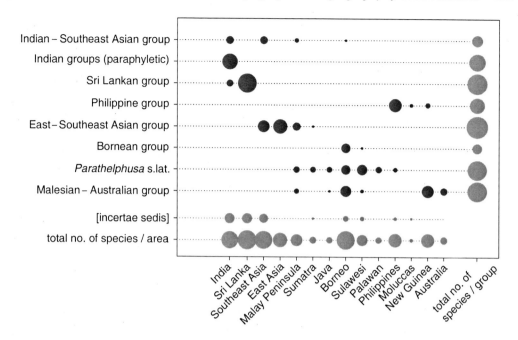

**Figure 6.** Diversity patterns of the Gecarcinucidae. The diameter of the circles is proportional to the species number within the respective gecarcinucid lineage (ordinate) and geographic area (abscissa). Species *incertae sedis* belong to genera not included in the phylogenetic analyses.

relatively high species diversity. In contrast, well-explored Sumatra and Java are depauperate in species number, even when considering cryptic speciation (unpublished data). Australia also shows a minor species and lineage diversity (Fig. 6), most likely due to a more recent dispersal of freshwater crabs from New Guinea across the Torres Strait, although there are still several species that need to be described (P.J.F. Davie, pers. comm.). The present analyses and the previous molecular phylogenies of the Old World freshwater crabs (Daniels et al. 2006; Klaus et al. 2006), as well as the fossil record (Klaus et al. 2006), argue against an origin of the Potamoidea predating the fragmentation of the former Gondwana continent. The fact that the Australian and New Guinean species nest deeply within the Gecarcinucidae, given the diversity pattern of the Australian region, excludes an Australian origin for the Gecarcinucidae.

Klaus et al. (2006) hypothesized that the Gecarcinucidae initially evolved on the Indian subcontinent, with subsequent dispersal to East and Southeast Asia. Based on the present data, this is difficult to resolve. The fact that the Indian groups (including the Sri Lankan group) branch off early within the gecarcinucid phylogeny could indicate an early radiation on the Indian subcontinent. But within the earliest separated Indian–Southeast Asian clade, taxa of both groups cluster together, with the species branching off first being *Lepidothelphusa cognetti* from Borneo.

For the sister group of the Sri Lankan clade (Fig. 2), an Indian origin seems to be most parsimonious with this phylogenetic split having already occurred on the Indian subcontinent (see also Bossuyt et al. 2004). As these non-Indian gecarcinucids are monophyletic, they are most likely the result of a single dispersal event eastward out of India. It was proposed by Klaus et al. (2006) that this dispersal event out of India could have occurred during the Miocene, when the climate became more humid again in northern India (Morley 2000), allowing the gecarcinucid crabs to expand their range.

Because only derived members of the East–Southeast Asian lineage occur in East Asia, the direction of gecarcinucid dispersal was probably first via the Malay Peninsula to the islands of the

Sunda Shelf. As regression events of variable magnitude were frequent during the glaciation periods of the Pliocene and Pleistocene, the resulting terrestrial connections could have allowed freshwater crabs to access the Greater Sunda Islands, although it is difficult to assign this initial and later dispersal events to defined periods of low sea level. The spreading of the Gecarcinucidae beyond the Sunda Shelf to the Philippines, Sulawesi, Halmahera, and further to New Guinea and Australia can be explained only by hypothetical dispersal via rafting, as there is no geological evidence to suggest terrestrial connections between these regions. The Philippine group contains the earliest separated descendants of such a gecarcinucid dispersal event with subsequent radiation on the Philippine islands. The distribution of this lineage covers a dispersal pathway following the Sangihe Island chain from the Philippine Islands to the Moluccas and with *Sendleria* onwards to New Guinea. This dispersal pathway was proposed as a track of general faunal exchange with New Guinea/Australia (Moss and Wilson 1998).

All of the four younger lineages (the Malesian–Australian group, the East–Southeast Asian group, the Bornean group, and *Parathelphusa*) probably evolved on the Sunda Shelf. *Sundathelphusa tenebrosa* from Borneo, sister group to these lineages, could represent an early clade within this radiation. A probable hypothesis is that the initial splits occurred on Borneo itself. Borneo certainly represents a biodiversity hotspot with respect to gecarcinucid diversity. Approximately 14% of the known gecarcinucid species occur on this island, as do representatives of four of the five lineages with Malesian representatives (Fig. 6). In addition, several new genera and species remain undescribed (unpublished data). The distribution pattern of the diverse Malesian–Australian group is congruent with this hypothesis, with an early differentiation of the Bornean genera *Balssiathelphusa*, *Bakousa*, *Perithelphusa*, and *Stygothelphusa*. Based on the present data, this lineage reached Sulawesi and New Guinea/Australia independently.

The East–Southeast Asian group successfully dispersed back into continental Asia. The species branching off first, *Niasathelphusa wirzi* (Nias island), *Salangathelphusa brevicarinata* (Phuket, Pulau Langkawi), and *Geithusa pulchra* (Pulau Redang Island, Malay Peninsula; 16S rRNA only), occur as relics on small islands off the coast of Sumatra and the Malay Peninsula. Therefore, it is probable that the East–Southeast Asian clade evolved in the area of Sumatra and the Malay Peninsula and spread to East Asia secondarily. During times of low sea level this could have occurred via the Siam palaeo-river system that drained the rivers of the Gulf of Thailand to the South China Sea (Voris 2000).

The genus *Parathelphusa* appears as sister group to the Malesian–Australian lineage. The relatively short branch lengths within *Parathelphusa* could indicate a more recent spreading of this genus, with high diversity on Borneo, and remarkable species radiations on Palawan (Ng & Takeda 1993; Freitag & Yeo 2004) and Sulawesi (Chia & Ng 2006; Schubart & Ng 2008). The genus reaches Mindoro and Balabac via Palawan (Ng & Takeda 1993) but is not reported from other Philippine islands. To the east, *Parathelphusa* has crossed Wallace's Line onto Sulawesi and Lombok and occurs in the west in the Malay Peninsula (Bott 1970b; Ng 1988, 1997).

## 6 CONCLUSIONS

Besides validation of gecarcinucid monophyly, this phylogenetic analysis increases profoundly our knowledge of the relationships within the Gecarcinucidae. In contrast to most previous approaches based on morphology alone, we can draw a much more detailed picture, identifying several lineages within the Gecarcinucidae.

Biogeographically, our phylogeny appears to support an early radiation of the Gecarcinucidae on the Indian subcontinent with subsequent dispersal to Southeast Asia. It allows the identification of diversity hotspots (Borneo and the Malay Peninsula) based on genetic diversity. It also provides insights to the historical freshwater crab biogeography of the Malesian (Malaysian) archipelago. Most conspicuously, the complex geography and palaeogeographical history of this region lead to reticulate area-lineage relationships, indicating: (1) independent colonization events at different time points, e.g., the Philippine group and *Parathelphusa* in the Philippines; the Philippine group and the Malesian–Australian group in New Guinea; or the Malesian–Australian group, the Bornean group, and Parathelphusa in Sulawesi; (2) recolonization events, e.g., the dispersal of the East–Southeast Asian group back to continental Asia; and (3) species radiations of related lineages on the same island, e.g., the Malesian–Australian group, the Bornean group, and *Parathelphusa* in Borneo. Although most of the gecarcinucid distribution patterns can be explained only by dispersal, vicariant events also contributed to the present distribution of gecarcinucid lineages, as sea level fluctuations both enabled isolation and faunal exchange on the Sunda Shelf.

## ACKNOWLEDGEMENTS

We thank Daisy Wowor (Bogor) and Peter Koller (Regensburg) for their companionship on various field trips; Agnes Lautenschlager (Warrnambool), Maria Schiwek, and Birgit Lautenschläger (Regensburg) for support with the histological work; Saskia Marijnissen (Amsterdam) for providing potamonautid tissue samples; and Aquarium Glaser (Offenbach) and Chris Lukhaup (Stuttgart) for supplying various specimens. We are deeply grateful to the following persons for access to material under their care and permission to take tissue and/or gonopod samples: Danièle Guinot and Regis Clever (MNHN), Ambros Hängi and Urs Wüest (NHMB), Paul Clark (NHML), Charles Fransen (RMNH), Michael Türkay and Kristin Pietratus (SMF), and Alireza Sari (ZUTEC). Richard von Sternberg, Joel Martin, and an anonymous reviewer kindly streamlined our English and gave valuable comments on the manuscript. We also thank the Deutsche Forschungsgemeinschaft for their financial support of this project (Br 2264/2-1, 2-2 and Schu 1460/6-1).

## REFERENCES

Alcock, A. 1910. *Catalogue of the Indian decapod crustacea in the collection of the Indian Museum. Part I. Brachyura. Fasciculus II The Indian fresh-water crabs — Potamonidae*. The trustees of the Indian Museum, Calcutta.

Bahir, M.M. & Yeo, D.C.J. 2007. The Gecarcinucid freshwater crabs of southern India (Crustacea: Decapoda: Brachyura. *Raffles Bull. Zool. Suppl.* 16: 309–354.

Bossuyt, F., Meegaskumbura, M., Beenaerts, N., Gower, D.J., Pethiyagoda, R., Roelants, K., Mannaert, A., Wilkinson, M., Bahir, M.M., Manamendra-Arachchi, K., Ng, P.K.L., Schneider, C.J., Oommen, O.V. & Milinkovitch, M.C. 2004. Local endemism within the Western Ghats-Sri Lanka biodiversity hotspot. *Science* 306: 479–481.

Bott, R. 1968. Parathelphusiden aus Hinterindien. *Senckenbergiana Biol.* 49: 403–422.

Bott, R. 1969. Die Flußkrabben aus Asien und ihre Klassifikation. *Senckenbergiana Biol.* 50: 339–366.

Bott, R. 1970a. Betrachtungen über die Entwicklungsgeschichte der Süßwasserkrabben nach der Sammlung des naturhistorischen Museums in Genf/Schweiz. *Rev. Suisse Zool.* 77: 327–344.

Bott, R. 1970b. Die Süßwasserkrabben von Europa, Asien und Australien und ihre Stammesgeschichte. *Abh. Senckenb. Naturforsch. Ges.* 526: 1–338.

Chia, O.K.S. & Ng, P.K.L. 2006. The freshwater crabs of Sulawesi, with descriptions of two new genera and four new species (Crustacea: Decapoda: Brachyura: Parathelphusidae). *Raffles Bull. Zool.* 54: 381–428.

Colgan, D.J., McLauchlan, C., Wilson, G.D.F., Livingston, S.P., Edgecombe, G.D., Macaranas, J., Cassis, G. & Gray, M.R. 1998. Histone H3 and U2 snRNA DNA sequences and arthropod molecular evolution. *Aust. J. Zool.* 46: 419–437.

Colosi, G. 1920. I Potamonidi. *Boll. Mus. Zool. Anat. comp. Univ. Torino.* 35: 1–27.

Cumberlidge, N., Daniels, S.R. & Sternberg, R.v. 2008. A revision of the higher taxonomy of the Afrotropical freshwater crabs (Decapoda: Brachyura) with a discussion of their biogeography. *Biol. J. Linn. Soc.* 93: 399–413.

Daniels, S.R., Cumberlidge, N., Pérez-Losada, M., Marijnissen, S.A.E. & Crandall, K.A. 2006. Evolution of Afrotropical freshwater crab lineages obscured by morphological convergence. *Mol. Phylogenet. Evol.* 40: 227–235.

Freitag, H. & Yeo, D.C.J. 2004. Two new species of *Parathelphusa* A. Milne Edwards, 1853, from the Philippines (Crustacea: Decapoda: Brachyura: Parathelphusidae). *Raffles Bull. Zool.* 52: 227–237.

Hall, R. 2003. An addiction to Southeast Asian biogeography. *J. Biogeogr.* 30: 161–163.

Hall, R. & Holloway, J.D. 1998. *Biogeography and Geological Evolution of SE Asia*. Leiden: Backhuys Publishers.

Hall, T.A. 1999. BioEdit: a user-friendly biological sequence alignment editor and analysis program for Windows 95/98/NT. *Nucleic Acids Symp. Ser.* 41: 95–98.

Holthuis, L.B. 1979. Cavernicolous and terrestrial decapod crustacea from northern Sarawak, Borneo. *Zool. Verh.* 171: 1–47.

Huelsenbeck, J.P. & Ronquist, F. 2001. MrBayes: Bayesian inference of phylogenetic trees. *Bioinformatics* 17: 754–755.

Klaus, S., Schubart, C.D. & Brandis, D. 2006. Phylogeny, biogeography and a new taxonomy for the Gecarcinucoidea Rathbun, 1904 (Decapoda: Brachyura). *Org. Divers. Evol.* 6: 199–217.

Klaus, S., Schubart, C.D. & Brandis D. 2008. Ultrastructure of spermatozoa and spermatophores of freshwater crabs (Brachyura: Potamoidea: Gecarcinucidae, Potamidae and Potamonautidae). *J. Morphol.* In press.

Martin, J.W. & Davis, G.E. 2001. An updated classification of the Recent Crustacea. *Natural History Museum of Los Angeles County, Science Series* 39 :1–124.

Morley, R.J. 2000. *Origin and Evolution of Tropical Rain Forests*. Chichester: Wiley.

Moss, S.J. & Wilson, M.E.J. 1998. Biogeographic implications of the Tertiary palaeogeographic evolution of Sulawesi and Borneo. In: Hall, R. & Holloway, J.D. (eds.), *Biogeography and Geological Evolution of SE Asia*: 133–163. Leiden: Backhuys.

Myers, N., Mittermeier, R.A., Mittermeier, C.G., da Fonseca, G.A.B. & Kent, J. 2000. Biodiversity hotspots for conservation priorities. *Nature* 403: 853–858.

Ng, P.K.L. 1988. *The Freshwater Crabs of Peninsular Malaysia and Singapore*: pp. 1–156. Singapore: Dept. Zool., Natn. Univ. Singapore: Shinglee Press.

Ng, P.K.L. 1997. On a new genus and four new species of freshwater crabs (Crustacea: Decapoda: Brachyura: Parathelphusidae) from Borneo and Java. *Raffles Bull. Zool.* 45: 105–121.

Ng, P.K.L. 2004. Crustacea: Decapoda, Brachyura. In: Yule C.M. & Yong H.S. (eds.), *The Freshwater Invertebrates of the Malaysian Region*: 311–336. Kuala Lumpur: National Academy of Sciences.

Ng, P.K.L. & Goh, R. 1987. Cavernicolous freshwater crabs (Crustacea, Decapoda, Brachyura) from Sabah, Borneo. *Stygologia* 3: 313–330.

Ng, P.K.L., Guinot, D. & Davie, P.J.F. 2008. Systema Brachyurorum: Part I. An annotated checklist of extant brachyuran crabs of the world. *Raffles Bul. Zool.* 17: 1–286.

Ng, P.K.L. & Rodríguez, G. 1995. Freshwater crabs as poor zoogeographical indicators: a critique of Bănărescu (1990). *Crustaceana* 68: 636–645.

Ng, P.K.L. & Sket, B. 1996. The freshwater crab fauna (Crustacea, Decapoda, Brachyura) of the Philippines. IV. On a collection of Parathelphusidae of Bohol. *Proc. Biol. Soc. Wash.* 109: 695–706.

Ng, P.K.L. & Takeda, M. 1993. The freshwater crab fauna (Crustacea, Brachyura) of the Philippines II. The genus *Parathelphusa* A. Milne Edwards, 1853 (family Parathelphusidae). *Bull. Natn. Sci. Mus., Tokyo, Ser. A.* 19: 1–19.

Ng, P.K.L. & Tay, F.W.M. 2001. The freshwater crabs of Sri Lanka (Decapoda: Brachyura: Parathelphusidae). *Zeylanica* 6:113–199.

Posada, D. & Crandall, K.A. 1998. MODELTEST: testing the model of DNA substitution. *Bioinformatics* 14: 817–818.

Rathbun, M.J. 1904–1906. Les crabes d'eau douce (Potamonidae). *Archives du Muséum d'Histoire naturelle Paris* 6(4), 7(4), 8(4): 225–312, 159–321, 33–122.

Schubart, C.D. & Ng, P.K.L. 2008. A new molluscivore crab from Lake Poso confirms multiple colonization of ancient lakes in Sulawesi by freshwater crabs (Decapoda: Brachyura). *Zool. J. Linn. Soc.* In press.

Schubart, C.D. (this volume). Mitochondrial DNA and decapod phylogenies. The importance of pseudogenes and primer optimization. In: Martin, J.W., Crandall, K.A., Felder, D.F. (eds.), *Crustacean Issues: Decapod Crustacean Phylogenetics*. Boca, Raton, Florida: Taylor & Francis/CRC Press.

Shih, H.T., Fang, S.H. & Ng, P.K.L. 2007. Phylogeny of the freshwater crab genus *Somanniathelphusa* Bott (Decapoda: Parathelphusidae) from Taiwan and the coastal regions of China, with notes on their biogeography. *Invertebr. Syst.* 21: 29–37.

Sternberg, R.v., Cumberlidge, N. & Rodríguez, G. 1999. On the marine sister groups of the freshwater crabs (Crustacea: Decapoda: Brachyura). *J. Zool. Syst. Evol. Research* 37: 19–38.

Sternberg, R.v. & Cumberlidge, N. 2001. Notes on the position of the true freshwater crabs within the brachyrhynchan Eubrachyura (Crustacea: Decapoda: Brachyura). *Hydrobiologia* 449: 21–39.

Voris, H.K. 2000. Maps of Pleistocene sea levels in Southeast Asia: shorelines, river systems and time durations. *J. Biogeogr.* 27: 1153–1167.

Wallace, A.R. 1869. *The Malay Archipelago*. pp. 515. Singapore: Graham Brash.

Yeo, D.C.J. & Ng, P.K.L. 1999. The state of freshwater crab taxonomy in Indochina (Decapoda, Brachyura). In: Schram, F.R. & von Vaupel Klein, J.C. (eds.), *Crustaceans and the Biodiversity Crisis*: 637–646. Leiden: Crustaceana, Special Volume.

Yeo, D.C.J., Ng, P.K.L., Cumberlidge, N., Magalhães, C., Daniels, S.R. & Campos, M.R. 2008. Global diversity of crabs (Crustacea: Decapoda: Brachyura) in freshwater. *Hydrobiologia* 595: 275–286.

# A Proposal for a New Classification of Portunoidea and Cancroidea (Brachyura: Heterotremata) Based on Two Independent Molecular Phylogenies

CHRISTOPH D. SCHUBART & SILKE REUSCHEL

*Biologie 1, Universität Regensburg, 93040 Regensburg, Germany*

ABSTRACT

Molecular methods are playing an increasingly important role in reconstructing phylogenetic relationships. Regardless of what source of DNA is used, the simple idea behind it is that the genetic distance (distinctness of DNA sequences) between any two taxa should be proportional to the time of their separation. Genetic markers with different degrees of variability appear appropriate for different taxonomic levels. The mitochondrial ribosomal RNA genes 12S and 16S have proven to be useful at the interspecific up to the interfamilial level in brachyuran crabs. Recent criticism has questioned the credibility of phylogenies based solely on mitochondrial DNA (mtDNA) as well as the specific value of commonly used mitochondrial markers such as 16S or Cox1. In this study, we present a molecular phylogeny of cancroid and portunoid crabs based on 1200 basepairs of mtDNA, which partly confirms and partly contradicts current morphology-based taxonomy. In order to test the reliability of mtDNA, we constructed a second phylogeny based on a nuclear gene corresponding to the histone H3. This phylogeny absolutely confirmed our initial results. Based on this independent evidence, we argue that mitochondrial DNA should still be considered a tool with high resolution power in decapod molecular phylogenies up to the interfamilial level. In view of the relatively unstable taxonomic classification of the two studied superfamilies, which are in the process of being revised (three new systems over the past three years), we propose a new taxonomy for the Cancroidea and Portunoidea that is based on significant evidence from two molecular markers and in part finds further support in larval morphology.

## 1 INTRODUCTION

The taxonomy of crabs included in the superfamilies Portunoidea and Cancroidea has been historically quite unstable (see Rathbun 1930; Karasawa et al. 2008). The swimming crabs of the genus *Portunus* and crabs of the genus *Cancer*, on which the superfamily names are based, clearly are different and easily separabale brachyuran heterotreme lineages. However, the establishment of higher taxonomic units in the form of subfamilies, families, and superfamilies, and the placement of different genera into those units based on sometimes convergent characters, has created a taxonomic system that is not necessarily composed of monophyletic units; it also has raised suspicions that members of the superfamilies Portunoidea and Cancroidea (as currently defined) would be better placed in the "other" superfamily or elsewhere (Schubart et al. 2000a; Flores & Paula 2000; Schubart & Reuschel 2005; Ng et al. 2008; Karasawa et al. 2008). Alternatively, genera or families classified elsewhere have been suggested to belong within the Portunoidea (Števčić 2005; Karasawa & Schweitzer 2006).

In order to obtain a stable and monophyletic taxonomic classification, corrections are often necessary at the superfamily, family, subfamily, and even genus level (e.g., Schubart et al. 2000b,

Table 1. Different arrangements of family (and subfamily) subdivisions of Portunoidea and Cancroidea, including extinct (†) and extant taxa.

| Martin & Davis (2001) | Ng et al. (2008) | Karasawa et al. (2008 |
|---|---|---|
| **PORTUNOIDEA** | | |
| **Portunidae** | **Portunidae** | **Portunidae** |
| (no subfamilies specified) | | Atoportuninae |
| | Caphyrinae | Caphyrinae |
| | Carupinae | Carupinae |
| | | Lupocyclinae |
| | | Necronectinae |
| | Podophthalminae | Podophtalminae |
| | Portuninae | Portuninae |
| | Thalamitinae | Thalamitinae |
| | | **Carcinidae** |
| | Carcininae | Carcininae |
| | Polybiinae | Polybiinae |
| | | **Macropipidae** |
| | | **Catoptridae** |
| | | **Mathildellidae** |
| | | **Carcineretidae** † |
| | | **Lithophylacidae** † |
| | | **Longusorbiidae** † |
| **Geryonidae** | **Geryonidae** | **Geryonidae** |
| **Trichodactylidae** | excluded | excluded |
| **CANCROIDEA** | | |
| **Cancridae** | **Cancridae** | |
| **Atelecyclidae** | **Atelecyclidae** | |
| **Pirimelidae** | **Pirimelidae** | |
| **Thiidae** | excluded | |
| **Corystidae** | excluded | |
| **Cheiragonidae** | excluded | |

2002, 2006 for the Grapsoidea). Therefore, it is necessary to understand the current taxonomy of Portunoidea and Cancroidea at different levels before contrasting it with our results based on two molecular phylogenies. Here, and in Table 1, we summarize the most important taxonomic revisions and conclusions at the family level for both superfamilies and at the subfamily level within the family Portunidae.

*Portunoid and cancroid families.* The composition of portunoid and cancroid crabs as used at the end of the 20th century was established by Bowman & Abele (1982). The history of classification of the Portunoidea previous to that has been summarized in detail by Karasawa et al. (2008: 83). Martin & Davis (2001) included the freshwater crab family Trichodactylidae within the Portunoidea based on findings by Rodríguez (1992), von Sternberg et al. (1999) and von Sternberg & Cumberlidge (2001). Števčić (2005) proposed his own explanation-free classification, in which he erected the Melybiidae as a portunoid family, moved the Geryonidae to the Goneplacoidea, and moved the Trichodactylidae to their own superfamily Trichodactyloidea. Ng et al. (2008) kept the Trichodactylidae removed from the Portunoidea (as also suggested by Schubart & Reuschel 2005), but left the Geryonidae within this superfamily. They also synonymized Števčić's (2005)

Melybiidae and kept the genus *Melybia* within the Xanthidae. That same year, Karasawa et al. (2008) published a taxonomic revision of the Portunoidea that emphasized fossil lineages and was based on a cladistic analysis of adult morphological characters. Their conclusion was that "the superfamily is much more diverse at the family level than has been previously recognized" (Karasawa et al. 2008: 82). Consequently, three subfamilies were elevated to family status (see below) and one new family, Longusorbiidae, and two new genera, exclusively composed of fossils, were described in their revision. According to Karasawa et al. (2008), and with inclusion of three additional fossil families (Carcineretidae, Lithophyllacidae, Longusorbiidae) and the extant Mathildellidae (which are Goneplacoidea according to Castro 2007 and Ng et al. 2008), the Portunoidea would consist of nine families (see Table 1; Karasawa et al. 2008: figs. 6-7).

Martin & Davis (2001) included six families within the superfamily Cancroidea (Table 1). In comparison to Bowman & Abele (1982), this meant the addition of the family Cheiragonidae Ortmann, 1893, with the two genera *Cheiragonus* and *Telmessus*, previously included within the Atelecyclidae. Ng et al. (2008) restricted the Cancroidea to the families Cancridae, Atelecyclidae, and Pirimelidae, separating the Cheiragonidae, Corystidae, and Thiidae into their own superfamilies: Cheiragonoidea, Corystoidea, and Thioidea (Table 1). Schweitzer & Feldmann (2000) redefined the family Cancridae with the inclusion of fossil taxa.

*Subfamilies of the Portunidae.* Ortmann (1893) included in his section Portuninea seven families, which later became subfamilies of the family Portunidae: Carupidae, Lissocarcinidae, Platyonychidae, Podophthalmidae, Polybiidae, Portunidae, and Thalamitidae. According to Davie (2002) and Ng et al. (2008), the Portunidae contains seven subfamilies: Caphyrinae Paul'son, 1875; Carcininae MacLeay, 1838; Carupinae Paul'son, 1875; Podophthalminae Dana, 1851; Polybiinae Ortmann, 1893; Portuninae Rafinesque, 1815; and Thalamitinae Paul'son, 1875. Števčić's (2005) system with eight subfamilies and 15 tribes will not be further discussed here, because it lacks supporting arguments and was not adopted in the more comprehensive revision by Ng et al. (2008). Most recently, previous taxonomies were challenged by the fossil work put forward by Karasawa et al. (2008). In addition to the inclusion of fossil taxa, Karasawa et al. (2008) elevated three subfamilies of the Portunidae, i.e., Catroptrinae, Carcininae, and Macropipinae, to full family level. Their results and conclusions will be discussed with our own later in this chapter.

The present study was initiated (Reuschel 2004; Schubart & Reuschel 2005) before the results of more recent revisions became available. Therefore, our taxon sampling was based on the classification by Martin & Davis (2001), with the goal to include taxa of all the portunoid and cancroid families listed in this monograph plus representatives of the seven subfamilies of the Portunidae as listed by Davie (2002). In this sense, our analysis is an independent revision to the ones by Ng et al. (2008) and Karasawa et al. (2008), which may also be said in terms of the methods used: adult morphology (Ng et al. 2008) and adult morphology plus fossils (Karasawa et al. 2008) versus DNA (present study). The goal of this study is to construct a phylogeny of cancroid and portunoid crabs (without claiming that these two superfamilies must represent sister taxa) and to propose a new taxonomy in which the taxa are classified according to their phylogenetic relationships based on two independent sources of DNA sequences. Based on these results, we propose a new taxonomic system, derived from two concordant phylogenetic hypotheses, that can be tested and ameliorated with additional morphological and molecular markers.

## 2 MATERIALS & METHODS

Samples for this study were obtained between 2000 and 2006, mostly from museum specimens and from colleagues (Table 2, Acknowledgements). All molecular studies were carried out at the University of Regensburg. DNA extractions and selective amplification of the mitochondrial complex, consisting of part of the large ribosomal subunit 16S rRNA, the tRNA$_{Leu}$, part of the NDH1

Table 2. List of crab species used for phylogenetic analyses with taxonomic classification following Martin & Davis (2001), locality of collection, museum catalogue number of voucher (if available), and genetic database accession numbers.

| Species | Taxonomy | Collection Locality | Voucher | mtDNA | nDNA |
| --- | --- | --- | --- | --- | --- |
| **PORTUNOIDEA** | | | | | |
| *Arenaeus cribrarius* | Portunidae: Portuninae | USA: North Carolina | SMF-32753 | FM208749 | FM208799 |
| *Callinectes sapidus* | Portunidae: Portuninae | GenBank: USA / USA: Lousiana | unknown/ULLZ3895 | AY363392 | FM208798 |
| *Laleonectes nipponensis* | Portunidae: Portuninae | French Polynesia | MNHN-B31434 | FM208753 | FM208792 |
| *Portunus hastatus* | Portunidae: Portuninae | Turkey: Beldibi | SMF-31989 | FM208780 | FM208796 |
| *Portunus inaequalis* | Portunidae: Portuninae | Ghana: Cape Coast | SMF-32754 | FM208752 | FM208795 |
| *Portunus ordwayi* | Portunidae: Portuninae | Jamaica: Priory | SMF-31988 | FM208751 | FM208794 |
| *Portunus pelagicus* | Portunidae: Portuninae | Australia | CSIRO uncatalogued | FM208750 | FM208797 |
| *Portunus trituberculatus* | Portunidae: Portuninae | GenBank: Japan | unknown | AB093006 | n.a. |
| *Scylla serrata* | Portunidae: Portuninae | Kenya: Lamu | MZUF 3657 | FM208779 | FM208793 |
| *Podophthalmus vigil* | Portunidae: Podophthalminae | Malaysia: Pontian | ZRC Y4821 | FM208760 | FM208787 |
| *Thalamita crenata* | Portunidae: Thalamitinae | Hawaii: Oahu | ULLZ 8664 | FM208754 | FM208800 |
| *Carupa ohashii* | Portunidae: Carupinae | Japan: Okinawa Island | SMF-32756 | FM208759 | FM208790 |
| *Carupa tenuipes* | Portunidae: Carupinae | New Caledonia | MNHN-B31436 | FM208758 | FM208789 |
| *Catoptrus nitidus* | Portunidae: Carupinae | New Caledonia | MNHN-B31435 | FM208755 | n.a. |
| *Libystes edwardsii* | Portunidae: Carupinae | New Caledonia | MNHN-B31437 | FM208761 | n.a. |
| *Libystes nitidus* | Portunidae: Carupinae | New Caledonia | MNHN-B31438 | FM208762 | n.a. |
| *Richerellus moosai* | Portunidae: Carupinae | New Caledonia (paratype) | MNHN-B22838 | FM208756 | FM208788 |
| *Lissocarcinus orbicularis* | Portunidae: Caphyrinae | Singapore: Southern Islands | no voucher, id. PKL Ng | FM208757 | FM208791 |
| *Carcinus maenas* | Portunidae: Carcininae | France: Le Havre | SMF-32757 | FM208763 | FM208811 |
| *Portumnus latipes* | Portunidae: Carcininae | UK: Hastings | SMF-32758 | FM208764 | FM208812 |
| *Polybius henslowii* | Portunidae: Polybiinae | Portugal | SMF-32759 | FM208765 | FM208816 |
| *Liocarcinus corrugatus* | Portunidae: Polybiinae | Spain: Ibiza | SMF-32760 | n.a. | FM208820 |
| *Liocarcinus depurator* | Portunidae: Polybiinae | Alborn Sea | MNHN-B31439 | FM208767 | FM208819 |
| *Liocarcinus holsatus* | Portunidae: Polybiinae | Germany: Helgoland | SMF-32750 | FM208766 | FM208817 |
| *Liocarcinus navigator* | Portunidae: Polybiinae | France: Normandie | SMF-32775 | n.a. | FM208821 |
| *Liocarcinus vernalis* | Portunidae: Polybiinae | Italy: Naples: Fusaro | SMF-32761 | FM208768 | FM208818 |
| *Necora puber* | Portunidae: Polybiinae | UK: Hastings | SMF-32749 | FM208771 | FM208813 |
| *Macropipus tuberculatus* | Portunidae: Polybiinae | Alborn Sea | MNHN-B31440 | FM208769 | FM208815 |
| *Bathynectes maravigna* | Portunidae: Polybiinae | Alborn Sea | MNHN-B31441 | FM208770 | FM208814 |
| *Benthochascon hemingi* | Portunidae: Polybiinae | New Caledonia | ZRC 2000.102 | FM208772 | FM208826 |

Table 2. continued.

| Species | Taxonomy | Collection Locality | Voucher | mtDNA | nDNA |
|---|---|---|---|---|---|
| Ovalipes trimaculatus | Portunidae: Polybiinae | Campagne MD50/Jasus | MNHN-B19785 | FM208773 | FM208823 |
| Ovalipes iridescens | Portunidae: Polybiinae | Taiwan: NE coast | ZRC 1995.855 | FM208774 | FM208825 |
| Ovalipes punctatus | Portunidae: Polybiinae | Taiwan | MNHN-B31442 | n.a. | FM208824 |
| Ovalipes australiensis | Portunidae: Polybiinae | Australia | CSIRO uncatalogued | n.a. | FM208822 |
| Geryon longipes | Geryonidae | Spain: Ibiza, fish market | SMF-32747 | FM208776 | FM208828 |
| Chaceon granulatus | Geryonidae | Japan | SMF-32762 | FM208775 | FM208827 |
| Trichodactylus dentatus | Trichodactylidae | Brazil: Bahia | SMF-32763 | FM208777 | FM208785 |
| **CANCROIDEA** | | | | | |
| Cancer pagurus | Cancridae | France: Le Havre | SMF-32764 | FM207653 | FM208806 |
| Cancer irroratus | Cancridae | USA: Maine | ULLZ 3843 | FM207654 | FM208807 |
| Atelecyclus rotundatus | Atelecyclidae | France: Bretagne | SMF-32765 | FM207652 | FM208804 |
| Atelecyclus undecimdentatus | Atelecyclidae | Portugal: Algarve | SMF-32766 | FM207651 | FM208805 |
| Pirimela denticulata | Pirimelidae | France: Guthary | SMF-32767 | FM208783 | FM208808 |
| Sirpus zariquieyi | Pirimelidae | Greece: Parga | SMF-32768 | FM208784 | FM208809 |
| Thia scutellata | Thiidae | France: Bretagne | SMF-32769 | FM208782 | FM208810 |
| Corystes cassivelaunus | Corystidae | France: Bretagne | SMF-32770 | FM208781 | FM208801 |
| Telmessus cheiragonus | Cheiragonidae | Japan: Hokkaido: Ozuchi | SMF-22475 | FM207656 | FM208802 |
| Erimacrus isenbeckii | Cheiragonidae | Japan | SMF-32752 | FM207657 | FM208803 |
| **PSEUDOTHELPHUSOIDEA** | | | | | |
| Epilobocera sinuatifrons | Pseudothelphusidae | Puerto Rico: Guilarte | SMF-32774 | FM208778 | FM208830 |
| **POTAMOIDEA** | | | | | |
| Geothelphusa dehaani/sp. | Potamidae | GenBank: Japan | unknown | NC007379 | DQ079677 |
| **CARPILIOIDEA** | | | | | |
| Carpilius sp. | Carpiliidae | French Polynesia | SMF-32771 | FM208748 | FM208786 |

CSIRO Marine Research, Invertebrate Museum, Hobart; MNHN: Muséum National d'Histoire Naturelle, Paris; MZUF: Museo Zoologico Universitá di Firenze 'La Specola', Florence; SMF: Senckenberg Museum, Frankfurt a.M.; ULLZ: University of Louisiana at Lafayette Zoological Collection, Lafayette.

**Table 3.** Primers used for amplification of approximately 1200 basepairs mtDNA (consisting of 16S rRNA, tRNA$_{Leu}$, NDH1) and exactly 328 basepairs nDNA corresponding to histone H3.

16S towards NDH1:
16L2: 5'–TGCCTGTTTATCAAAAACAT–3' (Schubart et al. 2002)
16L6: 5'–TTGCGACCTCGATGTTGAAT–3' (Schubart this volume)
16L11: 5'–AGCCAGGTYGGTTTCTATCT–3' (Schubart this volume)
16LLeu: 5'–CTATTTTGKCAGATDATATG–3' (Schubart this volume)
NDL8: 5'– TTA GTD GSR GTW GCY TTT GT–3' (new)

NDH1 towards 16S:
16H37: 5'–CCGGTYTGAACTCAAATCATGT–3' (Klaus et al. 2006)
16H11: 5'–AGATAGAAACCRACCTGG–3' (Schubart this volume)
16H10: 5'–AATCCTTTCGTACTAAA–3' (Schubart this volume)
16HLeu: 5'–CATATTATCTGCCAAAATAG–3' (Schubart this volume)
NDH1: 5'–TCCCTTACGAATTTGAATATATCC–3' (Schubart this volume)
NDH5: 5'–GCYAAYCTWACTTCATAWGAAAT–3' (Schubart this volume)

H3 forward and reverse:
H3af: 5'–ATGGCTCGTACCAAGCAGACVGC–3' (Colgan et al. 1998)
H3ar: 5'–ATATCCTTRGGCATRATRGTGAC–3' (Colgan et al. 1998)
H3H2: 5'–GGCATRATGGTGACRCGCTT–3' (new)

(16S-NDH1), in addition to amplification of part of the nuclear histone H3, were performed as reported in Schubart et al. (2006). The primers used to amplify an approximately 1200-bp unit of mtDNA (16S-NDH1 complex) and 328 bp of the nuclear histone H3 are listed in Table 3. PCR-amplifications were carried out with four minutes of denaturation at 94°C, 40 cycles with 45 s at 94°C, 1 min at 48°C, 1 min at 72°C, and 10 min final denaturation at 72°C. PCR products were purified with Microcon 100 filters (Microcon), ExoSAP-IT (Amersham Biosciences), or Quick-Clean (Bioline) and then sequenced with the ABI BigDye terminator mix followed by electrophoresis in an ABI Prism 310 Genetic Analyzer (Applied Biosystems, Foster City, USA). Forward and reverse strands were obtained as well as overlapping regions for larger DNA fragments. New sequence data were submitted to the European molecular database EMBL (see Table 2 for accession numbers). In addition, the following sequences archived in molecular databases were included in our analyses: mtDNA of *Portunus trituberculatus* (AB093006), *Callinectes sapidus* (AY363392), and *Geothelphusa dehaani* (NC007379), and nuclear DNA (nDNA) of *Geothelphusa* sp. (DQ079677).

Sequences were aligned with CLUSTAL W (Thompson et al. 1994) as implemented in the software BioEdit version 7.5.0.3 (Hall 1999) and corrected manually with BioEdit or xESEE version 3.2 (Cabot and Beckenbach 1989). The data for 16S-NDH1 and H3 were always analyzed as separate datasets for subsequent independent phylogenetic analyses. DNA sequence of *Carpilius* sp. (Carpiliidae) was included as an outgroup.

Phylogenetic congruence among mtDNA partitions was performed using the incongruence length difference (ILD) test (Farris et al. 1995) implemented in PAUP as the partition-homogeneity test (Swofford 1998). For this test, we used random taxon addition, TBR branch swapping, and heuristic searches with 1000 randomizations of the data. The model of DNA substitution that fit our data best was determined using the software MODELTEST 3.6 (Posada and Crandall 1998). This approach consists of successive pairwise comparisons of alternative substitution models using the hLRT and Akaike tests. Model selections were done separately for the mtDNA and nDNA. Two methods of phylogenetic inference were applied to our dataset: maximum parsimony (MP) using the software package PAUP (Swofford 1998) and Bayesian analysis (BI) as implemented in MrBayes v. 3.0b4 (Huelsenbeck & Ronquist 2001).

MP trees were obtained by a heuristic search with 100 replicates of random sequences addition and tree-bisection-reconnection as branch swapping options keeping multiple trees (MulTrees). Analyses were carried out by weighing transversions twice as much as transitions; gaps were always treated as missing. Subsequently, confidence values for the proposed groups within the inferred trees were calculated with the nonparametric bootstrap method (2000 pseudoreplicates, 10 replicates of sequence addition). Only minimal trees were retained and zero-length branches were collapsed. The BI trees were calculated using the suggested model of evolution. The Bayesian analysis was run with four MCMC (Markov chain Monte Carlo) chains for 2,000,000 generations, saving a tree every 500 generations (with a corresponding output of 4000 trees). The −lnL converged on a stable value between 20,000 and 60,000 generations ("burn-in phase"). The first 100,000 generations were thus excluded from the analysis to optimize the fit of the remaining trees. The posterior probabilities of the phylogeny were determined by constructing a 50% majority rule consensus of the remaining trees. Consensus trees were obtained using the "sumpt" option in MrBayes.

## 3 RESULTS

The total alignment of the sequenced portions of the 16S-NDH1 region consisted of 1497 bp, whereas the length of the sequenced region of the histone 3 gene consisted of 328 bp after removal of the primer regions. From the 1497-bp mtDNA, 671 were variable and 565 were parsimony-informative. The 328-bp nDNA had 111 variable positions and 100 parsimony-informative positions. The mtDNA fragment for most analyzed species was not longer than 1200 bp, but the sequence of the cancroid crab *Atelecyclus undecimdentatus* had an additional fragment of 284 bp inserted between the 16S rRNA and the tRNA$_{Leu}$ (explaining the high number of apparently constant characters). Comparing this fragment with sequences from the genetic database revealed that part of this DNA consists of a sequence corresponding to the tRNA$_{Val}$, whereas the rest of the sequence appears to be non-informative. Thus, we report a unique case of gene rearrangement, which appears to also occur in a similar fashion in other crabs of the genera *Cancer* and *Atelecylus*, based on the fact that we needed to amplify the apparently unconnected 16S rRNA and tRNA$_{Leu}$-NDH1 in separate PCRs (Schubart in preparation). Excluding this insertion in the DNA of *A. undecimdentatus*, we calculated a relatively high proportion of 46.6% parsimony-informative positions in the mtDNA as opposed to 30.5% parsimony-informative positions in the more conserved nDNA of histone 3.

The selected model of DNA substitution by hLRT and Akaike was the GTR + I + G model (Rodríguez et al. 1990) for the mitochondrial 16S-NDH1 as well as for the nuclear H3. This model was consequently used for the BI method. Character congruence between the 16S, tRNA$_{Leu}$, and the NDH1 gene fragments was not rejected according to the ILD test. We did not combine the mitochondrial and nuclear dataset, because one of the goals of this study was to compare results from the mitochondrial phylogeny with those from a nuclear dataset to address criticism concerning the credibility of phylogenies based on mtDNA (e.g., Mahon & Neigel 2008).

Both phylogenetic inference methods (BI and MP) resulted in trees that were surprisingly congruent in their overall topology for both sources of DNA, with most clusters showing consistently high confidence values. The results of the two methods are therefore shown together based on the topology of the BI tree, with all confidence values ≥ 50 plotted on the corresponding branches (figs. 1, 2). Posterior probabilities are expressed in a range from 0 to 100 (instead of from 0 to 1). In the case of H3, we also present the topology of the heuristic MP tree (Fig. 3), because the consensus tree of this relatively short gene fragment does not allow recognition of all branching patterns (without statistic support) at the base of the tree. The mtDNA MP heuristic search yielded

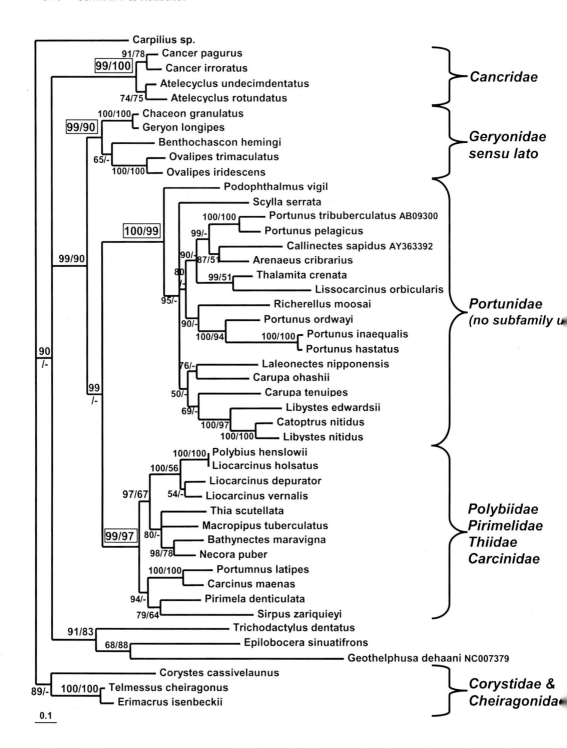

**Figure 1.** Phylogenetic consensus tree of 46 cancroid and portunoid crabs according to the classification of Martin & Davis (2001) based on 1497 basepairs of mtDNA (16S rRNA-NDH1); topology of a Bayesian Inference analysis with confidence values (only ≥ 50) corresponding to Bayesian posterior probabilities/maximum parsimony bootstrap values. *Carpilius* sp. was used as outgroup. The proposed taxonomic classification is given to the right.

# A Proposal for a New Classification of Portunoidea and Cancroidea 541

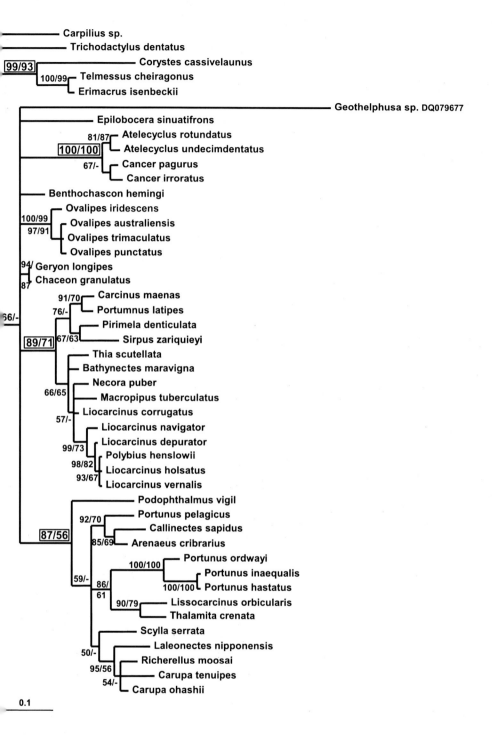

**Figure 2.** Phylogenetic consensus tree of 46 cancroid and portunoid crabs according to the classification of Martin & Davis (2001) based on 328 basepairs of nDNA (histone H3); topology of a Bayesian Inference analysis with confidence values (only ≥ 50) corresponding to Bayesian posterior probabilities/maximum parsimony bootstrap values. *Carpilius* sp. was used as outgroup.

**Figure 3.** Strict consensus of 45 shortest trees of maximum parsimony heuristic search of 46 cancroid and portunoid crabs; 328 basepairs of nDNA (histone H3). *Carpilius*, sp. was used as outgroup.

one shortest tree of length 6751 with tree scores CI = 0.30, RI = 0.51. The topology of this search was congruent with the consensus topology obtained after bootstrapping, with resulting bootstrap values shown in Figure 1. The nDNA MP heuristic search yielded 45 shortest trees of length 696 with tree scores CI = 0.42, RI = 0.69. The strict consensus topology of these 45 shortest trees is shown in Figure 3, whereas MP bootstrap values after 2000 bootstrap reiterations are included in Figure 2 for comparison with BI posterior probabilities.

Comparison of the phylogenetic results derived from the mtDNA dataset (Fig. 1) with the current classifications (Table 1) reveals striking differences. Most evident is that both superfamilies (Portunoidea and Cancroidea) cannot be recognized as monophyletic clades in the tree, regardless of which of the taxonomic systems of Table 1 is followed. Crabs that have been considered Portunoidea fall into three to four major subgroups, depending on whether freshwater crabs of the family Trichodactylidae are included. Without the trichodactylids, which cluster with freshwater crabs from two other families (Pseudothelphusidae and Potamidae), three strongly supported (confidence always ≥ 90) groups including portunoid crabs remain: 1) With a support of 100/99 (BI/MP), there is a clade that contains the core of the Portunidae, including the type genus *Portunus* and the type species *Portunus pelagicus*, and all included members of the subfamilies Portuninae, Thalamitinae, Carupinae, Caphyrinae, and Podophthalminae. However, whenever more than one species of the subfamilies (Portuninae and Carupinae) were available, they did not cluster together, casting some doubt on the validity of these taxonomic units. Additionally, the genera *Portunus*, *Carupa*, and *Libystes* do not appear as monophyletic units on this tree. 2) The second group of portunoid crabs clusters with a support of 99/97. This group includes the European representatives of the other two units previously treated as subfamilies (Polybiinae and Carcininae), but also three other European species that were considered to belong elsewhere: *Pirimela denticulata* and *Sirpus zariquieyi* (both Pirimelidae) and *Thia scutellata* (Thiidae). Interestingly, the genus *Liocarcinus* is not monophyletic, and its type species, *L. holsatus*, is genetically almost identical to the type species of the genus *Polybius*, *P. henslowii*. Two non-European genera that are commonly classified as Polybiinae, *Benthochascon* and *Ovalipes*, are not found in this group, but in 3) a cluster where they are united, with a support of 99/90, to the two deep water representatives of the family Geryonidae.

The allocation of the different members of the Cancroidea *sensu* Martin & Davis (2001) on the phylogenetic tree is equally fragmented. The core of the Cancroidea, with the type genus *Cancer* and type species *C. pagurus*, is found in a well-defined clade (88/100) together with members of the genus *Atelecyclus* (type genus of the family Atelecyclidae). However, the remaining "Cancroidea" have little phylogenetic affinity to these crabs. As mentioned above, the two families Pirimelidae and Thiidae are now embedded among the European Carcininae and Polybiinae. The Corystidae and Cheiragonidae cluster together, but without absolute support (89/-). Both families appear to hold a basal and unrelated position to all other crabs analyzed in this study. However, this study was not designed to discern (and the tree does not resolve) phylogenetic relationships at the root of the Heterotremata.

All of these groups could also be recovered with the much shorter and more conserved nuclear marker. The only exception is the cluster consisting of Geryonidae-*Benthochascon*-*Ovalipes*, which is unresolved at the level above 50% confidence (see Fig. 2). However, the heuristic search (Fig. 3) and additional analyses based on neighbor joining distances (not shown) also grouped these taxa together. Additional taxa and longer DNA fragments may be necessary to provide strong enough support from nuclear DNA to this potential clade. We did find support from nDNA for 1) the portunid group consisting of the subfamilies Portuninae, Thalamitinae, Carupinae, Caphyrinae, and Podophthalminae (87/56); 2) the second "portunid" group consisting of the European representatives of Carcininae and Polybiinae together with the "cancroid" families Pirimelidae and Thiidae (89/71); 3) the core group of Cancroidea restricted to the families Cancridae and Atelecyclidae (100/100); and 4) a clade uniting Corystidae and Cheiragonidae (99/93) in a potentially monophyletic assemblage.

According to this phylogenetic congruence of the two datasets, and with the goal to establish a taxonomic system that is in agreement with phylogenetic relationships, we propose a taxonomic classification as depicted in Figure 1 and Table 4.

**Table 4.** Proposed taxonomy of extant Portunoidea and Cancroidea, as well as taxa excluded from those superfamilies, based on the current molecular phylogenies and supporting evidence.

Superfamily Portunoidea Rafinesque, 1815
    Family Carcinidae MacLeay, 1838
    Family Geryonidae Colosi, 1923
    Family Pirimelidae Alcock, 1893
    Family Polybiidae Ortmann, 1893
    Family Portunidae Rafinesque, 1815
    Family Thiidae Dana, 1852

Superfamily Cancroidea Latreille, 1802
    Family Atelecyclidae Ortmann, 1892
    Family Cancridae Latreille, 1802

Superfamily Corystoidea Samouelle, 1819
    Family Corystidae Samouelle, 1819
    Family Cheiragonidae Ortmann, 1893

Superfamily Trichodactyloidea H. Milne Edwards, 1853
    Family Trichodactylidae H. Milne Edwards, 1853

## 4 DISCUSSION

The portunoid and cancroid taxonomic classifications as commonly used and summarized by Martin & Davis (2001) have been challenged by alternative classification schemes (Števčić 2005; Karasawa et al. 2008) and recently also by Ng et al. (2008, with the recognition of additional superfamilies). While Števčić's (2005) taxonomy was presented without further explanations, and evidently was based on subjective grouping according to adult morphology, Karasawa et al. (2008) used and listed adult morphological characters applied to extinct and extant portunoid crabs to support their classification. Adult morphology, especially carapace and chelar characters, is known to be influenced by convergent evolution. Therefore, we provide results from two molecular phylogenies (one mtDNA-based, the other nDNA-based) and use these to propose a new possible classification of portunoid and cancroid crabs. We do this realizing that all available classifications are still unsettled: "The composition of the superfamily Cancroidea has varied with different authors. The Portunoidea are sometimes included, and while there does appear(s) to be a link, we prefer to keep them apart until more compelling evidence surfaces" (Ng et al. 2008: 51). Nevertheless, we also propose a new taxonomy, because we are convinced that these molecular phylogenies correctly reflect the evolution of these groups and because we find independent confirmation of some of our conclusions in results from larval morphology (see below).

Our proposed taxonomy is summarized in Table 4 and with the labels of Figure 1. Most important is the recognition of six extant families within the superfamily Portunoidea instead of three (as in Martin & Davis 2001, Ng et al. 2008) or of a different six (Karasawa et al. 2008). In addition to the Geryonidae and the Portunidae *sensu novo*—which is now limited to members of the former subfamilies Carupinae, Caphyrinae, Podophthalminae, Portuninae, and Thalamitinae—we recognize the Carcinidae and Polybiidae as full families. We do not agree with Karasawa et al. (2008) in recognizing Mathildellidae Karasawa & Kato, 2003, as a portunoid family, based on preliminary DNA evidence that became available during revision of this manuscript (Schubart, in progress). This agrees with Ng & Manuel-Santos (2007), Castro (2007) and Ng et al. (2008), who also do not consider Mathildellidae to belong to Portunoidea. The Thiidae and Pirimelidae, which had been recognized as full families within the Cancroidea (according to Martin & Davis 2001) or placed in their own superfamily (Thioidea in Ng et al. 2008), are herewith moved into the vicinity of Polybiidae and Carcinidae (and into the Portunoidea, if superfamilies continue to be used). The close relationship of Thiidae and Pirimelidae to the European Polybiidae and Carcinidae (Figures 1, 2) not only justifies the removal of these two families and three genera from the Cancroidea and their inclusion into the Portunoidea, but also requires elevation of Polybiinae and Carcininae to family level, if Pirimelidae and Thiidae continue to be regarded as full families. Alternatively, Carcinidae, Pirimelidae, Polybiidae *sensu stricto*, and Thiidae would all need to be included within the family Carcinidae MacLeay, 1838.

Bourdillon-Casanova (1956) and Flores & Paula (2000) described the larval development of *Pirimela denticulata* and noticed a close morphological similarity to larvae of European Portunidae, especially Polybiinae and Carcininae. Based on larval morphology, Bourdillon-Casanova (1960) suggested a continuous evolutionary line from *Macropipus* to *Portumnus*, with *Pirimela* and *Sirpus* as intermediate forms. Flores & Paula (2000) concurred with Bourdillon-Casanova's opinion and pointed out that the latter two genera share most morphological characters with those of larvae of the European Carcininae, *Carcinus* and *Portumnus*. This is exactly where the molecular results would place these two genera, and it is an important confirmation that larval morphology is often congruent with molecular results, even if contrary to results from adult morphology (see Schubart et al. 2000b, 2002). Consequently, Flores & Paula (2000: 2139) concluded: "pirimelids could be regarded as non-swimming portunids between portunines and carcinines."

Karasawa et al. (2008) independently reached the conclusion that the Carcinidae and Macropipidae should be regarded as full families. That means that they also recognized differences important enough in the former Polybiinae and Carcininae to separate them from the remaining Portunidae at a family level. However, more drastically than in our classification, they modified the composition of these two families with respect to the composition of the subfamilies. According to their results, the European Carcinidae and Polybiidae are not monophyletic but consist of two lineages, with some genera falling into Karasawa et al.'s (2008) redefined Carcinidae (*Liocarcinus*, *Polybius*, *Portumnus*, *Xaiva*, *Carcinus*) and some into the redefined Macropipidae (*Bathynectes*, *Necora*, *Macropipus*), both of which are considered full families. Based on our results, we disagree with this classification. All our European Polybiidae and Carcinidae appear closely related. This includes the European representative of the genus *Macropipus*, *M. tuberculatus* Prestandrea, 1833. Our separation into two families (Carcinidae and Polybiidae) is justified by the fact that the morphologically derived Pirimelidae and Thiidae cluster among these crabs and by the fact that *Carcinus* and *Portumnus* cluster together as sister genera, whereas the Polybiidae form a second branch together with Thiidae. Karasawa et al. (2008) used only *Macropipus australis* Guinot, 1961, for material of that genus. If this species turns out to belong to a different lineage than the European *Macropipus tuberculatus*, it would have to be reclassified. However, the subfamily name Macropipinae Stephenson & Campbell (or the derived family name Macropipidae) remains with *M. tuberculatus*, and this species clearly belongs to the European Polybiidae Ortmann, 1893, which is the older family name and thus has preference (see also Holthuis 1968).

It is certainly true that our definition of the new Polybiidae and Carcinidae cannot be satisfactorily completed without including all members (at least all genera) of the former subfamilies in our analysis (currently in progress). The genera *Brusinia*, *Coenophthalmus*, *Echinolatus*, *Nectocarcinus*, *Parathranites*, *Raymanninus*, and *Xaiva* may belong to different evolutionary lineages and thus might require the definition of new taxa. The Polybiidae, however, is defined by the position of *Polybius henslowii* Leach, 1820, and for the moment includes the genera *Polybius*, *Liocarcinus* (for which a revision of all species is in progress), *Necora*, *Bathynectes*, and *Macropipus*. We realize, however, that according to our mtDNA tree, even the Polybiidae *sensu stricto* may be paraphyletic if the Thiidae keep their family status.

The heterogeneous character of the former Polybiidae is discernible the phylogenetic position of the genera *Ovalipes* and *Benthochascon* in our trees. They are clearly more closely related to Geryonidae than to Polybiidae. We therefore exclude them from the Polybiidae and place them provisionally in the Geryonidae *sensu lato* (Fig. 1). Morphologically, they are clearly distinct from *Geryon* and *Chaceon*, and they may deserve their own family. We anticipate placing these two genera in a new family, but we await further results on the phylogenetic position of the American members of *Ovalipes* and of *Raymanninus schmitti* (for long considered to be a member of the genus *Benthochascon*; see Ng 2000) and more conclusive confirmation from nuclear DNA (work in progress).

All representatives of the other former subfamilies of the Portunidae (Portuninae, Caphyrinae, Carupinae, Thalamitinae, and Podophtahlminae) appear in the same cluster and are not segregated by their subfamily status. This is also shown by Mantelatto et al. (this volume) for the subfamilies Portuninae, Thalamitinae, and Podophtalminae, a result that again differs from Karasawa et al. (2008), who considered the Catoptridae, consisting of the genera *Catoptrus* and *Libystes*, a separate family. The possible paraphyly of these subfamilies can be confirmed only if additional representatives of the Thalamitinae, Caphyrinae, and Podophthalminae are included. For the moment we can say that the subfamilies Portuninae and Carupinae, and also the genera *Portunus*, with the type species *P. pelagicus* Linnaeus, 1758 (see also Mantelatto et al. 2007), and *Carupa*, with the type species *Carupa tenuipes* Dana, 1852, are paraphyletic, and we suggest refraining from using these subfamilies before a redefinition at the genus level has been carried out.

The Cancroidea as a superfamily should now be limited to the families Cancridae and Atelecyclidae, the latter maybe in its restriction to the genus *Atelecyclus* (see Guinot et al. 2008). A similar conclusion was reached by Ng et al. (2008) when removing Thiidae, Corystidae, and Cheiragonidae from the Cancroidea and placing them in their own independent superfamilies; Ng et al. (2008) noted that these single-family taxa may be preliminary groupings. Upgrading families into monofamilial superfamilies, however, underscores that the phylogenetic position of the included species is unknown and only changes the taxonomic level of uncertainty. Based on our results, we now place the Pirimelidae and Thiidae within the Portunoidea in close relationship to Carcinidae and Polybiidae and confirm the separate status of Corystidae and Cheiragonidae. These last two families cluster together in the mtDNA as well as in the nDNA phylogenies and should constitute sister families in the same superfamily. In that case, the name Corystoidea Samouelle, 1819, has preference. However, also in this case, additional genera of both families and clarification of the phylogenetic relationships of some of the current Atelecyclidae will be necessary before confirming this taxonomic change.

Overall, we feel that this study serves as an example that molecular phylogenies based on mitochondrial DNA can provide new insights into evolutionary relationships among decapod Crustacea (and other animals), insights that then can be used to implement a more phylogenetically based taxonomic system. The obvious congruence with a second tree based on the independent nuclear marker H3 gives confidence that results from previously published phylogenies of brachyuran crabs based on mitochondrial DNA alone (e.g., Schubart et al. 2000b, 2006 and others) do not necessarily have to be questioned. However, it also remains true that only the combination of a maximum number of approaches will lead to the best possible understanding of often-unexpected phylogenetic relationships in the natural world.

ACKNOWLEDGEMENTS

This study was initiated following encouragement by Cédric d'Udekem d'Acoz, who also generously provided numerous crabs from his private collection. From an initial interest in European Polybiinae, the focus shifted to a more global perspective, thanks to an important loan from Alain Crosnier, including paratypes and then uncatalogued material from the Paris Museum. DNA aliquots were also made available by Joelle Lai and Peter K.L. Ng from the National University of

Singapore and by Amanda Sichter and Jawahar Patil from CSIRO Marine Research (Hobart, Australia). Tohru Naruse and Hiroaki Karasawa kindly helped with the inclusion of Japanese species. Without the important help of all these esteemed colleagues, we would not have been able to complete our taxon sampling for this study. Additional material (not all of it used) was contributed or collected by Klaus Anger, Gérard Breton, José A. Cuesta, Peter J.F. Davie, Darryl L. Felder, Michelle K. Harrison, Liu Hung-Chang, Rafael Robles, Clarissa Schubart, and Shih Hsi-Te. Cheryl Morrison suggested the use of an unpublished NDH1 primer in combination with 16S primers and thus initiated our extension into the NDH1 gene. Peter K.L. Ng gave valuable advice on an earlier version of this paper. The molecular work was carried out at the laboratory of Jürgen Heinze at the University of Regensburg, with the help of students. Christian Schütz and Sascha Barabas started the project in 2002, while Nicola Barabas was of invaluable help in filling the gaps between 2007 and 2008. Student salary came out of a grant from the Deutsche Forschungsgemeinschaft SCHU 1460-3/3. Our gratitude is extended to Jody Martin, Darryl L. Felder, and Keith Crandall for organizing a symposium on decapod phylogenetics in San Antonio (January 2008), for comments on an earlier draft, and for putting together this proceedings volume.

## REFERENCES

Bourdillon-Casanova, L. 1956. Le développement larvaire de *Pirimela denticulata* Montagu (Crustacea, Decapoda). *Bull. Inst. Océanogr. Monaco* 1073, 8 pp.

Bourdillon-Casanova, L. 1960. Le méroplancton du Golfe de Mraseille: les larves de crustacés decapodes. *Rec. Trav. Sta. Mar. d'Endoune* 30: 1–286.

Bowman, T.E. & Abele, L.G. 1982. Classification of the recent Crustacea. In: Abele, L.G. (ed.), *Systematics, the Fossil Record, and Biogeography*. Bliss, D.E. (series ed.), *The Biology of Crustacea*. Vol. 1: 1–27. New York: Academic Press.

Cabot, E.L. & Beckenbach, A.T. 1989: Simultaneous editing of multiple nucleic acid and protein sequences with ESEE. *Comp. Appl. Biosci.* 5: 233–234.

Castro, P. 2007. A reappraisal of the family Goneplacidae MacLeay, 1838 (Crustacea, Decapoda, Brachyura) and revision of the subfamily Goneplacinae, with the description of ten new genera and eighteen new species. *Zoosystema* 29: 609–773.

Colgan, D.J., McLauchlan, C., Wilson, G.D.F., Livingston, S.P., Edgecombe, G.D., Macaranas, J., Cassis, G. & Gray, M.R. 1998. Histone H3 and U2 snRNA DNA sequences and arthropod molecular evolution. *Aust. J. Zool.* 46: 419–437.

Davie, P.J.F. 2002. Crustacea: Malacostraca. Eucarida (Part 2): Decapoda—Anomura, Brachyura. In: Wells, A. & Houston, W.W.K. (eds.), *Zoological Catalogue of Australia*: 1–641. 19.3B. CSIRO Publications.

Farris, J.S., Källersjö, M., Kluge, A.G. & Bult, C. 1995. Testing significance of incongruence. *Cladistics* 10: 315–319.

Flores, A.A.V. & Paula, J. 2000. Larval and early juvenile stages of *Pirimela denticulata* (Montagu, 1808) (Crustacea, Brachyura, Pirimelidae) obtained in the laboratory. *J. Nat. Hist.* 34: 2123–2143.

Guinot, D., De Angeli, A. & Garassino, A. 2008. A new eubrachyuran family from the Upper Cretaceous (Cenomanian-Turonian) of Gara Sbaa, southeastern Morocco (Crustacea, Decapoda, Brachyura). *Atti Soc. ital. Sci. Natur. Mus. Civ. Stor. Natur. Milano* 149: in press.

Hall, T.A. 1999. BioEdit: a user-friendly biological sequence alignment editor and analysis program for Windows 95/98/NT. *Nucl. Acids Symp. Ser.* 41: 95–98.

Holthuis, L.B. 1968. Polybiinae, the correct name for the subfamily containing the genus *Macropipus* Prestandrea, 1833 (Decapoda, Brachyura, Portunidae). *Crustaceana* 14: 220–221.

Huelsenbeck, J.P. & Ronquist, F. 2001. MrBayes: Bayesian inference of phylogenetic trees. *Bioinformatics* 17, 754–755.

Karasawa, H. & Kato, H. 2003. The family Goneplacidae MacLeay, 1838 (Crustacea: Decapoda: Brachyura): systematics, phylogeny and fossil records. *Paleontol. Res.* 7: 129–151.

Karasawa, H. & Schweitzer, C.E. 2006. A new classification of the Xanthoidea sensu lato (Crustacea: Decapoda: Brachyura) based on phylogenetic analysis and traditional systematics and evaluation of all fossil Xanthoidea sensu lato. *Contrib. Zool.* 75: 23–73.

Karasawa, H., Schweitzer, C.E. & Feldmann, R.M. 2008. Revision of Portunoidea Rafinesque, 1815 (Decapoda: Brachyura) with emphasis on the fossil genera and families. *J. Crust. Biol.* 28: 82–127.

Klaus, S., Schubart, C.D. & Brandis, D. 2006. Phylogeny, biogeography and a new taxonomy for the Gecarcinucoidea Rathbun, 1904 (Decapoda: Brachyura). *Org. Div. Evol.* 6: 199–217.

Mahon, B.C. & Neigel, J.E. 2008. Utility of arginine kinase for resolution of phylogenetic relationships among brachyuran genera and families. *Mol. Phylogenet. Evol.* 48: 718–727.

Mantelatto, F.L., Robles, R. & Felder, D.L. 2007. Molecular phylogeny of the western Atlantic species of the genus *Portunus* (Crustacea, Brachyura, Portunidae). *Zool. J. Linn. Soc. London* 150: 211–220.

Mantelatto, F.L., Robles, R., Schubart, C.D. & Felder, D.L. (this volume). Geographic differentiation and molecular phylogeny of the genus *Cronius* Stimpson, 1860, with reclassification of *C. tumidulus* and several American species of *Portunus* within the genus *Achelous* De Haan, 1833 (Brachyura: Portunidae). In: Martin, J.W., Crandall, K.A. & Felder, D.L. (eds.), *Crustacean Issues: Decapod Crustacean Phylogenetics*. Boca Raton, Florida: Taylor & Francis/CRC Press.

Martin, J.W. & Davis, G.E. 2001. An updated classification of the recent Crustacea. *Nat. Hist. Mus. L.A. County, Sci. Ser.* 39: 1–124.

Ng, P.K.L. 2000. The deep-water swimming crabs of the genus *Benthochascon* (Decapoda: Brachyura: Portunidae) with description of a new genus for the American *B. schmitti*. *J. Crust. Biol.* 20 (special number 2): 310–324.

Ng, P.K.L., Guinot, D. & Davie, P.J.F. 2008. Systema Brachyurorum: Part I. An annotated checklist of extant brachyuran crabs of the world. *Raffles Bull. Zool. Suppl.* 17: 1–286.

Ng, P.K.L. & Manuel-Santos, M.R. 2007. Establishment of the Vultocinidae, a new family for an unusual new genus and new species of Indo-West Pacific crab (Crustacea: Decapoda: Brachyura: Goneplacoidea), with comments on the taxonomy of the Goneplacidae. *Zootaxa* 1558: 39–68.

Ortmann, A. 1893. Abtheilung: Brachyura (Brachyura genuina Boas), II. Unterabtheilung: Cancroidea, 2. Section: Cancrinea, 1. Gruppe: Cyclometopa. Die Decapoden Krebse des Strassburger Museums, mit besonderer Berücksichtigung der von Herrn Dr. Döderlein bei Japan und bei den Liu-Kiu-Inseln gesammelten und zur Zeit im Strassburger Museum aufbewahrten Formen, VII. Theil. *Zool. Jahrb., Abth. Syst. Geogr. Biol. Thiere* 7: 411–495, pl. 17.

Posada, D. & Crandall, K.A. 1998. MODELTEST: testing the model of DNA substitution. *Bioinformatics* 14: 817–818.

Rathbun, M.J. 1930. The cancroid crabs of America of the families Euryalidae, Portunidae, Atelecyclidae, Cancridae and Xanthidae. *U.S. Natl. Mus. Bull.* 152, 609 pp.

Reuschel, S. 2004. *Morphometrie, Ökologie und Molekulare Systematik bei europäischen Krabben der Familie Xanthidae und Portunidae, unter besonderer Berücksichtigung potentiell endemischer Formen.* Unpublished Diploma Thesis. University of Regensburg, 84 pp.

Rodríguez, F., Oliver, J.F., Marín, A. & Medina, J.R. 1990: The general stochastic model of nucleotide substitution. *J. Theor. Biol.* 142: 485–501.

Rodríguez, G. 1992. The freshwater crabs of America. Family Trichodactylidae and supplement to the family Pseudothelphusidae. *Faune Tropicale* 31: 1–189.

Schubart, C.D. (this volume). Mitochondrial DNA and decapod phylogenies. The importance of pseudogenes and primer optimization. In: Martin, J.W., Crandall, K.A. & Felder, D.L. (eds.), *Crustacean Issues: Decapod Crustacean Phylogenetics*. Boca Raton, Florida: Taylor & Francis/CRC Press.

Schubart, C.D., Cannicci, S., Vannini, M. & Fratini, S. 2006. Molecular phylogeny of grapsoid crabs and allies based on two mitochondrial genes and a proposal for refraining from current superfamily classification. *J. Zool. Syst. Evol. Res.* 44: 193–199.

Schubart, C.D., Cuesta, J.A., Diesel, R. & Felder, D.L. 2000b. Molecular phylogeny, taxonomy, and evolution of nonmarine lineages within the American grapsoid crabs (Crustacea: Brachyura). *Mol. Phylogenet. Evol.* 15: 179–190.

Schubart, C.D., Cuesta, J.A. & Felder, D.L. 2002. Glyptograpsidae, a new brachyuran family from Central America: larval and adult morphology, and a molecular phylogeny of the Grapsoidea. *J. Crust. Biol.* 22: 28–44.

Schubart, C.D., Neigel, J.E. & Felder, D.L. 2000a. Use of the mitochondrial 16S rRNA gene for phylogenetic and population studies of Crustacea. In: von Vaupel Klein, J.C. & Schram, F. (eds.), *Crustacean Issues 12: The Biodiversity Crisis and Crustacea. Proceedings of the Fourth International Crustacean Congress, Amsterdam, Netherlands, 20–24 July 1998*, vol. 2: 817–830. Rotterdam: Balkema.

Schubart, C.D. & Reuschel, S. 2005. Molecular phylogenetic relationships of cancroid and portunoid crabs (Decapoda: Brachyura) do not reflect current taxonomy. In: *Book of Abstracts, Sixth International Crustacean Congress, Glasgow, Scotland, UK 18–22 July, 2005*: 7.

Schweitzer, C.E. & Feldmann, R.M. 2000. Re-evaluation of the Cancridae Latreille, 1802 (Decapoda: Brachyura) including three new genera and three new species. *Contrib. Zool.* 69: 223–250.

von Sternberg, R., Cumberlidge, N. & Rodríquez, G. 1999. On the marine sister groups of the freshwater crabs (Crustacea: Decapoda: Brachyura). *J. Zool. Syst. Evol. Res.* 37: 19–38.

von Sternberg, R. & Cumberlidge, N. 2001. Notes on the positon of the true freshwater crabs within the brachyrynchan Eubrachyura (Crustacea: Decapoda: Brachyura). *Hydrobiologia* 449: 21–39.

Števčić, Z. 2005. The reclassification of brachyuran crabs (Crustacea: Decapoda: Brachyura). *Fauna Croatica* 14: 1–159.

Swofford, D.L. 1998. *PAUP*—Phylogenetic Analysis Using Parsimony (*and Other Methods). Version 4*. Sunderland, Massachusetts: Sinauer Associates.

Thompson J.D., Higgins, D.G. & Gibbson, T.J. 1994. CLUSTAL W: improving the sensitivity of progressive multiple sequence alignment through senquence weighting, position specific gap penalties and weight matrix choice. *Nucl. Acids Res.* 22: 4673–4680.

# Molecular Phylogeny of Western Atlantic Representatives of the Genus *Hexapanopeus* (Decapoda: Brachyura: Panopeidae)

BRENT P. THOMA[1], CHRISTOPH D. SCHUBART[2] & DARRYL L. FELDER[1]

[1] *University of Louisiana at Lafayette, Department of Biology and Laboratory for Crustacean Research, PO Box 42451, Lafayette, Louisiana 70504-2451, U.S.A.*
[2] *Universität Regensburg, Biologie I, 93040 Regensburg, Germany*

ABSTRACT

Species of the brachyuran crab genus *Hexapanopeus* Rathbun, 1898, are common benthic inhabitants in coastal and nearshore waters of the Americas. Despite the frequency with which they are encountered, they are taxonomically problematic and commonly misidentified by non-experts. Little previous work has been undertaken to explain relationships among the 13 nominal species of *Hexapanopeus* or their relationship to other phenotypically similar genera of the family Panopeidae. In the present study we examine partial sequences for 16S and 12S mitochondrial rDNA for 71 individuals representing 46 species of Panopeidae and related families of the Brachyura. Phylogenies inferred from both of these datasets are largely congruent and show, with one exception, the included genera and species of the Panopeidae to represent a monophyletic grouping. Within this group, *Hexapanopeus* is polyphyletic, being distributed among several separate major clades and clearly warranting taxonomic subdivision.

1 INTRODUCTION

As part of ongoing studies of the superfamily Xanthoidea *sensu* Martin & Davis (2001), we have undertaken a reexamination of phylogenetic relationships among genera assigned to the family Panopeidae Ortmann, 1893, on molecular and morphological bases. Early in the course of our morphological studies, we saw reason to conclude that the genus *Hexapanopeus* Rathbun, 1898, as currently defined, was polyphyletic. Differences in the characters of the carapace, chelipeds, and male first pleopod (gonopod) served to obscure what, if any, relationship existed among the species in the genus. The present study serves as the first step towards restricting species composition of the genus *Hexapanopeus s.s.* (*sensu stricto*) and defining its phylogenetic relationships.

Presently, the genus *Hexapanopeus* consists of 13 species distributed on both coasts of the Americas; six species are known from the western Atlantic ranging from Massachusetts to Uruguay, while seven more range in the eastern Pacific from Mexico to Ecuador (Table 1). Representatives of *Hexapanopeus* are commonly encountered in environmental studies and inhabit a variety of nearshore environments ranging from sand-shell bottoms to rubble and surface fouling accumulations, where they often reside amongst sponges and ascidians (Rathbun 1930; Felder 1973; Williams 1984; Sankarankutty & Manning 1997). Even so, available illustrations and morphological descriptions are of limited detail and quality for many species, and little can be deduced from present literature to clarify their phylogenetic relationships.

Herein, we provide evidence for polyphyly in the genus *Hexapanopeus* on the basis of two mitochondrial genes (16S rDNA and 12S rDNA). We also examine relationships among species

**Table 1.** Known species presently assigned to *Hexapanopeus* with authority and known distribution. Those preceded by an asterisk (*) are included in the present phylogenetic analyses, along with one putative new species of the genus from the western Gulf of Mexico, yet to be described.

| Taxon Name | Distribution |
| --- | --- |
| *\*Hexapanopeus angustifrons* (Benedict & Rathbun, 1891) | Western Atlantic; from Massachusetts to Brazil |
| *Hexapanopeus beebei* Garth, 1961 | Eastern Pacific; Nicaragua |
| *\*Hexapanopeus caribbaeus* (Stimpson, 1871) | Western Atlantic; southeast Florida to Brazil |
| *Hexapanopeus cartagoensis* Garth, 1939 | Eastern Pacific; Galapagos Islands, Ecuador |
| *Hexapanopeus costaricensis* Garth, 1940 | Eastern Pacific; Costa Rica |
| *\*Hexapanopeus lobipes* (A. Milne-Edwards, 1880) | Western Atlantic; Gulf of Mexico |
| *\*Hexapanopeus manningi* Sankarankutty & Ferreira, 2000 | Western Atlantic; Rio Grande do Norte, Brazil |
| *Hexapanopeus nicaraguensis* (Rathbun, 1904) | Eastern Pacific; Nicaragua |
| *Hexapanopeus orcutti* Rathbun, 1930 | Eastern Pacific; Mexico |
| *\*Hexapanopeus paulensis* Rathbun, 1930 | Western Atlantic; South Carolina to Uruguay |
| *Hexapanopeus quinquedentatus* Rathbun, 1901 | Western Atlantic; Puerto Rico |
| *Hexapanopeus rubicundus* Rathbun, 1933 | Eastern Pacific; Gulf of California |
| *Hexapanopeus sinaloensis* Rathbun, 1930 | Eastern Pacific; Mexico |

currently assigned to *Hexapanopeus* and relationships of this genus to other genera and species encompassed within the family Panopeidae. This serves to further clarify the species composition of *Hexapanopeus* s.s., and to confirm its phylogenetic proximity to other taxa constituting a putative panopeid lineage.

## 2 MATERIALS AND METHODS

### 2.1 *Taxon sampling*

Seventy-one individuals representing 46 species, 30 genera, and 10 families were subjected to molecular analyses. Of the 142 sequences used in this study, 132 were generated for this project, while the remaining 10 were obtained from GenBank (Table 2). Since the identity of the sister group to the family Panopeidae remains debatable (see Martin & Davis 2001, Karasawa & Schweitzer 2006, and Ng et al. 2008 for discussion), we included 22 taxa that represent the families Xanthidae MacLeay, 1838, Pseudorhombilidae Alcock, 1900, Pilumnidae Samouelle, 1819, Chasmocarcinidae Serène, 1964, Euryplacidae Stimpson, 1871, Goneplacidae MacLeay, 1838, Carpiliidae Ortmann, 1893, Eriphiidae MacLeay, 1838, and Portunidae Rafinesque, 1815.

Specimens used in this study were collected during research cruises and field expeditions and either directly preserved in 80% ethyl alcohol (EtOH) or first frozen in either seawater or glycerol at $-80°C$ before later being transferred to 80% EtOH. Additional materials were obtained on loan from the National Museum of Natural History—Smithsonian Institution (USNM). When possible, identifications of specimens were confirmed by two or more of the investigators to limit the chance of misidentifications.

**Table 2.** Crab species used for phylogeny reconstruction, showing catalog number, collection locality, and GenBank accession numbers for partial sequences of 16S and 12S, respectively (ULLZ = University of Louisiana at Lafayette Zoological Collection, Lafayette, Louisiana; USNM = United States National Museum of Natural History, Smithsonian Institution, Washington D.C.).

| Taxon | Catalog. No. | Collection Locality | 16S | 12S |
|---|---|---|---|---|
| **Carpiliidae Ortmann, 1893** | | | | |
| *Carpilius maculatus* (Linnaeus, 1758) | GenBank | | AF501732 | AF501705 |
| **Chasmocarcinidae Serène, 1964** | | | | |
| *Chasmocarcinus chacei* Felder & Rabalais, 1986 | ULLZ 8018 | Northern Gulf of Mexico; 2006 | EU863401 | EU863335 |
| *Chasmocarcinus mississippiensis* Rathbun, 1931 | ULLZ 7346 | Southwestern Gulf of Mexico; 2005 | EU863406 | EU863340 |
| **Eriphiidae MacLeay, 1838** | | | | |
| *Eriphia verrucosa* (Forskål, 1775) | ULLZ 4275 | Eastern Atlantic; Spain; Cadiz, 1998 | EU863398 | EU863332 |
| **Euryplacidae Stimpson, 1871** | | | | |
| *Frevillea barbata* A. Milne-Edwards, 1880 | ULLZ 8369 | Southeastern Gulf of Mexico; 2004 | EU863399 | EU863333 |
| *Sotoplax robertsi* Guinot, 1984 | ULLZ 7857 | Northern Gulf of Mexico; 2006 | EU863400 | EU863334 |
| **Goneplacidae MacLeay, 1838** | | | | |
| *Bathyplax typhlus* A. Milne-Edwards, 1880 | ULLZ 8032 | Northwestern Gulf of Mexico; 2006 | EU863397 | EU863331 |
| **Panopeidae Ortmann, 1893** | | | | |
| *Acantholobulus bermudensis* (Benedict & Rathbun, 1891) | ULLZ 5843 | Gulf of Mexico; Mexico; Campeche, 2002 | EU863355 | EU863289 |
| *Acantholobulus bermudensis* (Benedict & Rathbun, 1891) | ULLZ 6558 | Western Atlantic; Florida, Ft. Pierce, 2005 | EU863354 | EU863288 |
| *Acantholobulus bermudensis* (Benedict & Rathbun, 1891) | ULLZ 6924 | Western Atlantic; Florida, Ft. Pierce, 2006 | EU863372 | EU863306 |
| *Acantholobulus schmitti* (Rathbun, 1930) | ULLZ 6613 | Western Atlantic; Brazil; Sao Paulo, 1999 | EU863364 | EU863298 |
| *Acantholobulus schmitti* (Rathbun, 1930) | ULLZ 8367 | Western Atlantic; Brazil; Sao Paulo, 1999 | EU863357 | EU863291 |
| *Cyrtoplax* nr. *spinidentata* (Benedict, 1892) | ULLZ 8423 | Western Atlantic; Florida, Ft. Pierce, 2001 | EU863369 | EU863303 |
| *Dyspanopeus sayi* (Smith, 1869) | ULLZ 7227 | Western Atlantic; Florida, Ft. Pierce, 2006 | EU863395 | EU863329 |
| *Eucratopsis crassimanus* (Dana, 1851) | ULLZ 6427 | Western Atlantic; Florida, Ft. Pierce, 2006 | EU863392 | EU863326 |
| *Eurypanopeus abbreviatus* (Stimpson, 1860) | ULLZ 3753 | Western Atlantic; Florida, Ft. Pierce, 1998 | EU863388 | EU863322 |
| *Eurypanopeus depressus* (Smith, 1869) | ULLZ 3976 | Northern Gulf of Mexico; Mississippi, 1998 | EU863391 | EU863325 |
| *Eurypanopeus depressus* (Smith, 1869) | ULLZ 6077 | Eastern Gulf of Mexico; Tampa Bay, 2005 | EU863390 | EU863324 |
| *Eurypanopeus dissimilis* (Benedict & Rathbun, 1891) | ULLZ 5878 | Western Atlantic; Florida, Ft. Pierce, 1997 | EU863396 | EU863330 |
| *Eurypanopeus dissimilis* (Benedict & Rathbun, 1891) | ULLZ 8424 | Western Atlantic; Florida, Ft. Pierce, 1997 | EU863387 | EU863321 |
| *Eurypanopeus planissimus* (Stimpson, 1860) | ULLZ 4140 | Eastern Pacific; Mexico; Baja California, 1999 | EU863386 | EU863320 |
| *Glyptoplax smithii* A. Milne-Edwards, 1880 | ULLZ 6793 | Southwestern Gulf of Mexico; 2005 | EU863342 | EU863276 |
| *Glyptoplax smithii* A. Milne-Edwards, 1880 | ULLZ 7686 | Northern Gulf of Mexico; 2006 | EU863379 | EU863313 |
| *Glyptoplax smithii* A. Milne-Edwards, 1880 | ULLZ 8142 | Northern Gulf of Mexico; 2006 | EU863350 | EU863284 |
| *Glyptoplax smithii* A. Milne-Edwards, 1880 | ULLZ 8335 | Northern Gulf of Mexico; 2006 | EU863371 | EU863305 |
| *Glyptoplax smithii* A. Milne-Edwards, 1880 | ULLZ 9020 | Western Atlantic; Florida, Ft. Pierce, 2003 | EU863384 | EU863318 |

**Table 2.** continued.

| Taxon | Catalog. No. | Collection Locality | 16S | 12S |
|---|---|---|---|---|
| *Hexapanopeus angustifrons* (Benedict & Rathbun, 1891) | ULLZ 6943 | Western Atlantic; Florida, Ft. Pierce, 2006 | EU863343 | EU863277 |
| *Hexapanopeus angustifrons* (Benedict & Rathbun, 1891) | ULLZ 7174 | Western Atlantic; Florida, Ft. Pierce, 2003 | EU863368 | EU863302 |
| *Hexapanopeus angustifrons* (Benedict & Rathbun, 1891) | ULLZ 7757 | Western Atlantic; Florida, Ft. Pierce, 2006 | EU863351 | EU863285 |
| *Hexapanopeus angustifrons* (Benedict & Rathbun, 1891) | ULLZ 8368 | Eastern Gulf of Mexico; Florida, 2004 | EU863380 | EU863314 |
| *Hexapanopeus angustifrons* (Benedict & Rathbun, 1891) | ULLZ 9019 | Western Atlantic; Florida, Ft. Pierce, 2003 | EU863385 | EU863319 |
| *Hexapanopeus caribbaeus* (Stimpson, 1871) | ULLZ 6859 | Western Atlantic; Florida, Ft. Pierce, 2006 | EU863381 | EU863315 |
| *Hexapanopeus caribbaeus* (Stimpson, 1871) | ULLZ 6859 | Western Atlantic; Florida, Ft. Pierce, 2006 | EU863348 | EU863282 |
| *Hexapanopeus caribbaeus* (Stimpson, 1871) | ULLZ 7743 | Western Atlantic; Florida, Ft. Pierce, 2006 | EU863353 | EU863287 |
| *Hexapanopeus lobipes* (A. Milne-Edwards, 1880) | ULLZ 4731 | Northern Gulf of Mexico; Louisiana, 2001 | EU863356 | EU863290 |
| *Hexapanopeus lobipes* (A. Milne-Edwards, 1880) | ULLZ 6909 | Southeastern Gulf of Mexico; 2004 | EU863365 | EU863299 |
| *Hexapanopeus lobipes* (A. Milne-Edwards, 1880) | ULLZ 7828 | Northern Gulf of Mexico; 2006 | EU863352 | EU863286 |
| *Hexapanopeus manningi* Sankarankutty & Ferreira, 2000 | USNM 260923 | Western Atlantic; Brazil; Rio Grande do Norte, 1996 | EU863383 | EU863317 |
| *Hexapanopeus* nov. sp. | ULLZ 8646 | Northern Gulf of Mexico; Texas, 1998 | EU863361 | EU863295 |
| *Hexapanopeus paulensis* Rathbun, 1930 | ULLZ 3891 | Northern Gulf of Mexico; Texas, 1998 | EU863360 | EU863294 |
| *Hexapanopeus paulensis* Rathbun, 1930 | ULLZ 6608 | Western Atlantic; Brazil; Sao Paulo, 1996 | EU863373 | EU863307 |
| *Hexapanopeus paulensis* Rathbun, 1930 | ULLZ 6862 | Northern Gulf of Mexico; Texas, 2006 | EU863358 | EU863292 |
| *Hexapanopeus paulensis* Rathbun, 1930 | ULLZ 6870 | Northern Gulf of Mexico; Texas, 2006 | EU863374 | EU863308 |
| *Hexapanopeus paulensis* Rathbun, 1930 | ULLZ 6875 | Northern Gulf of Mexico; Texas, 2006 | EU863376 | EU863310 |
| *Hexapanopeus paulensis* Rathbun, 1930 | ULLZ 6882 | Northern Gulf of Mexico; Texas, 2006 | EU863375 | EU863309 |
| *Hexapanopeus paulensis* Rathbun, 1930 | ULLZ 8645 | Northern Gulf of Mexico; Panama City, 2007 | EU863377 | EU863311 |
| *Neopanope packardii* Kingsley, 1879 | ULLZ 3772 | United States; Florida, Ft. Pierce, 1998 | EU863349 | EU863283 |
| *Panopeus africanus* A. Milne-Edwards, 1867 | ULLZ 4273 | Eastern Atlantic; Spain; Cadiz, 1999 | EU863370 | EU863304 |
| *Panopeus americanus* Saussure, 1857 | ULLZ 8456 | Western Atlantic; Florida, Ft. Pierce, 1996 | EU863345 | EU863279 |
| *Panopeus herbstii* H. Milne Edwards, 1834 | ULLZ 8457 | Western Atlantic; South Carolina, 1997 | EU863362 | EU863296 |
| *Panopeus lacustris* Desbonne, 1867 | ULLZ 3818 | Western Atlantic; Florida, Ft. Pierce, 1997 | EU863363 | EU863297 |
| *Panopeus occidentalis* Saussure, 1857 | ULLZ 8640 | Northern Gulf of Mexico; Panama City, 2007 | EU863393 | EU863327 |
| *Panopeus occidentalis* Saussure, 1857 | ULLZ 8643 | Northern Gulf of Mexico; Panama City, 2007 | EU863394 | EU863328 |
| *Panoplax depressa* Stimpson, 1871 | ULLZ 8056 | Northern Gulf of Mexico; 2006 | EU863347 | EU863281 |
| *Rhithropanopeus harrisii* (Gould, 1841) | ULLZ 3995 | Northern Gulf of Mexico; Texas, 1998 | EU863346 | EU863280 |
| **Pilumnidae** Samouelle, 1819 | | | | |
| *Lobopilumnus agassizii* (Stimpson, 1871) | ULLZ 7121 | Southwestern Gulf of Mexico; 2005 | EU863402 | EU863336 |
| *Pilumnus floridanus* Stimpson 1871 | ULLZ 7343 | Southern Gulf of Mexico; 2005 | EU863403 | EU863337 |

Table 2. continued.

| Taxon | Catalog. No. | Collection Locality | 16S | 12S |
|---|---|---|---|---|
| **Portunidae** Rafinesque, 1815 | | | | |
| *Ovalipes punctatus* (De Haan, 1833) | GenBank | | DQ062733 | DQ060652 |
| **Pseudorhombilidae** Alcock, 1900 | | | | |
| *Trapezioplax tridentata* (A. Milne-Edwards, 1880) | ULLZ 8054 | Northern Gulf of Mexico; 2006 | EU863344 | EU863278 |
| **Xanthidae** MacLeay, 1838 | | | | |
| *Atergatis reticulatus* (De Haan, 1835) | GenBank | | DQ062726 | DQ060646 |
| *Batodaeus urinator* (A. Milne-Edwards, 1881) | ULLZ 8131 | Southern Gulf of Mexico; 2005 | EU863405 | EU863339 |
| *Eucratodes agassizii* A. Milne-Edwards, 1880 | ULLZ 8400 | Northern Gulf of Mexico; Louisiana, 1996 | EU863389 | EU863323 |
| *Garthiope barbadensis* (Rathbun, 1921) | ULLZ 8170 | Northern Gulf of Mexico; 2006 | EU863367 | EU863301 |
| *Garthiope barbadensis* (Rathbun, 1921) | ULLZ 8183 | Northern Gulf of Mexico; 2006 | EU863366 | EU863300 |
| *Liomera cinctimana* (White, 1847) | GenBank | | AF501736 | AF501708 |
| *Macromedaeus distinguendus* (De Haan, 1835) | GenBank | | DQ062731 | DQ060654 |
| *Micropanope sculptipes* Stimpson, 1871 | ULLZ 6603 | Southeastern Gulf of Mexico; 2004 | EU863404 | EU863338 |
| *Micropanope sculptipes* Stimpson, 1871 | ULLZ 8025 | Northern Gulf of Mexico; 2006 | EU863378 | EU863312 |
| *Speocarcinus lobatus* Guinot, 1969 | ULLZ 7820 | Northern Gulf of Mexico; 2006 | EU863407 | EU863341 |
| *Speocarcinus monotuberculatus* Felder & Rabalais, 1986 | ULLZ 7562 | Southwestern Gulf of Mexico; 2005 | EU863359 | EU863293 |
| *Xanthias canaliculatus* Rathbun, 1906 | ULLZ 4381 | Indian Ocean; South Africa; Sodwana Bay, 2001 | EU863382 | EU863316 |

**Table 3.** Primers used in this study.

| Gene | Primer | Sequence 5'→3' | Ref. |
|---|---|---|---|
| 16S | 16Sar | CGC CTG TTT ATC AAA AAC AT | (1) |
| 16S | 16Sbr | CCG GTC TGA ACT CAG ATC ACG T | (1) |
| 16S | 16L2 | TGC CTG TTT ATC AAA AAC AT | (2) |
| 16S | 1472 | AGA TAG AAA CCA ACC TGG | (3) |
| 12S | 12sf | GAA ACC AGG ATT AGA TAC CC | (4) |
| 12S | 12s1r | AGC GAC GGG CGA TAT GTA C | (4) |

*References:* (1) Palumbi et al. 1991, (2) Schubart et al. 2002, (3) Crandall & Fitzpatrick 1996, (4) Buhay et al. 2007.

## 2.2 DNA extraction, PCR, and sequencing

Genomic DNA was extracted from muscle tissue of the pereopods of a total of 66 specimens of the family Panopeidae and related taxa of the Xanthoidea *sensu* Martin & Davis (2001) utilizing one of the following extraction protocols: Genomic DNA Extraction Kit for Arthropod Samples (Cartagen Molecular Systems, Cat. No. 20810-050), Qiagen DNeasy® Blood and Tissue Kit (Qiagen, Cat. No. 69504), or isopropanol precipitation following Robles et al. (2007).

Two mitochondrial markers were selectively amplified using polymerase chain reaction (PCR). A fragment of the 16S large subunit rDNA approximately 550 basepairs (bp) in length was amplified using the primers 1472 or 16Sbr in combination with 16L2 and 16Sar and a fragment of the 12S small subunit rDNA approximately 310 bp in length was amplified using the primers 12sf and 12s1r (see Table 3 for complete primer information). PCR reactions were performed in 25-$\mu$l volumes containing: 0.5 $\mu$M forward and reverse primer, 200 $\mu$M each dNTP, 2.5 $\mu$l 10x PCR buffer, 3 mM $MgCl_2$, 1 M betaine, 1 unit NEB Standard Taq polymerase (New England Biolabs, Cat. No. M0273S), and 30–50 ng of genomic DNA. Reactions were carried out using the following cycling parameters: initial denaturation at 94°C for 2 min; 40 cycles at 94°C for 25 sec, 40°C (16S) or 52°C (12S) for 1 min, 72°C for 1 min; final extension at 72°C for 5 min. PCR products were purified using EPOCH GenCatch PCR Clean-up Kit (EPOCH BioLabs, Cat. No. 13-60250) and sequenced in both directions using ABI BigDye® Terminator v3.1 Cycle Sequencing Kit (Applied Biosystems, Foster City, CA, USA). Cycle sequencing products were purified using Sephadex G-50 columns (Sigma-Aldrich Chemicals, Cat. No. S6022). Sequencing products were run on an ABI PRISM® 3100 Genetic Analyzer (Applied Biosystems, Foster City, CA, USA).

## 2.3 Phylogenetic analyses

Sequences were assembled using Sequencher 4.7 (GeneCodes, Ann Arbor, MI, USA). Once assembled, sequences were aligned using MUSCLE (MUltiple Sequence Comparison by Log-Expectation), a computer program found to be more accurate and faster than other alignment algorithms (Edgar 2004). Alignments were further refined using GBlocks v0.91b (Castresana 2000) to omit poorly aligned or ambiguous positions. Default parameters were used for GBlocks except: 1) minimum length of a block = 4, 2) allowed gap positions = half. We conducted a partition heterogeneity test or incongruence length difference test (ILD) (Bull et al. 1993), as implemented in PAUP* v4b10 (Swofford 2003), to determine if the two gene regions could be combined.

The model of evolution that best fit each of the datasets was determined by likelihood tests as implemented in Modeltest version 3.6 (Posada & Crandall 1998) under the Akaike Information

Criterion (AIC). The maximum likelihood (ML) analyses were conducted using PhyML Online (Guindon et al. 2005) using the model parameters selected with free parameters estimated by PhyML. Confidence in the resulting topology was assessed using non-parametric bootstrap estimates (Felsenstein 1985) with 500 replicates.

The Bayesian (BAY) analyses were conducted in MrBayes (Huelsenbeck & Ronquist 2001) with computations performed on the computer cluster of the CyberInfrastructure for Phylogenetic RESearch project (CIPRES) at the San Diego Supercomputer Center, using parameters selected by Modeltest. A Markov Chain Monte Carlo (MCMC) algorithm with 4 chains and a temperature of 0.2 ran for 4,000,000 generations, sampling 1 tree every 1,000 generations. Preliminary analyses and observation of the log likelihood ($L$) values allowed us to determine burn-ins and stationary distributions for the data. Once the values reached a plateau, a 50% majority rule consensus tree was obtained from the remaining trees. Clade support was assessed with posterior probabilities (p$P$).

## 3 RESULTS

The initial sequence alignment of the 16S dataset, including gaps and primer regions, was 606 bp in length, while that of the 12S dataset was 384 bp in length. GBlocks was used to further refine the alignment, removing ambiguously aligned regions resulting in final alignments of 521 bp (86%) and 284 bp (74%) for 16S and 12S, respectively. Despite recent studies combining multiple loci into a single alignment (Ahyong & O'Meally 2004, Porter et al. 2005), we chose in this instance not to combine the datasets. The partition heterogeneity test or incongruence length difference test, as implemented in PAUP*, indicated that the combination of the two gene regions was significantly rejected (P = 0.0240). Furthermore, preliminary analysis of the combined dataset resulted in lower support for some of the tip branches than was the case in the single gene trees. This is due to different branching patterns (16S vs. 12S) at this level of the tree, which will be discussed later in this paper. This information would be lost in a combined tree.

Application of the likelihood tests as implemented in Modeltest revealed that the selected model of DNA substitution by AIC for the 16S dataset was HKY+I+G (Hasegawa et al. 1985) with an assumed proportion of invariable sites of 0.3957 and a gamma distribution shape parameter of 0.4975. The selected model for the 12S dataset was GTR+I+G (Rodríguez et al. 1990) with an assumed proportion of invariable sites of 0.3228 and a gamma distribution shape parameter of 0.6191.

Phylogenetic relationships among 71 individuals representing 46 species of the Xanthoidea *sensu* Martin & Davis (2001) were determined using Bayesian and ML approaches for both the 16S and 12S datasets. For the Bayesian analyses, the first 1,000 trees were discarded as burn-in and the consensus tree was estimated using the remaining 3,000 trees (= 3 million generations). Topologies resulting from the Bayesian analyses of both the 16S and 12S datasets were largely congruent (Figs. 1 and 2). A number of monophyletic clades are supported by both datasets, as follow: 1) *Acantholobulus bermudensis*, *Acantholobulus schmitti*, and *Hexapanopeus caribbaeus* with p$P$ (16S/12S) of 99/77, 2) *Hexapanopeus angustifrons* and *Hexapanopeus paulensis* with p$P$ of 100/99, 3) *Eurypanopeus depressus*, *Eurypanopeus dissimilis*, *Dyspanopeus sayi*, *Neopanope packardii*, and *Rhithropanopeus harrisii* with p$P$ of 97/99, 4) *Eurypanopeus abbreviatus* and *Eurypanopeus planissimus* with p$P$ of 99/87. In general, Bayesian posterior probabilities have been shown to be higher than the corresponding bootstrap values, but, in many cases, posterior probabilities tend to overrate confidence in a topology while bootstrap values based on neighbor joining, maximum parsimony, or ML methods tend to slightly underestimate support (Huelsenbeck et al. 2001, Huelsenbeck et al. 2002, Suzuki et al. 2002). With this in mind, it is not surprising to find that ML bootstrap supports for the same four clades are lower than the p$P$. The bootstrap values of the above clades are as follows: 1) <50/<50, 2) 72/51, 3) <50/<50, and 4) < 50/<50.

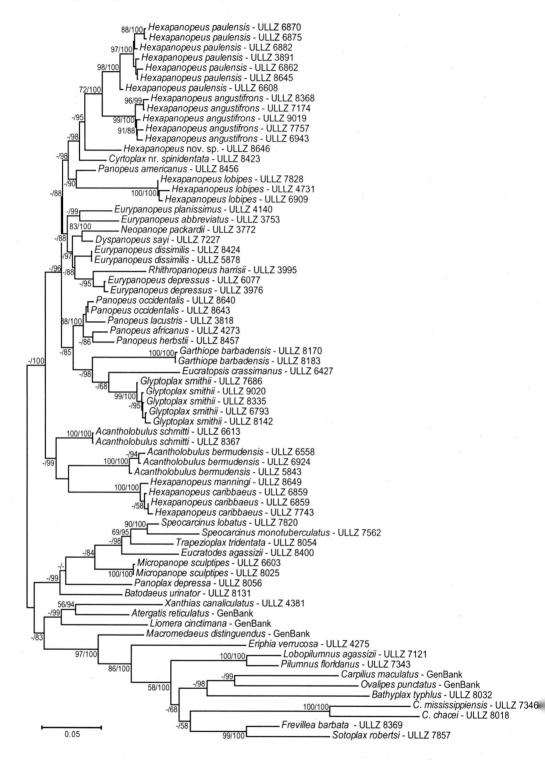

**Figure 1.** Phylogenetic relationships among panopeid crab species and selected representatives of the superfamily Xanthoidea *sensu* Martin & Davis (2001), inferred by Bayesian analysis from 521 basepairs of the 16S rDNA gene. Confidence intervals are from 500 bootstrap maximum likelihood analysis followed by Bayesian posterior probabilities. Genus shown as "*C.*" = *Chasmocarcinus*. Values below 50 are indicated by "-".

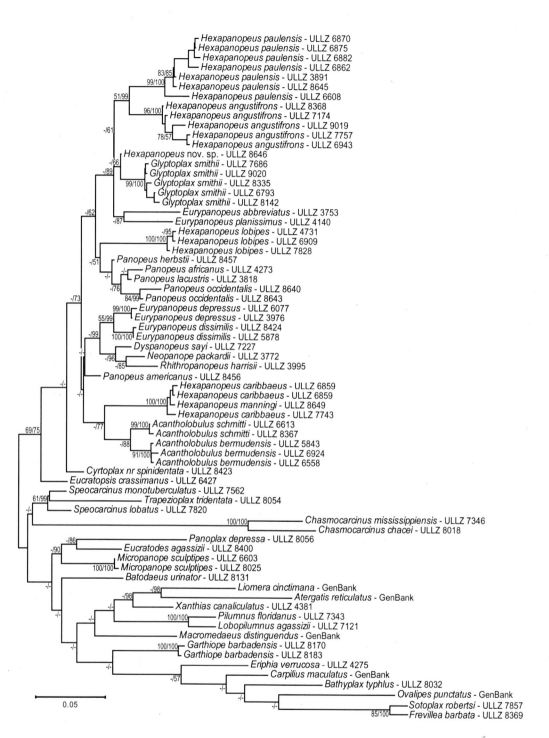

**Figure 2.** Phylogenetic relationships among panopeid crab species and selected representatives of the superfamily Xanthoidea *sensu* Martin & Davis (2001), inferred by Bayesian analysis from 284 basepairs of the 12S rDNA gene. Confidence intervals are from 500 bootstrap maximum likelihood analysis followed by Bayesian posterior probabilities. Values below 50 are indicated by "-".

## 4 DISCUSSION

Here we report two molecular phylogenies of the genus *Hexapanopeus* and related genera of the family Panopeidae. These phylogenies, which are based on partial sequences of the 16S and 12S rDNA, contain five of the 13 nominal species in *Hexapanopeus* and a single undescribed species that appears to be assignable to the genus. In addition, we have included representatives of 18 species of the family Panopeidae in order to better address both the monophyly of *Hexapanopeus* and the relationships of species currently assigned to *Hexapanopeus* to other panopeid taxa. Although only five species of *Hexapanopeus* are included in the dataset, these five species represent five of the six nominal species known from the western Atlantic. It is clear from our analyses that the genus *Hexapanopeus* is markedly polyphyletic and that further study of all its putative members is warranted, by both morphological and molecular methods.

### 4.1 Hexapanopeus angustifrons *and* Hexapanopeus paulensis

The phylogenies presented here lend support to a narrowed definition of *Hexapanopeus* that includes only the type-species of the genus *Hexapanopeus angustifrons* (Benedict & Rathbun, 1891) and *Hexapanopeus paulensis* Rathbun, 1930, pending results of morphological and molecular analyses for the remaining eight present congeners. It is interesting to note that in all analyses these taxa form a monophyletic clade and that within both species there is further evidence for genetic structure. It is unclear if the genetic divergence seen in these clades is the result of cryptic speciation or population differentiation, but the current analyses suggest some combination of the two might occur in each complex.

### 4.2 Hexapanopeus *nov. sp.*

In the analyses of the 16S dataset, the sister group to the *H. angustifrons*/*H. paulensis* clade is an undescribed species from intertidal waters of south Texas in the western Gulf of Mexico. This undescribed species resembles *H. paulensis* in general morphology, but it has a very distinctive gonopod, which most resembles that of *Acantholobulus schmitti* (Rathbun, 1930). In contrast to the results of the 16S dataset, the 12S dataset lends support to a clade that is composed of the undescribed species and *Glyptoplax smithii* A. Milne-Edwards, 1880, as the sister group to the *H. angustifrons*/*H. paulensis* clade. Unfortunately, suitable material of *Glyptoplax pugnax* Smith, 1870, the type species of the genus, has not to date been available for molecular analysis; therefore, it remains unclear whether this undescribed species is most appropriately treated as a member of the genus *Hexapanopeus*, the genus *Glyptoplax*, or a new monospecific genus.

### 4.3 Hexapanopeus lobipes

The species *Hexapanopeus lobipes* (A. Milne-Edwards, 1880) has had a very unsettled taxonomic history. After being described as a species of *Neopanope* A. Milne-Edwards, 1880, it was later transferred to the genus *Lophopanopeus* Rathbun, 1898, by Rathbun in 1898. In his 1948 revision of the genus *Lophopanopeus*, Menzies pointed out that *H. lobipes* does not fit the diagnosis of the genus *Lophopanopeus*. Upon transferring the species to the genus *Hexapanopeus*, he noted that "it seems to fit the diagnosis of that genus better than that of any other American genus." Only isolated records of *Hexapanopeus lobipes* have been reported since Menzies' 1948 work (Wicksten 2005, Felder et al. in press), and there has been no reassessment of its placement within the genus *Hexapanopeus*. The gonopod of *H. lobipes* is distinctive and has little resemblance to those in other members of the genus *Hexapanopeus*. Furthermore, unlike the carapaces of *H. angustifrons* and *H. paulensis*, which have five distinct anterolateral teeth, the 1st and 2nd antero-lateral teeth of

*H. lobipes* are generally fused, giving the appearance of four anterolateral teeth. On the basis of these and other morphological features, it is unclear whether *H. lobipes* is justifiably assignable to the genus *Hexapanopeus*. Whatever the case to be made on the basis of morphology alone, we cannot concur with Ng et al. (2008) in reassigning this species to *Lophopanopeus*.

Our analyses support removal of *H. lobipes* from the genus *Hexapanopeus* and appear to justify establishment of a new monospecific genus for *H. lobipes*. In both topologies, *H. lobipes* falls outside the clade formed by *H. angustifrons* and *H. paulensis*. In the phylogeny inferred from the 16S dataset, *H. lobipes* is the sister group to *Panopeus americanus* Saussure, 1857, with ML bootstrap and p$P$ values of <50/90, respectively. The phylogeny inferred from the 12S dataset presents *H. lobipes* as a sister group to *Panopeus s.s.* H. Milne Edwards, 1834, with ML bootstrap and p$P$ values of <50/51, respectively. Despite low support values, both topologies lend support to the removal of *H. lobipes* from the genus *Hexapanopeus* and the erection of a new genus for the species, as is currently in progress.

## 4.4 Hexapanopeus manningi

*Hexapanopeus manningi* Sankarankutty and Ferreira, 2000, was described on the basis of material from Rio Grande do Norte, Brazil. This species was distinguished from *Hexapanopeus caribbaeus* (Stimpson, 1871) by characters of the frontal margin, the 3rd anterolateral tooth of the carapace, and the apical process of the gonopod; however, upon the basis of synoptic comparisons of the male paratype (USNM 260923) to material of *H. caribbaeus* from eastern Florida, it appears that there is considerable morphological overlap between these two taxa, raising the question as to whether *H. manningi* might be a junior synonym of *H. caribbaeus*. The topology inferred from the 16S dataset places *H. manningi* in very close proximity to *H. caribbaeus*; distance between these taxa is very short and comparable to that within other accepted single-species clades in our tree. The clade containing both *H. manningi* and *H. caribbaeus* has high support values, with ML bootstrap and p$P$ values of 100/100, respectively. The strongest support for a synonymy of the two taxa comes from the topology inferred from the 12S dataset, with *H. manningi* positioned within the clade of *H. caribbaeus*. Our molecular phylogenies support synonymy of *H. manningi* with *H. caribbaeus*, and we herewith recommend that taxonomic revision, regardless of the eventual generic assignment to be accorded (see below).

## 4.5 Hexapanopeus caribbaeus

*Hexapanopeus caribbaeus* was originally described as a representative of the genus *Micropanope*; however, upon erection of the genus *Hexapanopeus*, Rathbun (1898) transferred this species to the genus *Hexapanopeus* apparently on the basis of carapace shape. It wasn't until the 1997 work by Sankarankutty and Manning that distinct differences between the gonopod of *H. caribbaeus* and that of the type-species *H. angustifrons* were noted. In the present analysis, this species is clearly separated from *Hexapanopeus s.s.*, and shown to be more closely allied to the genus *Acantholobulus*.

## 4.6 *Genus* Acantholobulus

Felder and Martin (2003) erected the genus *Acantholobulus* to accommodate a number of species from the genera *Panopeus* and *Hexapanopeus*, which included: 1) the type-species *Acantholobulus bermudensis* (Benedict & Rathbun, 1898), formerly *Panopeus bermudensis*; 2) *Acantholobulus miraflorensis* (Abele & Kim, 1989), formerly *Panopeus miraflorensis*; 3) *Acantholobulus pacificus* (Edmondson, 1931), formerly *Panopeus pacificus*; and 4) *Acantholobulus schmitti* (Rathbun, 1930), formerly *Hexapanopeus schmitti*. Despite similarities between *H. caribbaeus* and *A. schmitti* in both carapace and gonopod morphology, the possible relationship between *H. caribbaeus* and newly assigned members of the genus *Acantholobulus* was not addressed. The phylogenies inferred from both our datasets strongly support inclusion of *H. caribbaeus* within the genus *Acantholobulus*. While the phylogeny inferred from the 16S dataset shows *H. caribbaeus* nested with *Acantholobulus*, the topology inferred by analysis of the 12S datasets supports a sister group relationship between *H. caribbaeus* and both *A. bermudensis* and *A. schmitti*. Although both of these relationships are supported by p$P$ >75, the 16S dataset shows considerably higher p$P$ (99/77 for 16S/12S, respectively). As additional species of *Acantholobulus* become available for inclusion in our analysis, the relationship between *Acantholobulus* and its closest relatives should be more definitively resolved. Even so, it is by present findings established that *H. caribbaeus* is well separated from *Hexapanopeus* s.s., and we apply the new combination *Acantholobulus caribbaeus* (Stimpson, 1871).

### 4.7 Panopeus americanus

In a study of mud crabs from the northwestern Atlantic, Schubart et al. (2000) clearly showed polyphyly in the genus *Panopeus*, with both *Acantholobulus bermudensis* (as *Panopeus bermudensis*, see discussion above) and *Panopeus americanus* falling well outside *Panopeus* s.s. (Schubart et al. 2000, Fig. 1). In the present study, we find additional support for these findings with the topologies inferred from both datasets positioning *P. americanus* outside *Panopeus* s.s.; however, the topologies differ in where *P. americanus* is placed relative to species of other genera. In the topology inferred from the 16S dataset, *P. americanus* is a sister group to *H. lobipes*, while in the topology inferred from the 12S dataset, *P. americanus* is the sister group to the clade containing *E. depressus*, *E. dissimilis*, *N. packardii*, *D. sayi*, and *R. harrisii*. However, this arrangement is poorly supported with ML bootstrap and p$P$ values less than 50. Despite the differences in the topologies inferred from these two datasets, both provide evidence for the removal of *P. americanus* from *Panopeus*. Pending a thorough analysis of adult and larval morphology, data presented here support the establishment of a new genus for *P. americanus*.

### 4.8 *Genus* Eurypanopeus

Schubart et al. (2000, Fig. 1) also provided evidence for polyphyly among species presently assigned to the genus *Eurypanopeus* A. Milne-Edwards, 1880, with species of *Eurypanopeus* falling into three separate clades. In the present study, topologies inferred from both datasets support the polyphyletic nature of *Eurypanopeus*, with representatives found in three clades for 16S (Fig. 1) and two clades for 12S (Fig. 2). It is unclear what effect the addition of sequence data from other species of *Eurypanopeus* would have on the analyses; however, on the basis of evidence presented here and by Schubart et al. (2000), comprehensive study and taxonomic revision of the genus are needed.

### 4.9 Panoplax depressa

Despite a gonopod that shares little in common with that of the typical panopeid, *Panoplax depressa* Stimpson, 1871, has long been considered a member of the subfamily Eucratopsinae within the family Panopeidae (Martin & Abele 1986, McLaughlin et al. 2005, Ng et al. 2008). The analyses presented here provide no support for the inclusion of *Panoplax* within the family Panopeidae. In topologies inferred from both datasets, *Panoplax depressa* is well separated from remaining representatives of the family Panopeidae. In the phylogeny inferred from the 16S dataset, *Panoplax depressa* is found nested within a poorly supported clade containing representatives of the families Xanthidae and Pseudorhombilidae (ML/p$P$ <50/99). In the phylogeny inferred from the 12S dataset, *Panoplax depressa* is also excluded from the remaining representatives of the family Panopeidae, nested within a poorly supported clade containing representatives of the family Xanthidae (ML/p$P$ <50/90). Despite the low support values for the clades currently containing *Panoplax depressa*, there is little evidence to support the inclusion of *Panoplax* within the family Panopeidae.

## 4.10 Garthiope barbadensis

The genus *Garthiope* Guinot, 1990, was described to accommodate three small species formerly attributed to the genus *Micropanope*. Upon its erection, similarities between *Garthiope* and the family Trapeziidae were noted; however, in their recent review Ng et al. (2008) considered the genus to be a part of the family Xanthidae. In the present analyses the complex relationship of *Garthiope* to the remaining taxa of the Xanthoidea *sensu* Martin & Davis (2001) is shown in the conflict between the 16S dataset and 12S dataset in regards to the placement of *Garthiope*. In the phylogeny inferred from the 16S dataset, *Garthiope barbadensis* (Rathbun, 1921) is found within the family Panopeidae, where it is located within a clade containing representatives of the subfamily Eucratopsinae. However, this clade has support values with ML and p$P$ values of <50/98. To further confound our understanding, in the analyses of the 12S dataset, *Garthiope barbadensis* falls well outside the family Panopeidae in a clade containing representatives of the Eriphioidea, Carpilioidea, Goneplacoidea, and Portunoidea. As this arrangement also has poor support values (<50), the relationship of *Garthiope* to these groups remains unclear. The type-species of the genus *Garthiope spinipes* (A. Milne-Edwards, 1880) was not included in these analyses; as a result, it is unclear what effect its inclusion may have on the analyses. Further study of the group is needed to clarify how this genus is related to other representatives of the Xanthoidea *sensu* Martin & Davis (2001).

## 4.11 *Outgroup taxa*

Composition of the superfamily Xanthoidea *sensu* Martin & Davis (2001) is a subject of ongoing debate (Guinot 1978; Jamieson 1993; Coelho & Coelho Filho 1993; Schubart et al. 2000; Wetzer et al. 2003; Karasawa & Schweitzer 2006; Ng et al. 2008). In all of our analyses, the family Xanthidae is clearly shown to be polyphyletic. Analysis of the 16S dataset reveals a single clade containing representatives of Panopeidae, Pseudorhombilidae, and three subfamilies of Xanthidae; however, this clade is poorly supported with ML bootstrap values and p$P$ of <50/99 (Fig. 1). Furthermore, a second clade contains a single representative of the family Xanthidae as well as representatives of Eriphioidea, Pilumnoidea, Carpilioidea, Goneplacoidea, and Portunoidea. This clade is well supported with ML bootstrap values and p$P$ of 97/100. Within this clade we also find representatives of three families of Goneplacoidea, with two species of *Chasmocarcinus* representing Chasmocarcinidae, *Frevillea barbata* and *Sotoplax robertsi* representing Euryplacidae, and *Bathyplax typhlus* representing Goneplacidae. While Chasmocarcinidae and Euryplacidae form a poorly supported monophyletic clade, Goneplacidae is found in another clade with representatives of Portunoidea and Carpilioidea. Although neither of these clades is well supported (ML/p$P$ <50/58 & <50/98), they provide evidence for a polyphyletic Goneplacoidea. While the topology inferred from the 12S dataset (Fig. 2) still presents evidence for a polyphyletic Xanthidae and Goneplacoidea, the evidence differs from that inferred by the 16S dataset (Fig. 1). However, support values for the outgroup topology inferred by the 12S dataset are very low, making any conclusions drawn from this topology questionable. Regardless of differences between these two topologies, it is apparent that both Goneplacoidea and Xanthidae are polyphyletic and in need of revision.

## ACKNOWLEDGEMENTS

We thank H. Bracken, M. Brugler, J. Felder, S. France, E. Palacios-Theil, E. Pante, V. Paul, R. Robles, J. Thoma, and A. Windsor for assisting in obtaining specimens or with various aspects of data collection, analysis, and manuscript preparation. We are grateful to J. Martin and G. Davis for providing loans of materials from the Natural History Museum of Los Angeles County, R. Lemaitre for access to specimens at the National Museum of Natural History—Smithsonian Institution, and F. Mantelatto for loans of specimens from Brazil. This study was supported in part by U.S. National

Science Foundation grants NSF/BS&I DEB-0315995 and NSF/AToL EF-0531603 to D. Felder, as well as several small travel grants from the Smithsonian Marine Station, Ft. Pierce, Florida. Additional support to B. Thoma was provided under a Louisiana Board of Regents doctoral fellowship. This is University of Louisiana Laboratory for Crustacean Research contribution no. 128 and Smithsonian Marine Station contribution no. 737.

REFERENCES

Ahyong, S.T. & O'Meally, D. 2004. Phylogeny of the Decapoda Reptantia: resolution using three molecular loci and morphology. *Raff. Bull. Zool.* 52: 673–693.

Buhay, J.E., Moni G., Mann, N. & Crandall, K.A. 2007. Molecular taxonomy in the dark: evolutionary history, phylogeography, and diversity of cave crayfish in the subgenus *Aviticambarus*, genus *Cambarus*. *Mol. Phylogenet. Evol.* 42: 435–448.

Bull, J.J., Huelsenbeck, J.P., Cunningham, C.W., Swofford, D.L. & Waddell, P.J. 1993. Partitioning and combining data in phylogenetic analysis. *Syst. Biol.* 42: 384–397.

Castresana, J. 2000. Selection of conserved blocks from multiple alignments for their use in phylogenetic analysis. *Mol. Biol. Evol.* 17: 540–552.

Coelho P.A. & Coelho Filho, P.A. 1993. Proposta de classificação da família Xanthidae (Crustacea, Decapoda, Brachyura) através da taxonomia numérica. *Rev. Bras. Zool.* 10: 559–580.

Crandall, K.A. & Fitzpatrick Jr., J.F. 1996. Crayfish molecular systematics: using a combination of procedures to estimate phylogeny. *Syst. Biol.* 45: 1–26.

Edgar, R.C. 2004. MUSCLE: multiple sequence alignment with high accuracy and high throughput. *Nucleic Acids Res.* 32: 1792–1797.

Felder, D.L. 1973. *An annotated key to crabs and lobsters (Decapoda, Reptantia) from coastal waters of the northwestern Gulf of Mexico*. Center for Wetland Resources. Baton Rouge: Louisiana State University. Pp. 1–103.

Felder, D.L., Alvarez, F., Goy, J.W. & Lemaitre, R. (In press). Chapter 59, Decapod (Crustacea) of the Gulf of Mexico, with comments on the Amphionidacea. In: Felder, D.L. & Camp, D.K. (eds.), *Gulf of Mexico Origin, Waters, and Biota. Volume I, Biodiversity*: 1019–1104. College Station: Texas A&M University Press.

Felder, D.L. & Martin, J.W. 2003. Establishment of a new genus for *Panopeus bermudensis* Benedict and Rathbun, 1891 and several other xanthoid crabs from the Atlantic and Pacific oceans (Crustacea: Decapoda: Xanthoidea). *Proc. Biol. Soc. Wash.* 116: 438–452.

Felsenstein, J. 1985. Confidence limits on phylogenies: an approach using the bootstrap. *Evolution* 39: 783–791.

Guindon S., Lethiec, F., Duroux, P. & Gascuel, O. 2005. PHYML Online—a web server for fast maximum likelihood-based phylogenetic inference. *Nucleic Acids Res.* 33: 557–559.

Guinot, D. 1978. Principes d'une classification évolutive des Crustacés Décapodes Brachyoures. *Bull. Biol. Fr. Belg.* 112: 209–292.

Hasegawa, M., Kishino, H. & Yano, T. 1985. Dating of the human-ape splitting by a molecular clock of mitochondrial DNA. *J. Mol. Evol.* 21: 160–174.

Huelsenbeck, J.P. & Ronquist, F. 2001. MRBAYES: Bayesian inference of phylogeny. *Bioinformatics* 17: 754–755.

Huelsenbeck, J.P., Ronquist, F., Nielsen, R. & Bollback, J.P. 2001. Bayesian inference of phylogeny and its impact on evolutionary biology. *Science* 294: 2310–2314.

Huelsenbeck, J.P., Larget, B., Miller, R.E. & Ronquist, F. 2002. Potential applications and pitfalls of Bayesian inference of phylogeny. *Syst. Biol.* 51: 673–688.

Jamieson, B.G.M. 1993. Spermatological evidence for the taxonomic status of *Trapezia* (Crustacea: Brachyura: Heterotremata). *Mem. Qld. Mus.* 33: 225–234.

Karasawa, H. & Schweitzer, C.E. 2006. A new classification of the Xanthoidea *sensu lato* (Crustacea: Decapoda: Brachyura) based on phylogenetic analysis and traditional systematics and evaluation of all fossil Xanthoidea *sensu lato*. *Contr. Zool.* 75: 23–72.

Martin, J.W. & Abele, L.G. 1986. Notes on male pleopod morphology in the brachyuran crab family Panopeidae Ortmann, 1893, *sensu* Guinot (1978) (Decapoda). *Crustaceana*. 50: 182–198.

Martin, J.W. & Davis, G.E. 2001. *An updated classification of the Recent Crustacea*. Natural History Museum of Los Angeles County, Science Series 39: 1–124.

McLaughlin, P.A., Camp, D.K., Eldredge, L.G., Felder, D.L., Goy, J.W., Hobbs, III, H.H., Kensley, B., Lemaitre, R. & Martin, J.W. 2005. Order Decapoda. In: Turgeon, D. (ed.), *Common and Scientific Names of Aquatic Invertebrates of the United States and Canada. Names of Crustaceans Special Publications*. Vol. 31. Bethesda, Maryland: American Fisheries Society Special Publication. Pp. 209–326.

Menzies, R. J. 1948. A revision of the brachyuran genus *Lophopanopeus*. *Allan Hancock Occas. Pap.* 4: 1–27, figs. 1–3, pls. 1–6.

Ng, P.K.L., Guinot, D. & Davie, P.J.F. 2008. Systema Brachyurorum: Part I. An annotated checklist of extant brachyuran crabs of the world. *Raff. Bull. Zool.* 17: 1–286.

Palumbi, S., Martin, A., Romano, S., McMillan, W.O., Stice, L. & Grabowski, G. 1991. *The Simple Fool's Guide to PCR*. Honolulu, Department of Zoology and Kewalo Marine Laboratory.

Porter, M.L., Pérez-Losada, M. & Crandall, K.A. 2005. Model-based multi-locus estimation of decapod phylogeny and divergence times. *Mol. Phylogenet. Evol.* 37: 355–369.

Posada, D. & Crandall, K.A. 1998. Modeltest: testing the model of DNA substitution. *Bioinformatics* 14: 817–818.

Rathbun, M.J. 1898. The Brachyura of the biological expedition to the Florida Keys and the Bahamas in 1893. *Bull. Lab. Nat. Hist. Iowa* 4: 250–294.

Rathbun, M.J. 1930. The cancroid crabs of America of the families Euryalidae, Portunidae, Atelecyclidae, Cancridae, and Xanthidae. *Bull. U.S. Natl. Mus.* 152: 1–609.

Robles R., Schubart, C.D., Conde, J.E., Carmona-Suárez, C., Alvarez, F., Villalobos, J.L. & Felder, D.L. 2007. Molecular phylogeny of the American *Callinectes* Stimpson, 1860 (Brachyura: Portunidae), based on two partial mitochondrial genes. *Mar. Biol.* 150: 1265–1274.

Rodríguez, F., Oliver, J.L., Marín, A. & Medina, J.R. 1990. The general stochastic model of nucleotide substitution. *J. Theor. Biol.* 142: 485–501.

Sankarankutty, C. & Manning, R.B. 1997. Observations on *Hexapanopeus schmitti* Rathbun from Brazil (Crustacea: Decapoda: Xanthidae). *Proc. Biol. Soc. Wash.* 110: 249–255.

Schubart, C.D., Neigel, J.E. & Felder, D.L. 2000. A molecular phylogeny of mud crabs (Brachyura: Panopeidae) from the northwestern Atlantic and the role of morphological stasis and convergence. *Mar. Biol.* 137: 1167–1174.

Schubart, C.D., Cuesta, J.A. & Felder, D.L. 2002. Glyptograpsidae, a new brachyuran family from Central America: larval and adult morphology, and a molecular phylogeny of the Grapsoidea. *J. Crustac. Biol.* 22: 28–44.

Swofford, D. 2003. PAUP*. *Phylogenetic analysis using parsimony (*and other methods)*. Version 4. Sunderland, MA: Sinauer Assoc.

Suzuki, Y., Glazko, G.V. & Nei, M. 2002. Overcredibility of molecular phylogenies obtained by Bayesian phylogenetics. *Proc. Natl. Acad. Sci. U.S.A.* 99: 16138–16143.

Wetzer, R., Martin, J.W. & Trautwein, S.E. 2003. Phylogenetic relationships within the coral crab genus *Carpilius* (Brachyura, Xanthoidea, Carpiliidae) and of the Carpiliidae to other xanthoid crab families based on molecular sequence data. *Mol. Phylogenet. Evol.* 27: 410–421.

Wicksten, M. 2005. Decapod crustaceans of the Flower Garden Banks National Marine Sanctuary. *Gulf Mex. Sci.* 23: 30–37.

Williams, A.B. 1984. *Shrimps, Lobsters, and Crabs of the Atlantic Coast of the Eastern United States, Maine to Florida*. Washington, D.C.: Smithsonian Institution Press. 550 pp.

# Molecular Phylogeny of the Genus *Cronius* Stimpson, 1860, with Reassignment of *C. tumidulus* and Several American Species of *Portunus* to the Genus *Achelous* De Haan, 1833 (Brachyura: Portunidae)

FERNANDO L. MANTELATTO[1], RAFAEL ROBLES[2], CHRISTOPH D. SCHUBART[3] & DARRYL L. FELDER[2]

[1] *Laboratory of Bioecology and Crustacean Systematics, Department of Biology, FFCLRP, University of São Paulo, Brazil*
[2] *Department of Biology, Laboratory for Crustacean Research, University of Louisiana at Lafayette, Lafayette, Louisiana, U.S.A.*
[3] *Biologie I, Universität Regensburg, 93040 Regensburg, Germany*

## ABSTRACT

As currently recognized by most taxonomists, the genus *Cronius* Stimpson, 1860, encompasses only two species, both distributed in tropical and subtropical waters. *Cronius ruber* (Lamarck, 1818) is reported from both the Pacific and Atlantic American coastlines, as well as the eastern Atlantic, and *C. tumidulus* (Stimpson, 1871) is reported to occur exclusively in the tropical western Atlantic. We examine potential differences between allopatric populations assigned to *C. ruber*, test hypothesized monophyly of the genus, and resolve the phylogenetic position of its members within the Portunidae. In so doing, we also revisit taxonomic classification of American species currently assigned to the genus *Portunus*. New 16S mtDNA sequences were obtained from representatives of the genera *Charybdis*, *Cronius*, *Lupella*, *Lupocyclus*, *Polybius*, *Portunus*, and *Thalamita* for examination along with sequences from GenBank. Slight but consistent genetic differences were found among populations assigned to *Cronius ruber* from the Pacific American coastline, the Atlantic American coastline, and the eastern Atlantic coastline (West Africa). The name *C. edwardsii* (Lockington, 1877) is resurrected for specimens from the eastern Pacific, but further analyses are needed to determine if additional taxonomic revisions may be required to more narrowly restrict use of the name *C. ruber* among a complex of Atlantic populations. Presently assigned members of *Cronius* do not form a monophyletic group. The well-defined clade representing *C. ruber* (including the resurrected *C. edwardsii*) is placed in a weakly supported grouping with representatives of *Laleonectes*, *Thalamita*, and *Charybdis*. In contrast, *Cronius tumidulus* forms a well-supported cluster with several present American representatives of the genus *Portunus*, which themselves are well separated from *P. pelagicus*, type species of that genus. Thus, we propose a revised taxonomy with placement of *C. tumidulus* in the resurrected genus *Achelous* De Haan 1833, an assignment that we also propose for nine American species currently treated under *Portunus*.

## 1 INTRODUCTION

Portunoidea Rafinesque, 1815, *sensu* Martin & Davis (2001) is a highly diverse group that consists of three families: Geryonidae Colosi, 1923, Trichodactylidae H. Milne Edwards, 1853, and Portunidae Rafinesque, 1815. In the latter family, the subfamily Portuninae is the most diverse, containing 11 genera and more than 130 species. While this diverse group of marine and non-marine species shares clearly portunid adaptations, evolutionary lineages among the genera are poorly understood. Despite numerous studies on its classification (see Karasawa et al. 2008 for review), Portunidae is one of a few brachyuran families that have undergone little taxonomic revision in recent years. Systematic review is warranted to reflect current evidence of phylogenetic relationships among its constituent genera.

The genus *Cronius* was described by Stimpson (1860), being based upon "the *Lupa rubra* [= *Portunus ruber*] of M. Edwards, which forms the connecting link between the old genus *Lupa*, and *Charybdis*." Under current systematic treatments, the two species assigned to this genus are *Cronius ruber* (Lamarck, 1818) and *C. tumidulus* (Stimpson, 1871) (originally as *Acheloüs tumidulus*). However, another two species were once proposed but later synonymized. These are *C. millerii* (A. Milne-Edwards, 1868) from East Africa, which most authors consider a synonym of *C. ruber* (e.g., Rathbun 1930 and as discussed in Manning & Holthuis 1981), and *C. edwardsii* (Lockington, 1877) from the eastern Pacific.

The "blackpoint sculling crab" *Cronius ruber* is a typically shallow water species found among a variety of substrates, especially rock rubble in the sublittoral areas (including tide pools), but there are a few reports to depths near 100 m. Its reported distribution extends from New Jersey (USA) throughout the Gulf of Mexico and the Caribbean to Rio Grande do Sul (Brazil) in the western Atlantic; from California to Peru and the Galapagos Islands in the eastern Pacific (if accepting *C. edwardsii* as a synonym); and from Senegal to Angola along the eastern Atlantic (if accepting *C. millerii* as a synonym). However, recent translocation and rapid expansion of *Charybdis hellerii* (A. Milne-Edwards, 1867) into the western Atlantic (see Mantelatto & Dias 1999 for review), a species that also thrives in shallow rocky areas, seems to have a negative impact on native species (Mantelatto & Garcia 2001), and sympatric populations of *C. ruber* appear to be in decline along the Brazilian coast (FLM, personal observation). In contrast, the "crevice sculling crab" *C. tumidulus* is primarily resident on open areas of shallow waters, including seagrass bottoms, back-reef coral heads and flats, and coral reefs (FLM and DLF, personal observations). This species is found only in the western Atlantic and is currently reported only from Bermuda and Florida to Brazil (Rathbun 1930; Williams 1984; Melo 1996).

It is noteworthy that almost 150 years ago Stimpson (1860) considered *Cronius ruber* to potentially represent a link between *Portunus* Weber, 1795, and *Charybdis* De Haan, 1833. Given this potentially unique but uncertain phylogenetic position for *Cronius*, it was essential for us to include selected members of the subfamilies Portuninae and Thalamitinae in our analyses in order to test monophyly of the genus as well as its phylogenetic position within the Portunidae. At the same time, phylogeny and taxonomy of the widely distributed genus *Portunus* has long been a topic of debate (e.g., Stephenson & Campbell 1959), and polyphyly of the genus *Portunus* has been clearly demonstrated by Mantelatto et al. (2007). In this recently published molecular phylogeny, only the species *P. sayi* (Gibbes, 1850), among all included western Atlantic representatives of the genus, clustered with the Indo-West Pacific type species of the genus, *P. pelagicus* (Linnaeus, 1758). This lineage grouped with *Callinectes* Stimpson, 1860, and *Arenaeus* Dana, 1851, instead of other included species of *Portunus*. The other western Atlantic representatives of *Portunus* and *Laleonectes vocans* (A. Milne-Edwards, 1878) were instead consistently separated from this group and thus were noted to warrant eventual reclassification.

The current study aims to build on the molecular phylogeny of Mantelatto et al. (2007) by use of the same genetic marker, 16S mtDNA, but with inclusion of additional taxa representing the Portuninae and Thalamitinae. Special emphasis is given to the genus *Cronius* and constituent species in order to: 1) test intraspecific variability within *C. ruber* and the possible validity of *C. millerii* and *C. edwardsii*; 2) test monophyly of the genus *Cronius*; and 3) test the position of *Cronius* within the Portuninae and its postulated link to the subfamily Thalamitinae. On the basis of these results, we propose taxonomic reclassifications for the species and genera under study.

## 2 MATERIALS AND METHODS

### 2.1 Sample collection

Portunid crabs used in this study were newly collected or obtained as gifts or loans from museum collections (Table 1). Newly collected specimens for DNA analysis were preserved directly in 75 to 90% ethanol. Species identifications were confirmed on the basis of morphological characters from available references (Stimpson 1860; Rathbun 1930; Stephenson & Campbell 1959; Manning & Holthuis 1981; Williams 1984; Manning & Chace 1990). Voucher specimens from which tissue subsamples were taken have been deposited in permanent collections (Table 1). Tissues from paratype and holotype materials, excised by minimally destructive methods, were sequenced when possible (Table 1).

Along with populations of *Cronius* from both sides of the Atlantic Ocean and the eastern Pacific Ocean, we included several species representing *Portunus* and other genera of the family Portunidae for comparison, initially to more broadly root the analysis. It was essential to include other members of the subfamilies Portuninae and Thalamitinae in order to test monophyly of the genus *Cronius* and to determine its phylogenetic position within the Portunidae. Specifically, we used all sequences of 12 species of *Portunus* from the western Atlantic attained in the previous study on molecular phylogeny by Mantelatto et al. (2007); additional species of *Portunus* from the eastern Pacific (Mexico), eastern Atlantic (Mediterranean), and Indo-West Pacific; *Charybdis* from the Atlantic, and Indo-West Pacific; *Euphylax* Stimpson, 1860, from the eastern Pacific (Mexico); *Laleonectes* Manning & Chace, 1990, from the Atlantic; and species of *Lupocyclus* Adams & White, 1848, and *Thalamita* Latreille, 1829, from the Indo-Pacific. Additionally, specimens of the portunid crab genera *Ovalipes* Rathbun, 1898, and *Polybius* Leach, 1820, (Polybiinae) and *Carcinus* Leach, 1814, (Carcininae) were included in the analysis as outgroups because they putatively represented successively more distant lineages from the in-group taxa. Some of the comparative sequences included in the analysis were retrieved from GenBank (Table 1).

### 2.2 DNA analysis

We based our phylogenetic analysis exclusively on a partial fragment of the 16S rDNA gene, which has repeatedly shown its utility in both phylogenetic and population studies for more than a decade and is thus a common choice for use in phylogenetic studies on decapods (see Schubart et al. 2000 and Mantelatto et al. 2007 for literature review). DNA extraction, amplification, and sequencing protocols were implemented as per Schubart et al. (2000) with modifications as in Mantelatto et al. (2007) and Robles et al. (2007).

Total genomic DNA was extracted from muscle tissue of walking legs or chelipeds. Muscle was ground and incubated for 1–12 h in 600 $\mu$l lysis buffer at 65°C; protein was separated by addition of 200 $\mu$l 7.5 M ammonium acetate prior to centrifugation. DNA precipitation was made by addition of 600 $\mu$l cold isopropanol followed by centrifugation; the resultant pellet was washed with 70% ethanol, dried, and resuspended in 10–20 $\mu$l TE buffer.

An approximately 560-basepair region of the 16S rRNA gene was amplified from diluted DNA by means of polymerase chain reaction (PCR) (thermal cycles: initial denaturation for 10 min at 94°C; annealing for 38–42 cycles: 1 min at 94°C, 1 min at 45–48°C, 2 min at 72°C; final extension of 10 min at 72°C) with the following primers: 16Sar (5'–CGC CTG TTT ATC AAA AAC AT–3'), 16Sbr (5'–CCG GTC TGA ACT CAG ATC ACG T–3'), 16SH4 (5'–GTY GCC CCA ACC AAA

**Table 1.** Portunid crab species used for phylogenetic reconstructions, showing respective date and site of collection along with museum catalog number (ULLZ: University of Louisiana—Lafayette Zoological Collections; IVIC: Instituto Venezolano de Investigaciones Científicas—Laboratorio de Ecología y Genética de Poblaciones, Crustacean Collection "Dr. Gilberto Rodríguez;" CCDB: Crustacean Collection of Department of Biology, Faculty of Philosophy, Sciences and Letters of Ribeirão Preto, University of São Paulo; USNM: National Museum of Natural History, Smithsonian Institution, Washington D.C.; SMF: Senckenberg Forschungsinstitut und Museum, Frankfurt; MNHN: Muséum National d'Histoire Naturelle, Paris; CSIRO: Marine Research, Invertebrate Museum, Hobart) and GenBank accession number.

| Species | Collection site, date | Catalogue No. | GenBank accession number |
|---|---|---|---|
| *Arenaeus cribrarius* (Lamarck, 1818) | Venezuela: Falcón, 1999 | ULLZ 5173 | DQ407667[c] |
| *Callinectes bellicosus* Stimpson, 1859 | Mexico: Baja California, 1999 | ULLZ 4166 | DQ407670 |
| *Callinectes bocourti* A. Milne-Edwards, 1879 | Venezuela: Zulia, 1999 | ULLZ 4180 | AJ298170 |
| *Callinectes danae* Smith, 1869 | Venezuela: Falcón, 1998 | IVIC-LEGP-C-1 | AJ298184[a] |
| *Callinectes ornatus* Ordway, 1863 | Brazil: São Paulo, 1999 | ULLZ 4178 | AJ298186[a] |
| *Callinectes sapidus* Rathbun, 1896 | USA: Florida, 1998 | ULLZ 3766 | AJ298189 |
| *Carcinus maenas* Linnaeus, 1758 | USA: New Hampshire, 1998 | ULLZ 3840 | AJ130811 |
| *Charybdis hellerii* (A. Milne-Edwards, 1867) | Brazil: São Paulo, 1995 | CCDB 2038 | FJ152142 |
| *Charybdis feriatus* (Linnaeus, 1758) | China, 2005 | — | DQ062727 |
| *Cronius ruber* (Lamarck, 1818) | Ghana: Cape Coast, 2001 | SMF 31986 | FJ153143 |
| *Cronius ruber* (Lamarck, 1818) | Mexico: Veracruz, 2002 | ULLZ 6448 | FJ152144 |
| *Cronius ruber* (Lamarck, 1818) | Brazil: São Paulo, 1999 | ULLZ 4295: CCDB 138 | FJ152145 |
| *Cronius ruber* (Lamarck, 1818) | Brazil: São Paulo, 2000 | ULLZ 4772 | FJ152146 |
| "*Cronius ruber*" (Lamarck, 1818)*** | Panama: Pacific coast, 2007 | ULLZ 8673 | FJ152147 |
| "*Cronius ruber*" (Lamarck, 1818) | Panama: Pacific, Gulf of Chiriqui, 2007 | CCDB 1717 | FJ152148 |
| "*Cronius tumidulus*" (Stimpson, 1871) | Brazil: Ubatuba, 2000 | ULLZ 4770 | FJ152149 |
| "*Cronius tumidulus*" (Stimpson, 1871) | USA: Gulf of Mexico, 2005 | ULLZ 6838 | FJ152150 |
| "*Cronius tumidulus*" (Stimpson, 1871) | Providencia, Colombia, Caribbean, 1998 | ULLZ 9117 | FJ152151 |
| "*Cronius tumidulus*" (Stimpson, 1871) | Puerto Rico: Paguera, 1995 | USNM uncatalogued | FJ152152 |
| *Euphylax robustus* A. Milne-Edwards, 1874 | Costa Rica: Gulf of Nicoya, 2004 | CCDB 1122 | FJ152153 |
| *Laleonectes nipponensis* (Sakai, 1938) | French Polynesia, no date | MNHN-B 31434 | FJ152154 |
| *Laleonectes vocans* (A. Milne-Edwards, 1878) | USA: Louisiana, 2000 | ULLZ 4640 | DQ388051[d] |
| *Lupella forceps* (Fabricius, 1793) | R/V Oregon II, 1970 | USNM 284565 | FJ152155 |
| *Lupocyclus philippinensis* Semper, 1880 | China, 1998 | — | FJ152156 |
| *Ovalipes stephensoni* Williams, 1976 | USA: Florida, 2003 | ULLZ 5678 | DQ388050[d] |
| *Ovalipes trimaculatus* (De Haan, 1833) | Argentina: Mar del Plata, 2001 | ULLZ 4773 | DQ388049[d] |
| *Polybius henslowii* Leach, 1820 | Spain: Santander, 1992 | SMF 31991 | FJ152157 |
| *Portunus anceps* (Saussure, 1858) | Belize: Carrie Bow Cay, 1983 | ULLZ 4327 | DQ388054[d] |
| "*Portunus asper*" (A. Milne-Edwards, 1861) | Mexico: Sinaloa, 2004 | CCDB 1738 | FJ152158 |
| "*Portunus binoculus*" Holthuis, 1969** | USA: NW Atlantic, 1965 | USNM 113560 | DQ388062[d] |
| "*Portunus depressifrons*" (Stimpson, 1859)* | USA: Florida, 1996 | ULLZ 4442 | DQ388064[d] |
| *Portunus floridanus* Rathbun, 1930 | USA: Gulf of Mexico, 2000 | ULLZ 4695 | DQ388058[d] |
| "*Portunus gibbesii*" (Stimpson, 1859) | USA: Alabama, 2001 | ULLZ 4565 | DQ388057[d] |
| *Portunus hastatus* (Linnaeus, 1767) | Turkey: Beldibi, 2007 | SMF 31989 | FJ152159 |
| "*Portunus ordwayi*" (Stimpson, 1860)** | USA: Florida, 1915 | USNM 61174 | DQ388066[d] |
| "*Portunus ordwayi*" (Stimpson, 1860) | Jamaica: St. Ann – Priory, 2003 | SMF 31988 | FJ152160 |
| *Portunus pelagicus* (Linnaeus, 1758) | China, 2005 | — | DQ062734 |
| *Portunus pelagicus* (Linnaeus, 1758) | India: Gulf of Mainnar, 2003 | ULLZ 5682 | DQ388052[d] |
| *Portunus pelagicus* (Linnaeus, 1758) | Australia: Tasmania, no date | CSIRO uncatalogued | FJ152161 |
| "*Portunus rufiremus*" Holthuis, 1959** | French Guiana: Sinnamaryi, 1974 | USNM 151568 | DQ388063[d] |
| *Portunus sayi* (Gibbes, 1850) | USA: Louisiana, 2001 | ULLZ 4753 | DQ388053[d] |
| "*Portunus sebae*" (H. Milne Edwards, 1834) | USA: Florida, 2001 | ULLZ 4527 | DQ388067[d] |
| "*Portunus spinicarpus*" (Stimpson, 1871) | USA: Florida, 1996 | ULLZ 4618 | DQ388061[d] |
| "*Portunus spinimanus*" Latreille, 1819 | Jamaica: St. Ann – Priory, 2003 | SMF 31987 | FJ152162 |

**Table 1.** continued.

| Species | Collection site, date | Catalogue No. | GenBank accession number |
|---|---|---|---|
| *Portunus trituberculatus* (Miers, 1876) | Japan, 2002 | — | AB093006 |
| *Portunus ventralis* (A. Milne-Edwards, 1879) | Belize: Carrie Bow Cay, 1983 | ULLZ 4440 | DQ388060[d] |
| *Scylla olivacea* (Herbst, 1796) | Taiwan, 2003 | — | AF109321[b] |
| *Scylla paramamosain* Estampador, 1949 | Taiwan, 1998 | — | AF109319 |
| *Scylla serrata* (Forskål, 1775) | Taiwan, 2003 | — | AF109318[b] |
| *Scylla tranquebarica* (Fabricius, 1798) | Taiwan, 1998 | — | AF109320 |
| *Thalamita admete* Herbst, 1803 | South Africa, 2001 | ULLZ 4382 | FJ152163 |
| *Thalamita crenata* Latreille, 1829 | Hawaii, Oahu, 2003 | ULLZ 8664 | FJ152164 |
| *Thalamita danae* Stimpson, 1858 | Singapore: Labrador, 1999 | ULLZ 4760 | FJ152165 |
| *Thalamita sima* H. Milne Edwards, 1834 | Australia, 1980 | ULLZ 4761 | FJ152166 |

Specimens used for DNA analysis: * type; ** holotype.
[a] Schubart et al. 2001b; [b] Hideyuki et al. 2004; [c] Robles et al. 2007; [d] Mantelatto et al. 2007.
*** Quote marks (" ") are used to show commonly used present names that are proposed for revision in this paper.

TAA A–3'), 16SL2 (5'–TGC CTG TTT ATC AAA AAC AT–3'), 16SH2 (5'–AGA TAG AAA CCA ACC TGG–3'), 16SL15 (5'–GAC GATA AGA CCC TAT AAA GCT T–3') (for references on the primers, see Schubart et al. 2000 and Schubart et al. 2001a). We used 16SH4 and 16SL15 internal primers (in combination with 16SL2, 16Sar, and 16Sbr) for partial amplification of the possibly formalin-fixed specimens among museum materials. PCR products were purified using Microcon 100® filters (Millipore Corp.) and sequenced with the ABI PRISM® Big Dye™ Terminator Mix (Applied Biosystems) in an ABI PRISM® 3100 Genetic Analyzer (Applied Biosystems automated sequencer). All sequences were confirmed by sequencing both strands.

## 2.3 *Phylogenetic analyses*

A consensus sequence for the two strands was obtained and multiple alignments were performed using the Clustal W option as implemented in the sequence alignment editor BioEdit ver. 7 (Hall 1999). Phylogenetic and molecular evolutionary analyses were conducted using MRBAYES software for Bayesian analysis (BAY) and PAUP 4.0 b10 (Swofford 2000) for the maximum parsimony (MP) and neighbor joining (NJ) analyses. Sequences were first analyzed with the software MODELTEST (Posada & Crandall 1998) in order to find the model of evolution that best fit the data. The BAY analysis was performed sampling 1 tree every 500 generations for 2,000,000 generations, starting with a random tree using the model of evolution obtained with MODELTEST, thus obtaining 4,001 trees. Preliminary analysis showed that stasis was reached at approximately 25,000 generations; we discarded the first 30,000 generations and the initial random tree (= 61 trees) and obtained a majority rule consensus tree from the remaining 3,940 trees. NJ analysis was carried out with a maximum likelihood distance correction set, with the parameters obtained by MODELTEST. MP analysis was performed as a heuristic search with random sequence addition of 1000 random trees, including tree bisection and reconnection as a branch swapping option; ten trees were saved after every repetition; indels were treated as a fifth character. On molecular trees, bootstrap confidence values >50% were reported for both NJ (2000 bootstraps) and MP (2000 bootstraps). For the BAY analysis, values were shown for posterior probabilities of the nodes among the 3,940 saved trees. Sequences, as well as the complete alignment, have been deposited in GenBank (Table 1).

## 3 RESULTS

### 3.1 *Taxonomic account*

Morphological data, historical synonymies, and diagnoses for both species of *Cronius* have been gathered from descriptions in the references mentioned in the introduction, especially Stimpson

*Cronius ruber* (Lamarck, 1818) *sensu lato*

Material examined: 1 ♂ (81.77 mm CW), 2 ♀ (70.21, 78.95 mm CW), Brazil, São Paulo, Ubatuba, July 1998, CCDB 1445; 1 ♂ (82.81 mm CW), Brazil, São Paulo, Ubatuba, Ilha Anchieta, July 1999, ULLZ 4295 (only pereopods 4 and 5 as DNA voucher CDS) and CCDB 138; 1♂ (44.0 mm CW), Brazil, São Paulo, Ubatuba, Ilha Anchieta, June 2000, ULLZ 4772 (DNA voucher FLM); 1♂ (6.7 mm CW), Mexico, SW Gulf of Mexico, June 2005, ULLZ 7352; 1 ♀ (6.5 mm CW), USA, off Louisiana, Gulf of Mexico, June 2006, ULLZ 8180; 2 ♂ (43.2, 46.7 mm CW), 1 ♀ (50.5 mm CW), USA, Newfound Harbor Keys, Florida, June 1979, ULLZ 2288; 1 ♀ (50.1 mm CW), USA, Port Mansfield, Texas, August 1969, ULLZ 8662; 1 juvenile ♂ (7.6 mm CW), Mexico, Veracruz, Laguna La Mancha, July 2002, ULLZ 6448; 3 ♀ (18.3, 28.5, 33.4 mm CW), Mexico, Baja California, Isla del Carmen, January 1932, USNM 207834; 1 ♂ (17.4 mm CW), Panama, Pacific coast, 9 May 2005, CCDB 1717; 1 ♂ (14.6 mm CW), Panama, Pacific coast, 15 February 2007, ULLZ 8673; 1 ♂ (53.2 mm CW), Ecuador, September 1926, USNM 76854; 1 ♂ (23.7 mm CW), Venezuela, Cariaco Basin, NW of Barcelona, October 1963, USNM 152578; 1 ♀ (38.9 mm CW), Saint Lucia, Caribbean Sea, E of Saint Lucia, March 1966, USNM 180526; 1 ♂ (20.8 mm CW), 1 ♀ ovigerous (42.3 mm CW), USA, off Florida, Gulf of Mexico, SOFLA expedition, April 1981, USNM 242921; additional material examined labeled as *Cronius millerii* (A. Milne-Edwards, 1868): 1 ♂ (73.4 mm CW), 1 ♀ (71.8 mm CW); Ghana: Cape Coast, July 2001 (both DNA vouchers); 1 ♂ (not measurable mm CW), Senegal, Dakar, November 1950, USNM 173088.

*Cronius tumidulus* (Stimpson, 1871)

Material examined: 1 ♂ (24.5 mm CW), USA, Florida, Tortugas Isl., July 1924, USNM 61015; 1 ♂ (10.5 mm CW), 2 ♀ (8.80, 11.50 mm CW), USA, Florida, off Palm Beach, 1951, USNM 168055; 1 ♀ (11.40 mm CW), USA, Florida, off Palm Beach, April 1950, USNM 169257; 1 ♀ (26.3 mm CW), USA, Puerto Rico, Paguera, Lauri Reef, March 1995, USNM uncatalogued (DNA voucher); 1 ♂ (14.77 mm CW), 2 ♀ (11.62, 10.31 mm CW), Brazil, São Paulo, Ubatuba, February 1999, CCDB 2036; 1 ♀ (10.88 mm CW), Brazil, São Paulo, Ubatuba, March 1996, CCDB 131; 5 ♂ (19.65, 17.40, 15.26, 14.34, 8.82 mm CW), 2 ♀ (17.4, 13.54 mm CW), Brazil, São Paulo, Ubatuba, February 2000, CCDB 128; 1 ♂ (11.36 mm CW), Brazil, São Paulo, Ubatuba, February 1996, CCDB 127; 1 ♂ (15.01 mm CW), Brazil, São Paulo, Ubatuba, February 2000, ULLZ 4770 (DNA voucher FLM); 1 ♀ ovigerous (18.2 mm CW), Mexico, Gulf of Mexico, June 2005, ULLZ 6838; 1 ♀ (6.10 mm CW), Brazil, São Paulo, Ubatuba, Ubatumirim, February 2000, CCDB 2035.

### 3.2 *Molecular phylogeny*

In total, 545 positions of the 16S rRNA gene (not including primer regions) were aligned for 49 portunid species. The optimal model of evolution for the data set, selected under the Akaike information criterion (AIC) as implemented in Modeltest (Posada & Crandall 1998), was the TVM+I+G (Invariable sites + Gamma distribution) with the following parameters: assumed nucleotide frequencies A = 0.3821, C = 0.0820, G = 0.1446, T = 0.3913; substitution model A-C = 0.8814, A-G = 8.1643, A-T = 1.0082, C-G = 1.0959, C-T = 8.1643, G-T = 1.00; proportion of invariable sites I = 0.2746; variable sites follow a gamma distribution with shape parameter = 0.5018. Thus, posterior analyses are based on this evolutionary model.

The molecular tree (Figure 1) is based on three different algorithms (NJ, MP, BAY), which are mostly congruent. The resultant molecular phylogeny disagrees in several respects with the current

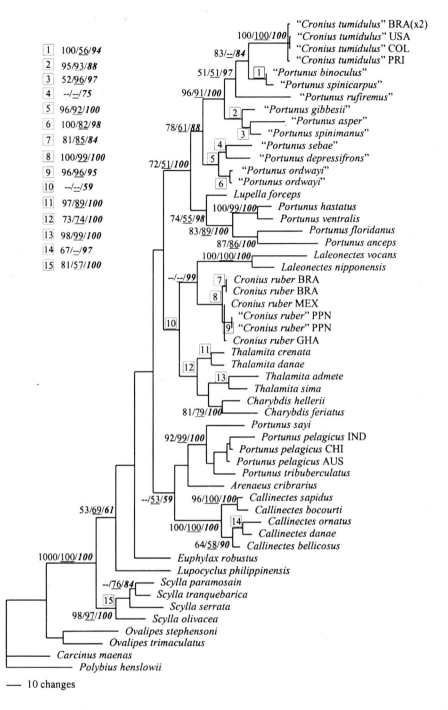

**Figure 1.** Evolutionary relationships of selected species of Portunidae based on a fragment of the 16S rDNA obtained with BAY analysis. Numbers on nodes are support values for that clade, NJ, MP, and BAY, respectively. Three-letter abbreviations are shown for species for which we obtained sequences from multiple populations (see Table 1 for details). BRA = Brazil, USA = United States of America, COL = Colombia, PRI = Puerto Rico, MEX = Mexico, PPN = Pacific Panama, GHA = Ghana, IND = India, CHI = China, AUS = Australia. Quote marks (" ") are used to show commonly used present names that are proposed for revision in this paper; (x2) indicates two identical sequences from the same locality. The name *C. edwardsii* is resurrected for specimens from the eastern Pacific. Even so, the genetic differences between this species and Atlantic populations are clearly less marked than in some trans-Panamic sister taxa of the genera *Alpheus* Fabricius, 1798 (see Knowlton et al. 1993), *Callinectes* (see Robles et al. 2007), and *Pachygrapsus* Randall, 1839 (see Schubart et al. 2005).

morphologically based classification of *Cronius*. Our analysis places *C. tumidulus* in a different clade from that of *C. ruber*, thus suggesting polyphyly of the genus. *Cronius tumidulus* appears derived from American representatives of the genus *Portunus sensu lato* with which it is clustered with high confidence values (78/61/88). On the other hand, all populations putatively assignable to *C. ruber* are found in a second clade, along with two species of *Laleonectes* and representatives of the Thalamitinae. However, the relationship between populations presently assigned to *C. ruber* and these other genera remains poorly resolved, as basal nodes are weakly supported. The genus *Portunus* is shown to be polyphyletic, with one clade encompassing two groups of species, among which are found all of the included American representatives except for *P. sayi*. Yet another clade contains *P. sayi* and the Indo-West Pacific species, which include *P. pelagicus*, type species of that genus.

Positional differences among putative populations of *Cronius ruber* were very limited. Even so, genetic divergences between Atlantic and Pacific populations are more pronounced (Gulf of Mexico vs. Pacific, 4 transitions [ts] and 1 transversion [tv]; Brazil vs. Pacific, 7 ts and 1 tv; Ghana vs. Pacific, 5 ts and 1 tv) than are divergences between Atlantic populations (Ghana vs. Gulf of Mexico, 1 tv; Ghana vs. Brazil, 4 ts).

## 4 DISCUSSION

### 4.1 *Paraphyly of* Cronius *and related taxonomic revisions*

Paraphyly of *Portunus* was reported previously by Mantelatto et al. (2007) and is corroborated here with treatment of additional taxa. According to our present molecular phylogeny, the genus *Cronius*, as currently defined, is also paraphyletic. We propose a new taxonomy, with restriction of the group defined as the genus *Portunus* and re-elevation of the subgenus *Achelous* De Haan, 1833, to full generic rank. Within *Achelous* (for the present) we include nine American species formerly assigned to the genus *Portunus* and *Cronius tumidulus* (see Table 1). The genus *Achelous* thus contains *A. asper* (A. Milne-Edwards, 1861), *A. binoculus* (Holthuis, 1969), *A. depressifrons* (Stimpson, 1859), *A. gibbesii* (Stimpson, 1859), *A. ordwayi* Stimpson, 1860, *A. rufiremus* (Holthuis, 1959), *A. sebae* (H. Milne Edwards, 1834), *A. spinicarpus* Stimpson, 1871, *A. spinimanus* (Latreille, 1819), and *A. tumidulus* Stimpson, 1871.

On the basis of our molecular genetic analyses of western Atlantic, eastern Atlantic, and eastern Pacific populations presently assigned to *Cronius ruber*, we for now continue to synonymize *C. millerii* with *C. ruber*. The small genetic differences in 16S mtDNA sequences, especially with our small sample size, are not deemed adequate for distinction of the African *C. millerii* as a separate species at this point, and its populations are thus treated under *C. ruber* provisionally. Similarly, Brazilian and Gulf of Mexico populations of *C. ruber* were not deemed to be adequately distinguished to justify separation, though analyses of additional samples and additional genes may warrant reconsideration of this issue in the future. Slightly more substantial genetic differences were found between populations of *C. ruber* from the Pacific American coastline and all of the populations in Atlantic waters. This divergence likely reflects historical separation of Atlantic and Pacific tropical waters by closure of the Panama Isthmus, as has been invoked to explain separations of other marine decapod species pairs or sister taxa (Knowlton & Weigt 1998).

As long as genetic homogeneity along both coastlines of the Atlantic remains unknown, it appears premature to recognize separate species for populations of *Cronius ruber* in the eastern and western Atlantic, and we elect to follow morphologically based conclusions (color pattern and ornamentation of the chelae) of Manning & Holthuis (1981). *Cronius ruber* thus has an amphi-Atlantic distribution and a closely related trans-Isthmian sister species. A similar distribution can be found for many other littoral decapod crustaceans, and questions remain whether such largely separated allopatric populations really belong to the same species. *Cronius tumidulus* shows clear genetic separation from *C. ruber* and clearly warrants treatment in a different genus. On the basis of its apomorphic morphological characters, one might assume it deserves treatment in its own genus.

However, its close genetic similarity to nine American representatives of the genus *Portunus*, a morphologically diverse group which is also in need of reclassification (see Mantelatto et al. 2007: fig. 1, clade C, plus *P. asper* from the eastern Pacific in the present work), prompts us to consolidate the taxonomy of this entirely American group by placing them together in one monophyletic genus. By elevating the available subgeneric name *Achelous* for this group, with the American *Portunus spinimanus* as type species of the genus, we alleviate the paraphyly of *Cronius* and partly address the polyphyly of *Portunus*. As treated here, the genus *Achelous* currently encompasses ten species listed above, but with high probability it will eventually include more eastern Pacific forms as studies progress (Mantelatto et al. in preparation). While *P. sayi* is positioned with strong support in a common clade with *P. pelagicus* (type species of the genus) and will thus remain within the genus *Portunus*, the taxonomic position and reclassification of other species of *Portunus* from the western Atlantic [*Portunus anceps* (de Saussure, 1858), *P. ventralis* (A. Milne-Edwards, 1879), and *P. floridanus* Rathbun, 1930] and Mediterranean [*Portunus hastatus* (Linnaeus, 1767)] must await further studies of additional American and western African representatives.

The western Atlantic *C. tumidulus* was originally described by Stimpson (1871) as *Achelous tumidulus*, even though he had also previously erected the genus *Cronius* for *C. ruber* in 1860. We can thus interpret that at least Stimpson did not see a close relationship between the two species. Later, *A. tumidulus* was reclassified under the genus *Neptunus*, as *N. tumidulus* (by A. Milne-Edwards 1879), as *Charybdella tumidula* (by Rathbun 1901), and finally within *Cronius* as *C. tumidulus* (by Rathbun 1920). Only after the present study does it again become a species of the genus to which it had been originally assigned.

Our molecular analysis agrees with recent results obtained from larval morphology (Fransozo et al. 2002). Important differences were noted between the larval morphological characters of *C. ruber* and *C. tumidulus*, which led those authors to cluster zoeae of Portuninae into two subgroups (see also Stuck & Truesdale 1988). Zoeae with relatively long antennal exopods were found typical of *C. tumidulus*, *Portunus gibbesii*, *P. spinicarpus*, and *Scylla serrata*, while those with short antennal exopods were found to represent *Cronius ruber*, *Arenaeus cribrarius*, *Callinectes danae*, *C. sapidus*, and *Charybdis hellerii*. With the exception of *Scylla serrata*, which holds a somewhat intermediate position in terms of larval morphology (see Fransozo et al. 2002: table 1) and a basal position in our molecular phylogeny, the zoeal subgroups correspond perfectly with those grouped by 16S mtDNA; only members of the newly defined *Achelous* have an antennal exopod length equal to or exceeding 1/3 the protopod length.

Rathbun (1930: 34–35) defined morphology of the subgenus *Achelous* in her keys as "Carapace narrow; antero-lateral margin the arc of a circle with short radius, whose center is near the center of the cardiac region," and for the subgenus *Portunus* as "Carapace wide; antero-lateral margin the arc of a circle with long radius, whose center is near the posterior margin of the carapace." She indicated *Cronius tumidulus* has a narrower carapace than *C. ruber*, which fits the description of *Achelous*. On the other hand, the defining characters of *Cronius* according to Rathbun (1930: 14) are "Movable portion of antenna excluded from orbit by a prolongation of its basal article. Antero-lateral teeth alternately large and small." Morphological studies of the representatives of *C. tumidulus* at our disposal did not reveal a clear exclusion of the movable portion of the antenna from the orbit (as opposed to the case in *Cronius*). The presence of alternately large and small anterolateral teeth, on the other hand, is not a character that excludes membership in the subgenera *Portunus* and *Achelous* as defined by Rathbun (1930). We therefore find no morphological contradictions for inclusion of *Cronius tumidulus* within *Achelous*.

The name *Portunus* was originally published by Weber (1795), used by Fabricius (1798), and included practically all the members of the Portunidae known at the time. The history of generic names for species assigned to the genus "*Portunus*" reflects a confused nomenclature, as was previously noted in an extensive revision and synonymy by Palmer (1927). Stephenson & Campbell (1959) built upon this earlier discussion and also gave arguments for and against the use of subgeneric definitions within this genus. *Achelous* (type species *Portunus spinimanus* Latreille, 1819) has previously been used as one of five valid subgenera within the genus *Portunus*, the others being *Lupocycloporus* Alcock, 1899 [type species *Achelous whitei* A. Milne-Edwards, 1861 = *Portunus gracilimanus* (Stimpson, 1858)], *Monomia* Gistel, 1848 [replacement name for *Amphitrite* De Haan, 1833; type species *Cancer gladiator* Fabricius, 1793], *Portunus* [type species *Cancer pelagicus* Linnaeus, 1758], and *Xiphonectes* A. Milne-Edwards, 1873 [type species *Amphitrite vigilans* Dana, 1852 = *Portunus longispinosus* (Dana, 1852)]. Stephenson & Campbell (1959) noted difficulties in placing four species of *Portunus* in any of the existing subgenera and discussed unresolved relationships with the genus *Callinectes*. They concluded (p. 88): "The real difficulties which arise over the four species above suggest that it is preferable at this stage to avoid the use of subgeneric categories while dealing with the Indo-West Pacific fauna." This suggestion has been followed from then on, not only for the Indo-West Pacific fauna, but also for the genus *Portunus* as a whole (e.g., Crosnier 1962; Türkay 1971; Stephenson 1972; Manning & Holthuis 1981; Williams 1984; Mantelatto et al. 2007). After almost fifty years, we break with this tradition by resurrecting one of the subgenera and elevating it to full generic status, similar to what Barnard (1950) did when using *Achelous*, *Hellenus* (= *Xiphonectes*), *Lupa* (= *Portunus*), and *Monomia* as full genera, into which he classified the South African swimming crabs. We are aware that this is but a first step that does not solve taxonomic issues for the entire genus. Future morphological and molecular systematic work must address whether other subgenera warrant elevation or whether other new genera need to be proposed (for example, as done by Manning 1989 for *Sanquerus* and Ng & Takeda 2003 for *Atoportunus*) in order to provide a natural classification based on monophyletic clades.

Six of the ten species we propose to include in the genus *Achelous* formerly belonged to that taxon as a subgenus (Rathbun 1930; Ng et al. 2008). It is noteworthy that three of them originally were described as species of *Achelous*: *A. ordwayi*, *A. spinicarpus*, and *A. tumidulus*. *Portunus vossi* Lemaitre, 1991, recently synonymized with *A. spinicarpus*, and *P. bahamensis* Rathbun, 1930, recently synonymized with *A. depressifrons* (see Mantelatto et al. 2007), would obviously also represent materials and descriptions now to be associated with *Achelous*. However, *A. asper*, *A. gibbesii*, and *A. rufiremus* have been treated recently as members of the subgenus *Portunus* (see Rathbun 1930; Ng et al. 2008), and their apparent morphological distinction from the other species of *Achelous* should be reexamined to confirm our proposition. The definition used by Stimpson (1860: 221) for *Achelous* differs somewhat from the later one by Rathbun (1930). Stimpson noted the genus to be "chiefly characterized by the shape of the merus-joint of the external maxillipeds, which is greatly produced anteriorly beyond the base of the palpus, with its outer margin usually straight, but sometimes little projecting at the antero-exterior angle." Perhaps this character, in addition to gonopod morphology, should be reconsidered in defining American members of *Portunus*, rather than depending upon vaguely defined differences in carapace shape. Stephenson & Campbell (1959) previously stressed the potential importance of gonopod morphology for subdivision of *Portunus* and provided examples of possible characters in gonopod structure that reflected subgeneric classifications among some Australian species of *Portunus*.

This study is an early step in revising taxonomy of the apparently polyphyletic genus *Portunus*. Not all western Atlantic species of *Portunus* dealt with in Mantelatto et al. (2007) have been addressed in this reclassification, which has focused primarily on those taxa potentially grouped with *Cronius tumidulus* and the resurrected type species of *Achelous*, *A. spinimanus* (those of clade C in Mantelatto et al. 2007). Our phylogeny adds evidence that the phylogenetic position of clade B in Mantelatto et al. (*Portunus anceps*, *P. ventralis*, and *P. floridanus*) requires future clarification, especially our adding of *P. hastatus* to this clade and revealing an apparent basal relationship of the entire clade to the genus *Lupella*. Additional taxa are currently being added to the analysis, with a special effort for broadened coverage of eastern Atlantic and Pacific genera.

## 4.2 Subfamily considerations and future work

The original description of *Cronius* by Stimpson (1860) suggested this new genus to occupy an intermediate position between *Portunus*, a member of the subfamily Portuninae Rafinesque, 1815, and *Charybdis*, a member of the subfamily Thalamitinae Paul'son, 1875. Most taxa of the subfamily Thalamitinae are representatives of two genera that we included in our analysis, *Charybdis* De Haan, 1833 (with approximately 50 species), and *Thalamita* Latreille, 1829 (with approximately 90 species); the remaining genera *Gonioinfradens* Leene, 1938 (one species), and *Thalamitoides* A. Milne-Edwards, 1869 (three species), apply to comparatively few representatives (Fig. 1). We included only six species of this putative subfamily and they resolved as a well-supported monophyletic clade, but it is positioned among different genera of the Portuninae rather than being separated from these at a basal node. At low support levels, *Cronius sensu stricto* and *Laleonectes* are positioned as a sister group to representatives of the Thalamitinae, *Charybdis* and *Thalamita*.

It is tempting to conclude that Thalamitinae simply represents a lineage within Portuninae that is characterized by broader fronts. That conclusion would be in agreement with Rathbun (1930), Stephenson & Campbell (1960), Türkay (1971), and Stephenson (1972), in which case the subfamilies would be synonymous and the name Portuninae would have priority. However, support levels for the basal nodes that position Thalamitinae in the present analysis remain too low for us to confidently draw this conclusion. We thus defer further consideration of this issue until we complete additional molecular analyses currently in progress.

Pending analyses include additional taxa of the aforementioned families, as well as an expanded subset of species representing Polybiinae and Carcininae. Topology of our present tree suggests that Polybiinae (represented by *Polybius* and *Ovalipes*) is polyphyletic, as von Sternberg & Cumberlidge (2001) have already indicated in their cladistic analysis, but again our present support values are low. The subfamily Polybiinae has been regarded as a basal group in the Portunoidea on the basis of morphological characters (Guinot 1978), zoeal evidence (Rice 1981), and molecular analysis (Mantelatto et al. 2007). Its potential monophyly and phylogenetic position within the family can be addressed only with broader representation of portunoid generic diversity in subsequent analyses.

## ACKNOWLEDGEMENTS

Major research support was provided by awards to FLM under FAPESP (grants 00/11220-1 and 02/08178-9) and under CNPq (Research Scholarships PQ 300279/95-7 and 301261/04-0; Research Grant 472746/04-9), both in Brazil and during residence on the University of Louisiana—Lafayette campus. Research support was provided by awards to DLF under U.S. National Science Foundation (grants DEB-BS&I 0315995 and EF-AToL 0531603) and the U.S. Department of Energy (grant DE-FG02-97ER1220). FLM and CDS thank Pró-Reitoria de Pós-Graduação of University of São Paulo (USP) (grant 2007.1.371.59.7) for support of CDS under the Brazil/Germany visiting program. We thank I. Wehrtmann, M. Hendrickx, R. Lemaitre, and G. Melo for making available fresh specimens and/or loans of materials from collections. This is contribution number 129 from the UL—Lafayette Laboratory for Crustacean Research.

## REFERENCES

Barnard, K.H. 1950. Descriptive catalogue of South African decapod Crustacea. *Ann. South Afr. Mus.* 38: 1–824.

Crosnier, A. 1962. Crustacés Décapodes, Portunidae. *Faune de Madagascar* 16: 1–154, pls.1–13.

Fabricius, J.C. 1798. *Supplementum Entomologiae Systematicae*: 572 pp. Hafniae (= Copenhagen): Proft & Storch.

Fransozo, A., Mantelatto, F.L. & Bertini, G. 2002. The first zoeal stage of the genus *Cronius* (Brachyura, Portunidae) from the Brazilian coast, hatched in the laboratory. *J. Plankton Res.* 24: 1237–1244.

Garth, J.S. & Stephenson, W. 1966. Brachyura of the Pacific Coast of America. Brachyrhyncha: Portunidae. *Allan Hanc. Monog. Mar. Biol.* 1: 1–154.

Guinot, D. 1978. Principes d'une classification évolutive des Crustacés Décapodes Brachyoures. *Bull. Biol. France Belg.* 112: 209–292.

Hall, T.A. 1999. BioEdit: a user-friendly biological sequence alignment editor and analysis program for Windows 95/98/NT. *Nucl. Acids Symp. Ser.* 41: 95–98.

Hideyuki, I., Cheng, J.H., Hamasaki, K. & Numachi, K.I. 2004. Identification of four mud crab species (genus *Scylla*) using ITS-1 and 16S rDNA markers. *Aquat. Liv. Res.* 17: 31–34.

Karasawa, H., Schweitzer, C.E. & Feldman, R.M. 2008. Revision of Portunoidea Rafinesque, 1815 (Decapoda: Brachyura) with emphasis of the fossil genera and families. *J. Crust. Biol.* 28: 82–127.

Knowlton, N. & Weigt, L.A. 1998. New dates and new rates for divergence across the Isthmus of Panama. *Proc. R. Soc. London B* 265: 2257–2263.

Knowlton, N., Weigt, L.A., Solórzano, L.A., Mills, D.K. & Bermingham, E. 1993. Divergence in proteins, mitochondrial DNA and reproductive compatibility across the Isthmus of Panama. *Science* 260: 1629–1632.

Manning, R.B. 1989. *Sanquerus*, a replacement name for *Posidon* Herklots, 1851 (Crustacea, Decapoda, Portunidae). *Proc. Biol. Soc. Wash.* 102: 698–700.

Manning, R.B. & Chace, F.A. Jr. 1990. Decapod and stomatopod Crustacea from Ascension Island, South Atlantic Ocean. *Smith. Contrib. Zool.* 503: 1–91.

Manning, R.B. & Holthuis, L.B. 1981. West African brachyuran crabs (Crustacea: Decapoda). *Smith. Contrib. Zool.* 306: 1–379.

Mantelatto, F.L. & Dias, L.L. 1999. Extension of the known distribution of *Charybdis hellerii* (A. Milne-Edwards, 1867) (Decapoda, Portunidae) along the western tropical South Atlantic. *Crustaceana* 72: 617–620.

Mantelatto, F.L. & Garcia, R.B. 2001. Biological aspects of the nonindigenous portunid crab (*Charybdis hellerii*) in the western tropical South Atlantic. *Bull. Mar. Sci.* 68: 469–477.

Mantelatto, F.L., Robles, R. & Felder, D.L. 2007. Molecular phylogeny of the western Atlantic species of the genus *Portunus* (Crustacea, Brachyura, Portunidae). *Zool. J. Linn. Soc.* 150: 211–220.

Martin, J.W. & Davis, G.E. 2001. An updated classification of the Recent Crustacea. *Natural History Museum of Los Angeles County, Science Series* 39: 1–124.

Melo, G.A.S. 1996. *Manual de identificação dos Brachyura (caranguejos e siris) do litoral brasileiro.* São Paulo, Brazil: Editora Plêiade/FAPESP.

Milne-Edwards, A. 1879 [1873–1881]. Études sur les Xiphosures et les Crustacés de la Région Mexicaine, Mission Scientifique au Mexique et dans l'Amerique centrale. *Recherches Zoologiques a l'Histoire de la Faune de l'Amerique Centrale et du Mexique (Paris)* part 5: 1–368, 61 pls.

Ng, P.K.L. & Takeda, M. 2003. *Atoportunus*, a remarkable new genus of cryptic swimming crab (Crustacea; Decapoda; Brachyura; Portunidae), with descriptions of two species from the Indo-West Pacific. *Micronesica* 35–36: 419–432.

Ng, P.K.L., Guinot, D. & Davie, P.J.F. 2008. Systema Brachyuororum: Part I. An annotated checklist of extant brachyuran crabs of the world. *Raffles Bull. Zool.* 17: 1–286.

Palmer, R. 1927. A revision of the genus "*Portunus*" (A. Milne Edwards, Bell, etc.). *J. Mar. Biol. Assoc. U. K.* 14: 877–908.

Posada, D. & Crandall, K.A. 1998. MODELTEST: testing the model of DNA substitution. *Bioinformatics* 14: 817–818.

Rathbun, M.J. 1901. The Brachyura and Macrura of Porto Rico. *Bull. U. S. Fish. Comm.* 20(2): 1–127, 2 pls.
Rathbun, M.J. 1920. Stalk-eyed crustaceans of the Dutch West Indies. *Rapport betreffende een voorloopig onderzoek naar den toestand van de Vischerij en de Industrie van Zeeproducten in de Kolonie Curaao, uitgebracht door Prof. Dr. Boeke, part 2, 's Gravenhage, 1919 (1920)*: 317–349 [internal page numbering, 1–33].
Rathbun, M.J. 1930. The cancroid crabs of America of the families Euryalidae, Portunidae, Atelecyclidae, Cancridae and Xanthidae. *U. S. Nat. Mus. Bull.* 152: 1–509, 230 pls.
Rice, A.L. 1981. Crab zoeae and brachyuran classification: a re-appraisal. *Bull. British Mus. Nat. Hist. (Zoology)* 40: 287–296.
Robles, R., Schubart, C.D., Conde, J.E., Carmona-Suárez, C., Alvarez, F., Villalobos, J.L. & Felder, D.L. 2007. Molecular phylogeny of the American *Callinectes* Stimpson, 1860 (Brachyura: Portunidae), based on two partial mitochondrial genes. *Mar. Biol.* 150: 1265–1274.
Schubart, C.D., Neigel, J.E. & Felder, D.L. 2000. Use of the mitochondrial 16S rRNA gene for phylogenetic and population studies of Crustacea. In: von Vaupel Klein, J.C. & Schram, F.R. (eds.), *Crustacean Issues 12, The Biodiversity Crisis and Crustacea*: 817–830. Rotterdam: Balkema.
Schubart, C.D., Cuesta, J.A. & Rodriguez, A. 2001a. Molecular phylogeny of the crab genus *Brachynotus* (Brachyura: Varunidae) based on the 16S rRNA. *Hydrobiologia* 449: 41–46.
Schubart, C.D., Conde, J.E., Carmona-Suárez, C., Robles, R. & Felder, D.L. 2001b. Lack of divergence between 16S mtDNA sequences of the swimming crabs *Callinectes bocourti* and *C. maracaiboensis* (Brachyura: Portunidae). *Fish. Bull.* 99: 475–481.
Schubart, C.D., Cuesta, J.A. & Felder, D.L. 2005. Phylogeography of *Pachygrapsus transversus* (Gibbes, 1850): the effect of the American continent and the Atlantic Ocean as gene flow barriers and recognition of *Pachygrapsus socius* Stimpson, 1871 as a valid species. *Nauplius* 13: 99–113.
Stephenson, W. 1972. An annotated check list and key of the Indo-West-Pacific swimming crabs (Crustacea: Decapoda: Portunidae). *R. Soc. N. Zeal. Bull.* 10: 1–64.
Stephenson, W. & Campbell, B. 1959. The Australian portunids (Crustacea: Portunidae) III. The genus *Portunus*. *Austr. J. Mar. Fresh. Res.* 10: 84–129.
Stephenson, W. & Campbell, B. 1960. The Australian portunids (Crustacea: Portunidae) IV. Remaining genera. *Austr. J. Mar. Fresh. Res.* 11: 73–122.
Stimpson, W. 1860. Notes on North American Crustacea, in the Museum of the Smithsonian Institution. No. II. *Ann. Lyc. Nat. Hist., New York* 7: 177–246, 5 pls.
Stimpson, W. 1871. Preliminary report on the Crustacea dredged in the Gulf Stream in the Straits of Florida, by L. F. de Pourtales, Assist. U. S. Coast Survey. Part I. Brachyura. *Mus. Comp. Zool.* 2(2): 109–160.
Stuck, K.C. & Truesdale, F.M. 1988. Larval development of the speckled swimming crab, *Arenaeus cribrarius* (Decapoda, Brachyura, Portunidae) reared in the laboratory. *Bull. Mar. Sci.* 42: 101–132.
Swofford, D.L. 2000. *PAUP\* Phylogenetic Analysis Using Parsimony (\*and Other Methods), Version 4*. Sunderland, Massachusetts: Sinauer Assoc.
Türkay, M. 1971. Die Portunidae des Naturhistorischen Museums Genf, mit einem Anhang über die Typen von *Ovalipes ocellatus floridanus* Hay & Shore 1918 (Crustacea, Decapoda). *Arch. Sci. Gen.* 24: 111–143.
von Sternberg, R. & Cumberlidge, N. 2001. Notes on the position of the freshwater crabs within the brachyrhynchan Eubrachyura (Crustacea: Decapoda: Brachyura). *Hydrobiologia* 449: 21–39.
Weber, F. 1795. *Nomenclator entomologicus entomologiam systematiam ill. Fabricii adjectis speciebus recens detectis et varietatibus.* Chilonii (Kiel) & Hamburgi. viii, 171 p.
Williams, A.B. 1984. *Shrimps, lobsters, and crabs of the Atlantic Coast of the eastern United States, Maine to Florida.* Washington, DC: Smith. Inst. Press.

# Index

## A

absence, *see* Presence/absence
*Acanthacaris*
   marine clawed lobster families, 357, 363–364
   morphological and molecular previous works, 358
   morphological convergence, 364
   *Oncopareia-Thaumastocheles*, 364
   *Palaeophoberus*, 364
   phylogenetic analysis, 362
   previous works comparison, 363
   Thaumastochelidae status, 364
*Acantholobulus*, 561–562
*Acantholobulus bermudensis*, 557, 561, 562
*Acantholobulus caribbaeus*, 562
*Acantholobulus schmitti*, 557, 560, 562
Acanthonychinae, 438
*Acanthonyx*, 444
*Acanthonyx petiverii*, 439
*Acetes*, 272
Achelata
   embryonic characters, 40
   fossil record, 5
   infraordinal relationships, 91
   phylogeny-based systematics, 23
   Polychelidan lobsters, 370
   Thalassinidea, mitochondrial and nuclear genes, 309, 312–313
   Thalassinidea, monophyly or paraphyly, 319
*Achelous*, 567, 574–576, *see Cronius* Stimpson, molecular phylogenies
*Achelous asper*, 574, 575, 576
*Achelous binoculus*, 574
*Achelous depressifrons*, 574, 576
*Achelous gibbesii*, 574, 576
*Achelous ordwayi*, 574, 576
*Achelous rufiremus*, 574, 576
*Achelous sabae*, 574
*Achelous spinicarpus*, 574, 576
*Achelous spinimanus*, 574, 576
*Achelous tumidulus*, 568, 574–576
*Achelous whitei*, 576
acrosome vesicle
   common sperm form, 102
   dimensions, 110
   internal complexity, 110
   presence/absence, 108–109
   shape, 109–110
   size, 110
   spermatozoal morphology, 105, 108–110
*Acrotheres*, 458
*Actaea areolatus*, 234, 236
*Actumnus setifer*, 226
*Adeleana*, 524

Aegeridae, 254
Aeglidae, 497
aflagellate sperm cells, 105
Afrotropical region, 499
Agostrocarididae, 294
AIC, *see* Akaike information criterion (AIC)
Akaike information criterion (AIC)
   Callianassidae, mitochondrial genes, 333
   *Cronius*, 572
   freshwater crayfishes, 345
   *Hexapanopeus*, 556–557, 557
   Majoidea, 443
   marine clawed lobster families, 361
   Portunoidea and Cancroidea, 538–539
   Thalassinidea, mitochondrial and nuclear genes, 313
*Alarconia seaholmi*, 467
Alarconiini, 467
Algae Tube, 151
*Aliaporcellana*, 111
Alloposidae, 134
Alpheidae
   basal lineages, 299
   Caridea, infraorder phylogeny, 281, 294, 296
   morphological characters, 252
   pair-bonding mating system, 127, 150–152
   superfamily Alpheoidea, 298
Alpheoidea, 298–299
*Alpheus*
   analysis, 486
   molecular markers, 253
   phylogenetic relationships, 252
   pseudogenes, 57
   superfamily Alpheoidea, 298
*Alpheus angulatus*, 127, 132
*Alpheus armatus*, 127
*Alpheus heterochaelis*, 127
*Alpheus heterochelis*, 132
*Alpheus roquensis*, 127
Alvinocarididae
   Caridea, infraorder phylogeny, 281, 294, 296
   phylogenetic relationships, 251
Amphionida, 36–37
Amphionidacea, 4
*Amphionides*, 7
Amphipoda, 35
*Amphitrite*, 576
*Amphitrite vigilans*, 576
*Anacalliax*, 337
analytical methods, 372–373
Anaspidacea, 35
ancestors, 9–11, 34–36
Anchistioididae, 296–297
anglerfish, 134

*Angustidontus*, 9
*Angustidontus seriatus*, 7
animals, Podotremata, 418
*Aniptumnus quadridentatus*, 223
Annelida, 108, 150
Anomala
    Brachyura, 417
    embryonic characters, 40
    eye design, 187
    historical developments, 399
    optical design diversity, 192
    specimens used, 418
    synapomorphies, 427
    Thalassinidea, monophyly or paraphyly, 319
Anomalifrontinae, 458, 466
Anomura
    arms, 104
    Bayesian influence, 405, 407
    Brachyura, 417
    carcinization, 410
    Caridea, infraorder phylogeny, 283
    classification implications, 410
    discussion, 407–410
    DNA extraction, 404
    fossil record, 5
    fundamentals, 399–400, 410–411, *ix*
    *Galatheoidea*, 407, 409–410
    guarding time duration, 146
    historical developments, 399
    infraordinal relationships, 91
    materials and methods, 400–404
    maximum parsimony, 405, 407
    molecular data, 401
    molecular taxonomy, 15
    neotenous male mating system, 159
    *Paguroidea*, 407, 409–410
    pair-bonding mating system, 154–155
    PCR, 404
    phylogenetic analysis, 404
    phylogeny-based systematics, 23
    podding mating system, 147
    precopulatory guarding mating system, 144
    results, 405–407
    sequences, 404, 405
    short courtship mating system, 141–142
    synapomorphies, 427
    taxon sampling, 400
    Thalassinidea, 309, 312, 319
*Antarcticheles*
    morphological trends, 378
    *Palaeopentacheles*, 375
    study results, 374
    *Willemoesia*, 377
*Antarcticheles antarcticus*, 369, 372
Anura, 108
apposition eyes, 185–186
*Arachnothelphusa rhadamanthysi*, 522
Archaeobrachyura, 418, 431
*Archangeliphausia*, 9, 10
*Archangeliphausia spinosa*, 4, 6

*Arenaeus*, 568
*Arenaeus cribrarius*, 575
Argonautidae, 134
Aristeidae
    acrosome vesicle, 109
    Caridea, Stenopodidea, and Nephropidae as outgoups, 269
    Caridea and Stenopodidea as outgoup, 267–268
    Caridea as outgoup, 265
    morphological character-based analysis, 261–262, 274
    Penaeoidea as natural group, 272
    protein-coding genes, 91, 94–95
    short courtship mating system, 124
Aristeinae, 261–262
*Armases miersii*, 233
*Artemesia longinaris*, 266, 268–269, 273
*Artemia*
    larval morphology, 222
    mitochondrial genome organization, 47
    scaphognathite origin, 37
Ascothoracica, 102, 105
Asian freshwater crabs *(Gecarcinucidae)*
    biogeography, 526–528
    Bornean group, 522
    Bott's system, similarities, 524, 526
    continental Indian species, 521
    discussion, 523–528
    East-Southeast Asian Group, 522
    frontal triangle, 523–524
    fundamentals, 509–510, 528–529
    historical systematic approaches, 510–511
    Indian-Southeast Asian group, 521
    lineages, 523–524
    Malesian-Australian group, 522–523
    materials and methods, 511–518
    molecular analysis, 511, 518
    monophyly, 523
    morphological analysis, 518, 523–524
    paraphyletic group, continental Indian species, 521
    *Parathelphusa*, 523
    Philippine group, 522
    results, 518, 521–523
    second gonopod, 523–524
    Sri Lankan group, 521
    *Sundathelphusa*, 526
    system of Bott similarities, 524, 526
assumptions, incorrect about ancestors, 9–11
Astacida, 38–41, 418, *see also* Freshwater crayfishes
Astacidae, *see also* Crayfishes
    embryonic characters, 38–41
    fossil calibration, 347
    freshwater crayfishes, 343–344, 348, 350
    short courtship mating system, 124
    Thalassinidea, mitochondrial and nuclear genes, 312
Astacidea
    Caridea, infraorder phylogeny, 283
    freshwater crayfishes, 344
    freshwater crayfishes diversification timing, 348
    infraordinal relationships, 91

microtubular arms, 111
mitochondrial genes, 18
nuclear genes, 19, 22
phylogeny-based systematics, 23
precopulatory guarding mating system, 144
short courtship mating system, 141–142
Thalassinidea, mitochondrial and nuclear genes, 309, 312
Astacoidea
  fossil calibration, 347
  freshwater crayfishes, 343, 345, 348, 350
  marine crab sister group, 497
*Astacoides*, 343
*Astacopsis*, 343
*Astacus*
  freshwater crayfishes, 343–344, 348
  ingroup topology insensitivity, 363
*Astacus licenti*, 351
*Astacus spinirostris*, 351
Asthenognathinae
  exclusions and exceptiona members, 466
  family development, 457
  specimens, 458, 461
*Asthenognathus atlanticus*, 463, 466
*Asthnosoma ijimai*, 236
Atelecyclidae, 535, 543, 546
*Atelecyclus*, 539, 543, 546
*Atelecyclus undecimdentatus*, 539
*Athanopsis*, 252
Atlantic Seamounts, 222
*Atoportunus*, 576
Atyidae
  basal lineages, 299
  fossil record, 6
  infraorder phylogeny, 281, 294, 296
  mitochondrial and nuclear genes, 298
  morphological characters, 252–253
  phylogenetic relationships, 250
  Thalassinidea, mitochondrial and nuclear genes, 312
*Atypopenaeus*, 249
*Austinixa*
  analysis, 485
  constituents of Pinnothereliinae, 468
  limited gene utility, 470
  Pinnothereliinae subfamily, 465
  Pinnotheridae restricted, 467–468
  pinnotherid subfamily definitions, 465
  specimens, 461
*Austinixa aidae*, 465, 468
*Austinixa behreae*, 465, 468
*Austinixa bragantina*, 465, 468
*Austinixa chacei*, 465, 468
*Austinixa cristata*, 465, 468
*Austinixa felipensis*, 465
*Austinixa gorei*, 465, 468
*Austinixa hardyi*, 465, 468
*Austinixa leptodactyla*, 465, 468
*Austinixa patagoniensis*, 465, 468
*Austinotheres*, 465, 469

*Austinotheres angelicus*, 466
Australasian region, 500–501
*Austrohelice*, 485
*Austropotamobius*, 343–344
*Austropotamobius torrentium*, 59
*Austrothelphusa*, 500–501, 522–523
Aves, 108
*Aviticambarus*, 344
Axianassidae
  Gebiidea families, 313, 318
  infraorder composition and relationships, 321
  mitochondrial and nuclear genes, 309–311
  monophyly or paraphyly, 319–320
  monophyly test, 313
  previously applied superfamilies, 321
Axiidae
  Axiidea families, 318
  Callianassidae, mitochondrial genes, 327–328
  infraorder composition and relationships, 321
  mitochondrial and nuclear genes, 309–311
  monophyly or paraphyly, 318–320
  monophyly test, 313
Axiidea
  Callianassidae, mitochondrial genes, 328
  infraorder composition and relationships, 321
  mitochondrial and nuclear genes, 318
  monophyly or paraphyly, 319
  pending analysis, 339
  previously applied superfamilies, 321
Axioidea
  mitochondrial and nuclear genes, 309, 311
  monophyly or paraphyly, 319
  monophyly test, 313
  previously applied superfamilies, 320–321

# B

*Bakousa*, 523, 528
*Balssiathelphusa*
  analysis, 522–523
  biogeography, 528
  Bott system similarities, 524
Banarescu, 494
banded shrimp, 127
*Barbicambarus*, 343
Barbouriidae, 283, 298
Barcode of Life initiative, 59
barnacles, 134
*Barytelphusa*, 521
basal lineages, 299
basally positioned groups, 334
basal parasites, 204
base exchangeability, 72
base frequency, 72
*Bathynectes*, 545
*Bathypalaemonella*, 299
Bathypalaemonellidae, 294, 296
*Bathyplax typhlus*, 563

Bayesian framework, methods, and properties
  Anomura, 405, 407
  Callianassidae, mitochondrial genes, 333
  Caridea, infraorder phylogeny, 292, 294
  *Cronius*, 571–572
  Cryptochiridae, 484–485
  freshwater crayfishes, 345–348, 350–351
  gaps, 71
  Gecarcinucidae, 511, 518
  *Hexapanopeus*, 557
  inference, 76, 82
  infraordinal relationships, 91
  Majoidea, 440, 443, 444
  marine clawed lobster families, 357, 361
  Pinnotheridae, 463
  Portunoidea and Cancroidea, 539
  Thalassinidea, mitochondrial and nuclear genes, 312–313
  tree topology statistical tests, 80
*Belotelson*, 10
Belotelsonidea, 4, 9
Benthesicymidae
  Caridea, Stenopodidea, and Nephropidae as outgoups, 269
  Caridea and Stenopodidea as outgoup, 266, 268
  Caridea as outgoup, 265
  eye design, 187
  morphological character-based analysis, 262
  optical design, 185
  Penaeoidea as natural group, 272
  protein-coding genes, 91, 94–95
  short courtship mating system, 124
  visual system components, 191
*Benthesicymidae*, 273
Benthesicyminae, 262
*Benthesicymus*
  Benthesicymidae as non-natural group, 273
  Caridea, Stenopodidea, and Nephropidae as outgoups, 269
  Caridea and Stenopodidea as outgoup, 268
  Caridea as outgoup, 265
*Benthesicymus bartletti*
  Caridea, Stenopodidea, and Nephropidae as outgoups, 269
  Caridea and Stenopodidea as outgoup, 268
  Caridea as outgoup, 265
*Benthochascon*, 543, 546
*Benthopanope indica*, 234, 236
*Biffarius*
  Axiidea families, 318
  Callianassinae subfamily relationships, 339
  Cheraminae and Callianassinae, 334, 336
  previously applied superfamilies, 321
BioEdit program
  *Cronius*, 571
  Gecarcinucidae, 511
  marine clawed lobster families, 360
  mitochondrial DNA, 49
  Pinnotheridae, 462

  Portunoidea and Cancroidea, 538
  Thalassinidea, mitochondrial and nuclear genes, 312
biogeography
  Asian freshwater crabs, 526–528
  freshwater crabs, 501–504
  parasites, 202–204
biology overview, parasites, 200–201
biparental care hypothesis, 132
blackpoint sculling crab, 568
blanket octopus, 134
*Blepharipoda*, 405
Blepharipodidae, 405
*Bonellia*, 134
*Bopyridae*
  coevolution example, 211, 213
  life cycles, 202
  parasites, 211
  phylogeny, 205–207
Bornean group, 522
Bott's system, similarities, 524, 526
*Bouchardina*, 343
*Brachycarpus*, 251
*Brachynotus*, 485
Brachyura
  Anomura, 400, 417
  arms, 104
  Caridea, infraorder phylogeny, 283
  embryonic characters, 40
  exclusions and exceptiona members, 466
  fossil record, 5
  fundamentals, *ix*
  guarding time duration, 146
  infraordinal relationships, 91
  marine crab sister group, 498
  microtubular arms, 111
  mitochondria, 108
  mitochondrial DNA, 49
  mitochondrial genes, 309, 312–313
  molecular taxonomy, 15
  monophyly or paraphyly, 319
  nuclear genes, 19, 309, 312–313
  optical design diversity, 192
  parasitism in, 203
  phylogeny-based systematics, 23
  podding mating system, 147
  Podotremata paraphyly, 431
  precopulatory guarding mating system, 144
  protein-coding genes, 95
  short courtship mating system, 142
  small groups, mating system, 158
  synapomorphies, 423
  taxonomy developments, 494
  taxon sampling, 400–401
Brachyura Heterotremata, *see* Portunoidea and Cancroidea, new classification proposal
Brachyura Panopeidae, *see Hexapanopeus*
Brachyura Podotremata, monophyletic group
  analysis, 419
  animals, 418
  Cyclodorippoidea, 422–429

discussion, 431–432
Dromiacea, 419, 429
*Eubrachyura*, 422–429
fundamentals, 417–418
Homoloidea, 419, 427, 429
materials and methods, 418–419
microscopy, 418
monophyly, 419–423, 432
phylogenetic relationships, 423–429
Podotremata, 431–432
Raninoidea, 421–422, 424–429
results, 419–429
subtaxa, 419–429
Brachyura Portunidae, *see Cronius* Stimpson, molecular phylogenies
Brachyura Potamoidea, *see* Gecarcinucidae
Branchiopoda, 37
Branchiura, 105
brand-and-bound algorithm, 74
Bremer index, 273–274
Bremer support, 79, 437
Bresiliidae, 283
Bresilioidea, 282
*Brusinia*, 545

## C

Calapidae, 95
*Calappa gallus*, 22
Calappidae, 22
*Calaxiopsis*, 318
*Calcinus*, 131
Calianassoidea, 321
*Callianassa*
   Cheraminae and Callianassinae, 336
   mitochondrial and nuclear genes, 311
   mitochondrial genes, 327
   subfamily relationships, 338–339
*Callianassa subterranea*
   Cheraminae and Callianassinae, 336
   mitochondrial genes, 328
   subfamily relationships, 339
Callianassidae
   Axiidea families, 318
   infraorder composition and relationships, 321–322
   mitochondrial and nuclear genes, 309–311
   monophyly or paraphyly, 318–320
   monophyly test, 313
   previously applied superfamilies, 321
Callianassidae, mitochondrial genes
   basally positioned groups, 334
   Callichirinae, 334, 337–338
   Cheraminae, 334, 336
   Ctenochelidae, 334
   dataset description, 333
   discussion, 336–339
   DNA extraction, 328
   fundamentals, 327–328
   methods, 328–333

model selection, 333
PCR, 328
pending analyses, 339
phylogenetic analyses, 328, 333
relationships, 336–339
results, 333–336
sequencing, 328
specimens, 328
tree topologies, 334
Callianassidea, 320–321
Callianassinae, 310, 337
Callianassoidea
   mitochondrial and nuclear genes, 309, 311
   monophyly test, 313
   previously applied superfamilies, 320
*Callianidea*, 311, 318
Callianideidae
   Axiidea families, 318
   infraorder composition and relationships, 321–322
   mitochondrial and nuclear genes, 309, 311
   monophyly or paraphyly, 319–320
   previously applied superfamilies, 321
*Callianopsis*, 337
*Calliax*, 327, 337
*Calliaxina*, 327, 334, 336–337
*Calliax punica*, 336
Callichirinae, 327
*Callichirinae*, 334, 337–338
*Callichirus*
   Callichirinae, 334
   monophyly or paraphyly, 319
   previously applied superfamilies, 321
   subfamily relationships, 337–338
*Callichirus garthi*, 337
*Callichirus islagrande*, 337
*Callichirus seilacheri*, 337
Callinassidae, 311
*Callinectes*, 568, 576
*Callinectes danae*, 575
*Callinectes sapidus*
   acrosome vesicle, 105
   sequencing, 538
   taxonomic revisions, 575
Calocarididae
   Axiidea families, 318
   infraorder composition and relationships, 321
   mitochondrial and nuclear genes, 309, 311
   monophyly or paraphyly, 320
   monophyly test, 313
*Calocaris*, 318
*Calyptraeotheres*
   constituents of Pinnotherinae, 469–470
   pinnotherid subfamily definitions, 465
   Pinnotherinae subfamily, 466
Cambarellinae, 343
*Cambarellus*, 343
Cambaridae, *see also* Crayfishes
   embryonic characters, 38–41
   fossil calibration, 347

freshwater crayfishes, 343–344, 348, 350
   mitochondrial and nuclear genes, 312
   short courtship mating system, 124
Cambarinae, 343–344
*Cambariodes*, 348
*Cambaroides*, 344
*Cambaroides japonicus*, 105, 350
*Cambarus*, 343, 363
Camin-Sokal Parsimony algorithm, 78
*Camposcia*, 438
Camptandridae, 484
Camptandriidae, 486
Campylonotidae, 283
*Cancer*
   acrosome vesicle, 105
   analysis, 539, 543
   nuclear genes, 19
   synapomorphies, 426
   taxonomy development, 533
*Cancer gladiator*, 576
*Cancer pagurus*, 543
*Cancer pelagicus*, 576
*Cancer pinnotheres*, 458
Cancridae
   analysis, 546
   mitochondrial and nuclear genes, 312
   precopulatory guarding mating system, 125
   sperm plugs, 147
Cancroidea, 543, 546, *see* Portunoidea and Cancroidea, new classification proposal
cape lobster, 365
Caphyrinae, 535, 543–544, 546
Carcineretidae, 535
Carcinidae, 545–546
Carcininae
   analysis, 543–545
   considerations and future work, 577
   taxonomy development, 535
carcinization, 410
*Carcinus*, 107, 545, 569
*Carcinus maenas*, 105, 108
*Cardisoma*, 49, 52, 56–58
*Cardisoma armatum*, 49, 52
*Cardisoma crassum*, 49, 53
*Cardisoma guanhumi*, 49, 52–53, 57
*Cardus*
   analysis, 374
   higher taxa diagnosis, 382
   morphological trends, 378–379
   Polychelidae, 376–377
   Polychelidan lobsters, 369, 372
*Cardus crucifer*, 375
Caridea
   Dendrobranchiata, monophyletic group, 272
   evolution and radiation, 245, 249–252
   fossil record, 6
   fossils, 254
   fundamentals, *ix*
   infraordinal relationships, 91
   large groups, mating system, 158
   mitochondrial and nuclear genes, 309, 312–313
   molecular taxonomy, 15
   monophyly or paraphyly, 320
   morphological character-based analysis, 262, 264–270
   morphological characters, 252
   phylogenetic relationships, 249
   phylogeny-based systematics, 23
   position, shrimp-like decapods, 247–248
   precopulatory guarding mating system, 144
   short courtship mating system, 141
   small groups, mating system, 157
Caridea, mitochondrial and nuclear genes
   *Alpheoidea*, 298
   Atyidae, 298
   basal lineages, 299
   *Crangonidae*, 299
   discussion, 296–300
   DNA extraction, 283, 291
   evolutionary history, 282–283
   fundamentals, 281, 300
   hypotheses testing, 300
   ingroup taxa, 283
   materials and methods, 283–292
   monophyly, 294, 296
   outgroup selection, 283
   *Palaemonoidea*, 297–298
   paraphyly, 294, 296
   PCR, 283, 291
   phylogenetic analyses, 292
   polyphyly, 294, 296
   Procaridoidea, 296–297
   Processidae, 299
   results, 294–296
   sequencing, 283, 291
   Thalassocarididae, 299
   Xiphocarididae, 298
Caridea mitochondrial and nuclear genes
   *Alpheoidea*, 298
   Atyidae, 298
   basal lineages, 299
   *Crangonidae*, 299
   discussion, 296–300
   DNA extraction, 283, 291
   evolutionary history, 282–283
   fundamentals, 281, 300
   hypotheses testing, 300
   ingroup taxa, 283
   materials and methods, 283–292
   monophyly, 294, 296
   outgroup selection, 283
   *Palaemonoidea*, 297–298
   paraphyly, 294, 296
   PCR, 283, 291
   phylogenetic analyses, 292
   polyphyly, 294, 296
   Procaridoidea, 296–297
   Processidae, 299
   results, 294–296
   sequencing, 283, 291

Thalassocarididae, 299
Xiphocarididae, 298
caridean shrimps, 123, 127
*Caridina*, 251
Caridoida, 34
*Carinus*, 545
Carpiliidae, 155, 552
Carpilioidea, 563
*Carpillus*, 538
Carpopenaeidae, 254
*Carupa*, 543, 546
*Carupa tenuipes*, 546
Carupidae, 535
Carupinae, 535, 543–544, 546
Catoptridae, 546
*Catoptrus*, 546
Catroptrinae, 535
*Ceylongthelphusa*, 521, 524
Ceylongthelphusinae, 510, 524
*Chaceon*, 52–54, 57, 546
*Chaceon affinis*, 49
*Chaceon fenneri*, 49, 54
*Chaceon granulatus*, 49
*Chaceon quinquedens*, 49, 54
*Chaetopterus*, 127
character homology, 70–71
*Charybdella tumidula*, 575
*Charybdis*, 567–569, 577
*Charybdis helleri*
 heterochrony, 226
 phylogenetics, 234, 236
 polarity, setal characters, 229–230, 232
 taxonomic revisions, 575
 taxonomy developments, 568
Chasmocarcinidae, 552, 563
*Chasmocarcinus*, 563
Cheiragonidae
 analysis, 543, 546
 precopulatory guarding mating system, 125
 sperm plugs, 147
 taxonomy development, 535
Cheiragonoidea, 535
*Cheiragonus*, 535
Cheraminae, 334, 336, 339
*Cheramus*, 339
*Cherax*, 343–344
*Cherax destructor*, 10, 40–41
*Chermus*, 334
Chilenophoberidae, 357
*Chimaerastacus pacifluvialis*, 350
Chimerastacidae, 357
*Chionoecetes*, 444
*Chionoecetes bairdi*, 126
*Chionoecetes japonicus*, 439
Chirostylidae, 399–400, 407
*Chlorodiella bidentata*, 223
*Chlorodiella nigra*
 phylogenetics, 234–236
 polarity, setal characters, 229–230, 232
 zoeal similarity, 223

Chlorodiellinae, 223, 234–235
Chlorodiinae, 223
*Chlorotocoides*, 299
Chordata, 149
CIPRES analysis
 Callianassidae, mitochondrial genes, 333
 Caridea, infraorder phylogeny, 291
 *Hexapanopeus*, 557
 phylogeny-based systematics, 23
 Pinnotheridae, 463
 Thalassinidea, mitochondrial and nuclear genes, 312
circular ommatidial facets, 186
Cirripedia, 102, 105
classification proposal, *Portunoidea* and *Cancroidea*
 discussion, 544–546
 families of, 534–535
 fundamentals, 533–534
 materials and methods, 535, 538–539
 results, 539, 542–543
 subfamilies of Portunidae, 535
cleavage pattern, 32–34
*Clibanarius*, 110
ClustalW program, 23
*Clypeasterophilus*
 constituents of Pinnotherinae, 469–470
 limited gene utility, 470
 pinnotherid subfamily definitions, 465
 Pinnotherinae subfamily, 466
*Clypeasterophilus rugatus*, 458
*Clypeasterophilus stebbingi*, 458, 470
Cnidaria, 148, 150
codon-based models, 73
*Coenobita*, 124
Coenobitidae, 399, 405
*Coenophthalmus*, 545
coevolution, 198–201, 207–213
COI gene, 58–59, 253, 440
COII gene, 319
*Coleia*, 374–375
*Coleia longipes*, 372, 378
Coleiidae
 higher taxa diagnosis, 380
 morphological trends, 378
 *Palaeopentacheles*, 375–376
 Polychelidan lobsters, 369
 study results, 374
 systematics, 379–380
 terminal taxa, 372
collection of larvae, 222
colonization, 502–503
combined mitochondrial analyses, 463
components as phylogenetic characters, 189–192
continental Indian species, 521
coral crabs, 127
coral gall crabs, 129
*Corallianassa*, 327, 334, 338
*Coralliocaris*, 297
*Corystes cassivelaunus*, 125
Corystidae, 125, 535, 543, 546
Corystoidea, 535

cospeciation, 205
crabs, *see also* Majoidea
    acrosome vesicle, 105, 110
    eye design, 187
    life history studies, 159
    mating systems study history, 123
    maxillipeds, 35
    mitochondria, 108
    mitochondrial DNA, 48
    neotenous male mating system, 130
    nucleus, 110
    opsin sequences, 188
    optical design, 185
    pair-bonding mating system, 155–156
    podding mating system, 126
    precopulatory guarding mating system, 125
    pseudogenes, 57
    reproductive swarm mating system, 129
    short courtship mating system, 124
    vision, 189
    visiting type mating system, 129, 134
    waving display mating system, 128, 129, 133
    zoeal similarity, 223
crabs, Asian freshwater
    biogeography, 526–528
    Bornean group, 522
    Bott's system, similarities, 524, 526
    continental Indian species, 521
    discussion, 523–528
    East-Southeast Asian Group, 522
    frontal triangle, 523–524
    fundamentals, 509–510, 528–529
    historical systematic approaches, 510–511
    Indian-Southeast Asian group, 521
    lineages, 523–524
    Malesian-Australian group, 522–523
    materials and methods, 511–518
    molecular analysis, 511, 518
    monophyly, 523
    morphological analysis, 518, 523–524
    paraphyletic group, continental Indian species, 521
    *Parathelphusa*, 523
    Philippine group, 522
    results, 518, 521–523
    second gonopod, 523–524
    Sri Lankan group, 521
    *Sundathelphusa*, 526
    system of Bott similarities, 524, 526
crabs, freshwater
    Afrotropical region, 499
    Australasian region, 500–501
    biogeography, 501–504
    colonization, 502–503
    distribution, 498–501
    diversity, 491
    fundamentals, 491
    gonopod 1 character evolution, 496
    higher taxonomy, 491–498
    mandibular palp character evolution, 494–496
    marine crab sister group, 497–498
    Neotropical region, 498–499
    Oriental region, 500
    origin theories, 503–504
    Palaearctic region, 499–500
    phylogeny, 491–492, 494–498
    Potamoidea, 496, 497–498
    potamoid type, 494–496
    Trichodactylidae, 497
crabs, gall
    discussion, 486–487
    fundamentals, 475, 482
    materials and methods, 482, 484
    results, 484–486
crabs, hermit
    acrosome vesicle, 105, 110
    ancestor assumptions, 10
    classification, 399
    eye design, 187
    mitochondrial DNA, 48
    optical design, 185
    pair-bonding mating system, 127
    precopulatory guarding mating system, 125, 131
    short courtship mating system, 124
Crangonidae
    basal lineages, 299
    evolutionary history of Caridea, 282
    infraorder phylogeny, 294, 296
    mitochondrial and nuclear genes, 299
    testing morphological hypotheses, 300
crayfishes
    acrosome vesicle, 105
    ancestor assumptions, 10
    embryonic characters, 38–41
    nuclear genes, 22
    opsin sequences, 188
    optical design, 185
    precopulatory guarding mating system, 131
    primer optimization, 59
    scaphognathite origin, 37
    short courtship mating system, 124, 131
    visual system components, 191
crayfishes, freshwater
    discussion, 350–351
    divergence time estimation, 347–348, 350
    DNA extraction, 345
    fossil calibrations, 346–347
    fundamentals, 343–345, 351
    interpreting results, 350–351
    methods, 345–348
    PCR, 345
    phylogenetic analyses, 345–346, 348
    phylogeny, 350
    results, 348, 350–351
    sequencing, 345
    taxon sampling, 345
crevice sculling crab, 568
*Crockerinus*, 344
*Cronius* Stimpson, molecular phylogenies
    *C. edwardsii*, 567, 569
    *C. millerii*, 569, 574

*C. ruber*, 567, 569, 572, 574–575
*C. tumidulus*, 572, 574–576
discussion, 574–577
DNA analysis, 569, 571
fundamentals, 567–569
future work, 577
materials and methods, 569–571
molecular phylogeny, 572, 574
paraphyly, 574–576
PCR, 569, 571
phylogenetic analysis, 571
results, 571–574
sample collection, 569
sequencing, 569, 571
subfamilies, 577
taxonomic account, 571–572
Crossotonotidae, 487
*Crossotonotus*, 487
Crustacea, 150
*Cryphiops*, 251, 297
Cryptochiridae
  discussion, 486–487
  fundamentals, 475, 482
  materials and methods, 482, 484
  results, 484–486
  visiting type mating system, 129
Cryptochiroidea, 486
*Ctenocheles*, 322, 337
Ctenochelidae
  Axiidea families, 318
  family relationships, 336–337
  infraorder composition and relationships, 321–322
  mitochondrial and nuclear genes, 309–311
  mitochondrial genes, 327–328, 334
  monophyly or paraphyly, 320
  monophyly test, 313
Cumacea, 35
*Currothelphusa*, 526
*Currothelphusa asserpes*, 522
*Cycleryon proinquus*, 372
*Cyclodius monticulosus*, 223
Cyclodorippidae
  Cyclodorippidea monophyly, 422
  nuclear genes, 22
  specimens used, 418
  synapomorphies, 423–424
Cyclodorippoidea
  Brachyura, 418
  Podotremata, 422–429, 431–432
  synapomorphies, 424–429
*Cylindrotelphusa*, 521
Cymonomidae, 418, 422
*Cymonomus aequilonius*, 418
*Cyrtograpsus*, 486

# D

*Dagnaudus petterdi*, 418
*Dardanus*, 185, 187
*Dardanus megistos*, 187, 191
data matrix, 388

data partition comparisons, 440, 443, 444
dataset description, 333
*Dawsonius*
  Axiidea families, 318
  family relationships, 337
  infraorder composition and relationships, 322
  mitochondrial genes, 327
Decapoda, origins
  ancestors, incorrect assumptions, 9–11
  fundamentals, 3, 11
  issues, 3–11
  paleobiogeography, 9
  paleoecology, 9
  Paleozoic fossils, 4–7, 9
  sister group, 4
Decapod phylogenetics, visual system characters of shrimp
  components as phylogenetic characters, 189–192
  evolutionary enigma, 187
  fundamentals, 183, 192
  molecular aspects, 187–189
  morphology, 183–186
  overview, 183–189
Decapod phylogeny, protein-coding genes
  Brachyura, 95
  fundamentals, 89, 97
  infraordinal relationships, 91
  molecular phylogeny, 90–91
  Palinuridae, 95–97
  Penaeoidea, 91, 94–95
  phylogenetic reconstruction, genera/species, 95–97
  super family/family level phylogeny studies, 91–95
Decapod phylogeny, spermatozoal morphology
  acrosome vesicle, 105, 108–110
  aflagellate sperm cells, 105
  fundamentals, 101–105, 112–113
  immotile sperm cells, 105
  microtubular arms, 111–112
  mitochondria, 108
  nucleus, 110–111
  sperm nuclear proteins, 106–107
  sperm nucleus, 107–108
  sperm uniqueness, 104–108
Decapods
  Ancestor, 3
  Body Plan, 3
  Fossils, 3
  natant, *ix*
  origin, 3
  Paleozoic, 3
  reptant, *ix*
  Sister Group, 3–4
  stem species, 32–34
Decapods, development, genes, and evolution
  ancestral development, 34–36
  cleavage pattern, 32–34
  decapod stem species, 32–34
  embryonic characters, 38–41
  freshwater crayfish monophyly, 38–41
  fundamentals, 31–32, 41

gastrulation, 32–34
scaphognathite origin, 37–38
Decapods, mitochondrial DNA and phylogenies
discussion, 54, 56–59
fundamentals, 47–49
material and methods, 49
primer optimization, 58–59
pseudogenes, 56–58
results, 52–54
Decapods, phylogenetics and molecular evolution
fundamentals, 15
genes and their diversity, 17–23
genetic marker development, 16–17
mitochondrial genes, 17–19
molecular taxonomy, 15
nuclear genes, 19, 22–23
phylogeny based systematics, 23
12S, 16S, COI, 17–19
decay index, 79
*Deckenia*, 495
Deckeniidae, 492, 494, 511
Deckeniinae, 523
decorator crabs, 437
Delta program, 263
Dendrobranchiata
Caridea, Stenopodidea, and Nephropidae as outgoups, 269–270
Caridea and Stenopodidea as outgoup, 266
Caridea as outgoup, 265
choice of outgroup and analyses, 270
early development, 32–34
eggs, 32
evolution and radiation, 245, 248–249
eye design, 187
fossil record, 6
fossils, 254
fundamentals, *ix*
maxillipeds, 35
molecular taxonomy, 15
morphological characters, 252
nuclear genes, 19, 22
phylogenetic relationships, 248–249
phylogeny-based systematics, 23
position, shrimp-like decapods, 247
short courtship mating system, 141–143
transformation types, 233
Dendrobranchiata, morphological character-based analysis
*Aristeidae*, 274
*Benthesicymidae*, 273
Caridea, 264–270
description of characters, 264
discussion, 270–274
fundamentals, 261–262
materials and methods, 262–263
monophyletic group, 271–272
morphological characters used, 276–278
natural group, 272–273
*Nephropidae*, 268–270
non-natural group, 273

optimization of characters, 264
outgroups, 264–271
*Penaeoidea*, 272–273
results, 264–270
*Sergestoidea*, 272
*Solenoceridae*, 273
species studied, 275
status of, 274
Stenopodidea, 266–270
dense perforatorial ring, 110
description of characters, 264
Desmocarididae, 296–297
development, Decapods
ancestral development, 34–36
cleavage pattern, 32–34
decapod stem species, 32–34
embryonic characters, 38–41
freshwater crayfish monophyly, 38–41
fundamentals, 31–32, 41
gastrulation, 32–34
scaphognathite origin, 37–38
*Diadema savignyi*, 236
*Diadema setosum*, 236
diagnosis of taxa, 379–386
DIALIGN-T program, 23
*Dicranodromia karubar*, 418
dimensions, acrosome vesicle, 110
*Diogenes*, 125, 131
Diogenidae
Anomura phylogeny, 405, 407
classification, 399
optical design, 185
precopulatory guarding mating system, 125, 131
direct optimization analysis, 443, 444–445
Disciadidae, 252, 296
*Discias*, 299
discussions
Anomura, 407–410
Asian freshwater crabs, 523–528
Callianassidae mitochondrial genes, 336–339
Caridea mitochondrial and nuclear genes, 296–300
freshwater crayfishes diversification timing, 350–351
gall crabs, 486–487
*Hexapanopeus*, 560–563
Majoidea, 445–448
marine clawed lobster families, 362–365
mitochondrial DNA, 54, 56–59
morphological character-based analysis, 270–274
Pinnotheridae, molecular genetic re-examination, 466–471
Podotremata, 431–432
Polychelidan lobsters, 375–379
Portunoidea and Cancroidea new classification proposal, 544–546
Thalassinidea mitochondrial and nuclear genes, 318–322
*Dissodactylus*
constituents of Pinnotherinae, 469–470
limited gene utility, 470

pinnotherid subfamily definitions, 465
Pinnotherinae subfamily, 466
distance methods, 77–78, 82
*Distocambarus*, 343
distribution, freshwater crabs, 498–501
divergence time estimation, 347–348, 350
diversification timing, freshwater crayfishes, *see also* Crayfishes
  discussion, 350–351
  divergence time estimation, 347–348, 350
  DNA extraction, 345
  fossil calibrations, 346–347
  fundamentals, 343–345, 351
  interpreting results, 350–351
  methods, 345–348
  PCR, 345
  phylogenetic analyses, 345–346, 348
  phylogeny, 350
  results, 348, 350–351
  sequencing, 345
  short courtship mating system, 124
  taxon sampling, 345
diversity, freshwater crabs, 491
DNA amplification, 360, 511
DNA extraction
  Anomura, 404
  Callianassidae mitochondrial genes, 328
  Caridea mitochondrial and nuclear genes, 283, 291
  freshwater crayfishes diversification timing, 345
  Gecarcinucidae, 511
  *Hexapanopeus*, 556
  marine clawed lobster families, 360
  mitochondrial DNA, 49
  Pinnotheridae, molecular genetic re-examination, 461–462
  Portunoidea and Cancroidea, 535
  Thalassinidea mitochondrial and nuclear genes, 312
Dollo parsimony algorithm, 78
Dorippidae, 418
Dotillidae, 128, 466
*Dottila*, 133
Dromiacea
  Brachyura, 417, 418, 432
  fossil record, 5
  optical design diversity, 192
  Podotremata paraphyly, 431–432
  synapomorphies, 423–427, 429
  Thalassinidea, monophyly or paraphyly, 319
*Dromiacea*, 419, 429
*Dromidiopsis*, 111
Dromiidae, 418–419
*Drosophila*, 189
*Drosophila melanogaster*, 22
dwarf male mating system, 121, 130, 134, 159
*Dynomene pilumnoides*, 418
Dynomenidae, 418–419
*Dyspanopeus sayi*, 557, 562

# E

East-Southeast Asian Group, 522
Echinodermata, 149, 151

*Echinoecus pentagonus*, 223, 234, 236
*Echinolatus*, 545
*Echinothrix calamarix*, 236
*Echinothrix diadema*, 236
Echiura, 151
ectoeloblasts, 31
Eiconaxiidae, 309–311, 321
*Eiconaxius*, 310
EMBL molecular database, 49, 538
embryonic characters, 38–41
*Emerita*, 134, 405
encounter rate competition, 123
*Engaeus*, 343–344
*Engaewa*, 344
Enoplometopidae
  infraordinal relationships, 91
  ingroup topology insensitivity, 363
  marine clawed lobster families, 357
  taxon sampling, 358–359
  Thalassinidea, mitochondrial and nuclear genes, 312
*Enoplometopus*, 358, 363
*Enoplometopus crosnieri*, 359
*Enoplometopus daumi*, 359
*Enoplometopus debelius*, 359
eocarid, 4
eocaridacea, 4
*Eocaris oervigi*, 4
Epialtidae, 438–439, 444, 447
Epialtinae, 438, 447
*Epilobocera sinuatifrons*, 511
*Eplumula australiensis*, 418
*Eriocheir*, 486
*Eriocheir japonicus*, 105
*Eriocheir sinensis*, 418
*Eriphia scabricula*, 230, 232
Eriphiidae, 552
Eriphiliidae, 147
Eriphioidea, 563
Erymidae, 357
*Eryon arctiformis*, 372
Eryonidae, 369, 372, 380
*Eryonidae*, 380
Essoidea, 10
*Euastacus*, 17, 22, 343–344
*Euastacus robertsi*, 22
Eubrachyura
  Brachyura, 417
  Cryptochiridae developments, 482
  Podotremata, 422–429
  Podotremata paraphyly, 431–432
  protein-coding genes, 95
  synapomorphies, 423–429
Eucalliacinae, 327, 334, 336
*Eucalliax*, 327, 336–337
Eucardia, 4, 283
eucarids, 6–7, 9
*Euchirograpsus*, 486–487
Eucratopsinae, 562
Eugonatonotidae, 296
Eumalacostraca, 11, 185
Eumedonidae, 223, 234
*Eumedonus niger*, 236

*Eunephrops*
  marine clawed lobster families, 357
  morphological convergence, 364
  phylogenetic analysis, 362
  Thaumastochelidae status, 364
Euphausiacea
  Caridea, infraorder phylogeny, 283
  maxillipeds, 35
  refracting superposition eyes, 185
  sister group to Decapoda, 4
*Euphylax*, 569
Eurynolambrinae, 438
*Eurypanopeus*, 562
*Eurypanopeus abbreviatus*, 557
*Eurypanopeus depressus*, 557, 562
*Eurypanopeus dissimilis*, 557, 562
*Eurypanopeus planissimus*, 557
Euryplacidae, 95, 552, 563
Euryrhynchidae, 296–297
eusociality mating system, 121–122, 128, 133, 157
evolution
  Caridea mitochondrial and nuclear genes, 282–283
  phylogenetic inference, molecular data, 72–73
  shrimp, visual system characters, 187
evolution, Decapods
  ancestral development, 34–36
  cleavage pattern, 32–34
  decapod stem species, 32–34
  embryonic characters, 38–41
  freshwater crayfish monophyly, 38–41
  fundamentals, 31–32, 41
  gastrulation, 32–34
  scaphognathite origin, 37–38
evolution and radiation, shrimp-like decapods
  Caridea, 249–252
  Dendrobranchiata, 248–249
  fossils, 254
  fundamentals, 245, 254–255
  molecular markers, 253
  morphological characters, 252–253
  phylogenetic relationships, 248–252
  position within decapoda, 247–248
  Stenopodidea, 248
exceptional members, 466–467
exclusions, 466–467
exhaustive search, 74
extended mate guarding hypothesis, 132
eye designs, *see* Visual system characters, shrimp

# F

*Fabia*
  constituents of Pinnotherinae, 470
  Pinnotheridae restricted, 467–468
  pinnotherid subfamily definitions, 465
  Pinnotherinae subfamily, 466
*Fallicambarus*, 343
false spider crabs, 498
families, 534–535
family development, 438

*Farfantepenaeus*, 124
*Farfantepenaeus paulensis*
  Caridea, Stenopodidea, and Nephropidae as outgoups, 269–270
  Caridea and Stenopodidea as outgoup, 268
  Caridea as outgoup, 266
  Penaeidae as non-natural group, 273
*Faxonella*, 343
Felsenstein zone, 82
female-centered competition, 123
*Fenneralpheus*, 298
*Fenneropenaeus*, 124
*Fenneropenaeus chinensis*, 253
fiddler crabs, 123, 128, 129
Fitch algorithm, 78
Folmer region, 59
football octopus, 134
fossils
  Decapoda, origins, 4–7, 9
  evolution and radiation, 254
  freshwater crayfishes diversification timing, 346–347, 350–351
  oldest known decapod, 34
Free Living mating system, 151
freshwater crabs
  Afrotropical region, 499
  Australasian region, 500–501
  biogeography, 501–504
  colonization, 502–503
  distribution, 498–501
  diversity, 491
  fundamentals, 491
  gonopod 1 character evolution, 496
  higher taxonomy, 491–498
  mandibular palp character evolution, 494–496
  marine crab sister group, 497–498
  Neotropical region, 498–499
  Oriental region, 500
  origin theories, 503–504
  Palaearctic region, 499–500
  phylogeny, 491–492, 494–498
  Potamoidea, 496, 497–498
  potamoid type, 494–496
  Trichodactylidae, 497
freshwater crabs, Asian
  biogeography, 526–528
  Bornean group, 522
  Bott's system, similarities, 524, 526
  continental Indian species, 521
  discussion, 523–528
  East-Southeast Asian Group, 522
  frontal triangle, 523–524
  fundamentals, 509–510, 528–529
  historical systematic approaches, 510–511
  Indian-Southeast Asian group, 521
  lineages, 523–524
  Malesian-Australian group, 522–523
  materials and methods, 511–518
  molecular analysis, 511, 518
  monophyly, 523

morphological analysis, 518, 523–524
paraphyletic group, continental Indian species, 521
*Parathelphusa*, 523
Philippine group, 522
results, 518, 521–523
second gonopod, 523–524
Sri Lankan group, 521
*Sundathelphusa*, 526
system of Bott similarities, 524, 526
freshwater crayfishes
  diversification timing, 343–345, 351
  monophyly, 38–41
  precopulatory guarding mating system, 131
  short courtship mating system, 131
freshwater crayfishes, timing of the diversification, *see also* Crayfishes
  discussion, 350–351
  divergence time estimation, 347–348, 350
  DNA extraction, 345
  fossil calibrations, 346–347
  fundamentals, 343–345, 351
  interpreting results, 350–351
  methods, 345–348
  PCR, 345
  phylogenetic analyses, 345–346, 348
  phylogeny, 350
  results, 348, 350–351
  sequencing, 345
  short courtship mating system, 124
  taxon sampling, 345
*Frevillea barbata*, 563
frontal triangle, 523–524
*Fungicola*, 475
fusion events, 81
future perspectives, 365–366, 471

## G

*Gaetice*, 486
*Galacantha*, 407
*Galathea*, 426, 429
Galatheacarididae, 283
*Galathea strigosa*, 418
Galatheidae, 312, 407, 497
Galatheiinae, 407
Galatheoidea
  Anomura, 405, 407, 409–410
  classification, 399
  eye design, 187
  historical developments, 399
  marine crab sister group, 497
  small groups, mating system, 157
  taxon sampling, 400
gall crabs *(Cryptochiridae)*, 16S rRNA sequence data, *see also* Freshwater crabs
  discussion, 486–487
  fundamentals, 475, 482
  materials and methods, 482, 484
  results, 484–486

gaps, dealing with, 70–71
GARLI, 482, 484
*Garthiope barbadensis*, 563
*Garthiope spinipes*, 563
gastrulation, 32–34
GBlocks program
  Caridea, infraorder phylogeny, 292, 294
  freshwater crayfishes, 345
  *Hexapanopeus*, 556, 557
  Majoidea, 440, 443
  phylogeny-based systematics, 23
  sequence alignment, 70
Gebiidea
  infraorder composition and relationships, 321
  mitochondrial and nuclear genes, 309
  monophyly or paraphyly, 319–320
  previously applied superfamilies, 321
*Gebiidea* families, 313, 318
Gecarcinidae
  analysis, 486
  mitochondrial DNA, 49
  short courtship mating system, 124
  taxon selection, 484
Gecarcinucidae
  Australasian region, 500
  biogeography, 526–528
  Bornean group, 522
  Bott's system, similarities, 524, 526
  continental Indian species, 521
  discussion, 523–528
  East-Southeast Asian Group, 522
  frontal triangle, 523–524
  fundamentals, 509–510, 528–529
  gonopod 1 characters, 496
  historical systematic approaches, 510–511
  Indian-Southeast Asian group, 521
  lineages, 523–524
  Malesian-Australian group, 522–523
  manibular palp characters, 494–495
  materials and methods, 511–518
  molecular analysis, 511, 518
  monophyly, 523
  morphological analysis, 518, 523–524
  Oriental region, 500
  paraphyletic group, continental Indian species, 521
  *Parathelphusa*, 523
  Philippine group, 522
  results, 518, 521–523
  second gonopod, 523–524
  Sri Lankan group, 521
  *Sundathelphusa*, 526
  system of Bott similarities, 524, 526
  taxonomy developments, 492, 494
  theories on origins, 503–504
Gecarcinucinae, 510, 518
Gecarcinucoidea, 494, 510
*Gecarcinucus*, 510
*Gecarcinucus jacquemonti*, 521
*Geelvinkia*, 522–523

*Geithusa pulchra*, 522, 528
General Time Reversible (GTR) model
    Callianassidae, mitochondrial genes, 333
    Caridea, infraorder phylogeny, 291, 294
    Cryptochiridae, 482
    freshwater crayfishes, 345
    Portunoidea and Cancroidea, 539
    Thalassinidea, mitochondrial and nuclear genes, 312
genes
    ancestral development, 34–36
    cleavage pattern, 32–34
    decapod stem species, 32–34
    embryonic characters, 38–41
    freshwater crayfish monophyly, 38–41
    fundamentals, 31–32, 41
    gastrulation, 32–34
    mitochondrial genes, 17–19
    nuclear genes, 19, 22–23
    12S, 16S, COI, 17–19
    scaphognathite origin, 37–38
genetic differences and saturation, 71
genetic marker development, 16–17
*Geocharax*, 343–344
*Geograpsus*, 484, 486
*Geosesarma*, 502
*Geosesarma notophorum*, 498
*Geothelphusa*, 538
*Geothelphusa dehaani*, 69, 538
*Geryon*, 52–57, 546
Geryonidae
    analysis, 543–544, 546
    sperm plugs, 147
    taxonomy development, 534
*Geryon longipes*, 49, 53, 57
*Geryon trispinosus*, 49
ghost crab, 57
ghost shrimps, 328
*Gilvossius*, 334
*Gilvossius setimanus*, 336
*Glabropilumnus edamensis*, 234
Glyphocrangonidae
    Caridea, infraorder phylogeny, 294, 296
    evolutionary history of Caridea, 282
    morphological characters, 252
    testing morphological hypotheses, 300
Glyptograpsidae, 485–487
*Glyptograpsus*, 485
*Glyptoplax*, 560
*Glyptoplax pugnax*, 560
*Glyptoplax smithii*, 560
*Glypturus*, 327, 334, 338
*Glypturus acanthochirus*, 338
*Gnathophausia*, 7
Gnathophyllidae
    basal lineages, 299
    Caridea, infraorder phylogeny, 281, 296
    pair-bonding mating system, 153
    phylogenetic relationships, 250–251
    superfamily Palaemonoidea, 297
*Gnathophylloides*, 297

*Gnathophyllum*, 297
Goneplacidae
    nuclear genes, 22
    protein-coding genes, 95
    taxonomy development, 534–535
    taxon sampling, 552
Goneplacoidea, 563
*Gonioinfradens*, 577
*Goniopsis*, 484, 486
gonopod 1 character evolution, 496
goodness of fit tests, 80
*Gourretia*
    Axiidea families, 318
    family relationships, 337
    infraorder composition and relationships, 322
    mitochondrial genes, 327
Gourretiidae, 318
Gourretiinae, 310
*Gramastacus*, 343–344
Grapsidae
    marine crab sister group, 497–498
    taxon selection, 484
    waving display mating system, 128
Grapsoidea
    analysis, 486–487
    exclusions and exceptiona members, 466
    homoplasy, 233
    pair-bonding mating system, 156
    short courtship mating system, 124
    taxonomy development, 534
    taxon selection, 484
*Grapsus*, 484, 486
greedy approach, 70
*Grynaminna*, 334, 338
*Grynaminna tamakii*, 338
guarding time, mating systems, 146
*Gubernatoriana*, 521

# H

*Haliphron atlanticus*, 134
*Halocaridina rubra*, 253
*Hapalocarcinus*, 484, 486
*Hapalocarcinus marsupialis*
    analysis, 484
    Cryptochiridae developments, 482
    developments, 475
    specimen, 482
    visiting type mating system, 129
*Hapalogaster dentata*, 125
harlequin shrimp, 127
*Harpilius*, 251
*Harrovia albolineata*, 223, 234, 236
heaps, *see* Podding mating system
*Hellenus*, 576
*Helograpsus*, 486
*Hemigrapsus*, 426, 486
*Hemigrapsus crenulatus*, 418
*Hemigrapsus sanguinensus*, 189
*Herbstia*, 444, 447
*Herbstia condyliata*, 439

hermit crabs
    acrosome vesicle, 105, 110
    ancestor assumptions, 10
    classification, 399
    eye design, 187
    mitochondrial DNA, 48
    optical design, 185
    pair-bonding mating system, 127
    precopulatory guarding mating system, 125, 131
    short courtship mating system, 124
Heterocarpodoidea, 282
heterochrony, 226–227
*Heterocrypta occidentalis*, 440
*Heterothelphusa*, 522
Heterotremata, see Portunoidea and Cancroidea, new classification proposal
    analysis, 543
    Cryptochiridae developments, 482
    Eubrachyura monophyly, 422
    protein-coding genes, 95
heuristic searches, 74
hexagonal ommatidial facets, 186
*Hexapanopeus*
    *Acantholobulus*, 561–562
    discussion, 560–563
    DNA extraction, 556
    *Eurypanopeus*, 562
    fundamentals, 551–552
    *Garthiope barbadensis*, 563
    *H. angustifrons*, 557, 560, 561
    *H. caribbaeus*, 557, 561, 561
    *H. lobipes*, 560–561
    *H. manningi*, 561
    *H. nov sp.*, 560
    *H. paulensis*, 557, 560, 561
    *H. schmitti*, 561
    materials and methods, 552–557
    outgroup taxa, 563
    *Panopeus americanus*, 562
    *Panoplax depressa*, 562
    PCR, 556
    phylogenetic analyses, 556–557
    results, 557
    sequencing, 556
    taxon sampling, 552, 556
Hexapoda, 11
hierarchical sequence pair alignment, 70
higher taxonomy, 491–498
Hippidae, 124, 312, 405
Hippoidea, 399, 405
*Hippolyte*, 298
Hippolytidae
    basal lineages, 299
    Caridea, infraorder phylogeny, 281, 294, 296
    nuclear genes, 22
    pair-bonding mating system, 153
    phylogenetic relationships, 250
    superfamily Alpheoidea, 298
    Thalassinidea, mitochondrial and nuclear genes, 312

history
    mating systems, 122–124
    parasites, 200
    systematic approaches, 510–511
*Hobbseus*, 343
*Holothuriophilus*, 465, 469–470
*Holothuriophilus pacificus*, 466
*Holthuisana*, 522–523
Homarida, 40
*Homarinus*
    marine clawed lobster families, 357, 364–365
    morphological and molecular previous works, 358
    morphological convergence, 364–365
    phylogenetic analysis, 362
    previous works comparison, 363
*Homarinus capensis*, 365
*Homarus*
    marine clawed lobster families, 357, 364–365
    morphological and molecular previous works, 358
    morphological convergence, 364–365
    phylogenetic analysis, 362
    precopulatory guarding mating system, 125
    previous works comparison, 363
    Thaumastochelidae status, 364
*Homarus americanus*
    freshwater crayfishes, 345
    morphological convergence, 365
    nuclear genes, 19
    precopulatory guarding mating system, 125
*Homarus capensis*, 365
*Homarus gammarus*, 365
*Homeryon*
    higher taxa diagnosis, 382
    morphological trends, 379
    Polychelidae, 376
    Polychelidan lobsters, 369, 372
    study results, 374
*Homola barbata*, 418
Homolidae
    Homoloidea monophyly, 419
    optical design diversity, 192
    protein-coding genes, 95
    specimans used, 418
Homolodromiidae, 418–419
Homolodromioidea, 431
Homoloidea
    Brachyura, 417, 418
    Dromiacea monophyly, 419
    Podotremata paraphyly, 431–432
    synapomorphies, 424–429
*Homoloidea*, 419, 427, 429
homoplasy, 233
Hoplocarida, 3, 9
*Hoploparia*, 357
host coevolution, 207–213
host specificity, 202–204
Hox gene ultrabithorax (UBX), 35–36
*Hyas*, 444
*Hyas araneus*, 439
*Hyas coarctatus alutaceus*, 439

*Hyas lyratus*, 126
*Hydrothelphusa*, 510
Hydrothelphusini, 510
*Hymenocera*, 297
*Hymenocera picta*, 127, 132
Hymenoceridae, 251, 296–297
Hymenosomatidae, 447, 498
Hymensomatidae, 438
*Hypoconcha*, 419
*Hypoconcha arcuata*, 418, 432
hypotheses testing, 300
Hyppolytidae, 124

## I

ILD, *see* Incongruence length difference (ILD)
immotile sperm cells, 105
*Imocaris*, 5, 9
*Imocaris colombiensis*, 5, 9
*Imocaris tuberculata*, 5, 9
Inachidae, 438–439
Inachoididae, 438–439, 447
*Inachus*, 123
*Inachus dorsettensis*, 226
*Inachus leptochirus*, 226
Inclusive Fitness Theory, 133
incongruence length difference (ILD)
    Callianassidae, mitochondrial genes, 333
    Caridea, infraorder phylogeny, 294
    *Hexapanopeus*, 556
    Majoidea, 440, 444
    marine clawed lobster families, 360–361
    multiple genes, 81
    Portunoidea and Cancroidea, 539
    Thalassinidea, mitochondrial and nuclear genes, 312
incorrect assumptions, ancestors, 9–11
INDELs, *see* Insertion/deletion events (INDELs)
Indian-Southeast Asian group, 521
inference methods, 75–79
infraorder composition, 321–322
infraordinal relationships, 91
ingroup taxa, 283
Insecta, 108
insertion/deletion events (INDELs), 443–444, 447
internal complexity, 110
internal family relationships, 321–322
interpreting results, 350–351
introns, 17, 22
*Irmengardia*, 522, 524
*Irmengardia johnsoni*, 523
*Irmengardia pilosimana*, 523
Isolapotamidae, 492, 494
Isopoda, 35
isopods, 134
issues, Decapoda, 3–11

## J

*Jagtia*, 357
Japanese freshwater crabs, 69
*Jaxea*, 311, 319, 321

*Jilinocaris chinensis*, 6
*Juxtafabia*, 465–468

## K

Kakaducarididae, 296–297
*Kakaducaris*, 251
Kashino-Hasegawa (KH) test, 80
*Kemponia*, 251, 297
KH, *see* Kashino-Hasegawa (KH) test
king crabs, 48, 399
*Kiwa*, 407
Kiwaidae, 399
*Knebelia bilobata*, 372
*Krangalanga*, 426
*Krangalangia spinosa*, 418

## L

*Laleonectes*
    considerations and future work, 577
    molecular phylogeny, 574
    sample collection, 569
    taxonomy developments, 567
*Laleonectes vocans*, 568
Lambrachaeidae, 438, 447
*Laomedia*, 311
Laomediidae
    Gebiidea families, 313, 318
    infraorder composition and relationships, 321
    mitochondrial and nuclear genes, 309–311
    monophyly or paraphyly, 318–320
    monophyly test, 313
    previously applied superfamilies, 321
large groups, mating system, 158
larval morphological characters, Majoidea
    Bayesian phylogenetic analysis, 443, 444
    data partition comparisons, 440, 443, 444
    direct optimization analysis, 443, 444–445
    discussion, 445–448
    fundamentals, 437–439
    larval morphology, 439–440, 449–451
    materials and methods, 439–444
    molecular data, 440, 443, 444
    morphological characters, 450–451
    morphological data comparison, 440, 443, 444
    phylogenetic analysis, 443, 444
    results, 444–445
larval morphology
    fundamentals, 221–222, 236–237
    Majoidea, 439–440, 449–451
larval morphology, *Brachyuran* phylogeny
    collection of larvae, 222
    fundamentals, 221–222, 236–237
    heterochrony, 226–227
    homoplasy, 233
    phylogenetics, 234–236
    polarity, setal characters, 228–230, 232
    setal observations, 222

transformation types, 232–233
zoeal similarity, 223
lateral transfer, 81
Latreilliidae, 95, 418–419
*Latreutes*, 298
*Lauridromia dehaani*, 401
*Leander*, 251
*Leandrites*, 251
*Leiogalathea*, 407
*Lepidopa*, 405
*Lepidophthalmus*, 334, 337, 339
Lepidopidae, 405
*Lepidothelphusa*, 521
*Lepidothelphusa cagnetti*, 527
*Lepidothelphusa cognetti*, 524
*Leptochela*, 299
*Leptodius exaratus*, 233
*Leptograpsus*, 484, 486
*Leptopalaemon*, 251
*Leptopalaemon gagadjui*, 297
*Leptopheus*, 298
Leptostraca, 35, 37
Leuciferinae, 261
Leucosiidae, 22, 111
*Libinia*, 444, 447
*Libinia dubia*, 439
*Libinia emarginata*, 22, 126, 439
*Libinia spinosa*, 226, 233
*Libystes*, 543, 546
life cycle, 201–202
life history studies, 159
likelihood-based methods, 82, *see also* Maximum likelihood (ML) method
likelihood framework, 73
*Limotheres*
constituents of Pinnotherinae, 469
limited gene utility, 470
pinnotherid subfamily definitions, 465
Pinnotherinae subfamily, 466
lineages, 523–524
*Linuparus*, 97
*Liocarcinus*, 543, 545
*Liocarcinus arcuatus*, 226
*Liocarcinus holsatus*, 543
*Liotelphusa*, 521
Liotelphusinae, 510, 524
Lissocarcinidae, 535
*Lithodes*, 405, 407
*Lithodes santolla*, 126
Lithodidae, 312
Lithophyllacidae, 535
*Litopenaeus*, 124
*Litopenaeus schmitti*
Caridea, Stenopodidea, and Nephropidae as outgoups, 269–270
Caridea and Stenopodidea as outgroup, 268
Caridea as outgroup, 266
Penaeidae as non-natural group, 273
*Litopenaeus vannamei*, 253
*Lobetelson*, 10

lobsters
maxillipeds, 35
opsin sequences, 188
precopulatory guarding mating system, 125
protein-coding genes, 95–97
visual system components, 191
lobsters, Nephropidae and Thaumastochelidae
*Acanthacaris*, 363–364
discussion, 362–365
DNA amplification, 360
DNA extraction, 360
fundamentals, 357–358, 365–366
future work, 365–366
*Homarinus*, 364–365
*Homarus*, 364–365
materials and methods, 358–361
*Metanephrops*, 365
morphological convergence, 364–365
*Nephrops*, 365
nucleotide composition, 360, 361
PCR, 360
phylogenetic analysis, 361–362
previous works, 358, 363
results, 361–362
sequencing, 360
taxon sampling, 358–359
*Thaumastochelidae*, 364
tissue sampling, 359
lobsters, Polychelidan
analytical methods, 372–373
*Coleiidae Van Straelen*, 380
data matrix, 388
diagnosis of taxa, 379–386
discussion, 375–379
*Eryonidae de Haan*, 380
fundamentals, 369–372
materials and methods, 372–373
morphological characters, 372, 389–392
morphological trends, 377–379
*Palaeopentacheles* position, 375–376
*Palaeopentachelidae*, 380–381
*Polychelida de Haan*, 379–380
*Polychelidae* genera, 376–377, 381–386
*Polychelidae Wood-Mason*, 381
polychelid sister group, 375–376
results, 373–374
systematics, 379–386
terminal taxa, 372, 387–388
*Tetrachelidae Beurlen*, 381
unambiguous character state changes, 393
*Willemoesia* position, 377
long-bodied decapods, 191
Longusorbiidae, 535
Lophiiformes, 134
*Lophopanopeus*, 560–561
*Lophozozymus pictor*, 226, 228
*Loxorhynchus grandis*, 126
*Lucifer*, 32, 248
Luciferidae

Caridea, Stenopodidea, and Nephropidae as outgoups, 269
Caridea and Stenopodidea as outgoup, 266
Caridea as outgoup, 265
Dendrobranchiata, monophyletic group, 272
morphological character-based analysis, 261
Sergestoidea as natural group, 272
*Lupa*, 568, 576
*Lupella*, 567, 576
*Lupocycloporus*, 576
*Lupocyclus*, 567, 569
*Lybia plumose*, 233
*Lyreidus tridentatus*, 418
Lysiosquillidae, 23
*Lysiosquillina maculata*, 23
*Lysmata debelius*, 127
Lystmatidae, 298

# M

*Macrobrachium*
  phylogenetic relationships, 251
  precopulatory guarding mating system, 125, 131
  superfamily Palaemonoidea, 297
*Macrobrachium australiense*, 125
*Macrobrachium rosenbergii*, 252–253
*Macrocheira*, 438–439
*Macrocheira kaempferi*, 226
Macrocheiridae, 438
Macrophthalmidae, 128, 466
*Macrophthalmus*, 133
Macropipidae, 545
Macropipinae, 535
*Macropipus*, 545
*Macropipus australis*, 545
*Macropipus tuberculatus*, 545
MAFFT program
  Cryptochiridae, 482
  freshwater crayfishes, 345
  phylogeny-based systematics, 23
*Mahatha*, 521
*Maia*, 438
*Maja*, 444
*Maja brachydactyla*, 439
*Maja squinado*, 126
Majidae
  arms, 104
  family development, 438–439, 447
  homoplasy, 233
  nuclear genes, 22
  precopulatory guarding mating system, 125
  specimens used, 418
  synapomorphies, 423
  zoeal similarity, 223
Majinae, 438, 447
Majoidea
  Bayesian phylogenetic analysis, 443, 444
  data partition comparisons, 440, 443, 444
  direct optimization analysis, 443, 444–445
  discussion, 445–448
  fundamentals, 437–439
  larval morphology, 439–440, 449–451
  marine crab sister group, 498
  materials and methods, 439–444
  molecular data, 440, 443, 444
  morphological characters, 450–451
  morphological data comparison, 440, 443, 444
  phylogenetic analysis, 443, 444
  results, 444–445
Malacostraca
  ancestor assumptions, 10, 11
  apomorphies, 34–35
  embryonic characters, 40
  Podotremata paraphyly, 431
*Maldivia*, 486
Malesian-Australian group, 522–523
mandibular palp character evolution, 494–496
marine clawed lobster families, mitochondrial gene-based phylogeny
  *Acanthacaris*, 363–364
  discussion, 362–365
  DNA amplification, 360
  DNA extraction, 360
  fundamentals, 357–358, 365–366
  future work, 365–366
  *Homarinus*, 364–365
  *Homarus*, 364–365
  materials and methods, 358–361
  *Metanephrops*, 365
  morphological convergence, 364–365
  *Nephrops*, 365
  nucleotide composition, 360, 361
  PCR, 360
  phylogenetic analysis, 361–362
  previous works, 358, 363
  results, 361–362
  sequencing, 360
  taxon sampling, 358–359
  *Thaumastochelidae*, 364
  tissue sampling, 359
marine crab sister group, 497–498
Markov chain Monte Carlo (MCMC), see also Monte Carlo (MC) simulation
  Anomura phylogeny, 404
  Bayesian methods, 76
  Caridea, infraorder phylogeny, 292
  Cryptochiridae, 484
  freshwater crayfishes diversification timing, 348
  gaps, 71
  Gecarcinucidae, 511, 518
  *Hexapanopeus*, 557
  multiple genes, 81
  Portunoidea and Cancroidea, 539
  ratchet approach, 75
Markov chains, 443
Markov models, 76
Marmorkrebs, 37
*Marsupenaeus*, 124
*Marsupenaeus japonicus*, 253

materials and methods
    Anomura, 400–404
    Asian freshwater crabs, 511–518
    Callianassidae mitochondrial genes, 328–333
    Caridea mitochondrial and nuclear genes, 283–292
    gall crabs, 482, 484
    *Hexapanopeus*, 552–557
    Majoidea, 439–444
    marine clawed lobster families, 358–361
    mitochondrial DNA, 49
    morphological character-based analysis, 262–263
    Pinnotheridae, molecular genetic re-examination, 458–463
    Podotremata, 418–419
    Polychelidan lobsters, 372–373
    Portunoidea and Cancroidea new classification proposal, 535, 538–539
    Thalassinidea mitochondrial and nuclear genes, 311–313
Mathildellidae, 535, 544–545
mating systems, evolution
    eusociality type, 128, 133, 157
    fundamentals, 121–122
    guarding time, 146
    history of study, 122–124
    large groups, 158
    life history studies, 159
    neotenous male type, 130, 134, 159
    pair-bonding type, 126–128, 132–133, 148–156
    podding, 126, 132, 147
    precopulatory guarding type, 125, 130–131, 144–145
    reproductive swarm type, 129–130, 134
    short courtship type, 124–125, 130–131, 141–143
    small groups, 157–158
    sperm plug, 143, 147
    types, 124–130
    visiting type, 129, 134
    waving display type, 128–129, 133–134
Matutidae, 95
maxillipeds, 35–36
Maxillopoda, 11
maxillopodans, 105
Maximum Composite Likelihood, 49
maximum likelihood (ML) method
    Bayesian methods, 76
    Callianassidae, mitochondrial genes, 333
    Caridea, infraorder phylogeny, 292, 294
    Cryptochiridae, 482, 485
    distance methods, 77
    freshwater crayfishes, 345
    fundamentals, 82
    *Hexapanopeus*, 557
    inference methods, 75–76
    marine clawed lobster families, 357, 361
    node support and tree comparison, 80
    phylogeny-based systematics, 23
    Pinnotheridae, 463
    searching for trees, 74
    Thalassinidea, mitochondrial and nuclear genes, 312–313
    tree topology statistical tests, 80

maximum parsimony (MP) method
    Anomura phylogeny, 404–405, 407
    Callianassidae, mitochondrial genes, 333
    *Cronius*, 572
    fundamentals, 81–82
    inference methods, 78–79
    Majoidea, 440
    Portunoidea and Cancroidea, 539
    previous works comparison, 363
    Thalassinidea, mitochondrial and nuclear genes, 312–313
*Maydelliathelphusa*, 521
MCMC, *see* Markov chain Monte Carlo (MCMC)
ME, *see* Minimum evolution (ME) method
*Medorippe lanata*, 418
MEGA4 program, 49
Meiura, 417
*Melicertus*, 124
*Melybia*, 535
Melybiidae, 534–535
membrane-bound feature, 110
membrane-bound nucleus, 110
*Menaethius*, 444
*Menaethius monoceros*, 439, 445
*Menippe adina*, 56
*Menippe mercenaria*, 56–57, 105, 125
*Menippe nodifrons*, 57
*Metanephrops*
    marine clawed lobster families, 357, 365
    morphological and molecular previous works, 358
    morphological convergence, 365
    phylogenetic analysis, 362
    previous works comparison, 363
    Thaumastochelidae status, 364
*Metapenaeopsis*, 249
*Metapenaeus*, 124
*Metaplax*, 128, 133
meta-sequence multiple genes, 80–81
methods, *see* Materials and methods
Micheleidae
    Axiidea families, 318
    infraorder composition and relationships, 321–322
    mitochondrial and nuclear genes, 309, 311
    monophyly or paraphyly, 320
    monophyly test, 313
*Micippa*
    Bayesian analysis, 444
    branching patterns, 447
    direct optimization analysis, 444
    family development, 447
*Micippa platipes*, 439
*Micippa thalia*, 439
Micromajinae, 438
*Micropanope*, 561, 563
*Microphrys*, 444, 447
*Microphrys bicornatus*, 440
microscopy, Podotremata, 418
microtubular arms
    number, 111
    origin, 111

presence/absence, 111
spermatozoal morphology, 111–112
Mictridae, 484
*Mictyris*, 484, 486
minimum evolution (ME) method, 74, 77
Mirocarididae, 250
Mithracidae, 438–439, 444, 447
Mithracinae, 438, 447
*Mithraculus*, 444, 447
*Mithraculus forceps*, 439
*Mithraculus sculptus*, 439
mitochondria, 108
mitochondrial analyses, 463–465
mitochondrial and nuclear genes, Caridea
   *Alpheoidea*, 298
   Atyidae, 298
   basal lineages, 299
   *Crangonidae*, 299
   discussion, 296–300
   DNA extraction, 283, 291
   evolutionary history, 282–283
   fundamentals, 281, 300
   hypotheses testing, 300
   ingroup taxa, 283
   materials and methods, 283–292
   monophyly, 294, 296
   outgroup selection, 283
   *Palaemonoidea*, 297–298
   paraphyly, 294, 296
   PCR, 283, 291
   phylogenetic analyses, 292
   polyphyly, 294, 296
   Procaridoidea, 296–297
   Processidae, 299
   results, 294–296
   sequencing, 283, 291
   Thalassocarididae, 299
   Xiphocarididae, 298
mitochondrial and nuclear genes, Thalassinidea
   *Axiidea* families, 318
   discussion, 318–322
   DNA extraction, 312
   fundamentals, 309–311, 322
   *Gebiidea* families, 313, 318
   infraorder composition, 321–322
   internal family relationships, 321–322
   materials and methods, 311–313
   monophyly, 313, 318–320
   paraphyly, 318–320
   PCR, 312
   phylogenetic analyses, 312–313
   previously applied superfamilies, 320–321
   results, 313–318
   sequencing, 312
   taxa included, 311–312
Mitochondrial DNA (mtDNA)
   amplification, 17
mitochondrial DNA (mtDNA), decapods
   advantages, 48–49
   disadvantages, 90
   discussion, 54, 56–59
   fundamentals, 47–49
   material and methods, 49
   phylogenetic reconstruction, 90
   primer optimization, 58–59
   pseudogenes, 56–58
   results, 52–54
mitochondrial gene-based phylogeny, marine clawed lobster families
   *Acanthacaris*, 363–364
   discussion, 362–365
   DNA amplification, 360
   DNA extraction, 360
   fundamentals, 357–358, 365–366
   future work, 365–366
   *Homarinus*, 364–365
   *Homarus*, 364–365
   materials and methods, 358–361
   *Metanephrops*, 365
   morphological convergence, 364–365
   *Nephrops*, 365
   nucleotide composition, 360, 361
   PCR, 360
   phylogenetic analysis, 361–362
   previous works, 358, 363
   results, 361–362
   sequencing, 360
   taxon sampling, 358–359
   *Thaumastochelidae*, 364
   tissue sampling, 359
mitochondrial genes, 17–19
mitochondrial genes, Callianassidae
   basally positioned groups, 334
   *Callichirinae*, 334, 337–338
   Cheraminae, 334, 336
   Ctenochelidae, 334
   dataset description, 333
   discussion, 336–339
   DNA extraction, 328
   fundamentals, 327–328
   methods, 328–333
   model selection, 333
   PCR, 328
   pending analyses, 339
   phylogenetic analyses, 328, 333
   relationships, 336–339
   results, 333–336
   sequencing, 328
   specimens, 328
   tree topologies, 334
*Miyadiella*, 249
ML, *see* Maximum likelihood (ML) method
modeling and model selection, 72–73, 333
ModelTest program
   Callianassidae, mitochondrial genes, 333
   Caridea, infraorder phylogeny, 292, 294
   *Cronius*, 571–572
   freshwater crayfishes, 345
   freshwater crayfishes diversification timing, 348
   Gecarcinucidae, 511, 518

*Hexapanopeus*, 556, 557
Majoidea, 443
marine clawed lobster families, 361
mitochondrial DNA, 49
Pinnotheridae, 463
Portunoidea and Cancroidea, 538
Thalassinidea, mitochondrial and nuclear genes, 312–313
molecular data
  Anomura, 401
  Majoidea, 440, 443, 444
molecular data, phylogenetic inference
  Bayesian methods, 76
  brand-and-bound algorithm, 74
  character homology, 70–71
  distance methods, 77–78
  evolution modeling and model selection, 72–73
  exhaustive search, 74
  fundamentals, 67–70, 75, 81–82
  gaps, dealing with, 70–71
  genetic differences and saturation, 71
  heuristic searches, 74
  inference methods, 75–79
  maximum likelihood, 75–76
  maximum parsimony, 78–79
  minimum evolution, 77
  morphology-based taxonomy comparison, 69–70
  multiple genes, 80–81
  nearest-neighbor interchange, 74
  neighbor joining, 77–78
  node support and tree comparison, 79–80
  ratchet, 75
  sequence alignment issues, 70–71
  statistical tests, tree topologies, 80
  subtree pruning and regrafting, 74
  tree bisection and reconnection, 74
  tree comparison and node support, 79–80
  trees, searching for, 73–75
molecular evolution, phylogenetics and
  fundamentals, 15
  genes and their diversity, 17–23
  genetic marker development, 16–17
  mitochondrial genes, 17–19
  molecular taxonomy, 15
  nuclear genes, 19, 22–23
  phylogeny based systematics, 23
  12S, 16S, COI, 17–19
molecular genetic re-examination, Pinnotheridae
  combined mitochondrial analyses, 463
  discussion, 466–471
  DNA extraction, 461–462
  exclusions and exceptional members, 466–467
  fundamentals, 457–458
  future perspectives, 471
  materials and methods, 458–463
  mitochondrial analyses, 463–465
  nuclear18S rRNA limited utility, 470
  PCR, 461–462
  phylogenetic analyses, 462–463
  Pinnothereliinae, 468–469
  Pinnotherinae, 465–466, 469–470
  restriction, 463, 465, 467
  results, 463–466
  specimens used, 458, 461
  subfamilies, 465–470
molecular markers, 253
molecular phylogenies, *Cronius* Stimpson
  *C. ruber sensu lato*, 572
  *C. tumidulus*, 572
  discussion, 574–577
  DNA analysis, 569, 571
  fundamentals, 567–569
  future work, 577
  materials and methods, 569–571
  molecular phylogeny, 572, 574
  paraphyly, 574–576
  PCR, 569, 571
  phylogenetic analysis, 571
  results, 571–574
  sample collection, 569
  sequencing, 569, 571
  subfamilies, 577
  taxonomic account, 571–572
molecular phylogenies, Portunoidea and Cancroidea classification proposal
  discussion, 544–546
  families of, 534–535
  fundamentals, 533–534
  materials and methods, 535, 538–539
  results, 539, 542–543
  subfamilies of Portunidae, 535
molecular phylogenies, protein-coding genes, 90–91
molecular phylogenies, Western Atlantic representatives, see *Hexapanopeus*
molecular taxonomy, 15
Mollusca, 148
monogamy, 122
*Monomia*, 576
monophyletic group, 271–272
monophyletic group, Brachyura Podotremata
  analysis, 419
  animals, 418
  *Cyclodorippoidea*, 422–429
  discussion, 431–432
  *Dromiacea*, 419, 429
  *Eubrachyura*, 422–429
  fundamentals, 417–418
  *Homoloidea*, 419, 427, 429
  materials and methods, 418–419
  microscopy, 418
  monophyly, 419–423, 432
  phylogenetic relationships, 423–429
  *Podotremata* paraphyly, 431–432
  *Raninoidea*, 421–422, 424–429
  results, 419–429
  subtaxa, 419–429
monophyly
  Asian freshwater crabs, 523
  Caridea mitochondrial and nuclear genes, 294, 296

compared to paraphyly, 318–320
Podotremata, 419–423, 432
testing for, 313
Thalassinidea mitochondrial and nuclear genes, 313, 318–320
Monte Carlo (MC) simulation, 73, *see also* Markov chain Monte Carlo (MCMC)
*Moreiradromia sarraburei*, 418
morphological analysis, 518, 523–524
morphological character-based analysis, Dendrobranchiata
    Aristeidae, 274
    Benthesicymidae, 273
    Caridea, 264–270
    description of characters, 264
    discussion, 270–274
    fundamentals, 261–262
    materials and methods, 262–263
    monophyletic group, 271–272
    morphological characters used, 276–278
    natural group, 272–273
    Nephropidae, 268–270
    non-natural group, 273
    optimization of characters, 264
    outgroups, 264–271
    Penaeoidea, 272–273
    results, 264–270
    Sergestoidea, 272
    Solenoceridae, 273
    species studied, 275
    status of, 274
    Stenopodidea, 266–270
morphological characters
    evolution and radiation, 252–253
    Majoidea, 450–451
    Polychelidan lobsters, 372, 389–392
morphological characters used, 276–278
morphological convergence, 364–365
morphological data comparison, 440, 443, 444
morphological trends, 377–379
morphology, 110–111, 183–186
morphology-based taxonomy comparison, 69–70
mounds, *see* Podding mating system
MP, *see* Maximum parsimony (MP) method
MrBayes program
    Callianassidae, mitochondrial genes, 333
    *Cronius*, 571
    Cryptochiridae, 484
    freshwater crayfishes, 345–346
    *Hexapanopeus*, 557
    Majoidea, 443
    marine clawed lobster families, 361
    Portunoidea and Cancroidea, 539
    Thalassinidea, mitochondrial and nuclear genes, 312–313
mtDNA, *see* Mitochondrial DNA (mtDNA)
Multidivtime package, 347–348
multiple genes, 80–81
Munidopsiane, 407
*Munidopsis*, 407

MUSCLE program
    Caridea, infraorder phylogeny, 292
    *Hexapanopeus*, 556
    Majoidea, 440, 443
    Pinnotheridae, 462
    sequence alignment, 70
mutation rates, 72
Mysida, 22, 185
Mysidacea, 35
Mystacocarida, 105

## N

*Nanocassiope melanodactyla*, 230, 232
Natantia
    molecular taxonomy, 15
    morphological character-based analysis, 261
    visual system components, 191
natural groups, 272–273
Nauplii, 233
*Naushonia*, 318
*Nautilothelphusa*, 524
nearest-neighbor interchange (NNI), 74
*Neaxius*, 321
*Nebalia*, 37
*Necora*, 545
*Nectocarcinus*, 426, 545
*Nectocarcinus antarcticus*, 418
neighbor joining (NJ)
    Callianassidae, mitochondrial genes, 333
    *Cronius*, 572
    distance methods, 77
    fundamentals, 82
    inference methods, 77–78
    previous works comparison, 363
    Thalassinidea, mitochondrial and nuclear genes, 312–313
Nematocarcinidae, 294, 296, 299
Nematocarcinoidea, 298
*Neocalichirus*, 334
Neocallichirus, 338
*Neoglyphea*, 358, 363
*Neopanope*, 560
*Neopanope packardii*, 557, 562
*Neophrops*, 364
neotenous male mating system, 121, 130, 134, 159
Neotropical region, 498–499
*Neotrypaea*, 319, 334, 339
*Neotrypaea califoriensis*, 336
*Neotrypaea gigas*, 336
*Neoxenphthalmus*, 466
Nephropidae, *see also* Marine clawed lobster families, mitochondrial gene-based phylogeny
    freshwater crayfishes, 343
    infraordinal relationships, 91
    marine clawed lobster families, 357
    mitochondrial and nuclear genes, 312
    monophyly or paraphyly, 319
    morphological character-based analysis, 268–270

Nephropidea, 262, 348, 350
*Nephropides*
  marine clawed lobster families, 357
  morphological convergence, 364
  phylogenetic analysis, 362
  Thaumastochelidae status, 364
Nephropinae, 364
Nephropoidea, 345, 497
*Nephrops*
  marine clawed lobster families, 357, 365
  morphological and molecular previous works, 358
  morphological convergence, 365
  phylogenetic analysis, 362
  previous works comparison, 363
  sperm nuclear proteins, 107
  Thaumastochelidae status, 364
Nephropsidae, 497
*Nephropsis*
  marine clawed lobster families, 357
  morphological and molecular previous works, 358
  phylogenetic analysis, 362
  previous works comparison, 363
  Thaumastochelidae status, 364
*Nephropsis agassizi*, 270
*Nepinnotheres*, 465, 469
*Nepinnotheres pinnotheres*, 458
*Nepinnotheres pinnotheroids*, 466
*Neptunus*, 575
*Neptunus tumidulus*, 575
*Niasathelphusa wirzi*, 522, 528
*Nihonotrypaea*, 336, 339
NJ, *see* Neighbor joining (NJ)
NNI, *see* Nearest-neighbor interchange (NNI)
node support and tree comparison, 79–80
non-natural groups, 273
nuclear and mitochondrial genes, Caridea
  *Alpheoidea*, 298
  Atyidae, 298
  basal lineages, 299
  *Crangonidae*, 299
  discussion, 296–300
  DNA extraction, 283, 291
  evolutionary history, 282–283
  fundamentals, 281, 300
  hypotheses testing, 300
  ingroup taxa, 283
  materials and methods, 283–292
  monophyly, 294, 296
  outgroup selection, 283
  *Palaemonoidea*, 297–298
  paraphyly, 294, 296
  PCR, 283, 291
  phylogenetic analyses, 292
  polyphyly, 294, 296
  Procaridoidea, 296–297
  Processidae, 299
  results, 294–296
  sequencing, 283, 291
  Thalassocarididae, 299
  Xiphocarididae, 298

nuclear and mitochondrial genes, Thalassinidea
  *Axiidea* families, 318
  discussion, 318–322
  DNA extraction, 312
  fundamentals, 309–311, 322
  *Gebiidea* families, 313, 318
  infraorder composition, 321–322
  internal family relationships, 321–322
  materials and methods, 311–313
  monophyly, 313, 318–320
  paraphyly, 318–320
  PCR, 312
  phylogenetic analyses, 312–313
  previously applied superfamilies, 320–321
  results, 313–318
  sequencing, 312
  taxa included, 311–312
nuclear genes, 19, 22–23
nuclear18S rRNA limited utility, 470
nucleotide composition, 360, 361
nucleus, 110–111

## O

Occam's Razor, 78
octopus, 134
*Ocypode*, 129
*Ocypode quadrata*, 57
Ocypodidae, 112, 128, 486
Ocypodoidae, 466
Ocypodoidea, 486–487
Ocythoidae, 134
*Oedignathus*, 405, 407
Ogyrididae, 281, 294, 296, 298
Oligochaeta, 108
*Ombrastacoides*, 343–344, 350
ommatidial facets, 186
*Oncopareia*, 364
*Opecarcinus hypostegus*, 129
Oplophoridae
  basal lineages, 299
  Caridea, infraorder phylogeny, 281, 294, 296
  phylogenetic relationships, 250
  testing morphological hypotheses, 300
Oplophoroidea, 282
optimization of characters, 264
*Orchestia cavimana*, 10
*Orconectes*, 343–344
Oregoniidae
  Bayesian analysis, 444
  branching patterns, 447
  direct optimization analysis, 444–445
  family development, 438–439
  larval morphology, 440
  Majoidea, 437
Oriental region, 500
origins, 111, 503–504
*Orthotheres*, 465, 469
*Orthotheres barbatus*, 466
*Ostrea puelchanas*, 134

outgroups, 264–271, 283, 563
*Ovalipes*
   analysis, 543, 546
   considerations and future work, 577
   sample collection, 569
   synapomorphies, 426
*Ovalipes catharus*, 418
oysters, 134
*Oziothelphusa*, 521, 524

## P

*Pachygrapsus*, 484
*Pachygrapsus marmoratus*, 484
*Pacifastacus*, 343–344, 348
Paguridae, 125, 131, 405, 407
Paguridea, 399
*Paguristes*, 131
Paguroidea, 399, 405–407, 409–410
*Pagurus*
   mitochondrial DNA, 48
   molecular data, 401
   precopulatory guarding mating system, 131
   sperm nuclear proteins, 107
*Pagurus bernhardus*, 105, 401
*Pagurus longicarpus*, 401
pair-bonding mating system, 121, 126–128, 132–133, 148–156
Palaearctic region, 499–500
*Palaemon*, 251
*Palaemonetes*, 251
*Palaemonetes paludosus*, 105
Palaemonidae
   basal lineages, 299
   Caridea, infraorder phylogeny, 281, 296
   fossil record, 6
   phylogenetic relationships, 250–251
   short courtship mating system, 124
   superfamily Palaemonoidea, 297
   Thalassinidea, mitochondrial and nuclear genes, 312
Palaemoninae, 252, 296
Palaemonoidea, 251, 297–298
*Palaeoechinastacus australianus*, 346
*Palaeonephrops*, 358
*Palaeopalaemon*, 9, 34
*Palaeopalaemon newberryi*, 4, 7
*Palaeopentacheles*
   morphological trends, 377
   Polychelidan lobsters, 369, 375–376
   study results, 374
   systematics, 379
*Palaeopentacheles roettenbacheri*, 372
Palaeopentachelidae, 369, 376, 379–381
paleobiogeography, 9
paleoecology, 9
Paleozoic fossils, 4–7, 9
Palicidae, 487
Palicoidea, 486
*Palicus*, 487
Palinura
   microtubular arms, 111
   molecular taxonomy, 15
   Polychelidan lobsters, 370
   Thalassinidea, monophyly or paraphyly, 319
Palinuridae, 95–97, 312
Palinuridea, 141–142
Palinuroidea, 283, 370
*Palinurus*, 97
*Palinurus longipes*, 97
*Palinurus ornatus*, 97
*Palinurus polyphagus*, 97
*Palinurus stimpsoni*, 97
*Palinurus versicolor*, 97
Palumbi region, 59
Pandalidae
   Caridea, infraorder phylogeny, 294, 296
   morphological characters, 252
   short courtship mating system, 124
   Thalassinidea, mitochondrial and nuclear genes, 312
Panopeidae, *see Hexapanopeus*
*Panopeus americanus*, 561–562
*Panopeus bermudensis*, 561, 562
*Panopeus miraflorensis*, 561
*Panopeus pacifus*, 561
*Panoplax depressa*, 562
*Panulirus*, 319
*Paracalliax*, 337
*Paracleistostoma*, 486
*Paraglypturus*, 336
*Paragrapsus*, 486
*Paralithodes brevipes*, 125
*Paralithodes camtschaticus*, 126
*Paramysis*, 22
*Paranephrops*, 343, 350
*Paranephrops zealandicus*, 418
Parapaguridae, 399, 405, 407
*Parapenaeus americans*
   Caridea, Stenopodidea, and Nephropidae as outgoups, 269–270
   Caridea and Stenopodidea as outgoup, 268
   Caridea as outgoup, 266
   Penaeidae as non-natural group, 273
*Parapenaeus longirostris*, 105
Parapeneini, 249
paraphyletic group, continental Indian species, 521
paraphyly
   Caridea mitochondrial and nuclear genes, 294, 296
   monophyly comparison, 318–320
   Thalassinidea mitochondrial and nuclear genes, 318–320
Parapthelphusidae, 510
parasites
   basal, 204
   biogeography, 202–204
   biology overview, 200–201
   *Bopyridae*, 202, 205–207, 211, 213
   coevolutionary theory, 198–200, 207–213
   coevolution example, 211, 213
   fundamentals, 197–200, 214–215
   history, 200
   host coevolution, 207–213
   host specificity, 202–204
   inferences, 209, 211

life cycles, 201–202
parasites, 211
phylogeny, 204–207
Rhizocephala, 201–205, 209, 211
taxonomy, 204–207
Parastacidae
embryonic characters, 39, 41
fossil calibration, 347
freshwater crayfishes, 343–344, 350
mitochondrial genes, 18, 312
nuclear genes, 22, 312
*Parastacidae*, 346
Parastacidea, 350
Parastacids, 348
Parastacoidea
fossil calibration, 347
freshwater crayfishes, 343–345, 348
marine crab sister group, 497
*Parastacus*, 343, 363
Parastasidae, 124
*Parathelphusa*
analysis, 522
Asian freshwater crabs, 523
biogeography, 528
Bott system similarities, 524
frontal triangle and second gonapod, 524
historical systematic approaches, 510
*Sundathelphusa*, 526
*Parathelphusa maculata*, 523
*Parathelphusa oxygona*, 523
*Parathelphusa pantherina*, 523
Parathelphusidae
frontal triangle and second gonapod, 524
historical systematic approaches, 511
Oriental region, 500
taxonomy developments, 492, 494
theories on origins, 504
Parathelphusinae
analysis, 518
Bott system similarities, 524
historical systematic approaches, 510–511
Parathelphusini, 510
Parathelphusoidea, 510
*Parathranites*, 545
*Paratrypaea*, 338–339
*Paratrypaea bouvieri*, 336
*Paratya*, 251
Paratymolidae, 438, 447
*Paratypton*, 486
Parental Manipulation Theory, 133
*Paromola japonica*, 401
parsimony analysis, *see also* Maximum parsimony (MP) method
fundamentals, 81–82
gaps, 71
inference methods, 78–79
modeling evolution and selection, 73
Parthenopidae, 223
partitioned branch support (PBS), 79
partitioned Bremer support (PBS), 443

partitioned likelihood support (PLS), 79
Partition Homogeneity Test, 81
*Pasiphaea*, 299
*Pasiphaea princeps*, 265
Pasiphaeidae
basal lineages, 299
Caridea, infraorder phylogeny, 281, 294, 296
evolutionary history of Caridea, 282
phylogenetic relationships, 250
*Pastilla*, 521
PAUP program
Anomura phylogeny, 404
Callianassidae, mitochondrial genes, 333
Caridea, infraorder phylogeny, 292
Cryptochiridae, 482, 484
Dendrobranchiata, morphological character-based analysis, 263
*Hexapanopeus*, 556, 557
Majoidea, 440, 443
marine clawed lobster families, 360–361
Pinnotheridae, 463
Polychelidan lobsters, 373
Portunoidea and Cancroidea, 538–539
Thalassinidea, mitochondrial and nuclear genes, 312
PBS, *see* Partitioned branch support (PBS); Partitioned Bremer support (PBS)
PCR, *see* Polymerase chain reaction (PCR)
*Peisos*, 272
Penaeidae
Caridea, Stenopodidea, and Nephropidae as outgoups, 269
Caridea and Stenopodidea as outgroup, 267–268
Caridea as outgroup, 265–266
morphological character-based analysis, 261–262
Penaeoidea as natural group, 272
phylogenetic relationships, 248–249
position, shrimp-like decapods, 247
short courtship mating system, 124
sperm plugs, 143
Penaeidea, 261, 272
Penaeinae, 261–262
Penaeoidea
Benthesicymidae as non-natural group, 273
Caridea, infraorder phylogeny, 283
Caridea, Stenopodidea, and Nephropidae as outgoups, 269
Caridea and Stenopodidea as outgroup, 266, 268
Caridea as outgroup, 265
Dendrobranchiata, monophyletic group, 272
morphological character-based analysis, 261–262, 272–273
protein-coding genes, 91, 94–95
*Penaeopsis serrata*
Caridea, Stenopodidea, and Nephropidae as outgoups, 269
Caridea and Stenopodidea as outgroup, 268
Caridea as outgroup, 266
Penaeidae as non-natural group, 273
*Penaeus*, 19, 22, 124
*Penaeus monodon*, 36, 189, 253

pending analyses, 339
Peneini, 249
*Pentacheles*
    higher taxa diagnosis, 382–383
    morphological trends, 379
    Polychelidae, 376
    Polychelidan lobsters, 369–372
    study results, 374
*Pentacheles laevis*, 370
Pentastomida, 105, 108
Peracarida, 9
*Perbrinckia*, 521, 524
*Percnon*, 484, 486–487
*Periclimenaeus*, 297
*Periclimenes*, 251
*Periclimenes sensu lato*, 251
*Perithelphusa*, 522–523, 528
*Permanotus purpureus*, 234, 236
*Permonotus purpureus*, 223
*Pestarella*, 334, 336
*Pestarella tyrrhena*, 336
*Petrolisthes*, 111, 425–427, 429
*Petrolisthes elongatus*, 418
*Philarius*, 251
Philippine group, 522
*Phoenice pasinii*, 6
*Phricotelphusa*, 521, 524
*Phricotelphusa gracilipes*, 521
*Phricotelphusa hockpingi*, 521
Phyllobranchiata, 261
Phyllotymolinidae, 422
phylogenetic analyses
    Callianassidae mitochondrial genes, 328, 333
    Caridea mitochondrial and nuclear genes, 292
    freshwater crayfishes diversification timing, 345–346
    *Hexapanopeus*, 556–557
    Pinnotheridae, molecular genetic re-examination, 462–463
    Thalassinidea mitochondrial and nuclear genes, 312–313
phylogenetic analysis
    Anomura, 404
    Majoidea, 443, 444
    marine clawed lobster families, 361–362
phylogenetic inference, molecular data
    Bayesian methods, 76
    brand-and-bound algorithm, 74
    character homology, 70–71
    distance methods, 77–78
    evolution modeling and model selection, 72–73
    exhaustive search, 74
    fundamentals, 67–70, 75, 81–82
    gaps, dealing with, 70–71
    genetic differences and saturation, 71
    heuristic searches, 74
    inference methods, 75–79
    maximum likelihood, 75–76
    maximum parsimony, 78–79
    minimum evolution, 77
    morphology-based taxonomy comparison, 69–70
    multiple genes, 80–81
    nearest-neighbor interchange, 74
    neighbor joining, 77–78
    node support and tree comparison, 79–80
    ratchet, 75
    sequence alignment issues, 70–71
    statistical tests, tree topologies, 80
    subtree pruning and regrafting, 74
    tree bisection and reconnection, 74
    tree comparison and node support, 79–80
    trees, searching for, 73–75
phylogenetic reconstruction, genera/species, 95–97
phylogenetic relationships, 248–252, 423–429
phylogenetics, 234–236, 348
phylogenetics, visual system characters of shrimp
    components as phylogenetic characters, 189–192
    evolutionary enigma, 187
    fundamentals, 183, 192
    molecular aspects, 187–189
    morphology, 183–186
    overview, 183–189
phylogenetics and molecular evolution
    fundamentals, 15
    genes and their diversity, 17–23
    genetic marker development, 16–17
    mitochondrial genes, 17–19
    molecular taxonomy, 15
    nuclear genes, 19, 22–23
    phylogeny based systematics, 23
    12S, 16S, COI, 17–19
phylogenies, mitochondrial DNA and
    discussion, 54, 56–59
    fundamentals, 47–49
    material and methods, 49
    primer optimization, 58–59
    pseudogenes, 56–58
    results, 52–54
phylogeny
    freshwater crabs, 491–492, 494–498
    freshwater crayfishes diversification timing, 350
    parasite, 204
    parasites, 204–207
phylogeny, bearing of larval morphology on
    collection of larvae, 222
    fundamentals, 221–222, 236–237
    heterochrony, 226–227
    homoplasy, 233
    phylogenetics, 234–236
    polarity, setal characters, 228–230, 232
    setal observations, 222
    transformation types, 232–233
    zoeal similarity, 223
phylogeny, protein-coding genes
    Brachyura, 95
    fundamentals, 89, 97
    infraordinal relationships, 91
    molecular phylogeny, 90–91
    Palinuridae, 95–97
    Penaeoidea, 91, 94–95
    phylogenetic reconstruction, genera/species, 95–97
    super family/family level phylogeny studies, 91–95

phylogeny, spermatozoal morphology
   acrosome vesicle, 105, 108–110
   aflagellate sperm cells, 105
   fundamentals, 101–105, 112–113
   immotile sperm cells, 105
   microtubular arms, 111–112
   mitochondria, 108
   nucleus, 110–111
   sperm nuclear proteins, 106–107
   sperm nucleus, 107–108
   sperm uniqueness, 104–108
phylogeny based systematics, 23
PhyML Online, 557
Physetocarididae, 283
*Pilodius areolatus*, 223
*Pilodius paumotensis*, 223
Pilumnidae
   homoplasy, 233
   outgroup taxa, 563
   taxon sampling, 552
   zoeal similarity, 223
Pilumnoidea
   larval morphology, 222
   phylogenetics, 234, 236
   zoeal similarity, 223
*Pilumnus hirtellus*, 230, 232, 234, 236
*Pilumnus vespertilio*, 234
*Pinnaxodes*, 465, 469–470
*Pinnaxodes chilensis*, 466
*Pinnixa*
   constituents of Pinnothereliinae, 468–469
   family development, 458
   limited gene utility, 470
   Pinnothereliinae subfamily, 465
   Pinnotheridae restriction, 465, 467–468
   pinnotherid subfamily definitions, 465
   specimens, 461
*Pinnixa chaetopterana*, 465, 468
*Pinnixa cristata*, 468
*Pinnixa cylindrica*, 465, 469–470
*Pinnixa monodactyla*, 465, 470
*Pinnixa rapax*, 468
*Pinnixa retinens*, 463, 465, 467
*Pinnixa sayana*, 465, 468
*Pinnixa valerii*
   exclusions and exceptiona members, 467
   limited gene utility, 470
   Pinnothereliinae subfamily, 465
   Pinnotheridae restriction, 463
Pinnothereliinae
   exclusions and exceptiona members, 467
   family development, 457
   Pinnotheridae, molecular genetic re-examination, 468–469
   Pinnotheridae restricted, 467
   pinnotherid subfamily definitions, 465
   specimens, 458, 461
*Pinnotheres*
   analysis, 485
   constituents of Pinnotherinae, 469
   family development, 458
   pinnotherid subfamily definitions, 465
*Pinnotheres pisum*, 466
Pinnotheridae
   analysis, 486
   Cryptochiridae developments, 482
   pair-bonding mating system, 156
   taxon selection, 484
   zoeal similarity, 223
Pinnotheridae, molecular genetic re-examination
   combined mitochondrial analyses, 463
   discussion, 466–471
   DNA extraction, 461–462
   exclusions and exceptional members, 466–467
   fundamentals, 457–458
   future perspectives, 471
   materials and methods, 458–463
   mitochondrial analyses, 463–465
   nuclear18S rRNA limited utility, 470
   PCR, 461–462
   phylogenetic analyses, 462–463
   Pinnothereliinae, 468–469
   Pinnotherinae, 465–466, 469–470
   restriction, 463, 465, 467
   results, 463–466
   specimens used, 458, 461
   subfamilies, 465–470
Pinnotherinae
   family development, 457
   Pinnotheridae, molecular genetic re-examination, 465–466, 469–470
   Pinnotheridae restricted, 467
   pinnotherid subfamily definitions, 465
   specimens, 458
Pinnotheroidea, 486
*Pirimela denticulata*, 543, 545
Pirimelidae, 535, 543, 545–546
Pisces, 151
Pisidae
   Bayesian analysis, 444
   branching patterns, 447
   direct optimization analysis, 445
   family development, 438–439
*Pisidia*, 111
Pisinae, 438, 447
*Pitho*, 444
*Pitho lherminieri*, 439, 444, 447
*Plagusia*, 484, 486–487
Plagusiidae, 484, 486
*Planes*, 484, 486
Planotergiinae, 438
*Platychirograpsus*, 485
Platyonychidae, 535
Platythelphusidae, 492, 494
*Pleistacantha*, 438
Pleocyemata
   choice of outgroup and analyses, 270
   early development, 32–34
   eggs, 32
   evolution and radiation, 245

infraordinal relationships, 91
   mitochondrial genes, 18
   molecular taxonomy, 15
   morphological character-based analysis, 262
   nuclear genes, 19, 22
   phylogeny-based systematics, 23
   Procaridoidea and Caridea, 297
   short courtship mating system, 141–143
   transformation types, 233
*Pleopteryx kuempeli*, 254
Pleopteryxoidea, 254
Pliosominae, 438
PLS, *see* Partitioned likelihood support (PLS)
podding mating system, 121, 126, 132, 147
*Podocallichirus*, 338
Podophthalmidae, 535
Podophthalminae, 535, 543–544, 546
Podotremata, 95, 431–432
Podotremata, monophyletic group
   analysis, 419
   animals, 418
   *Cyclodorippoidea*, 422–429
   discussion, 431–432
   *Dromiacea*, 419, 429
   *Eubrachyura*, 422–429
   fundamentals, 417–418
   *Homoloidea*, 419, 427, 429
   materials and methods, 418–419
   microscopy, 418
   monophyly, 419–423, 432
   phylogenetic relationships, 423–429
   *Podotremata* paraphyly, 431–432
   *Raninoidea*, 421–422, 424–429
   results, 419–429
   subtaxa, 419–429
polarity, setal characters, 228–230, 232
polyandry, 122, 123
Polybiidae, 535, 544–546
Polybiinae
   analysis, 543, 545
   considerations and future work, 577
   taxonomy development, 535
*Polybius*
   analysis, 543, 545
   considerations and future work, 577
   sample collection, 569
   taxonomy developments, 567
*Polybius henslowii*, 543
*Polycheles*
   higher taxa diagnosis, 383–384
   morphological trends, 379
   Polychelidae, 376
   Polychelidan lobsters, 369–372
   study results, 374
*Polycheles amemiyai*, 376
*Polycheles enthrix*, 376
*Polycheles kermadecensis*, 376
*Polycheles tanneri*, 376
*Polycheles tryphlops*, 370
Polychelida
   higher taxa diagnosis, 379–380
   molecular taxonomy, 15
   phylogeny-based systematics, 23
   Polychelidan lobsters, 370
*Polychelida*, 379–380
Polychelidae
   higher taxa diagnosis, 381
   infraordinal relationships, 91
   *Palaeopentacheles*, 376
   Polychelidan lobsters, 369
   terminal taxa, 372
*Polychelidae*, 376–377, 381–386
Polychelidan lobsters
   analytical methods, 372–373
   *Coleiidae Van Straelen*, 380
   data matrix, 388
   diagnosis of taxa, 379–386
   discussion, 375–379
   *Eryonidae de Haan*, 380
   fundamentals, 369–372
   materials and methods, 372–373
   morphological characters, 372, 389–392
   morphological trends, 377–379
   *Palaeopentacheles* position, 375–376
   *Palaeopentachelidae*, 380–381
   *Polychelida de Haan*, 379–380
   *Polychelidae* genera, 376–377, 381–386
   *Polychelidae Wood-Mason*, 381
   polychelid sister group, 375–376
   results, 373–374
   systematics, 379–386
   terminal taxa, 372, 387–388
   *Tetrachelidae Beurlen*, 381
   unambiguous character state changes, 393
   *Willemoesia* position, 377
polychelid sister group, 375–376
Polychilidae, 377
polygamy, 122
polygynandry, 122
polygyny, 122
polymerase chain reaction (PCR)
   Anomura, 404
   Anomura phylogeny, 404
   Callianassidae mitochondrial genes, 328
   Caridea mitochondrial and nuclear genes, 283, 291
   *Cronius*, 569, 571
   Cryptochiridae, 482
   freshwater crayfishes diversification timing, 345
   *Hexapanopeus*, 556
   marine clawed lobster families, 360
   mitochondrial DNA, 49
   Pinnotheridae, molecular genetic re-examination, 461–462
   Portunoidea and Cancroidea, 538
   Thalassinidea mitochondrial and nuclear genes, 312
*Polyonyx*, 111
*Polyonyx gibbesi*, 127
polyphyly, 294, 296
*Pontonia*, 297
*Pontonia margarita*, 127, 133

Pontoniinae
  Caridea, infraorder phylogeny, 296
  pair-bonding mating system, 127, 148–149
  phylogenetic relationships, 251
porcelain crab, 127, 399, *see also* Anomura
*Porcellana*, 111
Porcellanidae
  Anomura phylogeny, 407
  classification, 399
  nucleus, 111
  pair-bonding mating system, 127
  taxon sampling, 400
*Porcellanopagurus*, 10
*Porcellanopagurus nihonkaiensis*, 10
Porifera, 148, 150
*Portumnus*, 545
Portunidae, *see also Cronius* Stimpson, molecular phylogenies
  precopulatory guarding mating system, 125
  specimens used, 418
  sperm plugs, 147
  taxon sampling, 552
  Thalassinidea, mitochondrial and nuclear genes, 312
  zoeal similarity, 223
Portuninae, 543–544, 546
Portunoidea
  *Garthiope barbadensis*, 563
  marine crab sister group, 498
  outgroup taxa, 563
  phylogenetics, 234
  polarity, setal characters, 229
Portunoidea and Cancroidea, new classification proposal
  discussion, 544–546
  families of, 534–535
  fundamentals, 533–534
  materials and methods, 535, 538–539
  results, 539, 542–543
  subfamilies of Portunidae, 535
*Portunus*, *see Cronius* Stimpson, molecular phylogenies
  analysis, 543, 546
  sample collection, 569
  taxonomy development, 533
  taxonomy developments, 567–568
*Portunus anceps*, 575–576
*Portunus bahamensis*, 576
*Portunus floridanus*, 575–576
*Portunus gibbesii*, 575
*Portunus gracilimanus*, 576
*Portunus hastatus*, 575–576
*Portunus longispinosus*, 576
*Portunus pelagicus*
  analysis, 543, 546
  molecular phylogeny, 574
  taxonomic revisions, 575
  taxonomy developments, 567–568
  vision, 189
*Portunus sayi*, 574–575
*Portunus spinicarpus*, 575
*Portunus spinimanus*, 576
*Portunus tuberculatus*, 538

*Portunus ventralis*, 575–576
*Portunus vossi*, 576
position within decapoda, 247–248
Potamidae
  analysis, 518, 543
  frontal triangle and second gonapod, 524
  Gecarcinucidae monophyly, 523
  gonopod 1 characters, 496
  manibular palp characters, 494–495
  Oriental region, 500
  Palaearctic region, 499
  taxonomy developments, 492, 494
  theories on origins, 503
Potaminae, 499–500, 504
Potamiscinae, 499–500, 504
Potamocarcinidae, 492, 494
Potamoidea, *see also* Gecarcinucidae
  freshwater crabs, 496, 497–498
  fundamentals, 496
  marine crab sister group, 497–498
  microtubular arms, 112
  taxonomy developments, 494
potamoid freshwater crabs, 494–496
Potamonautidae
  Afrotropical region, 499
  distribution, 498
  frontal triangle and second gonapod, 524
  Gecarcinucidae monophyly, 523
  gonopod 1 characters, 496
  historical systematic approaches, 511
  manibular palp characters, 495
  taxonomy developments, 492, 494
Potamonautinae, 523
Poupiniidae, 419
precopulatory guarding mating system, 121, 125, 130–131, 144–145
presence/absence, 108–109, 111
previously applied superfamilies, 320–321
previous works, 358, 363
primer optimization, 58–59, *see also* Mitochondrial DNA (mtDNA)
*Prismatopus filholi*, 418
*Procambarus*, 343–344, 347
*Procambarus clarkii*, 105, 418
Procarididae
  basal lineages, 299
  Caridea, infraorder phylogeny, 281, 294, 296
  fossils, 254
  morphological characters, 252
  position, shrimp-like decapods, 247
Procaridoidea, 296–297
*Procaris*, 247, 252
*Procaris ascensionis*, 296
Processidae
  Caridea, infraorder phylogeny, 281, 294, 296
  Caridea mitochondrial and nuclear genes, 299
  morphological characters, 252
*Proeryon hartmanni*, 375
promiscuity, 122
proposal, classification of *Portunoidea* and *Cancroidea*

discussion, 544–546
families of, 534–535
fundamentals, 533–534
materials and methods, 535, 538–539
results, 539, 542–543
subfamilies of Portunidae, 535
protein-coding genes, decapod phylogeny
Brachyura, 95
fundamentals, 89, 97
infraordinal relationships, 91
molecular phylogeny, 90–91
Palinuridae, 95–97
Penaeoidea, 91, 94–95
phylogenetic reconstruction, genera/species, 95–97
super family/family level phylogeny studies, 91–95
Psalidopodidae, 252, 294, 296
Pseudochelidae, 249, 283
pseudogenes, 56–58, 70, *see also* Mitochondrial DNA (mtDNA)
*Pseudohapalocarcinus ransoni*, 475
*Pseudopalaemon*, 251
*Pseudopinnixa*, 461, 470
*Pseudopinnixa carinata*, 463, 466
Pseudorhombilidae, 552, 562, 563
*Pseudotealliocaris*, 5
Pseudothelphusidae
analysis, 543
distribution, 498
gonopod 1 characters, 496
manibular palp characters, 494–495
Neotropical region, 498–499
taxonomy developments, 492, 494
Pseudothelphusoidea, 494
*Puerulus*, 97
*Pugettia*, 444
*Pugettia quadridens*, 439
Pygocephalomorpha, 9
*Pylocheles*, 110, 405, 407
Pylojacquesidae, 400

# R

radiation and evolution, shrimp-like decapods
Caridea, 249–252
Dendrobranchiata, 248–249
fossils, 254
fundamentals, 245, 254–255
molecular markers, 253
morphological characters, 252–253
phylogenetic relationships, 248–252
position within decapoda, 247–248
Stenopodidea, 248
Raininoidea, 417
*Ranina*, 110
*Ranina ranina*, 418
Raninidae, 95, 418
Raninoidea
Brachyura, 418
Brachyura monophyly, 432
Podotremata, 421–422, 424–429, 431–432
synapomorphies, 424–429

rapid multiple clutch polygamy, 123
ratchet, 75
rate heterogeneity parameters, 72
RAxML
Callianassidae, mitochondrial genes, 333
Caridea, infraorder phylogeny, 292
freshwater crayfishes, 345–346
freshwater crayfishes diversification timing, 348
phylogeny-based systematics, 23
Pinnotheridae, 463
Thalassinidea, mitochondrial and nuclear genes, 312–313
*Raymanninus*, 545
*Raymanninus schmitti*, 546
recombination, 81
red king crab, 126
relationships, 334, 336–339
Remipedia, 11, 102, 105
reproductive swarm mating system, 121, 129–130, 134
Reptantia
fossil record, 5
higher taxa diagnosis, 379
infraordinal relationships, 91
molecular taxonomy, 15
morphological character-based analysis, 261
position, shrimp-like decapods, 248
Thalassinidea, monophyly or paraphyly, 319
resource-centered competition, 123
restriction, 463, 465, 467
results
Anomura, 405–407
Asian freshwater crabs, 518, 521–523
Callianassidae mitochondrial genes, 333–336
Caridea mitochondrial and nuclear genes, 294–296
freshwater crayfishes diversification timing, 348, 350–351
gall crabs, 484–486
*Hexapanopeus*, 557
Majoidea, 444–445
marine clawed lobster families, 361–362
mitochondrial DNA, 52–54
morphological character-based analysis, 264–270
Pinnotheridae, molecular genetic re-examination, 463–466
Podotremata, 419–429
Polychelidan lobsters, 373–374
Portunoidea and Cancroidea new classification proposal, 539, 542–543
Thalassinidea mitochondrial and nuclear genes, 313–318
reverse transcription-polymerase chain reaction (RT-PCR), 16
*Rhanbdonotus pictus*, 223, 234, 236
*Rhithropanopeus harrisii*, 557, 562
Rhizocephala
inferences, 209, 211
life cycle, 201–202
phylogeny, 204–205
Rhynchocinetidae, 294, 296, 299
*Rimapenaeus*, 124

*Rimapenaeus constrictus*
   Caridea, Stenopodidea, and Nephropidae as outgoups, 270
   Caridea and Stenopodidea as outgoup, 268
   Caridea as outgoup, 266
   Penaeidae as non-natural group, 273
river prawn, 125
RNA isolation, 16
rock shrimps, 123
*Rouxana*, 522–523
RT-PCR, *see* Reverse transcription-polymerase chain reaction (RT-PCR)

## S

12S, 16S, COI, 17–19
*Salangathelphusa*, 522, 524
*Salangathelphusa brevicarinata*, 522, 528
*Samastacus*, 343
sand crabs, 124, 130
Sankoff algorithm, 78
*Sanquerus*, 576
*Sartoriana*, 521, 524
*Sartoriana blandfordi*, 521
saturation, 71
*Sayamia*, 522
scaphognathite origin, 37–38
*Scapulicambarus*, 344
scarlet cleaner shrimp, 127
*Schizophrys*, 438
*Scleroplax*
   constituents of Pinnotherinae, 469
   Pinnotheridae restricted, 468
   Pinnotherinae subfamily, 465
*Scleroplax granulata*
   constituents of Pinnothereliinae, 468
   limited gene utility, 470
   Pinnothereliinae subfamily, 465
   Pinnotheridae restricted, 467
   pinnotherid subfamily definitions, 465
*Scopimera*, 133
Scyllaridae, 312
*Scylla serrata*, 575
sea shrimp, 185
second gonopod, 523–524
semi-terrestrial crabs, 124
semi-terrestrial grapsid crabs, 123
*Sendleria*
   analysis, 522
   biogeography, 528
   *Sundathelphusa*, 526
*Sendleria gloriosa*, 522
sensitivity to light, 187–188
sequence alignment issues, 70–71
Sequencher software, 360, 482
sequencing
   Anomura, 404, 405
   Callianassidae mitochondrial genes, 328
   Caridea mitochondrial and nuclear genes, 283, 291
   *Cronius*, 569, 571
   Cryptochiridae, 482
   freshwater crayfishes diversification timing, 345
   Gecarcinucidae, 511
   *Hexapanopeus*, 556
   marine clawed lobster families, 360
   mitochondrial DNA, 49
   Portunoidea and Cancroidea, 538
   Thalassinidea mitochondrial and nuclear genes, 312
Sergestidae
   acrosome vesicle, 109
   Caridea, Stenopodidea, and Nephropidae as outgoups, 269
   Caridea and Stenopodidea as outgoup, 266
   Caridea as outgoup, 265
   Dendrobranchiata, monophyletic group, 272
   morphological character-based analysis, 261–262
   position, shrimp-like decapods, 247
   Sergestoidea as natural group, 272
Sergestinae, 261
Sergestoidea
   Caridea, infraorder phylogeny, 283
   Caridea, Stenopodidea, and Nephropidae as outgoups, 269
   Caridea and Stenopodidea as outgoup, 266
   Caridea as outgoup, 265
   Dendrobranchiata, monophyletic group, 271–272
   morphological character-based analysis, 261–262, 272
   phylogenetic relationships, 248–249
*Sergio*
   Axiidea families, 318
   Callichirinae, 334
   mitochondrial genes, 327
   subfamily relationships, 337
*Sergio guassutinga*, 334, 338
*Sergio mericeae*, 334, 338, 345
*Sergio trilobata*, 334, 338
Sesarmidae
   analysis, 486
   colonization of freshwater, 502
   marine crab sister group, 497
   taxon selection, 484
setal observations, 222
*Seychellum*, 494–495
shape, 109–111
*Shinkaia*, 400, 407
Shinkaiinae, 407
short courtship mating system, 121, 124–125, 130–131, 141–143
shrimp
   acrosome vesicle, 105, 109
   embryonic characters, 40
   eusociality mating system, 128, 133
   mating systems study history, 123
   nucleus, 110
   opsin sequences, 188
   optical design, 185
   pair-bonding mating system, 127, 148–153
   precopulatory guarding mating system, 125, 131
   Procaridoidea and Caridea, 296–297

protein-coding genes, 91–95
pseudogenes, 57
short courtship mating system, 124, 131
visual system components, 191
shrimp, visual system characters
  components as phylogenetic characters, 189–192
  evolutionary enigma, 187
  fundamentals, 183, 192
  molecular aspects, 187–189
  morphology, 183–186
  overview, 183–189
shrimp-like decapods, evolution and radiation
  Caridea, 249–252
  Dendrobranchiata, 248–249
  fossils, 254
  fundamentals, 245, 254–255
  molecular markers, 253
  morphological characters, 252–253
  phylogenetic relationships, 248–252
  position within decapoda, 247–248
  Stenopodidea, 248
shrimp-like decapods, mitochondrial and nuclear genes
  *Alpheoidea*, 298
  Atyidae, 298
  basal lineages, 299
  *Crangonidae*, 299
  discussion, 296–300
  DNA extraction, 283, 291
  evolutionary history, 282–283
  fundamentals, 281, 300
  hypotheses testing, 300
  ingroup taxa, 283
  materials and methods, 283–292
  monophyly, 294, 296
  outgroup selection, 283
  *Palaemonoidea*, 297–298
  paraphyly, 294, 296
  PCR, 283, 291
  phylogenetic analyses, 292
  polyphyly, 294, 296
  Procaridoidea, 296–297
  *Processidae*, 299
  results, 294–296
  sequencing, 283, 291
  *Thalassocarididae*, 299
  *Xiphocarididae*, 298
shrimp-like decapods, morphological character-based analysis
  *Aristeidae*, 274
  *Benthesicymidae*, 273
  Caridea, 264–270
  description of characters, 264
  discussion, 270–274
  fundamentals, 261–262
  materials and methods, 262–263
  monophyletic group, 271–272
  morphological characters used, 276–278
  natural group, 272–273
  *Nephropidae*, 268–270

non-natural group, 273
optimization of characters, 264
outgroups, 264–271
*Penaeoidea*, 272–273
results, 264–270
*Sergestoidea*, 272
*Solenoceridae*, 273
species studied, 275
status of, 274
Stenopodidea, 266–270
*Siamthelphusa*, 522, 524
*Sicyonella*, 272
Sicyoniidae
  Caridea, Stenopodidea, and Nephropidae as outgoups, 269
  Caridea and Stenopodidea as outgroup, 266, 268
  Caridea as outgoup, 265
  morphological character-based analysis, 261–262
  Penaeoidea as natural group, 272
  protein-coding genes, 91, 94–95
  short courtship mating system, 124
Sicyoninae, 261–262
*Siderastrea stellata*, 129
Silentes, 96–97
Sinopotamidae, 492, 494
*Sirpus*, 545
*Sirpus Zariquieyi*, 543
sister group, Decapoda, 4
size, acrosome vesicle, 110
small groups, mating system, 157–158
*Snaha escheri*, 521
snapping shrimp, 127
Solenoceridae, 273
  Caridea, Stenopodidea, and Nephropidae as outgoups, 269
  Caridea and Stenopodidea as outgroup, 267
  Caridea as outgoup, 265
  morphological character-based analysis, 261–262
  Penaeoidea as natural group, 272
  protein-coding genes, 91, 94
  short courtship mating system, 124
Solenocerinae, 262
*Solitariopagurus*, 10
*Somanniathelphusa*, 522, 524
Somanniathelphusinae, 510, 524
*Sotoplax robertsi*, 563
southern king crab, 126
SOWH, *see* Swofford-Olsen-Waddell-Hillis (SOWH) test
spermatozoal morphology, decapod phylogeny
  acrosome vesicle, 105, 108–110
  aflagellate sperm cells, 105
  fundamentals, 101–105, 112–113
  immotile sperm cells, 105
  microtubular arms, 111–112
  mitochondria, 108
  nucleus, 110–111
  sperm nuclear proteins, 106–107
  sperm nucleus, 107–108
  sperm uniqueness, 104–108

sperm nuclear basic proteins (SNBPs), 106–108
sperm nuclear proteins, 106–107
sperm nucleus, 107–108
sperm plug mating system, 131, 143, 147
sperm plugs
    crab species, 147
    penacid shrimp, 143
    precopulatory guarding mating system, 125
sperm uniqueness, 104–108
spider crabs, 123, 437
*Spinastacoides*, 343
spiny lobsters, 95–97, *see also* Lobsters
*Spiralothelphusa*, 521, 524
Spiralothelphusinae, 510, 524
sponge-dwelling shrimp, 127–128
*Spongicola japonica*, 127
Spongicolidae, 127, 153
SPR, *see* Subtree pruning and regrafting (SPR)
square ommatidial facets, 186
squat lobsters, 399, *see also* Anomura
28S rDNA, 296
Sri Lankan group, 521
16S rRNA
    molecular markers, 253
    primer optimization, 58–59
    protein-coding genes, 95
18S rRNA, 95
16S rRNA sequence data, gall crabs
    discussion, 486–487
    fundamentals, 475, 482
    materials and methods, 482, 484
    results, 484–486
statistical tests, tree topologies, 80
Stenopodidae
    pair-bonding mating system, 127, 153
    phylogenetic relationships, 248
    position, shrimp-like decapods, 247–248
Stenopodidea
    Caridea, infraorder phylogeny, 283
    evolution and radiation, 245, 248
    fossil record, 6
    fossils, 254
    fundamentals, *ix*
    infraordinal relationships, 91
    molecular taxonomy, 15
    morphological character-based analysis, 262, 266–270
    morphological characters, 252
*Stenopus*, 109, 363
*Stenopus hispidus*
    Caridea and Stenopodidea as outgroup, 268
    Dendrobranchiata, monophyletic group, 272
    pair-bonding mating system, 127
*Stenorhynchus*, 444–445
*Stenorhynchus seticornis*, 439
*Stereomastis*
    higher taxa diagnosis, 384–385
    morphological trends, 379
    Polychelidae, 376
    Polychelidan lobsters, 369–372
    study results, 374

*Stereomastis aculeata*, 376
*Stereomastis alis*, 376
*Stereomastis auriculatus*, 376
*Stereomastis cerata*, 376
*Stereomastis galil*, 376, 386
*Stereomastis panglao*, 386
*Stereomastis phosphorus*, 376
*Stereomastis polita*, 376, 386
*Stereomastis suhmi*, 370
*Stereomastis surda*, 376
*Stereomastis trispinosus*, 376
Stomatopoda, 23, 35
*Stomatopods*, 3
stone crabs, 48
Strahlaxiidae
    Axiidea families, 318
    infraorder composition and relationships, 322
    mitochondrial and nuclear genes, 309, 311
    monophyly or paraphyly, 320
    monophyly test, 313
*Strahlaxius*, 318
strict parsimony analysis, 73
Stridentes, 96–97
*Stygothelphusa*, 522–523, 528
Stylodactylidae
    basal lineages, 299
    Caridea, infraorder phylogeny, 294, 296
    evolutionary history of Caridea, 282
subtree pruning and regrafting (SPR), 74
*Sundathelphusa*, 522, 526
*Sundathelphusa boex*, 522
*Sundathelphusa cassiope*, 523, 526
*Sundathelphusa cavernicola*, 522
*Sundathelphusa celer*, 522
*Sundathelphusa halmaherensis*, 522–523
*Sundathelphusa minahassae*, 526
*Sundathelphusa mindahassae*, 523
*Sundathelphusa picta*, 522
*Sundathelphusa rubra*, 522, 526
*Sundathelphusa tenebrosa*, 522, 528
Sundathelphusidae, 492, 494, 510
super family/family level phylogeny studies
    Brachyura, 95
    Penaeoidea, 91, 94–95
    protein-coding genes, 91–95
superposition eyes, 185–186
Swofford-Olsen-Waddell-Hillis (SOWH) test, 80
symbiosis, 122
symbiotic anomuran crabs, 123
*Synalpheus*, 252, 298
*Synalpheus neptunus neptunus*, 128
*Synalpheus regalis*, 128
Syncarida, 9
systematics, 379–386
system of Bott similarities, 524, 526

## T

*Taliepus*, 444
*Taliepus dentatus*, 439
Tanaocheleinae, 235

*Tanocheles bidenata*, 233–236
*Tanocheles stenochilus*, 235
TBR, *see* Tree bisection and reconnection (TBR)
*Tealliocaris*, 5
*Telmessus*, 535
Templeton test, 80
*Tennuibranchiurus*, 343
terminal taxa, 372, 387–388
*Terranthelphusa*, 522
*Terranthelphusa kuhli*, 523
territorial cooperation hypothesis, 132
*Tesnusocaris goldichi*, 11
*Tetrachela raiblana*, 372
Tetrachelidae
   higher taxa diagnosis, 381
   *Palaeopentacheles*, 375
   Polychelidan lobsters, 369
   terminal taxa, 372
Tetrachelidae Beurlen, 381
*Tetralia cavimana*, 234–236
Tetraliidae, 155
*Thaksinthelphusa*, 521
*Thalamita*
   considerations and future work, 577
   sample collection, 569
   taxonomy developments, 567
Thalamitidae, 535
Thalamitinae
   analysis, 543–544, 546
   considerations and future work, 577
   molecular phylogeny, 574
   sample collection, 569
   taxonomy development, 535
   taxonomy developments, 568–569
*Thalamitoides*, 577
*Thalassina*, 311, 321
Thalassinicea, 319
Thalassinida, 40
Thalassinidae
   Gebiidea families, 313, 318
   infraorder composition and relationships, 321
   mitochondrial and nuclear genes, 309–310
   monophyly or paraphyly, 318–320
   monophyly test, 313
   previously applied superfamilies, 321
Thalassinidea
   Caridea, infraorder phylogeny, 283
   fossil calibration, 347
   infraordinal relationships, 91
   microtubular arms, 111
   molecular taxonomy, 15
   pair-bonding mating system, 154
   phylogeny-based systematics, 23
   refracting superposition optics, 185
Thalassinidea, mitochondrial and nuclear genes
   *Axiidea* families, 318
   discussion, 318–322
   DNA extraction, 312
   fundamentals, 309–311, 322
   *Gebiidea* families, 313, 318

   infraorder composition, 321–322
   internal family relationships, 321–322
   materials and methods, 311–313
   monophyly, 313, 318–320
   paraphyly, 318–320
   PCR, 312
   phylogenetic analyses, 312–313
   previously applied superfamilies, 320–321
   results, 313–318
   sequencing, 312
   taxa included, 311–312
Thalassinoidea, 320
Thalassocarididae, 294, 296, 299
*Thaumastocheles*, 362, 364
Thaumastochelidae, 91, 357, *see also* Marine clawed lobster families, mitochondrial gene-based phylogeny
*Thaumastochelopsis*
   morphological and molecular previous works, 358
   phylogenetic analysis, 362
   previous works comparison, 363
   Thaumastochelidae status, 364
*Thelphusula*, 522, 524
*Thelphusula baramensis*, 522
*Thia scutellata*, 543
Thiidae, 535, 543, 545–546
Thioidea, 535
*Thomassinia*, 318
*Thomassinia gebioides*, 322
Thomassiniidae
   Axiidea families, 318
   infraorder composition and relationships, 321–322
   mitochondrial and nuclear genes, 309, 311
   monophyly or paraphyly, 320
   monophyly test, 313
Thoracotremata
   analysis, 484, 486
   Cryptochiridae developments, 482
   developments, 475
   Eubrachyura monophyly, 422
   mitochondrial DNA, 49
   protein-coding genes, 95
   taxon selection, 484
thoracotreme crabs, 110
*Thoralus*, 298
*Thymopides*
   marine clawed lobster families, 357
   morphological convergence, 364
   phylogenetic analysis, 362
Thymopinae, 364
*Thymops*
   marine clawed lobster families, 357
   morphological and molecular previous works, 358
   phylogenetic analysis, 362
   Thaumastochelidae status, 364
*Thymopsis*, 357, 364
tissue sampling, 359
TM9sf4, 22
total evidence approach, 81
*Tozeuma*, 298

Tracer program, 345–346
*Trachycaris*, 298
*Trachypenaeus*, 124
Trachypeneini, 249
transformation types, 232–233
Transition (TIM) model, 294
*Trapezia*, 127
*Trapezia richtersi*, 233
Trapeziidae
  pair-bonding mating system, 127, 155
  protein-coding genes, 95
  zoeal similarity, 223
Trapezioidea, 233–234, 236
*Travancoriana*, 524
*Travancoriana schirnerae*, 521
tree bisection and reconnection (TBR)
  Callianassidae, mitochondrial genes, 333
  phylogenetic inference, molecular data, 74
  Portunoidea and Cancroidea, 538
  Thalassinidea, mitochondrial and nuclear genes, 312
tree comparison and node support, 79–80
Tree of Life, 15, *ix*
trees, searching for, 73–75
tree topologies, 80, 334
Tremoctopodidae, 134
Trichobranchiata, 261
Trichodactylidae
  analysis, 543
  distribution, 498
  freshwater crabs, 497
  fundamentals, 497
  marine crab sister group, 497–498
  Neotropical region, 498–499
  taxonomy developments, 492, 494, 534
Trichodactyloidea, 494, 534
*Tritodynamia*, 466, 470
*Tritodynamia horvathi*, 463, 466
*Trizocheles*, 405, 407
*Troglocambarus*, 343
*Troglocarcinus corallicola*, 129
*Troglocaris*, 251, 253
Trophallaxis Theory, 133
true trees, 73–74
*Trypaea*, 339
*Tumidotheres*
  constituents of Pinnotherinae, 470
  limited gene utility, 470
  pinnotherid subfamily definitions, 465
  Pinnotherinae subfamily, 466
  specimens, 461
*Tunicotheres*
  constituents of Pinnotherinae, 469–470
  pinnotherid subfamily definitions, 465
  Pinnotherinae subfamily, 466
*Tunicotheres moseri*, 458, 461
Tychidae, 438
Tychinae, 438
*Tymolus*, 426
*Tymolus brucei*, 418
*Typhlatya*, 253
Typhlocarididae, 296–297

U

UBX, *see* Hox gene ultrabithorax (UBX)
*Uca*, 123, 128, 129, 133
*Uca tangeri*, 105
Udorellidae, 254
unambiguous character state changes, 393
*Uncina posidoniae*, 6
Uncinidae, 358
Unipeltata, 3
UPGMA method, 77
*Upogebia*
  Gebiidea families, 313
  pair-bonding mating system, 127
  previously applied superfamilies, 321
  Thalassinidea, mitochondrial and nuclear genes, 311
  Thalassinidea, monophyly or paraphyly, 319
Upogebiidae
  Gebiidea families, 313
  infraorder composition and relationships, 321
  previously applied superfamilies, 321
  Thalassinidea, mitochondrial and nuclear genes, 309–311
  Thalassinidea, monophyly or paraphyly, 318–320
  Thalassinidea monophyly test, 313
Upogebiinae, 310, 318
uropods, 201

V

*Vanni*, 521
*Varuna*, 486
*Varuna litterata*, 502
Varunidae
  analysis, 485–486
  marine crab sister group, 498
  specimans used, 418
  taxon selection, 484
  waving display mating system, 128
*Vetericaris*, 247
*Vetericaris chaceorum*, 296
*Virilastacus*, 343
visiting type mating system, 121, 129, 134
visual system characters, shrimp
  components as phylogenetic characters, 189–192
  evolutionary enigma, 187
  fundamentals, 183, 192
  molecular aspects, 187–189
  morphology, 183–186
  overview, 183–189
vraagteken effect, 254

W

Wallace's Line, 500, 509, 528
*Waterstonella*, 10
Waterstonellidea
  paleobiogeography and paleoecology, 9
  sister group to Decapoda, 4

waving display mating system, 121, 128–129, 133–134
Willemoesia
　higher taxa diagnosis, 385–386
　morphological trends, 378–379
　Polychelidae, 376
　Polychelidan lobsters, 369, 371–372, 377
　study results, 374
Willemoesia leptodactyla, 371

## X

Xaiva, 545
Xanthidae
　Garthiope barbadensis, 563
　homoplasy, 233
　outgroup taxa, 563
　pair-bonding mating system, 155
　Panoplax depressa, 562
　precopulatory guarding mating system, 125
　protein-coding genes, 95
　specimans used, 418
　taxonomy development, 535
　taxon sampling, 552
　zoeal similarity, 223
xanthid ring, 110
Xanthoidea
　analysis, 557
　DNA extraction, 556
　larval morphology, 222
　outgroup taxa, 563
　phylogenetics, 234, 236
　taxonomy developments, 551
Xantho poressa, 418
Xenograpsus testudinatus, 233
Xenommacarida, 4
Xenophthalminae
　exclusions and exceptiona members, 466
　family development, 457
　specimens, 458, 461

Xenophthalmus pinnotheroides
　exclusions and exceptiona members, 466
　limited gene utility, 470
　Pinnotheridae restriction, 463
XESEE 3.2 program, 49
Xiphocarididae, 294, 296, 298
Xiphocaris, 298
Xiphocaris elgonata, 298
Xiphoenaeus, 124
Xiphoenaeus kroyeri
　Caridea, Stenopodidea, and Nephropidae as outgoups, 269–270
　Caridea and Stenopodidea as outgroup, 268
　Caridea as outgoup, 266
　Penaeidae as non-natural group, 273
Xiphonectes, 576

## Y

Yagerocaris cozumel, 298

## Z

Zaops
　constituents of Pinnotherinae, 469
　limited gene utility, 470
　pinnotherid subfamily definitions, 465
　Pinnotherinae subfamily, 465–466
Zaops ostreum, 458
Zebrida adamsii, 223, 234, 236
zoeal larva, 32, 223
zoeal similarity, 223
Zuzalpheus, 133, 298
Zuzalpheus brooksi, 128
Zuzalpheus regalis, 128

# CRUSTACEAN ISSUES

General editor
## Prof. Dr. Stefan Koenemann
Institute for Animal Ecology and Cell Biology
University of Veterinary Medicine Hannover
Germany

1. Crustacean Phylogeny
   Schram, F.R. (ed.)
   1983    ISBN 90 6191 231 8 Sold out

2. Crustacean Growth: Larval Growth
   Wenner, A. (ed.)
   1985    ISBN 90 6191 294 6

3. Crustacean Growth: Factors in Adult Growth
   Wenner, A. (ed.)
   1985    ISBN 90 6191 535 X

4. Crustacean Biogeography
   Gore, R.H. & Heck, K.L. (eds.)
   1986    ISBN 90 6191 593 7

5. Barnacle Biology
   Southward, A.J. (ed.)
   1987    ISBN 90 6191 628 3

6. Functional Morphology of Feeding and Grooming in Crustacea
   Felgenhauer, B.E., Thistle, A.B. & Watling, L. (eds.)
   1989    ISBN 90 6191 777 8

7. Crustacean Egg Production
   Wenner, A. & Kuris, A. (eds.)
   1991    ISBN 90 6191 098 6

8. History of Carcinology
   Truesdale, F.M. (ed.)
   1993    ISBN 90 5410 137 7

9. Terrestrial Isopod Biology
   Alikhan, A.M.
   1995    ISBN 90 5410 193 8

10. New Frontiers in Barnacle Evolution
    Schram, F.R. & Hoeg, J.T. (eds.)
    1995    ISBN 90 5410 626 3

11  Crayfish in Europe as Alien Species - How to Make the Best of a Bad Situation
    Gherardi, F. & Holdich, D.M. (eds.)
    1999   ISBN 90 5410 469 4

12  The Biodiversity Crisis and Crustacea - Proceedings of the Fourth International Crustacean Congress, Amsterdam, Netherlands
    Vaupel Klein, J.C. von & Schram, F.R. (eds.)
    2000   ISBN 90 5410 478 3

13  Isopod Systematics and Evolution
    Kensley, B. & Brusca, R.C. (eds.)
    2001   ISBN 90 5809 327 1

14  The Biology of Decapod Crustacean Larvae
    Anger, K.
    2001   ISBN 90 2651 828 5

15  Evolutionary Developmental Biology of Crustacea
    Scholtz, G. (ed.)
    2004   ISBN 90 5809 637 8

16  Crustacea and Arthropod Relationships
    Koenemann, S. & Jenner, R.A. (eds.)
    2005   ISBN 0 8493 3498 5

17  The Biology and Fisheries of the Slipper Lobster
    Lavalli, K.L. & Spanier, E. (eds.)
    2007   ISBN 0 8493 3398 9

18  Decapod Crustacean Phylogenetics
    Martin, J.W., Crandall, K.A. & Felder, D.L. (eds.)
    2009   ISBN 1 4200 9258 8

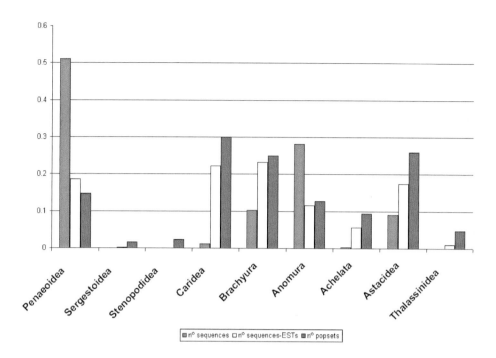

**Color Figure 1. (See Figure 2 in Palero & Crandall)** Decapod sequences in GenBank in April 2008, shown as a proportion of the sequences belonging to the different infraorders relative to the total number of sequences available (355,876), the total number of sequences available after excluding ESTs (337,603), and the relative proportion of population study datasets.

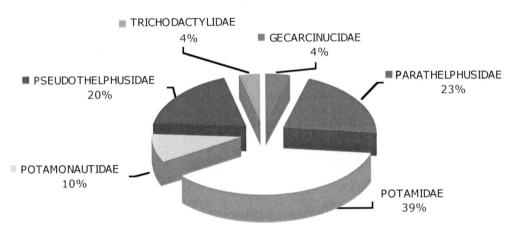

**Color Figure 2. (See Figure 1 in Cumberlidge & Ng)** Freshwater crab diversity.